Combustion Theory

Combustion Science and Engineering Series

Series Editor
Forman A. Williams
Princeton University

Advising Editors
Ronald Hanson, Stanford University
David T. Pratt, University of Washington
Daniel J. Seery, United Technologies

To Th. von Kármán,

the founder of aerothermochemistry

Combustion Theory

*The Fundamental Theory of
Chemically Reacting Flow Systems*

Second Edition

Forman A. Williams

Princeton University

ADDISON-WESLEY PUBLISHING COMPANY
The Advanced Book Program
Redwood City, California • Menlo Park, California • Reading, Massachusetts
New York • Don Mills, Ontario • Wokingham, United Kingdom • Amsterdam
Bonn • Sydney • Singapore • Tokyo • Madrid • San Juan

Sponsoring Editor: Richard W. Mixter
Production Coordinator: Kristina Montague
Copyeditor: Linda Thompson
Cover Design: Leigh McLellan

Library of Congress Cataloging in Publication Data

Williams, F. A. (Forman Arthur), 1934–
    Combustion theory.

    (Combustion, science, and engineering series)
    Bibliography: p.
    Includes index.
    1. Combustion.   I. Title.   II. Series.
QD516.W5   1985   541.3′61   85-3902
ISBN 0-8053-9801-5

3 4 5 6 7 8 9 10-MA-95 94 93 92 91

Preface to the Second Edition

Truly remarkable progress has been achieved in combustion theory during the 20 years since the completion of the first edition of this book. Because of this progress, when the first edition went out of print about 5 years ago, it was clear that a reprinting could no longer serve to bring readers to the frontiers of the subject. Revision of the entire text was needed, and I undertook the revision without realizing that it would require as much time as writing the first edition. Although it is regrettable that the book has been unavailable during the preparation of the second edition, it seemed best to try to complete as good a revision as possible instead of rushing into print. Thus consideration has been given to changes throughout; none of the chapters survived without modification, and outdated chapters have disappeared entirely and have been replaced by chapters devoted to new material of importance in the subject.

Nevertheless, the general format of the book and the philosophy of the presentation remain the same. The necessary background material still appears in the same five appendixes. These appendixes needed less revision than the rest of the volume; a few of the sections survived intact. Among the specific changes in the appendixes are updating of the discussion of thermochemical equilibrium calculations to take account of the revolutionary advances in computational power of electronic computers, presentation of clarifications that have been achieved in the concepts of steady-state and partial-equilibrium approximations in chemical kinetics, amplification of the discussions of branched-chain and thermal explosions on the basis of insights developed from ideas of asymptotic methods, presentation of newly available information on the chemical kinetics of hydrocarbon oxidation and of the production of oxides of nitrogen in flames, and augmentation of the treatment of transport processes to include additional aspects of radiant energy transfer. The first chapter, in which the basic mathematical formulations are presented, remains just as formidable to the faint-hearted as it was in the first edition. There are still 12 chapters, but the later ones, especially, exhibit modifications in coverage as a consequence of the progress made in the field.

The titles of the first six chapters remain essentially the same as in the first edition. Chapter 7 of the first edition, which was devoted mainly to a discussion of experiments on turbulent flames, has been deleted and replaced by a considerably more extensive chapter (Chapter 10) that develops turbulent-flame theory from first principles; noteworthy advances in the theory during the past 20 years have made this possible. Presentation of material on turbulent combustion, formerly in Chapter 7, is postponed until Chapter 10 because the new knowledge relies in part on various results that have been obtained in studies of laminar-flame instabilities (now covered in Chapter 9) and in investigations of ignition and extinction (included in Chapter 8). Monopropellant droplet burning, which comprised the subject of Chapter 10 of the first edition, has assumed a somewhat less prominent role in combustion in the intervening years and therefore is covered only briefly now, specifically at the end of Chapter 3, the chapter that includes detailed discussions of the burning of fuel droplets. Chapter 9 of the first edition addressed both the theory of solid-propellant deflagration and theories of combustion instability; such substantial advances have been made in these subjects that two chapters are now needed to approach the material properly, Chapter 7 on combustion of solid propellants and Chapter 9 on combustion instabilities. The general topic of Chapter 8, ignition, extinction, and flammability limits, is the same as in the first edition, as is the subject of Chapter 11, spray combustion. Also, the general thrust of Chapter 12 remains approximately the same, notwithstanding the title change, from "Chemical Reactions in Boundary Layers" to "Flame Attachment and Flame Spread," introduced to place emphasis on specific areas of application.

To delineate more thoroughly the modifications that have been introduced, it seems desirable to discuss briefly the content chapter by chapter. The basic formulations in Chapter 1 and the derivation in Chapter 2 of jump conditions across combustion waves are largely unchanged from the first edition. Therefore, the following discussion focuses on the remaining chapters.

Chapter 3, on diffusion flames and droplet burning, has been expanded considerably to include, for example, greater consideration of influences of buoyancy and turbulence on jet flames, discussion of kinetics of surface oxidation of carbon, consideration of combustion mechanisms of metal particles, and reference to treatment of a number of realities of droplet burning that were not addressed in the first edition, such as influences of radiant energy transfer, the occurrence of unsteady conditions in the gaseous and condensed phases, the role of momentum conservation, influences of variable properties and of Lewis numbers differing from unity, the selection of suitable reference states for the evaluation of effects of forced and natural convection, and identification of conditions for the occurrence of disruptive burning of droplets. In addition, sections on diffusion-flame structure and on the burning of monopropellant droplets have been added to Chapter 3, since these subjects, in which significant new progress in understanding has been made, fall within the realm of the title of the chapter; this more advanced material, which introduces the mixture fraction as an important variable and which makes use of asymptotic analysis for describing aspects such as flame-sheet broadening, ignition, and extinction, pedagogically should be covered only after study of Chapter 5, where the necessary background ideas are presented.

The material in Chapter 4, on finite-rate chemistry in flows with negligible molecular transport, has been amplified to a lesser extent beyond the presentation in the first edition. The discussions of ignition delay and of the well-stirred reactor have been expanded somewhat, and a treatment of two-phase nozzle flow has been added, while elsewhere in the chapter modifications in the writing have been restricted to updating of ideas.

The chapter on laminar-flame theory, where molecular transport and finite-rate chemistry first enter simultaneously, has been revised substantially in the light of advances in our conceptual foundations of the subject. Asymptotic methods in mathematical analysis, notably the so-called activation-energy asymptotics, have provided us with a greatly improved understanding of the elements of laminar-flame propagation. Most of the ad hoc approaches that were detailed in the first edition are deleted and replaced by a systematic development of activation-energy asymptotics. The results obtained thereby are employed to explain many previously perplexing properties of laminar-flame structure and flame speeds. In addition, new consideration is given to flames with multiple-step chemistry and to distributions of radicals in flames, on the basis of both asymptotic concepts and advances in computational abilities of electronic computers.

Because of rapid progress in detonation theory in the 1960s, Chapter 6 of the first edition on detonations was obsolete in certain respects even before it appeared in print. In contrast, many of the revolutionary advances in this subject seem now to have been brought to fruition and appraised in reviews. Therefore, the extensive revision of this chapter needed for preparation of the second edition, appears likely to remain adequate for a longer period of time. The major changes and additions begin with the discussion of effects of three-dimensional structures in detonation propagation. Consideration is given to the relevance of spin, to the mechanism of instability, to triple-shock interactions and to transverse spacings. In addition, new analyses of detonability limits and of failure of detonations are included, and criteria for direct initiation of detonations are presented along with a discussion of processes of transition from deflagration to detonation.

Chapter 7, on solid-propellant combustion, focuses mostly on the deflagration of homogeneous solids. Consideration is given to deflagrations controlled by interface, condensed-phase, gas-phase, and dispersed-phase reactions, now with utilization of concepts from asymptotics. Specific examples mentioned include the combustion of nitrocellulose and of ammonium perchlorate; attention is paid to both pressure and temperature sensitivities of burning rates of propellants. Appraisals are made of the current status of theories of combustion of heterogeneous propellants, and the combustion of black powder is addressed as an especially challenging theoretical problem involving the oldest propellant known to mankind. In addition, a discussion of erosive burning is given, including the equations for steady, frictionless, adiabatic flow in a constant-area channel with mass addition.

The presentation in Chapter 8 of the subject of ignition, extinction and flammability limits is a revision in nearly all respects. The new approach to this subject relies strongly on activation-energy asymptotics and thereby, I believe, achieves both enhanced understanding and improved accuracy in the results. Topics that were not present in the first edition include a discussion of physical aspects of spark ignition and an introduction to the use of activation-energy asymptotics in more complicated ignition problems, through the vehicle of the problem of ignition of a semi-infinite reactive solid by a constant energy flux applied to its surface.

Chapter 9 represents an effort to provide a unified and tutorial presentation of the broad field of the theory of combustion instabilities. The length of the chapter attests to the vastness of the field and to the progress that has been made therein in recent years. The final section of this chapter, on the theory of instabilities of premixed flames, is basic to analyses of premixed turbulent flame propagation and also has a bearing on aspects of flammability limits.

Chapter 10 on turbulent-flame theory also is long, as it must be because so many different viewpoints and approaches to the subject now are available. Included in this chapter are discussions of analyses of effects of strain on laminar flame sheets, topics of interest in themselves as well as in connection with turbulent combustion. Evolution equations for laminar flames in non-uniform flows also are given. The results outlined for turbulent burning velocities emphasize those aspects that have the strongest basic theoretical justifications. Since turbulent-flame theory is a subject of continuing development, improvements of results presented herein might be anticipated to be available in the not-too-distant future.

The presentation of the subject of spray combustion in Chapter 11 is not greatly different from that in the first edition. An updated outlook on the subject has been provided, and the formulation has been generalized to admit time dependences in the conservation equations. The analysis of spray deflagration has been abbreviated, and qualitative aspects of the results therefrom have been anticipated on the basis of simplified physical reasoning. In addition, brief discussions of the topics of spray penetration and of cloud combustion have been added.

The final chapter again is devoted largely to the theory of chemical reactions in boundary-layer flows. A general formulation of time-dependent, axisymmetric, boundary-layer equations has been added to provide the reader with a convenient point of departure for analyzing a variety of reacting boundary-layer flows. The problem of combustion of a fuel plate in an oxidizing stream then provides an illustration of the use of the general formulation, instead of standing independently. The analysis of the problem of premixed flame development in a mixing layer now is abbreviated and approached as part of the broader subject of mechanisms of flame stabilization by solid bodies and by fluid streams. The final section, addressing problems of flame spread through solid and liquid fuels from a simplified viewpoint, is new.

The present volume inevitably is appreciably longer than the first edition. The increase in length is due mainly to the increase in the breadth of the coverage. Details of the mathematical development have not been amplified; on the contrary, in a number of respects fewer intermediate steps have been included, so that on the average the reader will have to invest more time per page to extract the same depth of understanding. I think that this is an unavoidable consequence of the growth of combustion theory. Even the first edition contained too much material to be covered properly in a single course. At Princeton University portions of the contents have been used in a number of graduate courses. By the time students reach the last of these, combustion theory, they have been exposed to the subjects of the appendices, of Chapters 1, 2, and 4, and of most of Chapter 12, and they have studied the more qualitative aspects of Chapters 3 and 5; the students are then able to assimilate most of the rest of the book in one semester. Alternative orders of coverage would fit into differently designed curricula. It seems fair to say that at most universities courses do not cover all of the material in the text; in this respect, the presentation may serve as a reference for researchers specializing in the field.

In the period since the appearance of the first edition, a number of related books on combustion have been published. Many of these are referenced throughout the new text. They are quite helpful in providing alternative viewpoints toward much of the material. The progress in the field has resulted in significantly more extensive citations to references in the second edition, even though the bibliographical information certainly is not complete. Where I have forgotten to cite an important reference, I hope that the author will recognize the magnitude of my task and accordingly will forgive my oversight. In an effort to achieve some economy in printing, three oft-cited sources are abbreviated as follows: *C & F* for *Combustion and Flame*, *CST* for *Combustion Science and Technology*, and *n'th Symp.* for the currently biennial international symposia on combustion, the full citations of which are:

Proceedings of the First Symposium on Combustion and the Second Symposium on Combustion, The Combustion Institute, Pittsburg (1965), reprinted from *Industrial and Engineering Chemistry* (1928) and from *Chemical Reviews* (1937),

Third Symposium on Combustion and Flame and Explosion Phenomena, Williams and Wilkins Co., Baltimore (1949),

Fourth Symposium (International) on Combustion, Williams and Wilkins Co., Baltimore (1953),

Fifth Symposium (International) on Combustion, Rienhold Publishing Corp., New York (1955),

Sixth . . . , Reinhold Publishing Corp., New York (1957),

Seventh . . . , Butterworths Scientific Publications, London (1959),

Eighth . . . , Williams and Wilkins Co., Baltimore (1962),

Ninth . . . , Academic Press, New York (1963),

Tenth . . . , The Combustion Institute, Pittsburgh (1965),

Eleventh . . . (1967), *Twelfth . . .* (1969), *Thirteenth . . .* (1971), *Fourteenth . . .* (1973), *Fifteenth . . .* (1975), *Sixteenth . . .* (1977), *Seventeenth . . .* (1979), *Eighteenth . . .* (1981), *Nineteenth . . .* (1982).

I am indebted to many people for aid in preparation of the second edition. The colleagues and scientists who contributed significant discussions of technical material are too numerous to name, for if I were to attempt to acknowledge them all, I would surely slight someone by inadvertently omitting his or her name from the list. Nevertheless, I cannot refrain from stating here that Amable Liñán has done more than any other individual in improving my understanding of combustion theory. Even with respect to specific helpful recommendations for changes in the text, I do not dare to try to acknowledge individuals, although I would like to say that observations made by the team of Japanese combustion scientists, who translated the book from the typewritten manuscript as I completed the writing, proved to be notably useful in prompting clarifications in the English-language presentation. Nearly all of the original manuscript was typed by Peggy Berger, for whose patient help over a 5-year period I am greatly indebted. Finally, I wish to express my gratitude for the presence of mind encouraged by Elizabeth.

Forman A. Williams
Princeton, New Jersey
June 1984

Preface to the First Edition

This book contains a systematic development of the theory of chemical reactions in fluid flow. It is intended to serve either as an advanced graduate text or as a reference for workers in the field of combustion. An attempt has been made to cover the entire field, not just those parts of the field in which the author has done research. However, the book does emphasize the theoretical aspects of the subject and discusses experiments only where absolutely necessary. This type of approach appears to be justified in view of the existence of a number of excellent treatises on the experimental aspects of combustion, such as B. Lewis and G. von Elbe's classical work *Combustion, Flames and Explosions of Gases* (Academic Press, New York, 1st ed., 1951, 2nd ed. 1961), the book by A. G. Gaydon and H. G. Wolfhard on *Flames, Their Structure, Radiation and Temperature*, (Chapman and Hall, Ltd., London, 1960), and W. Jost, *Explosion and Combustion Processes in Gases* (McGraw-Hill, New York, 1946). A familiarity with the material in these books will provide the reader with a proper perspective for beginning a study of combustion theory. Relevant experimental work may also be found in the mammoth biennial volumes containing the proceedings of the *International Symposia on Combustion*, sponsored by the Combustion Institute, and in the journal *Combustion and Flame*.

The arrangement and the method of attack of the book require some explanation. In order to proceed with the development of the principal subject material as directly as possible, from the outset I assume that the reader has the scientific background necessary for understanding combustion theory. Thus, a knowledge of mathematics (principally, a thorough understanding of ordinary and partial differential equations), thermodynamics, statistical mechanics, chemical kinetics, and molecular transport theory is presupposed. To aid the reader who does not have a complete command of all of these subjects, and also as convenient places from which to draw specific background items used in the text, extended appendices are provided containing reviews on thermodynamics and statistical mechanics, on chemical kinetics, on the fundamental equations of fluid dynamics, and on transport properties. For courses in which students lack specific background topics, the appropriate appendices would be the best place to begin studying the book. However, some knowledge of elementary gas dynamics and thermodynamics and a thorough knowledge of applied mathematics are needed in any case, because no mathematical appendices are given, and topics such as heat engine cycle analysis, the meaning of Reynolds and Mach numbers, and the concept of turbulence are omitted.

Although pedagogical considerations did go into the arrangement of the material, the subject matter itself dictates more of a what-has-been-done book than a how-to-do book. I hope that by solving some problems and by pointing out related problems that have and have not been solved, I will be able to indicate fruitful lines of research and to suggest useful methods of solution to the reader.

Although no homework problems are provided explicitly, statements such as "it can be shown" and "the reader may prove" appear often in the text and can be treated as problems for the student.

My style of presentation is generally concisely deductive rather than discursively inductive. Perhaps, by stripping the presentation to the barest essentials and by removing adornments commonly incorporated for pedagogical reasons, I may clarify some possible points of confusion for a few readers.

The writing of this treatise was undertaken through the encouragement of Dr. S. S. Penner. Practically all of the book was completed while I was an Assistant Professor at Harvard University holding a National Science Foundation Grant for Fundamental Studies of Sprays. During this time I spent six months at Imperial College, University of London, on a National Science Foundation Postdoctoral Fellowship and devoted most of my time there to the writing.

Thanks are due to many of my colleagues with whom I have talked about the book. Deserving of special mention are Dr. William Nachbar and Professor Howard W. Emmons for informative conversations. Professor D. B. Spalding contributed many helpful suggestions and kindly read the first draft of the manuscript, recommending many important changes. But, most of all, I wish to thank my teacher, Dr. S. S. Penner, who wrote the original drafts of a few of the chapters and provided constant encouragement and help thereafter.

My family also deserves thanks for their patience.

F. A. Williams
La Jolla, California
June 1964

Contents

Summary of Relevant Aspects of Fluid Dynamics and Chemical Kinetics

Since combustion problems requiring theoretical analysis are primarily concerned with the flow of reacting and diffusing gases, the student of combustion theory must understand—in addition to chemical thermodynamics —the conservation equations of fluid dynamics, including transport properties and chemical kinetics. Condensed but reasonably complete treatments of these background topics are given in the appendixes. Readers unfamiliar with any of the background subjects are advised to read the appropriate appendixes before beginning Chapter 1.

The differential forms of the conservation equations derived in the appendixes for reacting mixtures of ideal gases are summarized in Section 1.1. From the macroscopic viewpoint (Appendix C), the governing equations (excluding the equation of state and the caloric equation of state) are not restricted to ideal gases. Most of the topics considered in this book involve the solutions of these equations for special flows. The forms that the equations assume for (steady-state and unsteady) one-dimensional flows in orthogonal, curvilinear coordinate systems are derived in Section 1.2, where specializations accurate for a number of combustion problems are developed. Simplified forms of the conservation equations applicable to steady-state problems in three dimensions are discussed in Section 1.3. The specialized equations given in this chapter describe the flow for most of the combustion processes that have been analyzed satisfactorily.

Boundary and interface conditions must be known if solutions to the conservation equations are to be obtained. Since these conditions depend strongly on the model of the particular system under study, it is difficult to give general rules for stating them; for example, they may require consideration of surface equilibria (discussed in Appendix A) or of surface rate processes (discussed in Appendix B). However, simple mass, momentum, and energy balances at an interface often are of importance. For this reason, interface conditions are derived through introduction of integral forms of the conservation equations in Section 1.4.

The organization of the remaining chapters of the book is discussed in Section 1.5.

1.1. THE CONSERVATION EQUATIONS FOR MULTICOMPONENT, REACTING, IDEAL-GAS MIXTURES

The conservation equations, to be used repeatedly, are derived in Appendixes C and D and may be summarized as follows. Overall continuity, equation (D-33), is

$$\partial \rho / \partial t + \nabla \cdot (\rho \mathbf{v}) = 0. \tag{1}$$

Momentum conservation, (D-35), is

$$\partial \mathbf{v}/\partial t + \mathbf{v} \cdot \nabla \mathbf{v} = -(\nabla \cdot \mathbf{P})/\rho + \sum_{i=1}^{N} Y_i \mathbf{f}_i. \tag{2}$$

Energy conservation, equation (D-37), is

$$\rho \, \partial u / \partial t + \rho \mathbf{v} \cdot \nabla u = -\nabla \cdot \mathbf{q} - \mathbf{P} : (\nabla \mathbf{v}) + \rho \sum_{i=1}^{N} Y_i \mathbf{f}_i \cdot \mathbf{V}_i. \tag{3}$$

Conservation of species, equation (D-40), is

$$\partial Y_i / \partial t + \mathbf{v} \cdot \nabla Y_i = w_i / \rho - [\nabla \cdot (\rho Y_i \mathbf{V}_i)] / \rho, \quad i = 1, \ldots, N. \tag{4}$$

In equation (2), p is obtained from equations (E-38)–(E-40), which may be combined to give

$$\mathbf{P} = [p + (\tfrac{2}{3}\mu - \kappa)(\nabla \cdot \mathbf{v})]\mathbf{U} - \mu[(\nabla \mathbf{v}) + (\nabla \mathbf{v})^T]. \tag{5}$$

In equation (3), $\mathbf{q}$ is given by equation (E-46), which may be written as

$$\mathbf{q} = -\lambda \nabla T + \rho \sum_{i=1}^{N} h_i Y_i \mathbf{V}_i + R^0 T \sum_{i=1}^{N} \sum_{j=1}^{N} \left(\frac{X_j D_{T,i}}{W_i D_{ij}} \right) (\mathbf{V}_i - \mathbf{V}_j) + \mathbf{q}_R. \tag{6}$$

In equations (3) and (4), $\mathbf{V}_i$ is determined by equation (E-18) which is

$$\nabla X_i = \sum_{j=1}^{N} \left(\frac{X_i X_j}{D_{ij}}\right)(\mathbf{V}_j - \mathbf{V}_i) + (Y_i - X_i)\left(\frac{\nabla p}{p}\right) + \frac{\rho}{p}\sum_{j=1}^{N} Y_i Y_j(\mathbf{f}_i - \mathbf{f}_j)$$

$$+ \sum_{j=1}^{N} \left[\left(\frac{X_i X_j}{\rho D_{ij}}\right)\left(\frac{D_{T,j}}{Y_j} - \frac{D_{T,i}}{Y_i}\right)\right]\left(\frac{\nabla T}{T}\right), \quad i = 1, \ldots, N. \tag{7}$$

The transport coefficients appearing in equations (5)–(7) are given in Appendix E. The external forces $\mathbf{f}_i$ are specified (not derived). The radiant flux, $\mathbf{q}_R$, is also viewed here as specified; it is found fundamentally through the integro-differential equation of radiation transport (see Appendix E). The reaction rates w_i in equation (4) are determined by the phenomonological expressions of chemical kinetics,

$$w_i = W_i \sum_{k=1}^{M} (v''_{i,k} - v'_{i,k})B_k T^{\alpha_k} e^{-(E_k/R^0 T)} \prod_{j=1}^{N} \left(\frac{X_j p}{R^0 T}\right)^{v'_{j,k}}, \quad i = 1, \ldots, N, \tag{8}$$

which are combinations of equations (B-6)–(B-8), (B-58), and (B-59).

The $N + 5$ dependent variables in equations (1)–(4) may be taken to be Y_i, ρ, T, and $\mathbf{v}$, in which case the other variables may be related to these through the ideal-gas equation of state,

$$p = \rho R^0 T \sum_{i=1}^{N} (Y_i/W_i), \tag{9}$$

the thermodynamic identity,

$$u = \sum_{i=1}^{N} h_i Y_i - p/\rho, \tag{10}$$

the caloric equation of state,

$$h_i = h_i^0 + \int_{T^0}^{T} c_{p,i} \, dT, \quad i = 1, \ldots, N, \tag{11}$$

and the identity

$$X_i = \frac{(Y_i/W_i)}{\sum_{j=1}^{N} (Y_j/W_j)}, \quad i = 1, \ldots, N. \tag{12}$$

In equations (1)–(7), vector notation is employed; $\mathbf{V}$ is the gradient operator, $\mathbf{U}$ is the unit tensor, two dots (:) imply that the tensors are to be contracted twice, and the superscript T denotes the transpose of the tensor. The symbols appearing here are summarized in Table 1.1.

TABLE 1.1

Symbols	Meaning
B_k	A constant in the frequency factor for the kth reaction*
$c_{p,i}$	Specific heat at constant pressure for species i*
D_{ij}	Binary diffusion coefficient for species i and j*
$D_{T,i}$	Thermal diffusion coefficient for species i*
E_k	Activation energy for the kth reaction*
$\mathbf{f}_i$	External force per unit mass on species i*
h_i	Specific enthalpy of species $i^\dagger$
h_i^0	Standard heat of formation per unit mass for species i at temperature T^0*
M	Total number of chemical reactions occurring*
N	Total number of chemical species present*
p	Hydrostatic pressure†
$\mathbf{P}$	Stress tensor†
$\mathbf{q}$	Heat-flux vector†
$\mathbf{q}_R$	Radiant heat-flux vector*
R^0	Universal gas constant*
T	Temperature†
T^0	A fixed, standard reference temperature*
u	Internal energy per unit mass for the gas mixture†
$\mathbf{V}_i$	Diffusion velocity of species $i^\dagger$
$\mathbf{v}$	Mass-average velocity of the gas mixture†
W_i	Molecular weight of species i*
w_i	Rate of production of species i by chemical reactions (mass per unit volume per unit time)†
X_i	Mole fraction of species $i^\dagger$
Y_i	Mass fraction of species $i^\dagger$
α_k	Exponent determining the temperature dependence of the frequency factor for the kth reaction*
κ	Bulk viscosity coefficient*
λ	Thermal conductivity*
μ	Coefficient of (shear) viscosity*
$v'_{i,k}$	Stoichiometric coefficient for species i appearing as a reactant in reaction k*
$v''_{i,k}$	Stoichiometric coefficient for species i appearing as a product in reaction k*
ρ	Density†

* Parameters that must be given in order to solve the conservation equations.
† Quantities determined by the equations given in this section.

1.2. ONE-DIMENSIONAL FLOW

1.2.1. Unsteady flow

The conservation equations in orthogonal, curvilinear coordinate-systems may easily be derived from equations (1)–(4). Since the complete form is complicated and will not be required in subsequent problems, the reader is referred to the literature for these equations.* Of special interest, however, are flows in which the radiant heat-flux vector, the (mass-average) velocity vector and the body-force vectors all are parallel to one of the three coordinates, and all properties are uniform along surfaces in the other two orthogonal directions. These conditions appear to define the most general type of one-dimensional flow.

The flow will be assumed to be in the x_1 direction, and the relation between the physical distances (s_1, s_2, s_3) and the coordinates (x_1, x_2, x_3) will be taken as

$$ds_1 = g_1 \, dx_1, \qquad ds_2 = g_2 \, dx_2, \qquad ds_3 = g_3 \, dx_3, \tag{13}$$

which essentially defines (g_1, g_2, g_3). For brevity we shall assume that $g_1 \, dx_1 = dx$ and use x in place of x_1 in the equations; in other words, the coordinate in the flow direction will be taken to be the actual physical distance in that direction ($g_1 = 1$). One-dimensional flow is then defined by the hypotheses that $\partial/\partial x_2 = 0$ and $\partial/\partial x_3 = 0$ for all flow properties† and that $v_2 = 0, v_3 = 0, f_{i2} = 0$, and $f_{i3} = 0$. Here the subscripts 2 and 3 denote the corresponding components of the vectors; the components of the vectors in the 1 (x) direction will be identified by italic type without a subscript.

Equation (7) now reduces to

$$\frac{\partial X_i}{\partial x} = \sum_{j=1}^{N} \left(\frac{X_i X_j}{D_{ij}}\right)(V_j - V_i) + (Y_i - X_i)\frac{1}{p}\frac{\partial p}{\partial x} + \frac{\rho}{p}\sum_{j=1}^{N} Y_i Y_j (f_i - f_j)$$
$$+ \sum_{j=1}^{N} \left[\left(\frac{X_i X_j}{\rho D_{ij}}\right)\left(\frac{D_{T,j}}{Y_j} - \frac{D_{T,i}}{Y_i}\right)\right]\frac{1}{T}\frac{\partial T}{\partial x}, \qquad i = 1, \dots, N, \tag{14}$$

and also shows that $V_{i2} = 0$ and $V_{i3} = 0$. Equation (6) then yields $q_2 = 0$, $q_3 = 0$, and

$$q = -\lambda\frac{\partial T}{\partial x} + \rho\sum_{i=1}^{N} h_i Y_i V_i + R^0 T \sum_{i=1}^{N}\sum_{j=1}^{N} \left(\frac{X_j D_{T,i}}{W_i D_{ij}}\right)(V_i - V_j) + q_R. \tag{15}$$

From equation (5) it is found that the off-diagonal elements of **P** vanish ($p_{12} = p_{23} = p_{31} = 0$) and that the diagonal elements of **P** are given by

$$p_{11} = p + (\tfrac{2}{3}\mu - \kappa)\frac{1}{g_2 g_3}\frac{\partial}{\partial x}(vg_2 g_3) - 2\mu\frac{\partial v}{\partial x}, \tag{16}$$

$$p_{22} = p_{11} + 2\mu\left(\frac{\partial v}{\partial x} - \frac{v}{g_2}\frac{\partial g_2}{\partial x}\right), \tag{17}$$

---

* See, for example, [1], where a form valid in arbitrary (nonorthogonal) coordinate systems also is given. (Note that the numbers in brackets refer to references at the end of each chapter.)
† Of course, g_2 and g_3 may depend upon x_2 and x_3.

and

$$p_{33} = p_{11} + 2\mu\left(\frac{\partial v}{\partial x} - \frac{v}{g_3}\frac{\partial g_3}{\partial x}\right), \tag{18}$$

where use has been made of expressions for the divergence and the dyadic gradient in orthogonal, curvilinear coordinate systems,

$$(\nabla \mathbf{v}) = \text{diagonal}\left(\frac{\partial v}{\partial x}, \frac{v}{g_2}\frac{\partial g_2}{\partial x}, \frac{v}{g_3}\frac{\partial g_3}{\partial x}\right)$$

in the present case.

The simplified form of equation (1) is

$$\frac{\partial \rho}{\partial t} + \frac{1}{g_2 g_3}\frac{\partial}{\partial x}(\rho v g_2 g_3) = 0. \tag{19}$$

By utilizing the expression for the divergence of a diagonal tensor in orthogonal, curvilinear coordinate systems, one can show that equation (2) reduces to

$$\frac{\partial v}{\partial t} + v\frac{\partial v}{\partial x} = -\frac{1}{\rho}\frac{\partial p_{11}}{\partial x} + \left(\frac{p_{22} - p_{11}}{\rho}\right)\frac{1}{g_2}\frac{\partial g_2}{\partial x}$$
$$+ \left(\frac{p_{33} - p_{11}}{\rho}\right)\frac{1}{g_3}\frac{\partial g_3}{\partial x} + \sum_{i=1}^{N} Y_i f_i. \tag{20}$$

Equation (3) becomes

$$\frac{\partial u}{\partial t} + v\frac{\partial u}{\partial x} = -\frac{1}{\rho}\frac{1}{g_2 g_3}\frac{\partial}{\partial x}(qg_2 g_3) - \frac{p_{11}}{\rho}\frac{\partial v}{\partial x}$$
$$- \frac{v}{\rho}\left(\frac{p_{22}}{g_2}\frac{\partial g_2}{\partial x} + \frac{p_{33}}{g_3}\frac{\partial g_3}{\partial x}\right) + \sum_{i=1}^{N} Y_i f_i V_i, \tag{21}$$

and equation (4) may be written as

$$\frac{\partial Y_i}{\partial t} + v\frac{\partial Y_i}{\partial x} = \frac{w_i}{\rho} - \frac{1}{\rho}\frac{1}{g_2 g_3}\frac{\partial}{\partial x}(\rho Y_i V_i g_2 g_3), \quad i = 1, \ldots, N. \tag{22}$$

Equations (19)–(22) may also be obtained by specializing the conservation equations given in [1]* to the case in which $\partial/\partial x_2 = 0$, $\partial/\partial x_3 = 0$, the 2 and 3 components of all vectors are zero, the pressure tensor is diagonal, and $g_1 = 1$. In Cartesian coordinate systems $g_2 = g_3 = 1$.

In combustion problems it is usually more convenient to work with the enthalpy $h = u + p/\rho$ instead of the internal energy u. A modified form of the energy equation may be derived from equation (21) by making the

---

* There is an error in equation (14) of [1]; (p_{jj}/h_j) should be replaced by $(p_{ii}u_j/h_iu_i)$.

substitution $u = h - p/\rho$ in equation (21), multiplying equation (20) by v, substituting the result into equation (21), and utilizing equation (19) to eliminate $\partial(1/\rho)/\partial t$. After recombining terms in the appropriate manner, the result is found to be

$$\frac{\partial}{\partial t}\left(h + \frac{v^2}{2}\right) + v\frac{\partial}{\partial x}\left(h + \frac{v^2}{2}\right) = -\frac{1}{\rho}\frac{1}{g_2 g_3}\frac{\partial}{\partial x}(q g_2 g_3) + \frac{1}{\rho}\frac{\partial p}{\partial t}$$

$$+ \frac{1}{\rho}\frac{1}{g_2 g_3}\frac{\partial}{\partial x}[(p - p_{11})v g_2 g_3]$$

$$+ \sum_{i=1}^{N} Y_i f_i(v + V_i). \tag{21a}$$

Actually, a relationship quite similar to equation (21a) may be derived from equations (1)–(3) for general flows.

1.2.2. Steady flow

In certain steady-flow systems, considerable simplification in equations (19)–(22) may be effected. Here $\partial/\partial t = 0$, the leading term on the left-hand side in equations (19)–(22) disappears, and $\partial/\partial x \to d/dx$ because only one independent variable remains. Equation (19) may be integrated immediately, yielding the simplified continuity equation

$$\rho v g_2 g_3 = \text{constant}. \tag{23}$$

In many steady-state combustion problems, body forces are negligible and a consequent simplification in the energy equation is possible. When $\partial/\partial t = 0$ and $f_i = 0$, multiplying equation (21a) by $\rho g_2 g_3$ and utilizing equation (23) shows that equation (21a) may be integrated once, giving

$$g_2 g_3\left[\rho v\left(h + \frac{v^2}{2}\right) + q - (p - p_{11})v\right] = \text{constant}. \tag{24}$$

Equation (24) states that the sum* of the enthalpy flux and the ordered kinetic-energy flux may change only because of heat flux ($g_2 g_3 q$) or viscous dissipation [$g_2 g_3(p - p_{11})v$]. Equation (24) also illustrates that the enthalpy is of more importance than the internal energy in steady-flow problems; the additional term p/ρ, which appears when equation (24) is written in terms of u, may be explained as "displacement work."

An integrated form of the momentum equation may be derived only under additional restrictive conditions. In steady-state flow with $f_i = 0$, if viscous forces are negligible, then equation (20) reduces to $\rho v\,dv/dx + dp/dx = 0$. If, in addition, $v^2/(p/\rho) \ll 1$, then $|d\ln p/dx| \ll |d\ln v/dx|$ (that is,

---

* This sum often is called the *stagnation* enthalpy flux and includes thermal enthalpy, chemical enthalpy, and kinetic energy.

fractional changes in the pressure are negligibly small) and the solution of the momentum equation is

$$p = \text{constant.} \tag{25}$$

The condition $v^2/(p/\rho) \ll 1$ is valid for low-speed flow; specifically, the square of the Mach number $[v^2/(\partial p/\partial \rho)_s, s \equiv \text{entropy}]$—a measure of the ratio of the ordered kinetic energy to the random, thermal kinetic energy of the molecules—must be small. When equation (25) is an acceptable approximation, equations (22) (with $\partial Y_i/\partial t = 0$) through (24) constitute an appropriate set of conservation equations. In these cases it is not necessary to use equation (20) (except to compute the actual small pressure change, after the rest of the problem has been solved with $p = \text{constant}$).

In (steady-state) plane, one-dimensional flow ($g_2 = g_3 = 1$), the additional restrictions imposed in the preceding paragraph are not needed in order to integrate the momentum equation. When $\partial v/\partial t = 0, f_i = 0$, and $g_2 = g_3 = 1$, equation (20) reduces to $\rho v\, dv/dx + dp_{11}/dx = 0$, the integral of which is

$$\rho v^2 + p_{11} = \text{constant,} \tag{26}$$

since equation (23) implies that

$$\rho v = \text{constant} \tag{27}$$

when $g_2 = g_3 = 1$. In equation (26), $p_{11} = p - (\tfrac{4}{3}\mu + \kappa)\, dv/dx$ according to equation (16). Equation (26) therefore expresses a balance between dynamic (inertial) forces, pressure forces, and viscous forces. Utilizing $g_2 = g_3 = 1$, $\partial/\partial t = 0$ and the above results in equations (22) and (24), we find that

$$\rho v\left(h + \frac{v^2}{2}\right) + q - \left(\frac{4}{3}\mu + \kappa\right)v\frac{dv}{dx} = \text{constant} \tag{28}$$

and

$$\frac{d}{dx}\left[\rho Y_i(v + V_i)\right] = w_i, \quad i = 1, \ldots, N. \tag{29}$$

Equations (26)–(29) are the conservation equations for steady, plane, one-dimensional flow with no body forces.

In steady one-dimensional flows it is often useful to introduce the **mass-flux fraction** of chemical species i,

$$\epsilon_i = \rho Y_i(v + V_i)/\rho v, \quad i = 1, \ldots, N. \tag{30}$$

Although ϵ_i need not lie between 0 and 1 (it may be negative or greater than unity), the definitions of Y_i and V_i imply that $\sum_{i=1}^{N} \epsilon_i = 1$. Multiplication of equation (22) by $\rho g_2 g_3$ and utilization of equation (23) shows that when $\partial/\partial t = 0$, the species conservation equation is

$$d\epsilon_i/dx = w_i/(\rho v g_2 g_3), \quad i = 1, \ldots, N. \tag{31}$$

In terms of ϵ_i, when $f_i = 0$ and thermal diffusion is neglected (for brevity), equation (14) becomes

$$\frac{dX_i}{dx} = \sum_{j=1}^{N} \left(\frac{X_i X_j}{D_{ij}}\right)\left(\frac{\epsilon_j}{Y_j} - \frac{\epsilon_i}{Y_i}\right)v + (Y_i - X_i)\frac{1}{p}\frac{dp}{dx}, \quad i = 1, \ldots, N. \quad (32)$$

Using these same assumptions and the (ideal-gas) condition $h = \sum_{i=1}^{N} h_i Y_i$, one can show by substituting equation (15) into equation (24) that if radiant transfer is neglected, then

$$g_2 g_3 \left[\rho v \left(\sum_{i=1}^{N} h_i \epsilon_i + \frac{v^2}{2}\right) - \lambda\frac{dT}{dx} - (p - p_{11})v\right] = \text{constant}. \quad (33)$$

In this form of equation (24), the effect of diffusion on the enthalpy flux is included implicitly in the first term, and the only term from the heat-flux vector that remains explicitly is that of heat conduction. The above equations serve to eliminate completely the diffusion velocities from the governing equations, replacing them by the flux fractions. The fact that equations (31) and (33) appear to be less complicated than equations (22) (with $\partial Y_i/\partial t = 0$) and (24) often justifies this transformation.

1.3. COUPLING FUNCTIONS

A number of effects contained in equations (1)-(7) are unimportant in a large class of combustion problems and will be neglected in nearly all of the applications to be considered subsequently and also in the equations to be derived in this section. These include:

1. Body forces* ($\mathbf{f}_i$ terms).
2. The Soret and Dufour effects (terms involving $D_{T,i}$).
3. Pressure-gradient diffusion [$(Y_i - X_i)\,\nabla p/p$].
4. Bulk viscosity (κ).
5. Radiant heat flux ($\mathbf{q}_R$).[†]

Reasons for neglecting these phenomena are discussed in Appendix E.

---

* In some low-speed combustion problems natural convection is of importance, and consequently $\mathbf{f}_i$ is not negligible. Since the acceleration due to gravity is the same for all species, $\mathbf{f}_i$ is the same for all i in these problems. Body forces disappear from equations (3) and (7) and remain only in the momentum equation.

† There are a number of combustion problems in which the neglect of $\mathbf{q}_R$ is invalid. Other radiative influences, radiation pressure and radiation-energy density, always appear to be negligible in combustion. Although the general form of the equation of radiation transport, from which $\mathbf{q}_R$ is obtained, is quite complicated, some combustion problems remain tractable when radiation is included because the effects of scattering and absorption often are negligible except at interfaces.

Introduction of additional restrictive assumptions is useful in analysis of many problems. The objective of the following manipulations is to express the equations for conservation of energy and of chemical species in a common form. This enables the chemical-source term to be removed from many equations by considering suitable linear combinations of dependent variables. These linear combinations often are descriptively called **coupling functions**. There are many approaches to the derivation of equations for coupling functions. One involves equations given in early papers of Shvab [2] and Zel'dovich [3], and the resulting formulation often is called the **Shvab-Zel'dovich formulation**. The assumptions needed limit the range of applicability of the formulation to a considerably greater extent than do the approximations given in the preceding paragraph. Nevertheless, the resulting equations remain valid, in a reasonable approximation, for many of the problems considered in later chapters. In addition to providing significant simplifications in analysis, the use of coupling functions often enhances understanding of the physical processes that occur.

For *steady flow* overall continuity, equation (1), reduces to

$$\mathbf{V} \cdot (\rho \mathbf{v}) = 0. \tag{34}$$

For *low-speed*, steady-flow problems, viscous effects are often negligible and, as pointed out in Section 1.2, the momentum equation implies that, approximately, p is a constant. The substitution of equation (6) into equation (3) yields, in this approximation,

$$\rho \mathbf{v} \cdot \mathbf{V}u = \mathbf{V} \cdot (\lambda \, \mathbf{V}T) - \mathbf{V} \cdot \left(\rho \sum_{i=1}^{N} h_i Y_i \mathbf{V}_i \right) - p(\mathbf{V} \cdot \mathbf{v}).$$

Utilizing equations (10) and (34) in this expression, we obtain

$$\mathbf{V} \cdot \left[\rho \sum_{i=1}^{N} Y_i h_i(\mathbf{v} + \mathbf{V}_i) - \lambda \, \mathbf{V}T \right] = 0 \tag{35}$$

for the energy equation. In view of equation (34), species conservation, equation (4), may be rewritten as

$$\mathbf{V} \cdot [\rho Y_i(\mathbf{v} + \mathbf{V}_i)] = w_i, \quad i = 1, \ldots, N. \tag{36}$$

When the *binary diffusion coefficients of all pairs of species are equal*, equation (7) reduces to

$$D \, \mathbf{V}X_i = X_i \sum_{j=1}^{N} X_j \mathbf{V}_j - X_i \mathbf{V}_i, \quad i = 1, \ldots, N$$

in the present approximation, where $D \equiv D_{ij}$. Since $\sum_{i=1}^{N} Y_i \mathbf{V}_i = 0$, $\sum_{i=1}^{N} Y_i = 1$, and $\sum_{i=1}^{N} X_i = 1$ by definition, multiplying this equation by Y_i/X_i and summing over i yields a relation for $\sum_{j=1}^{N} X_j \mathbf{V}_j$ that may be substituted back into the equation, showing that

$$D\left[\mathbf{V}\ln X_i - \sum_{j=1}^{N} Y_j \mathbf{V}\ln X_j\right] = -\mathbf{V}_i, \quad i = 1, \ldots, N.$$

Substituting equation (12) for X_i into all terms on the left-hand side of this expression, we obtain Fick's law,

$$\mathbf{V}_i = -D\,\mathbf{V}\ln Y_i, \quad i = 1, \ldots, N, \tag{37}$$

a simple and well-known diffusion equation that often is merely postulated, not derived from fundamentals through explicit hypotheses.

By substituting equations (11) and (36) into equation (35), we find

$$\mathbf{V}\cdot\left[\rho\mathbf{v}\left(\int_{T^0}^{T} c_p\,dT\right) + \rho\sum_{i=1}^{N} Y_i\mathbf{V}_i\left(\int_{T^0}^{T} c_{p,i}\,dT\right) - \lambda\,\mathbf{V}T\right] = -\sum_{i=1}^{N} h_i^0 w_i,$$

where

$$c_p \equiv \sum_{i=1}^{N} Y_i c_{p,i} \tag{38}$$

is the specific heat of the mixture.* Assuming that the *Lewis number (Le) is unity,*

$$\text{Le} \equiv \lambda/\rho D c_p = 1, \tag{39}$$

we may substitute equation (37) into the above form of the energy equation to show that

$$\mathbf{V}\cdot\left[\rho\mathbf{v}\left(\int_{T^0}^{T} c_p\,dT\right) - \frac{\lambda}{c_p}\,\mathbf{V}\left(\int_{T^0}^{T} c_p\,dT\right)\right] = -\sum_{i=1}^{N} h_i^0 w_i, \tag{40}$$

since $\mathbf{V}\int_{T_0}^{T} c_p\,dT = c_p\,\mathbf{V}T + \sum_{i=1}^{N}[(\mathbf{V}Y_i)\int_{T^0}^{T} c_{p,i}\,dT]$ according to equation (38). Equations (36) and (37) imply that

$$\mathbf{V}\cdot(\rho\mathbf{v}Y_i - \rho D\,\mathbf{V}Y_i) = w_i, \quad i = 1, \ldots, N. \tag{41}$$

In view of equation (39), the similarity in the forms of equation (40) (for the thermal enthalpy $\int_{T_0}^{T} c_p\,dT$) and equation (41) (for the mass fractions Y_i) is striking. Equations (40) and (41) are the energy- and species-conservation equations of Shvab and Zel'dovich. The derivation given for these equations required neither that any transport coefficient or the specific heat of the mixture is constant nor that the specific heats of all species are equal. Coupling functions may now be identified from equations (40) and (41).

---

* Here, precisely $\int_{T^0}^{T} c_p\,dT \equiv \sum_{i=1}^{N} Y_i \int_{T^0}^{T} c_{p,i}\,dT$; that is, the integration is performed at constant composition. This integral represents the thermal enthalpy, obtained from the total enthalpy, h, by subtraction of the chemical enthalpy, $\sum_{i=1}^{N} Y_i h_i^0$.

To emphasize more explicitly the similarity of equations (40) and (41), assume further that *chemical changes occur by a single reaction step*,

$$\sum_{i=1}^{N} v_i' \mathfrak{M}_i \rightarrow \sum_{i=1}^{N} v_i'' \mathfrak{M}_i, \tag{42}$$

where $\mathfrak{M}_i$ is the symbol for chemical species i. In this case, the quantity

$$w_i/W_i(v_i'' - v_i') \equiv \omega, \quad i = 1, \ldots, N, \tag{43}$$

is the same for all species (see Appendixes A and B), and equations (40) and (41) may both be expressed as

$$L(\alpha) = \omega, \tag{44}$$

where the linear operator L is defined by

$$L(\alpha) \equiv \mathbf{V} \cdot [\rho \mathbf{v} \alpha - \rho D \, \mathbf{V} \alpha], \tag{45}$$

and

$$\alpha = \frac{\displaystyle\int_{T^0}^{T} c_p \, dT}{\displaystyle\sum_{i=1}^{N} h_i^0 W_i(v_i' - v_i'')} \equiv \alpha_T \tag{46}$$

for equation (40), while

$$\alpha = Y_i/W_i(v_i'' - v_i') \equiv \alpha_i, \quad i = 1, \ldots, N, \tag{47}$$

for equation (41). The nonlinear rate term may be eliminated from all except one of the relations corresponding to equation (44) by subtraction; for example, if

$$L(\alpha_1) = \omega \tag{48}$$

can be solved for α_1, then the other flow variables are determined by the *linear* equations for the coupling functions β

$$L(\beta) = 0, \tag{49}$$

with $\beta = \alpha_T - \alpha_1 \equiv \beta_T$ and $\beta = \alpha_i - \alpha_1 \equiv \beta_i \, (i \neq 1)$.

In the operator L, the first term represents convection and the second diffusion. Equation (44) therefore describes a balance of convective, diffusive, and reactive effects. Such balances are very common in combustion and often are employed as points of departure in theories that do not begin with derivations of conservation equations. If the steady-flow approximation is relaxed, then an additional term, $\partial(\rho\alpha)/\partial t$, appears in L; this term represents accumulation of thermal energy or chemical species. For species conservation, equations (48) and (49) may be derived with this generalized definition of L, in the absence of the assumptions of low-speed flow and of a Lewis number of unity. Only the restrictions 1, 2, and 3, the equality of diffusion coefficients, and the assumption of a one-step reaction are needed; therefore, the formulation is of reasonably wide applicability for species conservation. To obtain equation (44) for energy conservation in time-dependent problems necessitates, in addition, the retention of assumption 5, equation (39) and the approximation of low-speed flow, as well as replacement of the steady-flow approximation by an additional assumption, to the effect that $\partial p/\partial t$ is negligible in an expression like equation (21a); thus the formulation is appreciably less general for energy conservation than for species conservation. Nevertheless, a result like equation (49) often can be derived for the total enthalpy h through reasonable approximations (see Section 3.4.2).

Although the one-reaction approximation is sufficiently accurate for many purposes in combustion, it is often important in detailed investigations to replace equation (42) by a kinetic scheme of greater complexity. In such cases, coupling functions obeying equation (49) still may be derived through use of atom conservation, equation (A-36). However, more than one rate equation of the type given in equation (48) remain to be solved. For systems in which reactants initially are unmixed, burning rates often may be determined by solving equation (49), without attacking equation (48) [4]. It should, however, be pointed out that equation (49) is deceptively simple in appearance; unless additional approximations are made, $\rho \mathbf{v}$ [which is determined by equation (34), an equation of state with $p = $ constant and in some cases an approximate form of the momentum equation] and ρD may depend on β_i or β_T, so that L depends implicitly on β [see equation (45)], and equation (49) actually becomes nonlinear.

1.4. CONSERVATION CONDITIONS AT AN INTERFACE

Relations expressing the conservation of mass, momentum, and energy at an interface may be derived by writing equations (1)–(4) in integral form and then passing to the limit in which the volume of integration approaches a surface.

The time rate of change of the integral of an arbitrary function $f(x, t)$ over a given volume $\mathscr{V}$ is

$$\frac{d}{dt}\left(\int_{\mathscr{V}} f \, d\mathscr{V}\right) = \int_{\mathscr{V}} \frac{\partial f}{\partial t} \, d\mathscr{V} + \int_{\mathscr{A}} f\left(\frac{d\mathbf{x}}{dt} \cdot \mathbf{n}\right) d\mathscr{A}, \qquad (50)$$

where $\mathscr{A}$ is the surface bounding $\mathscr{V}$, $\mathbf{n}$ is the outward-pointing normal to this surface, and the term involving $d\mathbf{x}/dt$ accounts for the changes associated with motion of the control volume. Integration of equation (1) over $\mathscr{V}$ and utilization of the divergence theorem and of equation (50) for $f = \rho$ yields

$$\frac{d}{dt}\left(\int_{\mathscr{V}} \rho \, d\mathscr{V}\right) + \int_{\mathscr{A}} \rho\left(\mathbf{v} - \frac{d\mathbf{x}}{dt}\right) \cdot \mathbf{n} \, d\mathscr{A} = 0. \qquad (51)$$

By multiplying equation (1) by $\mathbf{v}$ and adding the result to equation (2), one obtains equation (D-34), which may be integrated over $\mathscr{V}$ to show that

$$\frac{d}{dt}\left(\int_{\mathscr{V}} \rho \mathbf{v}\, d\mathscr{V}\right) + \int_{\mathscr{A}} \rho \mathbf{v}\left[\left(\mathbf{v} - \frac{d\mathbf{x}}{dt}\right)\cdot \mathbf{n}\right] d\mathscr{A}$$

$$= -\int_{\mathscr{A}} \mathbf{P}\cdot \mathbf{n}\, d\mathscr{A} + \int_{\mathscr{V}} \rho \sum_{i=1}^{N} Y_i \mathbf{f}_i\, d\mathscr{V} \tag{52}$$

by the procedures described above. If equation (2) is dotted into $\mathbf{v}$ and substituted into equation (3), then equation (1) may be used to obtain an equation which is essentially identical with equation (D-36). Integrating this result over $\mathscr{V}$ we find that

$$\frac{d}{dt}\left[\int_{\mathscr{V}} \rho\left(u + \frac{v^2}{2}\right) d\mathscr{V}\right] + \int_{\mathscr{A}} \rho\left(u + \frac{v^2}{2}\right)\left[\left(\mathbf{v} - \frac{d\mathbf{x}}{dt}\right)\cdot \mathbf{n}\right] d\mathscr{A}$$

$$= -\int_{\mathscr{A}} \mathbf{q}\cdot \mathbf{n}\, d\mathscr{A} - \int_{\mathscr{A}} \mathbf{v}\cdot \mathbf{P}\cdot \mathbf{n}\, d\mathscr{A} + \int_{\mathscr{V}} \rho \sum_{i=1}^{N} Y_i \mathbf{f}_i\cdot (\mathbf{v} + \mathbf{V}_i)\, d\mathscr{V}. \tag{53}$$

The substitution of equation (1) into equation (4) yields equation (D-39), which—upon integration over $\mathscr{V}$—reduces to

$$\frac{d}{dt}\left(\int_{\mathscr{V}} \rho Y_i\, d\mathscr{V}\right) + \int_{\mathscr{A}} \rho Y_i\left(\mathbf{v} + \mathbf{V}_i - \frac{d\mathbf{x}}{dt}\right)\cdot \mathbf{n}\, d\mathscr{A} = \int_{\mathscr{V}} w_i\, d\mathscr{V}, \quad i = 1, \ldots, N. \tag{54}$$

Equations (51)–(54) are the integral forms of the conservation equations.

We now let the control volume $\mathscr{V}$ be a thin slab, the thickness of which is allowed to approach zero (see Figure 1.1). In the limit, the integral over $\mathscr{A}$ is composed of two parts, namely, integrals over each face of the slab. Let us arbitrarily choose one face as the "positive" side of the slab and identify quantities on this side by the subscript $+$ and those on the other

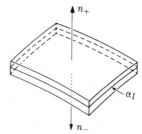

FIGURE 1.1.   Control volume for the derivation of interface conditions.

side by the subscript $-$. It is clear that $\mathbf{n}_- \to -\mathbf{n}_+$ at corresponding points on the two faces, and in the limit the integral over the surface of $\mathscr{V}$ may be replaced by an integral over the interface area $\mathscr{A}_I$, yielding

$$\int_{\mathscr{A}_I} \left[\rho_+ \left(\mathbf{v}_+ - \frac{d\mathbf{x}}{dt} \right) - \rho_- \left(\mathbf{v}_- - \frac{d\mathbf{x}}{dt} \right) \right] \cdot \mathbf{n}_+ \, d\mathscr{A} = - \lim_{\mathscr{V} \to 0} \left[\frac{d}{dt} \int_{\mathscr{V}} \rho \, d\mathscr{V} \right], \quad (55)$$

$$\int_{\mathscr{A}_I} \left\{ \rho_+ \mathbf{v}_+ \left[\left(\mathbf{v}_+ - \frac{d\mathbf{x}}{dt} \right) \cdot \mathbf{n}_+ \right] - \rho_- \mathbf{v}_- \left[\left(\mathbf{v}_- - \frac{d\mathbf{x}}{dt} \right) \cdot \mathbf{n}_+ \right] \right.$$

$$\left. + (\mathbf{P}_+ - \mathbf{P}_-) \cdot \mathbf{n}_+ \right\} d\mathscr{A}$$

$$= \lim_{\mathscr{V} \to 0} \left[\int_{\mathscr{V}} \rho \sum_{i=1}^{N} Y_i \mathbf{f}_i \, d\mathscr{V} - \frac{d}{dt} \int_{\mathscr{V}} \rho \mathbf{v} \, d\mathscr{V} \right], \quad (56)$$

$$\int_{\mathscr{A}_I} \left[\rho_+ \left(u_+ + \frac{v_+^2}{2} \right) \left(\mathbf{v}_+ - \frac{d\mathbf{x}}{dt} \right) - \rho_- \left(u_- + \frac{v_-^2}{2} \right) \left(\mathbf{v}_- - \frac{d\mathbf{x}}{dt} \right) \right.$$

$$\left. + \mathbf{q}_+ - \mathbf{q}_- + \mathbf{v}_+ \cdot \mathbf{P}_+ - \mathbf{v}_- \cdot \mathbf{P}_- \right] \cdot \mathbf{n}_+ \, d\mathscr{A}$$

$$= \lim_{\mathscr{V} \to 0} \left[\int_{\mathscr{V}} \rho \sum_{i=1}^{N} Y_i \mathbf{f}_i \cdot (\mathbf{v} + \mathbf{V}_i) \, d\mathscr{V} - \frac{d}{dt} \int_{\mathscr{V}} \rho \left(u + \frac{v^2}{2} \right) d\mathscr{V} \right], \quad (57)$$

and

$$\int_{\mathscr{A}_I} \left[\rho_+ Y_{i+} \left(\mathbf{v}_+ + \mathbf{V}_{i+} - \frac{d\mathbf{x}}{dt} \right) - \rho_- Y_{i-} \left(\mathbf{v}_- + \mathbf{V}_{i-} - \frac{d\mathbf{x}}{dt} \right) \right] \cdot \mathbf{n}_+ \, d\mathscr{A}$$

$$= \lim_{\mathscr{V} \to 0} \left(\int_{\mathscr{V}} w_i \, d\mathscr{V} - \frac{d}{dt} \int_{\mathscr{V}} \rho Y_i \, d\mathscr{V} \right), \quad i = 1, \ldots, N, \quad (58)$$

from equations (51)–(54), respectively. In equations (55)–(58), $\lim_{\mathscr{V} \to 0}$ is to be interpreted as the limit in which the thickness of the slab becomes infinitesimal (that is, the volume integral essentially approaches a surface integral).

The terms involving $\lim_{\mathscr{V} \to 0}$ of volume integrals remaining in equations (55)–(58), account for the possibility of surface forces, surface reactions, or a nonzero time rate of accumulation of mass, momentum, energy, or a chemical species at the interface. For example, if the interface is taken to be a solid surface at which chemical reactions are proceeding at a finite rate, then $w_i \to w_i' \, \delta(y - y_I)$, where w_i' is the surface reaction rate (mass of species i produced per unit area per unit time), y is the coordinate normal to the surface, $y = y_I$ at the surface, and δ is the Dirac δ-function. By substitution, we then find that in equation (58) $\lim_{\mathscr{V} \to 0} \int_{\mathscr{V}} w_i \, d\mathscr{V} = \int_{\mathscr{A}_I} w_i' \, d\mathscr{A}$, which is the mass rate of production of species i on the surface. If ρ_i' is the mass of species i per unit surface area, then $\rho Y_i \to \rho_i' \, \delta(y - y_I)$, and the remaining term on the right-hand side of equation (58) is

$$- \lim_{\mathscr{V} \to 0} \frac{d}{dt} \int_{\mathscr{V}} \rho Y_i \, d\mathscr{V} = - \frac{d}{dt} \lim_{\mathscr{V} \to 0} \int_{\mathscr{V}} \rho_i' \, \delta(y - y_I) \, d\mathscr{V} = - \frac{d}{dt} \int_{\mathscr{A}_I} \rho_i' \, d\mathscr{A},$$

which is the negative of the rate of accumulation of mass of species i on the surface. Equation (58) therefore expresses the following physically obvious conservation condition. The difference between the mass flux of species i leaving the surface and that entering the surface equals the mass rate of production of species i at the surface minus the mass rate of accumulation of species i on the surface. It is apparent that corresponding interpretations may also be given for equations (55)–(57).

In many problems, all terms in equations (55)–(58) involving $\lim_{\mathscr{V} \to 0}$ are zero; for example, there is usually no excess mass at the surface so that $\lim_{\mathscr{V} \to 0} \int \rho \, d\mathscr{V} = 0$ in equation (55). If it is also assumed that the interface is not moving ($dx/dt = 0$), that viscosity is negligible, and that approximations 1–4 of Section 1.3 are valid, then equations (55)–(58) simplify considerably. Using equations (5) and (6) and the identity $h = u + (p/\rho)$, we find by passing to the limit of small $\mathscr{A}_I$ that the differential forms of equations (55)–(58) reduce to

$$\rho_+ \mathbf{v}_+ \cdot \mathbf{n}_+ = \rho_- \mathbf{v}_- \cdot \mathbf{n}_+, \tag{59}$$

$$\rho_+ \mathbf{v}_+ (\mathbf{v}_+ \cdot \mathbf{n}_+) + p_+ \mathbf{n}_+ = \rho_- \mathbf{v}_- (\mathbf{v}_- \cdot \mathbf{n}_+) + p_- \mathbf{n}_+, \tag{60}$$

$$\left\{ \rho_+ \left[\left(h_+ + \frac{v_+^2}{2} \right) \mathbf{v}_+ + \sum_{i=1}^{N} h_{i+} Y_{i+} \mathbf{V}_{i+} \right] - \lambda_+ (\nabla T)_+ + \mathbf{q}_{R+} \right\} \cdot \mathbf{n}_+$$

$$= \left\{ \rho_- \left[\left(h_- + \frac{v_-^2}{2} \right) \mathbf{v}_- + \sum_{i=1}^{N} h_{i-} Y_{i-} \mathbf{V}_{i-} \right] - \lambda_- (\nabla T)_- + \mathbf{q}_{R-} \right\} \cdot \mathbf{n}_+, \tag{61}$$

and

$$\rho_+ Y_{i+} (\mathbf{v}_+ + \mathbf{V}_{i+}) \cdot \mathbf{n}_+ = \rho_- Y_{i-} (\mathbf{v}_- + \mathbf{V}_{i-}) \cdot \mathbf{n}_+, \tag{62}$$

respectively. In equations (61) and (62), the $\mathbf{V}_i$ values are determined in terms of concentration gradients at the interface either by the form of equation (7) in which the last three terms are omitted or, with additional approximations (see Section 1.3), by Fick's law. In equation (61), $\mathbf{q}_R$ has been retained because radiative heat losses from surfaces often are substantial. Although equation (60) will not be used explicitly in the subsequent applications, equations (59), (61), and (62) often will be needed as boundary conditions for equations (1)–(4). Detailed discussions of the importance of various terms in equations (59)–(62) for particular systems may be found in the literature.*

1.5. DISCUSSION OF THE APPROACH ADOPTED IN THE FOLLOWING DEVELOPMENT OF COMBUSTION THEORY

In the following chapters we shall begin with problems that are, fundamentally, the least complex and proceed to successively more-complicated problems. None of the differential conservation equations discussed in this chapter are needed in Chapter 2, where relations between the properties on the upstream and downstream sides of combustion waves are developed from overall (algebraic) conservation equations. Chapter 3 is concerned with problems in which transport effects are of importance but the chemical reaction-rate terms appearing in equation (4) and given by equation (8) need not be considered in obtaining first approximations for most of the information of interest; the formulation developed in Section 1.3 is useful in these problems. Chapter 4 deals with problems in which finite chemical reaction-rates must be considered, but the transport phenomena [the most highly differentiated terms appearing in equations (2)–(4) after use is made of equations (5)–(7)] all are negligible. Processes in which both transport phenomena and finite chemical reaction-rates must be considered simultaneously even in the most elementary descriptions are encountered in Chapter 5 on laminar flame theory and also appear in the discussions of detonation phenomena given in Chapter 6.

The arrangement of the remaining chapters is less strongly dependent on pedagogic considerations. The study of combustion of solid propellants, presented in Chapter 7, addresses a problem of engineering importance and also provides a vehicle for systematic amplification of basic concepts. The analysis in Chapter 8 of classical aspects of flammability emphasizes the influences of heat losses and of finite-rate chemistry. The next three chapters address from a basic viewpoint three classes of combustion problems that are of high practical interest, namely, combustion instability, turbulent combustion, and spray combustion. The bases of theoretical descriptions of unsteady burning in a wide variety of combustion problems are given in Chapter 9. The current status of our theoretical understanding of turbulent flames is reviewed in Chapter 10. Results of Chapter 3 are employed in Chapter 11 to apply the spray-combustion theory, developed therein, to problems of steady-state, one-dimensional burning in liquid-propellant combustors and of deflagrations in sprays. Chapter 12 concerns basic aspects

---

* See, for example, [5] for an analysis of equation (61), [6] and [7] for applications and generalizations of equations (61) and (62), and [8] for an amplification relevant to equation (62).

of problems of flame attachment in combustion devices and of flame-spread processes such as those occurring in fires. Since a thorough presentation of every aspect of combustion theory cannot be included in a volume of this size, an attempt has been made to cite references that can be pursued in advancing to any frontier of the subject.

REFERENCES

1. F. A. Williams, *J. Aero. Sci* **25**, 343 (1958).
2. V. A. Shvab, *Gos. Energ. izd.*, Moscow-Leningrad (1948).
3. Y. B. Zel'dovich, *Zhur. Tekhn. Fiz.* **19**, 1199 (1949), English translation, NACA Tech. Memo. No. 1296 (1950).
4. S. S. Penner and F. A. Williams, *Tech. Rept. No. 19, Contract No. DA-04-495-Ord-446*, California Institute of Technology, Pasadena, (1957).
5. S. M. Scala and G. W. Sutton, *ARS Journal* **29**, 141 (1959).
6. D. B. Spalding, *4th Symp.* (1953), 847–864.
7. F. A. Williams, "Condensed-Phase Mass and Energy Balances", chapter 3 of *Heat Transfer in Fires: Thermophysics, Social Aspects, Economic Impact*, P. L. Blackshear, ed., Scripta Book Co., New York: Wiley, 1974, 180–182.
8. P. A. Libby and F. A. Williams, *AIAA Journal* **3**, 1152 (1965).

CHAPTER 2

Rankine-Hugoniot Relations

In this chapter we shall classify the various types of infinite, plane, steady-state, one-dimensional flows involving exothermic chemical reactions, in which the properties become uniform as $x \to \pm \infty$. Such a classification provides a framework within which plane deflagration and detonation waves may be investigated. The experimental conditions under which these waves appear are described in Chapters 5 and 6, where detailed analyses of each type of wave are presented.

The Rankine-Hugoniot relations are the equations relating the properties on the upstream and downstream sides of these combustion waves. In this chapter, general Rankine-Hugoniot equations are derived and discussed first; then the Hugoniot curve for a simplified system is studied in detail in order to delineate explicitly the various burning regimes.

Since the major changes in the values of the flow variables in these systems usually take place over a very short distance (in nearly all cases less than a few centimeters; see Sections 5.1.2 and 6.1), in many problems deflagration and detonation waves can be treated as discontinuities (at which heat addition occurs) in the flow equations for an "ideal" (inviscid, nondiffusive, non-heat-conducting, nonreacting) fluid. In such problems, the equations derived in this chapter provide all of the information that is required concerning these waves, except their speed of propagation.

Some other texts discussing the Rankine-Hugoniot equations are listed in [1]–[5].

2.1. GENERAL RANKINE-HUGONIOT EQUATIONS

2.1.1. Derivation of the equations

A coordinate system that is stationary with respect to the wave will be adopted, and it will be assumed that the flow is in the $+x$ direction and that properties are uniform in planes normal to the x axis. Upstream conditions (at $x = -\infty$) will be identified by the subscript 0, while downstream conditions (at $x = +\infty$) are denoted by the subscript ∞. The flow is illustrated schematically in Figure 2.1. Equations (1-26)–(1-29) govern the system.

The continuity equation [equation (1-27)] implies that

$$\rho_0 v_0 = \rho_\infty v_\infty \equiv m, \tag{1}$$

where m is the mass flow rate per unit area. Since dv/dx approaches zero as $x \to \pm\infty$, equation (1-26) reduces to

$$\rho_0 v_0^2 + p_0 = \rho_\infty v_\infty^2 + p_\infty \tag{2}$$

(momentum conservation). The sequence of states at $x = \infty$ (with fixed parameters at $x = -\infty$) for which equations (1) and (2) are satisfied is often referred to as the **Rayleigh line**. Since q is proportional to temperature gradients and concentration gradients, $q \to 0$ as $x \to \pm\infty$, and equation (1-28) (the energy equation) implies

$$h_0 + v_0^2/2 = h_\infty + v_\infty^2/2, \tag{3}$$

where use has been made of equation (1). The species-conservation equations, equation (1-29), provide the additional requirements

$$w_{i,\infty} = 0, \quad i = 1, \ldots, N, \tag{4}$$

and

$$w_{i,0} = 0, \quad i = 1, \ldots, N. \tag{5}$$

The equation of state,

$$f(p_0, \rho_0, T_0, Y_{i,0}) = f(p_\infty, \rho_\infty, T_\infty, Y_{i,\infty}), \tag{6}$$

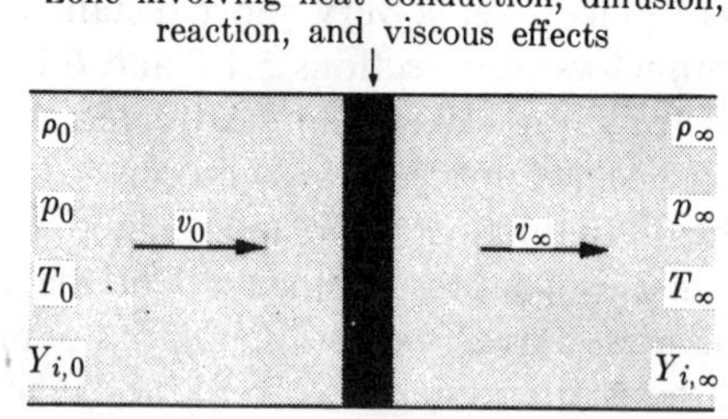

Zone involving heat conduction, diffusion, reaction, and viscous effects

FIGURE 2.1. Schematic diagram of a deflagration or detonation wave.

and the caloric equation of state,

$$g(h_0, p_0, T_0, Y_{i,0}) = g(h_\infty, p_\infty, T_\infty, Y_{i,\infty}), \tag{7}$$

constitute further relationships between the initial- and final-flow properties; for ideal-gas mixtures equations (6) and (7), respectively, reduce to equation (1-9) and the relation obtained by substituting equation (1-11) into equation (1-10) ($h \equiv u + p/\rho$).

In future work it will be convenient to employ modified forms of equations (2) and (3). Since equation (1) implies that

$$\rho_\infty v_\infty^2 - \rho_0 v_0^2 = m^2\left(\frac{1}{\rho_\infty} - \frac{1}{\rho_0}\right),$$

equation (2) may be written as

$$\frac{p_\infty - p_0}{(1/\rho_\infty) - (1/\rho_0)} = -m^2, \tag{8}$$

which defines a straight line (the Rayleigh line) with a negative slope in the $(p_\infty - p_0) - (1/\rho_\infty - 1/\rho_0)$ plane. Using equation (1) to express v in terms of m and ρ in equation (3) yields

$$h_\infty - h_0 = -\frac{1}{2}m^2\left(\frac{1}{\rho_\infty^2} - \frac{1}{\rho_0^2}\right),$$

which reduces to

$$h_\infty - h_0 = \frac{1}{2}\left(\frac{1}{\rho_\infty} + \frac{1}{\rho_0}\right)(p_\infty - p_0) \tag{9}$$

when equation (8) is used to eliminate m^2. Equation (9), which is called the **Hugoniot equation**, is a relationship among thermodynamic properties alone, as velocities have been eliminated.

The species-conservation equations require further discussion because equations (4) and (5) may not all be independent, and all of the information that can be obtained from equation (1-29) is not contained in equations (4) and (5). From equation (1-8) we see that the mass rate of production of species i by chemical reactions may be expressed as

$$w_i = \sum_{k=1}^{M} w_{i,k}, \quad i = 1, \dots, N, \tag{10}$$

where $w_{i,k}$ is the mass rate of production of species i in the kth reaction and there are M independent reactions. The convention to be adopted here is slightly different from that for equation (1-8); forward and backward reactions will not be considered separately as is implied by equation (1-8), so that the present M is half of the previous M, and $w_{i,k}$ is the difference of two terms of the type appearing inside the summation in equation (1-8). From the stoichiometry of the kth reaction, it follows that

$$w_{i,k} = W_i(v_{i,k}'' - v_{i,k}')\omega_k', \quad i = 1, \dots, N, \quad k = 1, \dots, M, \tag{11}$$

where ω'_k (which is independent of i) is the net rate of the kth reaction (that is, the difference between the forward rate and the backward rate; see Section B.1.2). Equations (10) and (11) imply that equations (4) and (5) are satisfied when

$$\omega'_{k,\,\infty} = 0, \quad k = 1, \ldots, M, \tag{12}$$

and

$$\omega'_{k,\,0} = 0, \quad k = 1, \ldots, M, \tag{13}$$

which may constitute fewer relations than equations (4) and (5), since $M \leq N$ in general.

However, when $M < N$ it is possible to derive additional (atom-conservation) conditions from equation (1-29). If $v_i^{(j)}$ denotes the number of atoms of kind j in a molecule of kind i, then the fact that atoms are neither created nor destroyed by chemical reactions is expressed by the equation

$$\sum_{i=1}^{N} v_i^{(j)} w_i / W_i = 0, \quad j = 1, \ldots, L, \tag{14}$$

where L is the total number of atoms in the system. Substituting equation (1-29) for w_i into equation (14) and integrating from $x = -\infty$ to $x = +\infty$ yields

$$\sum_{i=1}^{N} v_i^{(j)} Y_{i,\,0} / W_i = \sum_{i=1}^{N} v_i^{(j)} Y_{i,\,\infty} / W_i, \quad j = 1, \ldots, L, \tag{15}$$

since $\rho_0 v_0 = \rho_\infty v_\infty$ [equation (1)] and $V_i \to 0$ as $x \to \pm\infty$ (because diffusion velocities are proportional to gradients of the flow properties). Although it may be found that not all of the L additional relations between the properties at $x = +\infty$ and those at $x = -\infty$, given by equation (15), are linearly independent, it will always be true that $N - M$ of these relations are independent (compare Section A.3.7). Hence equations (12) and (15) provide a total of N independent relations among the flow properties at $x = \infty$; there are M equilibrium conditions and $N - M$ atom-conservation conditions, just as in the problem of computing equilibrium compositions in closed systems (Appendix A).

2.1.2. The cold-boundary difficulty

The restrictions given by equation (13) [or equation (5)] are not fulfilled exactly in deflagration or detonation waves. The combustible mixture approaching from $x = -\infty$ in Figure 2.1 is not in stable chemical equilibrium; the stable condition is that in which the mixture is composed of products of combustion. The initial mixture is in a metastable state and therefore actually reacts at a small but finite rate. Hence $\omega'_{k,\,0} \neq 0$, but $\omega'_{k,\,0}$

is very small compared with its value in the reaction zone so that, in some sense, equation (13) is satisfied approximately. Furthermore, equation (13) will be satisfied approximately for any initial mixture composition for which the mixture can be contained long enough to support a detonation or deflagration wave. Therefore, equation (13) actually provides no reasonable restriction on the initial conditions and will be omitted from the governing equations. If the problem were formulated in terms of wave propagation in a given combustible mixture, equation (13) would not even arise. However, it appears to be desirable to introduce this "cold-boundary difficulty" at an early stage, thus emphasizing how elementary it is, because the same trouble necessarily arises in studies of detonation structure (Section 6.1.2.2) and of deflagration propagation velocities (Section 5.3.2).

2.1.3. Use of the Rankine-Hugoniot equations

In summary, equations (1), (6)–(9), (12), and (15) comprise the independent relations between upstream and downstream conditions. If all upstream conditions (including v_0) were specified, then these $N + 5$ equations would determine the downstream flow variables v_∞, ρ_∞, p_∞, T_∞, h_∞, and $Y_{i,\infty}$. Although the initial composition and thermodynamic properties are adjustable experimentally, the propagation velocity v_0 is not controlled directly but instead is determined by other experimental parameters; in a typical experiment a given mixture in a tube will be ignited, and the velocity at which the combustion wave propagates down the tube will be measured. Therefore, study of combustion waves is facilitated by eliminating v_0 from as many equations as possible. After the equations that do not contain v_0 are solved for relations among flow properties downstream, the conditions that fix v_0 (to be defined in later sections) may be introduced to complete the solution. Equations (1), (6)–(9), (12), and (15) constitute the required forms since velocities appear only in equations (1) and (8) of this set. With ρ_0, p_0, T_0, h_0, and $Y_{i,0}$ given, the values of T_∞ and the chemical equilibrium composition ($Y_{i,\infty}$) can be calculated for various values of ρ_∞ and p_∞ from equations (6), (12), and (15),* the enthalpy h_∞ may then be evaluated for these same values of ρ_∞ and p_∞ from equation (7), and the result may finally be substituted into the Hugoniot equation, equation (9), in order to determine a curve in the ($p_\infty, 1/\rho_\infty$) plane along which the solution must lie regardless of the value of v_0. The nature of this curve, called the Hugoniot curve, is investigated in the following section. As illustrated in Figure 2.2, the intersection of the Hugoniot curve with the Rayleigh line defined by equation (8) determines the final thermodynamic state, after m has been obtained from v_0 for the particular experiment. The value of v_∞ may then

---

* This calculation is similar to the equilibrium-composition computations discussed in Appendix A.

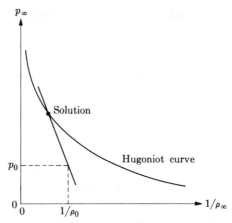

FIGURE 2.2. The determination of downstream flow properties from the Hugoniot curve.

be calculated from equation (1). Variations of this procedure have been programmed in recent years for solution by electronic computers (for example, [6]), which provide accurate Hugoniot curves based on the best values available for thermochemical properties.

2.2. ANALYSIS OF A SIMPLIFIED SYSTEM

2.2.1. Simplification of the Rankine-Hugoniot equations

The essential characteristics of the Hugoniot curve are most easily illustrated by studying a particular simple system. We shall consider an ideal gas mixture in order to assure that equations (1-9)–(1-11) are valid. It will be assumed that for the given initial thermodynamic properties (ρ_0, p_0, $Y_{i,0}$, and so on), (1) the final equilibrium composition along the Hugoniot curve is constant (that is, $Y_{i,\infty}$ are all fixed), (2) the final average molecular weight equals the initial average molecular weight—that is,

$$\left(\sum_{i=1}^{N} Y_{i,\infty}/W_i \right)^{-1} = \left(\sum_{i=1}^{N} Y_{i,0}/W_i \right)^{-1} \equiv \overline{W} \tag{16}$$

—and (3) between the initial temperature T_0 and any final temperature T_∞ along the Hugoniot curve, the specific heat at constant pressure at the final composition equals the specific heat at constant pressure of the initial mixture—that is,

$$\sum_{i=1}^{N} Y_{i,\infty} c_{p,i} = \sum_{i=1}^{N} Y_{i,0} c_{p,i,0} \equiv c_p \quad (= \text{constant}) \tag{17}$$

for $c_{p,i}$ in some T range including T_0 and T_∞. The simplest system for which equations (16) and (17) are valid is one in which all species have equal molecular weights ($W_i = \overline{W}$ for all i) and constant and equal heat capacities ($c_{p,i} = c_p$ = constant for all i).

In view of equation (16), equation (1-9) reduces to

$$\frac{p_\infty}{\rho_\infty T_\infty} = \frac{p_0}{\rho_0 T_0} = \frac{R^0}{\overline{W}}. \tag{18}$$

Substituting equation (1-11) into equation (1-10), solving for $h = u + p/\rho$, and evaluating the result at $x = -\infty$ and at $x = +\infty$ yields

$$h_\infty - h_0 = \sum_{i=1}^{N} (Y_{i,\infty} - Y_{i,0}) h_i^0 + \int_{T^0}^{T_\infty} \left(\sum_{i=1}^{N} Y_{i,\infty} c_{p,i} \right) dT$$

$$- \int_{T^0}^{T_0} \left(\sum_{i=1}^{N} Y_{i,0} c_{p,i} \right) dT \tag{19}$$

after subtraction. The heat of reaction of the mixture may be defined as

$$q \equiv \sum_{i=1}^{N} \left[\left(Y_{i,0} - Y_{i,\infty} \right) \left(h_i^0 + \int_{T^0}^{T_0} c_{p,i} \, dT \right) \right], \tag{20}$$

which, in view of equation (17), is the total chemical heat released per unit mass of combustible mixture. Substituting equation (20) into equation (19), we obtain

$$h_\infty - h_0 = -q + \int_{T_0}^{T_\infty} \left(\sum_{i=1}^{N} Y_{i,\infty} c_{p,i} \right) dT = -q + c_p(T_\infty - T_0), \tag{21}$$

where use has been made of equation (17) in obtaining the final equality.

The temperatures T_0 and T_∞ may be eliminated from equation (21) by using equation (18). It can be seen that the ratio $c_p/(R^0/\overline{W})$ will enter into the result. Since the specific heat at constant volume is given by $c_v = c_p - R^0/\overline{W}$ for an ideal-gas mixture, it follows that $c_p/(R^0/\overline{W}) = c_p/(c_p - c_v) = \gamma/(\gamma - 1)$, where $\gamma \equiv c_p/c_v$ is the specific-heat ratio for the final mixture. Solving equation (18) for T_∞ and T_0 and substituting these results into equation (21) therefore yields

$$h_\infty - h_0 = -q + \frac{\gamma}{\gamma - 1} \left(\frac{p_\infty}{\rho_\infty} - \frac{p_0}{\rho_0} \right). \tag{22}$$

Equations (9) and (22) show that the Hugoniot equation may be written in the form

$$\left(\frac{\gamma}{\gamma - 1} \right) \left(\frac{p_\infty}{\rho_\infty} - \frac{p_0}{\rho_0} \right) - \frac{1}{2} \left(\frac{1}{\rho_\infty} + \frac{1}{\rho_0} \right) (p_\infty - p_0) = q, \tag{23}$$

which contains only the known constants γ, q, p_0, and ρ_0 in addition to the variables p_∞ and ρ_∞. Equation (23) may also be obtained by considering a one-component ideal gas to which an amount of heat q (per unit mass) is added externally.

From equation (23) an explicit algebraic expression for p_∞ as a function of ρ_∞ along the Hugoniot curve may be obtained. Equations (8) and (23) therefore determine the complete solution for the simplified system.

2.2.2. Dimensionless form

The nature of the results will become more transparent if dimensionless variables are introduced. A dimensionless final pressure is the pressure ratio

$$p \equiv p_\infty/p_0.$$

A dimensionless final specific volume is the density ratio

$$v \equiv \rho_0/\rho_\infty,$$

which is also equal to the velocity ratio v_∞/v_0 according to equation (1). A dimensionless heat of reaction is

$$\alpha \equiv q\rho_0/p_0,$$

and a dimensionless mass-flow rate is

$$\mu = m^2/p_0\rho_0.$$

It will be observed that p, v, and μ must all be positive for physically acceptable solutions. Multiplying equation (23) by ρ_0/p_0 yields

$$\left(\frac{\gamma}{\gamma-1}\right)(pv-1) - \frac{1}{2}(v+1)(p-1) = \alpha,$$

the solution of which is

$$p = \frac{[2\alpha + (\gamma+1)/(\gamma-1)] - v}{[(\gamma+1)/(\gamma-1)]v - 1}. \tag{24}$$

Dividing equation (8) by $p_0\rho_0$ gives

$$\frac{p-1}{v-1} = -\mu. \tag{25}$$

2.2.3. Properties of the Hugoniot curve

The Hugoniot curves given by equation (24) are plotted in Figures 2.3 and 2.4 for $\gamma = 1.4$ and $\gamma = 1.2$, respectively.* All curves asymptotically approach the lines $v = (\gamma-1)/(\gamma+1)$ and $p = -(\gamma-1)/(\gamma+1)$. Therefore,

---

* The curves for $\alpha = 0$ are of significance in connection with ordinary shock waves.

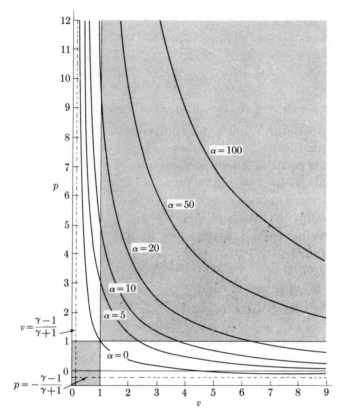

FIGURE 2.3. Hugoniot diagram from (24) with $\gamma = 1.4$.

the entire range of pressure ratio $(0 \le p \le \infty)$ may occur, but the specific-volume (or velocity) ratio is restricted to the interval

$$(\gamma - 1)/(\gamma + 1) \le v \le 2\alpha + [(\gamma + 1)/(\gamma - 1)],$$

with the upper bound corresponding to the limit $p \to 0$ [see equation (24)].

The intersection between the appropriate Hugoniot curve and the straight line through the point $(1, 1)$ with slope $-\mu$ [equation (25)] establishes the final state of the system. Since $\mu > 0$, the slope of this straight line is negative, and end states lying in the two shaded quadrants in Figures 2.3 and 2.4 are physically meaningless. Each Hugoniot curve is therefore divided into two nonoverlapping branches, an upper branch $[1 + (\gamma - 1)\alpha \le p \le \infty,$ $(\gamma - 1)/(\gamma + 1) \le v \le 1$; see equation (24)] called the **detonation branch** and a lower branch $[0 \le p \le 1, 1 + (\gamma - 1)\alpha/\gamma \le v \le 2\alpha + (\gamma + 1)/(\gamma - 1)$; see equation (24)] called the **deflagration branch**. Acceptable end states must lie on one of these two branches. Combustion waves are termed detonation waves or deflagration waves according to the branch of the Hugoniot curve upon which the final condition falls.

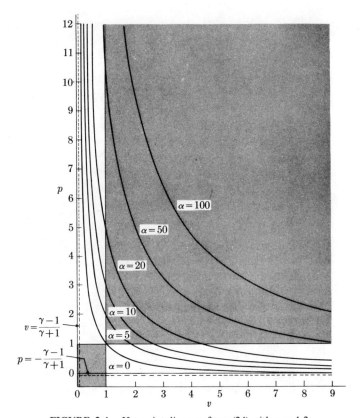

FIGURE 2.4. Hugoniot diagram from (24) with $\gamma = 1.2$.

Although this formal distinction is invaluable because of its unambiguity, an *understanding* of the difference between detonation and deflagration waves is best obtained by contrasting the characteristics of each. Thus (see Figures 2.3 and 2.4), in passing through a detonation wave the gas is slowed down, and its pressure and density increase; however, in going through a deflagration the gas speeds up and expands, and its pressure decreases. Other striking differences between the two types of waves will appear when we discuss wave structures in subsequent chapters (see Sections 5.3 and 6.1).

2.2.4. Analysis of the detonation branch

It is easily seen from Figure 2.3 or Figure 2.4 that a unique straight line through the point $(1, 1)$ is tangent to the detonation branch of the Hugoniot curve. This can be proven formally by considering the equation

$$dp/dv = (p - 1)/(v - 1), \tag{26}$$

where $p(v)$ and dp/dv are evaluated from the Hugoniot equation, equation (24). Equation (26), which states that the slope of the Hugoniot curve equals the slope of the straight line from the Hugoniot curve to the point $(1, 1)$, can be shown to possess only one solution, (p_+, v_+), lying on the detonation branch. This point (p_+, v_+) on the detonation branch is said to be the **upper Chapman-Jouguet point,** and the wave with this end condition (point B on the Hugoniot curve illustrated in Figure 2.5) is a **Chapman-Jouguet detonation.**

A straight line through the point $(1, 1)$ will not intersect the detonation branch anywhere if the magnitude of its slope is less than that of the tangent line, while it intersects the detonation branch at two points if the magnitude of its slope exceeds that of the tangent line. This is illustrated in Figure 2.5, where typical solutions lying on the detonation branch ABC are indicated. Equation (25) therefore implies that there is a minimum value of μ (that is, a minimum wave speed) for detonations, corresponding to the Chapman-Jouguet end conditions (at point B in Figure 2.5). When μ exceeds its minimum value μ_+ (that is, at higher wave speeds), there are two possible end states for the detonation, one lying on the line AB (a **strong detonation**) and the other lying on the line BC (a **weak detonation**).

The experimental conditions determine whether a strong detonation, a Chapman-Jouguet wave, or a weak detonation will be observed at a given value of μ. Under most experimental conditions, detonations are Chapman-Jouguet waves; this topic will be discussed more fully in Section 6.2, since the reasoning involves concepts of the structure of the wave.

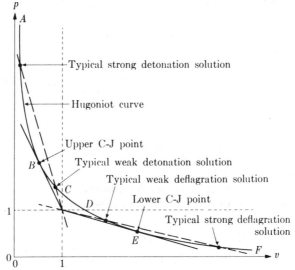

FIGURE 2.5. Schematic diagram depicting the various sections of the Hugoniot curve.

Figure 2.5 shows that as the nomenclature implies, the pressure ratio and the velocity change across a strong detonation exceed those across a weak detonation. The strong detonation with $p = \infty$ and the isochoric weak detonation ($v = 1$), both of which propagate at infinite velocity ($\mu = \infty$), represent unattainable limiting cases.

2.2.5. Analysis of the deflagration branch

Remarks precisely analogous to those given above for the detonation branch of the Hugoniot curve also apply to the deflagration branch. Figure 2.3, Figure 2.4, or equation (26), shows that a unique straight line through the point (1, 1) is tangent to the deflagration branch (*DEF* in Figure 2.5) at a point (p_-, v_-), which is called the **lower Chapman-Jouguet point** (point E in Figure 2.5). A wave with endpoint E is a **Chapman-Jouguet deflagration**.

A straight line through the point (1, 1) fails to intersect the deflagration branch if the magnitude of its slope exceeds that of the tangent line, and it intersects the deflagration branch at two points (once on *DE* and once on *EF*) if its slope is less than that of the tangent line. Hence the Chapman-Jouguet deflagration has the maximum wave speed ($\mu = \mu_-$) of all deflagrations. Figure 2.5 clearly shows that this maximum deflagration velocity is less than the minimum detonation velocity ($\mu_- < \mu_+$); this result can also be derived from equation (26).

Waves with end states lying along line *DE* are **weak deflagrations** and those with end states on line *EF* are **strong deflagrations**. The pressure and velocity changes across strong deflagrations exceed those across weak deflagrations. Limiting cases are the strong deflagration with $p = 0$, which has a finite (nonzero) propagation speed, and the isobaric weak deflagration ($p = 1$), which propagates at zero velocity ($\mu = 0$).

From considerations of combustion-wave structure, it will be indicated in Section 6.1.3 that strong deflagrations do not occur; hence the physically meaningful section of the deflagration branch of the Hugoniot curve is *DE*. Most deflagrations are, in fact, nearly isobaric.

2.2.6. Properties of Chapman-Jouguet waves

Chapman-Jouguet waves assume a special significance in many systems, particularly those involving detonations. It is therefore of interest to investigate the properties of the Chapman-Jouguet points in greater detail.

Differentiating equation (24) yields

$$\frac{dp}{dv} = -\frac{[(\gamma + 1)/(\gamma - 1)]p + 1}{[(\gamma + 1)/(\gamma - 1)]v - 1},$$

which may be substituted into equation (26) to obtain

$$p = v/[(\gamma + 1)v - \gamma] \tag{27}$$

at either Chapman-Jouguet point. By using this relation in equation (25), it can be shown that

$$\mu = \gamma p/v \tag{28}$$

at the Chapman-Jouguet points. From the definitions of the dimensionless variables μ, p, and v, the dimensional form of equation (28) is seen to be $m^2 = \gamma p_\infty \rho_\infty$, which implies that

$$v_\infty = \sqrt{\gamma p_\infty/\rho_\infty} \tag{29}$$

[see equation (1)]. Recalling that the speed of sound in an ideal gas is

$$a_\infty \equiv \sqrt{(\partial p/\partial \rho)_s}\big|_\infty = \sqrt{\gamma p_\infty/\rho_\infty}$$

(s = entropy per unit mass), we see that equation (29) implies that the downstream Mach number $M_\infty \equiv v_\infty/a_\infty$ is unity at both Chapman-Jouguet points. A more thorough analysis shows that the downstream gas velocity relative to the wave exceeds the sound velocity ($M_\infty > 1$) for weak detonations and strong deflagrations and is less than the sound velocity ($M_\infty < 1$) for strong detonations and weak deflagrations.

It is also possible to prove that along the Hugoniot curve, the entropy assumes a local minimum value only at the upper Chapman-Jouguet point and a local maximum value only at the lower Chapman-Jouguet point [7] and that along each tangent line given by equation (25) (the Rayleigh line), the entropy assumes a local maximum value at the Chapman-Jouguet point [8].

Explicit expressions for the dimensionless downstream properties (in terms of α and γ) are easily derived for Chapman-Jouguet waves. For example, if equation (27) is solved for v, giving

$$v = \gamma p/[(\gamma + 1)p - 1], \tag{30}$$

and this result is substituted into equation (24), then a quadratic equation for p is obtained, the solution to which is

$$p_\pm = 1 + \alpha(\gamma - 1)\left\{1 \pm \left[1 + \frac{2\gamma}{\alpha(\gamma^2 - 1)}\right]^{1/2}\right\}. \tag{31}$$

Upper signs correspond to the detonation branch, and lower signs correspond to the deflagration branch. Substitution of equation (31) into equation (30) yields

$$v_\pm = 1 + \alpha\left(\frac{\gamma - 1}{\gamma}\right)\left\{1 \mp \left[1 + \frac{2\gamma}{\alpha(\gamma^2 - 1)}\right]^{1/2}\right\}. \tag{32}$$

By using equations (31) and (32) in equation (25), we find that

$$\mu_\pm = \gamma + \alpha(\gamma^2 - 1)\left\{1 \pm \left[1 + \frac{2\gamma}{\alpha(\gamma^2 - 1)}\right]^{1/2}\right\}. \tag{33}$$

The temperature ratio may be computed from $T_\infty/T_0 = pv$ [see equation (18)]; the downstream temperature is found to exceed the upstream temperature at both Chapman-Jouguet points.

Equation (33) may be used to compute the initial Mach number for Chapman-Jouguet waves. The result is

$$M_{0\pm} = \left[1 + \frac{\alpha(\gamma^2 - 1)}{2\gamma}\right]^{1/2} \pm \left[\frac{\alpha(\gamma^2 - 1)}{2\gamma}\right]^{1/2}, \tag{34}$$

since $M_0 \equiv v_0/a_0 = v_0/\sqrt{\gamma p_0/\rho_0} = \sqrt{\mu/\gamma}$ from equation (1) and the definition of μ. Equation (34) shows that both M_{0+} and M_{0-} approach unity as the heat-release parameter α goes to zero, while $M_{0+} \to \infty$ and $M_{0-} \to 0$ as $\alpha \to \infty$. According to equation (34), the initial Mach number always exceeds unity for Chapman-Jouguet detonations and lies between zero and unity for Chapman-Jouguet deflagrations. Since the Chapman-Jouguet waves correspond to the minimum propagation speed for detonations and the maximum propagation speed for deflagrations, we conclude that all detonations propagate at supersonic velocities and all deflagrations propagate at subsonic velocities.

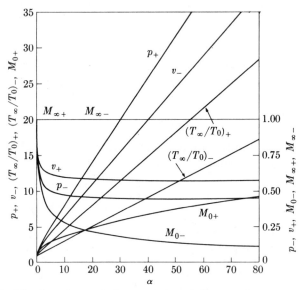

FIGURE 2.6. The dependence of the properties of Chapman-Jouguet waves upon the dimensionless heat-release parameter α, as determined by equations (31)–(34) with $\gamma = 1.3$.

Numerical values of the parameters for Chapman-Jouguet waves are plotted in Figure 2.6 as functions of the dimensionless heat-release parameter α for $\gamma = 1.3$. A representative value of α is $\alpha = 30$, for which Figure 2.6 shows that, approximately, $p_+ = 20$, $v_+ = 0.58$, $(T_\infty/T_0)_+ = 12.5$, and $M_{0+} = 5.8$ (giving $v_0 \approx 1000$ m/s) at the upper Chapman-Jouguet point, while $p_- = 0.46$, $v_- = 15.4$, $(T_\infty/T_0)_- = 7$, and $M_{0-} = 0.17$ at the lower Chapman-Jouguet point. In accordance with our earlier discussion, these numbers are typical of experimentally observed detonations but unrealistic for deflagrations.

2.3. EXTENSION OF THE RESULTS TO ARBITRARY SYSTEMS

2.3.1. Range of validity of the results of Section 2.2

More-general analyses than those given in the preceding section show that the qualitative properties of the Hugoniot curve derived in Section 2.2 remain valid for all the systems that fall within the framework of the equations presented in Section 2.1. Many quantitative results obtained in Section 2.2 are also of general validity; only some equations and certain limits on inequalities will require modification. The expressions which must be changed are those that involve α explicitly, since the single parameter α is not sufficient to characterize the behavior of arbitrary multicomponent mixtures. We shall briefly discuss the most important complication that arises in attempting to extend our results, and then we shall summarize most of the properties of arbitrary Hugoniot curves by means of a table.

2.3.2. Frozen versus equilibrium sound speeds

It was shown in Section 2.2 that the downstream Mach number is unity for Chapman-Jouguet waves. When interpreted correctly, this result applies to any combustible gas mixture. However, a possible source of ambiguity for multicomponent systems is the fact that more than one sound speed a can be defined.* Since there are $N + 2$ independent thermodynamic variables in an N-component gas mixture, N parameters besides s must be specified as constants in computing $\partial p/\partial \rho$ to evaluate a^2.†

A **frozen** (constant-composition) **sound speed** may be defined as

$$a_f = [(\partial p/\partial \rho)_{s, Y_i(i=1,\dots,N)}]^{1/2}$$

---

* The subscript ∞ is suppressed in this section for the sake of clarity in notation. It is understood that we are always referring to downstream conditions here.

† If s is the entropy per unit mass, then actually only $N + 1$ independent intensive parameters must be specified because p, ρ, and a are intensive properties that are independent of the size of the system.

($Y_i \equiv$ mass fraction of·chemical species i), and an **equilibrium sound speed** may be defined as

$$a_e = [(\partial p/\partial \rho)_{s, Y_i = Y_{i,e}(\rho, s)(i=1, \dots, N)}]^{1/2},$$

where $Y_{i,e}(\rho, s)$ represents the equilibrium mass fraction of species i at the given ρ and s in a mixture at the specified overall atomic composition [equation (15)]. The subscript notation $Y_i = Y_{i,e}(\rho, s)$ means that Y_i is to be set equal to its equilibrium function of ρ and s [which is determined by equation (12)] in computing the partial derivative. Since $a_f \geq a_e$ in general,* the problem is to determine which (if either) of these appears in the definition of M for which $M = 1$ at the Chapman-Jouguet points.

Since the Hugoniot curve is an equilibrium curve, it might be expected that the Chapman-Jouguet tangency condition would lead to $M = 1$ with M defined in terms of a_e. The rigorous analysis given in the following paragraph supports this guess; the downstream gas velocity relative to the wave equals the equilibrium sound speed at the Chapman-Jouguet points. Using $M = 1$ to define the Chapman-Jouguet point, with the M based on a_f instead of a_e, may lead to overestimates of the computed Chapman-Jouguet detonation-propagation velocities, which are comparable with the errors in the most accurate experimental measurements of this parameter (about one part in 10^3).† Further information on this topic may be found in [9] and in some of the references quoted therein (see also Section 6.2.4).

2.3.3. Proof that $v_\infty = a_{e,\infty}$ at the Chapman-Jouguet points

For a small change along the general Hugoniot curve [equation (9)], we find by differentiation that

$$dh_\infty = \frac{1}{2}\left(\frac{1}{\rho_\infty} + \frac{1}{\rho_0}\right) dp_\infty + \frac{1}{2}(p_\infty - p_0)\, d\left(\frac{1}{\rho_\infty}\right). \tag{35}$$

---

* This is a consequence of the fact that thermodynamically unstable equilibria are not attainable in natural processes. For example, if only one chemical reaction can occur in the system under consideration, then straightforward thermodynamic calculations show that

$$a_f^2 - a_e^2 = \left(\frac{\rho}{V}\right)\left[\left(\frac{\partial \epsilon}{\partial \rho}\right)_{s, Y_i = Y_{i,e}(\rho, s)(i=1, \dots, N)}\right]^2 \left\{\frac{\partial}{\partial \epsilon}\left[\sum_{i=1}^N \mu_i(v_i'' - v_i')\right]\right\}_{S, V},$$

where ϵ is the reaction progress variable defined in equation (A-19), other symbols are defined in Appendix A, and the partial derivatives are taken in a closed system (of entropy S and volume V). Since the factors preceding the brace on the right-hand side of this relation are intrinsically positive, the sign of $a_f^2 - a_e^2$ is the same as the sign of the last factor, [which equals $(\partial^2 U/\partial \epsilon^2)_{S, V}$ ($U \equiv$ internal energy)]. The reasoning following equation (A-20) implies that a small disturbance will cause the system to move away from the equilibrium point if this last factor is negative.

† Methods for computing Chapman-Jouguet detonation velocities are given in [1], [2], [6], [10], and [11], for example.

The tangent to the Hugoniot curve is a straight line through the point $(p_0, 1/\rho_0)$ provided that, in equation (35),

$$\frac{dp_\infty}{d(1/\rho_\infty)} = \frac{p_\infty - p_0}{1/\rho_\infty - 1/\rho_0}. \tag{36}$$

Using equation (36) to eliminate $d(1/\rho_\infty)$ from equation (35), we obtain the expression

$$dh_\infty = \frac{1}{\rho_\infty} dp_\infty. \tag{37}$$

Since the entropy change along the Hugoniot curve [at $Y_{i,\infty} = Y_{i,e,\infty}(\rho_\infty, s_\infty)$ for $i = 1, \ldots, N$] is given by the thermodynamic relationship*

$$T_\infty ds_\infty = dh_\infty - \left(\frac{1}{\rho_\infty}\right) dp_\infty, \tag{38}$$

equation (37) implies that

$$ds_\infty = 0 \tag{39}$$

at the Chapman-Jouguet points along the Hugoniot curve. Hence, in equation (36),

$$\frac{dp_\infty}{d(1/\rho_\infty)} = \left[\frac{\partial p_\infty}{\partial(1/\rho_\infty)}\right]_{s_\infty, Y_{i,\infty} = Y_{i,e,\infty}(\rho_\infty, s_\infty)(i=1,\cdots,N)}. \tag{40}$$

Equations (8) and (36) show that

$$\frac{dp_\infty}{d(1/\rho_\infty)} = -m^2 = -(\rho_\infty v_\infty)^2, \tag{41}$$

where the last equality follows from equation (1). Equations (40) and (41) finally yield

$$v_\infty^2 = (\partial p_\infty/\partial \rho_\infty)_{s_\infty, Y_{i,\infty} = Y_{i,e,\infty}(\rho_\infty, s_\infty)(i=1,\ldots,N)}, \tag{42}$$

which states that $v_\infty = a_{e,\infty}$ at the Chapman-Jouguet points as desired.

This analysis is illustrative of the type of reasoning required when the assumptions of Section 2.2 are inapplicable.

2.3.4. Summary of the properties of Hugoniot curves

Table 2.1 summarizes many of the properties of Hugoniot curves. In the table, quantities identified by the subscripts $+$, $-$, 1, max, and min are fixed constants, determined by the initial state of the mixture. The quantities

---

* Terms corresponding to $\sum_{i=1}^{N} \mu_i \, dN_i$ in equation (A-5) do not appear in equation (38) because the condition of chemical equilibrium [equation (A-16) with the equality sign] implies that these terms are zero in the present case.

TABLE 2.1. Properties of Hugoniot Curves

	Section in Fig. 2-5	Pressure ratio $p \equiv (p_\infty/p_0)$	Velocity and density ratios $v \equiv (v_\infty/v_0) = (\rho_0/\rho_\infty)$	Propagation Mach number $M_0 \equiv (v_0/a_{f,0})$	Downstream Mach number $M_\infty \equiv (v_\infty/a_{e,\infty})$	Remarks
Strong detonations	Line A-B	$p_+ < p < \infty$	$v_{min} < v < v_+$ ($v_{min} > 0$)	$M_{0+} < M_0 < \infty$	$M_\infty < 1$	Seldom observed; requires special experimental arrangement.
Upper Chapman-Jouguet point	Point B	$p = p_+$ ($p_+ > 1$)	$v = v_+$ ($v_+ < 1$)	$M_0 = M_{0+}$ ($M_{0+} > 1$)	$M_\infty = 1$	Usually observed for waves propagating in tubes.
Weak detonations	Line B-C	$p_1 < p < p_+$ ($p_1 > 1$)	$v_+ < v < 1$	$M_{0+} < M_0 < \infty$	$M_\infty > 1$	Seldom observed; requires very special gas mixtures.
Weak deflagrations	Line D-E	$p_- < p < 1$	$v_1 < v < v_-$ ($v_1 > 1$)	$0 < M_0 < M_{0-}$	$M_\infty < 1$	Often observed; $p \approx 1$ in most experiments.
Lower Chapman-Jouguet point	Point E	$p = p_-$ ($p_- < 1$)	$v = v_-$ ($v_- > 1$)	$M_0 = M_{0-}$ ($M_{0-} < 1$)	$M_\infty = 1$	Not observed.
Strong deflagrations	Line E-F	$0 < p < p_-$	$v_- < v < v_{max}$ ($v_{max} < \infty$)	$M_{0min} < M_0 < M_{0-}$ ($M_{0min} > 0$)	$M_\infty > 1$	Not observed; forbidden by considerations of wave structure.

with subscripts $+$ and $-$ are given by equations (31)–(34) only under the restrictive conditions imposed in Section 2.2. The quantity M_0 is based on the chemically frozen sound speed in the initial mixture, and M_∞ is based on the equilibrium sound speed in the final mixture. The limits for all of the inequalities except those in parentheses are intended to represent greatest lower bounds or least upper bounds. With this information, Table 2.1 should become self-explanatory.

REFERENCES

1. S. S. Penner, *Chemistry Problems in Jet Propulsion*, New York: Pergamon Press, 1957, 258–266.
2. B. P. Mullins and S. S. Penner, *Explosions, Detonations, Flammability and Ignition*, New York: Pergamon Press, 1959, 41–72.
3. J. O. Hirschfelder, C. F. Curtiss, and R. B. Bird, *Molecular Theory of Gases and Liquids*, New York: Wiley, 1954, 797–814.
4. R. A. Strehlow, *Fundamentals of Combustion*, Scranton, Penn.: International Textbook, 1968, 149–155.
5. Y. B. Zel'dovich and A. S. Kompaneets, *Theory of Detonation*, New York: Academic Press, 1960, 68–108.
6. S. Gordon and B. J. McBride, NASA SP-273 (1971).
7. R. L. Scorah, *J. Chem. Phys.* **3**, 425 (1935).
8. R. Becker, *Z. Physik.* **8**, 321 (1922); *Z. Elektrochem.* **42**, 457 (1936).
9. R. Gross and A. K. Oppenheim, *ARS Journal* **29**, 173 (1959).
10. R. Edse, "Calculation of Detonation Velocities in Gases," *Tech. Rept. No. 54-416*, Wright Air Development Center, Dayton (March 1956).
11. F. J. Zeleznik and S. Gordon, *ARS Journal* **32**, 606 (1962).

CHAPTER 3

Diffusion Flames and Droplet Burning

In the broadest sense, a **diffusion flame** may be defined as any flame in which the fuel and oxidizer initially are separated. With this usage, the term is synonymous with nonpremixed combustion. The broad definition encompasses a wide variety of processes—for example, a pan of oil burning in air, an aluminum sheet burning in a supersonic air stream, a lighted candle, a forest fire, and a fuel droplet burning in oxygen in a rocket engine. Included here are processes involving unsteady flow, high-speed flow, and highly turbulent motion. Therefore, it is unreasonable to attempt to discuss all of these processes from a unified viewpoint.

In a restricted sense, a diffusion flame may be defined as a nonpremixed, quasisteady, nearly isobaric flame in which most of the reaction occurs in a narrow zone that can be approximated as a surface. This is the class of flame problems that will be discussed in the present chapter. The coupling-function formulation (Section 1.3) provides a convenient framework within which these problems can be studied. Coupling functions usually are more useful for initially unmixed systems than for initially premixed systems because, as we shall see, the chemical reaction rate is often of negligible importance in answering certain questions raised for diffusion flames.

Even within our restricted definition, so many diffusion-flame problems exist that we do not have enough space to consider them all. Therefore, we shall merely illustrate the analytical procedure by means of three examples (Sections 3.1–3.3). The problem of droplet burning (Section

3.3) is of importance in other contexts (for example, liquid-fueled engine combustion) and therefore will be considered in somewhat greater detail. Most of these topics are also treated in [1]–[3].

In recent years appreciable clarification of structures of the thin sheets at which reactions occur has been obtained through asymptotic methods. Although study of these structures necessitates simultaneous consideration of both finite-rate chemistry and diffusion processes and therefore pedagogically belongs after Chapter 4, the relevance to diffusion flames dictates coverage here. For this reason, the reaction-sheet structure is analyzed in Section 3.4. Monopropellant droplet burning, treated in Section 3.5, also involves consideration of finite-rate chemistry but is related to the burning of a fuel droplet in an oxidizing atmosphere (Section 3.3). In a pedagogic approach, the reader may return to Sections 3.4 and 3.5 after completing Chapter 5.

3.1. THE FLAME AT THE MOUTH OF A TUBE IN A DUCT

3.1.1. Definition of the problem

The first successful detailed analysis of a diffusion-flame problem was given by Burke and Schumann in 1928 [4]. The Burke-Schumann problem is illustrated in Figure 3.1; fuel (or oxidizer) issues from a cylindrical

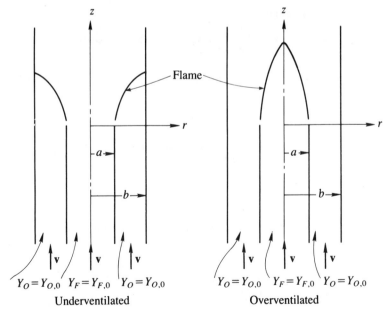

FIGURE 3.1. Diagram of the Burke-Schumann problem.

tube into a concentric cylindrical duct through which oxidizer (or fuel) is flowing. A flame is established at the mouth of the tube and either extends to the wall of the duct (the underventilated case) or converges to the axis (the overventilated case). The flame height, defined as the axial distance from the mouth of the tube to the point at which the flame reaches either the wall of the duct or the axis, is required for the design of burners of this type. Burke and Schumann obtained fairly accurate predictions of the flame height and the shape of the flame for laminar flow.

To investigate this problem, let the radii of the inner and outer tubes be a and b, respectively. We shall denote the species in the inner and outer tubes by the subscripts F and O, respectively, and shall consider the one-step overall chemical reaction

$$v_F \mathfrak{M}_F + v_O \mathfrak{M}_O \rightarrow \sum_{i=1}^{N-2} v_i \mathfrak{M}_i, \tag{1}$$

where the subscript i ($i = 1, 2, \ldots, N - 2$) identifies the various product species [compare equation (1-42)]. Let us employ a cylindrical coordinate system ($z \equiv$ distance along the axis, $r \equiv$ radial distance from the axis, $\theta \equiv$ azimuthal angle) and set the origin on the tube axis at the mouth of the inner tube (see Figure 3.1). Conditions at $z = 0$ will be identified by the subscript 0. The notation will be the same as in Chapter 1, particularly, Section 1.3.

3.1.2. Assumptions

In addition to all of the assumptions introduced in Section 1.3, the following approximations will be made.

1. The mass-average velocity is parallel to the axis of the tube everywhere, that is,

$$\mathbf{v} = v\mathbf{e}_z, \tag{2}$$

where $\mathbf{e}_z$ is a unit vector in the z direction.

2. The total mass flux in the z direction is constant everywhere in the duct, that is,

$$\rho v = \text{constant}. \tag{3}$$

3. The product ρD is constant throughout the duct, that is,

$$\rho D = \text{constant}. \tag{4}$$

4. Axial diffusion is negligible in comparison with radial diffusion, that is,

$$\partial Y_i / \partial z \ll \partial Y_i / \partial r. \tag{5}$$

5. The entire reaction occurs at a surface called the **flame sheet**, or **reaction sheet**.

The way in which approximation 5 is used will be explained in Section 3.1.4, and the validity of the approximations will be discussed after the solution is obtained.

3.1.3. Solution of the species conservation equation for the coupling function β

The only differential equation that we shall be obliged to consider is equation (1-49) with

$$\beta = \alpha_O - \alpha_F, \tag{6}$$

where, according to equations (1-47) and (1),

$$\alpha_F = -Y_F/W_F v_F \tag{7}$$

and

$$\alpha_O = -Y_O/W_O v_O. \tag{8}$$

By using equations (2)–(5) in the definition given in equation (1-45), we find that equation (1-49) becomes

$$\left(\frac{v}{D}\right)\frac{\partial \beta}{\partial z} - \frac{1}{r}\frac{\partial}{\partial r}\left(r\frac{\partial \beta}{\partial r}\right) = 0, \tag{9}$$

where the Laplacian has been written in cylindrical coordinates and terms involving $\partial/\partial\theta$ have been set equal to zero because of the symmetry of the problem. The boundary conditions for equation (9) are that β is bounded everywhere and that

$$\beta = Y_{F,0}/W_F v_F \qquad \text{at} \quad z = 0, \quad 0 \le r < a, \tag{10}$$

$$\beta = -Y_{O,0}/W_O v_O \quad \text{at} \quad z = 0, \quad a < r < b, \tag{11}$$

and

$$\partial \beta/\partial r = 0 \qquad \text{at} \quad r = b, \quad z > 0. \tag{12}$$

Equations (10) and (11) follow from equations (6)–(8), since $Y_F = Y_{F,0}$ and $Y_O = 0$ in the central pipe, while $Y_F = 0$ and $Y_O = Y_{O,0}$ in the outer duct. Equation (12) is a consequence of the physical requirement that neither fuel nor oxidizer can diffuse into the outer wall (at $r = b$):

$$(Y_F \mathbf{V}_F \cdot \mathbf{e}_r)_{r=b} = -[D(\partial Y_F/\partial r)]_{r=b} = 0$$

and

$$(Y_O \mathbf{V}_O \cdot \mathbf{e}_r)_{r=b} = -[D(\partial Y_O/\partial r)]_{r=b} = 0,$$

where $\mathbf{e}_r$ is a unit vector in the r direction.

It is convenient to introduce the dimensionless coordinates

$$\xi \equiv r/b \tag{13}$$

and

$$\eta \equiv zD/vb^2 \tag{14}$$

and to define the reduced parameters

$$c \equiv a/b \tag{15}$$

and

$$v \equiv Y_{O,0} W_F v_F / Y_{F,0} W_O v_O \tag{16}$$

and the reduced variable

$$\gamma \equiv \beta W_F v_F / Y_{F,0}. \tag{17}$$

In terms of these quantities, equations (9)–(12) become

$$\frac{\partial \gamma}{\partial \eta} = \frac{1}{\xi} \frac{\partial}{\partial \xi} \left(\xi \frac{\partial \gamma}{\partial \xi} \right), \tag{18}$$

$$\gamma = 1 \quad \text{at} \quad \eta = 0, \quad 0 \le \xi < c, \tag{19}$$

$$\gamma = -v \quad \text{at} \quad \eta = 0, \quad c < \xi < 1, \tag{20}$$

and

$$\partial \gamma / \partial \xi = 0 \quad \text{at} \quad \xi = 1, \quad \eta > 0. \tag{21}$$

To solve equation (18), subject to the boundary conditions given in equations (19)–(21) and to the additional requirement that γ be bounded, is a straightforward mathematical problem. By the method of separation of variables, for example, it can be shown (through the use of recurrence formulae and other known properties of Bessel functions of the first kind) that

$$\gamma = (1 + v)c^2 - v + 2(1 + v)c \sum_{n=1}^{\infty} \frac{1}{\varphi_n} \frac{J_1(c\varphi_n)}{[J_0(\varphi_n)]^2} J_0(\varphi_n \xi) e^{-\varphi_n^2 \eta}, \tag{22}$$

where J_0 and J_1 are Bessel functions of the first kind—of order 0 and 1, respectively—and the φ_n represent successive roots of the equation $J_1(\varphi) = 0$ (with ordering convention $\varphi_n > \varphi_{n-1}$, $\varphi_0 = 0$). Equation (22) provides the solution to equation (1-49) for β in the present problem.

3.1.4. The flame shape and the flame height

In order to obtain the shape of the flame and the flame height from equation (22), we must introduce approximation 5. If the entire reaction is to occur at a flame sheet, then this surface must divide the flow field into two regions. In one region (which will be identified by the subscript +), no oxidizer is present ($Y_{O+} = 0$), and therefore

$$\beta_+ = Y_F / W_F v_F \tag{23}$$

from equations (6)–(8); in the other region (identified by the subscript $-$),
no fuel is present ($Y_{F-} = 0$), and therefore

$$\beta_- = -Y_O/W_O\nu_O. \tag{24}$$

Equations (23) and (24) actually imply that equation (22) gives the concentration fields for both the fuel and the oxidizer, in their respective regions.

In view of the fact that the mass fractions Y_F and Y_O are nonnegative, equation (23) shows that $\beta_+ \geq 0$ in the fuel region, and equation (24) shows that $\beta_- \leq 0$ in the oxidizer region. Since equation (22) implies that β is continuous in the interior of the duct [see equation (17)],

$$\beta_+ = \beta_-$$

at the flame surface. This condition is compatible with the preceding inequalities only if

$$\beta = 0 \tag{25}$$

at the flame sheet. Hence, setting $\gamma = 0$ in equation (22) provides a relation between ξ and η that defines the locus of the flame. The shape of the surface obtained in this manner is shown in Figure 3.2 for $c = \frac{1}{2}$ and for two different values of ν.

The flame height is obtained by solving equation (22) for η after setting $\xi = 0$ for overventilated flames or $\xi = 1$ for underventilated flames

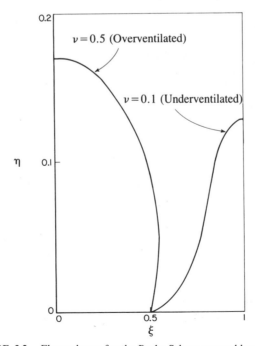

FIGURE 3.2.   Flame shapes for the Burke-Schumann problem ($c = \frac{1}{2}$).

(and, of course, $\gamma = 0$ in either case). Since flame heights are generally large enough to cause the factor $e^{-\varphi_n^2\eta}$ to decrease rapidly as n increases at these values of η, it usually suffices to retain only the first few terms of the sum in equation (22) for this calculation. Neglecting all the terms except $n = 1$, we obtain the rough approximation

$$\eta = \frac{1}{\varphi_1^2} \ln\left\{ \frac{2(1 + v)cJ_1(c\varphi_1)}{[v - (1 + v)c^2]\varphi_1 J_0(\varphi_1)} \right\} \tag{26}$$

for the dimensionless flame height of an underventilated flame. The first zero of $J_1(\varphi)$ is $\varphi_1 = 3.83$.

The flame shapes and flame heights obtained from equation (26) by Burke and Schumann are in fairly good agreement with experiments; the agreement is surprisingly good, considering the drastic nature of some of the approximations, which we shall now discuss.

3.1.5. The flame-sheet approximation

Since the flame-sheet approximation is common to all of the problems considered in Sections 3.1–3.3, it is of interest to study the character of this approximation in greater detail. The significance of the approximation can be addressed from the viewpoint of the general interface conservation condition given in equation (1-58).

In view of equation (1), equation (1-43) implies $w_F/W_F v_F = w_O/W_O v_O$. Hence, dividing equation (1-58) with $i = O$ by $W_O v_O$, dividing equation (1-58) with $i = F$ by $W_F v_F$, and subtracting the second result from the first,* we find that

$$\int_{\mathscr{A}_I} \left[\frac{\rho_+ Y_{F+}}{W_F v_F} (\mathbf{v}_+ + \mathbf{V}_{F+}) - \frac{\rho_- Y_{F-}}{W_F v_F} (\mathbf{v}_- + \mathbf{V}_{F-}) \right.$$
$$\left. - \frac{\rho_+ Y_{O+}}{W_O v_O} (\mathbf{v}_+ + \mathbf{V}_{O+}) + \frac{\rho_- Y_{O-}}{W_O v_O} (\mathbf{v}_- + \mathbf{V}_{O-}) \right] \cdot \mathbf{n}_+ \, d\mathscr{A} = 0$$

in the steady state. Here the unit vector normal to the flame sheet ($\mathbf{n}_+$), as well as all of the other symbols, have the same meaning as in Section 1.4. When there is no fuel on one side of the surface ($Y_{F-} \equiv 0$) and no oxidizer on the other side ($Y_{O+} \equiv 0$), the differential form of this equation is

$$\frac{\rho_+ Y_{F+}}{W_F v_F} (\mathbf{v}_+ + \mathbf{V}_{F+}) \cdot \mathbf{n}_+ = -\frac{\rho_- Y_{O-}}{W_O v_O} (\mathbf{v}_- + \mathbf{V}_{O-}) \cdot \mathbf{n}_+. \tag{27}$$

Equation (27) states formally that fuel and oxidizer must flow into the flame sheet from opposite sides in stoichiometric proportions.

---

* These operations are necessary because if all of the reactants are to be consumed at the interface, then $\lim_{\mathscr{V} \to 0} \int_{\mathscr{V}} w_i \, d\mathscr{V} \neq 0$ [that is, $w_i \to w_i' \, \delta(y - y_I)$ in the notation of Section 1.4].

Since equations (23)–(25) imply that $Y_{F+} = 0$ and $Y_{O-} = 0$, there is no fuel or oxidizer present at the flame sheet. This implies that the terms involving $\mathbf{v}_+$ and $\mathbf{v}_-$ (the convective terms) in equation (27) vanish, but it does not imply that the terms involving $\mathbf{V}_{F+}$ and $\mathbf{V}_{O-}$ vanish because

$$Y_i \mathbf{V}_i = -D\nabla Y_i, \quad i = F+, O-$$

[see equation (1-37)], and the concentration gradients remain nonzero at the flame sheet. Thus, for a diffusion flame, equation (27) reduces to the condition that the diffusion rates (which are now equal to the total rates) of transport of fuel and oxidizer to the flame sheet must be in stoichiometric proportions. In the coupling-function formulation, this condition is an automatic consequence of the continuity of the gradient of β at the flame.

Criteria for the validity of the flame-sheet approximation may be developed by analyzing the structure of the sheet (see Section 3.4). For calculation of flame shapes in the Burke-Schumann problem, the approximation usually is well justified, although uncertainties arise for strongly sooting flames.

3.1.6. The validity of the other approximations

Let us finally consider the validity of assumptions 1–4 of Section 3.1.2.

Assumptions 1 and 2 are mutually inconsistent in this problem. The approach flow in the inner and outer pipes will be parallel to the axis (assumption 1) and will have constant density, but the initial velocity profile will not be uniform (unless special experimental precautions are taken) because the velocity must approach zero at the walls. A velocity profile corresponding to fully developed pipe flow would be more reasonable, and this would invalidate assumption 2 in the approach stream; some consideration has been given to influences of inlet velocity profiles [5–8]. Furthermore, the heat release in the flame increases the temperature and therefore decreases the density of the isobaric stream, thus causing the velocity to increase according to equation (3). However, since the heat release is not uniform in planes perpendicular to the axis of the tube, the density decrease will also cause the flow to expand, and the streamlines will diverge from the hot regions (compare Figure 5.2). Therefore, the flame produces non-one-dimensional flow, violating assumption 1. Assumptions 1 and 2 appear to constitute the most unrealistic approximations in the present analysis; they are best justified by the analytical simplifications that they produce.

Assumption 3 has previously been shown to constitute a reasonably accurate approximation. However, uncertainties in the temperature and composition at which ρD is best evaluated arise from the difficulties associated with assumptions 1 and 2 and are particularly severe if the values of ρD for fuel and oxidizer differ appreciably. Assumption 4 is acceptable when the flame height is large enough (in comparison with the tube diameter) for the transport of fuel and oxidant to the flame to occur principally through radial diffusion; this can be shown to be true if the dimensionless parameter vb/D is sufficiently large. Assumption 4 can be eliminated at the expense of additional analytical complexity. Other refinements can also be introduced, further complicating the algebra; for example, the product ρD may be assigned different values in the fuel and oxidizer regions.

3.1.7. Comments on the formulation and the analysis

All the assumptions of Section 1.3 underlie the formulation. Assumption 1 deserves special mention because buoyancy forces on the hotter, less dense gas in the region of the flame often distort the shape of the flame sheet. A Froude number, $\mathrm{Fr} = v^2/(ag)$ (where g denotes the acceleration of gravity), measures the relative importance of inertial and buoyant forces. If Fr is sufficiently small (for example, $\mathrm{Fr} \ll \Delta\rho/\bar{\rho}$, where $\Delta\rho$ is a representative density difference and $\bar{\rho}$ a mean density), then the flame heights are controlled by buoyancy. Correlations of measured flame heights in the form $h/a \sim \mathrm{Fr}^n$, where h is the flame height and $\frac{1}{5} \leq n \leq \frac{1}{3}$, are available for $b \to \infty$ under buoyancy-influenced conditions with negligible viscous forces [9]–[11].* The result of equation (26), namely, $h/a \sim va/D \equiv \mathrm{Pe}$, is a Peclet-number (Pe) dependence that cannot be correlated with Fr and that holds in a "momentum-controlled" regime that occurs for sufficiently large values of Fr—for example, at sufficiently large v.

---

* If t is the time for a fuel element to travel from the burner port to the flame tip, then dimensionally $D \sim a^2/t$ and $h \sim \bar{v}t$, where $\bar{v}$ is an average velocity of the element. The analysis leading to equation (26) corresponds to setting $\bar{v}$ equal to the exit velocity v. If, instead, an estimate for domination by buoyancy is applied, $\bar{v} \sim gt \Delta\rho/\bar{\rho} \sim gt$ (roughly), then we obtain $t \sim \sqrt{h/g}$ and $h \sim ga^4/D^2$ from the starting expressions for h and D. This estimate is not immediately applicable for deriving the correlations [9]–[11] because they pertain to large-scale regimes, typically strongly turbulent, in which D here must be viewed as an effective diffusivity that depends on v, a, and Fr. The estimate might be thought to be usable directly for small-scale, laminar, buoyancy-controlled flames. For these situations, however, it has been suggested by Roper [see F. G. Roper, C & F **29**, 219 (1977) and F. G. Roper, C. Smith, and A. C. Cunningham, C & F **29**, 227 (1977)] that one further aspect of the physics must be considered. Associated with the change in v is a change in flame width, so that the characteristic transverse length for diffusion becomes an average value $\bar{a}$, different from a, giving $D \sim \bar{a}^2/t$. Imposition of a condition for conservation of fuel mass in axisymmetric flow, $\bar{a}^2\bar{v} \sim a^2 v$, then surprisingly results in cancellation of g from the estimate for h and therefore fortuitously produces a result in agreement with equation (26) concerning the dependence of h on v, a, and D. In other configurations the cancellation does not occur. For example, for a planar, two-dimensional slot burner of width a in the open atmosphere, the reader may use these methods to derive $h \sim va^2/D$ for the planar analog of the analysis underlying equation (22) but $h \sim (va)^{4/3}/(gD^2)^{1/3}$ for the buoyancy-dominated flow. There is evidence that the latter estimate is better under typical laboratory conditions and that buoyancy ceases to dominate only at burner dimensions below about 1 mm if $v \sim 10$ cm/s (that is, roughly, $\mathrm{Fr} \gtrsim 1$). Thus buoyancy usually plays a significant role in coflow laminar diffusion flames in the laboratory.

If v becomes too large, then the ratio of inertial to viscous forces (characterized by a Reynolds number, Re $= \bar{\rho}va/\bar{\mu}$, where $\bar{\mu}$ is a representative average value of the coefficient of viscosity) becomes large enough to cause the flow to be turbulent. Turbulence occurs in most practical burners and introduces qualitative differences from the present predictions; the flame fluctuates rapidly and on the average is thicker, with h/a approximately independent of v or a [12]. This flame-height behavior may be obtained from the present analysis by replacing D by a turbulent diffusivity that is assumed to be proportional to the product va. However, there are many properties of turbulent diffusion flames that cannot be predicted well by a simple approach of this kind (see Section 10.2).

Since increasing a increases Re and decreases Fr, there is a critical value of a above which the flow in the momentum-controlled regime always is turbulent. If Re_c is the critical value of Re above which the flow becomes turbulent (a geometry-dependent value perhaps on the order of 2000), then the analysis that has been given here is justified roughly for $\sqrt{ag\Delta\rho/\bar{\rho}} \lesssim v \lesssim \text{Re}_c\bar{\mu}/\bar{\rho}a$, a range that vanishes if $a \gtrsim (\text{Re}_c^2\bar{\mu}^2/g\bar{\rho}\Delta\rho)^{1/3}$. Large flames often encounter turbulence already in the buoyancy-controlled regime and pass into the momentum-controlled regime under fully turbulent conditions. The laminar analysis is justified best for relatively small laboratory flames.

The burning of a fuel jet issuing into an infinite oxidizing atmosphere may be viewed as the limit as $b \to \infty$ of the problem formulated here. Many experiments and some analyses [13] apply to this limit. Although the problem may be formulated directly with $b = \infty$, it is an interesting exercise to derive the solution from equation (22) by investigating the limit in which $c \to 0$ with ξ/c and η/c^2 held fixed. The result,

$$(\gamma + v)/(1 + v) = \int_0^\infty J_1(\omega)J_0(\omega\xi/c)e^{-\omega^2\eta/c^2}\, d\omega,$$

obtained with $\omega = c\varphi_n$ and $c(\varphi_{n+1} - \varphi_n) \to \pi c \to d\omega$, is the same representation that would be found by applying a Hankel transform to the directly formulated problem with $b = \infty$. By putting $\gamma = 0$ and $\xi = 0$ in this formula, we obtain an integral expression for the flame height, which becomes $h/a = (va/D)(1 + v)/4v$ if hD/va^2 is large (that is, if $v \ll 1$, a situation that often is encountered in practice).

It may be observed that only the fuel and oxidizer concentration fields have been considered in finding the flame shape. The nature of the boundary conditions makes it unnecessary to study the temperature- and product-concentration fields when the stated assumptions are adopted. If temperature or product concentrations are desired, they may be calculated a posteriori, in terms of the known fuel or oxidizer fields, by solving equation (1-49) for β_T and β_i with $\alpha_1 = \alpha_F$, for example. Temperatures at the flame sheet calculated in this way usually are too high (see Section 3.4).

3.2. THE OXIDATION OF CARBON AT THE WALLS OF A DUCT

3.2.1 Definition of the problem

There are many chemically reacting flow processes in which, under certain circumstances, the reaction takes place only at interfaces and no flame appears in the gas. The entire field of heterogeneous catalysis (see Section B.2.6), as well as all heterogeneous processes whose basic mechanisms are treated in Section B.4, fall within this category. Specific examples include the oxidation of some solid fuels (for example, carbon) and the surface recombination of various dissociated species (for example, oxygen). In one sense, processes of this type represent a degenerate kind of diffusion "flame" that is comparatively easy to analyze because the location of the "flame sheet" is known in advance. For illustrative purposes, we shall briefly discuss a very simple example. Similar problems are treated in [14]–[23], the last four of which constitute fairly comprehensive reviews of the subject. Present research is directed toward enlarging the class of soluble flow configurations, analyzing more complex surface reactions, including effects of simple homogeneous reactions and exploring peculiarities of convective-diffusive interactions.

Let us consider a rectangular duct (with sides of lengths a and b in the x and y directions, respectively), the walls of which are coated with carbon, and through which oxygen is passed with velocity v (in the z direction), as illustrated in Figure 3.3. The oxygen will be assumed to react instantaneously at the walls according to the process

$$2C + O_2 \rightarrow 2CO. \tag{28}$$

We wish to determine the distribution of the mass fraction of oxygen (Y) through the tube when the mass fraction at the inlet ($z = 0$) is $Y = Y_0$.

3.2.2. The nature of carbon combustion

Carbon combustion is an interesting illustration of gas-solid reactions [24] and also is relevant to the practical subject of the combustion of coal [25], since later stages of coal combustion, following devolatilization, mainly involve carbon burning. Detailed studies of the burning of solid fuels often reveal complexities peculiar to the fuel. For carbon there are a number of complexities that affect the simplifications applicable in models and, for example, limit the range of utility of equation (28).

The nature of carbon combustion depends strongly on the temperature of the fuel. At temperatures above 4000K, sublimation of solid carbon becomes nonnegligible at atmospheric pressure, and gas-phase combustion processes may be important. At lower temperatures, oxygen

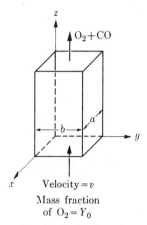

FIGURE 3.3. Diagram of a rectangular duct with carbon oxidation occurring at its walls.

may diffuse to the carbon surface, be adsorbed, and react there. The primary product of this surface reaction is CO_2 at temperatures below about 800K and CO at higher temperatures [26]. Thus equation (28) can be a good approximation only if the surface temperature of the carbon is between about 800K and 4000K (we shall see that there are additional restrictions).

The surface chemistry responsible for the interchange between CO_2 and CO as products is uncertain. A number of different theories have been qualitatively successful [27]–[31], but it is unclear that any of them are fundamentally correct. The Strickland-Constable scheme [29], [32], [33] seems popular; it involves two types of adsorption sites, S_1 and S_2, on which equilibrium adsorption and finite-rate reaction of oxygen occur (at different rates), with the sites being interconvertible ($S_1 \rightleftharpoons S_2$) at a finite rate. One viable alternative (developed by Ubhayakar in unpublished work) involves only one type of adsorption site, S_1, but two types of equilibrium adsorption of oxygen, one corresponding to $S_1 + O_2$ and the other to $2S_1 + O_2$, with a different finite-rate reaction occurring subsequent to each type of adsorption; one of these reactions produces CO_2 and the other CO. There are other alternatives (for example [30] and [31]). The rate constants and activation energies for the elementary steps differ for the different schemes; the overall rates obtained are not of the simple Arrhenius form but instead usually exhibit a maximum in the rate of carbon removal at a surface temperature between 1200K and 2500K, followed by a minimum at a temperature a few hundred degrees above that of the maximum. Each of the predictive schemes agrees with at least one set of experimental data, but data obtained by different authors differ.

A number of complications lead to differences in experimental results. Aside from possible catalytic influences of impurities in the carbon, there are different structural types of carbon that have been studied. These include amorphous carbon (typically randomly oriented graphite crystals) and pyrolytic graphite (a high-density, anisotropic carbon with distinctly different characteristics of base and edge planes). Amorphous carbon often tends to be porous; rates of diffusion of oxygen into pores can influence burning rates of porous materials. There are models that show how pore diffusion tends to reduce measured values of overall activation energies [26], [34]. Pore diffusion is seldom ever significant for pyrolytic graphite, but oxidation rates differ appreciably on the base and edge surfaces [35], [36]. An average of these rates would be relevant for most combustion processes.

As is seen in Section B.4, if the reaction rate at the surface and the gas pressure are high enough, then the burning rate is controlled by the rate of diffusion in the gas. The occurrence of this diffusion-controlled regime is well established for carbon combustion [31], [37]–[39]. The following analysis will be restricted to this limit, which ceases to apply if the dimensions of the carbon materials become too small [39], [40], [41].

Even when the reaction rate at the surface is effectively infinite, chemical kinetics in the gas can influence the carbon-burning rate. Equation (28) requires that residence times in the gas are short enough to prevent chemical depletion of O_2 from occurring before oxygen reaches the solid surface. At sufficiently high temperatures—for example, above about 3500K—a substantial amount of dissociation of O_2 may occur [39]; O atoms are appreciably more reactive than O_2 molecules on carbon surfaces [42]. At lower temperatures, oxygen dissociation is unimportant, but O_2 may react with CO in the gas to form CO_2 if the residence time is long enough [43], [44]. An alternative to equation (28) is a model in which $CO + \frac{1}{2}O_2 \rightarrow CO_2$ at a flame sheet in the gas and $CO_2 + C \rightarrow 2CO$ at the carbon surface [39], [41], [43], [45]. The rate of oxidation of carbon by CO_2 is so slow that this alternative seems reasonable only for fairly high surface temperatures and fairly low oxygen concentrations in the gas [41]. Uncertainties remain concerning influences of gas-phase chemistry [44].

From this discussion it may be seen that, depending on the conditions of temperature, pressure, composition, geometrical configuration, and gas velocity, there are a variety of different regimes of carbon combustion. The following analysis applies in a particular regime, typically with surface temperatures on the order of 2000K, appreciably lower gas temperatures, and oxygen-rich atmospheres at pressures on the order of atmospheric, for laminar flows with relatively short residence times in small ducts with dimensions on the order of centimeters.

3.2.3. Analysis

For simplicity, we shall again employ assumptions 1–5 of Section 1.3, the hypotheses of steady, low-speed flow, and assumptions 1–4 of Section 3.1.2. The other assumptions in Section 1.3 are not necessary here because we are studying a binary, nonreacting gas mixture and are not considering the temperature field. It is also found that the most severe approximations of Section 3.1.2 (that is, assumptions 1 and 2) can be avoided in these relatively elementary problems [16], [20], [23];* however, we shall not introduce such refinements in the present illustrative treatment.

Under the stated assumptions, equation (1-44) is valid for $\alpha = Y$ and with $\omega = 0$. In the Cartesian coordinate system shown in Figure 3.3 (and with assumptions 1–4 of Section 3.1.2), this governing equation reduces to

$$\left(\frac{v}{D}\right)\frac{\partial Y}{\partial z} = \frac{\partial^2 Y}{\partial x^2} + \frac{\partial^2 Y}{\partial y^2}. \tag{29}$$

The boundary conditions for equation (29) are that Y is bounded, that $Y = Y_0$ at $z = 0$, and that $Y = 0$ at $x = 0$, $x = a$, $y = 0$, and $y = b$ (that is, at the walls of the duct). This last condition follows directly from an analysis in the spirit of Section B.4; in the steady state, as the specific rate constant for the surface reaction approaches infinity, the concentrations of the reactant on the surface and in the gas phase adjacent to the surface both approach zero, and steps 1 and 5 of Section B.4 (the diffusion transport steps) become rate controlling. For recombination processes, the limit of an infinite surface rate constant is usually referred to as the case of a *perfectly catalytic wall*. Systems with finite surface reaction rates have been studied in [19] and [20], for example.†

By the method of separation of variables, the solution to equation (29), subject to the given boundary conditions, is found to be

$$Y = \left(\frac{4}{\pi}\right)^2 Y_0 \sum_{m=0}^{\infty} \sum_{n=0}^{\infty} \frac{1}{(2m+1)(2n+1)} \sin\left[\frac{(2m+1)\pi x}{a}\right]$$

$$\times \sin\left[\frac{(2n+1)\pi y}{b}\right]\exp\left\{-\left[\left(\frac{2m+1}{a}\right)^2 + \left(\frac{2n+1}{b}\right)^2\right]\pi^2\left(\frac{D}{v}\right)z\right\}. \tag{30}$$

Equation (30) gives the desired oxygen concentration field. At large distances from the entrance to the duct ($z \to \infty$), equation (30) becomes

$$Y = \left(\frac{4}{\pi}\right)^2 Y_0 \sin\left(\frac{\pi x}{a}\right)\sin\left(\frac{\pi y}{b}\right)\exp\left[-\left(\frac{1}{a^2} + \frac{1}{b^2}\right)\pi^2\left(\frac{D}{v}\right)z\right], \tag{31}$$

which shows that the initially uniform concentration profile assumes the shape of a half of a cycle of a sine wave and decays exponentially with the decay length

$$\left(\frac{v}{D}\right)\frac{1}{\pi^2}\left(\frac{1}{a^2} + \frac{1}{b^2}\right)^{-1}.$$

---

* Boundary-layer theory (see Chapter 12) can profitably be utilized to provide more accurate results in many of these problems. The reader is referred to cited reviews for interesting fluid-mechanical ramifications.

† In general, the boundary conditions for equation (29) become nonlinear in Y when the surface rates are finite.

3.3. THE BURNING OF A FUEL PARTICLE IN AN OXIDIZING ATMOSPHERE

3.3.1. Background and definition of the problem

A variety of phenomena are exhibited by the burning of a spherical fuel particle in an infinite oxidizing atmosphere. Here we shall consider one of the simplest situations, the quasisteady, spherically symmetrical burning of a liquid fuel that vaporizes and reacts in a gas-phase flame, producing gaseous products that flow and diffuse to infinity. However, at the outset it is of interest to indicate some of the complexities that may arise in other situations. Because of the diversity of high-temperature oxidation mechanisms of solids [24], [46], these complexities often are associated with the burning of solid fuels.

The relative volatility of the fuel and its oxide has an important influence on the combustion mechanism of the particle [47]. Carbon is an example of a nonvolatile solid fuel with volatile—in fact gaseous—reaction products; therefore, as indicated in the previous section, reactions tend to occur on the surface of a carbon particle during combustion. Among the recent theoretical studies on the burning of carbon particles that go beyond the computation of steady rates of burning [39], [41], [48]–[53] are analyses of extinctions that occur in cool ambient atmospheres when the particle becomes sufficiently small [49], [51], [52], investigations of conditions for ignition of a particle by a hot gas [51], [53] and calculation of combustion oscillations associated with diffusion of oxidizing gas into the interior of the particle, followed by internal chemical reactions [48]. To present analyses of these phenomena is beyond the scope of the discussion here.

If reaction products are less volatile, then their condensation can influence the combustion mechanism. For example, although boron is less volatile than B_2O_3, this oxide is sufficiently nonvolatile for its liquid phase to play a role in the combustion of boron particles under many circumstances [54]. Relatively volatile fuels with nonvolatile combustion products, such as magnesium and aluminum, practically always exhibit burning mechanisms influenced by product condensation. In the presence of product condensation, there are a number of possible modes of quasisteady burning. Condensed products may accumulate on the surface of the particle, may accumulate in a shell at a reaction sheet located at some distance from the surface of the particle, may accumulate in a shell at a condensation sheet located outside a thin primary gas-phase reaction sheet, or may flow and diffuse to infinity in the form of fine particles. The last of these processes may be enhanced by thermophoretic motion (see Section E.2.5) of fine particles away from the hottest reaction zone under the influence of the temperature gradient [55]. Many theoretical analyses of the various types of combustion processes have been published [55]–[62]. Law's models [60], [61] of different classes of combustion mechanisms provide a simplified but relatively accurate and useful framework within which many processes of combustion of metal particles can be placed and from which burning rates and flame locations may be obtained. Nevertheless, there is room for much additional work, directed, for example, toward ascertaining the physical reason for the occurrence of one burning mode rather than another or toward inclusion of thermophoretic effects, which have not yet been studied in the context of particle burning.

Liquid fuels with gaseous products do not share these complexities in their combustion mechanisms. The most ideal experiment in liquid-droplet combustion would be the steady-state, spherically symmetrical burning of a liquid sphere, internally supplied with fuel—for example, through a porous, spherical support in a gravity-free environment. The absence of gravity is significant because buoyant forces distort the shape of the flame significantly, unless the fuel particle is very small. A method for performing this most ideal experiment has not yet been devised. A conceptual design that most closely approximates this ideal involves the quasisteady burning of a free droplet in a laboratory in space [63]. Although this experiment has not yet been implemented, droplet burning in freely falling chambers on the earth has been studied [64]–[69]. The maximum test times achievable in these experiments are on the order of 2 s, insufficient for the complete burning of droplets with diameters on the order of 1 mm or larger; although spherical flames have been established, their diameters do not attain quasisteady values during the testing time [67].

Irrespective of whether a freely falling chamber is used, three different experimental techniques have been devised for studying the burning of a fuel droplet in an oxidizing atmosphere (see Figure 3.4). In part (a), fuel has been

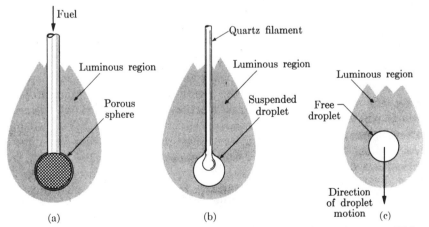

FIGURE 3.4. Diagrams of experiments on the burning of a fuel droplet in an oxidizing atmosphere. (a) Porous sphere. (b) Suspended droplet. (c) Falling droplet.

supplied internally to a stationary porous sphere at a steady rate, which is just sufficient to maintain a liquid film on the surface of the sphere during burning [70]–[80]. In part (b), fuel droplets have been suspended on stationary quartz fibers, and the rate of decrease of the diameter of the droplet has been measured after ignition [65], [81]–[92]. In part (c), small, freely falling fuel droplets have been ignited and observed during burning [82], [93]–[103]. Each of the techniques possesses certain intrinsic advantages over the others; for example, (a) is steady, (b) yields accurate results for the diameter as a function of time, and (c) is applicable to very small droplets.

In experiment (a) it is observed that the total mass per second that must be supplied to the sphere in order to maintain steady burning is

$$\dot{m} = K'r_l, \tag{32}$$

where r_l is the radius of the sphere and K' is independent of r_l. In experiments (b) and (c) it is found that after an initial unsteady period, the square of the droplet diameter decreases linearly with time; namely,

$$d_{l,0}^2 - d_l^2 = K(t - t_0), \tag{33}$$

where d_l is the droplet diameter, t is time, $d_{l,0}$ is the value of d_l at $t = t_0$, and K, which is called the **evaporation constant**, is independent of t. The *d-square law* in equation (33) also expresses the relationship between diameter and time for vaporizing droplets in the absence of combustion, as was observed by Sreznevsky [104] and explained by Langmuir [105]. Some of the objectives of theoretical analyses of droplet combustion have been to derive equations (32) and (33) and to predict observed flame shapes and temperature and concentration profiles. Successful analyses were first presented in [71], [72], [81], [106], and [107]; these analyses are discussed in textbooks [1], [108], and comprehensive reviews are available [109]–[115].

3.3.2 Assumptions

Although the shape of the reaction zone surrounding a burning droplet is not spherical in most experiments, salient features other than the flame shape [for example, equations (32) and (33)] are found to be predicted with good accuracy by analyses that postulate spherical symmetry. Spherical symmetry is the primary simplifying hypothesis that will be adopted in the following discussion. The theoretical model of the burning droplet is illustrated in Figure 3.5. The chemical reaction will be assumed to be described by equation (1), and all the assumptions of Section 1.3 will be adopted. Later it will also be assumed for simplicity that the flow variables (in particular, temperature) are constant inside the droplet. Departures from the main assumptions will be discussed subsequent to the analysis.

An assumption of Section 1.3 that merits special mention for this problem at the outset is the hypothesis that steady conditions prevail. There

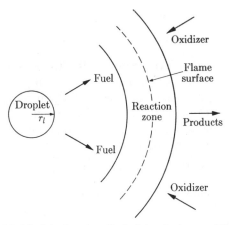

FIGURE 3.5.   Model of the burning of a fuel droplet in an oxidizing atmosphere.

is little doubt of the validity of this assumption for experiment (a); however, in experiments (b) and (c), there are two reasons for questioning the assumption. First, the initial unsteady ignition period may conceivably occupy the entire lifetime of the droplet. Second, since the size of the droplet is continually decreasing with time, at best quasisteady conditions can be attained, in which steady-state conservation equations determine the mass per second leaving the droplet ($\dot{m}$), after which the simple, overall mass-conservation condition

$$\dot{m} = -\frac{d}{dt}\left(\frac{4}{3}\pi r_l^3 \rho_d\right) = -4\pi r_l^2 \rho_d \frac{dr_l}{dt} \qquad (34)$$

($\rho_d \equiv$ liquid density = constant) determines the rate of change of diameter ($d_l = 2r_l$) of the droplet. A theoretical study of these two effects is presented in [116]. The ignition period usually is predicted to occupy less than 20% of the total burning time, in agreement with a number of experimental observations [93], [117], [118]. During the remainder of the burning time, the quasi-steady-state approximation [employing equation (34)] leads to a slight overestimate of $|dr_l/dt|$; the fractional error is somewhat less than the ratio of the gas density to ρ_d, which becomes appreciable only as the liquid approaches its critical point. An entirely different treatment, one which is basically unsteady, is required when the liquid is above the critical point [119]–[124].*

_____

* In highly supercritical regimes, the combustion process resembles diffusion of a gas from a point or distributed source; near critical conditions the strong dependence of the diffusion coefficient of the fuel on temperature can aid in establishing an approach to quasisteady conditions in the fuel-rich zone for situations under which quasisteadiness otherwise would not be expected. The theoretical predictions of [119]–[124] are in qualitative agreement with experimental observations in [77], [88], and [92] made near the critical point. Conditions of criticality may be attained in diesel and rocket combustion.

Equation (34) can be written in the form

$$d(d_l^2)/dt = -2\dot{m}/\pi\rho_d r_l, \tag{35}$$

which explicitly shows that if equation (32) is satisfied (that is, $\dot{m} \sim r_l$), then equation (33) will also be satisfied (that is, d_l^2 will vary linearly with t). Therefore, a steady-state theory that leads to equation (32) automatically yields equation (33) through the quasisteady hypothesis. Equations (32), (33), and (35) can be used to express the evaporation constant K in terms of the proportionality factor K', namely,

$$K = 2\dot{m}/\pi\rho_d r_l = 2K'/\pi\rho_d. \tag{36}$$

Equation (36) determines K from the results of the following steady-state analysis.

3.3.3. Analysis predicting the burning rate

In a one-dimensional, spherically symmetrical system, the overall continuity equation, equation (1-23), can be written as

$$\dot{m} = 4\pi r^2 \rho v = \text{constant}, \tag{37}$$

where r is the radial coordinate and v is the radial velocity. The burning rate $\dot{m}$ is found to appear as an eigenvalue of equation (1-49) and will be obtained by applying equation (1-49) to β_T with $\alpha \equiv \alpha_O$.

Employing the expression for the divergence in spherical coordinates, we find that equations (1-45) and (1-49) give

$$\frac{d}{dr}(r^2 \rho v \beta) = \frac{d}{dr}\left(r^2 \rho D \frac{d\beta}{dr}\right), \tag{38}$$

which reduces to

$$\frac{d\beta}{dr} = \frac{d}{dr}\left[\left(\frac{4\pi r^2 \rho D}{\dot{m}}\right)\frac{d\beta}{dr}\right] \tag{39}$$

when use is made of equation (37). In terms of the dimensionless independent variable

$$\xi \equiv \dot{m}\int_r^\infty (4\pi r^2 \rho D)^{-1}\, dr, \tag{40}$$

equation (39) becomes

$$d\beta/d\xi = -d^2\beta/d\xi^2, \tag{41}$$

the solution to which is

$$\beta = A + Be^{-\xi}, \tag{42}$$

where A and B are constants. Equation (42) is the solution for the profile of any coupling function β. The values of A and B depend on the boundary conditions, which must be discussed separately for each coupling function.

The coupling between the temperature and oxidizer concentration profiles is determined by the function

$$\beta_T \equiv \alpha_T - \alpha_O = \int_{T^0}^{T} c_p \, dT/q^0 + Y_O/W_O \nu_O, \tag{43}$$

where the standard heat of reaction at temperature T^0 is

$$q^0 \equiv h_F^0 W_F \nu_F + h_O^0 W_O \nu_O - \sum_{i=1}^{N-2} h_i^0 W_i \nu_i \tag{44}$$

[see equations (1-46), (1-47), and (1)]. The corresponding constants A_T and B_T in equation (42) with $\beta = \beta_T$ are determined by the boundary conditions for the temperature and for the mass fraction of oxidizer.

Equations (1-61) and (1-62) provide conditions on T and Y_O that must be satisfied at the surface of the droplet. If the subscripts + and − refer to conditions outside and inside the droplet, respectively, then equations (1-37) and (1-62) give

$$[\rho v Y_O - \rho D(dY_O/dr)]_+ = 0, \tag{45}$$

since there is no oxidizer inside the droplet (that is, $Y_{O-} \equiv 0$). Equation (45), which merely states that the net mass flow rate of oxidizer at the surface of the droplet is zero, yields the boundary condition

$$(dY_O/d\xi)_l = -Y_{O,l}, \tag{46}$$

where use has been made of equations (37) and (40) and the subscript l identifies conditions in the gas at the droplet surface.

The explicit derivation of the interface condition on T from equation (1-61) is somewhat more lengthy. It is clear on physical grounds that when the temperature of the droplet is constant and uniform, the amount of heat conducted into the droplet must be just sufficient to vaporize the fuel leaving the droplet; that is,

$$[4\pi r^2 \lambda (dT/dr)]_+ = \dot{m}L, \tag{47}$$

where L is the heat of vaporization per unit mass of the fuel at temperature T_l. In order to obtain equation (47) formally from equation (1-61), one must neglect the radiation* and kinetic-energy terms, use $(\nabla T)_- = 0$ and $\mathbf{V}_{i-} = 0$, and employ the mass-conservation conditions, equation (1-62), for the fuel, oxidizer, and product species. Using equations (1-39) and (40) in equation

---

* Radiative heat transfer is not negligible for liquids with high boiling points or for opaque droplets with highly luminous flames. The analysis requires modification for these systems (see Section 3.3.6).

(47), and taking the arbitrary standard reference temperature to be $T^0 = T_l$ for convenience,* we obtain the boundary condition

$$\left[d\left(\int_{T^0}^{T} c_p \, dT \right) \middle/ d\xi \right]_l = -L. \tag{48}$$

Equations (46) and (48) determine the gradient of β_T [equation (43)] at the surface of the droplet in the gas in terms of L and $Y_{O,l}$; thus,

$$\left(\frac{d\beta_T}{d\xi} \right)_l = -\frac{L}{q^0} - \frac{Y_{O,l}}{W_O v_O}. \tag{49}$$

In view of equation (42), equation (49) shows that

$$B_T e^{-\xi_l} = \frac{L}{q^0} + \frac{Y_{O,l}}{W_O v_O}, \tag{50}$$

where a dimensionless burning rate is

$$\xi_l = \dot{m} \int_{r_l}^{\infty} (4\pi r^2 \rho D)^{-1} \, dr \tag{51}$$

[see equation (40)].

An expression for the burning rate can now be obtained from equation (42). When equation (42) is applied to β_T and evaluated at $r = r_l$, we obtain

$$\beta_{T,l} = A_T + B_T e^{-\xi_l}. \tag{52}$$

Evaluating the same equation at $r = \infty$, where $\xi = 0$ [see equation (40)], we find

$$\beta_{T,\infty} = A_T + B_T. \tag{53}$$

By eliminating A_T between equations (52) and (53), it is found that

$$B_T = (\beta_{T,\infty} - \beta_{T,l}) + B_T e^{-\xi_l}. \tag{54}$$

In view of the meaning of $\int_{T^0}^{T} c_p \, dT$ (see Section 1.3) and our convention $T^0 = T_l$, equation (43) implies that

$$\beta_{T,\infty} - \beta_{T,l} = \frac{\sum\limits_{i=1}^{N} Y_{i,\infty} \int_{T_l}^{T_\infty} c_{p,i} \, dT}{q^0} + \frac{(Y_{O,\infty} - Y_{O,l})}{W_O v_O}. \tag{55}$$

Hence, utilizing equation (50) in equation (54) shows that

$$\xi_l = \ln \left[1 + \frac{\sum\limits_{i=1}^{N} \left(Y_{i,\infty} \int_{T_l}^{T_\infty} c_{p,i} \, dT \right) \middle/ q^0 + (Y_{O,\infty} - Y_{O,l})/W_O v_O}{L/q^0 + Y_{O,l}/W_O v_O} \right]. \tag{56}$$

---

* If this convention is not adopted, then additional terms involving $(dY_i/dr)_l$ appear in equation (48).

The fact that equation (56) is an expression for the burning rate can be seen from equation (51); thus, in dimensional notation,

$$\dot{m} = \frac{1}{\displaystyle\int_{r_l}^{\infty} (4\pi r^2 \rho D)^{-1}\, dr}$$

$$\times \ln \left[1 + \frac{\displaystyle\sum_{i=1}^{N} \left(Y_{i,\infty} \int_{T_l}^{T_\infty} c_{p,i}\, dT \right) + (Y_{O,\infty} - Y_{O,l})(q^0/W_O\nu_O)}{L + Y_{O,l}(q^0/W_O\nu_O)} \right]. \quad (57)$$

3.3.4. Discussion of the burning-rate formula

In many respects, equation (57) is of more general validity than the burning rate expressions usually quoted because arbitrary temperature dependences of (ρD) and c_p are permitted in equation (57), and the flame-sheet approximation has not been introduced. However, equation (57) will be useful for computing $\dot{m}$ only if all quantities appearing on the right-hand side of this equation are known. The specific heat at constant pressure for each pure species i ($c_{p,i}$), the heat of vaporization per unit mass for the fuel at temperature $T_l(L)$ and the quantity $q^0/W_O\nu_O$, which is the heat of reaction at temperature T_l for gaseous reactants *per unit mass of oxidizer consumed*, are all fundamental thermodynamic properties of the system and therefore are presumed to be known. In droplet-burning experiments, $Y_{F,\infty} = 0$ (there is no fuel in the ambient atmosphere), and an experimenter who controls the atmosphere in which the droplet burns can adjust the quantities $Y_{i,\infty}$ for the oxidizer and products and the temperature T_∞ to prescribed values. The remaining parameters, $Y_{O,l}$, T_l, and the factor preceding the logarithm, require more detailed discussion.

The value of $Y_{O,l}$ depends on the reaction rate and can be determined, in general, through numerical integration of the conservation equations or through asymptotic analyses of the type indicated in Section 3.4. In most droplet-burning studies it is assumed that $Y_{O,l} = 0$, since nearly all of the oxidizer is expected to be consumed before diffusing to the surface of the droplet. The validity of this assumption has been investigated by numerical integrations [107] and by asymptotic expansions for small values of the ratio of a reaction time to a residence time [125], [126]; values of $Y_{O,l}$ are found to be less than 10^{-2}. In the flame-sheet approximation, all the oxidizer is consumed instantaneously at the spherical surface where the reaction rate (per unit volume) is infinite, and $Y_O \equiv 0$ inside this surface (hence, in particular, $Y_{O,l} = 0$). It is generally quite accurate to put $Y_{O,l} = 0$ except very near extinction conditions (Section 3.4).

The value of T_l in the quasisteady state depends on the nature of the surface gasification process. The surface process may conceivably be a rate process, in which case an analysis analogous to that given in Section B.4 would be required in order to determine T_l, and T_l would be found to depend explicitly on $\dot{m}$. However, except for fuels that undergo surface or condensed-phase chemical reactions, such as depolymerization of polymeric fuels, the rates at which molecules of fuel enter and leave the liquid are sufficiently rapid to maintain surface equilibrium for the low values of $\dot{m}$ commonly encountered in droplet burning. Therefore, T_l is determined by the thermodynamic (phase equilibrium) condition that the partial pressure of the fuel at the surface of the droplet must be equal to the equilibrium vapor pressure of the fuel.* The application of this equilibrium condition entails obtaining the coupling between the fuel and oxidizer concentration profiles (i.e., solving for $\beta_F \equiv \alpha_F - \alpha_O$). However, unless the heat of reaction is extremely low or the fuel is extremely nonvolatile, the heat flux to the droplet surface is large enough to lead to equilibrium droplet surface temperatures that are only slightly below the boiling temperature of the liquid (see, for example, [116]). Therefore, $T_l = T_b$ ($T_b \equiv$ boiling point of the fuel) is a good approximation. See Section 7.3 for a more complete discussion of surface conditions.

Since the factor preceding the logarithm in equation (57) is influenced by the dependence of ρD on r, the variation of $\dot{m}$ with r_l, predicted by equation (57), is not immediately apparent. When ρD is independent of r in the vicinity of $r = r_l$, it can be shown from equation (57) that for small changes in r_l, $\dot{m} \sim r_l/(1 + ar_l)$, where $a =$ constant and usually $|ar_l| \ll 1$. Often $\rho D =$ constant is an acceptable approximation, and the factor preceding the logarithm becomes $4\pi\rho D r_l$, showing that $\dot{m} \sim r_l$. The quantity ρD can be replaced by λ/c_p, since Le $= 1$ [see equation (1-39)], and for many systems, the approximation $c_{p,i} = c_p =$ constant is reasonably accurate. These approximations will be considered further in subsequent discussion.

The results of the preceding discussion can be summarized by means of the approximate formula

$$\dot{m} = \left(\frac{4\pi\lambda r_l}{c_p}\right)\ln\left\{1 + \frac{1}{L}\left[c_p(T_\infty - T_b) + \frac{q^0 Y_{O,\infty}}{W_O \nu_O}\right]\right\}. \qquad (58)$$

This expression agrees with the empirical relationship given in equations (32) and (33) and can be used in equation (36) to determine the evaporation constant. Equation (58) shows that the principal dependence of $\dot{m}$ on the properties of the fluid is expressed by $\dot{m} \sim \lambda/c_p$, that $\dot{m}$ depends logarithmically on the heat of reaction and the heat of vaporization and that $\dot{m}$ is insensitive to the pressure and is not strongly influenced by the temperature.

---

* The droplets under consideration throughout Section 3.3 are so large (for example, greater than $10\,\mu$ in diameter) that effects of surface tension on phase equilibria are negligible. Experimental observations of the burning of single droplets small enough to exhibit this type of influence of surface tension would be difficult to obtain and have not been reported. Thus the surface-tension effects are purely mechanical, aiding maintenance of spherical symmetry.

These results are in agreement with experiment. Theoretical and experimental values of the evaporation constant K are roughly of the order of 10^{-2} cm^2/s.

3.3.5. Predictions of other characteristics of burning droplets

Since equation (42) is valid for any coupling function β, other results can be derived without explicitly invoking the flame-sheet approximation. For example, by considering

$$\beta_F \equiv \alpha_F - \alpha_O = -Y_F/W_F v_F + Y_O/W_O v_O, \tag{59}$$

and by using equations (45) and (56) and the surface conservation condition for fuel,

$$[\rho v Y_F - \rho D(dY_F/dr)]_+ = (\rho v)_+,$$

it can be shown that the mass fraction of fuel in the gas at the surface of the droplet is

$$Y_{F,l} = 1 - \frac{L[1 + Y_{O,\infty} W_F v_F/W_O v_O]}{L + \sum_{i=1}^{N} \left(Y_{i,\infty} \int_{T_l}^{T_\infty} c_{p,i}\, dT \right) + (q^0 Y_{O,\infty}/W_O v_O)}, \tag{60}$$

provided that $Y_{F,\infty} = 0$ and $Y_{O,l} = 0$. However, in order to obtain explicit temperature and concentration profiles and to locate the reaction zone, the flame-sheet approximation must generally be employed as a first approximation if numerical integrations are to be avoided.

In the flame-sheet approximation, the position of the flame is determined by

$$\beta_F = 0, \tag{61}$$

just as in Section 3.1. If quantities at the flame surface are identified by the subscript c, then $\alpha_F = \beta_F$, $\alpha_O = 0$ and $\alpha_T = \beta_T$ for $r < r_c$, while $\alpha_F = 0$, $\alpha_O = -\beta_F$ and $\alpha_T = \beta_T + \alpha_O = \beta_T - \beta_F$ for $r > r_c$. These observations serve to determine the fuel and oxidizer mass-fraction profiles and the temperature profile through equation (42). Representative profiles are shown in Figure 3.6, where Y_P denotes the total mass fraction of products. In many respects curves in Figure 3.6 are in qualitative agreement with the results of numerical integrations for reasonable chemical reaction rates [107] and with experimental results [80].

From equation (61), it can be shown that the value of ξ at the flame sheet is determined by

$$e^{\xi_c} = 1 + Y_{O,\infty} W_F v_F/W_O v_O,$$

which can be combined with equation (56) to show that

$$\frac{r_c}{r_l} = \frac{\ln\left[1 + (1/L)\left(\sum_{i=1}^{N} Y_{i,\infty} \int_{T_l}^{T_\infty} c_{p,i}\, dT + q^0 Y_{O,\infty}/W_O v_O\right)\right]}{\ln(1 + Y_{O,\infty} W_F v_F/W_O v_O)} \tag{62}$$

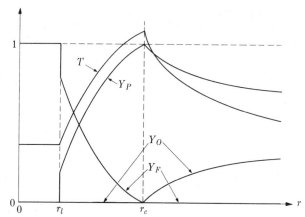

FIGURE 3.6. Temperature and mass-fraction profiles for a burning fuel droplet in the flame-sheet approximation.

when $Y_{O,1} = 0$ and $\rho D = $ constant. Because of the effects of natural convection in droplet-burning experiments, the value of the flame radius r_c predicted by equation (62) is larger than that typically observed by roughly a factor of 2, and, empirically, $(r_c - r_l)$ is more nearly independent of r_l than is r_c/r_l for supported droplets with diameters on the order of a millimeter or more.

3.3.6. Further realities of droplet burning

A number of phenomena excluded by the assumptions given in Section 3.3.2 can modify the burning rate of a droplet. These phenomena often are too complex to be included in accurate theoretical burning-rate analyses and necessitate semiempirical methods in deriving burning-rate formulas. The most important of these additional phenomena will now be considered.

First, it is of interest to write the results that have been obtained in a simplified notation. This may be accomplished by introducing the **transfer number** B, qualitatively representing the ratio of an impetus for interphase transfer to a resistance opposing the transfer. From the viewpoint of energetics, the sum of the heat released per unit mass of oxidizer consumed and the difference in the thermal enthalpy per unit mass between the ambient gas and the gas at the surface of the droplet comprises the impetus for transfer, while the heat of vaporization L per unit mass of fuel vaporized is the resistance. Hence, in equation (58) the logarithmic term is $\ln(1 + B)$. Generalizations in the definition of B (for example, [127]) lead to the logarithmic term in equation (57) being identified as $\ln(1 + B)$; in fact, in droplet burning the factor $\ln(1 + B)$ always appears in the expression for the burning rate. The introduction of B can be useful not only in droplet burning but also for all problems involving interphase transfer, such as for the problem discussed in Section 3.2. For a vaporizing droplet, gas-phase diffusion of the fuel rather than heat conduction is the main driving force for vaporization, and a better definition of B is $B = (Y_{F,l} - Y_{F,\infty})/(Y_{F,d} - Y_{F,l})$, where $Y_{F,d}$ is the mass fraction of fuel deep in the interior of the droplet (usually unity).

To write simple formulas it is useful to define $v \equiv (v_F W_F/v_O W_O)Y_{O,\infty}$, the product of the stoichiometric mass ratio of fuel to oxidizer with the oxidizer mass fraction at infinity. Then equation (62) for the flame-standoff ratio becomes

$$d_c/d_l = \ln(1 + B)/\ln(1 + v), \tag{63}$$

and it is found from equation (60) that the mass fraction of fuel at the droplet surface is

$$Y_{F,l} = (B - v)/(1 + B). \tag{64}$$

Substitution of equation (58) into equation (36) gives

$$K = (8/\rho_d)(\lambda/c_p)\ln(1 + B) \tag{65}$$

for the evaporation constant.

If it is of interest to account for the difference between T_l and T_b, then equation (A-23) may be employed with p representing the partial pressure of the fuel vapor, $X_{F,l}$ times the specified hydrostatic pressure; the mole fraction $X_{F,l}$ may be obtained from equation (1-12)—namely, $X_{F,l} = Y_{F,l}\overline{W}_l/W_F$—where $\overline{W}$ is the average molecular weight $[\sum_{i=1}^{N}(Y_i/W_i)]^{-1}$, and $Y_{F,l}$ is given by equation (64). The value of T_l obtained in this manner is a *wet-bulb* temperature somewhat below T_b.

In the application to the burning of a porous sphere, usually the fuel is not fed into the sphere at its boiling point, or wet-bulb temperature. Under these conditions gasification requires an energy $L + Q$ instead of L, where Q is the additional energy per unit mass of fuel needed for gasification. In the definition of B, for use in equations (63)–(65), the quantity L in the denominator must then be replaced by $L + Q$, as is readily verified by a suitably modified energy balance [compare equation (47), for example]. If the specific heat of the liquid fuel has the constant value c_l and the feed temperature is T_d, then the additional requirement in thermal enthalpy is $Q = c_l(T_l - T_d)$. If the fuel surface is at a sufficiently high temperature for the radiant energy lost from it to be of significance, then in view of equation (E-51), an additional term $\epsilon\sigma T_l^4/(\dot{m}/4\pi r_l^2)$ appears in Q; this causes equation (58) to become nonlinear in $\dot{m}$ and to cease to possess a solution if r_l is too large. A practical concern with porous-sphere experiments is that a significant amount of energy may be transferred to the liquid fuel from the flame that bathes the fine feed line [90]; this effectively decreases Q and may cause it to become negative in extreme circumstances.

For suspended or free droplets that are not initially at the boiling point, there is a contribution to Q given by $(\frac{4}{3})\pi r_l^3 \rho_d c_l(d\overline{T}_d/dt)/\dot{m}$, where $\overline{T}_d$

is the average temperature of the liquid. This contribution decreases with increasing time at a rate that increases as the thermal conductivity of the liquid increases. There have been a number of studies of the heat-up process [116], [128]–[130], including consideration of influences of convective motions of the liquid within the droplet [131]–[133]. Except under somewhat special conditions [133]—for example, for fuels with high boiling points—the heat-up typically occurs sufficiently rapidly to be unimportant during most of the burning, as was indicated in Section 3.3.2.

The possible occurrence of unsteady conditions in the gas phase for free droplets has also received further consideration [134]–[136]. It is evident physically that since a burning droplet exists only for a finite amount of time, there must be an outer region in which quasi-steady conditions are not attained. If the droplet is moving with respect to the gas, then this outer zone is swept away from the droplet relatively quickly, and quasi-steady conditions may prevail throughout the region of importance during most of the burning history. However, in spherically symmetrical burning, the gas-phase unsteadiness may be significant [134]–[136]. Typically, when ignition occurs there is relatively little fuel vapor in the gas, and the flame sheet is initially established closer to the droplet surface than a quasi-steady analysis such as equation (63) would predict. Thus, in addition to the time-dependent outer zone, there is an initial period of unsteady behavior throughout the gas, subsequent to liquid heat-up [135]. This period is usually relatively short, and the main influence of gas-phase unsteadiness on the burning time and on the flame-sheet location typically arises from the later period, during which there is a quasi-steady inner zone and a time-dependent outer zone. For burning in oxidizer-rich atmospheres, equation (63) usually predicts relatively small values of d_c/d_l such that the flame sheet lies in the quasisteady zone, and effects of gas-phase unsteadiness on d_c are relatively small. However, for burning in dilute atmospheres, the flame may lie in the outer zone, causing d_c to increase with time and equation (63) to be inapplicable. Theory predicts that d_c/d_l increases with time [135]; under suitable circumstances it achieves a quasi-steady value given by equation (63) during most of the burning history, but there always is a short period, as the droplet disappears, during which d_c/d_l approaches infinity (unless the flame is first extinguished; see Section 3.4). The predicted tendency for d_c/d_l to increase with time is in agreement with experiments [67] on the spherically symmetrical burning of free droplets in the absence of gravity. A correction factor to equation (65), accounting for aspects of gas-phase unsteadiness [135] is $1 + [(2/\pi)(\rho_\infty/\rho_d)\ln(1 + B)]^{1/2}$, where ρ_∞ is the density of the ambient atmosphere.*

The correction to K just cited remains consistent with the d-square law for the total burning time. If radiant transfer from the flame or from hot

---

* Further considerations may be found in a more recent publication, M. Matalon and C. K. Law, *C & F* **50**, 219 (1983).

surroundings, such as hot chamber walls, becomes a dominant source of energy input to the droplet, then the d-square law becomes a d-law for the usual situation in which the Planck-mean absorption length of the gas (Section E.5.2) is large compared with the droplet diameter. This is evident from the fact that the radiant energy input per unit area at the droplet surface, say q_R, is independent of d_l. A simplified energy balance for the droplet is $4\pi r_l^2 q_R = \dot{m}(L + Q)$ in a radiation-dominated situation, and in view of equation (34), this gives $dr_l/dt = -(q_R/\rho_d)/(L + Q)$, a d-law. The d-square law may be viewed as arising from the fact that the conductive input flux, $[\lambda(dT/dr)]_+$ in equation (47), is inversely proportional to r_c or r_l; gas-phase gradients steepen as the droplet size decreases.

An interesting paradox is posed by considering momentum conservation in connection with the flame-sheet approximation. Since Figure 3.6 shows that the temperature gradient is discontinuous at the flame sheet, it follows from the ideal-gas law with constant pressure p that the density gradient is discontinuous; therefore, according to equation (37) the velocity gradient is discontinuous. From equation (1-16) we see that this introduces a δ-function in the divergence of the shear-stress tensor at the flame sheet, and from equation (1-20) it is evident that there is nothing in momentum conservation that can balance this divergence, except possibly a discontinuity in p at the flame sheet, which would contradict the original hypothesis of constant p. The paradox has been resolved independently by Sepri [137] and by Cooper and Clarke [138], who showed that there is an adjustment region with thickness of order $r_c M^2$ (where M is a Mach number) just outside the flame sheet; in fact, p experiences a discontinuity in its gradient at the flame sheet and then decreases by a fractional amount of order M^2 through the adjustment region. Since M is small, the paradox concerns a small effect, which may, however, influence fluid flows appreciably in configurations that are more complex (for example, in turbulent flows). The phenomenon provides one mechanism for combustion to affect the fluid dynamics.

The analysis that has been given here relies on the assumption that the Lewis number is unity [equation (1-39)]. Although there have been a number of theoretical studies that consider Le $\neq$ 1 and that take into account variable transport properties and specific heats [106], [139]–[142], questions remain concerning these effects. The more general theories also are more complex and therefore offer greater resistance to attempts at extracting generally valid conclusions. Effects of variable properties and of Le $\neq$ 1 are significant in that they can introduce uncertainties in excess of a factor of 2 in K and in d_c/d_l. A few relevant principles have been identified and deserve to be discussed here.

If Le $\neq$ 1, then a formulation on the basis of coupling functions is less convenient. Equations (1-31), (1-32) and (1-33) become good starting equations for an analysis, especially if the flame-sheet approximation is adopted, since then the flux fractions ϵ_i remain constant on each side of the flame. These equations, or simplifications thereof, have in fact been employed in a number of the theories [106], [142]. A result, evident from equation (1-33) and also physically clear, is that a species i must be transported ($\epsilon_i \neq 0$) for its specific heat $c_{p,i}$ to play any role in the quasisteady combustion. In this respect the coupling-function theory can tend to be misleading; equation (58), for example, would suggest that an average specific heat of all species present in the gas is more relevant. This conclusion would be correct only for Le = 1 and would tempt one—for example, for burning in air—to employ c_p for nitrogen in λ/c_p, a procedure that clearly is entirely unjustified if Le $\neq$ 1. It has been reasoned [90] that since processes occurring between the flame and the droplet surface are of greatest importance in controlling burning, the c_p in equation (65) should be that for the fuel vapors (and possibly their gas-phase pyrolysis products), the only species transported in that region. Although the resulting value of c_p typically is larger by a factor of 2 than that obtained from other prescriptions, reasonable agreement with experiment is achieved by use of the c_p for fuel if suitable procedures are adopted for evaluation of transport properties [90].

When Le $\neq$ 1 it is found [140] that the main transport properties affecting the burning rate are (1) the thermal conductivity in the region between the flame and the droplet surface and (2) the diffusion coefficient of oxidizer in the region outside the flame. The thermal conductivity outside the flame is of secondary importance and the diffusion coefficient of the fuel vapor is practically irrelevant. These influences can be reasoned physically. Inside the flame, heat conduction is a dominant process and fuel diffusion is self-adjusting; outside oxidizer diffusion is dominant, and heat conduction merely provides a loss mechanism that decreases (or increases) the flame temperature, depending on whether the Lewis number outside is greater (or less) than unity. As a first approximation, the largest effect, that of heat conduction between the flame and the droplet, may be employed for evaluating the burning rate through use of equation (65).

In contrast to the relevant heat capacity, all species present contribute to the thermal conductivity. For burning in air, the λ in equation (65) may be taken as an average of that of the fuel vapor and air, 0.4 times that of fuel plus 0.6 times that of air [90]. Since λ and c_p vary with temperature, a temperature must be selected at which these properties are to be evaluated for use in equation (65). The arithmetic mean of the flame temperature and the boiling temperature may be used for this purpose [90], [114], [142] with good results. In calculating the flame temperature for this evaluation, equilibrium dissociation should be taken into account, since the unrealistically high temperatures often obtained when dissociation is neglected are clearly irrelevant to realistic property evaluations. On the other hand, if a formula for B is employed that explicitly contains a flame temperature instead of a standard heat of reaction q^0, then in this formula the theoretical undissociated flame temperature is best used in calculating B; the energy of dissociation is largely recovered in the cooler parts of the flame and therefore is available for driving vaporization.

The prescriptions that have now been given enable equations (63) and (65) to be employed with reasonable accuracy for spherically symmetrical burning. Forced and free convection introduce departures from spherical symmetry that modify burning rates and flame shapes. Thorough theoretical analyses for nonspherical situations are not yet available because of the complexity of the problem. Effects of forced convection at small convection velocities have been studied [143], [144], and ideas for corrections extending to large convective velocities have been considered [72], [96], [113], [145]. Nevertheless, results adapted from the early analysis of Frössling [146], [147] appear to remain as good as any currently available for forced-convection corrections.

Frössling studied vaporizing droplets without combustion and found that $K/K^0 = 1 + 0.276 \, \mathrm{Re}^{1/2} \mathrm{Sc}^{1/3}$, where Sc is the Schmidt number (Appendix E), Re is the Reynolds number, and K^0 is the value of K for evaporation in the same atmosphere without forced convection. Here

$$\mathrm{Re} = \rho u_\infty \, d_l / \mu \qquad (66)$$

where ρ and μ are the density and viscosity of the gas and u_∞ is the gas velocity far upstream, relative to the droplet. Since heat transfer is usually more important than mass transfer in droplet burning, it seems reasonable [148] to replace Sc by the Prandtl number Pr (see Appendix E) in the Frössling formula to obtain

$$K = K^0 (1 + 0.276 \mathrm{Re}^{1/2} \mathrm{Pr}^{1/3}) \qquad (67)$$

for droplet burning in forced convection, where K^0 is given by equation (65). In equation (67), Pr and the properties ρ and μ in Re may be evaluated at average conditions [148], which may be taken to be the ambient pressure, the ambient gas composition, and the arithmetic mean of the ambient temperature and the flame temperature (with dissociation taken into account) [90]. Within the accuracy of the correlation thereby obtained, Pr = 1 may be employed in equation (67). With these prescriptions, equation (67) provides reasonable agreement with data on flame-enveloped drops for Reynolds numbers from 1 to 2000, the range of practical interest.

Burning-rate corrections for free convection depend on the Grashof number,

$$\mathrm{Gr} = (g\rho^2 d_l^3 / \mu^2)(\Delta T / T), \qquad (68)$$

where g denotes the acceleration of gravity and ΔT denotes a temperature difference. With the prescription that has just been given for evaluation of the average temperature T and the average properties ρ and μ, equation (68) may be used with ΔT taken as the difference between the flame temperature (with dissociation) and the ambient temperature to calculate K, in the presence of free convection with Gr < 1, from the empirical formula [90], [149]:

$$K = K^0(1 + 0.533\text{Gr}^{0.52}),   (69)$$

where K^0 is obtained from equation (65). Formulas that differ somewhat from equation (69) may be found in the literature, based, for example, on heat-transfer analogies; good theoretical bases for specific functional forms are lacking.

Free or forced convection strongly influences the flame shape. Below or ahead of the droplet, the distance to the flame is reduced, and above or behind, it is increased significantly. Estimates are available of the distance to the flame (for example, [65] and [66]), but accurate theoretical results are unavailable. For hydrocarbon droplets burning in air, the elongated downstream flame yellows, and, under suitable conditions, opens and emits soot. At sufficiently high forced-convective velocities, the flame ahead of the droplet is extinguished; wake burning may ensue, which in turn is extinguished at a higher velocity. The forced convection may induce circulation within the droplet [131]–[133]. Fluid-mechanical details may become complex.

The phenomena of flame extinction are associated with finite-rate gas-phase chemistry, a topic considered in the following section. Just prior to extinction there is a measurable decrease in the burning rate [90]. A few correlations for corrections to K caused by finite-rate combustion are available [90].

For fuels with high boiling points, finite-rate chemistry may occur in the liquid phase. A pure fuel undergoing this type of fuel pyrolysis in the liquid phase gradually may become a multicomponent liquid during burning. Often the burning of initially multicomponent liquids is of interest; most practical fuels, for example, are not pure compounds. In the combustion of droplets composed of binary or multicomponent miscible liquids, the fuel constituents need not enter the gas in the same proportion in which they are present in the liquid. If rapid mixing is maintained throughout the droplet, then the burning may involve a batch distillation process, with the lighter (more volatile) fuels vaporizing first and the heavier ends later. Liquid-phase diffusion coefficients typically are sufficiently small that thorough mixing of the liquid is not anticipated theoretically. After vaporization of the lighter species from a thin outer layer, a quasi-steady concentration field may be established within the liquid, with a diffusion layer at the surface, through which species are transported at relative rates proportional to their relative bulk concentrations in the interior. Development of the diffusion layer is favored when liquid-phase diffusion times are long compared with the droplet lifetime.

Under these conditions, the surface temperature of the droplet consistent with interphase equilibrium may exceed the boiling point of the interior fuel mixture, since the concentration of the low-boiling constituent has been depleted to a large extent at the surface. Since thermal diffusivities of liquids are typically much higher than their diffusion coefficients, conduction causes the interior temperature of the droplet to increase toward the boiling point of the heavier constituent. This interior temperature may exceed a nucleation temperature for the interior mixture*, thereby inducing a disruptive burning process in which internal vaporization of light fuel causes the liquid to expand rapidly and then disintegrate into many tiny droplets. Disruptive burning has been observed experimentally [91], [101]–[103] for both miscible fuels and emulsified fuels. Theoretical models for predicting the occurrence of the phenomenon have been developed [114], [150]. The topic has been considered to be of practical importance because the performance of combustors can be enhanced in various ways through disruptive burning [91].

Since practical combustors burn clouds of droplets, interactions between quasi-steadily burning droplets have been subjected to continuing study, both experimental [98], [100], [151]–[155] and theoretical [154], [156]–[161]. Depending on the fuel, on the convective environment, and on the overall stoichiometry, flame sheets may surround either each individual droplet or entire clouds. In the latter case droplet vaporization in a hot environment without combustion is relevant. This process, of course, shares many attributes of droplet combustion (for example, the possibility of disruptive vaporization of multicomponent droplets). Droplet interactions can be complex and can produce increases or decreases in burning or vaporization rates, depending on the circumstances [151], [159]—for example, increases by bringing flames closer to droplets (at large separations) in convective situations [151] and decreases by producing larger effective diameters [155]. Further ideas on cloud combustion appear in Section 11.7.

3.4. STRUCTURE OF THE FLAME SHEET

3.4.1. Approaches to structure questions

Concentration and temperature profiles like those shown in Figure 3.6 are idealizations. In recent years a number of profile measurements have been made in laminar diffusion flames (for instance, [80], [162]–[168]). Configurations studied include porous spheres [80] and cylinders [163],

---

* The nucleation temperature, which exceeds the boiling point of the species, is the temperature at which bubbles spontaneously appear in the liquid. Bubble nucleation is a rate process, and its description on the basis of a nucleation temperature is a simplification. Homogeneous nucleation temperatures are substantially above the boiling point; heterogeneous nucleation—aided, for example, by impurities like dust—may occur at somewhat lower temperatures that nevertheless still exceed the boiling point.

through which fuel passes into initially stagnant or uniformly flowing oxidizer; counterflow diffusion flames with gaseous [162], liquid [165], or solid [166] fuels; the flame in the mixing layer of two parallel streams [164]; and the Burke-Schumann flame [167]. Results obtained are qualitatively similar, although behaviors are more complex in configurations that exhibit variations of profiles with distance along the flame sheet. A recent review [168] summarizes observations made for various counterflow configurations. For simplicity, attention here will be focused on the axisymmetric counterflow diffusion flame with a planar flame sheet. A photograph of one such flame is shown in Figure 3.7.

In Figure 3.7 a pool of liquid fuel is supplied with liquid from below, and an oxidizing gas with oxygen mass fraction Y_{O_2} flows out of a duct from above at a constant exit velocity U. The streamlines are made visible by seeding the gas with fine particles and illuminating it from the side. The particles are seen to pass through the luminous flame; there is a stagnation point very near the liquid surface. Concentration and temperature profiles measured in this flow are shown in Figure 3.8. These profiles are invariant with distance along the flame sheet except near the edge of the liquid.

The similarity of the profiles of temperature T, fuel (heptane, C_7H_{16}) and oxidizer O_2 with the profiles of Figure 3.6 may be seen in Figure 3.8. A rounding of the temperature profile in the reaction sheet is evident. Also, lighter fuel species, such as C_2H_2, H_2, and CH_4, are found on the fuel side of the reaction region; these are produced by finite-rate pyrolysis of the fuel vapor as it is heated. A small amount of oxygen penetrates the flame sheet and survives on the fuel side; some fuel species also penetrate to the oxidizer side, but this is less evident because the downward convection quickly

FIGURE 3.7. Photograph of a typical diffusion flame for heptane burning in air with streamlines revealed by a particle-track method [165].

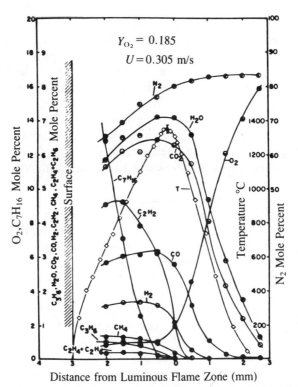

Distance from Luminous Flame Zone (mm)

FIGURE 3.8. Concentration and temperature profiles above a flat surface of liquid heptane burning in a downwardly directed stream of a mixture of oxygen and nitrogen [165].

reduces their concentrations in a convective-diffusive balance. Products such as CO_2 and H_2O tend to peak in concentration at the flame sheet and to decrease in concentration more rapidly on the oxidizer side than on the fuel side, again because of convective effects. The questions to be addressed here concern the rounding of the profiles of temperature and of product concentrations at the reaction sheet and the leakage of fuel and oxidizer through the flame.

The flame is a sheet of finite thickness, not precisely a surface. There are two phenomena that can lead to this reaction-zone broadening: equilibrium dissociation and finite-rate chemistry. Rounding of the temperature profile may occur by the former effect as a consequence of energy extracted chemically by dissociation. If all Lewis numbers are unity, then in principle it is straightforward to account for equilibrium dissociation; a mixture fraction may be introduced as a coupling function, and an equilibrium chemistry program (Section A.3.7) may be employed to calculate the temperature and composition at each mixture fraction. The formulations of Sections 3.1 and 3.3 in effect remain entirely applicable. Analyses by perturbation methods for equilibrium broadening of diffusion flames have been developed (for example, [169] and [170]), leading to formulas for flame thicknesses in terms of equilibrium constants, but the types of dissociation thus far included in the simplified formulas are not entirely realistic, being restricted mainly to dissociation of products into reactants rather than into radicals. It would be relatively straightforward to extend the perturbation methods to include realistic types of dissociation,* and the results could be of interest for large-scale, low-velocity, high-temperature, laminar diffusion flames. However, they would not be relevant for many of the applications— for example, for the profiles appearing in Figure 3.8—since the broadening shown is caused mainly by finite-rate chemistry. Leakage of reactants through the flame is a finite-rate effect that cannot be found from analyses of equilibrium broadening.

There are a number of possible approaches to the calculation of influences of finite-rate chemistry on diffusion flames. Known rates of elementary reaction steps may be employed in the full set of conservation equations, with solutions sought by numerical integration (for example, [171]). Complexities of diffusion-flame problems cause this approach to be difficult to pursue and motivate searches for simplifications of the chemical kinetics [172]. Numerical integrations that have been performed mainly employ one-step (first in [107]) or two-step [173] approximations to the kinetics. Appropriate one-step approximations are realistic† for limited purposes over restricted ranges of conditions. However, there are important aspects of flame structure (for example, soot-concentration profiles) that cannot be described by one-step, overall, kinetic schemes, and one of the major currently outstanding diffusion-flame problems is to develop better simplified kinetic models for hydrocarbon diffusion flames that are capable of predicting results such as observed correlations [172] for concentration profiles of nonequilibrium species.

With simplified chemical kinetics, perturbation methods are attractive for improving understanding and also for seeking quantitative comparisons with experimental results. Two types of perturbation approaches have been developed, Damköhler-number asymptotics and activation-energy asymptotics. In the former the ratio of a diffusion time to a reaction time, one of the similarity groups introduced by Damköhler [174], is treated as a large parameter, and in the latter the ratio of the energy of activation to the thermal

---

* Recently some progress has been made along these lines (N. Peters and F. A. Williams, "Effects of Chemical Equilibrium on the Structure and Extinction of Laminar Diffusion Flames," *Dynamics of Flames and Reactive Systems*, J. R. Bowen, N. Manson, A. K. Oppenheim and R. I. Soloukhin, eds. vol. 95 of *Progress in Astronautics and Aeronautics*, New York: American Institute of Aeronautics and Astronautics, 1984, 37–60).

† Among the earliest works employing simplified kinetics and obtaining analytical approximations are those of Y. B. Zel'dovich, *Zhur. Tekhn. Fiz.* **19**, 1199 (1949) [English translation, NACA Tech. Memo. No. 1296 (1950)] and of D. B. Spalding, *Fuel* **33**, 255 (1954). Chapter 7 of *The Mathematical Theory of Combustion and Explosion* by Y. B. Zel'dovich, G. I. Barenblatt, V. B. Librovich, and G. M. Makhviladze, Moscow: Nauka, 1980, treats the question of diffusion-flame structure for a one-step, second-order, Arrhenius reaction in a clear fashion.

energy, emphasized as important by Frank-Kamenetskii [37], is taken to be large.

Damköhler-number asymptotics can provide estimates of reaction-zone broadening in near-equilibrium situations [125], [126], [175]–[181]. They also afford possibilities of investigating other reaction regimes [176], [182]—for example, regimes of small Damköhler numbers—and they are extendable beyond one-step chemistry [177]–[181] more easily than are activation-energy asymptotics. However analyses of phenomena such as sharp ignition and extinction events cannot be performed on the basis of Damköhler-number asymptotics, but they can be treated by activation-energy asymptotics. Moreover, activation-energy asymptotics may lead to results valid for all Damköhler numbers, and therefore results of Damköhler-number asymptotics may be extracted from those of activation-energy asymptotics. Activation-energy asymptotics appear, in general, to be the more useful of the two types of perturbation approaches. A book is available devoted largely to activation-energy asymptotics [183]; the application to diffusion-flame structure was made by Liñán [184].

3.4.2. The mixture-fraction variable

Since the reaction zone is thin, most of the analysis of its structure can be performed without reference to a particular configuration. To introduce a general approach of this type, consider a two-stream problem having uniform properties over one portion of the boundary, called the **fuel stream** (subscript F, 0, possibly at infinity), and different uniform properties over the rest of the boundary, called the **oxidizer stream** (subscript O, 0, also possibly at infinity); assume that there is no oxidizer in the fuel stream and no fuel in the oxidizer stream. For a one-step reaction, the form given in equation (1) may be adopted, and in terms of the oxidizer-fuel coupling function β, appearing in equation (6), the mixture fraction may be defined as

$$Z = (\beta - \beta_{O,0})/(\beta_{F,0} - \beta_{O,0}), \tag{70}$$

so that $Z = 1$ in the fuel stream and $Z = 0$ in the oxidizer stream. Alternatively, for general chemistry the local mass fraction of an atom of type j may be defined as $Z_j = \sum_{i=1}^{N} (W_j/W_i)v_i^{(j)} Y_i$, where W_i is the molecular weight of molecule i, W_j is the atomic weight of atom j, and $v_i^{(j)}$, which appears in equation (A-36), is the number of atoms of kind j in molecule i; the mixture fraction may then be defined as in equation (70) with β replaced by Z_j for a selected atom j that does not have the same value of Z_j in each stream.

A conservation equation for Z may be derived from the general conservation equations. Under assumptions 1, 2, and 3 of Section 1.3, if the binary diffusion coefficients are equal then equation (1-37) is obtained, and from equation (1-4) it is readily found that

$$\rho \frac{\partial Z}{\partial t} + \rho \mathbf{v} \cdot \nabla Z = \nabla \cdot (\rho D \nabla Z). \tag{71}$$

The chemical-source term does not appear in equation (71) because (like β or Z_j) Z is a *conserved scalar*, a scalar quantity that is neither created nor destroyed by chemical reactions.

Under additional assumptions, all conserved scalars are expressible in terms of Z. For example, if initial conditions and boundary conditions are appropriate, then formulas for all Z_j in terms of Z are obtained by solving equation (70) for β, replaced by Z_j. Energy conservation warrants special consideration. The sum of the thermal and chemical enthalpies is $h = \sum_{i=1}^{N} h_i Y_i$. Substitution of this into equation (1-10) and the result into equation (1-3) may be shown by use of equations (1-1) and (1-2) to provide as a general form of energy conservation

$$\rho \frac{\partial}{\partial t}\left(h + \frac{1}{2}v^2\right) + \rho \mathbf{v} \cdot \mathbf{V}\left(h + \frac{1}{2}v^2\right) = \frac{\partial p}{\partial t}$$

$$+ \mathbf{V} \cdot [(p\mathbf{U} - \mathbf{P}) \cdot \mathbf{v}] + \rho \sum_{i=1}^{N} Y_i \mathbf{f}_i \cdot (\mathbf{v} + \mathbf{V}_i) - \mathbf{V} \cdot \mathbf{q}. \tag{72}$$

For flows at low Mach numbers with negligible radiant heat flux, it may be shown from equation (1-6) under the assumptions introduced above, that equation (72) may be written as

$$\rho \frac{\partial h}{\partial t} + \rho \mathbf{v} \cdot \mathbf{V} h = \frac{\partial p}{\partial t} + \mathbf{V} \cdot \left[\frac{\lambda}{c_p} \mathbf{V} h + \left(\rho D - \frac{\lambda}{c_p}\right) \sum_{i=1}^{N} h_i \mathbf{V} Y_i\right]. \tag{73}$$

Evidently, the further assumptions of a Lewis number of unity ($\lambda/c_p = \rho D$) and negligible $\partial p/\partial t$ reduce equation (73) to the same form as equation (71), thereby giving

$$h = h_{O,0} + (h_{F,0} - h_{O,0})Z \tag{74}$$

as a solution for h in terms of Z, provided that this form is consistent with initial and boundary conditions. This development has the important attribute of not invoking the assumptions of steady flow and of one-step chemistry, which were needed in Section 1.3.

Simplified, explicit expressions for concentrations and temperature in terms of Z and one other variable can be written if $c_{p,i} = c_p = $ constant and if the chemistry is described by equation (1). For the coupling functions $\beta_i \equiv \alpha_i - \alpha_F = Y_i/W_i v_i + Y_F/W_F v_F$, it is readily shown that

$$\rho \frac{\partial \beta_i}{\partial t} + \rho \mathbf{v} \cdot \mathbf{V} \beta_i = \mathbf{V} \cdot (\rho D \mathbf{V} \beta_i), \tag{75}$$

whence in view of equation (71), solutions in a form like equation (74) exist. From these solutions, the mass fractions Y_i may be expressed explicitly in terms of Z and Y_F, namely,

$$Y_i = Y_{iO,0} + (Y_{iF,0} - Y_{iO,0})Z - \frac{W_i v_i}{W_F v_F}(Y_F - Y_{F,0}Z), \quad i = 1, \ldots, N-2. \tag{76}$$

The similar expression

$$Y_O = Y_{O,0}(1 - Z) + \frac{W_O \nu_O}{W_F \nu_F}(Y_F - Y_{F,0} Z) \tag{77}$$

is obtained directly from equation (70). The temperature is expressed in terms of Z and Y_F through use of equations (74), (76), and (77) in the formula

$$h = h_F^0 Y_F + h_O^0 Y_O + \sum_{i=1}^{N-2} h_i^0 Y_i + c_p(T - T^0), \tag{78}$$

obtained from equation (1-11) with $c_{p,i} = c_p = $ constant. After algebraic reduction it is found that

$$T = T_{O,0} + (T_{F,0} - T_{O,0})Z - \frac{q^0/c_p}{W_F \nu_F}(Y_F - Y_{F,0} Z), \tag{79}$$

where q^0 is defined by equation (44). In equation (79), the quantity $q^0/W_F \nu_F$ is the heat released per unit mass of fuel consumed, and the first two terms,

$$T_f = T_{O,0} + (T_{F,0} - T_{O,0})Z, \tag{80}$$

give the "frozen" temperature profile that would be obtained for pure mixing, without chemical reactions. The frozen concentration profiles of fuel and oxidizer would be

$$Y_{F,f} = Y_{F,0} Z, \qquad Y_{O,f} = Y_{O,0}(1 - Z).$$

In the Burke-Schumann limit, a flame surface exists where $Y_F = Y_O = 0$, and all concentrations and temperatures may be written explicitly in terms of Z. From equation (77) it is seen that at $Y_F = Y_O = 0$ the corresponding (stoichiometric) value of the mixture fraction is

$$Z_c = (1 + Y_{F,0} W_O \nu_O / Y_{O,0} W_F \nu_F)^{-1}. \tag{81}$$

For $Z > Z_c$, no oxidizer is present, and according to equation (77),

$$Y_F = Y_{F,0}(Z - Z_c)/(1 - Z_c); \tag{82}$$

for $Z < Z_c$, no fuel is present, and equation (77) gives

$$Y_O = Y_{O,0}(1 - Z/Z_c). \tag{83}$$

The temperature may be obtained as a function of Z in the Burke-Schumann limit by use of these results in equation (79); it is found that

$$T = \begin{cases} T_f + \dfrac{q^0/c_p}{W_F \nu_F} Y_{F,0} Z, & Z < Z_c, \\[4mm] T_f + \dfrac{q^0/c_p}{W_O \nu_O} Y_{O,0}(1 - Z), & Z > Z_c, \end{cases} \tag{84}$$

where T_f is given by equation (80). Note that by putting $Z = Z_c$ in equation (84) the appropriate adiabatic flame temperature is obtained:

$$T_c = \frac{Y_{O,0} W_F \nu_F T_{F,0} + Y_{F,0} W_O \nu_O T_{O,0} + Y_{F,0} Y_{O,0} q^0/c_p}{Y_{O,0} W_F \nu_F + Y_{F,0} W_O \nu_O}. \tag{85}$$

In studying the structure of the flame sheet, equations (82), (83), and (84) cannot be taken as generally valid relationships, but equations (76), (77), and (79) may be employed.

The mixture fraction Z could have been introduced and employed directly in earlier examples. Since it is a coupling function, it could have been used in place of β in equations (9) and (38), with equation (70) employed to recover β from the solution for Z. In fact, it could have been introduced in Section 1.3, to replace β in equation (1-49). Although such selections of variables basically are matters of personal taste, the replacement of β by Z achieves a convenient normalization and also can help to clarify aspects of physical interpretations. For example, in equation (25) the flame-sheet condition, $\beta = 0$, becomes $Z = Z_c$, a condition of mixture-fraction stoichiometry. For the droplet-burning problem, when all the assumptions that underlie equation (58) are introduced, it is found that equation (42), interpreted for Z, becomes simply $Z = 1 - (1 + B)^{-r_l/r}$, where B is defined at the beginning of Section 3.3.6. The equivalence of equation (63) with the expression for r_c/r_l that is obtained from this solution can be derived from the fact that $1 + v = (1 - Z_c)^{-1}$ with Z_c given by equation (81); the two-stream problem most simply relevant to droplet burning has pure liquid fuel in one of the streams ($Y_{F,0} = 1$), even though $Y_{F,l} < 1$ according to equation (64).

3.4.3. Activation-energy asymptotics

Equation (71) is a conservation equation for the mixture fraction, and equations (76), (77), and (79) relate the other variables to this quantity and Y_F. To study the structure of the reaction sheet, we therefore need an additional conservation equation—for example, that for fuel. From equations (1-4), (1-8), (1-9) and (1-12) we find that, under the present assumptions, this additional equation can be written as

$$\rho \frac{\partial Y_F}{\partial t} + \rho \mathbf{v} \cdot \nabla Y_F = \nabla \cdot (\rho D \nabla Y_F) + w_F, \tag{86}$$

where

$$w_F = -\left(\frac{W_F v_F B T^\alpha}{W_F^{n_F} W_O^{n_O}}\right) \rho^{n_F + n_O} Y_F^{n_F} Y_O^{n_O} e^{-E/R^0 T}. \tag{87}$$

In this one-step approximation, the empirical reaction orders n_F and n_O have been introduced, as discussed in Section B.1.3, since the overall stoichiometry seldom provides the correct dependence of the rate on concentrations. For simplicity here we put $n_F = n_O = 1$, obtaining a qualitatively correct variation of the overall rate with the mass fractions of reactants. The overall activation energy is E, and the overall frequency factor for the rate of fuel consumption is $A_F = v_F B T^\alpha$.

Since T appears inside the exponential, it is convenient to employ T instead of Y_F as the variable to be considered in addition to Z. From equations (71), (79), and (86) it is readily shown that

$$\rho \frac{\partial T}{\partial t} + \rho \mathbf{v} \cdot \nabla T = \nabla \cdot (\rho D \nabla T) + w_T, \tag{88}$$

where

$$w_T = \left(\frac{q^0/c_p}{W_F \nu_F}\right)\left(\frac{A_F}{W_O}\right) \rho^2 Y_F Y_O e^{-E/R^0 T}. \tag{89}$$

The quantities Y_F and Y_O in equation (89) may be expressed in terms of Z and T by use of equations (77) and (79), namely,

$$Y_F = Y_{F,0} Z - \frac{W_F \nu_F}{q^0/c_p} (T - T_f) \tag{90}$$

and

$$Y_O = Y_{O,0}(1 - Z) - \frac{W_O \nu_O}{q^0/c_p} (T - T_f), \tag{91}$$

where T_f is given by equation (80). Equations (71) and (88)–(91), along with boundary and initial conditions, define the flame-structure problem to be investigated. Since $\mathbf{v}$ appears in these equations, in general it is necessary to consider the fluid dynamics as well, through equations (1-1) and (1-2), in seeking solutions.

We consider that equation (71) is solved first, prior to the study of equation (88). This may be done in principle if ρD is independent of T and if changes in the solution T have a negligible influence on the fluid dynamics. Otherwise it is only a conceptual aid, and we cannot investigate directly the variation of T with space and time. We can, however, investigate the variation of T with Z and study the influence of finite-rate chemistry on this variation. This type of investigation is facilitated by introducing into (88) the variable Z as an independent variable, in a manner analogous to that of Crocco [185].

Consider an orthogonal coordinate system in which Z is one of the coordinates and the other two, x and y, are distances along surfaces of constant Z. Let u and v be velocity components in the x and y directions, respectively, in this new reference frame. When use is made of equation (71) it is found that a formal transformation of (88) into the new coordinate system yields

$$\rho \frac{\partial T}{\partial t} + \rho \mathbf{v}_\perp \cdot \nabla_\perp T = w_T + \rho D |\nabla Z|^2 \frac{\partial^2 T}{\partial Z^2} + \nabla_\perp \cdot (\rho D \nabla_\perp T)$$
$$- \rho D \nabla_\perp (\ln |\nabla Z|) \cdot \nabla_\perp T \tag{92}$$

where $\mathbf{v}_\perp$ is a two-dimensional vector with x and y components being u and v, and where $\mathbf{\nabla}_\perp$ is the two-dimensional gradient operator in x and y.

In activation-energy asymptotics, $E/R^0 T_c$ is a large parameter. Equation (89) then suggests that w_T will be largest near the maximum temperature T_c and that w_T will decrease rapidly, because of the Arrhenius factor, as T decreases appreciably below T_c. Therefore values of the reaction rate some distance away from the stoichiometric surface ($Z = Z_c$) will be very small in comparison with the values near $Z = Z_c$. This implies that to analyze the effect of w_T, it is helpful to stretch the Z variable in equation (92) about $Z = Z_c$. A stretching of this kind causes the terms involving the highest Z derivative to be dominant, and therefore equation (92) becomes, approximately,

$$\partial^2 T/\partial Z^2 = -w_T/(\rho D|\mathbf{\nabla} Z|^2). \tag{93}$$

If $A_F \rho/(D|\mathbf{\nabla} Z|^2)$ is independent of x, y, and t, then—in view of equations (89), (90), and (91)—equation (93) possesses solutions in which T varies only with Z, and those solutions will, in fact, be exact solutions of equation (92) as well. Thus equation (93) viewed as an ordinary differential equation describes $T(Z)$ exactly for many problems.

To carry out an analysis of the reaction zone by activation-energy asymptotics, Z may be stretched about Z_c and T about T_c. As an independent variable, introduce $\eta = (Z - Z_c)/\epsilon$ and as a dependent variable, $\varphi = (T_c - T)E/R^0 T_c^2 - \gamma\eta$, where

$$\epsilon = 2R^0 T_c^2 Z_c (1 - Z_c)/[E(T_c - T_{f,c})] \tag{94}$$

and

$$\gamma = 2Z_c - 1 - 2Z_c(1 - Z_c)(T_{F,0} - T_{O,0})/(T_c - T_{f,c}), \tag{95}$$

in which $T_{f,c}$ is T_f evaluated from equation (80) at $Z = Z_c$. It can be shown that ϵ is the formal small parameter of expansion for activation-energy asymptotics applied to the diffusion flame and that $1 - \gamma$ is twice the ratio of the heat conducted from the flame toward negative values of $Z - Z_c$ to the total chemical heat released at the flame. The quantity $1/\epsilon$ is closely related to the Zel'dovich number, defined in Section 5.3.6. By introducing φ and η into equation (93), treating these variables as being of order unity, and expanding in ϵ—with use made of equations (89) (which has $n_F = n_O = 1$), (90), and (91)—in the first approximation the differential equation

$$d^2\varphi/d\eta^2 = \delta(\varphi^2 - \eta^2)e^{-(\varphi + \gamma\eta)} \tag{96}$$

is obtained, where a reduced Damköhler number is

$$\delta = \left(\frac{Z_c v_O Y_{F,0} \rho A_F e^{-E/R^0 T_c}}{W_F v_F D|\mathbf{\nabla} Z|^2}\right)\left[\frac{R^0 T_c^2}{E(T_c - T_{f,c})}\right]^3 [2Z_c(1 - Z_c)]^2, \tag{97}$$

with $A_F \rho/(D|\mathbf{\nabla} Z|^2)$ understood to be evaluated at $Z = Z_c$. The details of deriving equation (96) are not shown here but parallel the development of activation-energy asymptotics given in Chapter 5. Liñán's paper [184] may be consulted for methodology.

To obtain boundary conditions for equation (96), matching to the Burke-Schumann solution given by equation (84) may be imposed for $\eta \to \pm\infty$. As is usually true (Chapter 5), in the first approximation this matching requires the slopes dT/dZ to agree. From equation (84) and the definitions of φ and η, it may be shown that the matching requires $d\varphi/d\eta \to \pm 1$ as $\eta \to \pm\infty$. Solutions to equation (96) that satisfy these boundary conditions exist for a range of values of δ [184]. These solutions show that the maximum temperature is reduced below T_c by a fractional amount, proportional to ϵ, that increases as δ decreases. They also show, through evaluation of $\lim_{\eta \to \pm\infty}(\varphi \mp \eta)$ from the numerical integration of equation (96), that either fuel or oxidizer or both leak through the reaction zone in a fractional amount of order ϵ that increases as δ decreases [184]. The thickness of the reaction zone is seen to correspond to $Z - Z_c$ being of order ϵ.

As δ becomes large, a rescaling of variables in equation (96), with $\zeta \equiv \varphi \delta^{1/3}$ and $\xi \equiv \eta \delta^{1/3}$ treated as being of order unity, results in the parameter-free problem $d^2\zeta/d\xi^2 = \zeta^2 - \xi^2$, $d\zeta/d\xi \to \pm 1$ as $\xi \to \pm\infty$, encountered in Damköhler-number asymptotics and integrated numerically [175], [176]. This development shows that Damköhler-number asymptotics can be obtained as a limiting case of activation-energy asymptotics. In the large-δ limit, the fractional decrease of the maximum temperature below T_c is of order $\epsilon/\delta^{1/3}$, and the range of $Z - Z_c$ over which departures from the Burke-Schumann solution occur also is of order $\epsilon/\delta^{1/3}$. These scalings vary with the reaction order [125], [126]. The large-δ solution to the above equation has $\zeta \mp \xi \to 0$ as $\xi \to \pm\infty$ and therefore no leakage of fuel or oxidizer in the first approximation; a reactant having an overall reaction order greater than unity experiences leakage in the first approximation in Damköhler-number asymptotics [125], [126].

It is significant that ϵ, defined in equation (94), is the best formal parameter of expansion instead of simply RT_c/E. The difference $T_c - T_{f,c}$ is the local temperature increase, over its frozen value, associated with complete combustion occurring at the flame sheet ($Z = Z_c$). If this temperature increase is too small—that is, if the heat released in complete combustion is too small—then ϵ is not small enough for w_T to vary rapidly enough with temperature near the flame sheet to justify use of activation-energy asymptotics. To have ϵ small, the heat release need not be very large if E is large, but there must be some heat release; if the heat release is large, then smaller values of E still yield a small value of ϵ. In typical combustion applications, ϵ lies between 0.01 and 0.1, and results of expansions to the lowest order in activation-energy asymptotics are reasonably accurate.

The advent of activation-energy asymptotics has helped greatly in clarifying criteria for the validity of the flame-sheet approximation. The strong dependence of w_T on T for large E helps to confine the combustion reactions to a thin sheet, where $Z - Z_c$ is of order ϵ, by turning off the chemistry outside this sheet, where T is lower. For problems in which the peak temperature occurs at a boundary instead of at the flame sheet, the analysis shows that equilibrium prevails to all algebraic orders in ϵ between the hot boundary and the flame sheet, and the chemistry is frozen between the flame sheet and the cold boundary. For the more common problems in which the peak temperature occurs at the flame sheet, the chemistry is frozen to all algebraic orders in ϵ on both sides of the reaction zone. This suggests that equation (84), the Burke-Schumann solution, defines only one of many possible solutions that may apply outside the flame sheet, and in fact there are reaction regimes in which these equations do not apply [184]. These regimes will be discussed in the Section 3.4.4. Large values of the magnitude of the reaction rate in the reaction zone help to improve the validity of the flame-sheet approximation. For large values of δ, Damköhler-number asymptotics apply, and the thickness of the reaction zone decreases with increasing δ. The nonexistence of solutions to equation (96) when δ is too small means that if reaction rates are not sufficiently high, then extinction of the diffusion flame occurs. The extinction phenomenon will be addressed next.

3.4.4. Ignition and extinction

Analysis [184] shows that the minimum value of δ below which solutions to equation (96) cease to exist is

$$\delta_E = [(1 - |\gamma|) - (1 - |\gamma|)^2 + 0.26(1 - |\gamma|)^3 + 0.055(1 - |\gamma|)^4]e \quad (98)$$

within 1 % for $|\gamma| < 1$. For $|\gamma| > 1$, one of the two boundary temperatures exceeds the flame temperature, and an abrupt extinction event does not exist for the diffusion flame. Use of equation (98) in equation (97) provides an explicit expression for conditions of extinction of the diffusion flame.

As extinction is approached (through decreasing δ), leakage of one of the reactants through the reaction zone increases. The extent of leakage at extinction is proportional to ϵ. The reactant that leaks is described outside the reaction zone by an equation like equation (93) with $w_T = 0$, and solutions linear in Z are obtained. The result is a modification to equation (82) or equation (83), and also to equation (84), by a correction term proportional to ϵ. As $|\gamma|$ approaches unity, the proportionality factor at extinction between the leakage correction to the reactant mass fraction and ϵ becomes large. Therefore, the extinction formula can be inaccurate if $|\gamma|$ is too close to unity. This situation sometimes tends to be encountered in practice for the burning of hydrocarbon fuels in air; the value of Z_c in equation (95) can be small, typically 0.05, so that γ is near -1, and appreciable leakage of fuel into the oxidizer stream is calculated to occur at extinction.

When leakage of one reactant becomes sufficiently large, the situation in the vicinity of the reaction zone tends to begin to resemble that for a premixed flame, and modification of scaling for the asymptotic analysis becomes warranted. Qualitatively, in equation (96) the factor $\varphi^2 - \eta^2$ is the scaled product $(\varphi - \eta)(\varphi + \eta)$ of fuel and oxidizer concentrations in the reaction zone. One of these two factors tends to become constant; hence, depletion of only one reactant occurs, as in a premixed flame. A modification to the Burke-Schumann solution of order unity is needed on the side of the flame sheet to which reactant has leaked, and even in lowest order the flame temperature becomes a *premixed-flame temperature* less than T_c. The asymptotic analysis of this regime is due to Liñán [184], who terms it the **premixed-flame regime** of the diffusion flame, to distinguish it from the **diffusion-flame**, or **near-equilibrium regime** considered in the previous section. If $|\gamma|$ is near enough to unity and ϵ is not too small, then as δ decreases, extinction of the diffusion flame occurs subsequent to its passage into the premixed-flame region.

As the premixed-flame temperature decreases, conditions may be calculated for which the second reactant also begins to leak through the reaction zone in appreciable quantities. An asymptotic analysis may be developed for which, in the reaction zone, the factor $\varphi^2 - \eta^2$ remains effectively constant [184]. Departures from equations (82), (83), and (84) then occur even in lowest order on both sides of the flame sheet. This regime has been termed the **partial-burning regime** [184].

There is a fourth regime in which the first approximation to $Y_F(Z)$, $Y_O(Z)$, and $T(Z)$ is given by the frozen solutions, $Y_{F,f}$, $Y_{O,f}$, and T_f [see equation (80) and the sentence following it]. This has been termed the **nearly frozen**, or **ignition regime** [184]. In this regime, activation-energy asymptotics show that the reaction tends to occur in a region near the hotter of the two boundaries [184]. The method of analysis bears some relationship to the methods discussed in Section B.2.5.4, although there are significant differences because of the absence of spatial homogeneity. Provided that T_c exceeds both boundary temperatures, solutions are found to exist only for values of a Damköhler number, related to δ, that are *less* than a maximum value δ_I. Since nearly frozen solutions do not exist for $\delta > \delta_I$, the value δ_I may be interpreted as a critical Damköhler number for ignition; if $\delta > \delta_I$, then the solution must lie in one of the other regimes, all of which have a thin reaction zone of elevated temperature.

When T_c is less than a boundary temperature ($|\gamma| > 1$), nearly frozen solutions exist for all values of δ, but they tend to become inaccurate if δ is too large. In this case, as δ increases from zero with ϵ fixed, nearly frozen solutions lose accuracy and premixed-flame solutions gain accuracy, then these, in turn, lose accuracy at larger δ, and diffusion-flame solutions become accurate. There are no partial-burning solutions when T_c is less than a boundary temperature.

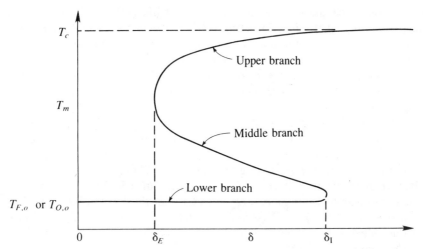

FIGURE 3.9. Schematic illustration of the dependence of the maximum temperature on the reduced Damköhler number for diffusion flames.

The most common situation is summarized qualitatively in Figure 3.9, which is a sketch of the maximum temperature T_m as a function of the Damköhler parameter δ, as obtained both by numerical integration of the full equations [186] and by activation-energy asymptotics [184]. This figure applies for small ϵ and only if T_c exceeds the boundary temperatures, $T_{F,0}$ and $T_{O,0}$. The upper branch and part of the middle branch are predicted by analysis of the diffusion-flame regime, much of the middle branch and part of the upper branch by analysis of the premixed-flame regime, and the lower branch and part of the middle branch by analysis of the ignition regime. A suitable partial-burning analysis can give the middle branch and both bends, but it is likely that the entire partial-burning regime is unstable and therefore not significant for diffusion flames.

If δ is gradually increased for a mixing flow with negligible chemical reaction, then the system may be expected to move along the lower branch until it reaches δ_I, at which point the solution must jump discontinuously to a vigorously burning state on the upper branch, usually in the diffusion-flame regime. If δ is gradually decreased for a vigorously burning diffusion flame on the upper branch, then at least when δ_E is passed, the solution must jump discontinuously to the lower branch, where generally there is negligible reaction. It is believed that the middle branch is unstable; flames on this branch have not been obtained experimentally. It is conceivable that the upper or lower branches experience instability, leading to an ignition or extinction jump, prior to the vertical tangent. Although entirely thorough stability analyses have not been completed, the few studies that have been performed (for example, [187] and [188]) indicate that instability begins at or very near the extrema. Of course, an ignition or extinction jump itself is not instantaneous; it is a very rapid, time-dependent process that would not be described by Figure 3.9 or equation (93) but instead would require retention of $\rho \, \partial T/\partial t$ in equation (92).

The stability of solutions in the premixed-flame regime for systems having T_c greater than the boundary temperatures has been questioned [189]. If this regime is unstable, then only solutions in the diffusion-flame and ignition regimes are significant with respect to Figure 3.9. It seems more likely (A. Liñán, personal communication) that in the premixed-flame regime, instability occurs near (not precisely at) the extremum; in this event the premixed-flame regime is relevant to Figure 3.9 in that, if $|\gamma|$ is near unity and ϵ is not too small, as δ decreases solutions on the upper branch in the diffusion-flame regime become less accurate than those in the premixed-flame regime prior to extinction.

The presentation here has not been cast within the framework of any particular configuration for the diffusion flame. The possibility of using the Crocco transformation to achieve this generality was suggested by Peters [190], [191]; it will prove useful later, notably in connection with turbulent flames (Chapter 10). The simplest configuration to which the formulation may be applied is the steady, stagnant diffusion layer. For this problem with ρD constant, Z equals the normal distance (measured from the oxidizer side of the layer) divided by the thickness of the layer, and Z therefore may be interpreted directly as a normalized physical coordinate; $|\nabla Z|$ is the reciprocal of the thickness of the layer. The analysis exposing activation-energy asymptotics originally was developed within the context of the problem of the steady counterflow diffusion flame [184]. It also has been applied to the quasisteady, spherically symmetrical burning of a droplet [192]. Analysis of one-dimensional, time-dependent mixing and reaction of initially separated fuel and oxidizer by activation-energy asymptotics [193] encounters many of the aspects that have been discussed here.

For steady-state diffusion flames with thin reaction sheets, it is evident that outside the reaction zone there must be a balance between diffusion and convection, since no other terms occur in the equation for species conservation. Thus these flames consist of *convective-diffusive zones* separated by thin reaction zones. Since the stretching needed to describe the reaction zone by activation-energy asymptotics increases the magnitude of the diffusion terms with respect to the (less highly differentiated) convection terms, in the first approximation these reaction zones maintain a balance between diffusion and reaction and may be more descriptively termed *reactive-diffusive zones*. Thus the Burke-Schumann flame consists of two convective-diffusive zones separated by a reactive-diffusive zone.

Since the possible simultaneous presence of three phenomena— convection, diffusion, and reaction—complicate analyses, it can be helpful to eliminate one of the three formally by means of a transformation. This is achieved in the analysis of the counterflow flame [184] by writing the equations in a convection-free form; that is, by finding a transformation to a new spatial variable such that the convective and diffusive terms coalesce into a single term that resembles a purely diffusive term in the new variable. The transformation to convection-free form is especially useful for analyzing the ignition regime, for which the reaction zone is not thin in the original spatial variable but instead extends an infinite distance into the hotter of the two streams, causing convection, diffusion, and reaction all to be important in the reaction zone. For any diffusion flame having a symmetry that eliminates the transverse gradient from equation (92), it is evident from equation (92) that the transformation to the mixture fraction as an independent variable achieves a convection-free form. Thus a diffusion-flame prescription for transformation to convection-free form is to introduce a mixture fraction as an independent variable.

Overall rate parameters for diffusion flames, needed in applying the analysis given here, often are not available. The analysis itself may be employed to design an experimental technique for obtaining these parameters from measurements of diffusion-flame extinction [194]. The counterflow configuration is convenient; the analysis addresses experiments with liquid fuels, shown in Figure 3.7. The quantity $D|\nabla Z|^2$ is proportional to the exit velocity U of gases leaving the oxidizer duct [195] if the experiment is carefully designed to establish a radial channel flow with a uniform rate of injection from above. To achieve reasonable accuracy in calculating $D|\nabla Z|^2$, it is necessary to take into account variable properties (ρ and D) and the influence of the presence of the condensed phase (through parameters such as the transfer number B) [194]. On the basis of the theory, an Arrhenius graph can be defined from equations (97) and (98), in which a semilog plot of a quantity proportional to the measured value of U at flame extinction, as a function of $1/T_c$, provides E from the slope and A_F from a value. Rate parameters obtained in this manner may be employed for calculating diffusion-flame structure and extinction in other configurations. Overall rate parameters for a number of fuels have been found in this way; a critical review is available [196].

3.5. MONOPROPELLANT DROPLET BURNING

A monopropellant is a fuel that is capable of reacting exothermically without an oxidizer. Numerous such fuels are known; many are liquids. A droplet of liquid monopropellant can burn in an inert atmosphere without an oxidizer. There are rockets that employ liquid monopropellants as fuels for propulsion; in some applications these are better suited than bipropellant rockets that employ a liquid fuel and a liquid oxidizer. The relevance of monopropellant droplet combustion to the operation of monopropellant rockets and fundamental curiosity have prompted studies of the subject.

Burning monopropellant droplets have the spherical configuration shared by burning fuel droplets in an oxidizing atmosphere and the premixedness of ordinary laminar flames (Chapter 5). Consequently, it may be expected that monopropellant droplets exhibit some characteristics of each of these systems. As is often true in intermediate cases, the analysis of monopropellant droplet combustion in certain respects is more complicated than that of either the planar premixed flame or the fuel droplet.

A number of theoretical studies of the quasisteady, spherically symmetrical burning of monopropellant droplets have been published. These include work involving numerical integration of the conservation equations or of approximations thereto [197], [198], analytical approaches involving ad hoc approximations [85], [199]–[201], methods employing Damköhler-number asymptotics [202]–[206] and procedures based on activation-energy asymptotics [207]–[210]. Reviews may be found in the literature [111], [183]. Consideration has been given to burning in inert atmospheres under both adiabatic and nonadiabatic conditions, as well as to burning in reactive atmospheres containing an oxidizer that can react with the fuel or its decomposition products [85], [205], [206].

The general character of results of these analyses may be understood on the basis of physical reasoning. Consider first the burning in an inert atmosphere, the temperature of which is at or above the adiabatic flame temperature for decomposition of the monopropellant. If the reaction rate is sufficiently low, then the combustion will have a negligible influence on the rate of mass loss of the droplet, and the dominant process will be inert vaporization in a hot atmosphere. We have seen that this process has a d-square law, and we know how to calculate the burning rate (for example, from equation (58) with $Y_{O, \infty} = 0$; see Section 3.3.6). The reaction occurs far from the droplet surface, in a region of large extent, whose location depends on a nondimensional activation energy and/or on a Damköhler number, $D_{II} = r_l^2 \bar{w}/\rho D$, where $\bar{w}$ is a representative value for the mass rate of consumption of fuel. A decrease in D_{II} or an increase in the activation energy moves the reaction region farther from the droplet. The burning-rate constant, K, is given by equation (65) and does not depend on D_{II}.

If the reaction rate is larger, then the combustion influences the rate of mass loss. For sufficiently large values of D_{II}, the reaction occurs in a narrow zone whose thickness is small enough for curvature effects to be negligible within it. If a nondimensional activation energy is large, then this zone is separated from the droplet surface by a convective-diffusive zone. In this case the structure of the reaction zone resembles that of the reaction zone of a premixed laminar flame (Chapter 5), with heat addition from superadiabatic products if the ambient temperature exceeds the adiabatic flame temperature. Qualitatively, the speed of propagation of the planar laminar flame begins to control the rate of mass loss of the droplet, and K becomes proportional to $\sqrt{D_{II}}$ (see Chapter 5). Since $K \sim dr_l^2/dt$ and $\sqrt{D_{II}} \sim r_l$, the d-square law is replaced by a d-law in which the first power of the droplet diameter decreases linearly with time.

The transition between these two limiting regimes occurs roughly where K, as given by equation (65) is of the same order as $\sqrt{D_{II}}$. Asymptotic analyses give sharper criteria for the transition [207]. Over any limited range of conditions, dr^k/dt is approximately constant, where $1 \leq k \leq 2$. In view of the dependence of D_{II} on r_l, for large droplets the d-law tends to hold; for small droplets the d-square law holds.

Additional phenomena occur if the temperature of the inert atmosphere is less than the adiabatic flame temperature. As might be expected from the discussion in the previous section, ignition and extinction events may be encountered [207]. Formulas for extinction conditions, obtained on the basis of activation-energy asymptotics, are available [207]. Within the context of activation-energy asymptotics, a partial-burning regime of the type discussed in the preceding section is found to arise [207]. A diffusion-flame regime, of course, does not exist for burning in an inert atmosphere; in this respect the problem of the monopropellant droplet is simpler than that of a fuel droplet in an oxidizing atmosphere.

For burning in a reactive atmosphere, conditions may be encountered for which the analysis of the diffusion-flame regime is relevant. Under suitable conditions there are two thin reaction sheets, an inner one at which premixed decomposition of the monopropellant occurs and an outer one having a diffusion-flame character. Categorizations are available for potential limiting behaviors at large Damköhler numbers in reactive environments [206]. There are many flame-structure possibilities, not all of which have been analyzed thoroughly.

REFERENCES

1. S. S. Penner, *Chemistry Problems in Jet Propulsion,* New York: Pergamon Press, 1957, Chapter XXII.
2. A. M. Kanury, *Introduction to Combustion Phenomena,* New York: Gordon and Breach, 1975, 142–269.
3. I. Glassman, *Combustion,* New York: Academic Press, 1977, 158–193.
4. S. P. Burke and T. E. W. Schumann, *Ind. Eng. Chem.* **20**, 998 (1928).
5. J. A. Fay, *J. Aero. Sci.* **21**, 681 (1954).
6. A. Goldburg and S. I. Cheng, *C & F* **9**, 259 (1965).
7. J. B. Haggard, Jr. and T. H. Cochran, *CST* **5**, 291 (1972).
8. T. M. Liu, *CST* **7**, 283 (1973).
9. P. H. Thomas, *9th Symp.* (1963), 844–859.
10. F. J. Kosdon, F. A. Williams and C. Buman, *12th Symp.* (1969), 253–264.
11. F. A. Williams, *Prog. Energy Combust. Sci.* **8**, 317 (1982).
12. H. C. Hottel and W. R. Hawthorne, *3rd Symp.* (1949), 254–266.
13. M. Klajn and A. K. Oppenheim, *19th Symp.* (1982), 223–235.
14. O. A. Tsukhanova, *6th Symp.* (1957), 573–577.

15. L. N. Khitrin and L. S. Solovyeva, *7th Symp.* (1959), 532–538.
16. H. A. Lauwerier, *Appl. Sci. Research* **A8**, 366 (1959).
17. D. E. Rosner, *A.I.Ch.E. Journal* **9**, 321 (1963).
18. D. E. Rosner, *Chem. Eng. Sci.* **19**, 1 (1964).
19. P. A. Libby and F. A. Williams, *AIAA Journal* **3**, 1152 (1965).
20. D. E. Rosner, *AIAA Journal* **2**, 593 (1964).
21. D. E. Rosner, *Int. J. Heat Mass Transfer* **9**, 1233 (1966).
22. D. E. Rossner, *11th Symp.* (1967), 181–196.
23. R. Aris, "Hierarchies of Models in Reactive Systems," in *Dynamics and Modeling of Reactive Systems*, W. E. Stewart, W. H. Ray, and C. C. Conley, eds., New York: Academic Press, 1980, 1–35.
24. D. E. Rosner, *Annual Review of Materials Science* **2**, 573 (1972).
25. I. Glassman, *Combustion*, New York: Academic Press, 1977, 258–269.
26. M. F. R. Mulcahy and I. W. Smith, *Rev. Pure and Appl. Chem.* **19**, 81 (1969).
27. G. Blyholder, J. S. Binford, Jr., and H. Eyring, *J. Phys. Chem.* **62**, 263 (1958).
28. R. H. Essenhigh, *8th Symp.* (1962), 813–814.
29. J. Nagle and R. F. Strickland-Constable, *Proc. Fifth Conf. on Carbon*, vol. 1, New York: Macmillan, 1962, 154–164.
30. J. N. Ong, Jr., *Carbon* **2**, 281 (1964).
31. H. G. Maahs, NASA Tech. Note No. D-6310 (1971).
32. J. R. Walls and R. F. Strickland-Constable, *Carbon* **1**, 333 (1964).
33. R. F. Strickland-Constable, *Second Conf. on Industrial Carbon and Graphite, Soc. Chem. Ind. London*, 235 (1966).
34. P. L. Walker, Jr., F. Rusinko, Jr., and L. G. Austin, *Advances in Catalysis*, vol. 11, New York: Academic Press, 1959, 133–221.
35. D. R. Olander, W. Seikhaus, R. Jones, and J. A. Schwarz, *J. Chem. Phys.* **57**, 408 (1972).
36. T. R. Acharya and D. R. Olander, *Carbon* **11**, 7 (1973).
37. D. A. Frank-Kamenetskii, *Diffusion and Heat Transfer in Chemical Kinetics*, New York: Plenum Press, 1969, 64–72.
38. K. Matsui, A. Koyama, and K. Uehara, *C & F* **25**, 57 (1975).
39. S. K. Ubhayakar and F. A. Williams, *J. Electrochem. Soc.* **123**, 747 (1976).
40. C. Park and J. P. Appleton, *C & F* **20**, 369 (1973).
41. P. A. Libby and T. R. Blake, *C & F* **36**, 139 (1979).
42. D. E. Rosner and H. D. Allendorf, *Carbon* **3**, 153 (1965); **7**, 515 (1969).
43. H. Tsuji and K. Matsui, *C & F* **26**, 283 (1976).
44. G. Adomeit, G. Mohiuddin, and N. Peters, *16th Symp.* (1977), 731–743.
45. D. B. Spalding, *Fuel* **30**, 6, 121 (1951).
46. D. E. Rosner, *Oxidation of Metals* **4**, 1 (1972).
47. I. Glassman, "Combustion of Metals. Physical Considerations," in *Solid Propellant Rocket Research*, vol. 1 of *Progress in Astronautics and Rocketry*, M. Summerfield, ed., New York: Academic Press, 1960, 253–258.
48. N. Peters, *15th Symp.* (1975), 363–375.
49. S. K. Ubhayakar, *C & F* **26**, 23 (1976).
50. N. R. Amundson and E. Mon, "Diffusion and Reaction in Carbon Burning," in *Dynamics and Modeling of Reactive Systems*, W. E. Stewart, W. H. Ray, and C. C. Conley, eds., New York: Academic Press, 1980, 353–374.
51. P. A. Libby, *C & F* **38**, 285 (1980).

52. M. Matalon, *CST* **24**, 115 (1980).
53. M. Matalon, *CST* **25**, 43 (1981).
54. G. Mohan and F. A. Williams, *AIAA Journal* **10**, 776 (1972).
55. R. P. Wilson, Jr., "Studies on Combustion of Aluminum Particles," Paper No. 69-3, Spring Meeting, Western States Section, The Combustion Institute, China Lake, CA (1969).
56. T. A. Brzustowski and I. Glassman, "Vapor-Phase Diffusion Flames in the Combustion of Magnesium and Aluminum: I Analytical Developments," in *Heterogeneous Combustion*, vol. 15 of *Progress in Astronautics and Aeronautics*, H. G. Wolfhard, I. Glassman, and L. Green Jr., eds., New York: Academic Press, 1964, 75–116.
57. D. K. Kuehl, *AIAA Journal* **3**, 2239 (1965).
58. L. A. Klyachko, *Fiz. Gor. Vzr.* **5**, 404 (1969).
59. P. F. Pokhil, A. F. Belaev, Yu. V. Frolov, V. S. Logachev, and A. I. Korotkov, *Combustion of Metal Powder Formulations in Reactive Environments*, Moscow: Nauka, 1972.
60. C. K. Law, *CST* **7**, 197 (1973).
61. C. K. Law, *CST* **12**, 113 (1976).
62. C. K. Law and F. A. Williams, *C & F* **22**, 383 (1974).
63. F. A. Williams, "Droplet Burning," Chapter II of *Combustion Experiments in a Zero-Gravity Laboratory*, T. H. Cochran, ed., vol. 73 of *Progress in Astronautics and Aeronautics*, New York: American Institute of Aeronautics and Astronautics, 1981, 31–60.
64. S. Kumagai, *Jet Propulsion* **26**, 786 (1956).
65. S. Kumagai and H. Isoda, *6th Symp.* (1957), 726–731.
66. H. Isoda and S. Kumagai, *7th Symp.* (1959), 523–531.
67. S. Kumagai, T. Sakai, and S. Okajima, *13th Symp.* (1971), 779–785.
68. S. Okajima and S. Kumagai, *15th Symp.* (1975), 401–407.
69. B. Knight and F. A. Williams, *C & F* **38**, 111 (1980).
70. G. N. Khudyakov, *Izvest. Akad. Nauk. S.S.S.R., Otdel. Tech. Nauk.* **4**, 508 (1949).
71. D. B. Spalding, *Fuel* **29**, 25 (1950); **32**, 169 (1953).
72. D. B. Spalding, *4th Symp.* (1953), 847–864.
73. G. A. Agoston, H. Wise, and W. A. Rosser, *6th Symp.* (1957), 708–717.
74. G. A. Agoston, B. J. Wood, and H. Wise, *Jet Propulsion* **28**, 181 (1958).
75. D. G. Udelson, *C & F* **6**, 93 (1962).
76. J. W. Aldred, J. C. Patel, and A. Williams, *C & F* **17**, 139 (1971).
77. G. S. Canada and G. M. Faeth, *14th Symp.* (1973), 1345–1354.
78. S. R. Gollahalli and T. A. Brzustowski, *14th Symp.* (1973), 1333–1344.
79. D. D. Ludwig, F. V. Bracco, and D. T. Harrje, *C & F* **25**, 107 (1975).
80. S. I. Abdul-Khalik, T. Tamaru, and M. M. El-Wakil, *15th Symp.* (1975), 389–399.
81. G. A. E. Godsave, *4th Symp.* (1953), 818–830.
82. A. R. Hall and J. Diederichsen, *4th Symp.* (1953), 837–846.
83. M. Goldsmith, *Jet Propulsion*, **26**, 172 (1956).
84. C. S. Tarifa, "On the Influence of Chemical Kinetics on the Combustion of Fuel Droplets," *Congres Internacional des Machines a Combustion*, Paper B9, Copenhagen: A. Busk, 1962.
85. C. S. Tarifa, P. P. del Notario, and F. Garcia Moreno, *8th Symp.* (1962), 1035–1056.
86. J. W. Aldred and A. Williams, *C & F* **10**, 396 (1966).

87. M. T. Monaghan, R. G. Siddal, and M. W. Thring, *C & F* **12**, 45 (1968).
88. G. M. Faeth, D. P. Dominicis, J. F. Tulpinsky, and D. R. Olson, *12th Symp.* (1969), 9–18.
89. G. M. Faeth and R. S. Lazar, *AIAA Journal* **9**, 2165 (1971).
90. C. K. Law and F. A. Williams, *C & F* **19**, 393 (1972).
91. F. L. Dryer, *16th Symp.* (1979), 279–295.
92. T. Kadota and H. Hiroyasu, *18 Symp.* (1981), 275–282.
93. H. C. Hottel, G. C. Williams, and H. C. Simpson, *5th Symp.* (1955), 101–129.
94. J. A. Bolt and M. A. Saad, *6th Symp.* (1957), 717–725.
95. B. J. Wood, W. A. Rosser, and H. Wise, *AIAA Journal* **1**, 1076 (1963).
96. P. Eisenklam, S. A. Arunachalam, and J. A. Weston, *11th Symp.* (1967), 715–728.
97. B. J. Wood and W. A. Rosser, *AIAA Journal* **7**, 2288 (1969).
98. A. S. M. Nuruzzaman, A. B. Hedley, and J. M. Beér, *13th Symp.* (1971), 787–799.
99. J. J. Sangiovanni and A. S. Kestin, *CST* **16**, 59 (1977).
100. J. J. Sangiovanni and L. G. Dodge, *17th Symp.* (1979), 455–465.
101. J. C. Lasheras, A. C. Fernández-Pello, and F. L. Dryer, *CST* **21**, 1 (1979).
102. J. C. Lasheras, A. C. Fernández-Pello, and F. L. Dryer, *CST* **22**, 195 (1980).
103. J. C. Lasheras, A. C. Fernández-Pello, and F. L. Dryer, *18th Symp.* (1981), 293–305.
104. B. Sreznevsky, *Zh. R. F. Kho.* **14**, 420, 483 (1882).
105. I. Langmuir, *Phys. Rev.* **12**, 368 (1918).
106. M. Goldsmith and S. S. Penner, *Jet Propulsion* **24**, 245 (1954).
107. J. Lorell, H. Wise, and R. E. Carr, *J. Chem. Phys.* **25**, 325, (1956).
108. B. P. Mullins and S. S. Penner, *Explosions, Detonations, Flammability and Ignition*, New York: Pergamon Press, 1959, 101–107.
109. H. Wise and G. A. Agoston, "Burning of a Liquid Droplet," in *Advances in Chemistry Series*, no. 20, Am. Chem. Soc. (1958), 116–135.
110. A. Williams, *Oxidation and Combustion Reviews* **1**, 1 (1968).
111. D. E. Rosner, "Liquid Droplet Vaporization and Combustion," Section 2.4 of *Liquid Propellant Rocket Combustion Instability*, D. T. Harrje and F. H. Reardon, eds., NASA SP-194, Washington (1972), 74–100.
112. A. Williams, *C & F* **21**, 1 (1973).
113. G. M. Faeth, *Prog. Energy Combust. Sci.* **3**, 191 (1977).
114. C. K. Law, *Prog. Energy Combust. Sci.* **8**, 171 (1982).
115. C. K. Law, "Mechanisms of Droplet Combustion," in *Proceedings of the Second International Colloquium on Drops and Bubbles*, D. H. Le Croisette, ed., JPL Publication 82-7, Jet Propulsion Laboratory, Pasadena (1982), 39–53.
116. F. A. Williams, *J. Chem. Phys.* **33**, 133 (1960).
117. K. Kobayasi, *5th Symp.* (1955), 141–148.
118. N. Nishiwaki, *5th Symp.* (1955), 148–151.
119. D. B. Spalding, *ARS Journal* **29**, 828 (1959).
120. D. E. Rosner, *AIAA Journal* **5**, 163 (1967).
121. A. Chervinsky, *AIAA Journal* **7**, 1815 (1969).
122. C. S. Tarifa, A. Crespo, and E. Fraga, *Astronautica Acta* **17**, 685 (1972).
123. C. E. Polymeropoulos and R. L. Peskin, *CST* **5**, 165 (1972).
124. D. E. Rosner and W. S. Chang, *CST* **7**, 145 (1973).
125. D. R. Kassoy and F. A. Williams, *Phys. Fluids* **11**, 1343 (1968).
126. D. R. Kassoy, M. K. Liu, and F. A. Williams, *Phys. Fluids* **12**, 265 (1969).

127. S. H. Sohrab, A. Liñán, and F. A. Williams, *CST* **27**, 143 (1982).
128. H. Wise and C. M. Ablow, *J. Chem. Phys.* **27**, 389 (1957).
129. C. K. Law, *C & F* **26**, 17 (1976).
130. C. K. Law and W. A. Sirignano, *C & F* **28**, 175 (1977).
131. S. Prakash and W. A. Sirignano, *Int. J. Heat Mass Transfer* **21**, 885 (1978).
132. S. Prakash and W. A. Sirignano, *Int. J. Heat Mass Transfer* **23**, 253 (1980).
133. P. Lara-Urbaneja and W. A. Sirignano, *18th Symp.* (1981), 1365–1374.
134. C. H. Waldman, *15th Symp.* (1975), 429–442.
135. A. Crespo and A. Liñán, *CST* **11**, 9 (1975).
136. C. K. Law, S. H. Chung, and H. Srinivasan, *C & F* **38**, 173 (1980).
137. P. Sepri, *C & F* **26**, 179 (1976).
138. C. A. Cooper and J. F. Clarke, *J. of Eng. Math.* **11**, 193 (1977).
139. H. Wise, J. Lorell, and B. J. Wood, *5th Symp.* (1955), 132–141.
140. D. R. Kassoy and F. A. Williams, *AIAA Journal* **6**, 1961 (1968).
141. B. N. Raghunandan and H. S. Mukunda, "Analysis of Bipropellant Droplet Combustion with Variable Thermodynamic and Transport Properties," *Rept. No. AE C&P-4*, Department of Aeronautical Engineering, Indian Institute of Science, Bangalore (1974); *C & F* **30**, 71 (1977).
142. C. K. Law and H. K. Law, *CST* **12**, 207 (1976); *C & F* **29**, 269 (1977).
143. F. E. Fendell, M. L. Sprankle, and D. A. Dodson, *J. Fluid Mech.* **26**, 267 (1966).
144. F. E. Fendell, D. E. Coats, and E. B. Smith, *AIAA Journal* **6**, 1953 (1968).
145. R. Natarajan and T. A. Brzustowski, *CST* **2**, 259 (1970).
146. N. Frössling, *Gerlands Beitr. Geophys.* **52**, 170 (1938).
147. W. E. Ranz and W. R. Marshall, *Chem. Eng. Prog.* **48**, 141, 173 (1952).
148. G. A. Agoston, H. Wise, and W. A. Rosser, *6th Symp.* (1957), 708–717.
149. S. Okajima and S. Kumagai, *19th Symp.* (1982), 1021–1027.
150. C. K. Law, *CST* **17**, 29 (1977); *AIChE Journal* **24**, 626 (1978).
151. J. F. Rex, A. E. Fuhs, and S. S. Penner, *Jet Propulsion* **26**, 179 (1956).
152. N. V. Feedoseeva, *Advances in Aerosol Physics* **2**, 110 (1972).
153. T. A. Brzustowski, E. M. Twardus, S. Wojcicki, and A. Sobiesiak, *AIAA Journal* **17**, 1234 (1979).
154. K. Miyasaka and C. K. Law, *18th Symp.* (1981), 283–292.
155. J. J. Sangiovanni and M. Labowsky, *C & F* **47**, 15 (1982).
156. H. H. Chiu and T. M. Liu, *CST* **17**, 127 (1977).
157. R. Samson, D. Bedeaux, M. J. Saxton, and J. M. Deutch, *C & F* **31**, 215 (1978).
158. M. Labowsky, *CST* **18**, 145 (1978).
159. M. Labowsky, *CST* **22**, 217 (1980).
160. A. Umemura, S. Ogawa, and N. Oshima, *C & F* **41**, 45 (1981).
161. A. Umemura, *18th Symp.* (1981), 1355–1363.
162. T. P. Pandya and F. J. Weinberg, *Proc. Roy. Soc. London* **279A**, 544 (1964).
163. H. Tsuji and I. Yamaoka, *12th Symp.* (1969), 997–1005; *13th Symp.* (1971), 723–731.
164. A. Melvin, J. B. Moss, and J. F. Clarke, *CST* **4**, 17 (1971).
165. J. H. Kent and F. A. Williams, *15th Symp.* (1975), 315–325.
166. K. Seshadri and F. A. Williams, *J. Polymer Sci.: Polymer Chem. Edition* **16**, 1755 (1978).
167. R. E. Mitchell, A. F. Sarofim, and L. A. Clomburg, *C & F* **37**, 227 (1980).
168. H. Tsuji, *Prog. Energy Combust. Sci.* **8**, 93 (1982).

169. P. M. Chung and V. D. Blankenship, *Phys. Fluids* **9**, 1569 (1966).
170. F. E. Fendell, *Astronautica Acta* **13**, 183 (1967).
171. T. M. Liu and P. A. Libby, *CST* **2**, 131 (1970).
172. R. W. Bilger, *C & F* **30**, 227 (1977).
173. V. K. Jain and H. S. Mukunda, *CST* **1**, 105 (1969).
174. G. Damköhler, *Zeit. Elektrochem.* **42**, 846 (1936).
175. S. K. Friedlander and K. H. Keller, *Chem. Eng. Sci.* **18**, 365 (1963).
176. F. E. Fendell, *J. Fluid Mech.* **21**, 281 (1965).
177. J. F. Clarke, *Proc. Roy. Soc. London* **307A**, 283 (1968); **312A**, 65 (1969).
178. J. F. Clarke and J. B. Moss, *CST* **2**, 115 (1970).
179. A. Melvin and J. B. Moss, *15th Symp.* (1975), 625–636.
180. R. W. Bilger, *CST* **22**, 251 (1980).
181. R. A. Allison and J. F. Clarke, *CST* **23**, 113 (1980).
182. J. D. Buckmaster, *C & F* **24**, 79 (1975).
183. J. D. Buckmaster and G. S. S. Ludford, *Theory of Laminar Flames*, Cambridge University Press, 1982.
184. A. Liñán, *Acta Astronautica* **1**, 1007 (1974).
185. L. Crocco, *Atti di Guidonia* **17**, 118 (1939).
186. F. E. Fendell, *Chem. Eng. Sci.* **22**, 1829 (1967).
187. L. L. Kirkby and R. A. Schmitz, *C & F* **10**, 205 (1966).
188. R. A. Schmitz, *C & F* **11**, 49 (1967).
189. N. Peters, *C & F* **33**, 315 (1978).
190. N. Peters, *CST* **30**, 1 (1982).
191. N. Peters and F. A. Williams, *AIAA Journal*, **21**, 423 (1983).
192. C. K. Law, *C & F* **24**, 89 (1975).
193. A. Liñán and A. Crespo, *CST* **14**, 95 (1976).
194. L. Krishnamurthy, F. A. Williams, and K. Seshadri, *C & F* **26**, 363 (1976).
195. K. Seshadri and F. A. Williams, *Int. J. Heat Mass Transfer* **21**, 251 (1978).
196. F. A. Williams, *Fire Safety Journal* **3**, 163 (1981).
197. J. Lorell and H. Wise, *J. Chem. Phys.* **23**, 1928 (1955).
198. V. K. Jain and N. Ramani, *CST* **1**, 1 (1970).
199. D. B. Spalding and V. K. Jain, "Theory of the Burning of Monopropellant Droplets," *Aero. Res. Council Tech. Rept. No. 20-176, Current Paper No. 447* (1959).
200. F. A. Williams, *C & F* **3**, 529 (1959).
201. V. K. Jain, *C & F* **7**, 17 (1963).
202. F. E. Fendell, *Astronautica Acta* **11**, 418 (1965).
203. F. E. Fendell, *AIAA Journal* **6**, 1946 (1968).
204. F. E. Fendell, *Int. J. Heat Mass Transfer* **12**, 223 (1969).
205. F. E. Fendell, *CST* **1**, 131 (1970).
206. J. D. Buckmaster, A. K. Kapila, and G. S. S. Ludford, *CST* **17**, 227 (1978).
207. A. Liñán, *Acta Astronautica* **2**, 1009 (1975).
208. A. K. Kapila, G. S. S. Ludford, and J. D. Buckmaster, *C & F* **25**, 361 (1975).
209. G. S. S. Ludford, D. W. Yannitell, and J. D. Buckmaster, *CST* **14**, 125 (1976).
210. G. S. S. Ludford, D. W. Yannitell, and J. D. Buckmaster, *CST* **14**, 133 (1976).

CHAPTER 4

Reactions in Flows with Negligible Molecular Transport

There are a number of combustion problems in which transport phenomena (which were of dominant importance in the preceding chapter) are negligible. Since the transport effects give rise to the most highly differentiated terms in the conservation equations, these effects will generally be unimportant provided that the gradients of the flow variables are sufficiently small. How small these gradients must be depends, of course, on the sizes of the other terms appearing in the conservation equations. Large values of the convective and time-dependent terms help to make the transport processes negligible. For example, in flows at high subsonic or supersonic speeds, transport effects are usually unimportant except in thin regions, such as shock waves and boundary layers, where the flow variables change very rapidly with position. The present chapter deals with systems involving negligible transport but finite reaction rates and focuses on methods of analysis, giving a few illustrative examples; textbooks [1] and [2] delve more deeply into details of some of the material to be covered here.

4.1. IGNITION DELAY AND THE WELL-STIRRED REACTOR

One simple example of an experiment in which transport effects are considered to be negligible is illustrated in Figure 4.1 and is described in detail in [3]. An oxidizing stream at an elevated temperature and traveling with

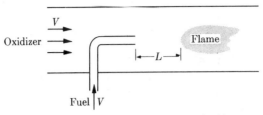

FIGURE 4.1.   Schematic diagram of spontaneous ignition experiment.

velocity V is brought into contact with a small amount of fuel which rapidly achieves this same temperature and velocity. An **ignition delay** τ may be defined as the ratio of the distance L downstream at which detectable luminous emissions begin to the convective velocity V. If it can be assumed that mixing occurs very rapidly in comparison with the rate of the gaseous chemical reaction, then transport terms can be neglected, and equation (1-4) for the fuel implies that, approximately,

$$1/\tau = -w_F/(\rho\Delta Y_F), \tag{1}$$

where ρ is the density of the reacting mixture at the temperature and pressure of the approaching stream, w_F is the rate of production of fuel (mass per unit volume per second), evaluated at the temperature and pressure of the oxidizing stream and at a mean composition of the ignition zone, and ΔY_F is the change in the fuel mass fraction that occurs prior to ignition. An estimate of the overall reaction rate during ignition can therefore be obtained by measuring V and L. Values of the ignition delay, obtained in experiments of this type that were performed at different temperatures, have been used in conjunction with equation (1) to deduce overall activation energies for the ignition of numerous fuels in air [3].

Various aspects of these ignition-delay experiments deserve further discussion. As explained in Section B.2.5, the chemical kinetics occurring during the ignition delay are complex and generally involve significant chain branching. Implicit in the writing of equation (1) is the idea that however complex the chemistry is, it can be related approximately to the disappearance of fuel, possibly to a small extent, prior to ignition. This interpretation is not essential to the use of τ; for example, it would be equally possible to interpret $1/\tau$ as the ratio of the rate of production of a critical chain carrier to the concentration of that carrier needed for ignition. The key aspect is that it must be possible to identify, in principle, a single variable that describes the progress of the ignition reactions. When this variable is normalized so that it begins at zero and achieves a value of unity at ignition, then the time rate of increase of the normalized progress variable Y is

$$dY/dt = 1/\tau. \tag{2}$$

Use of $\tau = L/V$ in equation (2) involves an approximation of relatively small changes in properties throughout the ignition region in Figure 4.1.

Equation (2) affords the possibility of employing ignition-delay data to calculate ignition processes in nonuniform flows [4]. Elements of gas containing fuel are identified and followed in a Lagrangian sense. For any such element, integration of equation (2) implies that ignition will occur at a time t_I defined by

$$\int_0^{t_I} (dt/\tau) = 1. \tag{3}$$

Application of equation (3) to different elements provides a value of t_I for each element; the minimum of these values is the shortest time for ignition to occur in the nonuniform system. Approaches of this type have been employed for estimating the chamber length required for combustion of a liquid injected into a supersonic stream [4] and for identifying conditions for ignition of streams of droplets of liquid fuels in oxidizing gases [5]. These approaches inherently lack accuracy because of their neglect of interdiffusional effects but often are convenient approximations, particularly for configurations that are too complex to be amenable to analyses that are more precise.

Another example of a system in which attention is focused primarily upon the rate of the chemical reaction is the conceptual **well-stirred reactor** [6]. It is supposed that by introducing extremely high-intensity turbulence into a reactor of volume $\mathscr{V}$, the thermodynamic properties can be made to be practically uniform throughout the reactor. The rate of consumption (mass per second) of fuel in the reactor will then be given by $-w_F \mathscr{V}$, where the thermodynamic variables appearing in the expression for w_F [equation (1-8)] are the average values of these variables in the reactor. For steady flow, the mass per second $\dot{m}_F$ of fuel entering the reactor would therefore have to be

$$\dot{m}_F = -w_F \mathscr{V}/\Delta Y_F, \tag{4}$$

where ΔY_F is the difference between the inlet and outlet mass fractions of fuel. This is the fundamental equation of the well-stirred reactor; it is sometimes coupled with an overall energy balance in order to determine the heat-release rate. The difficult problem in the analysis of any given system of this type is to ascertain the appropriate average properties within the reactor for computing w_F. A considerable amount of attention has been devoted to this problem, and a number of results have been obtained [6], [7], [8].*

---

* The earliest detailed theoretical analysis of the behavior of the well-stirred reactor appears to be that of Y. B. Zel'dovich and Y. A. Zysin, *Zhur. Tekhn. Fiz.* **11**, 493 (1941); see [7].

An experimental apparatus that under certain conditions approximates closely a well-stirred reactor is a spherical chamber into which premixed reactants are injected centrally through a perforated tube and out of which the reaction products flow through a number of ports located at various positions in the outer shell. The mixing is driven by the fuel-injection jets that are formed at the tube perforations, and the apparatus has been called the *jet-stirred reactor*. The average residence time of an element of gas in the reactor is

$$\tau_R = \rho \mathscr{V}/\dot{m}_F, \tag{5}$$

where ρ is the average density inside. Measurement of the average temperature and pressure inside and of the flow rate and inlet and outlet compositions for various values of $\dot{m}_F$ enables τ_R and w_F to be obtained as functions of temperature, for example. In this manner, overall rates of reaction are derived.

The overall rate of consumption of fuel obtained in a well-stirred reactor differs from that of an ignition-delay experiment because the combustion rather than the ignition of the fuel is investigated. Overall rates of other processes, such as pollutant production, also have been obtained in experiments with jet-stirred reactors by measuring levels of the appropriate products in the outlet streams. Results of some experiments on jet-stirred reactors are consistent with the concept of the well-stirred reactor [6], [9]–[17]. Some of the more recent work cited has endeavored to account for influences of imperfect mixing. Theoretical aspects of the well-stirred reactor have been presented in a recent book.*

Questions of the degree of perfection of mixing have plagued the use of the concept of the well-stirred reactor. In general, high turbulence intensities, small turbulence scales, slow rates of reaction, high reactor temperatures, small amounts of heat release, and relatively weak dependences of rates on temperature favor achievement of experimental results to which the concept can be applied, since under these conditions mixing rates are enhanced in comparison with reaction rates, and influences of nonuniformities within the reactor are minimized. Further information on favorable conditions may be found in Chapter 10.

In the applications that have been discussed here, high rates of transport have, somewhat paradoxically, favored attainment of conditions under which analyses neglecting transport effects can be applied. The rapid transport helps to achieve conditions of uniformity, under which transport no longer is significant, and effects of finite-rate chemistry can be studied. This same kind of situation prevails in various other experiments, such as those employing a suitably designed turbulent-flow reactor [18], [19], [20]. In

---

* Y. B. Zel'dovich, G. I. Barenblatt, V. B. Librovich and G. M. Makhviladze, *The Mathematical Theory of Combustion and Explosion*, Moscow: Nauka, 1980.

contrast, for the problems considered in the rest of this chapter, low rates of transport (in comparison with rates of convection and of reaction) favor attainment of conditions under which transport phenomena are negligible.

4.2. REACTIONS IN STEADY, QUASI-ONE-DIMENSIONAL FLOW

4.2.1. Steady-state, quasi-one-dimensional conservation equations

The equations governing the steady-state, quasi-one-dimensional flow of a reacting gas with negligible transport properties can easily be obtained from equations (1-19)–(1-22). When transport by diffusion is negligible ($D_{ij} \to 0$ and $D_{T,i} \to 0$ for $i, j = 1, \ldots, N$), the diffusion velocities, of course, vanish [$V_i \to 0$ for $i = 1, \ldots, N$, see equation (1-14)]. If, in addition, transport by heat conduction is negligible ($\lambda \to 0$) and $q_R \equiv 0$, then the heat flux q vanishes [see equation (1-15)]. Finally, in inviscid flow ($\mu \to 0$ and $\kappa \to 0$), equations (1-16)–(1-18) show that all diagonal elements of the pressure tensor reduce to the hydrostatic pressure, $p_{11} = p_{22} = p_{33} = p$. The steady-state forms of equations (1-20), (1-21a), and (1-22) then become

$$v\, dv/dx = -(1/\rho)\, dp/dx, \tag{6}$$

$$\frac{d}{dx}\left(h + \frac{v^2}{2}\right) = 0, \tag{7}$$

and

$$v\frac{dY_i}{dx} = \frac{w_i}{\rho}, \quad i = 1, \ldots, N, \tag{8}$$

respectively. In obtaining equations (6) and (7) body forces have been neglected ($f_i = 0$); this assumption, which will be retained throughout this chapter, is an excellent approximation for practically all problems of the present type. On the other hand, the use of $q_R = 0$ excludes many problems in radiation gas dynamics [2], [21].

The form assumed by the continuity equation, equation (1-19), can be derived formally by integration over the total cross-sectional area of the flow. The limits of the coordinates x_2 and x_3 [which appear in equation (1-13)] in such an integration must be independent of x because the boundaries of the cross section are streamlines and must therefore be parallel to the local x coordinate (that is, parallel to the local velocity vector). Thus, since the flow variables are independent of the coordinates x_2 and x_3, multiplication of equation (1-19) by $g_2 g_3$ followed by integration over the cross-sectional area shows that

$$\frac{d}{dx}\left(\rho v \iint g_2 g_3\, dx_2\, dx_3\right) = 0$$

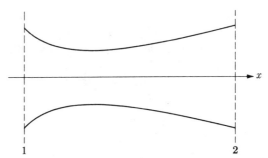

FIGURE 4.2   Schematic diagram of nozzle flow.

in steady flow. In view of equation (1-13), the double integral appearing here is simply $\iint ds_2\,ds_3 \equiv A(x)$, which is the total cross-sectional area of the flow. Hence the continuity equation is

$$d(\rho v A)/dx = 0. \tag{9}$$

Since the equation of state [for example, equation (1-9)] and the caloric equation of state [for example, equations (1-10) and (1-11) combined with the definition $h = u + p/\rho$] relate the thermodynamic properties h and p to ρ, T, and Y_i ($i = 1, \dots, N$), equations (6)–(9) constitute $N + 3$ first-order ordinary differential equations for the $N + 3$ unknowns Y_i ($i = 1, \dots, N$), T, ρ, and v. Therefore, if all flow variables are specified at some point in the flow (for example, at point 1 in Figure 4.2) and if the area $A(x)$ is given as a function of x, then—in principle—a stepwise forward integration of equations (6)–(9) [employing the subsidiary relations given in equations (1-8)–(1-12)] will yield all flow variables at any other point (x) in the channel. Although computational difficulties arise in forward integration procedures if some of the reactions approach equilibrium too closely,* numerical solutions for property profiles have been obtained (for example, [22]–[26]). These solutions are restricted primarily to supersonic, diverging portions of nozzles, the regions where effects of finite-rate chemistry are greatest; finite-rate processes influence the choking behavior in the vicinity of the throat in a manner that is explained in [1] and [2].

For a number of purposes it is convenient to replace equation (6) by an alternative expression. If equation (A-5) is applied to a closed system consisting of a unit mass of the reacting fluid, then we obtain

$$dh = T\,ds + (1/\rho)\,dp + \sum_{i=1}^{N} (\mu_i/W_i)\,dY_i,$$

where s is the entropy per unit mass, μ_i is the chemical potential of species i, and W_i is the molecular weight of species i. The formalism involved in

---

* These problems and approaches to their resolution are discussed in Section B.2.5.2.

deriving the form given for the last term in this expression is as follows:
$dN_i/M = d(N_i/M) = d[(N_i/N)/(M/N)] = d(X_i/\overline{W}) = d(Y_i/W_i) = dY_i/W_i$,
where M is the mass of the closed element, $N \equiv \sum_{i=1}^{N} N_i$ is the total number
of moles in this closed element, $\overline{W} \equiv M/N = \sum_{i=1}^{N} X_i W_i = (\sum_{i=1}^{N} Y_i/W_i)^{-1}$
is the average molecular weight, and use is made of Equation (1-12) in
obtaining the next-to-last equality. Since

$$v\frac{dv}{dx} = \frac{d}{dx}\left(\frac{v^2}{2}\right),$$

equations (6) and (7) and the preceding thermodynamic relationship yield

$$T\,ds/dx + \sum_{i=1}^{N} (\mu_i/W_i)\,dY_i/dx = 0,$$

which can be written as

$$v\,ds/dx = -(\rho T)^{-1} \sum_{i=1}^{N} (\mu_i/W_i)w_i \tag{10}$$

in view of equation (8). The right-hand side of equation (10) vanishes in
two important limiting cases: (1) **frozen flow**, in which the chemical reaction
rates w_i all vanish, and (2) **equilibrium flow**, in which $\sum_{i=1}^{N} \mu_i\,dN_i = 0$ in a
closed system by definition [see equation (A-16)], impling that

$$\sum_{i=1}^{N} (\mu_i/W_i)\,dY_i = 0$$

by the argument indicated above.

Equations (7) and (9) can always be written in integrated forms; for
example,

$$h + v^2/2 = h_0 \tag{11}$$

($h_0 \equiv$ total stagnation enthalpy) and

$$\rho v A = \dot{m} \tag{12}$$

($\dot{m} \equiv$ mass per second flowing through the channel). For either frozen or
equilibrium flow, Equation (10) can also be expressed in an integrated form,

$$s = s_0 \tag{13}$$

($s_0 =$ initial entropy). For frozen flow, equation (8) simply reduces to

$$Y_i = Y_{i,0},$$

while for equilibrium flow, equation (8) can be replaced by an appropriate
set of atom conservation and chemical equilibrium equations (see Section
A.3.7 and the last part of Section 2.1.1). Therefore, for either frozen or equili-

brium flow, equations (6)–(9) reduce to entirely algebraic relations,* and all property profiles can be determined without solving any differential equations. Useful methods for computing the flow variables in these limiting cases are given in [27] and [28], for example.

4.2.2. Specific impulse of rockets

One practical reason for studying chemical reactions in nozzle flow is to obtain accurate predictions of the thrust of a rocket motor. A rocket motor consists of a chamber in which propellants burn, followed by a nozzle (see Figure 2) in which the hot products of combustion are expanded to a high velocity. A force balance on the rocket motor[†] shows that provided the gas is expanded to the local atmospheric pressure at the nozzle exit (station 2), the thrust (force in the $-x$ direction) acting on the motor is $\dot{m}v_2$. An important rocket performance parameter is the **specific impulse** I_{sp}, that is, the impulse delivered to the rocket per unit weight of propellant ejected, which is seen to be given by $I_{sp} = \dot{m}v_2/\dot{m} = v_2$, apart from a dimensional constant.[‡] Propellant combinations with large values of I_{sp} are desirable. Hence, the value of v_2 is relevant; this value can be related to h_2 through equation (11) which may be rewritten as

$$v_2 = \sqrt{2(h_0 - h_2)}. \tag{14}$$

From equation (14) a simple and explicit formula for v_2 (and therefore for I_{sp}) can be derived if a one-component gas with a constant specific heat flows through the nozzle. Since $h = \text{constant} + c_p T$ in this case, $h_0 - h_2 = c_p T_0(1 - T_2/T_0)$. In place of the parameter T_2, it is generally more convenient to use p_2 because the atmospheric pressure (to which p_2 has been equated above) is usually specified in discussing propellant performance. For isentropic flow [equation (13)] of a one-component ideal gas with $c_p = \text{constant}$, it is well known and readily derived from equations (6), (7), and (1-9) that $T_2/T_0 = (p_2/p_0)^{(\gamma-1)/\gamma}$, where the ratio of specific heats is $\gamma \equiv c_p/c_v$ ($c_v = $ specific heat at constant volume). Since

$$c_p = [\gamma/(\gamma - 1)](R^0/W)$$

(from $c_p = c_v + R^0/W$), where R^0 is the universal gas constant and W is the molecular weight, equation (14) yields

$$I_{sp} = \sqrt{\left(\frac{2\gamma}{\gamma - 1}\right)\left(\frac{R^0 T_0}{W}\right)\left[1 - \left(\frac{p_2}{p_0}\right)^{(\gamma-1)/\gamma}\right]} \tag{15}$$

---

* This same simplification also occurs, of course, when some of the reactions are frozen and the rest are in equilibrium.

† See, for example, [29].

‡ The quantity I_{sp} is conventionally expressed in units of seconds, and therefore v_2 is divided by the standard gravitational acceleration in obtaining I_{sp}.

(aside from conversion factors), which shows that a high stagnation temperature T_0 ($T_0 \approx$ the adiabatic flame temperature of the propellants), a low molecular weight W, a high pressure ratio p_0/p_2 ($p_0 \approx$ the pressure in the combustion chamber of the rocket) and a low ratio of specific heats γ all favor high motor performance.

It is important to discover how equation (15) is modified when the many simplifying assumptions introduced in its derivation are removed. For frozen or equilibrium nozzle flow of a multicomponent ideal-gas mixture with variable heat capacities, procedures such as those discussed in [27], [28], and [30] are available for determining I_{sp}.* If reactions proceed at finite, nonzero rates in the nozzle, then differential equations must be solved, in general, to find I_{sp}. Since the computational procedures then become more complex, there has been interest in deriving simple criteria for determining whether frozen or equilibrium flow is a good approximation for any given reaction† and in developing approximate methods to account for small departures from either of these limiting cases. This topic will be considered in the rest of Section 4.2. It may be worth emphasizing first that variable specific heats and chemical reactions in the nozzle are not the only causes of departures from equation (15) in real rocket motors. We have not considered "off-design performance" (that is, formulas for I_{sp} when p_2 differs from the local atmospheric pressure). The occurrence and consequences of shock waves in the nozzle under certain conditions has been neglected. We have not mentioned the modification to I_{sp} caused by the divergence of the flow streamlines at the nozzle exit. We have also neglected the effects of incomplete combustion in the chamber (a nonequilibrium, burning mixture entering the nozzle), wall friction in the nozzle flow, heat transfer to the walls of the nozzle, regenerative cooling of the walls by liquid propellants, and so on. For background material and for discussions of these additional phenomena, the reader is referred to textbooks [32–36].

4.2.3. Near-equilibrium and near-frozen flows

Let f_j represent any flow variable (for example, T, ρ, v, Y_i), and consider situations in which the initial conditions at a selected position in the diverging

---

* In deriving these procedures, attention may be focused on obtaining effective mean molecular weights and specific-heat ratios that will again make equation (15) approximately valid [27], or effort may be placed on developing computer programs for solving the appropriate set of algebraic equations. The latter approach has given the results that can be used with greatest accuracy [28], [30].

† Clearly, these criteria must depend on the reaction-rate constants, which are unknown for many reactions. In such cases, the comparison of measured composition profiles with profiles calculated for frozen and equilibrium flow provides information concerning the reaction-rate constants through the use of these criteria. Thus a tool for measuring chemical reaction rates emerges [31].

portion of the nozzle shown in Figure 4.2 are given, are held fixed, and correspond to chemical equilibrium, while the area function $A(x)$ is specified.* The solutions to the equilibrium-flow equations are denoted by $f_{j,e}(x)$ and are presumed known. The magnitude of the difference $\Delta f_j \equiv f_j - f_{j,e}$ is assumed to be small compared with the magnitude of $f_{j,e}$ in order to define near-equilibrium flow. From the conservation equations, expressions of the form

$$\frac{d}{dx}\left(\frac{\Delta f_j}{f_{j,e}}\right) = \left(\frac{1}{A}\frac{dA}{dx}F_j\right) - G_j \tag{16}$$

may then be derived, where F_j is a function of the local equilibrium and nonequilibrium flow variables (but not of the rates), and G_j is a function of these properties and of the reaction rates as well. Qualitatively, the first term on the right-hand side of this equation results from the rate of change of $f_{j,e}$ caused by the area change, and the last term accounts for the tendency of f_j to approach $f_{j,e}$ under the influence of the finite reaction rate.

The general equation that has been given symbolically here can be expanded for $|\Delta f_j/f_{j,e}| \ll 1$. The form of the expanded equation depends on the flow variable selected. It is most convenient to employ equations (8), (10), (11), and (12) (along with state and rate equations) for describing the flow because the expansion of the entropy s, as given by equation (10), possesses special properties. Since s is constant in equilibrium, $F_j = 0$ for the entropy, and the corresponding G_j is found to be of higher order than G_j for other variables when an expansion is made for small values of $\Delta f_j/f_{j,e}$. Therefore, s is constant in the first approximation; the only remaining differential equations are those in equation (8), and the deviations in all other flow variables are related algebraically to ΔY_i. The one-term expansions $F_j = F_{e,j}$ and $G_j = \sum_i G_{e,ij}(\Delta Y_i/Y_{i,e})$ may now be introduced in equation (16), consistent with the fact that each $G_j = 0$ if all $\Delta f_j = 0$. Let M be the number of independent linear differential equations thereby obtained, and construct linear combinations of the ΔY_i that serve to diagonalize the coefficient matrix, thereby deriving for the new variables

$$\frac{d}{dx}\left(\frac{\Delta f_j}{f_{j,e}}\right) = \frac{F_{e,j}}{A}\frac{dA}{dx} - G_{e,j}\left(\frac{\Delta f_j}{f_{j,e}}\right), \quad j = 1,\ldots,M, \tag{17}$$

where $F_{e,j}$ and $G_{e,j}$ are evaluated from the local equilibrium-flow variables. General thermodynamic stability conditions require that $G_{e,j} > 0$ in this form.

For brevity of notation let us now suppress the subscript j; often only one independent reaction is neither frozen nor in equilibrium. The differential

---

* With upstream stagnation conditions and $A(x)$ given, finite-rate chemistry may influence m by modifying throat conditions. This effect is not addressed here.

equation for $\Delta f/f_e$ given in equation (17) is linear and possesses the solution

$$\frac{\Delta f}{f_e} = \exp\left(-\int_0^x G_e \, dx\right) \int_0^x \frac{F_e}{A} \frac{dA}{dx} \exp\left(\int_0^x G_e \, dx'\right) dx \qquad (18)$$

if $\Delta f/f_e = 0$ at $x = 0$. This equation explicitly gives the fractional deviation of a property f from its local equilibrium value f_e for near-equilibrium flow. However, even for simple reactions, to find the functions F_e and G_e from the prescription indicated above is complicated. Equation (18) shows that given a small positive number ϵ, $|\Delta f/f_e| \leq \epsilon$ if and only if the magnitude of the right-hand side is $\leq \epsilon$, but unfortunately this criterion for near-equilibrium flow depends on the history, since integrals over x appear.

An approximate condition for the occurrence of near-equilibrium flow, dependent only on local properties, can be obtained from the differential equation for $\Delta f/f_e$, equation (17). A reasonable condition is the requirement that $d|\Delta f/f_e|/dx$ be negative whenever $|\Delta f/f_e|$ exceeds its maximum permissible value ϵ. Thus at $\Delta f/f_e = \epsilon$, we require $d(\Delta f/f_e)/dx \leq 0$, and at $\Delta f/f_e = -\epsilon$, we require $d(\Delta f/f_e)/dx \geq 0$. These conditions can be written as $(F_e/A)dA/dx \leq G_e\epsilon$ at $\Delta f/f_e = \epsilon$ and $(F_e/A) dA/dx \geq -G_e\epsilon$ at $\Delta f/f_e = -\epsilon$. Since $G_e > 0$ and $\epsilon > 0$, a sufficient condition for these two inequalities to hold is that

$$\left| \frac{F_e}{G_e\epsilon} \frac{1}{A} \frac{dA}{dx} \right| \leq 1,$$

that is,

$$\left| \frac{1}{A} \frac{dA}{dx} \right| \leq \left| \frac{G_e\epsilon}{F_e} \right|. \qquad (19)$$

Equation (19) constitutes an upper bound on the fractional rate of change of area, which—if satisfied everywhere—is sufficient (but not, in general, necessary) to ensure that $|\Delta f/f_e| \leq \epsilon$ everywhere. It may be emphasized that the functional forms of F_e and G_e, and therefore the criterion given in equation (19), depend on the flow variable that f is taken to represent.

A development for near-frozen flows, parallel to that just discussed for near-equilibrium flows, is much simpler. One obtains

$$d(\Delta f/f_f)/dx = G_f \qquad (20)$$

and

$$\Delta f/f_f = \int_0^x G_f \, dx, \qquad (21)$$

where f is again an arbitrary flow variable, G is proportional to the reaction rate, the subscript f identifies quantities in the frozen flow, and $\Delta f \equiv f - f_f$. Thus the deviation from frozen conditions does not depend principally on how rapidly the frozen flow variables change with x, but instead is roughly proportional to the distance (or time) available for a departure from frozen conditions to occur.

4.2.4 Application to the reaction $A \rightleftharpoons B$ with species A and B present in only trace amounts

The calculations discussed in the preceding section are particularly simple for reactions involving species that are present only in trace amounts, because the reaction then has a negligible effect on the pressure, density, velocity, enthalpy, and other variables. Only the concentrations of the trace species involved in the reaction are affected appreciably by the rate of the reaction. The other flow variables may then be considered to be known functions of x in investigating the effect of the reaction rates on the trace concentrations. As an explicit illustration, let us consider the reaction $A \rightleftharpoons B$, identifying species A and B by the subscripts 1 and 2, respectively.

For the reaction under consideration, equations (8) and (2-11) yield

$$dY_1/dx = -W_1 \omega'/\rho v. \tag{22}$$

We need not consider dY_2/dx, since

$$Y_2 = 1 - Y_1 - \sum_{i=3}^{N} Y_i \tag{23}$$

and Y_i are known for the species not entering the reaction. The equilibrium condition is

$$X_2/X_1 = K_p(T) = Y_2/Y_1 \tag{24}$$

for this reaction [see equation (A-27)]. Here K_p is the equilibrium constant for partial pressures, and the last equality follows from the fact that $W_1 = W_2$ for this reaction. From rate formulas [for example, equations (1-8), (B-11), and (B-13)], it can be inferred that

$$\omega' = BT^{\alpha} e^{-E/R^0 T} \left(\frac{\rho}{W_1} \right) \left(Y_1 - \frac{Y_2}{K_p} \right), \tag{25}$$

where the rate constants refer to the forward reaction $A \to B$.

For equilibrium flow, the value of Y_1 is given by equations (23) and (24):

$$Y_{1,e}(x) = \left(1 - \sum_{i=3}^{N} Y_i \right) \bigg/ (1 + K_p). \tag{26}$$

Here Y_i ($i = 3, \dots, N$) and $K_p(T)$ are known functions of x and are the same in equilibrium and nonequilibrium flows (since only trace amounts of A and B are present). Equations (22), (23), (25), and (26) yield

$$dY_1/dx = -G_e(x)[Y_1 - Y_{1,e}(x)], \tag{27}$$

for nonequilibrium flow, where

$$G_e(x) \equiv \left(\frac{BT^\alpha}{v}\right) e^{-E/R^0 T} \left(1 + \frac{1}{K_p}\right) \tag{28}$$

is a known function of x and is the same in equilibrium and in non-equilibrium flows. For near-equilibrium conditions, we are interested in $\Delta Y_1/Y_{1,e}$, where $\Delta Y_1 \equiv Y_1 - Y_{1,e}$. Equation (27) shows that

$$\frac{d}{dx}\left(\frac{\Delta Y_1}{Y_{1,e}}\right) = -\frac{Y_1}{Y_{1,e}^2}\frac{dY_{1,e}}{dx} - G_e\left(\frac{\Delta Y_1}{Y_{1,e}}\right)$$

$$= -\frac{1}{Y_{1,e}}\frac{dY_{1,e}}{dx} - \left(G_e + \frac{1}{Y_{1,e}}\frac{dY_{1,e}}{dx}\right)\left(\frac{\Delta Y_1}{Y_{1,e}}\right). \tag{29}$$

When $|\Delta Y_1/Y_{1,e}| \ll 1$, the terms on the right-hand side of equation (29) will be of the same order if

$$\left|\frac{1}{Y_{1,e}}\frac{dY_{1,e}}{dx}\right| \ll G_e,$$

whence equation (29) becomes approximately

$$\frac{d}{dx}\left(\frac{\Delta Y_1}{Y_{1,e}}\right) = -\frac{1}{Y_{1,e}}\frac{dY_{1,e}}{dx} - G_e\left(\frac{\Delta Y_1}{Y_{1,e}}\right) \tag{30}$$

for near-equilibrium flow. The equivalence of equations (29) and (16) is apparent; to retrieve the previous notation, in equation (30) define $F_e \equiv -d \ln Y_{1,e}/dx/d \ln A/dx$. Thus the transcription of the results of the preceding section for near-equilibrium flow becomes obvious. In particular, the criterion for near equilibrium [equation (19)] is

$$\left|\frac{1}{Y_{1,e}}\frac{dY_{1,e}}{dx}\right| \leq \left(\frac{BT^\alpha}{v}\right) e^{-E/R^0 T} \left(1 + \frac{1}{K_p}\right) \epsilon \tag{31}$$

for the present reaction.

For frozen flow, $Y_1 = Y_{1,f} = $ constant, where $Y_{1,f}$ is the initial value of Y_1. For near-frozen flow, we are interested in $\Delta Y_1/Y_{1,f}$, where now $\Delta Y_1 \equiv Y_1 - Y_{1,f}$. In terms of this variable, equation (27) becomes

$$\frac{d}{dx}\left(\frac{\Delta Y_1}{Y_{1,f}}\right) = -G_e\left(\frac{\Delta Y_1}{Y_{1,f}} - \frac{Y_{1,e} - Y_{1,f}}{Y_{1,f}}\right). \tag{32}$$

Since $|\Delta Y_1/Y_{1,f}| \ll |(Y_{1,e} - Y_{1,f})/Y_{1,f}|$ for near-frozen flow, (32) can be written as

$$d(\Delta Y_1/Y_{1,f})/dx = G_f(x) \tag{33}$$

in this limit, where

$$G_f(x) \equiv G_e(Y_{1,e} - Y_{1,f})/Y_{1,f} \tag{34}$$

is a known function of x. Thus equation (20) and the corresponding results of the preceding section are obtained. A criterion for near-frozen flow can be written as

$$\left| \int_0^x \left(\frac{BT^\alpha}{v} \right) e^{-E/R^0 T} \left(1 + \frac{1}{K_p} \right) \left[\frac{(Y_{1,e} - Y_{1,f})}{Y_{1,f}} \right] dx \right| \leq \epsilon \qquad (35)$$

for this reaction.

4.2.5. Freezing of reactions

The reasoning outlined in Section 4.2.3 and illustrated in Section 4.2.4 was preceded by many approximate criteria that are less precise for the attainment of near-equilibrium flow [37]–[40]. These criteria play roles in design criteria to maximize nozzle performance. It has been found that for three-body atomic recombination reactions (primarily due to the cubic pressure dependence of the recombination rate), the composition profiles in a supersonic nozzle follow equilibrium curves very closely until a point is reached at which a relatively rapid transition to frozen flow occurs, after which the remaining portions of the composition profiles are practically frozen [24]. In order to determine the flow variables for flows that exhibit this "sudden-freezing" character, it is sufficient to obtain a condition specifying the location of the freezing point; near-equilibrium and near-frozen criteria are unimportant. When rate data are available, it is straightforward to estimate freezing points of this type [24].

When only one reaction occurs, combining equations (8) and (2-11) yields

$$v \, dY_i/dx = W_i(v_i'' - v_i')\omega'/\rho, \quad i = 1, \ldots, N, \qquad (36)$$

where ω' is the difference between the forward and backward rates of the reaction. Letting species 1 represent a product of the reaction, we can write equation (36) as

$$v \, dY_1/dx = \omega_F - \omega_B, \qquad (37)$$

where $\omega_F \equiv W_1(v_1'' - v_1')\omega_F'/\rho$, $\omega_B \equiv W_1(v_1'' - v_1')\omega_B'/\rho$, and ω_F' and ω_B' are the forward and backward rate terms in ω', respectively [for the forms of these terms, see equation (B-11)]. When the flow is near equilibrium, $\omega_F \approx \omega_B$; that is, $|v \, dY_1/dx| \ll \omega_F$ and $|v \, dY_1/dx| \ll \omega_B$ according to equation (37). On the other hand, when the flow is near frozen, either $\omega_F \ll \omega_B$ or $\omega_B \ll \omega_F$, and $|v \, dY_1/dx|$ is of the same order of magnitude as the larger of the two rate terms according to equation (37); that is, either $|v \, dY_1/dx| \gg \omega_F$ or $|v \, dY_1/dx| \gg \omega_B$. The intermediate freezing-point condition is therefore determined by either $|v \, dY_1/dx| \approx \omega_F$ or $|v \, dY_1/dx| \approx \omega_B$, depending on whether ω_F or ω_B becomes small in frozen flow. When freezing occurs suddenly, at the freezing point, the quantities may all be evaluated in near-equilibrium flow, and the freezing condition then becomes, approximately,

$$|v_e(dY_1 dx)_e| \approx \omega_{F,e} \approx \omega_{B,e}, \tag{38}$$

where the subscript e identifies equilibrium flow. All the quantities appearing in equation (38) can be evaluated from the equilibrium flow variables and the chemical kinetic formulas for ω_F and ω_B; in the first approximation sudden freezing occurs at the point where equilibrium calculations predict that the first term in equation (38) equals the second or the third. Good agreement of sudden-freezing predictions with experiment has been obtained for NO_2 recombination [41] but not for ion recombination [25].

A curious consequence of the sudden-freezing approximation is that the flow remains isentropic. This is evident from the fact that equation (13) applies in both equilibrium and frozen flow and is a further manifestation of the exceptional attributes of entropy in nozzle flow. There is a continuum of isentropic nozzle flows, ranging from fully equilibrium to fully frozen flows; the location of the freezing point parameterizes the continuum. This observation suggests an approach to obtaining an improved description of influences of finite-rate chemistry on nozzle flow. In the vicinity of the freezing point a perturbation may be sought in the time interval needed for freezing to occur, directed toward predicting the increase in entropy that occurs during freezing. The result would be a more accurate description of the property profiles, with regions of equilibrium and frozen flow separated by a transition region. This approach has not yet been pursued, although detailed analyses that treat the ratio of the thermal energy to the heat of dissociation as a small parameter provide important ideas that would be helpful in beginning the study.*

4.2.6. Two-phase nozzle flow

Many rocket propellants form solid or liquid products of combustion either in the combustion chamber or in the nozzle. This condensation tends to improve rocket performance by releasing heat but introduces the complications of two-phase flow in the nozzle. Gas-particle flows have much in common with reacting flows; for example, aerodynamic drag on the particles and heat transfer from the particles increase entropy, in precise analogy with the effects of finite-rate chemistry. Moreover, the two-phase effects are numerically larger and often of greater practical significance in contributing to losses in performance. Therefore, it appears to be of interest to discuss two-phase flows briefly.

As a first step consider a one-component ideal gas with constant specific heats, c_p and c_v, containing solid or liquid particles with constant

---

* See P. A. Blythe, *J. Fluid Mech.* **17**, 126 (1963); **20**, 243 (1964) and H. K. Cheng and R. S. Lee, *AIAA Journal* **6**, 823, 831 (1968).

specific heat, c_s, that are sufficiently small to maintain the same velocity and temperature as the gas and that do not exchange mass with the gas through condensation or evaporation [28], [33], [36]. The enthalpy per unit mass of this gas-particle system is $h = \text{constant} + Z c_p T + (1 - Z) c_s T$, where Z is the mass of gas per unit mass of the system. This relationship may be used in equation (14) for v_2, but the altered equation of state modifies equation (15). Neglecting the contribution of Brownian motion of the particles to the pressure, from equation (1-9) we may write $p = \rho_g T R^0 / W$, where ρ_g is the gas density, which is related to the total mass per unit volume ρ by

$$(1 - Z)/\rho_s + Z/\rho_g = 1/\rho, \tag{39}$$

where ρ_s is the specific gravity of the condensed phase. Equation (39) states that the total volume per unit mass is the sum of that of the condensed and gaseous phases; the gas is excluded from the volume occupied by the solid or liquid. Generally $(1 - Z)\rho/\rho_s \ll 1$, so that $\rho_g \approx Z\rho$, and the equation of state in terms of ρ becomes the same as that for an ideal gas with molecular weight W/Z. Since equations (9), (6), and (7) express appropriate mass, momentum, and energy balances, respectively, for the two-phase system, in this approximation the derivation of equation (15) remains valid, subject to the indicated replacement of molecular weight and to the use of $Z c_p + (1 - Z)c_s$ as the effective specific heat at constant pressure, in accordance with the expression for h. Thus in equation (15) W is replaced by W/Z and γ by $[Z\gamma + (1 - Z)(c_s/c_v)]/[Z + (1 - Z)(c_s/c_v)]$, as may be seen by algebraic manipulation. This effective γ decreases as $1 - Z$ increases, but the consequent increase in I_{sp} generally is less than the decrease associated with the factor $\sqrt{Z}$ that arises from the replacement of W. Therefore, even in this most ideal situation, a performance penalty is associated with the two-phase flow.

The penalty is greater if the particles cannot follow the changes in gas properties. For illustrative purposes let us assume that thermal equilibration occurs but allow for a particle velocity, v_s, different from that of the gas, v_g, while retaining all the other approximations indicated above. If the particles are spherical, all with radius r_s, and the Stokes law of drag applies, then a force balance on a particle gives

$$\tfrac{4}{3}\pi r_s^3 \rho_s v_s \, dv_s/dx = 6\pi r_s \mu(v_g - v_s), \tag{40}$$

where μ is the gas viscosity. In terms of the number density of particles, n_s, the mass flow rate of particles is

$$n_s(\tfrac{4}{3}\pi r_s^3 \rho_s)v_s A = \dot{m}_s = \text{constant}. \tag{41}$$

The mass flow rate of the gas is

$$(pW/R^0 T)v_g A = \dot{m}_g = \text{constant}, \tag{42}$$

and the total rate of mass flow is $\dot{m} = \dot{m}_s + \dot{m}_g$. The generalizations of equations (6) and (7) for overall momentum and energy conservation are

$$\dot{m}_s \, dv_s/dx + \dot{m}_g \, dv_g/dx = -A \, dp/dx \qquad (43)$$

and

$$\dot{m}_s(c_s T + v_s^2/2) + \dot{m}_g(c_p T + v_g^2/2) = \text{constant}. \qquad (44)$$

Equations (40)–(44) serve to determine the five unknown variables v_s, v_g, n_s, p, and T.

From equation (40) a characteristic time lag or response time for velocity equilibration can be identified as $2r_s^2 \rho_s/9\mu$. This time lag can be compared with a representative residence time in the nozzle, such as $\dot{m}/(A \, dp/dx)$. If the former is much larger,—for example, if r_s is sufficiently large—then in the first approximation v_s is approximately constant throughout the nozzle, and I_{sp} is low [28]. Perturbations about this large-lag limit may be developed [28]. If the former is much smaller, then $v_s = v_g$ in the first approximation, and the results described after equation (39) apply. Perturbations about this small-lag limit have been studied because of potential practical importance [42], [43] and have been considered in connection with optimized nozzle design [44], [28]. If the lag is neither large nor small, then numerical integration is generally needed [45]. Effects of temperature lags are similar to those of velocity lags. Condensation and agglomeration processes also may be important [46]. Particles may fail to follow streamlines and produce undesirable effects by impinging on walls; axisymmetric flows must be considered in analyzing these processes [47]–[50], and generalizations of the characteristic methods discussed below become useful.

Among additional topics of interest in two-phase flows are questions of sound propagation in two-phase media [28], [51], [52] (see Section 9.1.4) and many other problems [53]–[55]. We shall return to two-phase flows in connection with spray combustion (Chapter 11) to give a more systematic development.

4.3. REACTIONS IN UNSTEADY, THREE-DIMENSIONAL FLOW

4.3.1. Conservation equations; characteristic surfaces

Let us now consider the unsteady, three-dimensional flow of a homogeneous reacting fluid with negligible transport properties. In view of the simplifications that arise when transport effects are absent (see Section 4.2.1), equations (1-1)–(1-4) become

$$D\rho/Dt + \rho \mathbf{V} \cdot \mathbf{v} = 0, \qquad (45)$$

$$\rho D\mathbf{v}/Dt = -\mathbf{V}p, \qquad (46)$$

$$\rho Du/Dt = -p(\mathbf{V} \cdot \mathbf{v}), \qquad (47)$$

and

$$\rho D Y_i/Dt = w_i, \quad i = 1, \ldots, N, \tag{48}$$

respectively, where the substantial derivative is defined as

$$D/Dt \equiv \partial/\partial t + \mathbf{v} \cdot \nabla.$$

In order to illustrate most easily the mathematical properties of (45)–(48), it is convenient to write equations (45) and (47) in alternative forms.

When the same kind of reasoning used in deriving the expression for dh, employed in obtaining equation (10) in Section 4.2, is applied to equation (A-3), we find

$$du = T \, ds - p \, d(1/\rho) + \sum_{i=1}^{N} (\mu_i/W_i) \, dY_i.$$

Substituting this relationship and equation (45) into equation (47) gives

$$Ds/Dt = -(1/T) \sum_{i=1}^{N} (\mu_i/W_i)(D Y_i/Dt)$$

$$= -(1/\rho T) \sum_{i=1}^{N} (\mu_i/W_i) w_i, \tag{49}$$

where use has been made of equation (48) in the last equality. The right-hand member of equation (49) is the rate of production of entropy by the irreversible chemical reaction. Equation (49) is a useful replacement for equation (47).

By substituting equation (49) into the mathematical identity*

$$\frac{Dp}{Dt} = \left(\frac{\partial p}{\partial \rho}\right)_{s, Y_i} \frac{D\rho}{Dt} + \left(\frac{\partial p}{\partial s}\right)_{\rho, Y_i} \frac{Ds}{Dt} + \sum_{i=1}^{N} \left(\frac{\partial p}{\partial Y_i}\right)_{s, \rho, Y_{j(j \neq i)}} \frac{D Y_i}{Dt},$$

one can show through straightforward thermodynamic arguments that

$$\frac{D\rho}{Dt} = \frac{1}{a_f^2} \frac{Dp}{Dt} + \frac{\rho \beta_f}{c_{p, f}} \sum_{i=1}^{N} \left(\frac{\partial h}{\partial Y_i}\right)_{\rho, p, Y_{j(j \neq i)}} \frac{D Y_i}{Dt}, \tag{50}$$

where the frozen sound speed a_f is determined by

$$a_f^2 \equiv (\partial p/\partial \rho)_{s, Y_i}, \tag{51}$$

the frozen volumetric thermal expansion coefficient (which equals $1/T$ for an ideal gas mixture) is given by

$$\beta_f \equiv \rho [\partial(1/\rho)/\partial T]_{p, Y_i}, \tag{52}$$

---

* Here and in what follows, derivatives with respect to Y_i at constant $Y_j (j \neq i)$ are meant to be proportional to derivatives with respect to the mass of species i in the system at constant values of the masses of all other species in the system. With this interpretation, all N of the Y_i's are independently variable in the differentiation, and the restriction $\sum_{i=1}^{N} Y_i = 1$ is *not* to be imposed in evaluating such derivatives.

and the frozen specific heat at constant pressure is

$$c_{p,f} \equiv (\partial h/\partial T)_{p,Y_i}. \tag{53}$$

Substituting equations (45) and (48) into equation (50) yields the expression

$$\frac{1}{a_f^2}\frac{Dp}{Dt} = -\rho\nabla\cdot\mathbf{v} - \sum_{i=1}^{N}\sigma_i w_i, \tag{54}$$

where

$$\sigma_i \equiv (\beta_f/c_{p,f})(\partial h/\partial Y_i)_{\rho,p,Y_{j(j\neq i)}}. \tag{55}$$

For an ideal-gas mixture with a constant average molecular weight, the term $(\partial h/\partial Y_i)_{\rho,p,Y_{j(j\neq i)}}$ appearing in equation (55) is simply the (total) specific enthalpy of species i [that is, the quantity h_i given in equation (1-11)], and the dimensionless parameter σ_i becomes $\sigma_i = h_i/(c_{p,f}T)$; thus the term $-\sum_{i=1}^{N}\sigma_i w_i$ is a measure of the rate of liberation of heat by the chemical reaction. Equation (54) is a useful replacement for equation (45) in studies of the characteristic surfaces of the governing equations.

Equations (46), (48), (49), and (54) may be taken to be the governing differential conservation equations. Equations (48) and (49) are already in characteristic form;* they show that composition and entropy changes propagate only along streamlines. Equations (46) and (54) involve derivatives only of p and $\mathbf{v}$;[†] the characteristics of these equations require further study.

It is known [56] that an arbitrary system of quasi-linear partial differential equations which may be written in the form

$$\sum_{j,k_0,k_1,\ldots,k_n} A_{ij}^{(k_0,k_1,\ldots,k_n)}(x_0,x_1,\ldots,x_n;u_1,u_2,\ldots,u_N)\frac{\partial^{n_j}u_j}{\partial x_0^{k_0}\partial x_1^{k_1}\ldots\partial x_n^{k_n}}$$

$$+ \text{ terms in lower-order derivatives}$$

$$+ f_i(x_0,x_1,\ldots,x_n;u_1,u_2,\ldots,u_N) = 0, \quad i = 1,2,\ldots,N, \tag{56}$$

has a characteristic surface,

$$\Psi(x_0,x_1,x_n) = 0, \tag{57}$$

defined by the equation

$$\left| \sum_{\text{All }k_0,k_1,\ldots,k_n\,(k_0+k_1+\cdots+k_n=n_j)} A_{ij}^{(k_0,k_1,\ldots,k_n)}(x_0,x_1,\ldots,x_n;u_1,u_2,\ldots,u_N) \right.$$

$$\left. \times \left(\frac{\partial\Psi}{\partial x_0}\right)^{k_0}\left(\frac{\partial\Psi}{\partial x_1}\right)^{k_1}\cdots\left(\frac{\partial\Psi}{\partial x_n}\right)^{k_n} \right| = 0, \tag{58}$$

---

* For a general discussion of characteristics and for the definition of characteristic surfaces with an arbitrary number of dependent and independent variables, see, for example, [56]. The use of the method of characteristics in solving equations is considered in [57].

   [†] This is the reason that equation (54) is more useful than equation (45) in the present context.

where vertical lines represent the determinant of the matrix within them. In equation (56) there are $n + 1$ independent variables denoted by x_l ($l = 0, 1, \ldots, n$) and N dependent variables denoted by u_j. The quantities

$$A_{ij}^{(k_0, k_1, \ldots, k_n)}(x_0, x_1, \ldots, x_n; u_1, u_2, \ldots, u_N)$$

are the coefficients of the highest order (n_j) derivatives of u_j in the ith partial differential equation, while $f_i(x_0, x_1, \ldots, x_n; u_1, u_2, \ldots, u_N)$ represents the terms in the ith equation that involve no derivatives at all.

When there are no derivatives of order higher than the first ($n_j \leq 1$ for all j), the integers k_l may assume only the values 0 or 1. In this case, if one k_l is unity in a term of the summation in equation (56), then every other k_l in that term must be zero. We may then use a new notation which is obtained by letting

$$(k_0, k_1, \ldots, k_n) \to k,$$

where k takes on the values $0, 1, \ldots, n$ and is equal to the l value of the only nonzero k_l. In this simple case, equation (56) may be written as

$$\sum_{j=1}^{N} \sum_{k=0}^{n} A_{ij}^k(x_0, x_1, \ldots, x_n; u_1, u_2, \ldots, u_N) \frac{\partial u_j}{\partial x_k}$$

$$+ f_i(x_0, x_1, \ldots, x_n; u_1, u_2, \ldots, u_N) = 0, \quad i = 1, 2, \ldots, N, \quad (59)$$

and equation (58) becomes

$$\left| \sum_{k=0}^{n} A_{ij}^k(x_0, x_1, \ldots, x_n; u_1, u_2, \ldots, u_N) \frac{\partial \Psi}{\partial x_k} \right| = 0, \quad (60)$$

which is a first-order partial differential equation for the characteristic surface.

In equations (46) and (54), the highest derivatives are all of the first order ($n_j \leq 1$). Equations (46) and (54) may be written in full in Cartesian coordinates as follows:

$$\frac{\partial v_1}{\partial t} + v_1 \frac{\partial v_1}{\partial x_1} + v_2 \frac{\partial v_1}{\partial x_2} + v_3 \frac{\partial v_1}{\partial x_3} + \frac{1}{\rho} \frac{\partial p}{\partial x_1} = 0, \quad (61)$$

$$\frac{\partial v_2}{\partial t} + v_1 \frac{\partial v_2}{\partial x_1} + v_2 \frac{\partial v_2}{\partial x_2} + v_3 \frac{\partial v_2}{\partial x_3} + \frac{1}{\rho} \frac{\partial p}{\partial x_2} = 0, \quad (62)$$

$$\frac{\partial v_3}{\partial t} + v_1 \frac{\partial v_3}{\partial x_1} + v_2 \frac{\partial v_3}{\partial x_2} + v_3 \frac{\partial v_3}{\partial x_3} + \frac{1}{\rho} \frac{\partial p}{\partial x_3} = 0, \quad (63)$$

and

$$\frac{1}{\rho a_f^2} \left(\frac{\partial p}{\partial t} + v_1 \frac{\partial p}{\partial x_1} + v_2 \frac{\partial p}{\partial x_2} + v_3 \frac{\partial p}{\partial x_3} \right) + \frac{\partial v_1}{\partial x_1} + \frac{\partial v_2}{\partial x_2} + \frac{\partial v_3}{\partial x_3} + f = 0. \quad (64)$$

TABLE 4.1

	$j = 1$				$j = 2$			
	$k = 0$	$k = 1$	$k = 2$	$k = 3$	$k = 0$	$k = 1$	$k = 2$	$k = 3$
$i = 1$	1	v_1	v_2	v_3	0	0	0	0
$i = 2$	0	0	0	0	1	v_1	v_2	v_3
$i = 3$	0	0	0	0	0	0	0	0
$i = 4$	0	1	0	0	0	0	1	0

	$j = 3$				$j = 4$			
	$k = 0$	$k = 1$	$k = 2$	$k = 3$	$k = 0$	$k = 1$	$k = 2$	$k = 3$
$i = 1$	0	0	0	0	0	$\dfrac{1}{\rho}$	0	0
$i = 2$	0	0	0	0	0	0	$\dfrac{1}{\rho}$	0
$i = 3$	1	v_1	v_2	v_3	0	0	0	$\dfrac{1}{\rho}$
$i = 4$	0	0	0	1	$\dfrac{1}{\rho a_f^2}$	$\dfrac{v_1}{\rho a_f^2}$	$\dfrac{v_2}{\rho a_f^2}$	$\dfrac{v_3}{\rho a_f^2}$

Here the subscripts 1, 2, and 3 on x represent the various coordinates, and the subscripts on v denote the components of v in the corresponding directions. It is seen from equations (64) and (54) that $f \equiv \sum_{i=1}^{N} \sigma_i w_i / \rho$. By comparing equations (61)–(64) with equation (59), the A_{ij}^k may easily be determined. If $x_0 \equiv t$, $i = 1$ through 4 is allowed to correspond to equations (61)–(64), respectively, and $u_j \equiv v_j$ for $j = 1, 2, 3$, while $u_4 \equiv p$ then a table of the A_{ij}^k may be constructed (Table 4.1).

By substituting the values of $A_{ij}^k(x_0, x_1, \ldots, x_n; u_1, u_2, \ldots, u_N)$ given in Table 4.1 into equation (60), we obtain the following relationship defining the characteristic surface, $\Psi(t, x_1, x_2, x_3)$:

$$
\begin{vmatrix}
\dfrac{D\Psi}{Dt} & 0 & 0 & \dfrac{1}{\rho}\dfrac{\partial\Psi}{\partial x_1} \\[2ex]
0 & \dfrac{D\Psi}{Dt} & 0 & \dfrac{1}{\rho}\dfrac{\partial\Psi}{\partial x_2} \\[2ex]
0 & 0 & \dfrac{D\Psi}{Dt} & \dfrac{1}{\rho}\dfrac{\partial\Psi}{\partial x_3} \\[2ex]
\dfrac{\partial\Psi}{\partial x_1} & \dfrac{\partial\Psi}{\partial x_2} & \dfrac{\partial\Psi}{\partial x_3} & \dfrac{1}{\rho a_f^2}\dfrac{D\Psi}{Dt}
\end{vmatrix} = 0.
\tag{65}
$$

Evaluation of the determinant yields

$$\frac{1}{\rho}\left(\frac{D\Psi}{Dt}\right)^2\left[\frac{1}{a_f^2}\left(\frac{D\Psi}{Dt}\right)^2 - \left(\frac{\partial\Psi}{\partial x_1}\right)^2 - \left(\frac{\partial\Psi}{\partial x_2}\right)^2 - \left(\frac{\partial\Psi}{\partial x_3}\right)^2\right] = 0. \quad (66)$$

One solution of equation (66) is

$$D\Psi/Dt = 0, \quad (67)$$

which shows that streamlines are characteristic lines of the equations. The other solution to equation (66) may be written as

$$(D\Psi/Dt)^2/a_f^2 - (\nabla\Psi)\cdot(\nabla\Psi) = 0, \quad (68)$$

which defines characteristic surfaces that propagate normal to themselves with the frozen sound velocity a_f relative to the local mass-average velocity of the fluid.* This last result has been emphasized by Wood and Kirkwood [58] and by Chu [59].

4.3.2. The method of characteristics for steady, two-dimensional (axially symmetrical and plane) flows

For steady flows, equation (49) may be replaced by an expression that is both simpler and generally more useful. By substituting equation (45) and the identity $h = u + p/\rho$ into equation (47), we obtain the expression

$$Dh/Dt - (1/\rho)Dp/Dt = 0,$$

in which equation (46) may be employed to show that

$$D(h + v^2/2)/Dt - (1/\rho)\partial p/\partial t = 0. \quad (69)$$

For steady flow, equation (69) reduces to the relationship

$$\mathbf{v}\cdot\nabla(h + v^2/2) = 0, \quad (70)$$

which states that the total stagnation enthalpy $(h + v^2/2)$ is constant along streamlines. Since equation (70) is simpler than the steady-state form of equation (49), and since enthalpies are usually easier to compute than are entropies, it is generally convenient to employ $h + v^2/2$ in place of s as a new dependent variable, adopting equation (70) and the steady-state forms of equations (46), (48), and (54) as the governing differential conservation equations. That this choice is not convenient for unsteady flow is apparent from the fact that equation (69) involves $\partial p/\partial t$ and therefore would have to be

---

* This result becomes more transparent if the coordinates are transformed in such a way that the fluid is brought locally to rest; equation (68) then reduces to

$$\left(\frac{\partial\Psi}{\partial t}\right)^2 - a_f^2\left[\left(\frac{\partial\Psi}{\partial x_1}\right)^2 + \left(\frac{\partial\Psi}{\partial x_2}\right)^2 + \left(\frac{\partial\Psi}{\partial x_3}\right)^2\right] = 0.$$

included along with equations (46) and (54) in calculating the characteristic surfaces for the pressure and velocity equations.

For steady flow with axial symmetry, let x be the axis of symmetry, r be the radial coordinate normal to the axis of symmetry, v_1 be the component of velocity in the x direction, and v_2 be the component of velocity in the r direction. Then equation (70) and the steady-state forms of equations (46), (48), and (54) become, respectively,

$$v_1 \frac{\partial}{\partial x}\left(h + \frac{v_1^2 + v_2^2}{2}\right) + v_2 \frac{\partial}{\partial r}\left(h + \frac{v_1^2 + v_2^2}{2}\right) = 0, \tag{71}$$

$$\left. \begin{array}{l} \rho v_1 \dfrac{\partial v_1}{\partial x} + \rho v_2 \dfrac{\partial v_1}{\partial r} = -\dfrac{\partial p}{\partial x}, \\[3mm] \rho v_1 \dfrac{\partial v_2}{\partial x} + \rho v_2 \dfrac{\partial v_2}{\partial r} = -\dfrac{\partial p}{\partial r}, \end{array} \right\} \tag{72}$$

$$v_1 \frac{\partial Y_i}{\partial x} + v_2 \frac{\partial Y_i}{\partial r} = \frac{w_i}{\rho}, \quad i = 1, \ldots, N, \tag{73}$$

and

$$\frac{v_1}{a_f^2} \frac{\partial p}{\partial x} + \frac{v_2}{a_f^2} \frac{\partial p}{\partial r} + \rho \frac{\partial v_1}{\partial x} + \rho \frac{\partial v_2}{\partial r} + \frac{\rho v_2}{r} + \sum_{i=1}^{N} \sigma_i w_i = 0. \tag{74}$$

It is convenient to use the distance in the streamline direction, ξ, and the distance normal to streamlines, η, in place of x and r as independent variables. In terms of the magnitude of the velocity, $v = (v_1^2 + v_2^2)^{1/2}$, and the angle between the flow direction and the x axis, $\theta = \tan^{-1}(v_2/v_1)$, we find

$$d\xi = dx \cos \theta + dr \sin \theta, \qquad d\eta = -dx \sin \theta + dr \cos \theta, \tag{75}$$

and

$$v_1 = v \cos \theta, \qquad v_2 = v \sin \theta \tag{76}$$

(see Figure 4.3). In view of equations (75) and (76), equations (71)–(74) can be shown to reduce to

$$v \frac{\partial}{\partial \xi}\left(h + \frac{v^2}{2}\right) = 0, \tag{77}$$

$$v \, \partial v/\partial \xi = -(1/\rho) \, \partial p/\partial \xi, \qquad v^2 \, \partial \theta/\partial \xi = -(1/\rho) \, \partial p/\partial \eta, \tag{78}$$

$$v \, \partial Y_i/\partial \xi = w_i/\rho, \quad i = 1, \ldots, N, \tag{79}$$

and

$$\left(\frac{v}{\rho a_f^2}\right) \frac{\partial p}{\partial \xi} + \frac{\partial v}{\partial \xi} + v \frac{\partial \theta}{\partial \eta} + \frac{v \sin \theta}{r} + \frac{1}{\rho} \sum_{i=1}^{N} \sigma_i w_i = 0, \tag{80}$$

respectively. The first relation in equation (78) may be used to eliminate the velocity derivative from equation (80); the result is

$$\left(\frac{M^2 - 1}{\rho v^2}\right) \frac{\partial p}{\partial \xi} + \frac{\partial \theta}{\partial \eta} = -\frac{\sin \theta}{r} - \left(\frac{1}{\rho v}\right) \sum_{i=1}^{N} \sigma_i w_i, \tag{81}$$

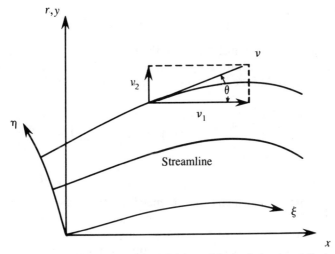

FIGURE 4.3. Schematic illustration of the definitions of the (x, r), (x, y), and (ξ, η) coordinate systems.

where $M \equiv v/a_f$ is the Mach number referred to the frozen speed of sound. Equations (77), (78), (79), and (81) may be taken as the governing set of differential conservation equations for steady, axially symmetrical flow. Their characteristic lines may easily be obtained by the methods of the preceding section.

The finite-difference forms of equations (77), (78), (79), and (81) are useful for obtaining numerical solutions by the method of characteristics for supersonic flows. Along a streamline ($\Delta\eta = 0$), equations (77) and (79) and the first relation in equation (78) reduce to total differential expressions and become, respectively,

$$\Delta(h + v^2/2) = 0 \quad \text{for} \quad \Delta\eta = 0, \tag{82}$$

$$\Delta Y_i = [w_i/(\rho v)] \, \Delta\xi, \quad i = 1, \ldots, N, \quad \text{for} \quad \Delta\eta = 0, \tag{83}$$

and

$$\Delta v = -[1/(\rho v)] \, \Delta p \quad \text{for} \quad \Delta\eta = 0 \tag{84}$$

(where the symbol Δ indicates a small increment). Equation (81) and the last relation in equation (78) constitute two independent equations for the derivatives of θ and p and have characteristic lines [57] given by

$$
\begin{vmatrix}
1 & 0 & 0 & \dfrac{1}{\rho v^2} \\[2mm]
0 & 1 & \dfrac{M^2 - 1}{\rho v^2} & 0 \\[2mm]
\Delta\xi & \Delta\eta & 0 & 0 \\[2mm]
0 & 0 & \Delta\xi & \Delta\eta
\end{vmatrix} = 0,
$$

that is, by

$$\Delta\eta = \pm \, \Delta\xi/\sqrt{M^2 - 1}, \tag{85}$$

which defines the Mach lines based on the frozen sound speed. Along these lines, one finds by substituting equation (85) into the finite-difference forms of equation (81) and of the last expression in equation (78), that the relationship

$$\left(\frac{\sqrt{M^2 - 1}}{\rho v^2}\right) \Delta p \pm \Delta\theta = \frac{\Delta l_\pm}{M}\left[-\frac{\sin\theta}{r} - \left(\frac{1}{\rho v}\right)\sum_{i=1}^{N} \sigma_i w_i\right]$$

$$\text{for} \quad \Delta\eta = \frac{\pm\Delta\xi}{\sqrt{M^2 - 1}} \tag{86}$$

must hold.* Here $\Delta l_\pm$, which is the geometric distance along the Mach lines, is given by $\Delta l_\pm = \pm M \, \Delta\eta$ along the lines $\Delta\eta = \pm \, \Delta\xi/\sqrt{M^2 - 1}$. When $M > 1$ [so that the values of $\Delta\eta$ given by equation (85) are real], equations (82), (83), (84), and (86) can be used to obtain a stepwise solution to the conservation equations.

The explicit numerical calculation procedure for solving equations (82), (83), (84), and (86) is very similar to the method of characteristics for nonreacting rotational flows [60]; the principal difference is that here entropy varies along streamlines as well as normal to them. Briefly, the basic step in the method of computation is as follows. All properties must be known at two neighboring points, A and B in Figure 4.4, which cannot be connected by a single streamline or Mach line. One may then construct Mach lines of opposite families at points A and B. These Mach lines will intersect at a point C, the new point at which all properties are to be computed. Since the distances AC and BC are known, equation (86) determines the values of p_C and θ_C (the pressure and the flow angle at point C). The extension of the streamline through the point C in the upstream direction (at angle θ_C) is then found to cross the line AB at a point D. By a linear interpolation along the line AB, the properties p_D, v_D, h_D, and so on may be found approximately at point D. By computing the length $\Delta\xi_C$ of the line DC, one may then determine v_C from equation (84), Y_{iC} from equation (83), and h_C from equation (82). All properties at the point C are now known, and one may proceed to the next step. In this manner, properties throughout a region of the shape indicated in Figure 4.5(a) may be obtained. While many refinements of the procedure outlined above yield better accuracy [60], there are no novel basic differences in the refined methods.

It is worthwhile to consider what kinds of boundary-value problems are correctly set for equations (77), (78), (79), and (81). A total of $N + 4$ dependent variables (p, θ, v, h, and Y_i) appear in these equations; subsidiary

---

* Expressions such as equation (86) are intended to represent two equations, one obtained by reading the upper signs and the other obtained by reading the lower signs.

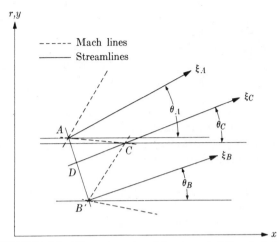

FIGURE 4.4.   Elementary step of the characteristic calculation.

relations (such as the equation of state) enable us to express all other quantities (namely, M, ρ, σ_i, and w_i) in terms of these dependent variables. We have
shown [equations (82), (83), (84), and (86)] that relationships between the
differentials of p and θ exist along Mach lines and that the differentials of
v, h, and Y_i are determined along streamlines. It follows that a correctly set
Cauchy initial value problem [57] is one in which the $N + 4$ dependent
variables are given along a noncharacteristic curve [that is, neither a Mach
line nor a streamline; see Figure 4.5(a)]. Unfortunately, in many practical

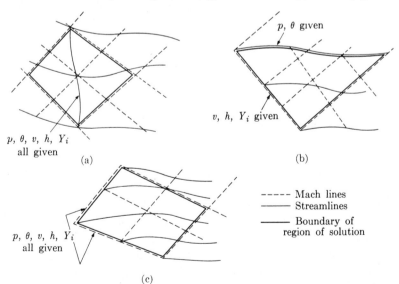

FIGURE 4.5.   (a) Variables specified along a noncharacteristic line. (b) Variables specified
along streamlines and Mach lines. (c) Variables specified along intersecting Mach lines.

problems data are given along characteristics, and we do not have a simple Cauchy problem. Data of this last type can, of course, also determine solutions in finite regions, as is illustrated in Figure 4.5(b) and (c). Procedures slightly different from that described in the preceding paragraph are required in solving such problems numerically.

Throughout this section we have been discussing steady, axially symmetrical flow. With only a few minor modifications in the formulas, our equations and results can be made to apply to steady, plane, two-dimensional flows. Thus if the radial coordinate r everywhere is replaced by y, the Cartesian coordinate perpendicular to x (see Figure 4.3), then we need only neglect the curvature terms in order to obtain equations that are valid in plane flows. Explicitly, we delete the terms $\rho v_2/r$ in equation (74), $v \sin \theta/r$ in equation (80), and $-\sin \theta/r$ in equations (81) and (86). The discussion is not modified in any way by these changes.

4.3.3. The method of characteristics for one-dimensional, unsteady flows

Let us next consider one-dimensional, unsteady flows with velocity v in the x direction. In this case, equations (46), (48), (49), and (54) reduce to

$$\rho \, \partial v/\partial t + \rho v \, \partial v/\partial x + \partial p/\partial x = 0, \tag{87}$$

$$\partial Y_i/\partial t + v \, \partial Y_i/\partial x = w_i/\rho, \quad i = 1, \ldots, N, \tag{88}$$

$$\partial s/\partial t + v \, \partial s/\partial x = -(\rho T)^{-1} \sum_{i=1}^{N} (\mu_i/W_i)w_i, \tag{89}$$

and

$$\rho \frac{\partial v}{\partial x} + \frac{1}{a_f^2} \frac{\partial p}{\partial t} + \frac{v}{a_f^2} \frac{\partial p}{\partial x} = -\sum_{i=1}^{N} \sigma_i w_i, \tag{90}$$

respectively. If we multiply equation (90) by a_f^2 and equation (87) by a_f and then add and subtract the resulting equations, we obtain the following pair of differential equations:

$$\frac{\partial p}{\partial t} + (v \pm a_f)\frac{\partial p}{\partial x} \pm \rho a_f \left[\frac{\partial v}{\partial t} + (v \pm a_f)\frac{\partial v}{\partial x}\right] = -a_f^2 \sum_{i=1}^{N} \sigma_i w_i. \tag{91}$$

Equations (88), (89), and (91) constitute a useful set of governing conservation equations for characteristic analyses of one-dimensional, unsteady flows.

The finite-difference forms of equations (88), (89), and (91) are used in obtaining numerical solutions by applying the method of characteristics in the x, t plane. Along the streamlines, $\Delta x = v \, \Delta t$, and equations (88) and (89) become

$$\Delta Y_i = (w_i/\rho) \, \Delta t/2, \quad i = 1, \ldots, N, \quad \text{for} \quad \Delta x = v \, \Delta t, \tag{92}$$

and

$$\Delta s = -(1/\rho T) \sum_{i=1}^{N} (\mu_i/W_i) w_i \, \Delta t/2 \quad \text{for} \quad \Delta x = v \, \Delta t. \tag{93}$$

Similarly, along the Mach lines, which are defined by $\Delta x = (v \pm a_f) \Delta t$, equation (91) assumes the total-differential form

$$\Delta p \pm \rho a_f \, \Delta v = -a_f^2 \sum_{i=1}^{N} \sigma_i w_i \, \Delta t/2, \quad \text{for} \quad \Delta x = (v \pm a_f) \Delta t. \tag{94}$$

Thus the streamlines and Mach lines again take on special physical significance; changes in entropy and composition propagate only along streamlines, and changes in pressure and velocity propagate only along Mach lines. Analyses are greatly facilitated by utilizing the special properties [equations (92)–(94)] of the differential equations along the characteristic lines [61]. The reader may consult [61] for details of procedures in which equations (92)–(94) are employed in obtaining numerical solutions and for discussions of correctly set initial-value problems.

4.3.4. One-dimensional, unsteady sound propagation in a binary, reacting ideal-gas mixture for the reaction $A \rightleftharpoons B$

4.3.4.1. Preliminary relationships

Much information can be gleaned from the linearized forms of equations (87)–(90) that are applicable for small departures from a uniform quiescent state of chemical equilibrium. For simplicity, we shall consider here only a binary ideal gas mixture experiencing the reaction $A \rightleftharpoons B$. All the essential physical phenomena can be exhibited in this special case. Corresponding analyses for a simple dissociating diatomic gas [62] and for a gas experiencing an arbitrary one-step reaction [59] may be found in the literature.

Letting the subscripts 1 and 2 identify species A and B, respectively, we then have $N = 2$, $W_1 = W_2 \equiv W$ (equal molecular weights), $X_i = Y_i$ $(i = 1, 2)$,

$$p = \rho(R^0/W)T \tag{95}$$

[from equation (1-9)], and [from equation (1-8)]

$$w_2 = -w_1 = WB_1 T^{\alpha_1 - 1}(p/R^0) e^{-E_1/R^0 T} [(1 - Y) - (Y/K_p)], \tag{96}$$

where $Y \equiv Y_2 = X_2$, $1 - Y = X_1$, and the equilibrium constant for partial pressures is given by

$$K_p = p_{2,e}/p_{1,e} = X_{2,e}/X_{1,e} = Y_e/(1 - Y_e)$$

[see equation (A-26)], that is,

$$Y_e = K_p/(1 + K_p), \tag{97}$$

in which the subscript e identifies chemical equilibrium conditions. Since $\sigma_i = h_i/(c_{p,f}T)$ for ideal gases with constant molecular weights [see equation (55)ff], the right-hand side of equation (90) becomes

$$-\sum_{i=1}^{N} \sigma_i w_i = (q/c_{p,f}T)w_2, \qquad (98)$$

where use has been made of equation (96) and the heat of reaction (per unit mass) has been defined as $q \equiv h_1 - h_2$. Equations (95)–(98) comprise subsidiary relationships that are required in the analysis.

4.3.4.2. Linearization

In order to linearize equations (87)–(90), for any dependent variable f appearing in these equations, we shall set $f = f_0 + f'$, where f_0 is a constant (independent of x and t) that corresponds to quiescent chemical equilibrium conditions and f' is small. Since the quiescent value v_0 of the velocity v is $v_0 = 0$, we have $v = v'$, and we shall therefore omit the prime on the small quantity v. Neglecting terms of order higher than the first in f' and in each of its derivatives, we then find that equations (87), (88) for $i = 2$, (89), and (90) become, respectively,

$$\rho_0 \, \partial v/\partial t + \partial p'/\partial x = 0, \qquad (99)$$

$$\partial Y'/\partial t = (w_Y Y' + w_p p' + w_s s')/\rho_0, \qquad (100)$$

$$\partial s'/\partial t = 0, \qquad (101)$$

and

$$\rho_0 \, \partial v/\partial x + (1/a_{f_0}^2) \partial p'/\partial t = (q/c_{p,f}T)_0(w_Y Y' + w_p p' + w_s s'), \qquad (102)$$

where

$$w_Y \equiv [(\partial w_2/\partial Y)_{p,s}]_0, \qquad w_p \equiv [(\partial w_2/\partial p)_{Y,s}]_0, \quad \text{and} \quad w_s \equiv [(\partial w_2/\partial s)_{p,Y}]_0.$$

In obtaining equations (100) and (102), use has been made of the fact that the unperturbed state is one of chemical equilibrium ($w_{20} = 0$). Equation (98) has also been employed in equation (102). In deriving equation (101), use is made of the fact that the right-hand side of equation (89) is of order higher than the first since $w_{20} = 0$ and $(\mu_2 - \mu_1)_0 = 0$ for chemical equilibrium [see equation (A-21)]; this result has already been emphasized in Section 4.2.3.

4.3.4.3. Reduction to a single partial differential equation

The solution to equation (101), subject to the condition $s' = 0$ at $t = 0$ for all x, is

$$s' = 0. \qquad (103)$$

Therefore, the only derivatives of w_2 that we need to evaluate for use in equations (100) and (102) are w_Y and w_p. Since the quantity in brackets in equation (96) is zero in equilibrium, we find from equation (96) that for small departures from equilibrium,

$$w_2 - w_{20} = [WB_1 T^{\alpha_1 - 1}(p/R^0)e^{-E_1/R^0 T}]_0$$
$$\times \{-[1 + (1/K_{p0})]Y' + (Y_0/K_{p0}^2)(dK_p/dT)_0 T'\}, \quad (104)$$

where use has been made of the fact that K_p depends only on temperature. Employing equation (97) and the van't Hoff equation [equation (A-49)], we find that the last term in the braces in this expression reduces to

$$-[1/(1 + K_{p0})](qW/R^0 T^2)_0 T'.$$

Since $h = Y_1 h_1 + Y_2 h_2$, we find by differentiation that

$$dh = -q\,dY + c_{p,f}\,dT,$$

and, therefore, when equation (A-5) is applied to a unit mass of the gas near equilibrium, we obtain

$$-q_0\,dY + c_{p,f0}\,dT = (1/\rho_0)\,dp, \quad (105)$$

where use has been made of equation (103) $(ds = 0)$ and of the fact that the coefficient of dY arising from the last term in equation (A-5) is zero at chemical equilibrium. The quantities dY, dT, and dp appearing in equation (105) may be identified with Y', T', and p', respectively, for our present purpose. By solving equation (105) for T' and substituting the result into equation (104), we can then easily identify the coefficients of Y' and p' and can thereby obtain w_Y and w_p. In this way, we find [in view of equation (95)], that

$$\frac{1}{\tau} \equiv -\frac{w_Y}{\rho_0} = [B_1 T^{\alpha_1} e^{-E_1/R^0 T}]_0 \left\{ \frac{1 + K_p}{K_p} + \left(\frac{1}{1 + K_p} \right) \left[\frac{q^2}{c_{p,f}(R^0/W)T^2} \right] \right\}_0$$

$$(106)$$

and that

$$\delta \equiv \rho_0 \left(\frac{q}{c_{p,f}T} \right)_0 \frac{w_p}{w_Y} = \left(\frac{1}{c_{p,f}T} \right)_0 \left\{ 1 + \left[\frac{(1 + K_p)^2}{K_p} \right] \left[\frac{c_{p,f}(R^0/W)T^2}{q^2} \right] \right\}_0^{-1}.$$

$$(107)$$

The quantity τ is clearly a representative reaction time, and δ will now be shown to be simply related to the difference between the frozen and equilibrium sound speeds.

In order to relate δ to the equilibrium sound speed a_e and the frozen sound speed a_f, let us compute $1/a_e^2 \equiv (\partial \rho/\partial p)_{s, Y = Y_e}$ for the present gas. In view of equation (51),

$$1/a_e^2 = 1/a_f^2 + (\partial Y/\partial p)_{s, Y = Y_e}(\partial \rho/\partial Y)_{s, p}. \quad (108)$$

Here

$$(\partial \rho/\partial Y)_{s,p} = (\partial T/\partial Y)_{s,p}(\partial \rho/\partial T)_{s,p} = -(\partial T/\partial Y)_{s,p}(pW/R_0 T^2), \quad (109)$$

where use has been made of equation (95). Since equation (105) is valid at constant s and at $Y = Y_e$,

$$(\partial Y/\partial p)_{s,Y=Y_e} = (1/\rho)/(c_{p,f} dT/dY_e - q)$$

$$= (R^0 T/Wpq)\{(c_{p,f}/q)(d[K_p/(1 + K_p)]/dT)^{-1} - 1\}^{-1}$$

$$= -(R^0 T/Wpq)\{1 + [(1 + K_p)^2/K_p][(c_{p,f} R^0 T^2)/(Wq^2)]\}^{-1}, \quad (110)$$

where use has been made of equations (95), (97), and (A-49) (the van't Hoff equation). Equation (105) may also be used to evaluate the remaining partial derivative in equation (109) at chemical equilibrium (the state of present interest); thus,

$$[(\partial T/\partial Y)_{s,p}]_0 = (q/c_{p,f})_0. \quad (111)$$

The substitution of equations (109)–(111) into equation (108) yields the required relationship

$$1/a_{e0}^2 = 1/a_{f0}^2 + \delta, \quad (112)$$

where δ is given by equation (107).

It is instructive [62] to reduce equations (99)–(102) to a single partial differential equation. In view of equation (103) after differentiating equation (102) with respect to t and subtracting the result (multiplied by ρ_0) from the undifferentiated form of equation (102) (multiplied by w_Y), we may make use of equation (100) to show that

$$\left[\frac{w_Y}{a_{f0}^2} + \rho_0 \left(\frac{q}{c_{p,f} T}\right)_0 w_p\right]\frac{\partial p'}{\partial t} + w_Y \rho_0 \frac{\partial v}{\partial x} - \left(\frac{\rho_0}{a_{f0}^2}\right)\frac{\partial^2 p'}{\partial t^2} - \rho_0^2 \frac{\partial^2 v}{\partial t\,\partial x} = 0.$$

By differentiating this expression with respect to x and then using the first and second t derivatives of equation (99) to eliminate $\partial^2 p'/\partial x\,\partial t$ and $\partial^3 p'/\partial x\,\partial t^2$, we obtain

$$\frac{\partial}{\partial t}\left[\frac{1}{a_{f0}^2}\frac{\partial^2 v}{\partial t^2} - \frac{\partial^2 v}{\partial x^2}\right] + \frac{1}{\tau}\left[\frac{1}{a_{e0}^2}\frac{\partial^2 v}{\partial t^2} - \frac{\partial^2 v}{\partial x^2}\right] = 0, \quad (113)$$

where use has been made of the definitions of τ and δ [equations (106) and (107)] and of equation (112). The only dependent variable appearing in equation (113) is the velocity v. Equation (113) implies that for very slow reactions $(1/\tau \to 0)$, v satisfies the simple wave equation

$$\frac{1}{a_{f0}^2}\frac{\partial^2 v}{\partial t^2} - \frac{\partial^2 v}{\partial x^2} = 0$$

with a propagation speed equal to the frozen sound velocity. On the other hand, for very fast reactions ($\tau \to 0$), v satisfies the wave equation

$$\frac{1}{a_{e0}^2}\frac{\partial^2 v}{\partial t^2} - \frac{\partial^2 v}{\partial x^2} = 0$$

with the propagation speed a_{e0} (the equilibrium sound speed).

4.3.4.4. Dispersion relations [62]

Solutions to equation (113) with a sinusoidal time dependence are of interest, since they illustrate sound dispersion resulting from finite reaction rates. Suppose one generates sound waves in a gas occupying the region $x > 0$ by oscillating a piston harmonically about the point $x = 0$ (see Figure 4.6). The velocity of the piston can be represented by the real part of

$$v(0, t) = V(0)e^{i\omega t},$$

where $V(0)$ is the constant (small) amplitude of the piston velocity perturbation and ω is the (angular) vibrational frequency. Harmonic solutions to equation (113) will then be given by (the real part of)

$$v(x, t) = V(x)e^{i\omega t},$$

where, by substitution into equation (113),

$$(1 + i\omega\tau)\,d^2V/dx^2 + (\omega/a_{f0})^2[(a_{f0}^2/a_{e0}^2) + i\omega\tau]V = 0.$$

From the solution to this equation, for which V remains bounded as $x \to \infty$, we obtain

$$v(x, t) = V(0)\exp[-(\omega\alpha/a_{f0})x \sin \varphi]\exp\{i\omega[t - (\alpha/a_{f0})x \cos \varphi]\},$$

(114)

where

$$\alpha \equiv \{[(a_{f0}/a_{e0})^4 + (\omega\tau)^2]/[1 + (\omega\tau)^2]\}^{1/4} \tag{115}$$

and

$$\varphi \equiv \tfrac{1}{2}\tan^{-1}\{[(a_{f0}/a_{e0})^2 - 1]\omega\tau/[(a_{f0}/a_{e0})^2 + (\omega\tau)^2]\}. \tag{116}$$

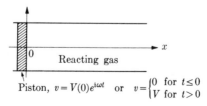

FIGURE 4.6. Generation of sound waves in a reacting gas.

This result shows that the phase velocity of the disturbance is $a = a_{f0}/(\alpha \cos \varphi)$ and that the disturbance attenuates as x increases with a characteristic decay length given by $l = a_{f0}/(\omega \alpha \sin \varphi)$. The consequent dependence of the phase velocity on the frequency [see equations (115) and (116)] represents sound dispersion, and the attenuation illustrates sound absorption. Various specific results follow directly from equations (114)–(116). As $\omega \to 0$, $a \to a_{e0}$ and $l \to \infty$ (no attenuation); as $\omega \to \infty$, $a \to a_{f0}$ and $l \to 2a_{f0}\tau/[(a_{f0}/a_{e0})^2 - 1]$; for intermediate values of ω, $a_{e0} < a < a_{f0}$. The product $\omega\tau$ appears in equations (115) and (116) as a dimensionless parameter, thus indicating that near-equilibrium sound propagation will occur for frequencies such that $\omega\tau \ll 1$ and that near-frozen conditions will prevail if $\omega\tau \gg 1$.

4.3.4.5. An initial-value problem [59]

A more thorough insight into the physical significance and implications of equation (113) is best obtained by solving an initial-value problem instead of by studying the dispersion relations further. Suppose again that the gas occupies the region $x > 0$ and that a piston is located at $x = 0$. For $t \leq 0$, the gas is entirely quiescent and is in chemical equilibrium. The piston is given an impulse at $t = 0$, causing the velocity at $x = 0$ to assume the constant value V for $t > 0$ (see Figure 4.6). Since no disturbance is imposed at $x = \infty$, we require $v \to 0$ as $x \to \infty$ for all (finite) values of t. These initial and boundary conditions determine a unique solution $v(x, t)$ to equation (113).

The solution may easily be obtained by the Laplace transform method. If the Laplace transform of the velocity is defined as

$$u(x, s) \equiv \int_0^\infty e^{-st} v(x, t) \, dt,$$

where s is now the Laplace transform variable, then by taking the transform of equation (113), we obtain

$$s[(s/a_{f0})^2 u - \partial^2 u/\partial x^2] + (1/\tau)[(s/a_{e0})^2 u - \partial^2 u/\partial x^2] = 0,$$

which, upon rearrangement, reduces to

$$a_{f0}^2 \{1 - [1 - (a_{e0}/a_{f0})^2]/(1 + s\tau')\} \partial^2 u/\partial x^2 - s^2 u = 0, \qquad (117)$$

where

$$\tau' \equiv (a_{e0}/a_{f0})^2 \tau$$

is a slightly modified representative reaction time. Since the boundary conditions on u become $u = V \int_0^\infty e^{-st} \, dt = V/s$ at $x = 0$ and $u \to 0$ as $x \to \infty$, the solution to equation (117) is

$$u = (V/s)\exp[-(sx/a_{f0})\{1 - [1 - (a_{e0}/a_{f0})^2]/(1 + s\tau')\}^{-1/2}]. \quad (118)$$

By taking the inverse transform, we then find that the velocity is given by

$$
v = \left(\frac{V}{2\pi i}\right) \int_{\gamma - i\infty}^{\gamma + i\infty}
$$

$$
\times \exp[s't' - s'x'\{1 - [1 - (a_{e0}/a_{f0})^2]/(1 + s')\}^{-1/2}] \, ds'/s', \quad (119)
$$

where $s' \equiv s\tau'$, the dimensionless time is $t' \equiv t/\tau'$, and the dimensionless distance is $x' \equiv x/a_{f0}\tau'$. Here γ is a positive constant such that there are no singularities of the integrand to the right of the line $s' = \gamma$ in the complex s' plane.

For $x' > t'$ (that is, for $x > a_{f0}t$), the sign of the real part of the argument of the exponential in equation (119) is negative when $|s'| \to \infty$ with $\text{Re}\{s'\} > 0$. The contour of integration may then be closed to the right, and the contribution from the semicircular arc along which $|s'| \to \infty$ vanishes, thus showing that $v = 0$ for $x' > t'$. Therefore, no disturbance propagates faster than the frozen sound speed $x/t = a_{f0}$.

When $x' < t'$, analogous reasoning shows that the contour of integration must be closed to the left, and the simple pole at $s' = 0$ and branch points at $s' = -1$ and at $s' = -(a_{e0}/a_{f0})^2$ will provide a nonzero contribution to v. Thus some disturbance always propagates with the velocity a_{f0}.

In order to find the shape of the wave front at large values of x and t, one may perform an asymptotic expansion of the integral in equation (119) for t' and x' approaching infinity with the ratio x'/t' fixed. By means of an interesting application of the method of steepest descents, the reader may show that

$$
v = (V/2)\{1 - \text{erf}[(t - x/a_{e0})/\{2[1 - (a_{e0}/a_{f0})^2]\tau t\}^{1/2}]\}, \quad (120)
$$

where erf denotes the error function

$$
\text{erf}(z) \equiv (2/\sqrt{\pi}) \int_0^z e^{-y^2} \, dy.
$$

Equation (120) shows that at large values of x' and t', the main part of the signal travels with the equilibrium sound speed $x/t = a_{e0}$, and the width of the signal broadens (as x and t increase) in proportion to $1/\sqrt{t}$. Since t' is the dimensionless time variable appearing in equation (119), the asymptotic shape given by equation (120) is attained earlier (at smaller values of t and x) for smaller values of the reaction time τ.

It is now clear that the time development of the sound wave behaves as illustrated in Figure 4.7. The magnitude of the pulse at $x = a_{f0}t$ decreases as t increases. At small values of t/τ, the sharp frozen wave is a close approximation to reality, while at large values of t/τ, the spreading equilibrium wave closely represents the true wave pattern. The intermediate transition

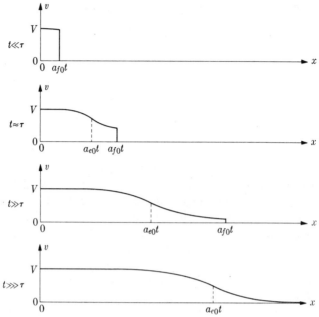

FIGURE 4.7. Schematic diagram of the one-dimensional, unsteady propagation of a sound pulse into a reacting gas.

region occurs when $t \approx \tau$. This behavior is consistent with the picture developed in Section 4.3.4.4.

4.3.4.6. Related problems

By analyses that are similar to those presented above, many related problems can be solved. For example, the structure of the steady, two-dimensional sound wave produced when a supersonic reacting gas stream is turned through an infinitesimally small angle (see Figure 4.8) can be obtained by the methods of Section 4.3.4.5. The essential characteristics of the results of such an analysis can be inferred by interpreting the time t of Section 4.3.4.5 as the distance measured along the sound wave in the steady, two-dimensional system. A decaying wave propagates at an angle determined by the frozen sound speed a_{f0}, and a spreading wave develops at some distance from the wall and propagates at an angle determined by the equilibrium sound speed a_{e0}.

The wave field produced in the steady, two-dimensional flow of a reacting gas past a wavy wall has been treated in [63] and [64]. Lick [65] has obtained solutions to the nonlinear, steady, two-dimensional conservation equations governing the flow of a reacting gas mixture about a blunt body. Reviews of these and other studies may be found in [1], [2], and [66]–[71].

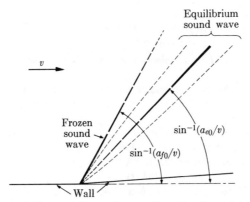

FIGURE 4.8. Schematic diagram of a steady, two-dimensional sound wave in a supersonic reacting gas stream.

REFERENCES

1. J. F. Clarke and M. McChesney, *The Dynamics of Real Gases*, London: Butterworths, 1964.
2. W. G. Vincenti and C. H. Kruger, Jr., *Introduction to Physical Gas Dynamics*, New York: Wiley, 1965.
3. B. P. Mullins, *Fuel* **32**, 327, 451 (1953).
4. F. A. Williams, *Astronautica Acta* **15**, 547 (1970).
5. J. J. Sangiovanni and A. S. Kestin, *16th Symp.* (1977), 577–592.
6. J. P. Longwell and M. A. Weiss, *Ind. Eng. Chem.* **47**, 1634 (1955).
7. L. A. Vulis, *Thermal Regimes of Combustion*, New York: McGraw-Hill, 1961, Chapter 2.
8. M. V. Herbert, *8th Symp.* (1962), 970–982.
9. H. C. Hottel, G. C. Williams, and M. L. Baker, *6th Symp.* (1957), 398–411.
10. M. A. Weiss, R. J. Lang, and J. P. Longwell, *Ind. Eng. Chem.* **50**, 257 (1958).
11. A. E. Clarke, J. Odgers, and P. Ryan, *8th Symp.* (1962), 982–994.
12. O. Blichner, *8th Symp.* (1962), 995–1002.
13. V. K. Jain and D. B. Spalding, *C & F* **9**, 37 (1965).
14. H. C. Hottel, G. C. Williams, and G. A. Miles, *11th Symp.* (1967), 771–778.
15. D. R. Jenkins, V. S. Yumlu, and D. B. Spalding, *11th Symp.* (1967), 779–790.
16. B. R. Bowman, D. T. Pratt, and C. T. Crowe, *14th Symp.* (1973), 819–830.
17. W. S. Blazowski, *CST* **21**, 87 (1980).
18. I. Eberstein and I. Glassman, *10th Symp.* (1965), 365–374.
19. F. L. Dryer and I. Glassman, *14th Symp.* (1973), 987–1003.
20. D. Aronowitz, R. J. Santoro, F. L. Dryer, and I. Glassman, *17th Symp.* (1979), 633–644.
21. S. S. Penner and D. B. Olfe, *Radiation and Reentry*, New York: Academic Press, 1968.
22. F. J. Krieger, *J. Am. Rocket Soc.* **21**, 179 (1951).
23. S. P. Heims, NACA Tech. Note No. 4144 (1958).

24. K. N. C. Bray, *J. Fluid Mech.* **6**, 1 (1959).
25. K. N. C. Bray, *9th Symp.* (1963), 770–784.
26. A. A. Westenberg and S. Favin, *9th Symp.* (1963), 785–798.
27. S. S. Penner, *Chemistry Problems in Jet Propulsion*, New York: Pergamon Press, 1957.
28. F. A. Williams, M. Barrère, and N. C. Huang, *Fundamental Aspects of Solid Propellant Rockets*, AGARDograph No. 116, Slough, England: Technivision Services, 1969.
29. P. G. Hill and C. R. Peterson, *Mechanics and Thermodynamics of Propulsion*, Reading, Mass.: Addison-Wesley, 1965.
30. S. Gordon and B. J. McBride, NASA SP-273 (1971).
31. P. P. Wegener, *Phys. Fluids* **2**, 264 (1959).
32. N. A. Hall, *Thermodynamics of Fluid Flow*, Englewood Cliffs, N.J.: Prentice-Hall, 1951.
33. D. Altman and J. M. Carter, "Expansion Processes," in *Combustion Processes*, vol. II of *High Speed Aerodynamics and Jet Propulsion*, B. Lewis, R. N. Pease, and H. S. Taylor, eds., Princeton: Princeton University Press, 1956, 26–63.
34. L. Crocco, "One-Dimensional Treatment of Steady Gas Dynamics," in *Fundamentals of Gas Dynamics*, vol. III of *High Speed Aerodynamics and Jet Propulsion*, H. W. Emmons, ed., Princeton: Princeton University Press, 1958, 64–349.
35. M. Summerfield, "The Liquid Propellant Rocket Engine," in *Jet Propulsion Engines*, vol. XII of *High Speed Aerodynamics and Jet Propulsion*, O. E. Lancaster, ed., Princeton: Princeton University Press, 1959, 439–520.
36. M. Barrère, A. Jaumotte, B. Fraeijs de Veubeke, and J. Vandenkerckhove, *Rocket Propulsion*, New York: Elsevier, 1960.
37. S. S. Penner and D. Altman, *J. Franklin Inst.* **245**, 421 (1948); D. Altman and S. S. Penner, *J. Chem. Phys.* **17**, 56 (1949); S. S. Penner, *J. Am. Chem. Soc.* **71**, 788 (1949); *J. Franklin Inst.* **249**, 441 (1950); *J. Chem. Phys.* **17**, 841 (1949), **19**, 877 (1951), **20**, 341 (1952).
38. M. Rudin, *Phys. Fluids* **1**, 384 (1958); *1958 Heat Transfer and Fluid Mechanics Institute*, Stanford: Stanford University Press, 1958, 91–103.
39. M. Rudin and H. Aroeste, *C & F* **3**, 273 (1959).
40. M. Lenard, "Gas Dynamics of Chemically Reacting Gas Mixtures Near Equilibrium," in *Dynamics of Manned Lifting Planetary Entry*, S. M. Scala, A. C. Harrison, and M. Rogers, eds., New York: Wiley, 1963, 841–865.
41. P. P. Wegener, *ARS Journal* **30**, 322 (1960).
42. W. D. Rannie, "Perturbation Analysis of One-Dimensional Heterogeneous Flow in Rocket Nozzles," in *Detonation and Two-Phase Flow*, vol. 6 of *Progress in Astronautics and Rocketry*, S. S. Penner and F. A. Williams, eds., New York: Academic Press, 1962, 117–144.
43. F. E. Marble, "Dynamics of a Gas Containing Small Solid Particles," in *Proceedings of the Fifth AGARD Combustion and Propulsion Colloquium*, New York: Pergamon Press, 1963, 175–215.
44. F. E. Marble, *AIAA Journal* **1**, 2793 (1963).
45. M. Gilbert, L. Davis, and D. Altman, *Jet Propulsion* **25**, 25 (1955).
46. W. G. Courtney, *9th Symp.* (1963), 799–810.
47. W. S. Bailey, E. N. Nilson, R. A. Serra, and T. F. Zupnik, *ARS Journal* **31**, 793 (1961).

48. J. R. Kliegel and G. R. Nickerson, "Flow of Gas-Particle Mixtures in Axially Symmetric Nozzles," in *Detonation and Two-Phase Flow*, vol. 6 of *Progress in Astronautics and Rocketry*, S. S. Penner and F. A. Williams, eds., New York: Academic Press, 1962, 173–194.

49. J. R. Kliegel, *9th Symp.* (1963), 811–826.

50. J. D. Hoffman and S. A. Lorenc, *AIAA Journal* **3**, 103 (1965); **4**, 169 (1966).

51. G. F. Carrier, *J. Fluid Mech.* **4**, 376 (1958).

52. R. A. Dobbins and S. Temkin, *AIAA Journal* **2**, 1106 (1964).

53. N. A. Fuchs, *The Mechanics of Aerosols*, New York: Macmillan, 1964.

54. S. L. Soo, *Fluid Dynamics of Multiphase Systems*, Waltham, Mass.: Blaisdell, 1967.

55. G. B. Wallis, *One-Dimensional Two-Phase Flow*, New York: McGraw-Hill, 1969.

56. I. G. Petrovskii, *Partial Differential Equations*, New York: Interscience, 1954, Chapter I, Section 3.

57. R. Courant and K. O. Friedrichs, *Supersonic Flow and Shock Waves*, New York: Interscience, 1948, Chapter II.

58. W. W. Wood and J. G. Kirkwood, *J. Appl. Phys.* **28**, 395 (1957); *J. Chem. Phys.* **27**, 596 (1957).

59. B. T. Chu, "Wave Propagation in a Reacting Mixture," *1958 Heat Transfer and Fluid Mechanics Institute*, Stanford: Stanford University Press, 1958, 80–90; "Wave Propagation and the Method of Characteristics in a Reacting Gas Mixture with Application to Hypersonic Flow," *Tech. Note No. 57–213*, Wright Air Development Center (1957).

60. A. Ferri, *Elements of Aerodynamics of Supersonic Flows*, New York: Macmillan, 1949.

61. G. Rudinger, *Wave Diagrams for Nonsteady Flow in Ducts*, New York: D. van Nostrand, 1955.

62. J. F. Clarke, *J. Fluid Mech.* **7**, 577 (1960).

63. W. G. Vincenti, *J. Fluid Mech.* **6**, 481 (1959).

64. F. K. Moore and W. E. Gibson, *J. Aero. Sci.* **27**, 117 (1960).

65. W. Lick, *J. Fluid Mech.* **7**, 128 (1960); "Inviscid Flow Around a Blunt Body of a Reacting Mixture of Gases, Part A, General Analysis; Part B, Numerical Solutions," *AFOSR Tech. Note No. 58–522 (R.P.I. Tech. Rept. No. AE 5810)* and *AFOSR Tech. Note No. 58–1124 (R.P.I. Tech. Rept. No. AE 5814)*, Rensselaer Polytechnic Institute, Troy (1958).

66. T. Y. Li, *ARS Journal* **31**, 170 (1961).

67. F. E. Marble, *Annual Review of Fluid Mechanics* **2**, 397 (1970).

68. E. Becker, *Annual Review of Fluid Mechanics* **4**, 155 (1972).

69. G. B. Whitham, *Linear and Nonlinear Waves*, New York: Wiley, 1974.

70. V. V. Rusanov, *Annual Review of Fluid Mechanics* **8**, 377 (1976).

71. D. G. Crighton, *Annual Review of Fuid Mechanics* **11**, 11 (1979).

CHAPTER 5

Theory of Laminar Flames

This chapter concerns the structures and propagation velocities of the deflagration waves defined in Chapter 2. Deflagrations, or laminar flames, constitute the central problem of combustion theory in at least two respects. First, the earliest combustion problem to require the simultaneous consideration of transport phenomena and of chemical kinetics was the deflagration problem. Second, knowledge of the concepts developed and results obtained in laminar-flame theory is essential for many other studies in combustion. Attention here is restricted to the steadily propagating, planar laminar flame. Time-dependent and multidimensional effects are considered in Chapter 9.

The problem of determining the propagation velocity of a deflagration wave was first studied by Mallard and le Chatelier [1], who considered heat loss to be of predominant importance and rates of chemical reactions to be secondary. The essential result that the burning velocity is proportional to the square root of the reaction rate and to the square root of the ratio of the thermal conductivity to the specific heat at constant pressure was first demonstrated by Mikhel'son [2], whose work has been discussed in more recent literature [3], [4]. Independent investigations in later generations by Taffanel [5] and by Daniell [6] based on simplified models of the combustion wave reached the same conclusion. Subsequently, improved basic equations became available for use in theoretical analyses.

In more recent years, the idea that the overall activation energy is large in comparison with the thermal enthalpy has led to the development of asymptotic analyses that lend further clarification to the wave structure and that provide formulas with improved accuracy for the burning velocity. The implications of this idea were first derived heuristically by Zel'dovich and Frank-Kamenetskii [7]. Many ad hoc approximations that rest in varying measure on the concept of large activation energy are reviewed in [8]. Systematization of the concept through formal introduction of the method of matched asymptotic expansions was achieved first by Bush and Fendell [9]. A number of recent studies of deflagration structure have been based on asymptotic methods of this type. The development of laminar-flame theory to be given in Section 5.3 emphasizes the utility of asymptotic concepts. Much of the recent work is covered in [10].

We shall begin with a brief discussion of experiments and a simple indication of the primary physical phenomena involved. Next, appropriate basic differential equations are derived for describing the structures of combustion waves. The essential features of the mathematical problem of predicting the propagation speed of a one-dimensional laminar deflagration wave will then be presented by examining in detail the unimolecular process $R \rightarrow P$ ($R \equiv$ reactant, $P \equiv$ product). This choice may be justified by the fact that more-complicated processes may be treated by extensions of the approaches developed. The necessary modifications caused by the presence of chain reactions are discussed in the final section; in particular, a criterion for the validity of the steady-state approximation for reaction intermediaries is described. It will be seen that although useful approaches to the theoretical analysis are available, many outstanding questions remain concerning deflagration structure.

5.1. DESCRIPTION OF LAMINAR FLAMES

5.1.1. Experiments

When a quiescent combustible gas mixture contained in an open tube is ignited by a spark at one end of the tube, a combustion wave spreads through the gas. Provided that the tube is not too short and the electrical energy released by the spark is not too much larger than the minimum energy required for ignition, the combustion wave often is observed to be nearly planar and to travel at approximately a constant speed between (very roughly) two and ten tube diameters from the spark. This constant speed is an empirical laminar burning velocity or flame speed, which is characteristic of the combustible mixture.*

---

* Recent efforts to distinguish between the terms *burning velocity* and *flame speed* on the basis of Eulerian and Lagrangian coordinate systems appear to introduce confusion. Therefore, the terms are used interchangeably here, as synonyms for such terms as *deflagration velocity*, *wave speed*, and *propagation velocity*. They all refer to velocities measured with respect to the gas ahead of the wave.

Because of various inherent difficulties, the experiment just described does not yield accurate burning velocities. An alternative apparatus that provides burning-velocity data of greater accuracy is the Bunsen-type burner schematically illustrated in Figure 5.1. The mixture composition is determined by adjusting the flow rates of the fuel and oxidizer (often air), and a steady, conical flame sits at the mouth of the cylindrical burner if the exit velocity exceeds the flame speed. The radius of curvature of the flame cone is generally so large compared with the width of the flame that the flame structure is the same as the structure of a plane flame except at the tip of the cone and at the burner port; the curvature can be eliminated by use of a slot burner, a Bunsen-type burner with an elongated rectangular outlet. The component of velocity tangent to the flame surface remains constant across the flame, but the component of velocity normal to the flame increases, as is illustrated by the vector diagrams in Figure 5.1. The component of velocity normal to the flame surface, just upstream from the flame, is the velocity of propagation (burning velocity) of the flame. This propagation velocity can be determined from measurements of the flow velocity at the exit of the burner tube, accompanied by measurements, typically photographic, of the geometry of the flame surface (for example, the apex angle of the flame cone). Numerous experiments of this type have been performed, and various

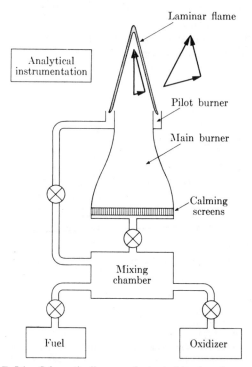

FIGURE 5.1.   Schematic diagram of a typical laminar-flame experiment.

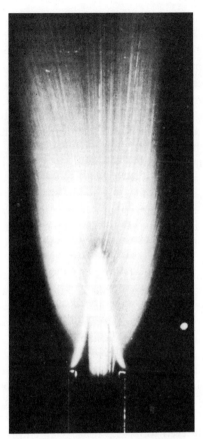

FIGURE 5.2. Photograph of a natural gas-air flame on a rectangular burner, with very small magnesium oxide particles introduced into the flow and illuminated stroboscopically in order to make the gas streamlines visible (from [11], courtesy of B. Lewis and of Academic Press).

procedures for calculating burning velocities from the measurements have been developed [11].* A photograph of a typical laminar flame produced in this way is shown in Figure 5.2.

Laminar deflagration waves may be produced in a number of other experiments. For example, a steady deflagration may be stabilized in a combustible mixture, flowing steadily at a velocity greater than the burning velocity†, by a wire or rod inserted perpendicular to the flow direction. In ramjet combustors, the steady flow is contained within a duct, and one or more stabilizing rods are employed; the propagation velocity of the wedge-shaped flame determines the minimum chamber length required for the

---

* See [11] for a discussion of the techniques and the accuracy of each.

† At a velocity below the burning velocity the flame would move upstream and *flash-back* would be said to occur.

combustion to be completed inside the duct, but since the flow in practical ramjets is turbulent, the turbulent burning velocity (Chapter 10) must be employed. A configuration useful for studying laminar flames in the laboratory is the **flat-flame burner**, in which a planar flame is stabilized above one or between two porous plates in a flow that is essentially one-dimensional. This arrangement affords a good research tool because it offers the experimenter a relatively large number of variable parameters and facilitates analysis and interpretation of measurements by virtue of the plane, one-dimensional geometry, but heat loss to the porous plate generally is significant and must be taken into account. Variants of the flat-flame burner that also are useful in the laboratory are stagnation-flow devices, in which a flow with a stagnation point is established by directing a uniform combustible stream either normally onto a flat plate or counter to the flow of a uniform, inert stream; the flame produced near the stagnation point is flat locally and can be arranged to have negligible heat loss but is "stretched" in that streamlines diverge in the vicinity of the flame.

Motions of propagating laminar flames and shapes of steady flames often are influenced by buoyancy and by various instabilities (Chapter 9). The curvature of the streamlines in the upper part of Figure 5.2 is produced by buoyancy. Advantages of some Bunsen, flat-flame, and stagnation-flow burners are their tendencies to minimize these effects. Flames propagating upward, downward, and horizontally in chambers usually behave differently from each other because of buoyant effects. Even in the absence of both gravity and instabilities, walls influence flame motions in chambers; to calculate histories of flame shapes for slow propagation of a flame sheet at a specified flame speed from an initially uniform and quiescent state in a closed chamber with fixed walls is a nontrivial, elliptic problem even with inviscid fluid mechanics if there is an appreciable change in density across the sheet.* Nevertheless, approximate values of flame speeds often are obtained by monitoring the pressure histories in a closed chamber, or "bomb" (typically spherical), with data reduction performed on the basis of an assumed spherical shape of a flame of negligible thickness. In using this technique, it is best to verify by observation that the flame is nearly spherical. The complication of pressure variation with time can be eliminated by the **soap-bubble method**, in which the combustible mixture, possibly enclosed within a soap bubble, is ignited in the open or in a very large chamber. In this experiment, if ρ_0 is the density of the unburnt mixture and v_0 is the flame speed, then the rate of increase of the mass of burnt gas inside a spherical flame of radius r is $4\pi r^2 \rho_0 v_0$, and since the mass of burnt gas is $\frac{4}{3}\pi r^3 \rho_\infty$, where ρ_∞ is the density of the burnt gas, the flame speed is $v_0 = (\rho_\infty/\rho_0)\, dr/dt$, a constant fraction of the observed growth rate of the spherical flame.

---

* A simple solution exists only for spherical symmetry.

In all these laminar-flame experiments, the combustion wave propagates at a definite velocity that empirically depends on the pressure, temperature, and composition of the initial combustible mixture. Our primary objective in this chapter is to predict this burning velocity. In Section 5.1.2 a simple physical picture of the deflagration wave is presented, leading to a crude estimate of the burning velocity. The discussion, which is similar to that of Landau and Lifshitz [3], illustrates the essential mechanism involved.

5.1.2. Phenomenological analysis of a deflagration wave

Consider a stationary plane deflagration wave of thickness δ, into which a combustible gas mixture flows at velocity v_0 (the laminar burning velocity). If q is the heat of reaction (energy released per unit mass of the reactant mixture) and w is the reaction rate (mass of the reactant mixture converted per unit volume per second), then $qw\delta$ is the energy released per unit area of the wave per second. The cold incoming gas is heated to a temperature at which it reacts at an appreciable rate by means of heat conduction from the reaction zone. The rate at which heat is conducted upstream is roughly $\lambda\, dT/dx \approx \lambda(T_\infty - T_0)/\delta$ (energy per unit area per second), where λ is the mean thermal conductivity of the gas, T is the temperature, x is the distance normal to the wave, and the subscripts 0 and ∞ identify conditions upstream and downstream of the wave, respectively. If the wave is adiabatic so that no energy is lost downstream or from the sides of the wave, then energy conservation implies that $q = c_p(T_\infty - T_0)$ (where c_p is an average specific heat of the mixture) and that the entire heat released must be conducted upstream, that is, $qw\delta \approx \lambda(T_\infty - T_0)/\delta$. These results lead to the equation $c_p(T_\infty - T_0)w\delta \approx \lambda(T_\infty - T_0)/\delta$, which implies that the thickness of the wave is

$$\delta \approx \sqrt{\lambda/c_p w}. \tag{1}$$

The burning velocity v_0 can be related to this wave thickness as follows. The mass of combustible material per unit area per second flowing into the wave is $\rho_0 v_0$, where ρ_0 is the density of the initial combustible gas mixture. The deflagration wave consumes these reactants at a rate $w\delta$ (mass per unit area per second). Hence mass conservation implies that $\rho_0 v_0 = w\delta$, which, in conjunction with equation (1), yields

$$v_0 \approx (1/\rho_0)\sqrt{(\lambda/c_p)w}. \tag{2}$$

Equation (2) shows that the burning velocity is proportional to the square root of a reaction rate and to the square root of the ratio of the thermal conductivity to the specific heat. The factor on the right-hand side of equation (2) with the strongest temperature dependence is usually w, which varies roughly as $e^{-E_1/R^0 T_\infty}$, where E_1 is an activation energy, and R^0 is the universal gas constant [see equation (1-8)]. At constant temperature,

λ and c_p are independent of the pressure p, $\rho_0 \sim p$ according to the ideal gas law [equation (1-9)], and $w \sim p^n$, where n is the order of the reaction [see equation (1-8)]. Hence the pressure and temperature dependences of v_0 are given approximately by

$$v_0 \sim p^{(n/2)-1} e^{-E_1/2R^0 T_\infty}. \qquad (3)$$

In principle, equation (2) should provide a numerical estimate for v_0. However, in practice, values of w are so uncertain (see Appendix B) that equation (2) is more useful for estimating w from experimental values of v_0. A useful way to deduce an overall order and an overall activation energy for a reaction is to measure the T and p dependences of v_0 and to correlate these empirically with equation (3) by adjusting n and E_1.

Empirical laminar burning velocities lie between 1 and 1000 cm/s. Since these velocities are small compared with the speed of sound, equation (1-25) is valid for laminar flames. Thus laminar flames are nearly isobaric, we were justified in not attaching a subscript to p in equation (3), and the quantity c_p appearing above is the specific heat at constant pressure.

Equations (1) and (2) enable us to estimate the flame thickness δ from measured values of v_0. Eliminating w between equations (1) and (2) leads to the relation

$$\delta \approx \lambda/c_p \rho_0 v_0. \qquad (4)$$

With representative values for λ, c_p, and ρ_0, and with $v_0 \sim 50$ cm/s, equation (4) gives $\delta \sim 10^{-2}$ cm. Therefore δ is large compared with a molecular mean free path (about 10^{-5} cm), and the continum equations of fluid dynamics are valid within the deflagration wave; but δ is small compared with typical dimensions of experimental equipment (for example, the diameter of the burner mouth, and hence the radius of curvature of the flame cone, for experiments with Bunsen-type burners), and laminar deflagration waves may be approximated as discontinuities in many experiments. Since equations (3) and (4) imply that $\delta \sim p^{-(n/2)}$ at constant temperature, experimental studies of the interiors of laminar flames are often performed at subatmospheric pressures.

The remainder of this chapter is devoted mainly to a mathematical amplification of the concepts introduced in this section. In particular, it will be seen that the burning velocity is an eigenvalue of the conservation equations. Many questionable aspects of the reasoning employed in this section will have been remedied before the end of this chapter.

5.2. MATHEMATICAL FORMULATION

5.2.1. Introductory remarks

In this section we shall develop equations capable of describing the structures of both deflagrations and detonations. Thus the approximation $p = $ constant, which was shown in the preceding section to be valid for ordinary deflagrations but which is invalid for detonations (see Chapter 2), will not be made here. Our principal reason for adopting this approach is to avoid repetition in Chapter 6. The approach is unnecessarily complicated for deflagrations only in that the momentum conservation equation must be considered and one additional approximation [assumption 8 below] must be made. These disadvantages appear to be insignificant in view of the greater generality of some of the results [for example, equation (24)]. For a detailed derivation of the basic relations for deflagrations alone, see [12] or [13].

5.2.2. Preliminary assumptions and equations

In order to analyze the structure of a combustion wave, it will obviously be expedient to adopt a coordinate system which is at rest with respect to the wave. The following assumptions will then be made:

1. The flow is steady, plane, and one-dimensional in the adopted frame of reference.
2. Radiative heat transfer is negligible.
3. The Soret and Dufour effects are negligible.
4. Diffusion caused by pressure gradients is negligible.
5. External forces are negligible.
6. The gas mixture is ideal.

The approximate validity of these assumptions and of others to be made subsequently can be justified by order-of-magnitude considerations for many systems.

In view of assumptions 1–6, if x is the coordinate normal to the combustion wave, then equations (1-26), (1-27), (1-31), (1-32), and (1-33) are valid with $g_2 = g_3 = 1$ and with the last term in equation (1-32) omitted. Thus

$$\rho v^2 + p - (\tfrac{4}{3}\mu + \kappa)\, dv/dx = \text{constant}, \tag{5}$$

$$\rho v \equiv m = \text{constant}, \tag{6}$$

$$d\epsilon_i/dx = w_i/m, \quad i = 1, \ldots, N, \tag{7}$$

$$\frac{dX_i}{dx} = \left(\frac{m}{\rho}\right) \sum_{j=1}^{N} \frac{X_i X_j}{D_{ij}} \left(\frac{\epsilon_j}{Y_j} - \frac{\epsilon_i}{Y_i}\right), \quad i = 1, \ldots, N, \tag{8}$$

and

$$m\left(\sum_{i=1}^{N} h_i \epsilon_i + \frac{v^2}{2}\right) - \lambda \frac{dT}{dx} - \left(\frac{4}{3}\mu + \kappa\right) v \frac{dv}{dx} = \text{constant}, \tag{9}$$

where the notation was defined in Chapter 1. Equations (1-8), (1-9), (1-11), and (1-12) are necessary subsidiary relationships among the variables appearing in equations (5)–(9).

The boundary conditions for these differential equations are to be applied at $x = -\infty$ (where properties will be identified by the subscript 0) and at $x = +\infty$ (where properties will be identified by the subscript ∞). At $x = -\infty$, all properties are uniform ($d/dx = 0$) (see, however, Section 2.1.2) and the values of all variables except v_0 are known (compare Section 2.1.3). At $x = +\infty$, properties again become uniform ($d/dx = 0$) and chemical equilibrium is reached. In Chapter 2 (Section 2.1) it was shown that these conditions imply that relationships exist between the properties at $x = +\infty$ and those at $x = -\infty$.

5.2.3.   Approximations that further simplify the energy equation

Equation (9) can be simplified by making two additional assumptions:

7. All chemical species may be considered to have constant and equal average specific heats at constant pressure c_p.
8. The effective coefficient of viscosity ($\frac{4}{3}\mu + \kappa$) equals the ratio of the thermal conductivity to the average specific heat (λ/c_p).

Assumption 7 is not essential because the important qualitative results to be obtained do not depend on this assumption. However, it enables the analysis to be simplified considerably and, in some respects, clarifies the physical significance of the results. In view of assumption 7, equation (1-11) reduces to

$$h_i = h_i^0 + c_p(T - T^0). \tag{10}$$

Assumption 8 is entirely unnecessary for ordinary deflagrations (that is, for $p \approx$ constant) because the terms involving the viscosity are negligibly small. However, this assumption leads to a considerable simplification in the analysis for detonations and for flames with nonnegligible kinetic energy. Assumption 8 states that

$$\tfrac{4}{3}\mu + \kappa = \lambda/c_p, \tag{11}$$

which is a relation between transport properties. When bulk viscosity is negligible ($\kappa = 0$), equation (11) implies that the Prandtl number is $\frac{3}{4}$ (see Appendix E), which is a good approximation for many gases (see, for example, Appendix E or [14], p. 16). Nevertheless, we reemphasize that assumption 8 will *not* be present in subsequent sections of this chapter.

As was pointed out by Spalding [15], some aspects of the analysis of combustion waves in which kinetic energy is of importance can be simplified by using the stagnation enthalpy ($c_p T + v^2/2$) as a variable. A normalized dimensionless measure of the stagnation enthalpy is

$$\tau \equiv \frac{(c_p T + v^2/2) - (c_p T_0 + v_0^2/2)}{(c_p T_\infty + v_\infty^2/2) - (c_p T_0 + v_0^2/2)}, \tag{12}$$

which varies (monotonically for purely exothermic reactions) from 0 at $x = -\infty$ to 1 at $x = +\infty$.

By using equations (10)–(12) (along with the boundary conditions $d/dx = 0$ at $x = \pm\infty$ and the identity $\sum_{i=1}^{N} \epsilon_i = 1$), the reader can prove that equation (9) reduces to

$$\frac{\lambda}{mc_p} \frac{d\tau}{dx} = \left[\frac{\sum_{i=1}^{N} h_i^0 (\epsilon_i - \epsilon_{i,\infty})}{\sum_{i=1}^{N} h_i^0 (\epsilon_{i,0} - \epsilon_{i,\infty})} \right] - (1 - \tau). \tag{13}$$

The denominator inside the brackets in equation (13) is the total heat release per unit mass of mixture, q. Thus

$$q \equiv \sum_{i=1}^{N} h_i^0 (\epsilon_{i,0} - \epsilon_{i,\infty}) = (c_p T_\infty + v_\infty^2/2) - (c_p T_0 + v_0^2/2), \tag{14}$$

where the last equality, which is simply the statement of overall energy conservation for the wave, follows directly from equations (9) and (10) and the boundary conditions. In view of equation (14), the definition of τ [equation (12)] reduces to

$$\tau = [(c_p T + v^2/2) - (c_p T_0 + v_0^2/2)]/q. \tag{12a}$$

5.2.4. Simplifications in the energy and diffusion equations for unimolecular reactions in binary mixtures

Since equations (7) and (8) constitute $2N$ differential equations, general conclusions regarding the wave structure are difficult to obtain. If only one chemical reaction occurs [$M = 1$ in equation (1-8)], then equation (7) provides $N - 1$ simple relations among the N flux fractions ϵ_i, and a single flux fraction is sufficient to characterize the progress of the reaction. This case has been treated in most analyses of combustion waves; equation (7) reduces to a single differential equation. But, even in this case, simplifying assumptions regarding the diffusion coefficients (usually that all binary diffusion coefficients are approximately equal) must be made in order to decrease the number of independent relations in equation (8). In order to avoid all these complications while retaining the essential physical aspects of the problem, we shall consider the following system:

9. The gas is a binary mixture in which the unimolecular reaction R → P occurs.

Assumption 9 automatically reduces the number of independent relations in each of the equations (7) and (8) to one. Furthermore, when the analysis is extended to include multicomponent systems in which any one-step reaction occurs ($M = 1$) and for which the binary diffusion coefficients

of the pairs of principal species do not differ appreciably, there is virtually
no change in the results. However, assumption 9 will have to be omitted in
the final section of this chapter because the distribution of reaction intermediaries obviously cannot be studied on the basis of a one-step process.

According to assumption 9, $M = 1$ and $N = 2$. Letting the subscripts
1 and 2 identify the properties of species R and P, respectively, we find that
$v''_{1,1} = v'_{2,1} = 0$, $v'_{1,1} = v''_{2,1} = 1$, $W_1 = W_2 = W$ [since mass is conserved in
chemical reactions, or formally by summing equation (7) over i], $X_i = Y_i$
[from equation (1-12)], $Y_1 = 1 - Y_2$, and $\epsilon_1 = 1 - \epsilon_2$. Equation (1-9) then
reduces to

$$p = \rho R^0 T/W, \tag{15}$$

and equation (1-8) becomes

$$w_1 = -w_2 = -\frac{WpB_1 T^{\alpha_1}}{R^0 T} X_1 e^{-E_1/R^0 T}$$

$$= -\rho B_1 T^{\alpha_1}(1 - Y_2)e^{-E_1/R^0 T}, \tag{16}$$

where use has been made of equation (15). The addition of assumption 9 to
assumptions 1–7 also renders valid all of the results of Section 2.2.

Assumption 9 enables us to simplify equation (13) further. Since
$\epsilon_1 = 1 - \epsilon_2$, the quantity in the brackets in equation (13) is

$$(\epsilon_2 - \epsilon_{2,\infty})/(\epsilon_{2,0} - \epsilon_{2,\infty}) = 1 - \epsilon,$$

where the normalized flux fraction (or, reaction progress variable) has been
defined as

$$\epsilon \equiv (\epsilon_2 - \epsilon_{2,0})/(\epsilon_{2,\infty} - \epsilon_{2,0}), \tag{17}$$

which goes from 0 at $x = -\infty$ to 1 at $x = +\infty$. If the dimensionless distance
variable

$$\xi \equiv \int_0^x (mc_p/\lambda)\, dx \tag{18}$$

is also introduced, then equation (13) reduces to

$$d\tau/d\xi = \tau - \epsilon. \tag{19}$$

For the present two-component mixture, equation (8) reduces to

$$dY_i/dx = (m/\rho D_{12})(Y_i - \epsilon_i), \quad i = 1, 2, \tag{20}$$

which is Fick's law. Since the boundary conditions require that $dY_i/dx = 0$
at $x = \pm\infty$, equation (20) implies that $Y_{i,0} = \epsilon_{i,0}$ and $Y_{i,\infty} = \epsilon_{i,\infty}$ (the
diffusion velocities vanish far upstream and downstream). Hence, in terms
of the normalized mass fraction

$$Y \equiv (Y_2 - Y_{2,0})/(Y_{2,\infty} - Y_{2,0}) \tag{21}$$

(which goes from 0 at $x = -\infty$ to 1 at $x = +\infty$), equation (20) becomes

$$(1/\text{Le})\, dY/d\xi = Y - \epsilon, \tag{22}$$

where use has been made of equations (17) and (18) and the Lewis number is

$$\text{Le} = \lambda/\rho D_{12} c_p. \tag{23}$$

Actually, since the reaction goes to completion at $x = +\infty$ $[d\epsilon_i/dx \to 0$; see equation (7)], equation (16) implies that $Y_{2,\infty} = 1$ $(Y_{1,\infty} = 0)$, and therefore also $\epsilon_{2,\infty} = 1$ $(\epsilon_{1,\infty} = 0)$.

5.2.5. Solution of the diffusion equation when Le $= 1$.

From equations (19) and (22) it can be seen that Y is simply related to τ provided that the next assumption is allowed:

10. The Lewis number is unity.

It is shown in Appendix E that Le $= 1$ is a reasonable approximation. The difference of equations (19) and (22) is $d\tau/d\xi - (1/\text{Le})(dY/d\xi) = \tau - Y$, which, when Le $= 1$, has the general solution

$$(\tau - Y) = Ae^{\xi},$$

where A is an arbitrary constant. Since $\tau - Y$ must remain bounded as $x \to \infty$, the only acceptable value of A is $A = 0$, implying that

$$Y = \tau, \tag{24}$$

which also satisfies the correct boundary conditions at $x = \pm\infty$ [see equations (12) and (21)]. Equation (24) provides an explicit relationship between the composition and the stagnation enthalpy throughout the wave.

5.2.6. Dimensionless forms for the momentum equation and the species-conservation equation

If the constant in equation (5) is evaluated at $x = -\infty$ and use is made of equations (6), (11), (15), and (18), then equation (5) becomes

$$\frac{dv}{d\xi} = (v - v_0) + \frac{R^0}{W}\left(\frac{T}{v} - \frac{T_0}{v_0}\right). \tag{25}$$

A convenient dimensionless velocity variable is

$$\varphi \equiv v/v_0, \tag{26}$$

which approaches unity as $x \to -\infty$. Equation (12a) can be used to express the temperature T in terms of φ and τ; the result is

$$\frac{T}{T_0} = 1 + \alpha'\tau - \left(\frac{\gamma - 1}{2}\right)M_0^2(\varphi^2 - 1), \tag{27}$$

where the ratio of specific heats is

$$\gamma \equiv c_p/(c_p - R^0/W), \tag{28}$$

the dimensionless heat release is

$$\alpha' \equiv q/c_p T_0, \tag{29}$$

and the initial Mach number is

$$M_0 \equiv v_0/\sqrt{\gamma(R^0/W)T_0}. \tag{30}$$

The relation between α' and the dimensionless heat release α used in Section 2.2.2 is

$$\alpha' \equiv \frac{q}{c_p T_0} = \left(\frac{\gamma - 1}{\gamma}\right)\frac{q}{(R^0/W)T_0} = \left(\frac{\gamma - 1}{\gamma}\right)\frac{q}{p_0/\rho_0} = \left(\frac{\gamma - 1}{\gamma}\right)\alpha.$$

By substituting equations (26), (27), and (30) into equation (25), we obtain the dimensionless form of the momentum conservation equation,

$$d\varphi/d\xi = F_\varphi(\tau, \varphi), \tag{31}$$

where

$$F_\varphi(\tau, \varphi) \equiv (\varphi - 1) + \frac{1}{\gamma M_0^2}\left\{\frac{1}{\varphi}\left[1 + \alpha'\tau - \left(\frac{\gamma - 1}{2}\right)M_0^2(\varphi^2 - 1)\right] - 1\right\}. \tag{32}$$

The function $F_\varphi(\tau, \varphi)$ depends only on τ, φ, and constants.

In equation (7) with $i = 2$, use may be made of $Y_{2\infty} = 1$ and equations (16)–(18), (21), and (24) to show that

$$d\epsilon/d\xi = F_\epsilon(\tau, \varphi), \tag{33}$$

where

$$F_\epsilon(\tau, \varphi) \equiv \left[\frac{\lambda\rho B_1 T^{\alpha_1}}{m^2 c_p}\right](1 - \tau)e^{-E_1/R^0 T}. \tag{34}$$

Equation (33) is a dimensionless form of the species-conservation equation. We are entitled to indicate that $F_\epsilon(\tau, \varphi)$, as defined in equation (34), is a function of τ and φ, because the additional variables ρ and T, which appear on the right-hand side of equation (34), are easily expressed in terms of τ, φ, and constants through equations (6) and (27). The function $F_\epsilon(\tau, \varphi)$ is non-negative over the entire range $0 \leq \tau \leq 1$, in the physically acceptable range of φ, and it equals zero only at $\tau = 1$, although it becomes very small (because of the exponential factor involving T) near $\tau = 0$ where T is small.

5.2.7. Summary of the simplified mathematical problem

Equations (19), (31), and (33) are three first-order differential equations, which determine the variations of the variables τ, φ, and ϵ (dimensionless measures of the stagnation temperature, the velocity, and the progress of the reaction, respectively) through the combustion wave. Since the distance variable ξ appears only as $d/d\xi$ in these equations, it is easily eliminated, thereby reducing the total number of differential equations from three to two. Through division of equations (31) and (33) by equation (19), it is seen that the two governing equations can be taken to be

$$d\varphi/d\tau = F_\varphi(\tau, \varphi)/(\tau - \epsilon) \tag{35}$$

and

$$d\epsilon/d\tau = F_\epsilon(\tau, \varphi)/(\tau - \epsilon), \tag{36}$$

over the range $0 \leq \tau \leq 1$. Boundary conditions for equations (35) and (36) are $\epsilon = 0$ and $\varphi = 1$ at $\tau = 0$, and $\epsilon = 1$ and $\varphi = \varphi_\infty$ [which is determined either by setting $F_\varphi(\tau, \varphi) = 0$ and $\tau = 1$ in equation (32) or by consulting Section 2.2] at $\tau = 1$. Since both of these endpoints are singular points of equations (35) and (36), the boundary conditions alone do not enable us to draw any conclusions concerning the existence or uniqueness of solutions. After solutions for $\varphi(\tau)$ and $\epsilon(\tau)$ have been obtained from equations (35) and (36), the profiles in the physical coordinate can be found by integrating equation (19); namely,

$$\xi = \int_{\tau_1}^\tau d\tau/(\tau - \epsilon), \tag{37}$$

where τ_1 is the value of τ at $x = 0$.

5.3. THE UNIMOLECULAR DECOMPOSITION FLAME WITH LEWIS NUMBER OF UNITY

5.3.1. Governing equations

Let us now restrict our attention to deflagrations with $M_0 \ll 1$, retaining all of the assumptions of the preceding section. If φ and $d\varphi/d\xi$ remain small compared with $1/M_0^2$ (as is, in fact, found to be true from the deflagration solution), then equations (27) and (31) [with F_φ given by equation (32)] imply that

$$T/T_0 = 1 + \alpha'\tau \tag{38}$$

and

$$\varphi = 1 + \alpha'\tau. \tag{39}$$

In view of equation (29), equation (38) states that

$$\tau = c_p(T - T_0)/q; \tag{40}$$

that is, τ is a measure of the static temperature, since ordered kinetic energy is negligible. Using equations (6), (26), and (38) in equation (39), one obtains

$$\rho T = \rho_0 T_0 = \text{constant}, \tag{41}$$

which implies that $p = $ constant, in view of equation (15). These formal results were anticipated in Section 5.1; they can be derived more directly by estimating the magnitudes of the terms in equations (5) and (9). This latter derivation rests only on the assumptions that the Mach number is small compared with unity throughout the wave and that the coefficient of viscosity is not abnormally large; thus the results do not depend on assumption 8.

Since equation (39) represents an approximate solution to equation (35), the only differential equation that remains to be solved is equation (36), in which $F_\epsilon(\tau, \varphi) = F_\epsilon(\tau, \varphi(\tau))$ depends only on τ. By utilizing equations (38) and (41) in equation (34), we can obtain the τ dependence of F_ϵ explicitly. When this result is substituted into equation (36), we find that

$$d\epsilon/d\tau = \Lambda\omega/(\tau - \epsilon), \qquad (42)$$

where the dimensionless reaction-rate function $\omega = \omega(\tau)$ is defined as

$$\omega \equiv (1 - \tau)\exp[-\beta'(1 - \tau)/(1/\alpha' + \tau)], \qquad (43)$$

the dimensionless activation energy is

$$\beta' \equiv E_1/R^0 T_\infty, \qquad (44)$$

and the dimensionless factor Λ has been defined as

$$\Lambda \equiv (\lambda\rho B_1 T^{\alpha_1}/m^2 c_p)e^{-\beta'}. \qquad (45)$$

So that Λ may be treated as a constant without complicating the definition of ω, we shall (solely for simplicity) employ a final assumption:

11. The temperature dependence of λ is given by

$$\lambda T^{\alpha_1 - 1} = \text{constant}.$$

Two realistic cases in which the nonessential assumption 11 is valid are $\alpha_1 = 0$, $\lambda \sim T$ and $\alpha_1 = \frac{1}{2}$, $\lambda \sim \sqrt{T}$.

The two-point boundary conditions for equation (42) are $\epsilon = 0$ at $\tau = 0$ and $\epsilon = 1$ at $\tau = 1$. Three constants α', β', and Λ, enter into equation (42). The first two of these constants are determined by the initial thermodynamic properties of the system, the total heat release, and the activation energy, all of which are presumed to be known. In addition to depending on known thermodynamic, kinetic, and transport properties, the third constant Λ depends on the mass burning velocity m, which, according to the discussion in Section 5.1, is an unknown parameter that is to be determined by the structure of the wave. Since equation (42) is a first-order equation with two boundary conditions, we may hope that a solution will exist only for a particular value of the constant Λ. Thus Λ is considered to be an eigenvalue of the nonlinear equation (42) with the boundary conditions stated above; Λ is called the **burning-rate eigenvalue**.

From equations (6), (16), (44), and (45) it can be seen that

$$v_0 = (1/\rho_0)\sqrt{(\lambda_\infty/c_p)w_2(1, T_\infty)}(1/\sqrt{\Lambda}),$$

where λ_∞ is the value of λ at $T = T_\infty$ and $w_2(1, T_\infty)$ denotes the reaction rate [appearing in equation (16)] evaluated at $X_1 = 1$ and $T = T_\infty$. The comparison of this expression with equation (2) shows that the expected primary dependence of v_0 on the properties of the system has been included in the definition of Λ. Thus Λ itself should not depend strongly on the properties of the system; indeed, it can depend only on the two dimensionless groups α' and β'. For typical activation energies, $(1/\sqrt{\Lambda}) \sim 0.1$.

5.3.2. The cold-boundary difficulty [16], [17], [18]

The cold-boundary condition is $\tau = 0$ at $\epsilon = 0$. If $\omega(\tau)$ does *not* go to zero at $\tau = 0$ [as is implied by equation (43)], then the point ($\tau = 0$, $\epsilon = 0$) is not a singular point of equation (42), and it is a simple matter to determine the behavior of the solution in the neighborhood of this point. With $\epsilon = 0$ at $\tau = 0$ and $\omega(0) \neq 0$, equation (42) implies that $(d\tau/d\epsilon)_0 = 0$, whence $\tau \ll \epsilon$ near $\tau = 0$, and hence τ may be neglected in comparison to ϵ on the right-hand side. Equation (42) is then easily integrated to show that

$$\tau = -\epsilon^2/2\Lambda\omega(0)$$

near $\tau = 0$, $\epsilon = 0$. Hence, for $\Lambda > 0$, τ becomes *negative* (the temperature drops) as ϵ increases near the cold boundary; the reaction proceeds rapidly and heat is conducted downstream. This can be seen more directly by differentiating the reciprocal of equation (42) with respect to ϵ; when $\omega(\tau)$ is given by equation (43), the result is $(d^2\tau/d\epsilon^2)_0 = -1/\Lambda\omega(0)$, which implies that τ becomes negative as ϵ increases [since $(d\tau/d\epsilon)_0 = 0$] for realistic* values of Λ. It is physically unacceptable for τ to decrease as ϵ increases near $\tau = 0$, $\epsilon = 0$; for example, in the case $Y_{2,0} = 0$, this implies that the mass fraction of products becomes negative [see equations (21) and (24)].

The problem identified here is the one in which the *cold-boundary difficulty*, introduced in Section 2.2.2, first arose. It is clear that the trouble lies not in the mathematics but instead in the mathematical model of the physical situation; the mathematical problem defined by equations (42) and (43) with the stated boundary conditions is ill-posed. A further indication of this fact may be obtained by differentiating equation (19) with respect to ξ and using the result $d\epsilon/d\xi = (d\epsilon/d\tau)(d\tau/d\xi) = \Lambda\omega$, derived from equations (42) and (19). We thereby see that

$$d^2\tau/d\xi^2 - d\tau/d\xi = -\Lambda\omega, \tag{46}$$

---

* If Λ were negative, then m would have to be imaginary according to equation (45).

the three terms of which nondimensionally represent diffusion, convection, and reaction, respectively. At the cold boundary, $\xi \to -\infty$, we expect τ to approach zero smoothly, so that $d\tau/d\xi$ and $d^2\tau/d\xi^2$ should both approach zero, but this is impossible according to equation (46) if ω does not approach zero as $\tau \to 0$.

In the formulation of the problem, the combustible mixture has been asked to flow from $x = -\infty$ at a finite velocity (that is, for an infinite time) without reacting appreciably, even though $\omega \neq 0$. It cannot do so; it will have to have reacted before it arrived. The equations attempt to demonstrate this result by suggesting that $\Lambda = 0$, so that both m and v_0 are infinite [see equations (45) and (6)] if $\omega(0) \neq 0$. This may be seen, for example, by starting an integration from the hot boundary and finding that, for reasonable functions $\omega(\tau)$, ϵ approaches zero much more rapidly than τ at the cold boundary (see the inset in Figure 5.3). In the vicinity of the cold boundary, it is therefore reasonable to neglect ϵ in comparison with τ in the denominator of equation (42), whence equation (42) may be integrated to show that

$$\Lambda = \int_0^{\epsilon'} d\epsilon \bigg/ \int_0^{\tau'} (\omega/\tau)\, d\tau,$$

where ϵ' is the value of ϵ at $\tau = \tau'$ (near zero). Since the integral in the denominator diverges at $\tau = 0$ if $\omega > 0$ and $\omega(0) \neq 0$, we then see that $\Lambda = 0$.

Avoidance of the appearance of the cold-boundary difficulty can be achieved only by revision of the physical model on which the mathematical formulation is based. There are many ways in which this can be done. In one approach, von Kármán and Millán [17] replace $\tau = 0$ by $\tau = \tau_i$ (where $0 < \tau_i < 1$) as the position at which to apply the cold-boundary condition $\epsilon = 0$. The introduction of this artifice is equivalent to employing the physical concept of an ignition temperature T_i, below which the chemical reaction rate vanishes [see equations (19) and (33)]. An alternative procedure for determining a finite, nonzero value for Λ involves the assumption that a flame holder serves as a weak heat sink [18], removing an amount of heat per unit area per second equal to $(\lambda\, dT/dx)_i = mc_p(T_\infty - T_0)(d\tau/d\xi)_i$. Since $\epsilon = 0$ at the flame holder (that is, there is no reaction upstream), equation (19) implies that $(d\tau/d\xi)_i = \tau_i$, whence the heat-sink concept is seen to be mathematically equivalent to the ignition-temperature concept at the cold boundary.

With either of these methods of remedying the cold-boundary difficulty, the value of Λ depends, of course, on the chosen value of τ_i. It is found that the calculated estimate for Λ assumes a pseudostationary value as τ_i, or the heat loss to the flame holder, is allowed to vary between reasonable limits [16]. This conclusion depends directly on the strong temperature dependence of the reaction-rate function. The pseudostationary behavior is illustrated in Figure 5.3, where the dashed line indicates the pseudostationary value of Λ. It is reasonable to identify the pseudostationary value as the

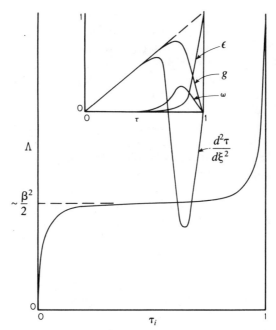

FIGURE 5.3. Schematic representation of the solution profiles and of the pseudostationary behavior of the burning-rate eigenvalue as a function of the nondimensional ignition temperature.

physically significant value, since values of τ_i very near zero or unity seem unrealistic. Thus a burning rate that depends only weakly on τ_i is obtained; this dependence may be eliminated entirely—for example, by arbitrarily defining Λ as the value corresponding to the inflection point of the curve in Figure 5.3.

These early approaches to the elimination of the cold-boundary difficulty are mathematically equivalent to reformulation of the physical model through modification of $\omega(\tau)$ by putting $\omega = 0$ for $\tau < \tau_i$. The nondimensional ignition temperature τ_i is small, and the corresponding T_i is not to be associated in any way with a physical ignition temperature to be found experimentally. Instead, τ_i is a purely computational device, one which can lead to somewhat laborious computations if solutions are to be generated for a range of values of τ_i in search of pseudostationarity. For this reason alternative modifications to $\omega(\tau)$ have been sought. One alternative that has enjoyed some popularity is that of Friedman and Burke [19] in which $1/\alpha'$ is set equal to zero in equation (43). This avoids the cold-boundary difficulty by causing $\omega(\tau)$ to approach zero rapidly, in an exponential fashion as $\tau \to 0$ and results in a graph of $\omega(\tau)$ that is not very different from that for reasonable values of $1/\alpha' \neq 0$. However, it is a modification that extends over the entire range of τ and, for example, changes the slope $d\omega/d\tau$ at $\tau = 1$, thereby affecting the value of Λ obtained and providing an incorrect dependence of Λ on α'. Therefore, this particular modification is not recommended; changes are permissible in $\omega(\tau)$ only near $\tau = 0$.

A certain amount of rigorous study has been given to the mathematical form of $\omega(\tau)$ near $\tau = 0$ that renders the problem well-posed [20], [21], [22]. As might be expected from the above discussion, the behavior of ω/τ as $\tau \to 0$ is relevant. If ω/τ approached infinity as $\tau \to 0$, then the problem remains ill-posed. Much of the work has been concerned with continuous functions $\omega(\tau)$ that are positive for $0 < \tau < 1$ and that vanish at $\tau = 0$ and at $\tau = 1$ [22]. For rate functions in this class, if $\lim_{\tau \to 0}(\omega/\tau)$ is finite (either positive or zero), then solutions satisfying the given boundary conditions exist for all values of Λ less than or equal to a certain maximum value [22]. Modifications to ω of the form $\omega = $ constant $(1 - \tau)^m \tau^n$, where m and n are positive integers, have been studied [23], [24], [25]; for these rate functions, the maximum value of Λ appears physically to be the most realistic [26] and is the value that has been obtained [23]. If ω vanishes in a finite neighborhood of the point $\tau = 0$, then Λ is unique [21]. This attribute of the ignition-temperature approach to the correction of the cold-boundary difficulty helps to make it more attractive than other approaches that involve modification of ω.

An understanding of the special characteristics of the ignition-temperature method may be obtained from a phase-plane analysis of equation (42). If such an analysis is pursued [26] by standard methods (for example, [27]) it is found that the singular point (or critical point) at $\tau = \epsilon = 1$ is a saddle point. For nonnegative functions $\omega(\tau)$, only one of the two integral curves from the saddle enters the physically acceptable region $\tau < 1, \epsilon < 1$, and it does so with $\epsilon < \tau$. If $\omega(\tau)$ is positive except at $\tau = 0$ and $\tau = 1$, then the only other singularity in the region of interest $(0 \le \tau \le 1, 0 \le \epsilon \le 1)$ is the cold boundary, $\tau = 0, \epsilon = 0$. Multiplying equation (42) by $(1 - \epsilon/\tau)$ and passing to the limit $\tau \to 0$, $\epsilon \to 0$, we obtain a quadratic equation for $(d\epsilon/d\tau)_0$, the solution to which is

$$\left(\frac{d\epsilon}{d\tau}\right)_0 = \frac{1 \pm \sqrt{1 - 4\Lambda \lim_{\tau \to 0}(\omega/\tau)}}{2}.$$

If ω/τ approaches a finite limit as $\tau \to 0$, then this formula gives two directions of integral curves at the cold boundary for Λ less than a maximum value, and moreover the cold boundary is found to exhibit a nodal type of behavior when Λ does not exceed the maximum. It is because of this nodal character that one integral curve from the saddle passes through the cold boundary for each value of Λ less than or equal to the maximum. In contrast, with an ignition temperature τ_i, the solution $\epsilon = 0$ for $\tau < \tau_i$ causes the relevant cold-boundary point to be $\tau = \tau_i$, $\epsilon = 0$, which is not a singularity of equation (42). In this case, the requirement that the integral curve must pass through both the nonsingular point $(\tau = \tau_i, \epsilon = 0)$ and the hot-boundary saddle serves to select a unique value of Λ.

Asymptotic analysis for large nondimensional activation energies (β' appreciably greater than unity; see Section 5.3.6) occupies a unique position with respect to the cold-boundary difficulty. The asymptotic approach in effect determines Λ through an expansion dominated by the behavior of the solution in the hotter portions of the flame. The specific form of $\omega(\tau)$ in the vicinity of $\tau = 0$ plays no role in generating any of the terms in the expansion. In this respect, the cold-boundary difficulty simply does not appear within the context of the asymptotic expansion, and artifices of ignition temperatures, flame holders, or rate-function modifications become superfluous. The asymptotic method thereby provides a physically correct solution to the ill-posed problem. If the problem is rendered well-posed by introduction of τ_i, then successive orders of the asymptotic expansion yield, in an asymptotic sense, successively improved approximations to the pseudostationary value of Λ. When the asymptotic method is adopted, it is permissible but unnecessary to state the problem in a well-posed fashion. In a very real sense, the idea of a steadily propagating premixed laminar flame is an asymptotic concept. If the asymptotic approach becomes inappropriate, then the initial metastable mixture of reactants is insufficiently stable for a flame to propagate through it at an identifiable quasisteady speed. Further clarification of the situation has been achieved recently [28] on the basis of analysis of an initial-value problem in which the initial rate of reaction is not negligible. Unlike the steady-flow problem in the infinite domain considered here, the initial-value problem in the infinite domain (or the steady-flow problem in the semi-infinite domain) is well-posed.

5.3.3. Bounds on the burning-rate eigenvalue

When the problem is formulated in a well-posed manner, upper and lower bounds may be established rigorously for Λ. This has been accomplished by Johnson and Nachbar [21], [29] for the formulation based on the ignition temperature. The analysis employs a form of equation (46) obtained by defining $g = d\tau/d\xi$ and by treating τ as the independent variable and g as the dependent variable. From equation (46) we see that

$$g - g \, dg/d\tau = \Lambda\omega. \tag{47}$$

The boundary condition at the hot boundary for equation (47) is $g = 0$ at $\tau = 1$; that at the cold boundary (originally $g = 0$ at $\tau = 0$) becomes $g = \tau_i$ at $\tau = \tau_i (0 < \tau_i < 1)$ in the ignition-temperature formulation.

Integrating equation (47) from τ to $\tau = 1$ gives

$$g^2/2 = \Lambda \int_\tau^1 \omega \, d\tau - \int_\tau^1 g \, d\tau, \tag{48}$$

which reduces to

$$\Lambda\left(\int_{\tau_i}^1 \omega \, d\tau\right) = \tau_i^2/2 + \int_{\tau_i}^1 g \, d\tau \tag{49}$$

when evaluated at $\tau = \tau_i$. Dividing equation (47) by g and integrating from $\tau = \tau_i$ to τ yields

$$g = \tau - \Lambda \int_{\tau_i}^{\tau} (\omega/g) \, d\tau. \tag{50}$$

By substituting equation (50) into the integral on the right-hand side of equation (49), it is found that

$$\Lambda \int_{\tau_i}^{1} \omega \, d\tau = \tfrac{1}{2} - \Lambda \int_{\tau_i}^{1} \int_{\tau_i}^{\tau} [\omega(\tau')/g(\tau')] \, d\tau' \, d\tau,$$

which becomes

$$2\Lambda \int_{\tau_i}^{1} \omega \left(1 + \frac{1 - \tau}{g} \right) d\tau = 1 \tag{51}$$

when the double integral is reduced to a single integral through integration by parts. Equation (51) will be used shortly to determine an upper bound for Λ.

Equation (47) may also be written in the form

$$d(g^2)/d\tau = 2(g^2)/g - 2\Lambda\omega, \tag{52}$$

which resembles a linear equation for g^2 and therefore suggests obtaining a "solution" by means of an "integrating factor." Multiplying equation (52) by $\exp[-2\int_{\tau_i}^{\tau} (1/g) \, d\tau]$ in fact shows that

$$\frac{d}{d\tau} \left\{ g^2 \exp\left[-2 \int_{\tau_i}^{\tau} \left(\frac{1}{g} \right) d\tau \right] \right\} = -2\Lambda\omega \exp\left[-2 \int_{\tau_i}^{\tau} \left(\frac{1}{g} \right) d\tau \right], \tag{53}$$

as can be seen explicitly by expanding the derivative on the left-hand side of equation (53) by parts. Integrating equation (53) from $\tau = \tau_i$ to $\tau = 1$ and utilizing the boundary conditions on g yields

$$2\Lambda \left(\frac{1}{\tau_i^2} \right) \int_{\tau_i}^{1} \omega \exp\left[-2 \int_{\tau_i}^{\tau} \left(\frac{1}{g} \right) d\tau \right] d\tau = 1 \tag{54}$$

after dividing by τ_i^2. Equation (54) will be used to determine a lower bound for Λ.

Since ω, $(1 - \tau)$, and g are all nonnegative, it is clear that by substituting an upper bound for $g(\tau)$ into equation (51) a lower bound for the integral is obtained, whence the resulting value of Λ given by equation (51) will be an upper bound.* On the other hand, substituting this same upper

---

* The observant reader may have noticed that by setting $\tau = 1$ (and therefore $g = 0$) in equation (50), one obtains $\Lambda \int_{\tau_i}^{1} (\omega/g) \, d\tau = 1$, which also yields an upper bound for Λ when an upper bound for $g(\tau)$ is used to evaluate the integral. Why, then, should we employ the more complicated expression given in equation (51)? The reason is that for most upper-bound trial functions $g(\tau)$, equation (51) yields a lower upper bound for Λ (a value closer to the correct value of Λ) than does the simpler equation. This is readily seen with the upper bound $g(\tau) = \tau$, for which the value of Λ given by the simpler equation is twice the value given by equation (51).

bound for $g(\tau)$ into equation (54) will yield a lower bound for Λ, because the inner integral will be a lower bound, implying that the exponential and therefore the outer integral assume upper bounds. Equations (51) and (54) can therefore be used to provide limits between which the eigenvalue must lie.

Since $\Lambda \int_{\tau_i}^{\tau} (\omega/g)\, d\tau \geq 0$, it follows from equation (50) that

$$g \leq \tau. \tag{55}$$

Hence τ may be used for the upper bound of g in equations (51) and (54). Equation (51) then implies that

$$\Lambda \leq \left[2 \int_{\tau_i}^{1} (\omega/\tau)\, d\tau \right]^{-1}, \tag{56}$$

and equation (54) shows that

$$\Lambda \geq \left[2 \int_{\tau_i}^{1} (\omega/\tau^2)\, d\tau \right]^{-1} \tag{57}$$

after the inner integration over τ is performed. Equations (56) and (57) constitute explicit, rigorous bounds for the burning rate eigenvalue Λ. These bounds typically are reasonably narrow, and equation (56) or the average of equations (56) and (57) provides a closer estimate for Λ than that obtained by many approximate methods.

5.3.4. Iterative procedures and variational methods

An iterative technique for obtaining successively narrower upper and lower bounds for the eigenvalue Λ and successively lower upper bounds for the function $g(\tau)$ has been developed by Johnson and Nachbar [21]. Since they have proven that this procedure converges and that $g(\tau)$ monotonically approaches the solution of equation (47), the oscillating divergence sometimes observed with other iterative schemes cannot appear here, and the present method constitutes a truly rigorous procedure for obtaining solutions of any desired accuracy. Integrating equation (50) from $\tau = \tau_i$ to τ yields

$$\int_{\tau_i}^{\tau} g\, d\tau = \tfrac{1}{2}(\tau^2 - \tau_i^2) - \Lambda \int_{\tau_i}^{\tau} \int_{\tau_i}^{\tau'} \frac{\omega(\tau'')}{g(\tau'')}\, d\tau''\, d\tau'$$

$$= \tfrac{1}{2}(\tau^2 - \tau_i^2) - \Lambda \int_{\tau_i}^{\tau} (\tau - \tau') \frac{\omega(\tau')}{g(\tau')}\, d\tau', \tag{58}$$

where the last equality is obtained through integrating the last term by parts. Integrating equation (47) from $\tau = \tau_i$ to τ gives

$$g^2/2 = \tau_i^2/2 - \Lambda \int_{\tau_i}^{\tau} \omega\, d\tau + \int_{\tau_i}^{\tau} g\, d\tau, \tag{59}$$

which becomes

$$g^2 = \tau^2 - 2\Lambda \int_{\tau_i}^{\tau} \omega(\tau')\{[(\tau - \tau')/g(\tau')] + 1\} \, d\tau' \tag{60}$$

when equation (58) is substituted into the last term. Equations (54) and (60) provide the basis for the iterative procedure.

An upper bound for $g(\tau)$ is substituted into equation (54) in order to obtain a lower bound for Λ. If this value of Λ and the upper bound for $g(\tau)$ are substituted into the right-hand side of equation (60), then, since the integral in equation (60) will clearly assume a lower bound, the right-hand side of equation (60) will assume an upper bound, and hence a new upper bound for $g^2(\tau)$ will be given by equation (60). The fact that the resulting new upper bound for $g(\tau)$ is lower than the previous upper bound requires proof by mathematical induction.*

The nth approximation for g is then given by [see equations (60) and (54)]

$$g_{[n]}^2 = \tau^2 - 2\Lambda_{[n-1]} \int_{\tau_i}^{\tau} \omega(\tau')\{[(\tau - \tau')/g_{[n-1]}(\tau')] + 1\} \, d\tau', \tag{61}$$

where

$$\Lambda_{[n-1]} = \left\{ \frac{2}{\tau_i^2} \int_{\tau_i}^{1} \omega(\tau) \exp\left[-2 \int_{\tau_i}^{\tau} \left(\frac{1}{g_{[n-1]}(\tau')} \right) d\tau' \right] d\tau \right\}^{-1}. \tag{62}$$

In accordance with the preceding discussion, as $n \to \infty$, $g_{[n]}(\tau)$ approaches the correct solution from above and $\Lambda_{[n]}$ approaches the correct eigenvalue from below. From the discussion in Section 5.3.3 it is seen that for each $g_{[n]}(\tau)$ a value of Λ may also be computed from equation (51), and that the resulting sequence of values will approach the correct eigenvalue from above. Thus successively narrower bounds for the burning rate and successively smaller upper bounds for $g = d\tau/d\xi$ are obtained.

Since the first estimates for the bounds of Λ given in equations (56) and (57) are so close together, it appears likely that few iterations would be

---

* It is obvious that if $g = \tau$ is used as the first approximation, then the second approximation obtained from equation (60) will be a lower upper bound. The induction proof is then completed by using the assumption that the nth approximation for g is a lower upper bound than is the $(n - 1)$st approximation to prove that the $(n + 1)$st approximation is a lower upper bound than the nth approximation. This last proof is obtained quite simply from equations (54) and (60), since the fact that the nth approximation for g is lower than the $(n - 1)$st approximation immediately shows [via equations (54) and (60)] that the right-hand side of equation (60) used in obtaining the $(n + 1)$st approximation is smaller than the right-hand side of equation (60) used in obtaining the nth approximation, whence equation (60) implies that the $(n + 1)$st approximation is smaller than the nth approximation. Throughout this reasoning, equation (60) could have been replaced by the simpler expression given in equation (50); however, convergence is usually slower when equation (50) is used instead of equation (60) (see the preceding footnote).

required to obtain an exceedingly high degree of accuracy. From equations (61) and (62) it is seen that the integrations that must be performed in carrying out the present technique are not excessively complicated. However, for most applications, either equations (56) and (57) or one of the other approximate formulas to be given in Sections 5.3.5 and 5.3.6 will be sufficiently accurate, and it will be unnecessary to employ the method described here. It may be noted that if the present methods are employed, then calculations must be performed for various values of τ_i to find conditions of pseudo-stationarity.

Other iterative methods have been proposed that are less well justified. For example, one approach [30] employs equations (48) and (49) directly, with the nth approximation for $g(\tau)$ substituted into the integrals to obtain the $(n + 1)$st approximations for Λ and $g(\tau)$. A variational approach also has been developed [25], based on the introduction of $\eta = e^{\xi}$ as a new variable, so that equation (46) can be written in a standard form, $d^2\tau/d\eta^2 = -\Lambda\omega/\eta^2$. None of these approaches circumvent the cold-boundary difficulty unless $\omega(\tau)$ is modified suitably, for example, by the introduction of $\tau_i \neq 0$. The approximations to be discussed next bypass the cold-boundary difficulty in a natural way.

5.3.5. The approximations of Zel'dovich, Frank-Kamenetskii, and von Kármán

Many approximate formulas for Λ have been developed on the basis of the physical idea that conditions in the vicinity of the hot boundary control the burning rate (for example, [8], [31]). One way to see how these formulas may be obtained is to follow the approach of von Kármán [32]. Multiplying equation (42) by $(1 - \epsilon) - (1 - \tau)$ and integrating from 0 to 1 yields

$$\Lambda I = \tfrac{1}{2} - \int_0^1 (1 - \tau)\, d\epsilon, \tag{63}$$

where

$$I = \int_0^1 \omega\, d\tau, \tag{64}$$

which is easily evaluated in terms of exponential integrals [33] when ω is given by equation (43). For the purpose of burning-velocity calculations, von Kármán suggests that an approximation to Λ can be obtained from equation (63) by approximating the integral therein on the basis of the behavior of $\tau(\epsilon)$ in the vicinity of the hot boundary. In this approximate evaluation, the integral is convergent, and the cold-boundary difficulty thereby disappears.

A linear relationship, $1 - \tau = c(1 - \epsilon)$, may be employed, with the constant c obtained in terms of Λ by evaluating the slope, $d\tau/d\epsilon = c$, from the limiting form taken by equation (42) at the hot boundary. This may be shown to give $\Lambda = (1 - \sqrt{2I})/(2I)$. A simpler approximation is to put $c = 0$, obtaining $\Lambda = (2I)^{-1}$ directly from equation (63). These formulas of von Kármán [32] become increasingly accurate as β' increases; if $\beta' > 2$, then the error in v_0 typically is less than 10% when the first formula is used and less than 30% for the second.

The formula of Zel'dovich and Frank-Kamenetskii [7], [34] may be obtained from the simplest result of von Kármán by performing an asymptotic expansion of the integral I for large values of β' and retaining only the first term in the expansion. In view of equation (43), by transforming from the variable τ to $z \equiv \beta'(1 - \tau)/(1/\alpha' + \tau)$ in equation (64), we find that

$$I = \left(\frac{1 + 1/\alpha'}{\beta'}\right)^2 \int_0^{\beta'\alpha'} \frac{ze^{-z}}{(1 + z/\beta')^3}\, dz.$$

The activation energy and the heat of reaction are usually sufficiently large that a negligible error in this integral is produced by extending the upper limit to infinity ($1/\beta'\alpha' \approx 0$). The factor $(1 + z/\beta')^{-3}$ in the integrand of the resulting expression for I may be expanded in powers z/β' to obtain an asymptotic expression for I useful for large values of β'. Keeping only the first term in the expansion, we obtain, with $\mathcal{O}$ meaning terms of order,

$$I \approx \left(\frac{1 + 1/\alpha'}{\beta'}\right)^2 \int_0^\infty ze^{-z}\, dz + \mathcal{O}\left(\frac{1}{\beta'^3}\right) = \left(\frac{1 + 1/\alpha'}{\beta'}\right)^2 + \mathcal{O}\left(\frac{1}{\beta'^3}\right).$$

The introduction of this result into the formula $1/\sqrt{\Lambda} = \sqrt{2I}$ yields

$$1/\sqrt{\Lambda} = \frac{\sqrt{2}(1 + 1/\alpha')}{\beta'}, \tag{65}$$

which is the expression of Zel'dovich and Frank-Kamentskii for the non-dimensional quantity $1/\sqrt{\Lambda}$, which is proportional to m or v_0. This formula is nearly as accurate as the von Kármán expression from which it was derived and is exactly the same as the first approximation obtained by the asymptotic analysis to be presented next. The significance of this result and of the appearance of the nonnegligible factor $1/\beta'$ on the right-hand side will become clear after the activation-energy asymptotics have been developed.

5.3.6. Asymptotic analysis for strongly temperature-dependent rates

Although the asymptotic analysis was developed first on the basis of a formulation in the variables appearing in equation (42) [9], it is more instructive to begin with equation (46). For later convenience in analysis, the combinations $\alpha = \alpha'/(1 + \alpha')$ (not to be confused with the α of Chapter 2) and $\beta = \alpha'\beta'/(1 + \alpha')$ are employed in place of α' and β'. In terms of α and β, the rate expression in equation (43) becomes

$$\omega = (1 - \tau)\exp\{-\beta(1 - \tau)/[1 - \alpha(1 - \tau)]\}. \tag{66}$$

Note that α is the temperature rise divided by the final temperature, $(T_\infty - T_0)/T_\infty$ $(0 < \alpha < 1)$, and β is the product of α with the nondimensional activation energy $E_1/R^0 T_\infty$. Thus $\beta < \beta'$, and large activation energy and large heat release both favor large β. It will be found that β, rather than β', is the appropriate large parameter of expansion. Therefore, in addition to β' being large, the heat release cannot be too small if the asymptotic results are to be accurate. Values of α typically are 0.8 and seldom are less than 0.5; thus β generally is large if β' is large. Representative values of β in fact lie between 5 and 15, numbers that are sufficiently large for asymptotic results to be reasonably accurate. Because of the many important contributions of Zel'dovich to our physical understanding of the theory, it seems appropriate to call β the **Zel'dovich number**.

For the sake of being definite, the analysis will be developed on the basis of the rate expression given specifically in equation (66) by considering the parametric limit in which $\beta \to \infty$ with α of order unity and held fixed and with Λ required to vary with β in such a way that a nontrivial result is obtained. It may be emphasized, however, that other rate expressions can be treated in a similar way. The Zel'dovich number is a nondimensional measure of the temperature sensitivity of the overall reaction rate, qualitatively a measure of the strength of the dependence of the reaction rate on the extent of reaction completed. The attribute that is essential to the asymptotic approach is that the nonnegative single-peaked rate function possesses a maximum that narrows and that occurs at a location progressively closer to the fully burnt condition as $\beta \to \infty$.

From equation (66) it is seen that as β becomes large the reaction term in equation (46) becomes very small ("exponentially small," since β appears inside the exponential) unless τ is near unity. Hence for $1 - \tau$ of order unity, there is a zone in which the reaction rate is negligible and in which the convective and diffusive terms in equation (46) must be in balance. In describing this **convective-diffusive zone**, an outer expansion of the form $\tau = \tau_0(\xi) + H_1(\beta)\tau_1(\xi) + H_2(\beta)\tau_2(\xi) + \cdots$ may be introduced, following the formalism of matched asymptotic expansions [35]. Here

$$\lim_{\beta \to \infty} [H_{n+1}(\beta)/H_n(\beta)] = 0, \text{ with } H_0(\beta) = 1.$$

Substitution of this expansion into equation (46), passing to the limit $\beta \to \infty$, subtracting the result from equation (46), dividing by $H_1(\beta)$, again passing to the limit $\beta \to \infty$, and so on, produces the sequence of outer equations

$$d^2\tau_n/d\xi^2 = d\tau_n/d\xi, \quad n = 0, 1, 2, \ldots, \tag{67}$$

so long as $\lim_{\beta \to \infty}[\Lambda e^{-a\beta}/H_n(\beta)] = 0$ for any fixed $a > 0$ of order unity. This last restriction is satisfied by all algebraic (power-type) ordering functions— for example, for $H_n(\beta) \sim \beta^{-n}$; the reaction-rate term can contribute to equation (67) only at exponentially small orders—for example, for $H_n(\beta) \sim e^{-a\beta}$. The solutions to equation (67) satisfying the boundary condition that $\tau = 0$ at $\xi = -\infty$ are

$$\tau_n = A_n e^\xi, \quad n = 0, 1, 2, \ldots, \tag{68}$$

where the A_n are constants to be determined by matching conditions.

Equation (68) must fail if ξ becomes sufficiently large; in fact, from equation (66) it is seen that when $1 - \tau$ becomes of order β^{-1}, the reaction-rate term is no longer exponentially small. This observation indicates the need for a downstream zone in which $1 - \tau$ is of order β^{-1}. An appropriate stretched dependent variable for this zone apparently is $y = \beta(1 - \tau)$. To ascertain formally a suitable independent variable for use with y necessitates investigating various possibilities. If this independent variable is denoted by η, then any selection having $\lim_{\beta \to \infty}(d\eta/d\xi) = 0$ produces in the lowest order a first-order equation that expresses a convective-reactive balance and that lacks the flexibility needed to satisfy the downstream boundary condition and simultaneously match with equation (68). Thus the downstream zone cannot be thicker than the upstream zone. Study also shows that if $d\eta/d\xi$ is of order unity in the limit, then matching again cannot be achieved. Therefore, $\lim_{\beta \to \infty}(d\eta/d\xi) = \infty$; the downstream zone is thinner than the upstream zone and may reasonably be viewed as an inner zone. Any stretching of this type causes the convective term in equation (46) to become small compared with the diffusive term, and since at least two terms are needed to satisfy the matching and boundary conditions successfully, there must be a reactive-diffusive balance in the inner zone to the lowest order. The inner zone therefore is reasonably termed a **reactive-diffusive zone**. With the stretching $\eta = (\xi - \xi_0)/s(\beta)$, where $s \to 0$ as $\beta \to \infty$, equation (46) written in inner variables is

$$d^2 y/d\eta^2 - s \, dy/d\eta = s^2 \Lambda y \exp[-y/(1 - \alpha y/\beta)], \tag{69}$$

where use has been made of equation (66).

In equation (69) an inner expansion of the form

$$y = y_0(\eta) + h_1(\beta)y_1(\eta) + h_2(\beta)y_2(\eta) + \cdots$$

may be introduced, where $\lim_{\beta \to \infty}[h_{n+1}(\beta)/h_n(\beta)] = 0$, with $h_0(\beta) = 1$. From equation (69) it is seen that in the limit $\beta \to \infty$ ($s \to 0$), the necessary reactive-diffusive balance with y and η of order unity will be maintained only if $s^2\Lambda$ is finite and nonzero in the limit. Therefore, the expansion

$$\Lambda = (1/s^2)[\Lambda_0 + s_1(\beta)\Lambda_1 + s_2(\beta)\Lambda_2 + \cdots]$$

is introduced as well, where $\lim_{\beta \to \infty}(s_{n+1}/s_n) = 0$, $s_0 = 1$. Following a procedure analogous to that described for the outer expansion, we then generate a sequence of inner equations, the first of which is

$$d^2 y_0/d\eta^2 = \Lambda_0 y_0 e^{-y_0}. \tag{70}$$

The downstream boundary conditions, $y_n \to 0$ as $\eta \to \infty$, may be applied to the inner equations. The first integral of equation (70) then is

$$\tfrac{1}{2}(dy_0/d\eta)^2 = \Lambda_0 \int_0^{y_0} y' e^{-y'}\, dy' = \Lambda_0[1 - (1 + y_0)e^{-y_0}]. \tag{71}$$

It is found, as may be anticipated from the fact that τ decreases (y increases) as η decreases, that to achieve matching the negative square root must be taken in equation (71) in obtaining $dy_0/d\eta$. A further integration then gives

$$\eta = -\frac{1}{\sqrt{2\Lambda_0}} \int^{y_0} \frac{dy'}{[1 - (1 + y')e^{-y'}]^{1/2}} + \text{constant.}$$

An expansion of this formula for large negative values of η produces

$$y_0 = -\sqrt{2\Lambda_0}\,\eta + \text{constant} + \text{e.s.t.}, \tag{72}$$

where e.s.t. stands for terms that go to zero exponentially in η as $\eta \to -\infty$. Unlike equation (70), the differential equations for y_n ($n = 1, 2, \ldots$) are linear, and methods for solving them are available (for example, [36]).

Completion of a solution by matched asymptotic expansions entails employing matching conditions. Although there are many ways to effect matching, the most infallible approach currently available is to investigate a parametric limit in an intermediate variable [35]. Thus we consider $\beta \to \infty$ with η_t held fixed, where $\eta_t = s(\beta)\eta/t(\beta) = (\xi - \xi_0)/t(\beta)$, with $t(\beta) \to 0$ and $t(\beta)/s(\beta) \to \infty$ in the limit. The general matching condition is then written as

$$\tau_0 + H_1\tau_1 + \cdots \underset{\widetilde{\tau}}{} 1 - \beta^{-1}(y_0 + h_1 y_1 + \cdots), \tag{73}$$

where $\underset{\widetilde{\tau}}{}$ implies that when η_t is introduced on both sides, the two asymptotic expansions must agree. From equations (68) and (72), equation (73) may be written as

$$A_0 e^{\xi_0}(1 + t\eta_t + \cdots) + H_1 A_1 e^{\xi_0} + \cdots \underset{\widetilde{\tau}}{} 1 + \beta^{-1}\sqrt{2\Lambda_0}(t/s)\eta_t + \cdots. \tag{74}$$

Equation (74) is used repeatedly at successively higher orders to evaluate constants and ordering functions. Thus we first find that $A_0 e^{\xi_0} = 1$ and next find that s must be of order β^{-1}; hence we may put $s = \beta^{-1}$, thereby obtaining the further matching requirement $\sqrt{2\Lambda_0} = 1$. These last results provide the first approximation to the burning-rate eigenvalue, $\Lambda \simeq \Lambda_0/s^2 = \beta^2/2$, in agreement with equation (65). By proceeding further with the expansion and matching, it may be shown that $H_n = \beta^{-n}$, $h_n = \beta^{-n}$, $s_n = \beta^{-n}$, and $\Lambda_1 = 3\alpha - \gamma$, where $\gamma = \int_0^\infty \{1 - [1 - (1 + y_0)e^{-y_0}]^{1/2}\}\, dy_0 = 1.3440$. With these results, a two-term expansion for Λ is available, and a composite expansion for $\tau(\xi)$ may be constructed by standard methods [35], if desired.

The extended discussion of activation-energy asymptotics that has been given here was included to illustrate details of the technique that sometimes are needed in analyzing difficult problems. In most publications, as well as elsewhere in this volume (for example, Section 3.4.3), abbreviated presentations are made; for example, matching conditions such as equation (73) seldom are written. Analyses at the depth indicated here underlie applications of activation-energy asymptotics, at least in principle. However, often physical understanding allows shortcuts to be taken. For example, the expansions $\Lambda = \beta^2 \Lambda_0 + \beta \Lambda_1 + \Lambda_2 + \cdots$ and $y = y_0 + \beta^{-1} y_1 + \cdots$ could have been guessed in advance and later verified to work. Because of the translational invariance of the original problem, the value of ξ_0, which specifies the first approximation to the location of the inner zone, cannot be calculated; simplification can be achieved by employing this fact to demand at the outset that the reaction occur at $\xi = 0$. Suitably imposing this stipulation at all algebraic orders allows the outer solution to be written in advance as $\tau = e^\xi$ for $\xi < 0$, $\tau = 1$ for $\xi > 0$; the expansion in terms of $H_n(\beta)$ then need not be introduced at all. A simpler matching criterion—namely, matching of the slopes of the inner and outer solutions [37]—may then be employed directly in equation (71) for $\eta \to -\infty$ ($y_0 \to \infty$) to obtain $\Lambda_0 = \frac{1}{2}$, giving the first approximation for the burning rate. With these simplifications, the entire analysis can be completed in a few equations [38], although much thought underlies the equations.

The flame structure that has been calculated here is illustrated schematically in Figure 5.4. There are two zones through which the temperature varies: a thicker upstream zone in which the reaction rate is negligible to all algebraic orders in β^{-1}, followed by a thinner reaction zone in which convection is negligible to the lowest order in β^{-1}. The reaction occurs in the thin downstream zone because the temperature is too low for it to occur appreciably upstream; where the reaction is occurring, the reactant concentration has been depleted by diffusion so that it is of order β^{-1} times the initial reactant concentration. In the formulation based on equation (42), where $\epsilon(\tau)$ is sought, the convective-diffusive zone need not be mentioned

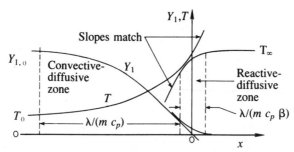

FIGURE 5.4. Schematic illustration of deflagration structure obtained by activation-energy asymptotics.

since the solution therein is $\epsilon = 0$; no change occurs in $\epsilon(\tau)$ until τ is within β^{-1} of unity (see Figure 5.3). The reaction zone is sufficiently narrow in the ξ coordinate to allow the reaction-rate function to be approximated, for many purposes, as a delta function in ξ that describes appropriate jump conditions across a reaction sheet [39]. This delta-function idea underlies the early development of equation (65) [7], [34].

In dimensional form [see equation (45)], the burning velocity given by the two-term expansion $\Lambda = \frac{1}{2}\beta^2 + \beta(3\alpha - \gamma)$, is*

$$v_0 = \left(\frac{2\lambda\rho B_1 T^{\alpha_1}}{\beta^2 \rho_0^2 c_p}\right)^{1/2} e^{-E_1/2R^0 T_\infty} \left[1 + \frac{1.344 - 3\alpha}{\beta} + \mathcal{O}\left(\frac{1}{\beta^2}\right)\right] \quad (75)$$

where $\alpha = (T_\infty - T_0)/T_\infty$ and $\beta = E_1(T_\infty - T_0)/R^0 T_\infty^2$. This result has come formally from the matching condition and represents the only burning velocity for which an internally consistent asymptotic structure for the deflagration can be obtained. The two-term result in equation (75) lies within a few percent of the pseudostationary burning velocity calculated numerically over the entire range of values of β that could be of practical interest ($\beta \geq 2$; see [9]). It is at least as accurate as the best results obtained by ad hoc methods prior to its derivation. Retention of the second term in equation (75) is especially important at the lower values of β, where it can produce corrections to v_0 approaching 50% in extreme cases. At representative values of β, the two-term asymptotic expansion for v_0 is significantly better than the corresponding one-term expansion of Zel'dovich and Frank-Kamenetskii obtained from equation (65) and given by the first term in equation (75). The value of v_0 obtained from the two-term expansion typically is 10% below that given by the first term. Thus Zel'dovich numbers for real flames lie in a range that is both appropriate for application of asymptotic methods and sufficiently far from the limit to require a formalism capable of generating the second term in the expansion, if accurate results are wanted.

It is of interest to compare equation (75) with equations (2) and (3). In making this comparison, for simplicity let us ignore the correction terms in equation (75) of order β^{-1} or higher. The principal functional dependences of v_0 agree in equations (2) and (75), notably the strongest dependences on $\lambda/c_p, B_1, p, T_\infty$, and E_1, the latter two controlled by the Arrhenius exponential. However, if w in equation (2) is evaluated at the initial concentration and at the final temperature, then the numerical value of v_0 obtained from equation (2) is roughly β times that of equation (75); the difference is an order of magnitude. The results could be made to agree by evaluating w in equation (2) at a suitable lower temperature, but no physical basis exists for selecting this temperature. Figure 5.4 shows the physics that have been overlooked in deriving equation (2). In the reaction zone, the reactant concentration has

---

* Here $\mathcal{O}$ means *terms of order*.

been depleted through diffusion by the factor β^{-1}; therefore, in equation (2) w should be evaluated at β^{-1} times the initial concentration (and at T_∞, since the reaction does indeed occur at the downstream side of the deflagration). Furthermore, because of the strong temperature dependence of w, the reaction proceeds not in a zone of thickness δ, as given by equation (4), but rather in a thinner zone of thickness δ/β. These two factors of β^{-1} produce the β^{-2} in equation (75). It would not be difficult to modify the physical reasoning given in Section 5.1.2 on the basis of considerations derived from a structure like that shown in Figure 5.4 to obtain correct β factors in equation (2). The reader may be interested in doing this to test his or her understanding. There are evident secondary dependences of v_0 on the overall activation energy and on the heat release, appearing through β in equation (75), that cannot be obtained from equation (2) without introducing these modifications.

5.3.7. Generalizations to other flames

Some additional aspects of the deflagration structure are illustrated qualitatively in Figure 5.5 in the physical (x) coordinate. All properties vary monotonically through the wave (except, of course, the reaction rate). The small pressure drop shown is not predicted by the equations in this section (since $p = $ constant here); however, since v increases through the wave, it is clear from equation (5), for example, that p must decrease. The actual pressure profile is easily computed from equation (5) by using the velocity profiles obtained from the equations in this section; the total pressure drop seldom exceeds 1 mm of mercury. Mach numbers are sufficiently small for this fundamentally perturbative approach for obtaining $p(x)$ to be well justified. Increasing the activation energy makes the reaction-rate curve more sharply peaked toward the hot boundary but has relatively little effect on the shapes

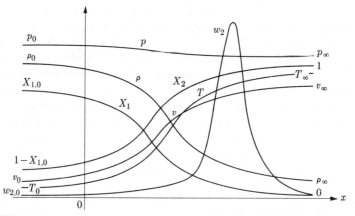

FIGURE 5.5. Representative variation of flow properties in a deflagration wave.

of the other curves. Increasing λ/c_p and ρD_{12}, keeping Le $= 1$, spreads all curves over a larger range of x.

If λ/c_p is increased with ρD_{12} held constant, then Le $\neq 1$, assumption 10 is violated, and the analysis in this section is invalid. However, the way in which the curves in Figure 5.5 will be modified can be inferred from reasoning like that in Section 5.1.2. Since thermal effects alone were considered in deriving equation (1), increasing λ/c_p should spread the T curve (and, consequently, the ρ, v, and p curves) over a somewhat larger x range, while causing less change in the composition curves, provided that there is no appreciable modification in w. If there were a single-zone structure to the flame—that is, if the Zel'dovich number β were not large—then no influence of λ/c_p on w would be anticipated. However, the structure illustrated in Figure 5.4 indicates that for reasonable values of β, there may be a change in the average reactant concentration in the reaction zone. An increase in λ/c_p at fixed ρD_{12} may be expected physically to spread the reaction region of elevated temperature over a larger range of reactant concentration, thereby effectively increasing the average reactant concentration in the reaction zone and hence increasing w.

A quantitative estimate of the magnitude of the effects of relative diffusion of reactant and heat can be obtained on the basis of the structure of the convective-diffusive zone of Figure 5.4. In this zone $\epsilon = 0$, and solutions to equations (19) and (22) are $\tau = e^{\xi}$ and $Y = e^{\text{Le}\,\xi}$, respectively, where the conditions $\tau = Y = 1$ have been imposed at the reaction zone ($\xi = 0$). Since the fuel mass fraction is $Y_1 = Y_{1,0}(1 - Y)$, the expansion of the convective-diffusive solution about $\xi = 0$ is $Y_1 \approx Y_{1,0}\,\text{Le}\,\xi$, which indicates that in the reaction zone, the fuel concentration is proportional to Le. Since w is proportional to the fuel concentration, equation (1) yields $\delta \sim \sqrt{\rho D_{12}}$, independent of λ/c_p, and equation (2) implies that $m \sim (\lambda/c_p)/\sqrt{\rho D_{12}}$. Thus an increase in λ/c_p at constant ρD_{12} increases the burning velocity proportionally but leaves the extent of the T profile in physical space unchanged; the spatial extent of the composition curves is narrowed, proportional to c_p/λ, because of the increase in m at constant ρD_{12}. Increasing ρD_{12} with λ/c_p held constant decreases m (physically by diffusively reducing the reactant concentration in the reaction zone, thereby decreasing w) and consequently increases the spatial extent of the T curve (in proportion to $\sqrt{\rho D_{12}}$) and of the concentration curves [in proportion to $(\rho D_{12})^{3/2}$]. The asymmetry in the effects of variations of λ/c_p and ρD_{12} for large β arises through the influence on the reactant concentration in the reaction zone. A symmetric behavior could be anticipated if the reaction extended throughout the entire thickness of the deflagration.

The type of reasoning that has been given here can be extended to take into account reaction orders different from unity and multiple-reactant, one-step chemistry. For a one-reactant process of reaction order n, w is proportional to the reactant concentration (in the reaction zone) to the nth power, and therefore from equation (1), the extent of the temperature profile is proportional to $\sqrt{(\lambda/c_p)/\mathrm{Le}^n}$; from equation (2) the burning velocity is proportional to $\sqrt{(\lambda/c_p)\mathrm{Le}^n}$; and from the Y solution in the convective-diffusive zone, the extent of the concentration profiles is seen to be proportional to $(\rho D_{12} c_p/\lambda)\sqrt{(\lambda/c_p)\mathrm{Le}^n}$. More precisely, the reactant concentration in the reaction zone is proportional to $\mathrm{Le}\,\beta^{-1}$, so that w is proportional to $\mathrm{Le}^n\beta^{-n}$, and powers of β appear that depend on the reaction order. For $\mathrm{Le} = 1$ a two-term expansion of the burning-rate eigenvalue for a one-reactant, one-step process of order n is [36]*

$$\Lambda = \{\beta^{n+1}/[2\Gamma(n + 1)]\}\{1 + 2\beta^{-1}[\alpha(n + 1)(n + 2)/2 - \gamma] + \mathcal{O}(\beta^{-2}) \quad (76)$$

where Γ denotes the gamma function $[\Gamma(n + 1) \equiv \int_0^\infty y^n e^{-y}\, dy = n!]$ and

$$\gamma = \int_0^\infty \left\{1 - \left[\int_0^y y'^n e^{-y'}\, dy'/\Gamma(n + 1)\right]^{1/2}\right\} dy.$$

Here γ, which depends on n, has been evaluated by numerical integration for various values of n [36] and is shown in Figure 5.6; representative values are $\gamma = 2(1 - \ln 2)$ for $n = 0$, $\gamma = 0.970$ for $n = \frac{1}{2}$, $\gamma = 1.3440$ for $n = 1$, and $\gamma = 2.1274$ for $n = 2$. For $n < 1$, the reaction goes to completion ($y = 0$) at a finite value of η in the reaction zone; for $n > 1$, there is a thick downstream region, where ξ is of order unity, in which a balance between convection, reaction, and diffusion develops at a very low concentration level (of order $\beta^{-(n+1)/(n-1)}$ times the initial concentration). Occurrence of these finer details of the reaction-zone structure is comprehensible qualitatively on the basis of the relative rapidity with which the reaction rate decreases with decreasing reactant concentration at very low reactant concentrations—this rapidity is low for small n and high for large n, as is evident from the functional form y^n. To the lowest order, equation (76) provides a flame-speed formula in which every factor except the constant $2\Gamma(n + 1)$ could have been obtained by physical reasoning based on Figure 5.4, without performing the asymptotic analysis.

There have been a number of analyses employing activation-energy asymptotics for two-reactant, one-step chemistry. Most of these have been summarized by Mitani [40]; reviews also may be found elsewhere [10], [13].

---

* In equation (76) Λ is defined as in equation (45) but with the additional factors $v'_{1,1}(Y_{1,0}\rho/W_1)^{n-1}$ on the right-hand side. This modification is needed because, although $M = 1$ and $N = 2$, assumption 9 and many of its consequences no longer are valid for $n \neq 1$ [for example, with the rate expression given by equation (1-8), equations (16) and (34) are revised]. In the asymptotic analysis to lowest order, approximations corresponding to assumption 11 are unnecessary because the reaction rate is relevant only near the burnt-gas state; it is sufficient to evaluate the quantities in the definition of Λ—for example, in equation (45)—at the hot boundary. At higher orders, expansions of these quantities about their hot-boundary values appear. See B. Rogg and F. A. Williams, *CST*, **42**, 301 (1985).

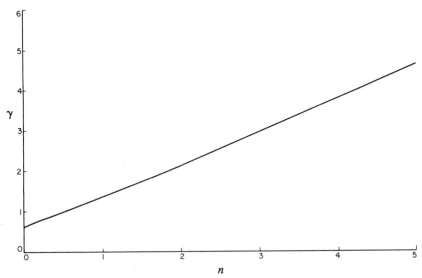

FIGURE 5.6. Dependence of the term γ in the formula for the burning-rate eigenvalue, equation (76), on the reaction order n.

With notable exceptions [41], the formulations have been based not on equation (8) but instead on diffusion equations of the type

$$dY_i/dx = (m/\rho D_i)(Y_i - \epsilon_i), \quad i = 1, \ldots, N-1, \tag{77}$$

where D_i is a diffusion coefficient for species i in the mixture. Equation (77) is derived formally by assuming that an inert species $i = N$ is present in great excess so that the other species effectively diffuse through this inert, whose concentration profile is calculated a posteriori from $\sum_{i=1}^{N} Y_i = 1$. Apparently results based on equation (77) can be surprisingly good [41] if the inert is taken to be an average product.

If $i = 1, 2$ are the two reactants and n_1 and n_2 identify the reaction orders with respect to these reactants, while Le_1 and Le_2 are the corresponding Lewis numbers ($Le_i = \lambda/\rho D_i c_p$), then to the lowest order in β^{-1} the asymptotic analysis yields [40]

$$v_0 = \left[\frac{2\lambda_\infty v'_{1,1}(v'_{2,1}/v'_{1,1})^{n_2} \rho_\infty^{n_1+n_2} B_{1,\infty} T_\infty^{\alpha_1} Y_{1,0}^{n_1+n_2-1}}{\rho_0^2 c_{p,\infty} W_1^{n_1+n_2-1} \beta^{n_1+n_2+1} Le_1^{-n_1} Le_2^{-n_2}} \right]^{1/2}$$
$$\times \sqrt{G(n_1, n_2, a)} e^{-E_1/2R^0 T_\infty}, \tag{78}$$

where

$$G(n_1, n_2, a) = \int_0^\infty y^{n_1}(y+a)^{n_2} e^{-y} \, dy,$$

in which $a = \beta(\phi - 1)/Le_2$, with $\phi = v_1 W_1 Y_{2,0}/v_2 W_2 Y_{1,0}$ being the **equivalence ratio**. Equation (78) holds for $\phi > 1$—that is, if species 2 is present in excess of the stoichiometric requirement or if the mixture is stoichiometric; if it is found that $\phi < 1$, then equation (78) may be employed merely by interchanging the definitions of species 1 and 2 throughout. For stoichiometric conditions ($\phi = 1$, or—more precisely—if $\phi - 1$ is small compared with β^{-1}), we have $G(n_1, n_2, a) = \Gamma(n_1 + n_2 + 1)$. If conditions are sufficiently far from stoichiometry for $\phi - 1$ to be large compared with β^{-1}, then $G(n_1, n_2, a) = a^{n_2}\Gamma(n_1 + 1)$, and equation (78) reduces to

$$v_0 = \left[\frac{2\lambda_\infty v'_{1,1}(v'_{2,1}/v'_{1,1})^{n_2}\rho_\infty^{n_1+n_2}B_{1,\infty} T_\infty^{\alpha_1} Y_{1,0}^{n_1+n_2-1}\Gamma(n_1 + 1)}{\rho_0^2 c_{p,\infty} W_1^{n_1+n_2-1}\beta^{n_1+1} \mathrm{Le}_1^{-n_1}(\phi - 1)^{-n_2}} \right]^{1/2} e^{-E_1/2R^0 T_\infty}.$$

(79)

The principal features of these flame-speed results are readily comprehensible physically from the previous discussion. Consider, in particular, the powers of β and of the Lewis numbers in equation (78). The extra 1 in $n_1 + n_2 + 1$ for β comes from the thinness of the reaction zone in comparison with the flame thickness, δ. The remaining powers for β (n_1 and n_2) and the Lewis numbers come from the rate expression w and reflect the fact that in the reaction zone the concentration of reactant i is of order Le_i/β times its initial concentration because of the concentration reduction through the convective-diffusive zone. For off-stoichiometric conditions, species 2 is not depleted appreciably, and therefore the rate-reduction factor $(\mathrm{Le}_2/\beta)^{n_2}$ does not appear in equation (79).

Formulas like equations (78) and (79) are useful for calculating the flame speed as a function of the equivalence ratio. It must, however, be recognized that many of the quantities appearing in these equations vary with ϕ. Usually the strongest variation occurs in the Arrhenius factor, since $E_1/R^0 T_\infty$ is large, and T_∞ peaks near $\phi = 1$, usually smoothly because of equilibrium dissociation of products. It is also quite significant that in one-step approximations, the overall rate parameters—such as E_1, n_1, and n_2—generally differ in stoichiometric, rich, and lean mixtures (see, for example, [42]). Equations (78) and (79) have the merit of exhibiting explicitly the influences of transport coefficients on flame speeds and of providing explanations for some initially counterintuitive results (for example, that the high diffusion coefficient of H_2 tends to reduce the burning velocity of a fuel-lean H_2–O_2 flame), but these formulas cannot be employed in the absence of appropriate chemical-kinetic information for overall rate parameters.

For one-step reactions the analyses that have been given here for flame structures and flame speeds have successfully removed many of the most restrictive assumptions that were introduced beginning in Section 5.2.2. Assumptions 8–11 have been largely eliminated, with some minor questions remaining, such as influences of variable Lewis numbers. Assumption 7 could be circumvented in an asymptotic analysis by use of the idea that the reaction occurs mainly near the hot boundary, and departures from assumptions 3 and 4 could be investigated readily. The physical effects that have been considered here refer specifically to steady, planar, adiabatic flames (assumptions 1 and 2); additional physical phenomena occur if these assumptions are relaxed (see Sections 8.2 and 9.5). For steady, planar, adiabatic deflagrations, the critical remaining approximation is that of a one-step reaction; it is unclear that much further work directed toward removal of other underlying assumptions would be warranted prior to study of the importance of the one-step approximation. Investigations of multiple-step kinetics are addressed in the following section.

5.4. FLAMES WITH MULTIPLE-STEP CHEMISTRY

5.4.1. Discussion of the background and the nature of the problem

5.4.1.1. Formulation

Let us now consider flames in which more than one chemical reaction occurs. If assumptions 1–7 of Section 5.2 and the isobaric approximation $p = $ constant are retained, then the flame equations may be taken as

$$d\epsilon_i/d\tau = (\lambda/m^2 c_p)(w_i/f), \quad i = 1, \ldots, N, \tag{80}$$

and

$$dX_i/d\tau = (\lambda R^0 T/c_p p)\left\{\sum_{j=1}^{N}(1/D_{ij})[X_i(\epsilon_j/W_j) - X_j(\epsilon_i/W_i)]\right\}\Bigg/f, \quad i = 1, \ldots, N, \tag{81}$$

where

$$f \equiv \left[\sum_{j=1}^{N} h_j^0(\epsilon_j - \epsilon_{j,\infty})/q\right] - (1 - \tau) \tag{82}$$

is a linear function of τ and ϵ_j. Equation (80) is the ratio of equation (7) to equation (13), and equation (81) can be derived from equation (8) by first expressing Y_i in terms of X_i, W_i, and the average molecular weight $\overline{W} = 1/\sum_{i=1}^{N} Y_i/W_i$, [namely, $Y_i = X_i W_i/\overline{W}$; compare equation (1-12)], then using equation (1-9) to eliminate $\overline{W}/\rho$, and finally dividing by equation (13). The independent variable employed here is the normalized dimensionless temperature τ, which is given by equation (40). The $2N$ dependent variables ϵ_i and X_i appear in the $2N$ expressions given in equations (80) and (81) (with pressure constant, the quantities w_i, λ, and so on are known functions of τ and X_i). If an ignition temperature is defined in such a way that $w_i = 0$ for

T below the ignition temperature and w_i is given by equation (1-8) for T above the ignition temperature, then a reasonable set of boundary conditions [for example, $Y_i = \epsilon_i$ at $\tau = 0$ and all ϵ_i are specified at $\tau = 0$, while $Y_i = \epsilon_i$ at $\tau = 1$ and chemical equilibrium exists ($w_i = 0$) at $\tau = 1$] is expected to determine a (unique) solution to equations (80) and (81) only for a particular value of the mass burning rate m.

The normalized temperature τ, given by equation (40), is a convenient independent variable only if it varies monotonically with distance through the flame. Although it is conceivable that nonmonotonic behavior of τ could occur—for example, for flames in which endothermic reactions become dominant at the highest temperatures—no practical examples of adiabatic flames with temperature peaks or valleys are known; the use of τ seems appropriate from the viewpoint of monotonicity. The main limitation to the formulation given is assumption 7 to the effect that $c_{p,i} = c_p = $ constant. This assumption can be removed, with $\tau = (T - T_0)/(T_\infty - T_0)$, merely at the expense of complicating the definition of f in equation (82), namely, $f = \sum_{j=1}^{N} (h_j \epsilon_j - h_{j,\infty} \epsilon_{j,\infty})/c_p(T_\infty - T_0)$, where each h_j is a known function of T and c_p is arbitrary since it cancels from the equations. It is also straightforward to include Soret diffusion by adding appropriate terms to equation (81). These modifications should not introduce appreciable additional complexity if solutions are to be sought by numerical integration; the formulation then becomes quite general, subject only to the steady, planar, adiabatic, and isobaric approximations, for ideal-gas mixtures with Dufour effects neglected.

However, with complicated chemistry the integration of equations (80) and (81) faces computational difficulties, associated not only with "stiffness" (Sections B.2.5.2 and 4.2.1) but also with complications from molecular transport and from eigenvalue aspects. For example, tendencies for stepwise integrations in the direction of increasing τ to diverge from the hot-boundary point motivate integrations starting at the hot boundary, but these are difficult to begin because, with complicated chemistry, formal expansions about the hot boundary demonstrate the necessity of selecting values of more than one parameter for starting a numerical integration, thereby requiring a search in a multidimensional space to satisfy suitable conditions at the cold boundary [43]. This "multiple-eigenvalue" aspect of the problem has led to questionable suggestions to the effect that flammability limits may be associated with the nonexistence of solutions to equations (80) and (81), and numerical integrations of the ordinary differential equations have been completed successfully only by the introduction of special devices [44], [45], [46] that seem better suited to some flames than to others. Thus, notwithstanding the significant advances made in recent years (see Section 5.4.1.3), reliable and general approaches to the accurate calculation of flame structures and flame speeds for arbitrary reaction kinetics do not yet appear to be available.

5.4.1.2. Objectives of Analyses

There are two broad classes of theoretical analyses of steady, planar, adiabatic, isobaric deflagrations with multiple-step reaction kinetics, namely, studies with model kinetics and studies with real kinetics. Studies in the former category employ relatively simple rate expressions designed to retain some essentials of the full chemistry in various flames but not based on accepted rate parameters for any specific flame; the objectives are to develop physical insight, not to obtain predictions for real flames that can be compared with experimental results. Studies in the latter category endeavor to employ correct kinetic data for particular flames to the extent that these data are known; their objectives are to improve knowledge of chemical processes in specific flames and to achieve agreement with experimental results, thereby possibly obtaining better values for poorly known rate parameters of certain elementary steps. The methods employed in studies with model kinetics often are analytical in character, while those for studies with real kinetics usually are numerical.

As indicated in Appendix B, the chemistry of flames generally involves chain reactions. Three specific objectives of studies with real kinetics may be identified. In the order of increasing difficulty, they are to predict the burning velocity, to predict the distributions of temperature and of major species (principally reactants and products), and to predict the distributions of species present in lower concentrations (principally chain carriers). Burning velocities often are influenced primarily by conditions near the hot boundary, as has been seen in the previous section; they depend on overall levels of chain-carrier concentrations but seldom are affected appreciably by distributions of radicals or atoms in the cooler portions of the flame. Concentration profiles of major species usually are controlled mainly by the burning velocity and transport coefficients. Studies have shown that burning velocities, temperature profiles, and concentration profiles of major species often can be obtained with reasonable accuracy even if predictions of chain-carrier distributions are poor. In particular, the burning velocity per se seldom is a sharp discriminant of proper kinetic details. Thus the most challenging problem in laminar-flame theory is to predict the distributions of chain carriers in flames. These predictions are of interest not only because of their relationships to fundamental transport and chemical-kinetic properties of chain carriers but also because observable properties of flames, such as emission of chemiluminescent radiation, are associated with chain-carrier distributions.

5.4.1.3. Literature

Investigations with model kinetics date from the early work of Zel'dovich and Semenov [47]. A number of different model mechanisms have been studied on the basis of analytical approximations designed to provide estimates of flame speeds (for example, [48]–[51]). More recently, the techniques of asymptotic analysis have been brought to bear on simple mechanisms, to search for the full range of possible behaviors, and to calculate flame speeds and concentration and temperature profiles. Mechanisms that have been studied include the sequential reaction A → B → C, [52]–[56]* and the competing reaction A → B, A → C, [57], [58]. The two-reactant, sequential process A + βB → αC, C + γB → δD has been considered for conditions under which species B is present in excess [59]; here α, β, γ, and δ are stoichiometric coefficients. Related two-reactant mechanisms also have been investigated [60], [61]. Additional mechanisms that have received consideration are the straight-chain model [10] A + B → 2C + B, A + C → D + C, in which A functions as a reactant, B as an inert, C as a chain carrier and D as a product; the mechanism A + B → 2C (at a negligible rate), A + C → 2C, and A + C → D + C (or 2C → D), which models a branched-chain reaction with carrier C [10], [34], [62]; and also related processes with varying extents of chain branching.

A brief consideration of the mechanism A → B → C may serve to illustrate the range of possibilities that occur. Let α_k and β_k represent non-dimensional heat releases and activation energies, respectively, for the first and second steps ($k = 1, 2$), and let δ be a measure of the ratio of the rate of the second step to that of the first (since the discussion is purely qualitative, these quantities need not be defined precisely here). For there to be a positive net heat release, $\alpha_1 + \alpha_2$ must be positive, and β_1 is large in all problems of interest. Most of the work has treated both β's as being large and both α's as being positive; in this case, if δ is sufficiently small, then two separate flames may occur, the first, A → B, traveling faster than the second, B → C. If δ is larger, then there usually is a single flame with two reaction zones separated by a transport zone, but these two reaction zones merge at sufficiently large values of δ. If $\beta_2 = 0$ with α_1 positive, then at sufficiently small values of δ, there is a thick convective-reactive zone for B → C, downstream from the A → B flame; however, as δ becomes sufficiently large the step B → C begins to occur in the reactive-diffusive zone of the flame and to influence the flame speed (it may be anticipated that even if $\beta_2 > 0$, there may be a range of values of β_2 and δ for which B → C occurs in a downstream region having a convective-reactive balance). If $\alpha_1 \leq 0$ and $\beta_2 = 0$ (a case that may model processes with highly reactive reaction intermediaries), then no matter how small δ is, the process A → B cannot proceed without energy input from B → C, and generally both steps occur in the same thin reactive-diffusive zone. When $\alpha_1 \leq 0$ and β_2 is large, in addition to this structure there appears to be the possibility (for sufficiently small values of δ) of a thin, endothermic, reactive-diffusive zone for A → B, embedded within the convective-diffusive

---

* Numerical methods rather than asymptotic analyses were employed in the first of these references.

zone that precedes the exothermic reactive-diffusive zone of $B \to C$. Thus it is seen that even for the simple example of $A \to B \to C$, many different flame structures may be encountered. For the competing reaction, $A \to B, A \to C$, there is even a range of parameters in which two stable steady-state solutions for the flame structure exist, with different flame temperatures, different flame speeds, and different percentages of reactant converted to B and C [58]; the ignition procedure may be presumed to select one of the multiplicity of solutions.

Investigations with real kinetics have employed expansions about the hot boundary for the purpose of calculating burning velocities [33] and various numerical methods of integration to obtain burning velocities and profiles of temperature and of species concentrations [30], [43]–[46], [63]–[87]*. The approaches to numerical integration fall into two broad categories, namely, the integration of ordinary differential equations that describe the steady flame and the integration of one-dimensional, time-dependent, partial differential equations that reduce to the equations for the steady flame when the transient terms vanish. The majority of the recent investigations [67]–[87] employ the latter technique; the earliest studies [63]–[66] and a small fraction of the more recent work [43]–[46] involve the former.

Although solving time-dependent partial differential equations numerically usually is thought to be more complicated than solving ordinary differential equations, the reverse is true for the laminar-flame problem with complex chemical kinetics because of the eigenvalue character of the problem. With the availability of large computers, the time-dependent approach offers many advantages. Standard routines for the integration of systems of parabolic equations may be employed, with known boundary conditions upstream and downstream from the flame and with guessed initial profiles of temperature and concentrations, selected to approximate those of the steadily propagating flame. Experience has shown that by marching forward in time, the solutions usually converge rapidly to a description of the steady-state flame. In principle, if the flame is unstable to one-dimensional disturbances, then the solutions will reveal the instability; there are examples in which this has occurred (for example, [88]). The time-dependent approach tends to establish the steadily propagating flame in the same way that nature does.

However, numerical integration of the ordinary differential equations also possesses advantages. Foremost among these are the inherently greater accuracies of the solutions, when they can be obtained; one-dimensional

---

* Concerning the work employing numerical methods, it should be emphasized that the extensive citation of literature given here is not complete and omits, in particular, many relevant papers co-authored by Dixon-Lewis; the last two references are general reviews that mention flame-structure studies by numerical methods.

programs achieve finer intervals in space than two-dimensional programs.*
In principle, this can be a significant benefit if different reactions are of
importance in different narrow zones of the flame. In addition, the ordinary
differential equations can provide multiple steady-state solutions when
they exist (a relatively difficult task for time-dependent methods) and
unstable solutions as well, which could aid in the understanding of flame
structures. Thus both classes of the numerical approaches have attractive
attributes.

Successful integrations of the ordinary differential equations have
been obtained for relatively few flames. Only the earliest work [43], [44],
[63]-[66] employed τ as the independent variable, using a formulation like
that in equations (80) and (81); later studies adopted instead a variable
related to ξ. In all these investigations, the integration was started first from
the hot boundary, a procedure that achieves simplification with one-step
chemistry and often yields good approximations to the flame speed at early
stages in the computation. However, with complex chemistry, these integrations generally exhibit saddlelike divergences as τ is decreased as a
consequence of the varying kinetics [44]; thus the numerical solutions tend
to move away from the correct integral curve whether the integration is
performed in the direction of increasing or decreasing τ. Criteria therefore
must be developed for modifying parameters in a manner designed to overcome these divergences [44]. In view of these difficulties, there may be merit
in trying to integrate first from the cold boundary with an ignition-temperature approximation, using a modern routine for "stiff" equations; the
advantage would be that—irrespective of the complexity of the kinetics—
only one parameter, the burning rate, would be available for adjustment, and
criteria similar to those previously developed for identifying divergent
solutions might be applicable for locating the correct integral curve. A
successful technique of this type would enable a pseudostationary value of
the burning velocity to be identified and the corresponding temperature and
concentration profiles to be calculated with good accuracy. From this
discussion it should be evident that further research on the numerical
integration of the laminar-flame equations seems warranted.

Many important conclusions concerning chemical kinetics in specific
flames have been obtained in the references cited (for example, [63]-[85])
and in related work. Since these are too numerous to be discussed here, the
reader is referred to the references.

5.4.1.4. Burning-velocity calculations

It has been seen from the viewpoint of asymptotic analysis that when
a one-step approximation is applicable for the chemical kinetics, the burning

---

* In this respect, there are one-dimensional methods involving collocation that are
better classified as two-dimensional methods.

velocity is determined by matching the solution in a downstream reactive-diffusive zone at the hot boundary to the inert solution in the upstream convective-diffusive zone. This conclusion remains true for multiple-step kinetics if the sequence of reactions can be reduced to an effective one-step process through introduction of a suitable combination of steady-state and partial-equilibrium approximations, as described in Section B.2.5.2. Throughout the convective-diffusive zone, the flux fractions ϵ_i remain equal to their initial values, $\epsilon_{i,0}$, and when the generally weak dependence of λ on composition is neglected, equation (81) (or the generalization thereof that eliminates assumption 7) reduces to a system of linear ordinary differential equations for $X_i(\tau)$ that may be solved subject to the known final equilibrium values $X_{i,\infty}$ at $\tau = 1$. If D_{ij} may be assumed to have the same temperature dependence for each pair of species, then by writing $D_{ij} = D_{ij,0}g(T)$ and $\lambda = \lambda_0 h(T)$, where $D_{ij,0}$ and λ_0 are constants and the functions g and h are nondimensional, and by defining $\text{Le}_{ij,0} = \lambda_0(T_\infty - T_0)/pD_{ij,0}$, one may derive a linear system with constant coefficients from equation (81):

$$\left.\begin{array}{l} d(X_i - X_{i,0})/d\theta = \sum_{j=1}^{N} C_{ij}(X_j - X_{j,0}), \quad i = 1, \ldots, N, \\[2mm] X_i - X_{i,0} = X_{i,\infty} - X_{i,0} \quad \text{at} \quad \theta = 0, \quad i = 1, \ldots, N, \end{array}\right\} \quad (83)$$

where $X_{i,0} = \epsilon_{i,0}\overline{W}_0/W_i$ [in which $\overline{W}_0 = (\sum_{j=1}^{N}\epsilon_{j,0}/W_j)^{-1}$ is the initial average molecular weight],

$$C_{ij} = \begin{cases} X_{i,0}\,\text{Le}_{ij,0}, & i \neq j \\[2mm] -\sum_{j=1,j\neq i}^{N} X_{j,0}\,\text{Le}_{ij,0}, & i = j \end{cases}$$

and

$$\theta = \int_{\tau}^{1} \frac{h(T)R^0 T/\overline{W}_0}{g(T)\sum_{j=1}^{N}(h_j\epsilon_{j,0} - h_{j,\infty}\epsilon_{j,\infty})}\, d\tau.$$

The problem in equation (83) is well-posed and solvable by matrix methods, with all eigenvalues being negative, and the derivatives $dX_i/d\tau$ at $\tau = 1$ are readily computed from equation (83). These derivatives provide linear relationships between X_i and τ in the reaction zone that enable the rates w_i there to be expressed explicitly in terms of τ. A progress variable ϵ for the effective one-step process may be defined, which changes from 0 to 1 across the reaction zone, and an equation of the form $d\epsilon/d\tau = \Lambda\omega(\tau)$ may be derived from equation (80), where Λ is inversely proportional to m^2 and $\omega(\tau)$ is known. Then

$$\Lambda = 1 \Big/ \int_{0}^{1} \omega(\tau)\, d\tau \qquad (84)$$

provides the first approximation to the burning-rate eigenvalue after a suitable expansion for $\omega(\tau)$, paralleling that of Section 5.3.6, has been introduced. In addition to giving the burning rate, this procedure may be extended to provide the concentration and temperature profiles from the solution to equation (83) and the integral of equation (13). The function $\omega(\tau)$ may be sufficiently complicated to necessitate a numerical evaluation of the integral for obtaining Λ.

This procedure fails if an insufficient number of justifiable steady-state and partial-equilibrium approximations is available. This situation often is encountered in real flames and is largely responsible for their individual structural features. Suitable knowledge of the kinetics may still enable asymptotic methods to be employed for calculating burning velocities and some structural aspects, even if the flame speed is not controlled by conditions in the hottest part of the flame. For example, three-body recombination processes may be sufficiently slow to occur in a downstream convective-reactive zone that does not influence the flame speed, and the upstream structure and propagation velocity may be calculated first by neglecting these processes; then the recombination region may be analyzed afterward. This situation appears to prevail for the ozone decomposition flame under many conditions. Particular species may be produced and consumed in relatively narrow upstream reaction zones in such a manner that they exert a minimal effect on the burning velocity and can be ignored at first and then calculated later in an analysis specifically addressing their reaction zone. This may occur under appropriate conditions for bromine molecules in hydrogen-oxygen flames to which HBr has been added. There may be a single reaction zone with a reactive-diffusive character at the hot boundary, but within which more than one independent reaction must be considered because of inapplicability of steady-state types of approximations. This necessitates development of special analyses for the structure of the reactive-diffusive zone. These and other unique characteristics of individual flames may motivate future theoretical studies. In the rest of this chapter we shall be concerned primarily with methods for testing the validity of the steady-state approximation for chain carriers.

5.4.2. The extended steady-state approximation

When the steady-state approximation of Section B.2.5.2 is applied to nonisothermal systems involving molecular transport, it is sometimes referred to as the **extended** steady-state approximation. The simplifications that have just been indicated to follow from the application of this approximation underscore the importance of having methods for ascertaining the validity of the approximation. Criteria for the applicability of the extended steady-state approximation have been developed by Giddings and Hirschfelder [64] and improved by Millán and Da Riva [89]. The discussion here largely follows [89]. Factors that may contribute to inapplicability of the steady-state approximation in flames [89], [90] are (1) too short a residence time in the flame for the chain-carrier concentration to reach its steady-state value and (2) the diffusion of chain carriers away from regions of maximum chain-carrier concentration.

A statement of the extended steady-state approximation is that throughout the flame, the net mass rate of production of any reaction intermediary r is negligibly small,

$$w_r = 0. \tag{85}$$

The integral of equation (80) for $i = r$ is then $\epsilon_r = 0$ because the initial (as well as final) chain-carrier concentration is zero. Since the production terms w_r are functions of τ and of the mole fractions X_i, relations such as equation (85) permit explicit representation of X_r in terms of τ and the remaining X_i. Thus if equation (85) is valid for all chain carriers, the mole fractions of all these reaction intermediaries may be eliminated from the flame equations, and the flux fractions of all of these species are zero. Since stoichiometry conditions relate the remaining ϵ_i, only one independent expression remains in equation (80), and the problem is reduced to that of a one-step reaction. The flame equations may then be solved explicitly to give all the mole fractions, including the X_r, in terms of τ.

5.4.3. Conservation equations for reaction intermediaries

For the sake of simplicity, consideration will be restricted to systems in which only one intermediary exists; this will be denoted by the subscript r, and the other species will be denoted by the subscripts j ($j = 1, \ldots, N - 1$). The discussion and results are applicable qualitatively for a given chain carrier in flames containing any number of chain carriers.

Since reaction intermediaries are present in low concentrations, equation (81) may be simplified for species r. It can be shown [91] that when $X_r \ll 1$, if the diffusion velocities are the same for all species other than r, then equation (81) reduces to

$$dX_r/d\tau = (\lambda R^0 T/c_p p \overline{W} D_r)[X_r - (\overline{W}/W_r)\epsilon_r]/f, \tag{86}$$

where the diffusion coefficient for species r in the multicomponent mixture (D_r) has been defined by

$$1/D_r \equiv \sum_{j=1}^{N-1} (X_j/D_{rj}).$$

Equation (86) may be taken as qualitatively true in flames even though the diffusion velocities of major species differ. For species r, equation (80) is

$$d\epsilon_r/d\tau = (\lambda/m^2 c_p)(w_r/f). \tag{87}$$

The sum in the quantity f [equation (82)] appearing in the denominators of equations (86) and (87) now extends only over the $N - 1$ species other than r because $X_r \ll 1$ implies that the term corresponding to $j = r$ will be small. Hence f depends only on τ and ϵ_j for $j \neq r$.

It is reasonable to define the Lewis number for the radical r as

$$\text{Le}_r \equiv \lambda R^0 T / c_p p \overline{W} D_r = \lambda / \rho D_r c_p,$$

which may depend on the composition (excluding X_r since $X_r \ll 1$) and τ. A useful form for equations (86) and (87) may then be obtained by multiplying equation (86) by $(f/\text{Le}_r)(W_r/\overline{W})$ and then differentiating the resulting equation and substituting for $d\epsilon_r/d\tau$ from equation (87). The result, in which ϵ_r does not appear, is

$$\mathscr{D}(X_r) \equiv f \frac{d}{d\tau}\left(\frac{f}{\text{Le}_r} \frac{W_r}{\overline{W}} \frac{dX_r}{d\tau}\right) - f \frac{d}{d\tau}\left(\frac{W_r}{\overline{W}} X_r\right) = -\left(\frac{\lambda}{m^2 c_p}\right) w_r, \quad (88)$$

a study of which will yield a criterion for the applicability of the extended steady-state approximation.*

5.4.4. A criterion for the applicability of the extended steady-state approximation

For illustrative purposes let us assume that chain breaking occurs through a process that is of the first order with respect to species r. Then w_r is of the form

$$w_r = b(a - X_r), \quad (89)$$

where a and b may depend on τ and all X_j except X_r. Substituting equation (89) into equation (88) yields

$$\mathscr{D}(X_r) = -\Omega(a - X_r), \quad (90)$$

in which

$$\Omega \equiv \lambda b / m^2 c_p \quad (91)$$

is a dimensionless function of τ and the X_j's excluding X_r. Since the steady-state approximation ($w_r \approx 0$) implies that

$$X_r \approx a \quad (92)$$

---

* It will be noted that the diffusion coefficient of the radical affects only the first term of the operator $\mathscr{D}$ in equation (88). This term, which goes to zero as $D_r \to 0$, produces the primary influence of diffusion upon the radical concentration profile. The three terms in equation (88) represent diffusion, convection, and reaction for the chain carriers. These same three physical effects must appear irrespective of the formulation and would also arise in testing criteria for partial equilibria.

from equation (89), equation (90) shows that this approximation will be valid if

$$|\mathscr{D}(X_r)|/\Omega \ll X_r. \tag{93}$$

Equation (93) will be satisfied if Ω is sufficiently large; hence a criterion for the applicability of the steady-state approximation is that Ω be large. Exactly how large Ω must be is, of course, not determined by these approximate considerations; according to the discussion in Section 5.4.1, we expect that accurate predictions of chain-carrier distributions will require a larger Ω than will accurate predictions of burning velocities.

The physical significance of the criterion in equation (93) may be seen by noting that equation (91) implies that

$$\Omega = (\delta/v)/(\rho/b), \tag{94}$$

where $\delta \equiv \lambda/mc_p$ is a characteristic flame thickness [see equation (4)]. It follows that $\delta/v \equiv t_{res}$ is a characteristic residence time in the flame, and [from equation (89)], $\rho/b \equiv t_{ch}$ is a characteristic chemical time for the destruction of chain carriers. Hence

$$\Omega = t_{res}/t_{ch}, \tag{95}$$

and the steady-state approximation will be valid if $t_{res} \gg t_{ch}$. The time ratio in equation (95) was first introduced by Damköhler [92], [93] in an entirely different context and is referred to as **Damköhler's first similarity group**.

The result given in equation (95) is easily shown to be independent of the special assumption introduced with regard to the order of the reaction causing removal of the chain carriers, and other forms for w_r, such as $w_r = b(aX_r - X_r^2)$ and $w_r = b(a - X_r^2)$, also lead essentially to equation (93).

5.4.5. Methods of analysis for testing steady-state approximations

Since the steady-state approximation applies if Ω is sufficiently large, one approach to solving equation (88) for small deviations from the steady state is to assume that

$$X_r = \sum_{n=0}^{\infty} X_{r,n}/\overline{\Omega}^n, \tag{96}$$

where $\overline{\Omega}$ is a constant representative of the magnitude of Ω. The value of $\overline{\Omega}$ may be chosen in such a way that

$$\Omega \equiv \overline{\Omega}b', \tag{97}$$

where $b' \leq 1$ and b' expresses the variation of Ω through the flame. Substituting equations (96) and (97) into equation (90) and collecting terms involving the same powers of $\overline{\Omega}$ gives

$$X_{r,0} = a \tag{98}$$

for the steady-state distribution and

$$X_{r,n} = \mathscr{D}(X_{r,n-1})/b', \quad n = 1, 2, \ldots, \tag{99}$$

for small deviations from the steady state. Thus the flame equations are first solved with $X_r = a(\tau, X_j)$, the operator $\mathscr{D}$ is computed, and $b'(\tau)$ and $\mathscr{D}(a)$ are evaluated; $X_{r,1}(\tau)$ is then computed from equation (99) and the flame equations are again solved with $X_r = a(\tau, X_j) + X_{r,1}(\tau)/\overline{\Omega}$. This process is continued until sufficient convergence is obtained.

Special situations exist for which this procedure simplifies considerably. If the intermediary under consideration is not a chain carrier but is merely produced and consumed through unimportant side reactions, then the burning velocity and the composition profiles of all other species in the flame are virtually unaffected by the presence of this intermediary. The structure of the flame (excluding the X_r profile) can therefore be determined completely by setting $X_r = 0$ in the flame equations. After this structure is determined, a, b', and the coefficients of the linear differential operator $\mathscr{D}(X_r)$ are known functions of τ. Therefore, equation (90) reduces to a linear nonhomogeneous differential equation with known variable coefficients, and in the solution procedure described above $\mathscr{D}$ and b' need not be recomputed (through the flame equations) in each step because they remain unchanged. For very small deviations from the steady state, it is often reasonable to omit the recomputation of $\mathscr{D}$ and b' in successive steps even for chain carriers.

As the hot boundary is approached ($\tau \to 1$), f goes to zero, causing $\mathscr{D}(X_r)$ to approach zero [see equation (88)]. Since equation (90) then implies that $X_{r,n}$ approaches zero at the hot boundary for $n \geq 1$, the series given in equation (96) should converge rapidly as the adiabatic flame temperature is approached. On the other hand, near the cold boundary b' is small and, therefore, the series tends to converge slowly upstream [see equation (99)].

The approach that has been described here is an example of a perturbation method for large Damköhler numbers and may be termed **Damköhler-number asymptotics**. It has been developed on the basis of an expansion that does not distinguish among special zones within the flame. It is possible that the Damköhler-number expansion will often be good in the hot reaction zone but poor elsewhere, while radical distributions away from the hot reaction zone have relatively little influence on the main characteristics of the flame. Under these circumstances, an approach based on matched asymptotic expansions, treating different zones differently, may be helpful. Sharper definitions of values of $\overline{\Omega}$ consistent with the steady-state approximation (in the zone where it is applied) might thereby be developed.

On the basis of the approach defined here, minimum values of $\overline{\Omega}$ consistent with an acceptable steady-state approximation may be estimated by comparing $X_{r,1}(\tau)/\overline{\Omega}$ with $a(\tau, X_j)$; the former must be sufficiently small. An alternative approach to the estimation of minimum acceptable values of $\overline{\Omega}$ is to introduce simplified functions $f(\tau)$, $a(\tau)$, $b'(\tau)$, $Le_r(\tau)$, and $W_r/\overline{W}$ that enable equation (88) to be solved analytically [89]. Comparison of the solutions for finite values of $\overline{\Omega}$ with those for $\overline{\Omega} = \infty$ then provides an indication of how small $\overline{\Omega}$ can be if the steady state is to be reasonable. In this manner, $\overline{\Omega} \geq 100$ has been obtained as an approximate criterion [89].

5.4.6. Observations on theories of flame structure

It would be of interest to employ results like those in the preceding section to ascertain the applicability of the steady-state approximation in specific flames. This has been done by different investigators, often with conflicting conclusions. The primary reason for the differences is uncertainty in values of reaction-rate parameters. Key specific reaction-rate constants sometimes are uncertain by an order of magnitude at representative flame temperatures. Better rate-constant data are needed to aid in the application of the methods discussed herein to specific flames.

In addition, the validity of simplifying approximations can vary strongly with the initial pressure, temperature, and composition of the reactant mixture. Therefore, these conditions must be specified in stating whether the steady-state assumption is applicable. For the ozone decomposition flame in an ozone-oxygen mixture initially at room temperature and atmospheric pressure, it appears that the steady-state approximation for oxygen atoms (with the mechanism $O_3 + M \to O_2 + O + M$, $O_3 + O \to 2O_2$) is reasonably accurate at sufficiently low initial concentrations of ozone (between 20% and 30% by volume, for example) but inaccurate at high ozone concentrations (above 30%). For the hydrogen-bromine flame, there appears to be a wide range of conditions under which the steady-state approximation for hydrogen atoms is excellent; the validity of the steady-state approximation for bromine atoms in this flame is less clear and is likely to be significantly dependent on initial conditions.

Few flames have rate parameters known as accurately as those for the hydrogen-oxygen flame. This flame affords a good testing ground for theoretical methods of prediction. Numerical integrations have been applied successfully to the hydrogen-oxygen flame. Because of the complexity of the chemical kinetics, analytical approaches have been less successful. For example, there are conditions under which introduction of the steady-state approximation for the intermediaries O, OH, and H leads to reasonable agreement between calculated and measured burning velocities, even though estimates indicate that the steady state is inapplicable, at least for the last two of these chain carriers [42].

There is motivation to pursue analytical methods based on asymptotic analysis for the hydrogen-oxygen flame, in spite of the availability of numerical methods. Aside from the fact that successful analytical methods provide formulas for the burning velocity and for temperature and concentration profiles, asymptotic methods may reveal interesting intricacies of the flame structure, such as combinations of reaction steps to form different overall effective reactions in different parts of the flame, that may be hidden by purely numerical results. Activation-energy asymptotics currently comprise the most promising analytical approaches for studying the hydrogen-oxygen flame and other flames. For the hydrogen-oxygen system, it appears that flame-structure simplifications appropriate to activation-energy asymptotics may vary appreciably with initial conditions; there may be many different hydrogen-oxygen flames. In general, it may be anticipated that in the future, asymptotic analysis may be restricted less to model kinetics and applied more to real kinetics.

The complexities of the hydrogen-oxygen flame underscore uncertainties associated with more complex systems, such as hydrocarbon-air flames. It is possible that, although the hydrogen-oxygen chain is involved in the kinetics of hydrocarbon combustion, the overall rates may be modified in such a way that fewer variations in flame structure occur as initial conditions are varied. A single primary reactive-diffusive zone with a rate that can be approximated as a one-step process may control the flame speed for typical hydrocarbon-air flames. Measured burning velocities are readily correlated by expressions like equation (78). However, recent numerical integrations for methane-air flames, for methanol-air flames, and for other flames with full kinetics (to the extent known at present) show numerous species participating in many reactions [86]. Identification of useful, fundamentally justified simplifications applicable for these more complex fuels is a continuing, worthwhile topic of research.

There are important chemical processes involving species present in low concentrations that exert little or no influence on the flame speed and on other aspects of the flame structure. Notable among these are the production of oxides of nitrogen in fuel-air combustion and the production of solid carbonaceous materials (soot) in fuel-rich hydrocarbon combustion. To the extent that these trace species do not influence the flame propagation, they may be ignored at first, and their production rates may be calculated after the rest of the flame structure has been obtained. Often this a posteriori calculation can be simplified by use of activation-energy asymptotics. This is especially true for the production of oxides of nitrogen by the Zel'dovich mechanism (Section B.2.5.5) because of the strong temperature dependence of the rate. So many elementary steps are involved in soot production in flames that accurate calculations of soot concentrations are difficult whether numerical or analytical methods are employed; simplified overall kinetic expressions for soot processes appear to be needed.

REFERENCES

1. E. Mallard and H. L. le Chatelier, *Ann. Mines* **4**, 379 (1883).
2. V. A. Mikhel'son, "On the Normal Ignition Rate of Fulminating Gas Mixtures," Ph.D. Thesis, Univ. Moscow (1889); see *Collected Works*, vol. 1, Moscow: Novyi Agronom Press, 1930.
3. L. D. Landau and E. M. Lifshitz, *Fluid Mechanics*, New York: Pergamon Press, 1959, Chapter XIV, pp. 474–476.
4. A. S. Predvoditelev, *Int. J. Heat Mass Transfer* **5**, 435 (1962).
5. J. Taffanel, *Compt. Rend. Acad. Sci., Paris* **157**, 714 (1913); **158**, 42 (1914).
6. P. J. Daniell, *Proc. Roy. Soc. London* **126A**, 393 (1930).
7. Y. B. Zel'dovich and D. A. Frank-Kamenetskii, *Zhur. Fiz. Khim.* **12**, 100 (1938).
8. S. S. Penner and F. A. Williams, *Astronautica Acta* **7**, 171 (1961).
9. W. B. Bush and F. E. Fendell, *CST* **1**, 421 (1970).
10. Y. B. Zel'dovich, G. I. Barenblatt, V. B. Librovich, and G. M. Makhviladze, *The Mathematical Theory of Combustion and Explosion*, Moscow: Nauka, 1980.
11. B. Lewis and G. von Elbe, *Combustion, Flames and Explosions of Gases*, 1st ed., New York: Academic Press, 1951, 213–220, 381–400.
12. S. S. Penner, *Chemistry Problems in Jet Propulsion*, London: Pergamon Press, 1957, Chapter XXIV.
13. J. D. Buckmaster and G. S. S. Ludford, *Theory of Laminar Flames*, Cambridge: Cambridge University Press, 1982.
14. J. O. Hirschfelder, C. F. Curtiss, and R. B. Bird, *The Molecular Theory of Gases and Liquids*, New York: Wiley, 1954, Chapter 11.
15. D. B. Spalding, "Contribution to the Theory of the Structure of Detonation Waves in Gases," report prepared under *N.A.S.A. Grant No. NSG-10-59*, University of California, Berkeley, July 1960; see also *9th Symp.* (1963), 417–423.
16. H. W. Emmons, *Harvard University Report* (1950).
17. Th. von Kármán and G. Millán, *Anniversary Volume on Applied Mechanics Dedicated to C. B. Biezeno*, N. V. de Techniche Uitgeverji H. Stam., Haarlem (Holland) (1953), 55–69.
18. J. O. Hirschfelder, C. F. Curtiss, and D. E. Campbell, *4th Symp.* (1953), 190–210.
19. R. Friedman and E. Burke, *J. Chem. Phys.* **21**, 710 (1953).
20. A. N. Kolmogorov, I. G. Petrovskii, and N. S. Piskunov, "Study of the diffusion equation with a source term and its application to a biological problem," *Bulletin of the State University of Moscow*, International series, vol. 1, Section A, 1–25 (1937).
21. W. E. Johnson and W. Nachbar, *Arch. Rational Mech. Anal.* **12**, 58 (1963).
22. W. E. Johnson, *Arch. Rational Mech. Anal.* **13**, 46 (1963).
23. D. B. Spalding, *C & F* **1**, 287, 296 (1957).
24. G. Rosen, *Jet Propulsion* **28**, 839 (1958).
25. G. Rosen, *7th Symp.* (1959), 339–341; *J. Chem. Phys.* **32**, 311 (1960).
26. D. Layzer, *J. Chem. Phys.* **22**, 222 (1954).
27. G. Birkhoff and G. C. Rota, *Ordinary Differential Equations*, New York: Wiley, 1959.
28. Y. B. Zel'dovich, *C & F* **39**, 219 (1980).
29. W. E. Johnson and W. Nachbar, *8th Symp.* (1962), 678–689.
30. G. Klein, University of Wisconsin Naval Research Laboratory Repts. (Sept. 1954, Feb. 1955, April 1955); *Phil. Trans. Roy. Soc. London* **249A**, 389 (1957).

31. J. M. de Sendagorta, *C & F* **5**, 305 (1961).
32. Th. von Kármán, *6th Symp.* (1957), 1–11.
33. Th. von Kármán and S. S. Penner, *Selected Combustion Problems, Fundamentals and Aeronautical Applications*, AGARD, London: Butterworths Scientific Publications, 1954, 5–41.
34. Y. B. Zel'dovich, *Zhur. Fiz. Khim.* **22**, 27 (1948), English translation, NACA Tech. Memo. No. 1282 (1951).
35. J. D. Cole, *Perturbation Methods in Applied Mathematics*, Waltham, Mass.: Blaisdell Publ. Co., 1968.
36. F. A. Williams, "A Review of Some Theoretical Considerations of Turbulent Flame Structure," in AGARD Conference Proceedings No. 164, *Analytical and Numerical Methods for Investigation of Flow Fields with Chemical Reactions, Especially Related to Combustion*, M. Barrère, ed., AGARD, NATO, Paris (1975), II 1–1 to II 1–25.
37. M. Van Dyke, *Perturbation Methods in Fluid Mechanics*, New York: Academic Press, 1964.
38. F. A. Williams, "Asymptotic Methods in Deflagrations," *Colloque International Berthelot-Vieille-Mallard-le Chatelier, First Specialists Meeting (International) of The Combustion Institute*, Pittsburgh: The Combustion Institute, 1981, LXX–LXXIV.
39. G. Joulin and P. Clavin, *C & F* **35**, 139 (1979).
40. T. Mitani, *CST* **21**, 175 (1980).
41. A. K. Sen and G. S. S. Ludford, *18th Symp.* (1981), 417–424.
42. T. Mitani and F. A. Williams, *C & F* **39**, 169 (1980).
43. E. S. Campbell, F. J. Heinen, and L. M. Schalit, *9th Symp.* (1963), 72–80.
44. E. S. Campbell, *C & F* **9**, 43 (1965).
45. K. A. Wilde, *C & F* **18**, 43 (1972).
46. G. Dixon-Lewis, J. B. Greenberg, and F. A. Goldsworthy, *Proc. Roy. Soc. London* **346A**, 261 (1975).
47. Y. B. Zel'dovich and N. N. Semenov, *Zhur. Eksp. Teor. Fiz.* **10**, 1116, 1427 (1940).
48. L. A. Lovachef, *Dokl. Akad. Nauk S.S.S.R.* **120**, 1287, **123**, 501 (1958); **124**, 1271, **125**, 129, **128**, 995 (1959); **131**, 876 (1960).
49. L. A. Lovachef, *C & F* **4**, 357 (1960).
50. J. Adler and D. B. Spalding, *C & F* **5**, 123 (1961).
51. L. A. Lovachef, *8th Symp.* (1962), 411–417, 418–426.
52. H. F. Korman, *CST* **2**, 149 (1970).
53. V. S. Berman and Y. S. Ryazantsev, *Prikl. Mat. Mekh.* **37**, 1049 (1973).
54. V. S. Berman and Y. S. Ryazantsev, *Zhur. Prikl. Mekh. Tekhn. Fiz.*, no. 1, 75 (1973).
55. G. Joulin and P. Clavin, *C & F* **25**, 389 (1975).
56. A. K. Kapila and G. S. S. Ludford, *C & F* **29**, 167 (1977).
57. V. S. Berman and Y. S. Ryazantsev, *Prikl. Mat. Mekh.* **39**, 306 (1975).
58. B. I. Khaikin and S. I. Khudyaev, *Dokl. Akad. Nauk. SSSR* **245**, 155 (1979).
59. S. B. Margolis and B. J. Matkowsky, *CST* **27**, 193 (1982).
60. S. B. Margolis, *CST* **28**, 107 (1982).
61. S. B. Margolis and B. J. Matkowsky, *SIAM J. Appl. Math.* **42**, 982 (1982).
62. K. Seshadri and N. Peters, *CST* **33**, 35 (1983).
63. J. O. Hirschfelder, C. F. Curtiss, and D. E. Campbell, *J. Phys. Chem.* **57**, 403 (1953).

64. J. C. Giddings and J. O. Hirschfelder, *6th Symp.* (1957), 199–212; *J. Phys. Chem.* **61**, 738 (1957).

65. E. S. Campbell, *6th Symp.* (1957), 213–221.

66. E. S. Campbell, J. O. Hirschfelder, and L. M. Schalit, *7th Symp.* (1959), 332–338.

67. D. B. Spalding, *Phil. Trans. Roy. Soc. London* **249A**, 1 (1957).

68. Y. B. Zel'dovich and G. I. Barenblatt, *C & F* **3**, 61 (1959).

69. G. K. Adams and G. B. Cook, *C & F* **4**, 9 (1960).

70. G. Dixon-Lewis, *Proc. Roy. Soc. London* **298A**, 495 (1967).

71. M. J. Day, D. V. Stamp, K. Thompson, and G. Dixon-Lewis, *13th Symp.* (1971), 705–721.

72. D. B. Spalding, P. L. Stephenson, and R. G. Taylor, *C & F* **17**, 55 (1971).

73. D. B. Spalding and P. L. Stephenson, *Proc. Roy. Soc. London* **324A**, 315 (1971).

74. P. L. Stephenson and R. G. Taylor, *C & F* **20**, 231 (1973).

75. L. Bledjian, *C & F* **20**, 5 (1973).

76. L. D. Smoot, W. C. Hecker, and G. A. Williams, *C & F* **26**, 323 (1976).

77. G. Dixon-Lewis and R. J. Simpson, *16th Symp.* (1977), 1111–1119.

78. G. Tsatsaronis, *C & F* **33**, 217 (1978).

79. J. Warnatz, *Ber. Bunsenges Phys. Chem.* **82**, 193, 643, 834 (1978).

80. C. K. Westbrook, *CST* **23**, 191 (1980).

81. J. M. Heimerl and T. P. Coffee, *C & F* **39**, 301 (1980).

82. T. P. Coffee and J. M. Heimerl, *C & F* **43**, 273 (1981).

83. J. Warnatz, *CST* **26**, 203 (1981).

84. J. Warnatz, *18th Symp.* (1981), 369–384.

85. M. A. Cherian, P. Rhodes, R. J. Simpson, and G. Dixon-Lewis, *18th Symp.* (1981), 385–396.

86. C. K. Westbrook and F. L. Dryer, *18th Symp.* (1981), 749–767.

87. E. S. Oran and J. P. Boris, *Prog. Energy Combust. Sci.* **7**, 1 (1981).

88. S. B. Margolis, *J. Comp. Phys.* **27**, 410 (1978).

89. G. Millán and I. Da Riva, *8th Symp.* (1962), 398–411.

90. M. Gilbert and D. Altman, *6th Symp.* (1957), 222–236.

91. C. F. Curtiss and J. O. Hirschfelder, *J. Chem. Phys.* **17**, 550 (1949).

92. G. Damköhler, *Zeit. Elektrochem.* **42**, 846 (1936).

93. S. S. Penner, *Combustion Researches and Reviews 1955*, London: Butterworths Scientific Publications, 1956, 140–162.

CHAPTER 6

Detonation Phenomena

Let us reconsider the experiment described at the beginning of Section 5.1.1. If the open tube containing the combustible gas mixture is sufficiently long, then after traveling a distance on the order of ten tube diameters, the deflagration wave begins to accelerate markedly. An unsteady transitional region develops, and a high-speed (about 3×10^5 cm/s) nearly planar combustion wave often emerges and travels at a constant velocity through the remainder of the combustible mixture to the end of the tube. This high-speed wave is a detonation, which is invariably found to propagate at a speed correspondingly closely to the upper Chapman-Jouguet point (Chapter 2). The rate of development of the detonation can be enhanced by closing the ends of the tube.

Detonations can also be initiated by passing a sufficiently strong shock wave through a combustible mixture in a shock tube; if the pressure ratio across the shock wave is very large, these are strong detonations that propagate faster than the Chapman-Jouguet velocity and that decay slowly toward the Chapman-Jouguet condition. Standing detonations (stationary in the laboratory) may be produced by adjusting air flow conditions in a converging-diverging nozzle so that a Mach-reflected normal shock sits downstream from the nozzle exit. When the air is preheated to a sufficiently high temperature and fuel (usually hydrogen) is added, then the shock wave ignites the mixture, and the ensuing combustion has been observed to transform the shock into a (steady, plane) strong detonation. We shall consider the structures and propagation velocities of detonations produced by the methods described above.

Detonations were first investigated scientifically by Berthelot and Vieille [1] and by Mallard and le Chatelier [2]. Hypotheses leading to mathematical predictions of the speed of propagation of these waves were given by Chapman [3] and Jouguet [4], and reasonable postulates for the wave structure were presented by Zel'dovich [5], von Neumann [6], and Döring [7]. Significant strides toward improved understanding of detonation structure began in the late 1950s with the gradual recognition of the prevalence of unsteady, nonplanar wave patterns. Background material and references to the large body of literature on detonations that existed prior to 1960 may be found in books [8]–[13] and in reviews [14], [15]; the wealth of information that has been obtained more recently is covered in reviews [16]–[22] and in newer books [23]–[28].

In the following section we shall discuss the structure of steady, plane, one-dimensional, gaseous detonations. Factors influencing detonation propagation velocities are considered in Section 6.2. Important unsteady and nonplanar aspects of detonation phenomena are treated in Section 6.3. Detonations in media that are not purely gaseous will be discussed in Section 6.4. Since the material on detonations is too extensive to be developed fully here, the reader is directed to the literature cited above for further details.

6.1. PLANAR DETONATION STRUCTURE

6.1.1. Governing equations

The structure of a steady, planar detonation wave must be understood before investigations of unsteady, nonplanar phenomena can properly be begun and is most easily illustrated by analyzing the simplest representative system. Therefore, we shall begin by adopting assumptions 1–10 of Section 5.2; that is, initially we shall consider the one-step first-order reaction R → P in a binary mixture with a Lewis number of unity and an effective Prandtl number of $\frac{3}{4}$. The two differential equations describing the wave structure in such a system were derived in Section 5.2 and are given by equations (5-35) and (5-36), in which $F_\varphi(\tau, \varphi)$ and $F_\epsilon(\tau, \varphi)$ are defined in equations (5-32) and (5-34), respectively. It will be recalled that φ is the ratio of the local velocity to the upstream ($x = -\infty$) velocity [equation (5-26)], τ is the normalized dimensionless stagnation temperature [equation (5-12)], and ϵ is the reaction progress variable [equation (5-17)]. The status of the mathematical problem is summarized in Section 5.2.7.

6.1.2. Properties of the governing equations

The solution to equations (5-35) and (5-36) is a line in the three-dimensional (φ, τ, ϵ) space, beginning at the singular point $\varphi = 1$, $\tau = 0$,

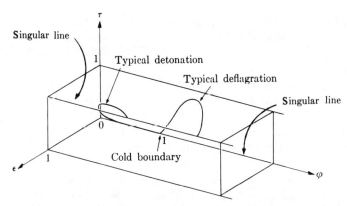

FIGURE 6.1. Schematic illustration of the $(\varphi, \tau, \epsilon)$ region in which the solution for the structure of a combustion wave must lie.

$\epsilon = 0$ and ending at the singular point $\varphi = \varphi_\infty$, $\tau = 1$, $\epsilon = 1$. Physically acceptable solutions must lie entirely within a region shaped like the interior of a semiinfinite beam of unit square cross section, as illustrated in Figure 6.1. In order to deduce the general shape of the solution curve, it is necessary to locate the singularities of equations (5-35) and (5-36) in the region of interest, to determine the behavior of the solutions in the neighborhoods of these singularities, and to sketch integral curves between the singularities using the slopes given by equations (5-35) and (5-36).

6.1.2.1. Location of Singularities

Since the numerator and the denominator on the right-hand side of equation (5-35) or equation (5-36) are bounded in the region of interest, the singularities will be located at the points or lines where both the numerator and the denominator vanish. The denominator for equation (5-35) and equation (5-36) vanishes when

$$\epsilon = \tau, \tag{1}$$

which defines a slanting plane containing the φ axis in Figure 6.1. In this plane, heat conduction precisely balances viscous dissipation, and diffusion vanishes [see Section 5.2; for example, equation (5-19)]. The numerator of equation (5-35) vanishes when

$$F_\varphi(\tau, \varphi) = 0, \tag{2}$$

and the numerator of equation (5-36) vanishes when

$$F_\epsilon(\tau, \varphi) = 0. \tag{3}$$

Since equations (2) and (3) are independent of ϵ, their solutions will define cylindrical surfaces, the generatrices of which are parallel to the ϵ axis.

From equation (5-34) it is clear that the solution to equation (3) in the region of interest is the plane

$$\tau = 1. \tag{4}$$

In view of the discussion in Section 5.3.2, we may also want to treat the plane

$$\tau = 0 \tag{5}$$

as a solution to equation (3) in order to remedy the cold-boundary difficulty. Equations (1), (4), and (5) imply that the lines $\epsilon = \tau = 1$ and $\epsilon = \tau = 0$ may be taken to be the singular lines of importance for equation (5-36).

Equation (5-32) shows that equation (2) defines a parabola in the (φ, τ) plane. Solving for τ, we find that this parabola is given by

$$\alpha'\tau = -\left(\frac{\gamma + 1}{2}\right)M_0^2\varphi^2 + (1 + \gamma M_0^2)\varphi - \left[1 + \left(\frac{\gamma - 1}{2}\right)M_0^2\right], \tag{6}$$

or [solving equation (6) for φ] by

$$\varphi = \varphi_\pm(\tau) \equiv \frac{(\gamma M_0^2 + 1) \pm [(M_0^2 - 1)^2 - 2(\gamma + 1)M_0^2\alpha'\tau]^{1/2}}{(\gamma + 1)M_0^2}, \tag{7}$$

where the notation implies that $\varphi_+(\tau) \geq \varphi_-(\tau)$. From equation (5-31) it can be seen by recalling the meaning of the variables that when equation (6) [that is, equation (2)] is valid, the viscous force term in equation (5-5) is zero. Hence the line in the (φ, τ) plane defined by equation (6) is the Rayleigh line [along which equation (2-2) relates the properties at every point]. The intersection of the parabolic surface [in $(\varphi, \tau, \epsilon)$ space] given by equation (6) with the plane given by equation (1) is the singular line of equation (5-35); along this line, viscous effects, heat conduction, and diffusion vanish.

Since the only ϵ-dependent condition above is $\epsilon = \tau$, it will be convenient to view the location of the singularities in a projection of the $(\varphi, \tau, \epsilon)$ space on the (φ, τ) plane. The construction of such a projection is aided by Figure 6.2, in which $\alpha'\tau$ is sketched as a function of φ for various values of M_0 as given by the Rayleigh-line condition [that is, by equation (6), which is the only condition with any degree of complexity]. The product $\alpha'\tau$ was chosen as the vertical coordinate in Figure 6.2, because in equation (6) the relationship between this product and φ depends only on γ and M_0 and mainly on M_0, since γ generally lies within a very narrow range of values. Pertinent facts, applicable to all of these curves and easily deduced from equations (6) and (7), are illustrated in Figure 6.2 for the curves labeled $M_0 = 5$ and $M_0 = 0.1$.

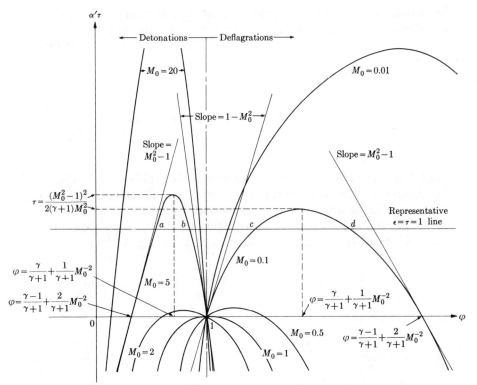

FIGURE 6.2. Schematic diagram illustrating the properties of various Rayleigh lines. End states at points a, b, c, d correspond to strong and weak detonations and weak and strong deflagrations, respectively.

The most important singular points are those common to both equations (5-35) and (5-36), namely;

$$(\varphi, \tau, \epsilon) = (\varphi_+(0), 0, 0), \tag{8}$$

$$(\varphi, \tau, \epsilon) = (\varphi_-(0), 0, 0), \tag{9}$$

$$(\varphi, \tau, \epsilon) = (\varphi_+(1), 1, 1), \tag{10}$$

and

$$(\varphi, \tau, \epsilon) = (\varphi_-(1), 1, 1). \tag{11}$$

Equation (7) shows that for a detonation (that is, for $M_0 > 1$), equation (8) identifies the cold-boundary point [$(\varphi, \tau, \epsilon) = (1, 0, 0)$], while for a deflagration (that is, for $M_0 < 1$), equation (9) represents the cold-boundary point. By using equation (7) and the results of Section 2.2 (for example, Figure 2.5), it can be seen that equation (10) is the hot-boundary point for a weak detonation or a strong deflagration and equation (11) is the hot-boundary point for a strong detonation or a weak deflagration. For Chapman-Jouguet waves, equation (2-34) [in which $\alpha = \alpha'\gamma/(\gamma - 1)$] implies that the quantity under the radical sign in equation (7) vanishes at $\tau = 1$, and therefore $\varphi_+(1) = \varphi_-(1)$, causing equations (10) and (11) to coalesce into a single point.

6.1.2.2. Solutions in the neighborhoods of singular points

Let us first consider the relationship between ϵ and τ near the points defined by equations (8)–(11). In the neighborhood of ($\tau = 0, \epsilon = 0$) [equations (8) and (9)], we must assume that $F_\epsilon(\tau, \varphi) = 0$ over a short range of τ in order to remedy the cold-boundary difficulty (see Section 5.3.2). Hence equation (5-36) implies that $\epsilon = 0$ as τ increases in the neighborhood of either of these points (that is, $\tau > \epsilon$ here).

The nature of the singularities ($\tau = 1, \epsilon = 1$) [equations (10) and (11)] can be deduced by expanding equation (5-36) about these points. In this manner, these singularities are found to be saddle points in the (τ, ϵ) plane; only one integral curve reaches ($\tau = 1, \epsilon = 1$) from the permissible region ($\tau \leq 1, \epsilon \leq 1$); all others bend away from the singularities. The slope of the curve that reaches ($\tau = 1, \epsilon = 1$) from ($\tau \leq 1, \epsilon \leq 1$) is

$$\frac{d\epsilon}{d\tau} = \frac{1}{2}\left\{1 + \sqrt{1 + 4 \lim_{\tau \to 1} [F_\epsilon(\tau, \varphi)/(1 - \tau)]}\right\}, \tag{12}$$

which implies that $\tau > \epsilon$ along this curve. Equation (12) shows that the nature of the singularity depends quantitatively but not qualitatively on the value of φ. The qualitative properties of the singularity do depend on our choice of a first-order reaction; however, for any reasonable reaction order, only one solution reaches the singularity from the permissible region and this solution has $\tau > \epsilon$ near the singularity.

Let us next investigate the relationship between τ and φ near the points defined by equations (8)–(11). For equations (8) and (9) we may assume (in view of the preceding discussion) that $\epsilon = 0$. Expanding equation (5-35) about $\tau = 0, \varphi = \varphi_\pm(0)$ then yields the equidimensional equation

$$\frac{d\varphi}{d\tau} = \frac{\alpha'}{\gamma M_0^2 \varphi_\pm(0)} + \frac{(-1)^a}{\gamma \varphi_\pm(0)}\left(1 - \frac{1}{M_0^2}\right)\left[\frac{\varphi - \varphi_\pm(0)}{\tau}\right], \tag{13}$$

where $a = 0$ for $\varphi_\pm(0) = 1$ (the cold boundary) and $a = 1$ for $\varphi_\pm(0) \neq 1$ (the other singularity at $\tau = 0$). The general solution to equation (13) is

$$\varphi - \varphi_\pm(0) = \left[\frac{\alpha'}{\gamma M_0^2 \varphi_\pm(0) - (-1)^a(M_0^2 - 1)}\right]\tau$$
$$+ c\tau^{[(-1)^a/\gamma\varphi_\pm(0)](1 - 1/M_0^2)}, \tag{14}$$

where c is an arbitrary constant when $(-1)^a(M_0^2 - 1) > 0$, but $c = 0$ when $(-1)^a(M_0^2 - 1) \leq 0$ [since φ must approach $\varphi_\pm(0)$ as τ goes to zero]. Thus an infinite number of integral curves, in fact, all integral curves (with $\epsilon = 0$) in the neighborhood of the singular point, pass through the singularity when $(-1)^a(M_0^2 - 1) > 0$, and such singularities are nodal points in the (φ, τ) plane. On the other hand, when $(-1)^a(M_0^2 - 1) < 0$ (that is, $c = 0$), it can be deduced from equation (13) or (14) that the singularity is a saddle point in the (τ, φ) plane. Since $M_0^2 > 1$ for detonations and $M_0^2 < 1$ for deflagrations, we conclude that $[\varphi = \varphi_-(0), \tau = 0]$ is a saddle point and $[\varphi = \varphi_+(0) = 1, \tau = 0]$ is a nodal point for detonations, and that $[\varphi = \varphi_-(0) = 1, \tau = 0]$ is a saddle point and $[\varphi = \varphi_+(0), \tau = 0]$ is a nodal point for deflagrations. Thus the cold boundary $(\varphi = 1, \tau = 0)$ is a saddle point in the (φ, τ) plane for deflagrations and a nodal point in the (φ, τ) plane for detonations; this is the primary reason that the burning velocity is an eigenvalue for deflagrations, while wave-structure solutions exist with any propagation velocity (greater than or equal to the Chapman-Jouguet wave speed) for detonations.

The singularities $[\varphi = \varphi_+(1), \tau = 1]$ and $[\varphi = \varphi_-(1), \tau = 1]$ [equations (10) and (11)] have been considered in detail by Hirschfelder and Curtiss [29]. Analysis of the integral curves in the (φ, τ) plane in the neighborhoods of these points is somewhat more complicated than analysis for the other singularities; the precise properties of the solutions differ for various ranges of values of the parameters (that is, M_0) and depend on the order of the reaction. Nevertheless, for all realistic reactions, the point $[\varphi = \varphi_+(1), \tau = 1]$ (the end state for weak detonations or strong deflagrations) always exhibits a saddle type of behavior in the (φ, τ) plane, and the point $[\varphi = \varphi_-(1), \tau = 1]$ (the end state for strong detonations or weak deflagrations) always exhibits a nodal type of behavior in the (φ, τ) plane.

6.1.2.3. General properties of the integral curves

The (τ, ϵ) relationship along the solution curves is relatively simple; equation (5-36) and the result that $\tau > \epsilon$ in the neighborhoods of the hot and cold boundaries suggest that $\tau > \epsilon$ throughout the range $0 < \tau < 1$ and that ϵ varies monotonically with τ in this region. These results enable us to use equation (5-35) and the previously deduced properties of the singularities in the (φ, τ) plane in order to construct the projection of the solution curves on the (φ, τ) plane. This projection is shown in Figure 6.3.

Figure 6.3 clearly illustrates the nodal type of behavior of solutions about $\varphi_+(0)$ and $\varphi_-(1)$ and the saddle type of behavior of solutions about $\varphi_-(0)$ and $\varphi_+(1)$. The cold boundary $(\varphi = 1)$ is located at $\varphi_-(0)$ for deflagrations and at $\varphi_+(0)$ for detonations. The hot boundary is located at $\varphi_-(1)$ or $\varphi_+(1)$, depending on the type of wave, as stated previously. The encircled signs are the signs of the slopes of the integral curves in regions bounded

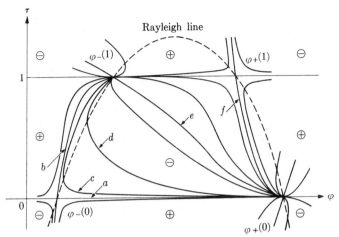

FIGURE 6.3. Schematic diagram of the projection of solution curves of (5-35) and (5-36) on the (φ, τ) plane.

by $\tau = 0$, $\tau = 1$, and the Rayleigh line; these signs depend only on the result that $\tau > \epsilon$ for $0 < \tau < 1$ and the assumption (adopted for clarity) that $\tau < \epsilon$ for $\tau > 1$ and $\tau < 0$.

It should be emphasized that for given parameters (M_0, E_1, B_1, and so on) the (ϵ, τ) behavior of the solution ensures that at most one integral curve from the cold-boundary singularity reaches the hot-boundary point in $(\varphi, \tau, \epsilon)$ space; all other integral curves from the cold boundary that pass through $[\varphi = \varphi_{\pm}(1), \tau = 1]$ do so with $\epsilon \neq 1$ [and hence with infinite slope in the (φ, τ) plane] and therefore do not satisfy the hot-boundary conditions. Thus, each solution curve shown in Figure 6.3 corresponds to a different value of some wave parameter (for example, the reaction-rate frequency factor B_1).

Since $\varphi_{+}(0)$ and $\varphi_{-}(1)$ are both nodes, an infinite number of integral curves connect these two points in Figure 6.3. Hence, acceptable solutions passing between these two points will exist for a range of values of the wave parameters; that is, regardless of the reaction rate, strong detonations exist for any propagation speed (above the upper Chapman-Jouguet wave speed). On the other hand, only one integral curve (the curve labeled f) passes between $\varphi_{+}(0)$ and $\varphi_{+}(1)$ in Figure 6.3; that is, weak detonations can exist only for particular reaction-rate functions at any given wave speed. Similarly, only one integral curve (the curve labeled b) passes between $\varphi_{-}(0)$ and $\varphi_{-}(1)$; that is, weak deflagrations propagate with a definite wave speed that depends on the reaction rate. Finally, no integral curves pass between $\varphi_{-}(0)$ and $\varphi_{+}(1)$ in Figure 6.3; that is, solutions to equations (5-35) and (5-36) for the structure of strong deflagrations do not exist. It is interesting that consideration of the basic (elliptic versus hyperbolic) character of the external flow, with the wave treated as a discontinuity, leads to roughly the opposite variation in the number of free parameters [30]; the results pertain to different questions and are not in conflict.

6.1.3. Remarks on deflagrations

A more detailed demonstration that strong deflagrations do not exist can be given. From equations (7) and (14) it is easily shown that, with $M_0 < 1$, the solution curve leaves the cold boundary ($\varphi = 1$, $\tau = 0$, $\epsilon = 0$) with $\tau > \epsilon$ and with a slope in the (φ, τ) plane which exceeds that of the Rayleigh line, as is illustrated at $\varphi_-(0)$ in Figure 6.3. Hence, the integral curve must cross the Rayleigh line at some point between $\tau = 0$ and $\tau = 1$ in order to reach the hot-boundary point [$\varphi = \varphi_+(1)$, $\tau = 1$, $\epsilon = 1$] of a strong deflagration. Since the slope in the (φ, τ) plane of the integral curve on the Rayleigh line is infinite (unless $\epsilon = \tau$) and the slope in the region to the left of the Rayleigh line is positive (unless $\epsilon \geq \tau$) the integral curve cannot reach the Rayleigh line unless ϵ becomes equal to τ at some point in the region $0 < \tau < 1$. At $\epsilon = \tau$, equation (5-19) shows that $d\tau/d\xi = 0$, while the ξ derivative of equation (5-19) shows that $d^2\tau/d\xi^2 = -d\epsilon/d\xi$. Since equations (5-33) and (5-34) imply that $d\epsilon/d\xi > 0$, it follows that τ begins to decrease as ξ (or ϵ) increases, and therefore τ becomes less than ϵ. Once ϵ exceeds τ, equation (5-36) shows that τ continues to decrease as ϵ (or ξ) increases. Hence τ will not approach the required hot-boundary condition $\tau = 1$ at $\epsilon = 1$. We may therefore conclude that solutions to equations (5-35) and (5-36) cannot satisfy the boundary conditions for strong deflagrations.

Since the above argument relies on the assumptions used to derive equations (5-35) and (5-36), we have shown only that strong deflagrations are impossible in binary systems with a first-order reaction, a Lewis number of unity, a Prandtl number of $\frac{3}{4}$, and so on. Since the result depends on the structure of the wave, it is impossible to invent a generally valid proof. However, for all combustion systems that have been studied, the result has been found to be true; all known deflagrations in the laboratory are weak deflagrations. Solutions for structures of strong deflagrations may be obtained if an endothermic process follows the exothermic reaction [31]. These solutions have been proposed for astrophysical problems of stellar and galactic winds involving ionization-recombination fronts [32]. Searches for strong deflagrations in combustion could focus on systems with both extraordinarily high burning velocities and endothermic attributes near the hot boundary. Currently it is unclear whether fundamental reaction-rate limitations will preclude the occurrence of strong deflagrations in combustion.

For ordinary deflagration waves, M_0 is very small compared with unity and the Rayleigh line is very nearly a straight line with slope $d(\alpha'\tau)/d\varphi = 1$ in the range $0 \leq \tau \leq 1$ (see Figure 6.2). Equation (14) shows that the initial slope of the integral curve at the cold boundary approaches this same value as $M_0 \to 0$. Since the (φ, τ) projection of the solution curve lies between the Rayleigh line and the straight line through the cold boundary with this initial slope, the solution itself must have $d(\alpha'\tau)/d\varphi = 1$ everywhere when M_0 is sufficiently small. In terms of dimensional variables, this condition implies that $T/T_0 = v/v_0$, which is equivalent to the assumption (introduced in Section 5.3.1) of an isobaric wave.

6.1.4. Approximate solution for the structure of a detonation

Since detonations propagate at supersonic velocities, while deflagrations propagate at low subsonic velocities, the mass flow rate m is much larger for detonations than for deflagrations. Consequently, the dimensionless reaction-rate function $F_\epsilon(\tau, \varphi)$, defined in equation (5-34), is much smaller for detonations; typically $F_\epsilon(\tau, \varphi) \lesssim 10$ in a deflagration wave and

$$F_\epsilon(\tau, \varphi) \lesssim 10^{-2} \ll 1 \tag{15}$$

in a detonation wave. This result has a number of important implications concerning detonation structure.

Since the change in ϵ through the detonation wave is $\Delta\epsilon = 1$, equations (15) and (5-33) imply that a representative dimensionless distance scale $\Delta\xi$ is large compared with unity for detonations (it is of the order of unity for deflagrations). An estimate of the order of magnitude of the left-hand side of equation (5-19) then shows that $d\tau/d\xi \ll 1$ (since the change in τ through the wave is $\Delta\tau = 1$), and therefore equation (5-19) implies that, approximately,

$$\epsilon = \tau \tag{16}$$

throughout the detonation wave. Equations (16) and (5-24) show that, approximately, $\epsilon = Y$ in the wave; that is, diffusion is negligible. Equations (16) and (5-33) lead to

$$d\tau/d\xi = F_\epsilon(\tau, \varphi), \tag{17}$$

which serves to eliminate ϵ from the problem; equations (5-31) and (17) determine the approximate structure of the detonation.

In view of equations (5-34) and (15), the temperature must become relatively large in the detonation wave before τ begins to change appreciably from its value ($\tau = 0$) at the cold boundary. Hence, the upstream portion of the detonation structure will lie very near the line $\tau = 0$ connecting the points $\varphi_+(0)$ and $\varphi_-(0)$ in Figure 6.3, and the detonation structure will be represented by a curve similar to the one labeled c in Figure 6.3. The solution curve c is closely approximated by curve a followed by curve b in Figure 6.3.

Since $\epsilon = \tau = 0$ along curve a, no reaction occurs along this curve, and the structure of the upstream part of the wave is obtained by integrating equation (5-31) with $\tau = 0$ between $\varphi_+(0)$ and $\varphi_-(0)$. The integration can be performed analytically; the result is

$$\xi - \xi_1 = \left(\frac{\gamma M_0^2}{M_0^2 - 1}\right)\left\{\ln(1 - \varphi) - \left[\frac{2 + (\gamma - 1)M_0^2}{(\gamma + 1)M_0^2}\right]\right.$$
$$\left. \times \ln\left[\varphi - \frac{2 + (\gamma - 1)M_0^2}{(\gamma + 1)M_0^2}\right]\right\}, \tag{18}$$

where ξ_1 is an arbitrary constant.

Since the characteristic distance over which properties change appreciably is $\Delta\xi \gg 1$, along curve b (the downstream part of the wave) the term $d\varphi/d\xi$ may be neglected in equation (5-31). The relationship between φ and τ along curve b is therefore given approximately by $F_\varphi(\tau, \varphi) = 0$, the solution to which is $\varphi = \varphi_-(\tau)$, where $\varphi_-(\tau)$ is given by equation (7). Hence, curve b lies very close to the Rayleigh line, and the structure of the downstream part of the wave is obtained by integrating equation (17) with $\varphi = \varphi_-(\tau)$; namely,

$$\xi = \int d\tau / F_\epsilon[\tau, \varphi_-(\tau)]. \tag{19}$$

Since equation (5-27) shows that the temperature T depends on both φ and τ and equation (7) shows that $\varphi_-(\tau)$ is a fairly complicated function, the integral in equation (19) must generally be evaluated numerically. However, equation (19) shows that the asymptotic ξ dependence of τ (as $\xi \to \infty$) is

$$\tau = 1 - e^{-\bar{F}(\xi - \xi_2)}, \tag{20}$$

where ξ_2 is a constant and the dimensionless specific reaction-rate constant at $\xi = \infty$ is

$$\bar{F} \equiv \lim_{\tau \to 1} \{F_\epsilon[\tau, \varphi_-(\tau)]/(1 - \tau)\} = \text{constant}.$$

For the highly unrealistic case in which the activation energy is zero ($E_1 = 0$) and the pressure and temperature dependences of λ and B_1 are such that the factor in brackets in equation (5-34) is a constant, equation (20) is the solution to equation (17) throughout the downstream portion of the wave.

6.1.5. Discussion of detonation structure

We have inferred that for planar detonations, curve c in Figure 6.3 represents the most reasonable structure. The relationship between pressure and volume along this curve is illustrated in Figure 6.4 (the labels a through f identify corresponding curves in Figures 6.3 and 6.4). Since curve c lies close to curve a followed by curve b, we concluded that these two curves afford a good approximation to the structure of most planar detonations. Since

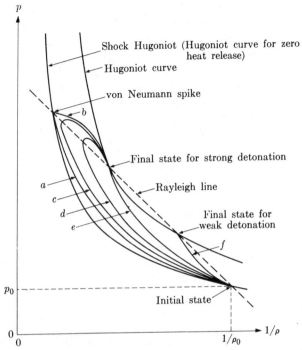

FIGURE 6.4. Schematic illustration of detonation structure in a pressure-volume plane.

no heat release occurs along curve a, this curve represents the structure of a normal shock wave in the reactant gas. Curve b represents the structure of a deflagration with upstream conditions given by the point $\varphi_-(0)$ (which is called the von Neumann spike in Figure 6.4). Hence a good approximation to the usual detonation structure is a shock wave followed by a deflagration. This is quite reasonable on physical grounds; the shock wave heats the reactants to a temperature at which they can react at a rate that is high enough for the ensuing deflagration to propagate as fast as the shock. The structure of a high-speed "deflagration" of this type necessarily differs greatly from the deflagration structure discussed in Chapter 5. The a-b (shock-deflagration) structure of detonations was first proposed by Zel'dovich [5], von Neumann [6], and Döring [7] and may be referred to as the **ZND wave structure**.

On the basis of numerical calculations of steady, planar detonation structure (for example, [33]) and of good experimental measurements performed mainly in the 1950s (for example, [34]–[37]; see review in [38]), it was widely believed that the ZND structure was representative of most real detonations. This structure excludes weak detonations, which require special consideration (see Section 6.2.2). It is likely to apply to sufficiently strong detonations (*over-driven waves*, see Sections 6.2.6 and 6.3.3). However, for the most common—Chapman-Jouguet waves—more recent studies,

to be discussed in Sections 6.2.5 and 6.3, established that this structure
usually is unstable and that, in fact, these detonations generally possess a
complex three-dimensional substructure involving transverse waves that
produce a steady, planar wave only on the average (for reviews see, for example, [16]–[18], [22]–[25], [38]). Knowledge of the ZND structure
nevertheless is essential to the understanding of the three-dimensional
structure (to be discussed in Section 6.3). Therefore, further deductions on
the basis of the ZND structure deserve to be pursued here.

There is a fundamental reason that large departures from the ZND
structure (such as curve e in Figures 6.3 and 6.4) are not to be expected for
steady, planar waves. The dimensionless distance over which φ changes
appreciably in the shock wave, as given by equation (18), is $\Delta\xi \sim \gamma M_0^2/$
$(M_0^2 - 1)$; that is, $\Delta\xi \sim \gamma$ for detonations with a reasonable amount of
heat release. In view of equations (5-11) and (5-18) and the fact that $\kappa \approx 0$,
this implies that the width of the shock wave in physical dimensions is
$\Delta x \sim 4\gamma\mu/3m = 4\gamma\mu/3\rho_0 a_0 M_0$, where a_0 is the initial velocity of sound.
Since a_0 is of the same order of magnitude as molecular velocities, the kinetic
theory expression for the viscosity μ [for example, equation (E-27) combined
with equation (E-34)] indicates that Δx is of the order of a few molecular
mean free paths. Hence, molecules suffer only two or three collisions in
passing through shocks of the strengths ordinarily encountered in detonations. Thus, since chemical reactions can proceed only through molecular
collisions, the chemical reaction rate would have to approach the maximum conceivable reaction rate (that is, the intermolecular collision rate)
before an appreciable amount of heat release could occur in the shock wave.
It appears probable that such highly reactive mixtures usually would explode
in a more or less homogeneous manner instead of supporting a propagating
detonation wave.

The result that Δx is of the order of a few mean free paths in the shock
wave casts doubt upon the validity of our continuum equations for the upstream portion of a detonation. Kinetic-theory analyses of strong shock
waves in nonreactive gases (for example, [39]) yield wave thicknesses that
are nearly twice as large as those obtained from continuum theory and
that are in better agreement with the few available experimental results
[40]. No kinetic-theory analyses of detonation structure have been reported,
but such analyses would be expected to show a similar increase in the width
of the upstream (shock) portion of the detonation wave. This increase
should not substantially modify our other conclusions regarding the wave
structure. In general, the shock waves that occur in detonations are strong
enough to be treated as discontinuities for nearly all purposes.

6.1.6. The structure of ZND detonations

The approximate solution given in equation (18) for the structure of
the upstream part of the detonation wave was first derived by Becker [41]

for shock waves in nonreactive gases. Equation (18) shows that as ξ increases from $-\infty$ to $+\infty$, φ decreases monotonically from $\varphi_+(0)$ to $\varphi_-(0)$. There is no solution for the structure of a wave in which φ increases from $\varphi_-(0)$ to $\varphi_+(0)$ as ξ increases; this would represent a finite-amplitude expansion wave, which can be shown to violate the second law of thermodynamics in ideal gases with constant specific heats. Since no reaction occurs in the shock wave, equation (18) is valid for arbitrary reaction mechanisms and diffusion coefficients, (that is, it does not depend on assumptions 9 and 10 of Section 5-2). On the other hand, simple analytical solutions for the shock-wave structure cannot be obtained from the continuum equations unless the effective Prandtl number is $\frac{3}{4}$ (assumption 8) [42], [43]. The spatial dependence of the flow variables for a shock wave in a gas is illustrated by the upstream (left-hand) portion of the curves in Figure 6.5.

The solution given in equation (19) for the downstream part of the detonation wave represents a high-speed deflagration in which the kinetic energy is appreciable but the transport processes (viscosity, heat conduction, and diffusion) are negligible. This deflagration wave is therefore radically different from those discussed in Chapter 5. The difference is primarily a consequence of the much larger mass flow rate (or convection velocity) in detonations; very large gradients would be necessary to generate transport fluxes comparable in magnitude to the convective fluxes, and the chemical reaction rate is not high enough to produce these large gradients. The absence of significant molecular transport removes the eigenvalue character of the deflagration problem, thereby making it essentially an initial-value problem; the chemical reaction rate influences the length of the reaction zone but not directly the speed of the wave. The spatial dependence of the flow

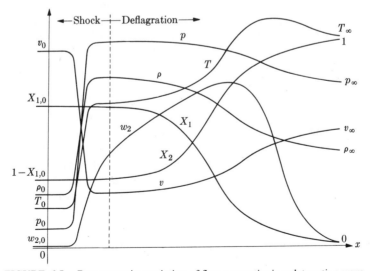

FIGURE 6.5. Representative variation of flow properties in a detonation wave.

variables in the deflagration is illustrated in the downstream (right-hand) part of Figure 6.5. Because of the high velocity, the pressure is not constant in the deflagration region of Figure 6.5. Figure 6.5 shows a small decrease in temperature as the hot boundary is approached. This effect is not present for most strong detonations; it occurs for Chapman-Jouguet waves because, as heat is added, the temperature decreases (and the Mach number, of course, increases) along the Rayleigh line when the Mach number is between $1/\sqrt{\gamma}$ and 1.

Since transport effects are neglected in equation (19), the equation does not depend on assumptions 8 and 10 of Section 5.2. Clearly, equation (19) can also be made to apply to a system involving any one-step reaction by suitably modifying the definition of $F_\epsilon(\tau, \varphi)$. In fact, since transport effects are negligible, the problem of computing the structure of the downstream portion of the detonation wave for any reaction mechanism merely involves a forward integration of a set of equations of the form

$$dX_i/dx = f_i(X_1, \ldots, X_N), \quad i = 1, \ldots, N \tag{21}$$

(x = distance, X_i = mole fraction of species i, N = number of chemical species), provided that the rate constants and heats of reaction are known for all elementary reaction steps; the continuity, momentum, energy, and state equations reduce to algebraic expressions relating velocity, density, temperature, and pressure to X_i. Equation (21) is comparable in complexity to the equations governing the time dependence of the concentrations of chemical species in a homogeneous, isothermal, reacting system (see Appendix B).

In view of the range of validity of equations (18) and (19) discussed above, these two equations describe the structure of a ZND detonation involving any one-step reaction and satisfying assumptions 1–8 of Section 5.2.* However, a small ambiguity arises in connecting the solutions given in equations (18) and (19). Mathematically, the shock wave extends from $\xi = -\infty$ to $\xi = +\infty$, while (for Arrhenius rate functions) the deflagration extends from a finite value of ξ to $\xi = +\infty$. Hence the two waves overlap, and in using equations (18) and (19) without modification, the shock wave must be cut off at an arbitrary point and tacked onto the beginning of the deflagration in order to obtain a single-valued solution for the detonation structure. The cutoff point may be defined as the point at which $[\varphi - \varphi_-(0)]/\varphi_-(0) = \delta$, where δ is a number that is small compared with unity. For typical detonations the width of the shock wave is so small in comparison with the width of the reaction zone that curves similar to those shown in Figure 6.5 are modified imperceptibly when δ is varied from 10^{-2}

---

* The restriction to one-step reactions may be removed by replacing equation (19) by equation (21); ZND detonations involving chain reactions can be analyzed more easily than the deflagrations with chain reactions discussed in Section 5.4.

to 10^{-4}. Wood [44] recognized that ambiguities of this kind are characteristic of problems of the boundary-layer type; by means of a boundary-layer expansion near the shock and an expansion in powers of $1/\Delta\xi$ near the Rayleigh line, he has obtained a natural connection between the shock and deflagration solutions, thereby providing an analytical approximation to curve c in Figures 6.3 and 6.4. A related expansion was developed independently [45].

For purposes of further analyses of detonation structure, the shock wave may be treated as a discontinuity. Both the viscous interaction between the shock and the reaction region and the molecular transport within the reaction region are small perturbations that do not appear to exert qualitatively significant influences on the wave structure. This conclusion appears to apply not only to steady, planar waves but also to unsteady, three-dimensional structures; it affords one helpful simplification in the complicated analyses of transverse wave structures. It also alters the interpretation of a detonation as a deflagration-supported shock; the "support" provided by the chemical reactions is of a nonplanar compressible gasdynamic character with negligible molecular transport.

Analysis of the reaction region of a steady, planar, ZND detonation is facilitated by the introduction of simplifications in the chemical kinetics. An asymptotic analysis with the reaction rate assumed to be zero below an ignition temperature and constant above it has been reported [46]. Activation-energy asymptotics with the chemistry approximated as a one-step process have been introduced on the basis of physical reasoning [13] and through formalisms of matched asymptotic expansions [47], [48]* to identify an induction zone followed by a shorter heat-release zone and to obtain asymptotic expressions for the length of the induction zone. More-complex approximations to the chemical kinetics generally have required numerical integrations [33], [49].

6.2. PROPAGATION VELOCITIES FOR DETONATIONS TRAVELING IN TUBES

6.2.1. Basic considerations for plane waves

We have seen that considerations of detonation structure do not restrict the detonation propagation velocity (at least for strong and Chapman-Jouguet waves). Therefore, wave speeds must depend on the experimental configuration and can be discussed only within the framework of a class of experiments. Here we shall first consider detonations produced by

---

* See also H. K. Cheng, "An Analytic, Asymptotic Theory of Shock Supported Arrhenius Reactions and Ignition Delay in Gas," unpublished (1970) for an early version of the study in [48]. The work in [47] involves a double limit that also describes slight merging of the shock with the reaction zone.

igniting a combustible mixture with a spark at the closed end of a tube. Many experimental studies have been concerned with systems of this kind. Empirically, detonations in such systems are found to travel at the Chapman-Jouguet wave speed after an unsteady period of development.

Steady, strong detonations are easily ruled out by dynamic considerations. For a strong detonation propagating into a quiescent gas mixture, the gas velocity behind the detonation is directed toward the wave (in the laboratory coordinate system) and is subsonic with respect to the wave (see Chapter 2). Since the gas velocity is necessarily zero at the closed end of the tube, there is a net increase in the gas velocity in the direction of propagation of the wave (in the laboratory frame) between the end of the tube and the detonation. When transformed to a coordinate system moving with the detonation velocity, this velocity change becomes an increase from a subsonic speed immediately behind the detonation to a supersonic speed at the closed end of the tube (see Figure 6.6). Such a velocity increase requires that a **rarefaction** wave follows the detonation. The presence of the rarefaction wave is well substantiated experimentally (for example, [34]). Since the Mach number behind a strong detonation is subsonic (relative to the detonation wave), this rarefaction wave will overtake the detonation and therefore weaken it, decreasing the detonation velocity until the Mach number behind the wave becomes sonic. When the Chapman-Jouguet condition is reached ($M_\infty = 1$), the rarefaction wave can no longer affect the detonation.

This same qualitative argument cannot be used to rule out weak detonations. Brinkley and Kirkwood [50] have formalized the dynamic argument by analyzing a model in which the detonation is of infinitesimal thickness and the pressure behind the wave initially decreases monotonically (with finite slope) toward the closed end of the tube. They find that the time rate of increase of the pressure immediately behind the detonation wave is negative for strong detonations, positive for weak detonations, and zero for Chapman-Jouguet waves. Hence (compare Figure 2.5) weak detonations

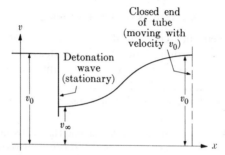

FIGURE 6.6. Schematic illustration of the velocity profile behind a detonation wave propagating in a tube; all velocities in the $+x$ direction.

(as well as strong detonations) approach the Chapman-Jouguet condition as time increases for the Brinkley-Kirkwood model.

6.2.2. Further comments on weak detonations

A very convincing argument against the existence of weak detonations utilizes the results on detonation structure presented in Section 6.1.2.3. A ZND structure is not possible for weak detonations, and chemical reaction rates generally are not high enough to produce a direct transition to the end state.

Weak detonations are believed to represent the "condensation shocks" observed in supersonic wind tunnels [12], [51]. Supercooled water vapor in a supersonic stream has been observed to condense rapidly through a narrow wave. The amount of liquid formed is so small that the equations for purely gaseous waves are expected to apply approximately. Since a normal shock wave would raise the temperature above the saturation point (thus ruling out the ZND structure, for example), and the flow is observed to be supersonic downstream from the condensation wave, it appears reasonable to assume that condensation shocks are weak detonations. This hypothesis may be supported by the fact that unlike chemical reaction rates, the rate of condensation increases as the temperature decreases. Proposals that weak detonations also represent various processes occurring in geological transformations have been presented [52].

Theoretical analyses of condensation waves for steady supersonic flows in nozzles are based primarily on the early work of Oswatitsch [53]. Numerical integrations of sets of ordinary differential equations are performed to describe homogeneous nucleation and droplet growth in supersonic flows with negligible molecular transport in the streamwise direction. The approach is discussed in [54] and [55], for example. Good agreement with the experimental measurements on the location and width of the condensation wave is obtained. However, this success does not produce simple formulas for predicting the onset of condensation. Simplified approaches to the analysis of condensation waves can provide useful results and improved understanding of the wave structure. One such approach has been reported recently [56]. Consideration of Figure 6.3 leads to the idea that a condensation wave can exist in a supersonic nozzle flow only at a definite location that depends on the wave structure. This idea of treating the condensation-wave problem as an eigenvalue problem reminiscent of the deflagration problem has not yet been exploited with the objective of obtaining simplified descriptions of condensation waves.

6.2.3. Effects of tube walls

As experiments on detonation waves in tubes were refined, it soon became apparent that these waves do not propagate precisely at the

Chapman-Jouguet speed. For example, detonation velocities decrease approximately linearly with the reciprocal of the tube diameter at fixed initial conditions; standard experimental practice is to plot the detonation velocity as a function of the reciprocal of the tube diameter and to extrapolate to zero to obtain the "true" experimental wave speed (for example, [57]). Also, detonation velocities have been observed to decrease approximately linearly with the reciprocal of the initial pressure in a tube of fixed diameter. The first satisfactory explanation of these effects was presented by Fay [58], who accounted for the influence of the boundary layer behind the shock front in a Chapman-Jouguet wave with the ZND structure.

Fay's model is illustrated in Figure 6.7, where a coordinate system that is stationary with respect to the detonation wave has again been adopted. Geometric considerations clearly imply that in this frame, the velocity in the boundary layer exceeds the velocity in the reaction zone. Furthermore, if it is assumed (quite reasonably) that the pressure is independent of the coordinate y (normal to the direction of propagation of the wave) and that the wall of the tube is maintained approximately at room temperature, then the ideal gas equation of state implies that the density in the boundary layer exceeds the density in the reaction zone. Hence, the mass flux ρv in the boundary layer exceeds the mass flux in the reaction zone, and therefore as the boundary layer grows, the streamlines in the reaction zone must diverge (at a rate determined by the rate of increase of the displacement thickness of the boundary layer). Conditions relating properties across oblique shock waves readily show that this divergence implies that the shock front must be slightly convex toward the unburned gas; this effect has been observed but, in itself, influences the detonation velocity to a negligible extent. On the other hand, the fact that the flow diverges throughout the reaction zone does affect the detonation velocity, since the equations of Chapter 2 must now be replaced by quasi-one-dimensional flow equations with a small rate of increase of area [that is, (4-6) through (4-9)] for the reaction zone (excluding the boundary layer).

The calculation of detonation velocities from the quasi-one-dimensional flow equations is considerably more difficult than the calculations discussed in Chapter 2. For example, the wave speed now depends on the

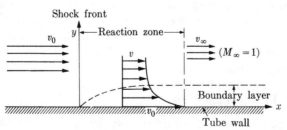

FIGURE 6.7.   Schematic diagram of a detonation near a tube wall [58].

profiles of the static and dynamic pressures in the reaction zone; that is, the propagation velocity is influenced by the wave structure. Another problem lies in determining where to apply the downstream Chapman-Jouguet condition $M_\infty = 1$; the condition cannot be applied at $x = \infty$ because the boundary layer would eventually occupy the entire cross section of the tube at some finite value of x. Fay solved this last problem by noting that the effects of area increase and of heat addition counteract each other in quasi-one-dimensional flow (in the subsonic regime, addition of heat increases M and an increase in area decreases M). A phenomenon similar to that occurring at the throat of a converging-diverging nozzle can therefore occur; at a point where the rate of increase of area is in the proper proportion to the rate of increase of stagnation enthalpy, the flow can experience a smooth transition through $M = 1$, from $M < 1$ to $M > 1$. The Chapman-Jouguet condition must therefore be applied at the finite value of x at which the rate of growth of the boundary layer is appropriately related to the reaction rate. Events occurring downstream from this $(M = 1)$ plane cannot affect the detonation because in the downstream region the velocity with respect to the wave is supersonic both inside and outside of the boundary layer.

On the basis of approximate calculations of the detonation velocity for representative systems containing stoichiometric hydrogen-oxygen mixtures [58], the fractional decrease in the detonation speed (below its classical value) is found to be $\Delta v_0/v_0 = 2.1 \delta^*/d$, where δ^* is the boundary-layer displacement thickness at the plane where $M = 1$, and d is the diameter of the tube. This result exhibits the observed inverse dependence on diameter, and (since δ^* should be approximately proportional to the width of the reaction zone, which is empirically [34] approximately inversely proportional to the pressure) is in rough agreement with the observed pressure dependence. By estimating the reaction rate from schlieren studies of detonations and by utilizing an empirical formula for the rate of growth of a supersonic turbulent boundary layer, Fay obtained numerical values of $\Delta v_0/v_0$ which varied from 1 % to 4 % and which were in good agreement with experimental results (see Table 6.1).

6.2.4. Ambiguities associated with frozen and equilibrium sound speeds

If a detonation is viewed as a plane Chapman-Jouguet wave traveling steadily in a tube in a direction away from its closed end, then we may question whether the result that $v_\infty = a_{e,\infty}$, obtained in Section 2.3.3, is consistent with the end-wall condition. From the standpoint of the method of characteristics for reacting gas mixtures (Section 4.3), the fact that the characteristics propagate at the frozen sound speed implies [59] that the upstream end of the rarefaction wave that follows the detonation (Section

TABLE 6.1. Comparison of calculated and observed fractional velocity deficit at 1-atm pressure in 1-cm radius tubes [58]

Composition	v_0, m/sec	$\dfrac{\Delta v_0}{v_0} \times 10^2$, calculated	$\dfrac{\Delta v_0}{v_0} \times 10^2$, observed
1. $2H_2 + O_2$	2830	1.0	1.2
2. $2H_2 + O_2 + 5He$	3660	2.2	2.5
3. $2H_2 + O_2 + 13He$	3823	3.3	4.4
4. $2H_2 + O_2 + 13A$	1522	1.6	2.7
5. $16\% \ H_2 + 84\% \ air$	1552	1.4	1.6
6. $53\% \ C_2H_2 + 47\% \ O_2$	2820	1.0	1.3

6.2.1) can smoothly join the downstream end of the detonation wave only at a point where the flow velocity relative to the detonation wave is the frozen sound speed a_f. From the characteristic analysis (specifically, from the requirement that the pressure and velocity gradients must remain finite where rarefaction joins the detonation), it is found that the condition which must be applied at this point is [compare equations (4-54), (4-55), and (4-91)]

$$\sum_{j=1}^{N} w_j (\partial h / \partial Y_j)_{\rho, \, p, \, Y_i(i \neq j)} = 0, \tag{22}$$

where the notation is that introduced in Chapter 1. Since equation (22) will be satisfied if $w_j = 0$ for all j, which is the statement of chemical equilibrium for the system, it appears that the detonation will propagate steadily if its downstream (chemical equilibrium) condition is applied at $v_\infty = a_{f, \infty}$, i.e., where the Mach number based on the *frozen* sound speed is unity. This result suggests that contrary to the conclusion reached in Section 2.3.3, the end states for detonations propagating in closed tubes correspond to the frozen Chapman-Jouguet condition indicated by point C in Figure 6.8 [59], a point situated slightly onto the weak-detonation branch of the equilibrium Hugoniot.

From Figure 6.8 it may be seen that this conclusion poses difficulties in connection with the wave structure. The gas mixture in a ZND wave would meet the equilibrium Hugoniot at point D in Figure 6.8 before reaching state C, and a transition from D to C is difficult to visualize. Possibilities that have been considered for resolving these difficulties include the investigation of "pathological" [59] systems in which equation (22) (which merely states that the enthalpy change is zero when a small amount of reaction occurs at constant pressure and density) is satisfied with some $w_j \neq 0$ and the analysis of systems with endothermic reactions near the hot boundary (to aid in a D-C transition) [60], [61]. Although it is clear that there are hot-boundary complexities associated with the behaviors of planar detonations having

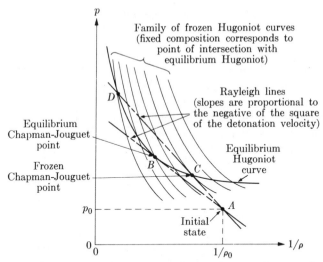

FIGURE 6.8. Schematic illustration of the difference between frozen and equilibrium Chapman-Jouguet detonations.

$a_{f,\infty} \neq a_{e,\infty}$, these effects are small (typically less than 1 % in the detonation velocity) and masked in the laboratory by multidimensional structures of Chapman-Jouguet waves.

6.2.5. Effects of three-dimensional structures

Observations of three-dimensional structures for Chapman-Jouquet detonations propagating in closed tubes range from spinning detonations (Section 6.3.1) in weak mixtures to "turbulent" detonations [62] in strong mixtures. The prevalence of the three-dimensional phenomenon, arising from instability of the planar wave and discussed in Section 6.3, necessitates its consideration in the development of accurate theories for propagation velocities. Appropriate analyses that couple the three-dimensional behavior to the wave field behind the detonation are difficult to develop with accuracies sufficient to distinguish between equilibrium and frozen sound speeds. It seems likely that when the characteristic dimensions of the nonplanar disturbances are small compared with the tube diameter, the three-dimensional effects will result in average propagation velocities (after removal of the wall effects discussed in Section 6.2.3) slightly in excess of the theoretical one-dimensional Chapman-Jouguet speed calculated on the basis of chemical equilibrium downstream. To substantiate this speculation remains a challenging theoretical problem. When the characteristic transverse dimensions are larger—for example, for near-limit or single-spin detonations—average sustained detonation velocities below the theoretical values by amounts up to about 20 % are observed, largely through the onset of a reduced average extent of combustion.

6.2.6. Chemical reactions behind a shock wave produced in a shock tube

Up until now in Section 6.2 we have been considering detonations in closed tubes filled with combustibles and ignited weakly at one end. It is possible instead to initiate the detonation by exploding a strong charge (of a solid explosive, for example) in the tube. If the shock wave (or "blast" wave) from the explosive is stronger than the shock that precedes the reaction zone in the ZND structure of the Chapman-Jouguet wave, then a strong detonation is established; this is termed an **over-driven wave**. Gasdynamic (inviscid) interactions with boundaries tend to weaken the strong detonation rapidly to a Chapman-Jouguet wave. This weakening is enhanced if the end of the tube at which the charge is exploded is open to a low-pressure reservoir, but provision of a sufficiently high-pressure reservoir at the initiation end retards the decay. With a sufficiently large reservoir at sufficiently high pressure, it is merely necessary to burst a diaphragm between the reservoir and the combustible to establish a strong detonation; the pressurized reservoir serves to maintain the downstream gas pressure above that of the Chapman-Jouguet state until the reflected wave from the end of the reservoir overtakes the detonation. This is the principle of operation of a **shock tube**. Shock tubes have often been used to investigate strong (and Chapman-Jouguet) detonations in a controlled fashion (see, for example, [12] and [62]-[64]).

Shock tubes have become standard tools for studying chemical kinetics at high temperatures [65]. Shock waves in suitably designed tubes are very planar and decay extremely slowly through viscous interactions. Often the reactant material is introduced in small amounts so that the reaction occurs nearly isothermally. If the overall reaction is exothermic at all, then—in principle—the shock followed by the reaction region together comprise a strong detonation wave. The evident stability of these waves clearly demonstrates that sufficiently strong planar detonations are stable; they propagate at the velocity of the shock wave initially established in the shock tube. If the amount of heat released in the reaction is high enough to generate pressure waves locally that can interact significantly with the leading shock, then the potential for development of instability arises. The transverse structures that are observed at or near Chapman-Jouguet conditions are consequences of this type of instability, as discussed in the following section.

6.3. TRANSVERSE STRUCTURES FOR DETONATIONS

6.3.1. Spinning detonations and stability considerations

It is curious that the basis for the discovery of transverse structures in detonations existed more than 50 years ago. Peculiar detonations that appeared to propagate along helical paths in tubes with round cross sections were observed [66], [67]. These "spinning detonations" leave a helical inscription on a smoke-covered foil that has been placed on the inside wall of the tube, and they may also be studied by many types of photographic techniques [23]–[25], [68]–[72]. The writings on the walls mark the trajectories of the strongest waves; the general properties of the writings have been described theoretically [73]–[76]. For 30 years after their discovery, spinning detonations were thought to be unusual phenomena associated only with gas mixtures near the limits of detonability (Section 6.3.4). Had it been realized that spin is merely a particular manifestation of the currently well-known transverse structures, clarifications of these structures could have been obtained much earlier (through use of the available experimental techniques to study mixtures progressively removed from limits of detonability).

If it is accepted that a spinning mode of propagation exists, then an acoustic theory may be applied to predict the spin frequency, the slope of the helical path on the tube wall, and other characteristics of spinning detonations [75]–[76]. For a planar Chapman-Jouguet wave in an infinitely long tube and in a frame of reference in which the burnt gas is at rest, the circumferential velocity of a traveling tangential acoustic wave at the wall is a known constant times a_∞, the sound speed of the burnt gas. In this frame of reference, the leading planar shock moves at the velocity $v_0 - a_\infty$. Therefore, if the markings on the wall occur where the tangential wave crest intersects the leading shock (thereby providing local maxima of pressure and temperature), the slope of the helical path will be proportional to $a_\infty/(v_0 - a_\infty)$. The resulting predictions are in reasonable agreement with experiment.

The acoustic aspect of the theory suggests that transverse modes other than the spinning (traveling-wave) mode may occur. In particular, in tubes with noncircular cross sections, purely spinning modes are forbidden. Experiments show that more-complex modes exist, but their identification is more difficult. The number of transverse nodes in the tube is observed to increase with an increase in the tube diameter or with an increasing departure from the limit of detonability. In this respect, with the tacit understanding that the transverse waves often do not involve rotation, detonations in mixtures far removed from the limit of detonability may be described roughly as multiple-headed spinning detonations. Depending on the experimental conditions, the resulting smoked-foil patterns on walls may be very regular or highly irregular.

The experimental prevalence of multiheaded spin has prompted categorical statements to the effect that all self-sustaining detonations exhibit significant three-dimensional structure [25]. The belief that no Chapman-Jouguet detonations are planar rests on the conclusion that all such planar waves are unstable to certain nonplanar disturbances. This conclusion is difficult to substantiate in general because of the complexity of the needed stability analysis. However, a sufficient number of stability studies have now been performed [77]–[92] to provide a good qualitative understanding of the instability question. The results indicate that few, if any, real combustible gas mixtures admit stable planar Chapman-Jouguet waves.

The stability investigations address the time-dependent, inviscid conservation equations having the ZND steady-state solution. The analytical work introduces linearizations about this steady-state solution and employs various different approximations for the rates of chemical reactions. More recent numerical studies have dealt with the nonlinear time-dependent equations, usually in a one-dimensional approximation (for example, [28], [87], and [92]). The general idea that emerges is that if the overall rate of heat release increases sufficiently rapidly with increasing temperature or extent of reaction, then a mechanism exists for amplification of disturbances. The amplification rate varies with the transverse wavelength and may achieve a maximum at a particular wavelength but apparently does not vanish at either large or small wavelengths. In particular, the ZND structure usually is unstable even to planar disturbances, thereby giving rise to pulsating detonations; however, these may not represent the most unstable mode and therefore may not be observed experimentally.

It should be understood that since the stability predictions involve reaction-rate properties, planar Chapman-Jouguet detonations are stable for suitable rate functions. For example, if the rate of heat release decreases monotonically with an increasing extent of reaction behind the shock, then the mechanism for the instability is absent. The failure to find Chapman-Jouguet detonations without transverse structures reflects the inability to encounter real chemical systems with reaction-rate properties suitable for stability.

To obtain a rough physical understanding of the mechanism of the instability, attention may be focused first on a planar detonation subjected to a one-dimensional, time-dependent perturbation. Since the instability depends on the wave structure, a model for the steady detonation structure is needed to proceed with a stability analysis. As the simplest structure model, assume that properties remain constant at their Neumann-spike values for an induction distance l_e, after which all of the heat of combustion is released instantaneously. If v is the gas velocity with respect to the shock at the Neumann condition, then l_e may be expressed approximately in terms of the explosion time t_e, given by equation (B-57) as $l_e = vt_e$. From normal-shock relations for an ideal gas with constant specific heats in the strong-shock limit, the Neumann-state conditions are expressible by $v/v_0 = \rho_0/\rho = (\gamma - 1)/(\gamma + 1)$, $p = p_0 M_0^2 2\gamma/(\gamma + 1)$, and $T = T_0 M_0^2 2\gamma(\gamma - 1)/(\gamma + 1)^2$. If v_0 is perturbed by an amount δv_0, then l_e is perturbed by an amount δl_e, where

$$\frac{\delta l_e}{l_e} = \frac{\delta v}{v} + \frac{\delta t_e}{t_e} = \frac{\delta v}{v} - \frac{E}{R^0 T}\frac{\delta T}{T} - \alpha\frac{\delta T}{T} - (n - 1)\frac{\delta p}{p},$$

in which n is the overall order of the reaction (the pressure exponent) and α measures the temperature exponent of the preexponential factor, the quantity B_0 in equation (B-57) having been made proportional to $p^n T^{\alpha+1}$ [see equation (B-51)]. Use of the strong-shock relations then gives

$$\delta l_e/l_e = (\delta v_0/v_0)[1 - 2E/(R^0 T) - 2\alpha - 2(n - 1)], \tag{23}$$

from which it may be seen that $\delta l_e/\delta v_0 < 0$ if $E/(R^0 T) + \alpha + (n - 1) > \frac{1}{2}$.

Assume that an external perturbation of the leading shock introduces $\delta v_0 > 0$. This locally causes an increase in p and T that is convected downstream at velocity v. The consequent decrease in the ignition time then causes an upstream displacement ($\delta l_e < 0$) of the heat-release front when the perturbation arrives at l_e if $E/(R^0 T) + \alpha + (n - 1) > \frac{1}{2}$. This local increase of heat release generates a compression wave, somewhat like a piston motion, which propagates upstream at velocity $a - v$ with respect to the leading shock (where a is the sound speed at the Neumann state) and overtakes the shock, thereby strengthening it and increasing v_0 further. A characteristic period for this instability mechanism, which involves convection in the downstream direction followed by propagation of an acoustic wave in the upstream direction, is

$$\frac{l_e}{v} + \frac{l_e}{a - v} = \frac{t_e}{(1 - M)} \approx \frac{l_e}{v_0}\left(\frac{\gamma + 1}{\gamma - 1}\right)\left[1 - \sqrt{\frac{\gamma - 1}{2\gamma}}\right]^{-1} \sim 10\frac{l_e}{v_0},$$

where $M \approx \sqrt{(\gamma - 1)/(2\gamma)}$ is the Mach number at the Neumann state.

The condition for occurrence of the one-dimensional instability mechanism is thus seen to be that $E/R^0 T$ be sufficiently large. For a sufficiently strong Chapman-Jouguet wave, it may be shown by use of equations (2-34), (5-29), and (5-30) that, in terms of the specific heat c_p and the heat release q per unit mass of the mixture, the temperature at the Neumann state is $T = (q/c_p)4\gamma(\gamma - 1)/(\gamma + 1)$; by use of equations (2-31) and (2-32), in terms of the final temperature it is $T = T_\infty 2(\gamma - 1)/\gamma$. Thus instability also requires that $E/R^0 T_\infty$ or $E c_p/R^0 q$ be sufficiently large for Chapman-Jouguet waves. It appears physically that for overdriven waves there will be a further requirement that the heat release be sufficiently large; otherwise, the strength of the compression that is generated may be insufficient to strengthen the leading shock appreciably. In general, if $E/R^0 T$ is so small that equation (23) gives $\delta l_e/\delta v_0 > 0$, generation of an expansion wave instead of a compression would be expected by $\delta v_0 > 0$, and the steady detonation would be anticipated to be stable to planar disturbances. From comparison with more complex results, it is found that the specific criterion on $E/R^0 T$ given above is quite conservative in that detonations are stable for a range of values of $E/R^0 T$ large enough to satisfy the inequality.

The discussion given here has been qualitative and does not constitute a correct stability analysis. Even for the simplified model adopted, consideration should be given to jump conditions for interactions at the shock and reaction fronts, to allowance for outgoing waves beyond these boundaries, and to inclusion of acoustic waves propagating in both directions in between; equation (4-3) may be taken as an integral condition for defining l_e, with the integral performed in a Lagrangian frame. A proper analysis of the simplified model would be too lengthy to be included here. Moreover, it is better to take into account distributed heat release, as is done in current analyses (for example, [92]). Even the linear stability analyses then become quite complicated. Simplifications in the stability investigations and greater generality in presentation of results might be achieved by formal application of asymptotic analysis for large values of $E/R^0 T$ with the distributed reaction. Although some consideration has been given to large values of this parameter, the full power of activation-energy asymptotics has not yet been brought to bear on the problem. Accurate analyses for real systems include such factors as multistep chemistry and variable heat capacities and become very complicated.

Thus far our stability discussion has been restricted to one-dimensional instabilities. Mathematical formulations of multidimensional instability problems are readily written [78]–[85], [88], and although the resulting mathematical problems are not appreciably more complicated than those for one-dimensional perturbations, it remains difficult to draw general conclusions from analyses. A stability approach in terms of ray trajectories for transverse acoustic waves helps to clarify physically the character of the multidimensional instability [25], [93]. The types of processes discussed above—generation of acoustic waves by heat release in hot, high-pressure regions and their subsequent propagation (now nonplanar) and interaction with the leading shock—are involved in the instability mechanism. Simplified physical models may be developed [23] based on ideas of nonplanar wave propagation. In one such model [77] a disturbance with a transverse wavelength of order l_e is considered within the context of the simplified structure discussed above, and a criterion for instability is proposed on the basis of isentropic propagation of the wave generated at the reaction front. Although simplified models may aid in physical understanding, their results [including that obtained from equation (23)] are inaccurate and only suggestive. There is need for development of a better understanding of the instability mechanism.

6.3.2. Theories of transverse structures

Instability analyses do not provide good indications of fully developed transverse structures of detonations because these structures correspond to highly nonlinear phenomena. A great deal of nonlinear evolution would occur between onset of instability and attainment of a mature multidimensional detonation structure. Intersections of oblique shocks are known to constitute a central element in transverse structures of detonations [69], [72]. Oblique-shock relations are therefore relevant to the nonplanar structure.

In a frame of reference fixed with respect to an oblique shock, let ψ be the angle between the approach velocity and the shock plane and δ be the deflection angle of the flow upon passage through the shock, as illustrated in Figure 6.9 for the wave labeled incident shock. For an ideal gas with constant specific heats, it is then readily shown [94] that in the adopted frame,

$$M_2^2 = \frac{2 + (\gamma - 1)M_1^2}{2\gamma M_1^2 \sin^2 \psi - (\gamma - 1)} + \frac{2M_1^2 \cos^2 \psi}{2 + (\gamma - 1)M_1^2 \sin^2 \psi}, \tag{24}$$

$$\cot \delta = \tan \psi \left[\frac{(\gamma + 1)M_1^2}{2(M_1^2 \sin^2 \psi - 1)} - 1 \right], \tag{25}$$

$$\frac{p_2}{p_1} = \frac{2\gamma M_1^2 \sin^2 \psi - (\gamma - 1)}{\gamma + 1} \tag{26}$$

and

$$\frac{\rho_2}{\rho_1} = \frac{p_2}{p_1}\frac{T_1}{T_2} = \frac{(\gamma + 1)M_1^2 \sin^2 \psi}{(\gamma - 1)M_1^2 \sin^2 \psi + 2}, \tag{27}$$

where the subscripts 1 and 2 identify conditions upstream and downstream from the shock, respectively. If two intersecting shocks are considered with different angles ψ in the same uniform approach flow, then it is found to be impossible for both δ and p_2, given by equations (25) and (26), to be the same for both shocks; that is, steady double-shock interactions do not exist. However, it may be shown [95] that steady triple-shock intersections do exist, as illustrated in Figure 6.9. These triple-shock interactions lie at the core of understanding of multidimensional detonation structures.

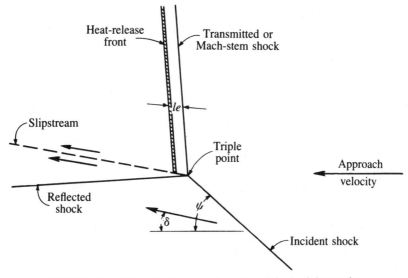

FIGURE 6.9. Schematic diagram of reactive triple-shock interaction.

The wave labeled incident shock in Figure 6.9 leads the other waves and differs from them in that in a frame fixed with respect to the wave pattern, the component of the approach velocity tangential to the shock is directed toward the point of intersection. It is in this respect that the leading shock is "incident," or incoming, and the others are outgoing. The incident shock must be sufficiently oblique to have $M_2 > 1$. The transmitted, or Mach-stem, shock that it intersects is more nearly normal, hence stronger, and it generally has a subsonic downstream Mach number ($M_2 < 1$). To satisfy continuity of the pressure and flow direction across the illustrated slipstream, the indicated reflected shock is needed. This reflected shock is the weakest of the three. The triple point is the point at which the three shocks meet. The pattern in Figure 6.9 is conveniently viewed as a limiting case of a problem in which a shock in one medium is incident on another medium [95], the limit being that in which the two media become identical. The two media are separated along a streamline through the triple point and the slipstream, and the Mach-stem shock is the shock transmitted into the second medium, the analogy with ray optics then being evident.

The steady wave pattern described here is well known in nonreactive gasdynamics. For given values of M_1 and ψ for the incident shock, the configuration is determined more or less uniquely (see [95], for example, for a discussion of uniqueness). In self-sustaining detonations the incident shock may not be sufficiently strong to initiate chemical heat release in the time available, but the strong Mach-stem shock usually is. In Figure 6.9 a heat-release front therefore is shown behind the Mach-stem shock. The streamwise distance from the Mach-stem shock to the heat-release front is the induction distance. For the conditions represented by Figure 6.9, the induction distance behind the incident or reflected shock has been taken to be too long for a reaction front to be shown in the figure. The heat release, of course, will modify the flow field and change the conditions from those calculated for nonreactive wave interactions.

A relatively simple manner in which triple-shock intersections may combine to produce transverse structures for detonations is illustrated in Figure 6.10. This figure represents wall traces left by a detonation with a two-dimensional structure having conditions that are uniform in the direction normal to the figure. The markings on the smoked wall are produced by triple points like the one illustrated in Figure 6.9.* The triple points are observed to propagate along curved paths and to collide periodically. The shock-wave pattern is indicated in Figure 6.10 at one point along one of the markings. At this point the incident and reflected shocks extend above the

---

* This has been demonstrated experimentally beyond doubt by A. K. Oppenheim and co-workers, who employed high-speed photographic techniques to photograph simultaneously in real time the wave pattern and the wall markings as they develop during detonation propagation.

Direction of propagation
of detonation

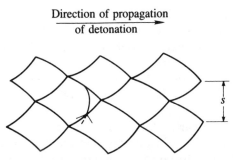

FIGURE 6.10. Sketch of a regular pattern of smoked-foil inscriptions on the wall of a detonation tube with a rectangular cross section.

trajectory of the triple point in the figure, and the Mach-stem shock and slipstream extend below. The figure indicates that the incident shock is slightly curved and extends to another symmetrically located triple point on an adjacent wall trace. All the shocks in the detonation in fact are curved and are intensified mainly where the triple points collide; they then tend to decay therafter. For the pattern in Figure 6.10, the incident and Mach-stem shocks both are strong enough to generate heat release at some distance behind them, but the rate of this heat release is considerably less than that occurring at the triple-point collisions. The markings in Figure 6.10 may be viewed as corresponding to a standing transverse mode in a tube.

A readily measurable characteristic of the detonation illustrated in Figure 6.10 is the transverse spacing s shown, often called the **cell size**, since detonations with transverse structures have been called cellular detonations. Calculation of s is a primary objective of theories of transverse structures. The spacing is regular in Figure 6.10; considerably more complex patterns with both regular and irregular spacings have been observed and explained at least partially on the basis of various types of triple-point trajectories. Transverse spacings in self-sustaining Chapman-Jouguet detonations are observed to decrease with an increasing pressure or with an increasing amount of heat release of the combustible mixture. For spinning detonations the transverse spacings are of the same order of magnitude as the tube diameter.

Theories for calculating s are based on simplifications that adopt ray acoustics for analyzing the propagation of trapped transverse waves [25], [93], [96]–[98] or on attempts to integrate numerically the full two-dimensional, time-dependent equations of gasdynamics with shock discontinuities and with finite rates of heat release [27], [49], [99], [100]. Even the former approach usually employs numerical integrations for calculating ray trajectories [97], and in spite of these complexities, the best theoretically predicted values of s differ from the observed values by amounts on the order of a factor of two. In principle, numerical integration of the full equations describes the process correctly, but although these calculations have produced interesting wave dynamics similar to those observed, the computations are time-consuming, even with the largest electronic computers [100]. Moreover, idealizations and simplifications providing analytical results usually lie at the heart of theories that develop impoved understanding. Thus far, analytical theories for transverse structures of detonations have been developed only on the basis of the acoustic approximation for transverse rays [98].

The spacing s varies with the initial pressure, temperature, and chemical composition in the same way that the induction length l_e varies. Thus the ratio s/l_e may be taken as constant in a first approximation. The value of s is appreciably larger than l_e; perhaps $s/l_e \approx 10$ or 30 [101]. Thus s/v_0 may be on the order of the period of the longitudinal disturbance, identified in the preceding subsection. The fundamental reasons for this are not entirely clear.

6.3.3. Standing detonations

Experimentally, the standing detonations, defined at the beginning of this chapter, do not exhibit the transverse structures discussed above [102]–[106]. For these strong waves the experimental environment is steady and three-dimensional (usually axially symmetric). The velocity is subsonic behind the normal shock existing in the flow before the detonation is formed, and the rarefaction discussed in Section 6.2.1 is not present. Therefore, the fact that these are strong detonations is entirely consistent with the boundary conditions. The length of the reaction zone downstream from the normal shock typically is comparable with the characteristic longitudinal length of the wave pattern downstream from the nozzle exit. Therefore, it is currently unclear whether transverse-wave patterns would develop if the reaction zone were very short in comparison with the external wave-pattern dimension. Based on the stability considerations of Section 6.3.1, it might be anticipated that a transverse structure would occur if the wave conditions were near enough to the Chapman-Jouguet conditions but that the ZND structure would be observed for sufficiently strongly overdriven waves. These standing detonations, or variants thereof, have been suggested for use in ramjet combustion chambers for supersonic propulsion at high Mach numbers [107]–[110].

6.3.4. Detonability limits and quenching thickness

A number of practically important phenomena associated with detonations pose fundamentally more difficult theoretical problems than those addressed above. These include the initiation of detonations, transmission of a detonation from confined to unconfined spaces, quenching of a detonation, and limits of detonability. Discussion of these phenomena may conveniently be introduced by considering quenching of detonations first.

It is found experimentally that self-supported propagation of a detonation through a combustible mixture contained in a tube with rigid walls cannot occur if the tube diameter is less than a critical value, a quenching diameter for the detonation. Similarly, self-supported propagation does not occur through a combustible mixture contained between two rigid, parallel, infinite plates if the distance between the plates is less than the critical quenching distance for the detonation. Replacement of one of the two plates by an infinite, quiescent, inert gas again results in failure of the detonation if the thickness of the combustible layer is less than the critical quenching thickness for the detonation, the value of which appreciably exceeds the detonation quenching distance of the same mixture [111]–[113]. For an infinite, planar slab of combustible bounded on both faces by an inert gas, the critical slab thickness for self-sustained detonation is approximately twice the quenching thickness. Bases have been laid for the development of a degree of theoretical understanding of these phenomena of quenching of detonations [111], [114]–[121].

Although the development of a description of detonation quenching necessitates consideration of detonation structure, an approximate physical understanding of the phenomenon may be obtained without addressing the complexities of cellular structures of detonations. The quenching typically occurs near conditions of spin, for example, with at most a small number of cells present, so that the wave that is quenched may be approximated as being planar and noncellular; experimentally, cellular structures often are not observed during quenching [111]–[113]. Therefore, we may consider a model like that illustrated in Figure 6.7, with a reaction zone of length l following a normal shock. As discussed in Section 6.2.3, for a detonation traveling in a tube the boundary-layer growth has the effect of an area expansion on the quasi-one-dimensional reaction zone and therefore tends to reduce the Mach number, while the heat release tends to increase the Mach number. It seems logical that for the detonation to exist the latter effect must exceed the former at the Neumann state; otherwise behind the leading shock the subsonic Mach number will begin to decrease further, with an associated temperature and reaction-rate decrease, and the system will not be likely to be able to recover to reach the Chapman-Jouguet state. Thus a quenching criterion is that the area-expansion effect exceeds the heat-release effect at the Neumann state.

If A is the effective cross-sectional area of the quasi-one-dimensional flow and T^0 is the stagnation temperature, then for an ideal gas with constant specific heats [122] in variable-area, isentropic flow,

$$\frac{dM}{M}\left[1 - \frac{\frac{\gamma+1}{2}M^2}{1 + \frac{\gamma-1}{2}M^2}\right] = -\frac{dA}{A}, \tag{28}$$

while in constant-area flow with heat addition,

$$2\frac{dM}{M}\left[1 - \frac{2\gamma M^2}{1 + \gamma M^2} + \frac{[(\gamma - 1)/2]M^2}{1 + [(\gamma - 1)/2]M^2}\right] = \frac{dT^0}{T^0}. \tag{29}$$

According to the specified criterion, the detonation exists only if at the Neumann state, dT^0/T^0 divided by the coefficient of dM/M on the left-hand side of equation (29) exceeds dA/A divided by the coefficient of dM/M on the left-hand side of equation (28). For sufficiently strong leading shocks, the value of M at the Neumann state is such that this criterion reduces approximately to

$$dT^0/T^0 > 2\, dA/A. \tag{30}$$

An equality may be used in equation (30) to define approximately a general condition for detonation quenching. If for order-of-magnitude purposes, dT^0/dx is approximated as T^0/l, then this equality becomes $dA/dx \approx A/(2l)$, in which l may be estimated from experimental results (for example, [123]–[125]).

In applying the general criterion of equation (30) to detonations confined by walls, available approximations [58], [117] may be employed for relating dA/dx to the growth rate of the displacement thickness δ^* of the boundary layer; roughly,

$$dA/dx \approx [\rho_w(v_0 - v)/(\rho v)](4A/H)\, d\delta^*/dx$$

$$\approx (\rho_w/\rho)\{8A/[H(\gamma - 1)]\}\, d\delta^*/dx,$$

where ρ_w is the density of the gas at the wall and H is the hydraulic diameter (four times the ratio of the area to the perimeter of the duct, which is the diameter for a circular tube and twice the height for a channel). In using this formula $d\delta^*/dx$ may be approximated as $\delta^*/2l$, where δ^* is the boundary-layer thickness at the Chapman-jouguet plane; the condition for quenching then becomes very roughly

$$\delta^* \approx (\gamma - 1)H/8. \tag{31}$$

Order-of-magnitude estimates of quenching diameters and of quenching distances may be obtained from equation (31) if boundary-layer theory is employed to estimate δ^*—namely, $\delta^* \approx l/\sqrt{\text{Re}}$—where the Reynolds number is $\text{Re} = \rho l(v_0 - v)/\mu$, in which μ is the coefficient of viscosity at the Neumann state (typically $\text{Re} \gtrsim 10^5$).

To apply equation (30) to the estimation of a quenching thickness for a combustible layer adjacent to an inert gas, we must consider the shock pattern in the inert. An approximate model for this shock pattern is illustrated in Figure 6.11. Although not perfectly consistent with the triple-point discussion related to Figure 6.9, in a first approximation it may be assumed that the detonation generates an oblique shock with deflection

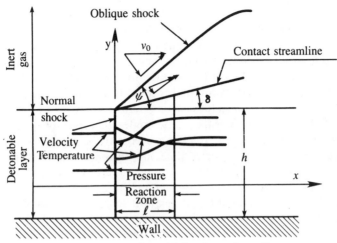

FIGURE 6.11. Schematic diagram of model for detonation adjacent to a compressible medium [121].

angle δ in the inert gas and that this deflection angle also occurs at the boundary of the combustible and thereby imposes a divergence in the cross-sectional area of the reacting flow. The geometry in Figure 6.11 shows that in terms of the thickness h of the detonable layer, $dA/dx = (A/h)\tan \delta$. Equation (25) may be used for δ, with the shock angle ψ obtained in terms of the pressure ratio and the upstream Mach number (M_1) from equation (26). To achieve an approximate lateral pressure balance, the pressure ratio p_2/p_1 may be taken to be the pressure ratio p_+ across the undisturbed detonation given by equation (2-31) for a Chapman-Jouguet wave. With $M_1 = v_0/\hat{a}$, where $\hat{a}$ is the speed of sound in the inert gas, the cited formulas suffice to determine ψ and δ. If $\hat{a} \approx a$, then algebraic manipulations show that for sufficiently strong waves, $\sin \psi \approx 1/\sqrt{2}$; that is, $\psi \approx \pi/4$ and $\tan \delta \approx 1/\gamma$. Thus the approximate condition for quenching, obtained from the order-of-magnitude version of equation (30), is

$$l \approx (\gamma/2)h. \tag{32}$$

Since δ^*/l is small, comparison of equations (31) and (32) shows that the hydraulic diameter for quenching under confinement is smaller than the quenching thickness for the combustible bounded by the compressible medium. The reason for the difference clearly is that the relief provided by the compressible boundary allows $d \ln A/dx$ behind the leading shock to be larger than it can be for a detonation under confinement.

There are procedures for making more accurate estimates of quenching conditions than those outlined here. For example, differences between properties of the inert gas and of the combustible may be taken into account [111]. It is interesting that very weak walls can often effectively provide complete confinement [112] because at the short time scales relevant to passage of the reaction zone, the inertia rather than the strength of the wall is relevant. A wall with mass m per unit area subjected to a pressure p moves a distance $(p/m)t^2/2$ in time t, and confinement will be provided if this is small compared with l during the passage time l/v_0. Thus $(p/m)(l/2v_0^2) \ll 1$ leads to confinement, a condition that is roughly $m \gg \rho_0 l$, since the pressure behind the shock is $p \approx \rho_0 v_0^2$ for a strong detonation, where ρ_0 is the initial density of the combustible. Since solid materials typically have densities on the order of $10^3 \rho_0$, the wall thickness need only be 10^{-3} times the length of the reaction zone to provide confinement in representative experiments. From these results it may be inferred that properties of bounding gaseous media may significantly influence quenching thicknesses.

Quenching of detonations is relevant to the subject of detonability limits. It is known experimentally that combustible gas mixtures diluted sufficiently with inerts, or fuel-oxidant mixtures that are too rich or too lean, cannot be made to support a self-sustained detonation; these mixtures are said to lie outside the detonability limits. In practice, detonability limits are narrower than the flammability limits to be discussed in Chapter 8, in that there are mixtures capable of supporting deflagrations but not detonations, although theoretically this need not always be true because the phenomena differ. Explanations of detonability limits were sought first solely on the basis of chemical-kinetic mechanisms and rates [126], [127] but later it was realized that quenching-thickness phenomena provide better bases for pursuing explanations [111], [116], [117], [120], [128]. According to these later views, detonability limits may be equated to quenching limits for detonations and therefore depend on the size and configuration of the system.

Even with these newer explanations of detonability limits, there tend to be mixtures for which these limits are insensitive to dimensions of the system at sufficiently large dimensions. If the induction time t_e or the reaction-zone length l increases greatly with a small amount of dilution for chemical-kinetic reasons, then in equation (31) or (32) the variation of δ^* or l with dilution will be so great that no matter how large the value of the dimension H or h is within practical bounds, the quenching condition will be achieved approximately at a fixed limit composition, calculable from the equation by use of a representative value for H or h. This type of behavior enables simplified criteria for the nonexistence of self-sustained detonations to be specified—for example, that the temperature at the Neumann state of the Chapman-Jouguet wave lie below a critical value. If chemical-kinetic mechanisms are such that there exists a critical dilution above which l increases strongly with increasing dilution and below which l is much less sensitive to dilution, then this critical dilution may represent a detonability limit determined by the chemistry and effectively independent of the configuration. In this sense, chemical kinetics sometimes may control detonability limits, with quenching details playing a very minor role. The relative importance of the chemical kinetics and of the system configuration to detonability limits for specific combustible mixtures currently is a topic of uncertainty.

6.3.5. The transition from deflagration to detonation

Ways to initiate detonations in detonable mixtures may be subdivided into direct initiation and indirect initiation. In **direct initiation** a sufficient amount of energy is deposited rapidly in the mixture, which results in the immediate emergence of a detonation after an initial period of adjustment, without the intervening establishment of a deflagration. In **indirect initiation** the ignition source is weaker and produces a deflagration, which subsequently undergoes a transition to detonation. Criteria for direct initiation involve the energy of the igniter, while the process of indirect initiation is independent of the ignition mechanism. Since direct initiation fundamentally is a simpler process, it deserves some comment prior to a discussion of the transition from deflagration to detonation.

On the basis of the ZND detonation structure, a simple criterion can be stated for direct initiation. It seems evident that the energy deposited in the combustible medium must be sufficient to maintain a shock at least as strong as the leading shock of a Chapman-Jouguet wave for a time period at least as long as the chemical induction time (or perhaps the overall chemical reaction time) at the Neumann state if the detonation structure is to be established directly by the initiator. This idea was developed by Zel'dovich in the 1940s [129]-[131] and in essence underlies more-recent theories for direct initiation [22], [132]-[140]. Use may be made of the similarity solutions for strong blast waves [141]-[143] to show that the energy (for spherical waves, $k = 3$), the energy per unit length (for cylindrical waves, $k = 2$), or the energy per unit area (for planar waves, $k = 1$) that must be deposited instantaneously in order to initiate a detonation directly is approximately

$$U_k = C_k \Gamma_k(\gamma) \rho_0 v_0^{k+2} t_e^k, \qquad (33)$$

where $C_1 = 1$, $C_2 = 2\pi$, $C_3 = 4\pi$, and the function $\Gamma_k(\gamma)$, dependent on both k and the ratio of specific heats, typically is roughly of order unity [22]. Numerical integrations of the time-dependent conservation equations [144]-[146] tend to provide a little additional support for equation (33). Recent work [139], [140] has focused on clarifying influences of the time history of energy deposition on the initiation conditions; in addition to a minimum critical energy, clearly there must exist a minimum critical power as well because an initiator with a power level that is too low cannot establish a shock wave as strong as the leading shock of the Chapman-Jouguet wave.

Indirect initiation involves a transition from deflagration to detonation. This process has been studied extensively for waves propagating in tubes, and many reviews are available [14]-[28]. The detonation develops through a complicated sequence of unsteady, non-one-dimensional events. The earliest studies of detonation development consisted of experimental measurements of detonation induction distances (the distance required for the steady detonation to form in the tube) [147]–[154]. Observations that the induction distance depends on the gas composition [148], [149] indicate that the chemical reaction rates are of importance, and observations that the induction distance depends on the tube diameter [150] suggest that wall-interaction effects are of importance. Wall roughness, turbulence-producing grids, and vents in the tube also affect the induction distance. Refined experimental techniques have given detailed pictures of the mechanism of transition from deflagration to detonation [16]–[24], [155]–[158]. The process is known to involve acceleration of the deflagration by the expanding hot gases behind the wave, generation of Mach waves by the flame, coalescence of the Mach waves to form shock waves, development of turbulence ahead of and within the deflagration caused by the increased flow velocities, and a complicated wave interaction in the multiwave, turbulent system to produce the final Chapman-Jouguet detonation.

Oppenheim [158]–[161] and others [162], [163] have developed simplified one-dimensional, laminar, multiwave models of the transition process that are capable of correlating many of the experimental observations. The simplest picture [159], [160], [163], involving only two wave discontinuities (a shock wave and a flame), predicts that both waves must propagate at velocities greater than the Chapman-Jouguet speed before they can coalesce to form the Chapman-Jouguet detonation. From a much more detailed graphical analysis [161], it was found to be possible to obtain a wave pattern that was in excellent agreement with the experimental results of [155]. More recent studies employ more-sophisticated computational techniques [164]. The fundamental problems posed by the transition process remain formidable. It is difficult to explain the high turbulent flame speeds needed for coalescence of a deflagration with a shock; localized initiations of explosions by focused pressure waves and a more nearly direct development of a detonation from these energy sources are likely to play an important role that causes deflagration-shock coalescence to be unnecessary. Studies of wave interactions in reacting media [165]–[168] therefore are relevant to the transition process, while one-dimensional time-dependent analyses of fast deflagrations propagating at velocities approaching sonic would not appear to be very relevant.

The importance of the buildup of pressure waves in promoting transition to detonation indicates that confining a combustible mixture by walls aids in the development of a detonation. Detonations are more difficult to initiate in unconfined combustibles [169]. If a transition from deflagration to detonation is to occur in the open, then unusually large flame accelerations are needed, and these are more difficult to achieve without confinement. Numerical integrations of the conservation equations demonstrate that transition to detonation is rare in the open [170]. There are similarity solutions for flame-supported pressure fields in the absence of confinement [171]–[177], and these solutions tend to be stable in that they resist coalescence of the flame with the pressure waves to produce a detonation [173], [177]. Therefore, provision of even partial confinement can significantly enhance the tendency for transition to occur. To obtain improved quantification of the dependence of transition on the degree of confinement is a topic of current research [178].

Since the initiation of a detonation is achieved more readily under confinement, there is interest in establishing the conditions that are necessary for a Chapman-Jouguet detonation traveling in a tube to be transmitted into an unconfined combustible at an open end of the tube [179]–[188]. It is found experimentally that the critical tube diameter needed is much greater than the quenching diameter for the detonation; the value is approximately 13 times the cell size s—that is, roughly on the order of $200l_e$ (see Section 6.3.2). A degree of theoretical understanding of this observation has been developed [183]. To ascertain influences on detonation behavior of the extent of confinement in various configurations remains a topic of active research.

6.4. DETONATIONS IN SOLIDS, LIQUIDS, AND SPRAYS

Our attention thus far has been focused on detonations in gaseous combustibles. There are many liquid and solid materials that are capable of experiencing exothermic decomposition in a detonative mode. These include many solid propellants as well as explosives such as trinitrotoluene (TNT). Detonative processes in liqud and solid combustibles are fundamentally like detonative processes in gases in many ways; the foregoing material is relevant. However, there are special complications that arise from the phase changes and especially from the extremely high pressures produced. The pressures achieved are so high (on the order of 10^5 atm) that an appropriate equation of state for the system is uncertain, and therefore Chapman-Jouguet detonation velocities cannot be calculated by the methods of Chapter 2. In fact, measurement of detonation velocities becomes an important way to obtain information about the equation of state through use of Chapman-Jouguet and Hugoniot relations. In addition, fundamentals of reaction-rate theory become hazy at these high pressures. Thus it is unclear whether the instability mechanism of gaseous detonations (Section 6.3.1) is generally operative for solid explosives; experimentally, the reaction zones of high explosives are usually found to be very thin, and local departures from planar waves often seem to be dominated by granular structures of the material. Over larger distances the wave front tends to be curved and appears to propagate in the charge as if it were unconfined because the pressures achieved not only exceed the yield strengths of wall materials but also typically are large enough in acting on the wall inertia to produce a significant displacement over a reaction-zone length. There are critical charge diameters below which detonation cannot occur, much like quenching diameters for detonations in gases, but usually corresponding to unconfined conditions (except for some liquid and dilute explosives). These often are called **failure diameters** because the detonation fails to propagate in charges of smaller diameters.

In solid or liquid explosives, reactive molecules are continually interacting, and limitations on detonation structures associated with molecular mean free paths no longer apply. It becomes entirely possible for significant release of chemical energy to occur within the structure of the leading shock. This fact motivates new approaches to studies of detonation structure on the basis of molecular dynamics [189], [190]. Although the fundamental complexities that are encountered make the problem difficult, further pursuit of these lines of investigation seems desirable.

Questions of practical importance concern failure diameters and conditions necessary for initiation of detonation. Theoretical descriptions of failure conditions in terms of chemical induction times have been developed [191]–[196] for solid and liquid explosives through reasoning that bears a relationship to that of Section 6.3.4 for gases. Susceptibility to initiation is usually measured by impact tests in standard configurations that effectively expose the material to one or more shock waves and result in sensitivity classifications for the material. With this type of initiation, the process of transition from shock to detonation is relevant. Initiation also may occur as a consequence of a deflagration propagating through the material, and the process of transition from deflagration to detonation then is of interest. Although there have been many investigations of these various transition processes for propellants as well as for high explosives (see, for example, [197]–[208]), numerous unknown aspects remain; representative questions concern the homogeneous and heterogeneous rate processes that may occur.

Background information concerning detonations in solid and liquid explosives is available in a number of books [11], [209]–[215]. Additional information may be found not only in the proceedings of the combustion symposia and of the international colloquia on gasdynamics of explosions and reactive systems but also in the proceedings of the international symposia on detonation, published by the U.S. Office of Naval Research. These last symposia have been held periodically, approximately once every 5 years.

Sprays of liquid fuels in gaseous oxidizers can also support detonations. Reasonably extensive theoretical [216]–[223] and experimental [219], [224]–[230] studies of detonations in sprays have been reported in the literature, and reviews are available [231]–[234]. Since the liquid fuels must vaporize and mix with the oxidizer before combustion can occur, it is generally more difficult to initiate detonations in sprays than in gas mixtures. If the fuel has a sufficiently high vapor pressure, then the detonation may propagate through the gas approximately independently of the liquid droplets. Otherwise, breakup and vaporization of fuel droplets is a dominant aspect of the propagation mechanism, and reaction zones therefore are much longer than the reaction zones of gaseous detonations. Incomplete combustion and losses to boundaries then tend to occur, and the detonation velocity correspondingly is less than the theoretical maximum Chapman-Jouguet value. Nevertheless, the existence of detonations in nonvolatile sprays under suitable conditions is now a well-established fact.

Detonations also may occur in solid-gas mixtures (for example, of aluminum powder [235] or coal dust [236]) and have been involved in dust explosions in grain elevators, for example. When the solid fuel particles are sufficiently fine, the heterogeneous reactions can occur rapidly enough to support nearly Chapman-Jouguet waves in systems of dimensions encountered in practice. The mechanisms of propagation of these detonations are not well understood. Among other configurations in which detonations have been observed and studied are thin films of liquid fuels on surfaces adjacent to oxidizing gases [237]-[242]; some degree of theoretical understanding of these detonations has been developed [243]-[247]. It may be said in general that for any system capable of experiencing combustion or exothermic chemical reactions, the possibility of detonation deserves consideration at least for safety reasons, if not for scientific reasons.

REFERENCES

1. M. Berthelot and P. Vieille, *Compt. Rend. Acad. Sci., Paris* **93**, 18 (1881).
2. E. Mallard and H. L. le Chatelier, *Compt. Rend. Acad. Sci., Paris* **93**, 145 (1881).
3. D. L. Chapman, *Phil. Mag.* **47**, 90 (1899).
4. E. Jouguet, *J. Mathématique* **6**, 347 (1905); *Méchanique des Explosifs*, O. Doin, Paris (1917).
5. Y. B. Zel'dovich, *Zhur. Eksp. Teor. Fiz.* **10**, 542 (1940), English translation, NACA Tech. Memo. No. 1261 (1950); see also "Theory of Combustion and Detonation of Gases," *Tech. Rept. No. F-TS-1226-IA (GDAM A9-T-45)*, Wright Patterson Air Force Base, Dayton (1949).
6. J. von Neumann, "Theory of Detonation Waves," *Prog. Rept. No. 238* (April 1942) *O.S.R.D. Rept. No. 549* (1942).
7. W. Döring, *Ann. Physik* **43**, 421 (1943); W. Döring and G. Burkhardt, "Contribution to the Theory of Detonation," *Tech. Rept. No. F-TS-1227-IA (GDAM A9-T-46)*, Wright-Patterson Air Force Base, Dayton (1949).
8. B. Lewis and G. von Elbe, *Combustion, Flames and Explosions of Gases*, 1st ed., Academic Press, New York, (1951), 579–627; 2nd ed. (1961), 511–554.
9. R. Courant and K. O. Friedrichs, *Supersonic Flow and Shock Waves*, New York: Interscience (1948), 204–234.

10. J. O. Hirschfelder, C. F. Curtiss, and R. B. Bird, *The Molecular Theory of Gases and Liquids*, New York: Wiley, 1954, 797–813.

11. A. R. Ubbelohde and J. Capp, "Detonation Processes in Gases, Liquids and Solids," in *Combustion Processes*, vol. II of *High Speed Aerodynamics and Jet Propulsion*, B. Lewis, R. N. Pease, and H. S. Taylor, eds., Princeton: Princeton University Press, 1956, 577–612.

12. B. P. Mullins and S. S. Penner, *Explosions, Detonations, Flammability and Ignition*, New York: Pergamon Press, 1959, 41–72.

13. Y. B. Zel'dovich and A. S. Kompaneets, *Theory of Detonation*, New York: Academic Press, 1960.

14. R. Gross and A. K. Oppenheim, *ARS Journal* **29**, 173 (1959).

15. J. A. Fay, *8th Symp.* (1962), 30–40.

16. A. K. Oppenheim, N. Manson, and H. G. Wagner, *AIAA Journal*, **1**, 2243 (1963).

17. R. A. Strehlow, *C & F* **12**, 81 (1968).

18. J. H. S. Lee, R. I. Soloukhin, and A. K. Oppenheim, *Astronautica Acta* **14**, 565 (1969).

19. J. H. S. Lee, *Astronautica Acta* **17**, 455 (1972).

20. A. K. Oppenheim and R. I. Soloukhin, *Annual Review of Fluid Mechanics* **5**, 31 (1973).

21. J. H. S. Lee, *Annual Review of Physical Chemistry* **28**, 75 (1977).

22. J. H. S. Lee, *Annual Review of Fluid Mechanics* **16**, 311 (1984).

23. K. I. Shchelkin and Y. K. Troshin, *Gasdynamics of Combustion*, Baltimore: Mono Book Corp., 1965.

24. R. I. Soloukhin, *Shock Waves and Detonations in Gases*, Baltimore: Mono Book Corp., 1966.

25. R. A. Strehlow, *Fundamentals of Combustion*, Scranton, Pa.: International Textbook Co., 1968, 287–352.

26. A. K. Oppenheim, *Introduction to Gasdynamics of Explosions*, New York: Springer-Verlag, 1972.

27. C. L. Mader, *Numerical Modeling of Detonations*, Berkeley: University of California Press, 1979.

28. W. Fickett and W. C. Davis, *Detonation*, Berkeley: University of California Press, 1979.

29. J. O. Hirschfelder and C. F. Curtiss, *J. Chem. Phys.* **28**, 1130 (1958); C. F. Curtis, J. O. Hirschfelder, and M. P. Barnett, "Theory of Detonations, III," *Tech. Rept. No. WIS-AF-9*, University of Wisconsin, Madison (1958).

30. W. D. Hayes, "The Basic Theory of Gasdynamic Discontinuities," in *Fundamentals of Gas Dynamics*, vol. III of *High Speed Aerodynamics and Jet Propulsion*, H. W. Emmons, ed., Princeton: Princeton University Press, 1958, 417–481.

31. W. I. Axford, *Phil. Trans. Roy. Soc. London* **253A**, 301 (1961).

32. R. C. Newman and W. I. Axford, *Astrophysical Journal* **151**, 1145 (1968).

33. A. K. Oppenheim and J. Rosciszewski, *9th Symp.* (1963), 424–441.

34. G. B. Kistiakowsky and P. H. Kydd, *J. Chem. Phys.* **22**, 1940 (1954); **23**, 271 (1955); **25**, 824 (1956).

35. D. H. Edwards, G. T. Williams, and J. C. Breeze, *J. Fluid Mech.* **6**, 947 (1959).

36. B. Levitt and D. F. Hornig, *J. Chem. Phys.* **36**, 219 (1962).

37. C. Guerraud, J. C. Leyer, and C. Brochet, *Compt. Rend.* **264**, 5 (1967).

38. D. H. Edwards, *12th Symp.* (1969), 819–828.

39. H. M. Mott-Smith, *Phys. Rev.* **82**, 885 (1951).
40. W. G. Vincenti and C. H. Kruger, Jr., *Introduction to Physical Gas Dynamics*, New York: Wiley, 1965, 412–424.
41. R. Becker, *Z. Physik* **8**, 321 (1922).
42. M. Morduchow and P. A. Libby, *J. Aero. Sci.* **16**, 674 (1949).
43. P. A. Libby, *J. Aero. Sci.* **18**, 286 (1951).
44. W. W. Wood, *Phys. Fluids* **4**, 46 (1961).
45. T. C. Adamson, *Phys. Fluids* **3**, 706 (1960).
46. J. R. Bowen, *Phys. Fluids* **10**, 290 (1967).
47. W. B. Bush and F. E. Fendell, *CST* **2**, 271 (1971).
48. D. A. Beckstead, "Theory of Shock-Supported Arrhenius Reactions and Ignition Delay," Ph.D. Thesis, Department of Mechanical Engineering, University of Southern California, Los Angeles (1973).
49. E. S. Oran, J. P. Boris, T. Young, M. Flanigan, T. Burks, and M. Picone, *18th Symp.* (1981), 1641–1649.
50. S. R. Brinkley, Jr. and J. G. Kirkwood, *3rd Symp.* (1949), 586–590.
51. S. G. Reed, Jr., *J. Chem. Phys.* **20**, 1823 (1952).
52. A. R. Ubbelohde, *4th Symp.* (1953), 464–467; see also the discussion by B. Lewis, p. 467.
53. K. Oswatitsch, *Z. Agnew. Math. Mech.* **22**, 1 (1942).
54. H. G. Stever, "Condensation Phenomena in High Speed Flows," in *Fundamentals of Gas Dynamics*, vol. III of *High Speed Aerodynamics and Jet Propulsion*, edited by H. W. Emmons, Princeton: Princeton University Press, 1958, 526–573.
55. P. P. Wegener, "Gasdynamics of Expansion Flows with Condensation and Homogeneous Nucleation of Water Vapor," Chapter 4 of *Nonequilibrium Flows*, vol. 1, part 1, P. P. Wegener, ed., New York: Marcel Dekker, 1969.
56. R. A. Dobbins, *Journal of Fluids Engineering*, **105**, 414 (1983).
57. H. M. Peek and R. G. Thrap, *J. Chem. Phys.* **26**, 740 (1957).
58. J. A. Fay, *Phys. Fluids* **2**, 283 (1959).
59. J. G. Kirkwood and W. W. Wood, *J. Chem. Phys.* **22**, 1915 (1954).
60. W. W. Wood and Z. W. Salsburg, *Phys. Fluids* **3**, 549 (1960).
61. J. J. Erpenbeck, *Phys. Fluids* **4**, 481 (1961).
62. D. R. White, *Phys. Fluids* **4**, 465 (1961).
63. R. B. Morrison, "A Shock Tube Investigation of Detonative Combustion," *Rept. UMM-97*, University of Michigan Engineering Research Institute, Ann Arbor (1952).
64. J. A. Fay, *4th Symp.* (1953), 501–507.
65. A. Lifshitz, editor, *Shock Waves in Chemistry*, New York: Marcel Dekker, 1981.
66. C. Campbell and D. W. Woodhead, *J. Chem. Soc.*, 3010, (1926), 1572 (1927).
67. W. A. Bone and R. P. Fraser, *Phil. Trans. Roy. Soc.*, London **288A**, 197 (1929); **230A**, 363 (1932).
68. R. E. Duff and H. T. Knight, *J. Chem. Phys.* **20**, 1493 (1952); R. E. Duff, H. T. Knight, and H. R. Wright, *J. Chem. Phys.* **22**, 1618 (1954).
69. R. E. Duff, *Phys. Fluids* **4**, 1427 (1961).
70. J. E. Dove and H. G. Wagner, *8th Symp.* (1962), 589–599.
71. B. V. Voitsekhovskii, V. V. Mitrofanov, and M. E. Topchain, *Zhur. Prikl. Mekh. Tekhn. Fiz.*, no. 3, 27 (1962).
72. G. L. Schott, *Phys. Fluids* **8**, 850 (1965).

73. N. Manson, *Compt. Rend. Acad. Sci., Paris* **222**, 46 (1946).
74. Y. B. Zel'dovich, *Dokl. Akad. Nauk SSSR* **52**, 147 (1946).
75. J. A. Fay, *J. Chem. Phys.* **20**, 942 (1952).
76. B. T. Chu, *Proceedings of the Gas Dynamics Symposium on Aerothermochemistry*, Northwestern University, Evanston (1956), 95–111.
77. K. I. Shchelkin, *Zhur. Eksp. Teor. Fiz.* **36**, 600 (1959).
78. R. M. Zaidel, *Dokl. Akad. Nauk. SSSR* **136**, 1142 (1961).
79. J. A. Fay, "Stability of Detonation Waves at Low Pressures," in *Detonation and Two-Phase Flow*, vol. 6 of *Progress in Astronautics and Rocketry*, S. S. Penner and F. A. Williams, eds., New York: Academic Press, 1962, 3–16.
80. J. J. Erpenbeck, *Phys. Fluids* **5**, 604 (1962).
81. R. M. Zaidel and Y. B. Zel'dovich, *Zhur. Prikl. Mekh. Tekhn. Fiz.*, no. 6, 59 (1963).
82. V. V. Pukhnachev, *Dokl. Akad. Nauk SSSR* **149**, 798 (1963).
83. J. J. Erpenbeck, *Phys. Fluids* **7**, 684 (1964).
84. J. J. Erpenbeck, *9th Symp.* (1963), 442–453.
85. J. J. Erpenbeck, *Phys. Fluids* **8**, 1192 (1965).
86. L. A. Il'kaeva and N. A. Popov, *Fiz. Gor. Vzr.* **1**, 20 (1965).
87. W. Fickett and W. W. Wood, *Phys. Fluids* **9**, 903 (1966).
88. J. J. Erpenbeck, *Phys. Fluids* **9**, 1293 (1966).
89. G. G. Chernyi and S. A. Medvedev, *Astronautica Acta* **15**, 371 (1970).
90. W. Fickett, J. D. Jacobson, and G. L. Schott, *AIAA Journal* **10**, 514 (1972).
91. P. Howe, R. Frey, and G. Melani, *CST* **14**, 63 (1976).
92. G. E. Abouseif and T. Y. Toong, *C & F* **45**, 67 (1982).
93. H. O. Barthel and R. A. Strehlow, *Phys. Fluids* **9**, 1896 (1966).
94. L. D. Landau and E. M. Lifshitz, *Fluid Mechanics*, London: Pergamon Press, 1959.
95. L. F. Henderson, *J. Fluid Mech.* **26**, 607 (1966).
96. H. O. Barthel, *Phys. Fluids* **15**, 43 (1972).
97. H. O. Barthel, *Phys. Fluids* **17**, 1547 (1974).
98. K. W. Chiu and J. H. S. Lee, *C & F* **26**, 353 (1976).
99. S. Taki and T. Fujiwara, *AIAA Journal* **16**, 73 (1978).
100. S. Taki and T. Fujiwara, *18th Symp.* (1981), 1671–1681.
101. C. K. Westbrook, "Chemical Kinetics of Gaseous Detonations," in *Fuel-Air Explosions*, J. H. S. Lee and C. M. Guirao, eds., Waterloo, Canada: University of Waterloo Press, 1982, 189–242.
102. R. A. Gross, *ARS Journal* **29**, 63 (1959).
103. J. A. Nicholls, E. K. Dabora, and R. L. Gealer, *7th Symp.* (1959), 766–772.
104. J. A. Nicholls and E. K. Dabora, *8th Symp.* (1962), 644–655.
105. J. A. Nicholls, *9th Symp.* (1963), 488–498.
106. P. M. Rubins and R. P. Rhodes Jr., *AIAA Journal* **1**, 2778 (1963).
107. M. Roy, *Compt. Rend. Acad. Sci., Paris* **222**, 31 (1946).
108. J. A. Nicholls, H. R. Wilkinson, and R. B. Morrison, *Jet Propulsion* **27**, 534 (1957).
109. R. Dunlap, R. L. Brehm, and J. A. Nicholls, *Jet Propulsion* **28**, 451 (1958).
110. R. A. Gross and W. Chinitz, *J. Aero. Sci.* **27**, 517 (1960).
111. E. K. Dabora, J. A. Nicholls, and R. B. Morrison, *10th Symp.* (1965), 817–830.
112. W. P. Sommers, *ARS Journal* **31**, 1780 (1961).
113. T. G. Adams, *AIAA Journal* **16**, 1035 (1978).
114. W. P. Sommers and R. B. Morrison, *Phys. Fluids* **5**, 241 (1962).
115. K. I. Shchelkin, *Fiz. Gor. Vzr.* **4**, 39 (1968).

116. S. Tsugé, H. Furukawa, M. Matsukawa, and T. Nakagawa, *Astronautica Acta* **15**, 377 (1970).
117. S. Tsugé, *CST* **3**, 195 (1971).
118. T. Fujiwara and S. Tsugé, *J. Phys. Soc. Japan* **33**, 237 (1972).
119. S. Tsugé and T. Fujiwara, *Zeit. Angew. Math. Mech.* **54**, 157 (1974).
120. J. E. Dove, B. J. Scroggie, and H. Semerjian, *Acta Astronautica* **1**, 345 (1974).
121. F. A. Williams, *C & F* **26**, 403 (1976).
122. N. A. Hall, *Thermodynamics of Fluid Flow*, Englewood Cliffs, N.J.: Prentice-Hall, 1951.
123. G. B. Kistiakowsky and P. H. Kydd, *J. Chem. Phys.* **25**, 824 (1956).
124. H. G. Wagner, *9th Symp.* (1963), 454–460.
125. P. L. Lu, E. K. Dabora, and J. A. Nicholls, *CST* **1**, 65 (1969).
126. F. E. Belles, *7th Symp.* (1959), 745–751.
127. R. W. Patch, *ARS Journal* **31**, 46 (1961).
128. K. I. Shchelkin, *Usp. Fiz. Nauk* **87**, 273 (1965).
129. Y. B. Zel'dovich, *Zhur. Eksp. Teor. Fiz.* **21**, 3 (1947).
130. Y. B. Zel'dovich, *Zhur. Fiz. Khim.* **23**, 1362 (1949).
131. Y. B. Zel'dovich, S. M. Kogarko, and N. N. Semenov, *Zhur. Tehkn. Fiz.* **26**, 1744 (1956).
132. G. G. Bach, R. A. Knystautas, and J. H. S. Lee, *13th Symp.* (1971), 1097–1110.
133. R. S. Fry and J. A. Nicholls, *15th Symp.* (1975), 43–52.
134. J. H. S. Lee, R. A. Knystautas, and C. M. Guirao, *15th Symp.* (1975), 53–67.
135. R. A. Knystautas and J. H. S. Lee, *C & F* **27**, 221 (1976).
136. J. H. S. Lee and K. Ramamurthi, *C & F* **27**, 331 (1976).
137. M. Sichel, *Acta Astronautica* **4**, 409 (1977).
138. C. W. Wilson and A. A. Boni, *CST* **21**, 183 (1980).
139. G. E. Abouseif and T. Y. Toong, *C & F* **45**, 39 (1982).
140. E. K. Dabora, "The Relationship between Energy and Power for Direct Initiation of Hydrogen-Air Detonations," in *Proceedings of the Second International Conference on the Impact of Hydrogen on Water Reactor Safety*, Sandia National Laboratories, Albuquerque, N.M. (1982), 931–952.
141. L. I. Sedov, *Dokl. Akad. Nauk SSSR* **52**, 17 (1946).
142. G. I. Taylor, *Proc. Roy. Soc., London* **186A**, 273 (1946); **201A**, 159 (1950).
143. V. P. Korobeinikov, *Annual Review of Fluid Mechanics* **3**, 317 (1971).
144. E. Bishimov, "Numerical Solution of a Problem of a Strong Point Explosion in a Detonating Gas," in *Differential Equations and Their Applications*, Alma-Alta: Nauka, (1968).
145. A. A. Boni and C. W. Wilson, *Acta Astronautica* **4**, 409 (1977).
146. A. A. Boni, C. W. Wilson, M. Chapman, and J. L. Cook, *Acta Astronautica* **5**, 1153 (1978).
147. M. Berthelot and P. Vieille, *Compt. Rend. Acad. Sci., Paris* **94**, 101 (1882).
148. H. L. le Chatelier, *Compt. Rend. Acad. Sci., Paris* **130**, 1775 (1900).
149. H. B. Dixon, *Phil. Trans. Roy. Soc., London* **200A**, 316 (1903).
150. P. Laffitte, *Compt. Rend. Acad. Sci., Paris* **176**, 1392 (1923).
151. P. Dumanois and P. Laffitte, *Compt. Rend. Acad. Sci., Paris* **183**, 284 (1926).
152. A. C. Egerton and S. F. Gates, *Proc. Roy. Soc., London* **114A**, 137, 152 (1927); **116A**, 516 (1927).
153. A. S. Sokolik and K. I. Shchelkin, *Fiz. Z.* **4**, 795 (1933); *Zhur. Fiz. Khim.* **5**, 1459 (1934).

154. M. W. Evans, F. I. Given, and W. E. Richeson, *J. Appl. Phys.* **26**, 1111 (1955).
155. E. H. W. Schmidt, H. Steinicke, and U. Neubert, *4th Symp.* (1953), 658–666.
156. F. J. Martin, *Phys. Fluids* **1**, 399 (1958).
157. F. J. Martin and D. R. White, *7th Symp.* (1959), 856–865.
158. A. J. Laderman and A. K. Oppenheim, *Phys. Fluids* **4**, 778 (1961).
159. A. K. Oppenheim, *J. Appl. Mech.* **19**, 63 (1952).
160. A. K. Oppenheim, *4th Symp.* (1953), 471–480.
161. A. K. Oppenheim and R. A. Stern, *7th Symp.* (1959), 837–850.
162. G. B. Kistiakowsky, *Ind. Eng. Chem.* **43**, 2794 (1955).
163. G. I. Taylor and R. S. Tankin, "Transformation from Deflagration to Detonation," in *Fundamentals of Gas Dynamics*, vol. III of *High Speed Aerodynamics and Jet Propulsion*, H. W. Emmons, ed., Princeton: Princeton University Press, 1958, 645–656.
164. E. S. Oran and J. P. Boris, *Prog. Energy and Combust. Sci.* **7**, 1 (1981).
165. J. F. Clarke, *J. Fluid Mech.* **81**, 257 (1977); **89**, 343 (1978); **94**, 195 (1979).
166. J. F. Clarke, *Acta Astronautica* **5**, 543 (1978).
167. P. A. Blythe, *17th Symp.* (1979), 909–916.
168. G. E. Abouseif and T. Y. Toong, *J. Fluid Mech.* **103**, 1 (1981).
169. R. A. Strehlow, *14th Symp.* (1973), 1189–1200.
170. A. A. Boni, M. Chapman, J. L. Cook, and G. P. Schneyer, "Transition to Detonation in an Unconfined Turbulent Medium," in *Turbulent Combustion*, vol. 28 of *Progress in Astronautics and Aeronautics*, L. A. Kennedy, ed., New York: American Institute of Aeronautics and Astronautics, 1978, 379–405.
171. A. L. Kuhl, M. M. Kamel, and A. K. Oppenheim, *14th Symp.* (1973), 1201–1215.
172. C. M. Guirao, G. G. Bach, and J. H. S. Lee, *C & F* **27**, 341 (1976).
173. F. A. Williams, *CST* **12**, 199 (1976).
174. K. W. Chiu, J. H. S. Lee, and R. A. Knystautas, *J. Fluid Mech.* **82**, 193 (1977).
175. J. Kurylo, H. A. Dwyer, and A. K. Oppenheim, *AIAA Journal* **18**, 302 (1980).
176. A. L. Kuhl, "Similarity Analysis of Flame-Driven Blast Waves with Real Equations of State," *Colloque International Berthelot-Vieille-Mallard-le Chatelier, First Specialists Meeting (International) of the Combustion Institute*, Pittsburgh: The Combustion Institute, 1981, 491–496.
177. F. A. Williams, *J. Fluid Mech.* **127**, 429 (1982).
178. R. A. Knystautas, J. H. S. Lee, I. O. Moen, and H. G. Wagner, *17th Symp.* (1979),, 1235–1245.
179. V. V. Mitrofanov and R. I. Soloukhin, *Dokl. Akad. Nauk SSSR* **159**, 1003 (1964).
180. R. I. Soloukhin and K. W. Ragland, *C & F* **13**, 295 (1969).
181. D. H. Edwards, M. A. Nettleton, and G. O. Thomas, *J. Fluid Mech.* **95**, 79 (1979).
182. H. Matsui and J. H. S. Lee, *17th Symp.* (1979), 1269–1280.
183. D. H. Edwards, G. O. Thomas, and M. A. Nettleton, "Diffraction of a Planar Detonation in Various Fuel-Oxygen Mixtures at an Area Change," in *Gasdynamics of Detonations and Explosions*, vol. 75 of *Progress in Astronautics and Aeronautics*, J. R. Bowen, N. Manson, A. K. Oppenheim, and R. I. Soloukhin, eds., New York: American Institute of Aeronautics and Astronautics, 1981, 341–357.
184. P. A. Urtiew and C. M. Tarver, "Effects of Cellular Structure on the Behavior of Gaseous Detonation Waves under Transient Conditions," in *Gasdynamics of Detonations and Explosions*, vol. 75 of *Progress in Astronautics and Aeronautics*, J. R. Bowen, N. Manson, A. K. Oppenheim, and R. I. Soloukhin, eds., New York: American Institute of Aeronautics and Astronautics, 1981, 370–384.

185. I. O. Moen, M. Donato, R. A. Knystautas, and J. H. S. Lee, *18th Symp.* (1981), 1615–1622.
186. R. A. Knystautas, J. H. S. Lee, and C. M. Guirao, *C & F* **48**, 63 (1982).
187. P. Thibault, Y. K. Liu, C. Chan, J. H. S. Lee, R. A. Knystautas, C. M. Guirao, B. Hjertager, and K. Fuhre, *19th Symp.* (1983), 599–606.
188. I. O. Moen, S. B. Murray, D. Bjerketvedt, A. Rinnan, R. A. Knystautas, and J. H. S. Lee, *19th Symp.* (1983), 635–644.
189. A. M. Karo, J. R. Hardy, and F. E. Walker, *Acta Astronautica* **5**, 1041 (1978).
190. A. N. Dremin and V. Y. Klimenko, "The Effect of the Shock-Wave Front on the Origin of Reaction," in *Gasdynamics of Detonations and Explosions*, vol. 75 of *Progress in Astronautics and Aeronautics*, J. R. Bowen, N. Manson, A. K. Oppenheim, and R. I. Soloukhin, eds., New York: American Institute of Aeronautics, 1981, 253–268.
191. H. Jones, *Proc. Roy. Soc., London* **189A**, 415 (1947).
192. H. Erying, R. F. Powell, G. H. Duffey, and R. B. Parlin, *Chem. Revs.* **45**, 69 (1949).
193. W. W. Wood and J. G. Kirkwood, *J. Chem. Phys.* **22**, 1920 (1954).
194. M. W. Evans, *J. Chem. Phys.* **36**, 193 (1962).
195. A. N. Dremin and V. S. Trofimov, *10th Symp.* (1965), 839–843.
196. H. N. Presles and C. Brochet, "Induction Delay and Detonation Failure Diameter of Nitromethane Mixtures," in *Gasdynamics of Detonations and Explosions*, vol. 75 of *Progress in Astronautics and Aeronautics*, J. R. Bowen, N. Manson, A. K. Oppenheim, and R. I. Soloukhin, eds., New York: American Institute of Aeronautics and Astronautics, 1981, 282–295.
197. M. A. Cook, D. H. Pack, and W. A. Gey, *7th Symp.* (1959), 820–836.
198. H. W. Hubbard and M. H. Johnson, *J. Appl. Phys.* **30**, 765 (1959).
199. A. W. Campbell, W. C. Davis, and J. R. Travis, *Phys. Fluids* **4**, 498 (1961).
200. A. W. Campbell, W. C. Davis, J. B. Ramsay, and J. R. Travis, *Phys. Fluids* **4**, 511 (1961).
201. M. H. Boyer and R. Grandey, "Theoretical Treatment of the Detonation Behavior of Composite Solid Propellants," in *Detonation and Two-Phase Flow*, vol. 6 of *Progress in Astronautics and Rocketry*, S. S. Penner and F. A. Williams, eds., New York: Academic Press, 1962, 75–98.
202. F. P. Bowden, *9th Symp.* (1963), 499–516.
203. S. J. Jacobs, T. P. Liddiard, Jr., and B. E. Drimmer, *9th Symp.* (1963), 517–529.
204. F. J. Warner, *9th Symp.* (1963), 536–544.
205. H. Krier and J. A. Kezerle, *17th Symp.* (1979), 23–34.
206. R. R. Bernecker and D. Price, *17th Symp.* (1979), 55–62.
207. M. Cowperthwaite, "Single-Shock Curve Buildup and a Hydrodynamic Criterion for Initiation of Detonation," in *Gasdynamics of Detonations and Explosions*, vol. 75 of *Progress in Astronautics and Aeronautics*, J. R. Bowen, N. Manson, A. K. Oppenheim, and R. I. Soloukhin, eds., New York: American Institute of Aeronautics and Astronautics, 1981, 269–281.
208. J. J. Dick, *18th Symp.* (1981), 1623–1629.
209. J. Taylor, *Detonation in Condensed Explosives*, Oxford: Oxford University Press, 1952.
210. F. P. Bowden and A. D. Yoffee, *Initiation and Growth of Explosions in Liquids and Solids*, Cambridge: Cambridge University Press, 1952.
211. F. P. Bowden and A. D. Yoffee, *Fast Reactions in Solids*, New York: Academic Press, 1958.

212. M. A. Cook, *The Science of High Explosives*, ACS Monograph No. 39, New York: Reinhold, 1958.
213. J. Berger and J. Viard, *Physique des Explosifs Solides*, Paris: Dunod, 1962.
214. A. N. Dremin, S. D. Savrov, V. S. Trofimov, and K. K. Schvedov, *Detonation Waves in Condensed Media*, Moscow: Nauka, 1970.
215. W. E. Baker, *Explosions in Air*, Austin: University of Texas Press, 1973.
216. F. A. Williams, *Phys. Fluids* **4**, 1434 (1961).
217. F. A. Williams, "Detonations in Dilute Sprays," in *Detonation and Two-Phase Flow*, vol. 6 of *Progress in Astronautics and Rocketry*, S. S. Penner and F. A. Williams, eds., New York: Academic Press, 1962, 99–114.
218. T. H. Pierce and J. A. Nicholls, *14th Symp.* (1973), 1277–1284.
219. J. A. Nicholls, M. Sichel, R. S. Fry, and D. R. Glass, *Acta Astronautica* **1**, 385 (1974).
220. S. A. Gubin and M. Sichel, *CST* **17**, 109 (1977).
221. V. V. Mitrofanov, A. V. Pinaev, and S. Zhdan, *Acta Astronautica* **6**, 218 (1979).
222. E. K. Dabora, *Acta Astronautica* **6**, 269 (1979).
223. S. Eidelman and A. Burcat, *18th Symp.* (1981), 1661–1670.
224. W. T. Webber, *8th Symp.* (1962), 1129–1140.
225. F. B. Cramer, *9th Symp.* (1963), 482–487.
226. E. K. Dabora, K. W. Ragland, and J. A. Nicholls, *Astronautica Acta* **12**, 9 (1966).
227. K. W. Ragland, E. K. Dabora, and J. A. Nicholls, *Phys. Fluids* **11**, 2377 (1968).
228. E. K. Dabora, K. W. Ragland, and J. A. Nicholls, *12th Symp.* (1969), 19–26.
229. Z. Gabrijel and J. A. Nicholls, *Acta Astronautica* **5**, 1051 (1979).
230. R. Bar-Or, M. Sichel, and J. A. Nicholls, *18th Symp.* (1981), 1599–1606.
231. E. K. Dabora and L. P. Weinberger, *Acta Astronautica* **1**, 361 (1974).
232. A. A. Borisov and B. E. Gel'fand, *Arch. Termodynamiki i Spalaia* **7**, 273 (1976).
233. E. K. Dabora, "Fundamental Mechanisms of Liquid Spray Detonations," in *Fuel-Air Explosions*, J. H. S. Lee and C. M. Guirao, eds., Waterloo, Canada: University of Waterloo Press, 1982, 245–264.
234. M. Sichel, "The Detonation of Sprays: Recent Results," in *Fuel-Air Explosions*, J. H. S. Lee and C. M. Guirao, eds., Waterloo, Canada: University of Waterloo Press, 1982, 265–302.
235. W. A. Strauss, *AIAA Journal* **6**, 1753 (1968).
236. K. N. Palmer and P. S. Tonkin, *C & F* **17**, 159 (1971).
237. R. Loison, *Compt. Rend. Acad. Sci.*, Paris **234**, 512 (1952).
238. V. E. Gordeev, V. F. Komov, and Y. K. Troshin, *Dokl. Akad. Nauk SSSR* **160**, 853 (1965).
239. V. F. Komov and Y. K. Troshin, *Dokl. Akad. Nauk SSSR* **175**, 109 (1967).
240. K. W. Ragland and J. A. Nicholls, *AIAA Journal* **7**, 859 (1969).
241. J. R. Bowen, K. W. Ragland, F. J. Steffes, and T. G. Loflin, *13th Symp.* (1971), 1131–1140.
242. K. W. Ragland and C. F. Garcia, *C & F* **18**, 53 (1972).
243. A. A. Borisov, S. M. Kogarko, and A. V. Lyubimov, *Dokl. Akad. Nauk SSSR* **164**, 125 (1965); *Fiz. Gor. Vzr.* **1**, 31 (1965).
244. M. Sichel, C. S. Rao, and J. A. Nicholls, *13th Symp.* (1971), 1141–1149.
245. C. S. Rao, M. Sichel, and J. A. Nicholls, *CST* **4**, 209 (1972).
246. Y. Fujitsuna and S. Tsugé, *14th Symp.* (1973), 1265–1275.
247. Y. Fujitsuna, *Acta Astronautica* **6**, 785 (1979).

CHAPTER 7

Combustion of Solid Propellants

Solid propellants are solid materials that are capable of experiencing exothermic reactions without the addition of any other reactants. They are employed mainly as propellants for rocket vehicles and as propelling charges for projectiles in guns. It is intended that the propellants will deflagrate, because a detonation would be damaging. Therefore, sensitive explosives generally are not employed as solid propellants; solid-propellant formulations must be sufficiently metastable to resist transition to detonation.

Solid propellants conveniently are subdivided into two categories, homogeneous and heterogeneous propellants. In **homogeneous propellants** the solid is homogeneous, the fuel and oxidizer being intimately mixed at the molecular level; in principle, the solid may consist of a single constituent such as nitrocellulose, but in practice it is a mixture of constituents. The most common homogeneous propellants are *double-base propellants*, so named because they are composed mostly of two exothermic compounds (usually nitrocellulose and nitroglycerin), but single-base and triple-base propellants also have been used. In **heterogeneous propellants** the solid is heterogeneous, usually consisting of a continuous fuel matrix or *binder* containing oxidizer particles, thus forming a composite material that is termed a *composite propellant*. There are many different composite propellants; a typical example is a polybutadiene acrylic-acid copolymer as the fuel with crystals of ammonium perchlorate as oxidizer. There are heterogeneous propellants in which one or both of the constituents could serve as a homogeneous propellant (for example, there are *composite-modified double-base propellants*, in which oxidizer crystals are added to the homogeneous propellant, and even the composite-propellant constituent ammonium perchlorate can deflagrate alone at sufficiently high ambient pressures). In addition, various other heterogeneous additives often are used. Thus both double-base and composite propellants may be *metalized*; metal particles (such as aluminum) are added and these burn in the gaseous reaction products of the propellant constituents and enhance performance.

When solid propellants deflagrate, they burn from the surface inward, and the surface is observed to regress normal to itself at a definite rate. An objective of theories of solid-propellant deflagration is to ascertain the dependence of this linear regression rate or burning rate on the ambient pressure, temperature, propellant composition, and so on. In addition to being of fundamental interest, the burning rate is of importance to the design of systems such as rocket motors that employ solid propellants. The present chapter is concerned with theories of deflagration of solid propellants. This subject is used here more as a means for introducing various fundamental concepts than as an accurate derivation of burning rates of real solid propellants. Therefore, most of the attention will be devoted to homogeneous propellants; different approaches to the analyses of their combustion processes will be indicated. Heterogeneous propellants will be considered briefly in Section 7.7, where various elements of their combustion mechanisms will be reviewed.

There are a number of useful background references on solid propellants and their combustion. Most books on rocket propulsion devote at least one chapter to solid propellants [1]–[5]. No less than three books [6]–[8] have been concerned solely with solid-propellant topics, the last [8] being a 700-page text on the subject. Readers may consult this literature for more detailed information.

7.1. DESCRIPTION OF STEADY DEFLAGRATION OF A HOMOGENEOUS SOLID

The simplest examples of deflagrating solids are those in which all reactants and products are in the solid state (for example, mixtures of molybdenum with boron) [9]. These materials may deflagrate in a process in which a single exothermic reaction front propagates through the solid. Although processes of this type are of fundamental interest and also of applied interest in materials processing, gaseous products of combustion are needed in propulsion applications; the simplest example of a solid propellant would be a solid that gasifies in an exothermic rate process at a planar solid-gas interface. Gasification steps more commonly tend to be endothermic. A simple model mechanism for propellant deflagration involves an endothermic sublimation of the solid at an interface, followed by an exothermic gas-phase reaction. A number of liquid monopropellants—for instance, ethyl nitrate— burn by a mechanism that differs from this only in that vaporization replaces the sublimation step [10].

At one time ammonium perchlorate was thought to deflagrate by the simple model mechanism just defined. However, by further study it was found that this solid melts and then vaporizes during burning instead of subliming directly; moreover, significant exothermic reactions were inferred to occur in the liquid phase [11]. A number of energetic crystals behave in this same general way, including materials currently subjected to continuing investigation for use in propellants, such as RDX (1,3,5-trinitrohexahydro-1,3,5-triazine, chemical formula $(NO_2N \cdot CH_2)_3$, one member of a class of materials, called *nitramines*, containing both fuel and oxidizing species in the same molecule) [12].* Unless the liquid layer on the surface of the solid is very thin, bubbles may develop within it in the course of the exothermic, gas producing reaction, and a **foam** reaction zone may exist. The foam zone may be observed, for example, in the burning of double-base propellants with sufficiently high contents of nitroglycerin. At low nitroglycerin fractions, the foam zone typically is less prominent than a **fizz** reaction zone, a gas-phase region in which condensed particles that leave the surface of the homogeneous solid experience exothermic gasification. Nitrocellulose burns with a subsurface reaction zone in which the solid softens appreciably, followed by a fizz zone, and then, after a dark zone, by a luminous gas-phase flame.

Because of these complex combustion mechanisms, analysis of the deflagration of homogeneous solid propellants is a challenging field of research. The problem becomes especially difficult if it is necessary to account for two-phase flow that occurs in foam or fizz zones. A two-phase region has been called a **dispersed phase**, and the combustion of double-base propellants then usually involves reactions occurring in the condensed phase, in the dispersed phase, and in the gas phase. Although analyses of the reacting dispersed phase have been developed [13]–[15], these theories will not be addressed in detail here. Instead, a simplified model will be considered in which there is a sharp planar boundary between condensed and gaseous phases. A model of this type can be useful for limited purposes, such as for calculating burning rates, if the neglected aspects influence the phenomenon of interest to a sufficiently small extent. Burning rates may be controlled primarily by chemical-reaction processes occurring in one phase. For example, for most double-base formulations, the burning rate appears

---

* There have been some recent clarifications of combustion mechanisms of propellants containing these materials [see N. Kubota, 18th Symp. (1981), 187–194; 19th Symp. (1982), 777–785].

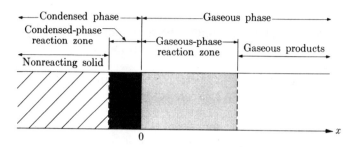

FIGURE 7.1   Model of a burning solid.

to depend mainly on reactions occurring in the condensed phase [16]–[18].*
Thus simplified models may be relevant to the burning of real propellants.
A simplified model that may provide a reasonable representation of the
steady-state, plane, one-dimensional burning of a homogeneous solid is
illustrated schematically in Figure 7.1. Heat conducted into the solid from
the reaction region raises its temperature to a point at which a phase transition or some other (endothermic or exothermic) chemical (equilibrium or
rate) process begins to occur. At a higher temperature, the condensed phase
gasifies either in a simple (endothermic) sublimation process or in a more
complex (endothermic or exothermic) chemical reaction. The gases evolved
react exothermically in the gas phase to produce the final (gaseous) reaction
products. Given the fundamental physical and chemical parameters describing each of these processes, one would like to be able to predict the
linear regression rate of the solid (that is, the velocity at which the interface
between the gaseous and condensed phases moves relative to the stationary
solid).

    As in most steady, planar flow problems, it is convenient here to
assume that the solid and the gas are infinite in extent and to adopt a coordinate system in which the flow is steady. Thus, in Figure 7.1, x is the coordinate normal to the surface, the interface is located at $x = 0$, the solid extends
to $x = -\infty$, and the gas extends to $x = +\infty$. We are interested in computing
the mass burning rate m (g/cm$^2 \cdot$ s) at which the solid must move in the
$+x$ direction so that the interface will remain at $x = 0$.

    Since the processes described above occur in series, the slowest
processes may be expected to govern the overall burning rate. A number
of different burning-velocity formulas may therefore be obtained, depending
on which processes are dominant. We shall first discuss interface processes
and then consider spatially distributed processes in different phases.

---

    * See also F. A. Williams, "Deflagration Mechanisms of Double-Base Propellants at
Moderate and High Pressures," unpublished (1976).

7.2. APPLICATIONS OF TRANSITION-STATE THEORY

Let us consider the case in which the burning rate is determined by the rate at which molecules from the condensed phase enter the gas. We shall focus our attention on the process occurring at the interface $x = 0$. Transition-state theory (Appendix B) provides information concerning the rate of this gasification process.

Solid propellants generally contain a number of different chemical species, each of which may gasify by processes of varying complexity. However, for illustrative purposes, we shall study the simplest case of a one-component condensed phase that sublimes in a one-step, unimolecular process:

$$\mathfrak{M}_1(s) \rightarrow \mathfrak{M}_1(g), \tag{1}$$

where $\mathfrak{M}_1$ is the chemical symbol for the subliming species, and s and g denote solid and gas phases, respectively. Related analyses of more realistic systems may be found in [19] and [20], for example.

If $\gamma_{\neq}$ denotes the number of moles of activated complexes (for the sublimation process) per unit surface area, and τ denotes the average time for an activated complex to sublime, then the number of moles per unit area per second leaving the surface is $\gamma_{\neq}/\tau$. Hence, the mass burning rate becomes

$$m = W_1 \gamma_{\neq}/\tau, \tag{2}$$

where W_1 is the molecular weight of species 1 (the species of which the solid is composed).

From the viewpoint of the transition-state theory, $\gamma_{\neq}$ is related to the actual number of moles of species 1 per unit area, γ_1, through the equilibrium condition $\gamma_{\neq}/\gamma_1 = K_{\gamma}^{\neq}$, where $K_{\gamma}^{\neq}$ is the equilibrium constant for surface concentrations for the process in which the complex is formed. Hence,

$$m = W_1 \gamma_1 (K_{\gamma}^{\neq}/\tau). \tag{3}$$

The equilibrium constant $K_{\gamma}^{\neq}$ can be expressed as a ratio of partition functions *per unit surface area* through reasoning similar to that leading to equation (A-24) for K_c. Proceeding as in Section B.3.4, we then find that

$$m = W_1 \gamma_1 (k^0 T_i/h)(q_{\neq}/q_1)e^{-E/R^0 T_i}, \tag{4}$$

where k^0 is Boltzmann's constant, R^0 is the universal gas constant, h is Planck's constant, E is the activation energy for the sublimation process, T_i is the interface temperature (at $x = 0$), q_1 is the partition function per unit surface area for species 1 (with the ground-state energy factor removed), and $q_{\neq}$ is the same kind of partition function for the activated complex without the degree of freedom corresponding to translational motion in the direction of the reaction coordinate. Since the same number of partition functions appear in the numerator and in the denominator of equation (4) (because of the assumption of a unimolecular process), the "per unit surface area" parts of $q_{\neq}$ and q_1 cancel and therefore can be neglected.

The surface concentration γ_1 appearing in equation (4) is easily related to the density and molecular weight of the solid. Geometrically, $\gamma_1^{3/2}$ is seen to be the number of moles per unit volume in the solid. Hence,

$$\gamma_1 = (\rho_s/W_1)^{2/3}, \tag{5}$$

where ρ_s is the specific gravity of the solid. In view of equation (5), equation (4) becomes

$$m = W_1^{1/3}\rho_s^{2/3}(k^0 T_i/h)(q_{\neq}/q_1)e^{-E/R^0 T_i}. \tag{6}$$

Equation (6) is the required mass-burning-rate formula, which (under our present assumptions) determines m in terms of the interface temperature T_i and the properties of both the solid and the activated complex.

In order to use equation (6), we must measure, estimate, or compute the value of T_i for the burning solid. The surface temperature usually depends on the amount and rate of heat release in the gas (and perhaps in the solid) as well as on the heat transfer rates in the gas and in the solid. Therefore, equation (6) should be supplemented by an analysis of the thermal processes occurring in the system. An example of such an analysis will be considered in Section 7.5. In the remaining part of this section we shall discuss a more basic difficulty encountered in attempting to use equation (6).

The structures of the solid and of the activated complex affect m through the parameters E and $q_{\neq}/q_1$ in equation (6). Generally, so many different possibilities exist for the structure of the complex that these parameters cannot be calculated from purely theoretical considerations. Semiempirical procedures must therefore be used to determine the burning rate from equation (6) even when T_i is known. In order to amplify these assertions, let us consider the ratio $q_{\neq}/q_1$.

The partition function for species 1 on the surface of the solid differs considerably from that for species 1 in the gas. The translational degrees of freedom of the gas molecule generally become vibrational degrees of freedom in the solid, and some (or all) of the rotational modes in the gas may also become vibrational modes (librations) for the solid. Furthermore, the activated complex on the surface of the solid may differ in more than one way from nonactivated surface molecules; for example, rotational motion may be less inhibited for the complex (leading to fewer librations and a greater number of fully excited rotational modes), or vibrational modes corresponding to translational motion of the center of mass may assume different frequencies in the complex. A very simple case is that in which $q_{\neq}$ differs from q_1 principally in having one less vibrational factor, as indicated in Section B.3.6 for the unimolecular gaseous reaction; that is,

$$q_{\neq}/q_1 = 1/q_v, \tag{7}$$

where q_v denotes this vibrational factor. However, equation (7) represents only one of many possibilities.

In view of these uncertainties, the actual structure of the complex is usually best determined by using all available data concerning the structure of the solid (to find q_1) in conjunction with *measured* gasification rates, in order to obtain $q_{\neq}$ through equation (6). These rate measurements may be performed for the burning solid, in which case equation (6) merely provides a formula for correlating data, or for a gasification process in which the gaseous combustion is supressed (for example, sublimation at pressures below the limit of flammability). With the latter strategy, equation (6) is first used in order to obtain $q_{\neq}$ from experiments that do not involve combustion, and afterward equation (6) also determines m for the burning solid. Actually, a measurement of m gives $q_{\neq} e^{-E/RT_i}$, not $q_{\neq}$ alone [see equation (6)]; since E is generally also unknown, the measurements must be performed at different interface temperatures in order to determine E (from the primary T_i dependence of m) and $q_{\neq}$ separately.

The program outlined above has not been attempted for conventional solid propellants, primarily because propellant compositions are so complex that a simple formula of the type given in equation (6) is not expected to be valid. However, a number of propellant ingredients have been studied extensively from our present viewpoint. Experiments on the rate of sublimation of a solid in a vacuum [21]–[26] [for which equation (6) is least questionable, since gas-phase processes cannot affect the measured sublimation rates] and experiments in which the solid is pressed against a heated plate at atmospheric pressure [27]–[31] have been performed. Hypotheses concerning the mechanism of the sublimation process and the structure of the complex have been invented in order to correlate the measured m values [26]–[28], [30]–[33]. The heated-plate, or *linear-pyrolysis*, experiments demand special consideration of gas-phase processes in their analysis [8], [30], [33]–[38] as a penalty that must be paid for their abilities to approach the higher surface temperatures achieved in solid-propellant deflagration. Alternative methods for reaching high temperatures, such as laser heating, also need gas-phase analyses for their full interpretation [39], [40]. The presence of the gaseous phase can cause a departure from equation (6) as a consequence of the possibility of attainment of interphase equilibrium.

7.3 APPROACH TO INTERFACIAL EQUILIBRIUM

At the surface of a burning solid, the reverse process

$$\mathfrak{M}_1(g) \rightarrow \mathfrak{M}_1(s) \tag{8}$$

may be fast enough to alter appreciably the formula for the burning rate given in equation (6). Equation (8) represents the process in which the gaseous species evolved from the solid returns to the surface. This process may be described from the viewpoint of molecular gas kinetics, which is presented in Section B.4.3. By reasoning analogous to that leading to equation (B-80), we find that the mass per unit area per second returning to the solid is $W_1\beta_1\alpha$, where β_1 [which is given by equation (B-78) when the vapor is an ideal gas] is the number of moles of species 1 (in the gas) per unit area per second striking the surface, and α is a (dimensionless) surface accommodation coefficient or evaporation coefficient for molecules of species 1. Thus, equation (6) should be replaced by the expression

$$m = W_1^{1/3}\rho_s^{2/3}(k^0 T_i/h)(q_{\neq}/q_1)e^{-E/R^0 T_i} - W_1\beta_1\alpha. \tag{9}$$

The factor β_1 introduces into equation (9) an explicit dependence of m on the concentration $c_{1,i}$ of species 1 in the gas adjacent to the interface [see equation (B-78)]. Except for this difference, equation (9) contains the same kinds of parameters as does equation (6), since the coefficient α can be analyzed from the viewpoint of transition-state theory. Although α may depend in general on T_i and the pressure and composition of the gas at the interface, a reasonable hypothesis, which enables us to express α in terms of kinetic parameters already introduced and *thermodynamic* properties of species 1, is that α is independent of the pressure and composition of the gas [$\alpha = \alpha(T_i)$]. Under this condition, at constant T_i the last term in equation (9) is proportional to the concentration $c_{1,i}$ and the first term on the right-hand side of equation (9) is independent of $c_{1,i}$. Therefore, by increasing the concentration (or partial pressure) of species 1 in the gas, the surface equilibrium condition for species 1—$m = 0$—can be reached. If $p_{1,e}(T_i)$ denotes the equilibrium partial pressure of species 1 at temperature T_i, then when $m = 0$, equation (9) reduces to

$$W_1[p_{1,e}(T_i)/R^0 T_i]\sqrt{k^0 T_i/2\pi m_1}\,\alpha = W_1^{1/3}\rho_s^{2/3}(k^0 T_i/h)(q_{\neq}/q_1)e^{-E/R^0 T_i},$$

where use has been made of equation (B-78) and of the ideal-gas equation for species 1 ($c_1 = p_1/R^0 T$). This relation shows that $\alpha(T_i)$ is given by

$$\alpha = [p_{1,e}(T_i)]^{-1}\sqrt{2\pi W_1 R^0 T_i}(\rho_s/W_1)^{2/3}(k^0 T_i/h)(q_{\neq}/q_1)e^{-E/R^0 T_i} \tag{10}$$

under our present assumptions. In view of equation (B-78) and the ideal-gas law, equation (10) implies that equation (9) can be written in the alternative form

$$m = W_1^{1/3}\rho_s^{2/3}(k^0 T_i/h)(q_{\neq}/q_1)e^{-E/R^0 T_i}[1 - p_{1,i}/p_{1,e}(T_i)], \tag{11}$$

where $p_{1,i}$ is the partial pressure of species 1 in the gas at $x = 0$.

The new parameter $p_{1,e}(T_i)$, appearing in equations (10) and (11), is the equilibrium vapor pressure, which is given by equation (A-23) and is approximately proportional to $e^{-L/R^0 T_i}$, where L is the heat of sublimation. If $p_{1,e}(T_i)$ cannot be calculated from known thermodynamic properties or measured by conventional techniques, its value may sometimes be deduced by measuring m with the solid pressed against a heated plate [34]. After $p_{1,e}(T_i)$ has been obtained, to compute the burning rate of the solid from equation (11) entails considering gas-phase reactions in order to determine $p_{1,i}$. Equation (11) is, therefore, somewhat more difficult to use than is equation (6).

If $p_{1,i}/p_{1,e}(T_i) \ll 1$, then equation (11) merely provides a small correction to equation (6). However, $p_{1,i}/p_{1,e}(T_i)$ may approach unity for some burning solids. In such cases, the gasification process, equation (1), no longer governs the overall burning rate; some other process becomes rate controlling. In extreme cases it is sometimes reasonable and convenient to replace equation (11) by the simpler requirement of surface equilibrium, namely,

$$p_{1,i} = p_{1,e}(T_i), \tag{12}$$

in computing the mass burning rate from an analysis of another process, such as finite-rate chemistry in the condensed or gaseous phase.

If surface equilibrium prevails, then it is relatively straightforward to generalize the interface condition to chemical processes that are more complex than equations (1) and (8). This is of interest, since propellant materials often experience processes of this type; for example, NH_4ClO_4 undergoes dissociative sublimation into NH_3 and $HClO_4$ [33]. For a general process in which the condensed material is transformed to $\sum_{j=1}^{N} v_j'' \mathfrak{M}_j$ in the gas, the surface equilibrium condition (for an ideal gas mixture and a solid whose thermodynamic properties are independent of pressure) is

$$\prod_{j=1}^{N} p_{j,i}^{v_j''} = K_p(T_i), \tag{13}$$

where $K_p(T_i)$ is an equilibrium constant for partial pressures [compare equation (A-26)]. When equation (13) is applied with $p_j = pX_j$, it is important to realize that the mole fractions X_j in the gas are not deducible directly from the composition of the solid. There are relationships for interface flux conservation of species, obtainable from equation (1-58), but these involve diffusion velocities in addition to the concentrations. A nearby source or sink for a species in the gas—for example, a reaction sheet—can influence the relative concentration of that species at the interface through differential diffusion.

It is desirable to be able to ascertain in advance whether equation (6) or equation (12) is the better approximation. A few general observations can be made concerning this question. Polymeric materials generally gasify in irreversible rate processes, effectively exhibiting an infinite value of $p_{1,e}(T_i)$; they may be considered to exist in metastable states because molecules entering the interface from the gas do not combine to add to the polymer chain in the solid. On the other hand, for most simple subliming or vaporizing materials, $W_1 \beta_1 \alpha$ is much larger than the values of m typically encountered in solid deflagration, and it may then be seen from equation (9) that surface equilibrium is a good approximation. The smaller α is, the less likely it will be that equation (12) can be used. An increasing extent of molecular rearrangement upon gasification is associated with a decreasing likelihood of achieving interfacial equilibrium. For a simple dissociative sublimation such as that of NH_4ClO_4, surface conditions may or may not correspond to equilibrium (they usually do for NH_4ClO_4 [33], but not necessarily for the dissociative vaporization of B_2O_3 [41]). For processes that are more complex than these, finite surface rates usually are encountered. In double-base propellants, the situation is more complicated in that the nitrocellulose component cannot achieve equilibrium, while equilibrium vaporization of nitroglycerin may occur. This complexity enables interfacial equilibrium to be unimportant for some processes (for example, steady deflagration [40]) but possibly of significance for others (for example, radiant ignition [42]). Use of the more general interface condition given by equation (11), instead of equation (6) or equation (12), has not been made in studies of deflagration of solids but could be warranted if α is in an appropriate intermediate range.

7.4. DEFLAGRATION CONTROLLED BY CONDENSED-PHASE REACTION RATES

If there is exothermicity in the condensed-phase reaction zone illustrated in Figure 7.1, then the burning rate may be controlled by reactions that occur in this zone. Deflagration analyses paralleling those discussed in Chapter 5 may be developed for condensed-phase combustion. For illustrative purposes, let us assume that the condensed-phase reaction is the one-step process

$$\mathfrak{M}_0(s) \rightarrow \mathfrak{M}_1(s), \tag{14}$$

which may be considered to precede equation (1). An Arrhenius reaction of zero order will be treated here for simplicity. The most important physical difference between condensed-phase and gas-phase reactions is that diffusion of species is much slower in the condensed phase; here we shall neglect molecular diffusion entirely to expose this difference most strongly.

A formulation like that of Chapter 5, with the assumption of low Mach number (p approximately a constant) introduced at the ouset, results in

$$dY_1/dx = w_1/m \tag{15}$$

and

$$\lambda \, dT/dx = m(h_0 Y_0 + h_1 Y_1) + \text{constant}, \tag{16}$$

in place of equations (5-7) and (5-9). With the heat released per unit mass of reactant consumed in the condensed phase defined as $q_s = h_0^0 - h_1^0$, equation (16) may be written as

$$\lambda \, dT/dx = m \left[\int_{T_0}^{T} c_p \, dT + q_s(Y_{1,0} - Y_1) \right], \tag{17}$$

where the subscript 0 identifies conditions at $x = -\infty$ (at which point $dT/dx = 0$) and where c_p is the specific heat at constant pressure for species 0 and 1, assumed equal. In equation (15) for a zero-order reaction, we write

$$w_1 = \rho B_1 T^{\alpha_1} e^{-E_1/R^0 T} \tag{18}$$

[compare equation (5-16)], in which all quantities now of course refer to the condensed phase. Boundary conditions for equations (15) and (17) are that $Y_1 = Y_{1,0}$ and $T = T_0$ at $x = -\infty$ and that $Y_1 = 1$ and $T = T_i$ at $x = 0$.

The problem that has been defined here possesses the cold-boundary difficulty discussed in Section 5.3.2 and can be approached by the same variety of methods presented in Section 5.3. The asymptotic approach of Section 5.3.6 has been seen to be most attractive and will be adopted here. Thus we shall treat the Zel'dovich number

$$\beta = E_1 \left(\int_{T_0}^{T_i} c_p \, dT \right) / (R^0 c_{p,i} T_i^2) \tag{19}$$

as a large parameter. Although a relationship between Y_1 and T like equation (5-24) cannot be derived because the Lewis number differs from unity, the variable Y_1 can be eliminated by differentiating equation (17) and using equation (15). From the solution to the resulting equation for T, the value of m is then found from the integral of equation (15), namely,

$$m = \int_{-\infty}^{0} w_1 \, dx/(1 - Y_{1,0}), \tag{20}$$

in which the asymptotic expansion of this divergent integral is intended.

There is an upstream *convective-diffusive* zone in which only convection and heat conduction occur. In the variable ξ defined in equation (5-18), the solution in this zone is $Y_1 = Y_{1,0}$ and

$$\int_{T_0}^{T} c_p \, dT = \left(\int_{T_0}^{T_i} c_p \, dT \right) e^{\xi}, \tag{21}$$

where a matching condition has been anticipated in putting T_i equal to the upper limit of the integral on the left-hand side. This solution allows for a variable heat capacity, and subsurface phase changes (for example, melting) are readily included by introducing delta functions in c_p. In the reactive-diffusive zone the stretched variables $\eta = -\beta\xi$ and

$$y = \beta c_{p,i}(T_i - T) \left/ \int_{T_0}^{T_i} c_p \, dT \right. \tag{22}$$

may be introduced to produce in lowest order the problem

$$d^2y/d\eta^2 = \Lambda e^{-y}, \tag{23}$$

with $y = 0$ at $\eta = 0$ and $dy/d\eta \to 1$ as $\eta \to \infty$ (through matching), where

$$\Lambda = \frac{R^0 q_s \rho_i B_1 T_i^{\alpha_1+2} \lambda_i}{E_1 \left(\displaystyle\int_{T_0}^{T_i} c_p \, dT \right)^2 m^2} e^{-E_1/R^0 T_i}, \tag{24}$$

which is assumed to be of order unity. The first integral of equation (23), subject to the matching condition, gives

$$dy/d\eta = (1 - 2\Lambda e^{-y})^{1/2}, \tag{25}$$

provided that $2\Lambda \leq 1$. In the inner-zone variables that have been introduced, the first approximation to equation (20) is

$$\hat{\alpha} = \Lambda \int_0^\infty e^{-y} \, d\eta, \tag{26}$$

where

$$\hat{\alpha} = (1 - Y_{1,0}) q_s \bigg/ \int_{T_0}^{T_i} c_p \, dT. \tag{27}$$

Use of equation (25) in equation (26) produces, after algebra,

$$\Lambda = \hat{\alpha} \, (1 - \tfrac{1}{2} \hat{\alpha}). \tag{28}$$

In dimensional terms, this result is

$$m = \left\{ \frac{R^0 \rho_i B_1 T_i^{\alpha_1+2} \lambda_i e^{-E_1/R^0 T_i}}{E_1(1 - Y_{1,0}) \left[\displaystyle\int_{T_0}^{T_i} c_p \, dT - \frac{1}{2} q_s(1 - Y_{1,0}) \right]} \right\}^{1/2}. \tag{29}$$

The subscripts i here on ρ and λ indicate that they are to be evaluated in the condensed phase at $T = T_i$.

The derivation of equation (29) has not employed conservation conditions at the interface with the gas; hence the value of T_i remains undetermined. Therefore, additional physical information must be introduced to calculate m. The simplest condition to apply at the interface is adiabaticity, $dT/dx = 0$, which serves to isolate the solid from the gas and shows from equation (17) that $\int_{T_0}^{T_i} c_p \, dT = q_s(1 - Y_{1,0})$, therefore giving T_i. In this case equation (29) reduces to a formula for m derived originally by Zel'dovich [43]. It is interesting to compare the resulting expression with equation (5-75) for a gaseous flame with a Lewis number of unity. In the first approximation the two formulas differ only by an additional factor of $\beta^{-1/2}$ on the right-hand side of equation (5-75). This difference is readily understandable on the basis of the physical reasoning given in Section 5.3.7. Diffusion of the reactant is responsible for a factor $\beta^{-1/2}$ in equation (5-75) through the consequent reduction in the reactant concentration (hence in the reaction rate) in the reaction zone. This diffusion is not present in the condensed-phase deflagration, and the corresponding factor is therefore absent. The reactant concentration in the condensed-phase reaction zone is of the same order of magnitude as it is in the cold propellant. This fact could have been used to scale equation (15), thereby producing, with equation (23), a pair of simultaneous differential equations for describing the reaction-zone structure [40]. The solutions obtained with the adiabatic boundary condition are unstable to one-dimensional disturbances (see Section 9.2).

The derivation of equation (29) is invalid if $q_s(1 - Y_{1,0}) > \int_{T_0}^{T_i} c_p \, dT$, since equation (17) then implies that $dT/dx < 0$ at $x = 0$, so that contrary to assumption, the peak temperature of the condensed phase no longer occurs at its surface. In a sense, the adiabatic condition represents a limiting condition at which the condensed-phase deflagration breaks away from the surface and ceases to be influenced by interface or gas-phase processes. From the viewpoint of the condensed-phase analysis, it is necessary that $q_s(1 - Y_{1,0}) \leq \int_{T_0}^{T_i} c_p \, dT$. When the inequality prevails in this relationship, heat is fed into the condensed phase at its surface. With T_0, q_s, and other condensed-phase properties fixed, this energy input increases m because the consequent increase in T_i exerts a larger influence on the Arrhenius factor in the numerator of equation (29) than on the integral in the denominator. Strictly speaking, whenever the inequality holds, the burning rate is not controlled entirely by condensed-phase processes because interface or gas-phase processes then influence m by modifying T_i. The simplest approach to the use of equation (29) in these nonadiabatic cases is to assign T_i a constant value, a "gasification temperature" that is presumed to be known, but this approach fails to address the question of the physics that determine the value of T_i.

If we consider T_i fixed and q_s decreasing, then we find that the derivation of equation (29) remains valid as q_s vanishes or becomes negative. The requirements of the analysis are that $T_i > T_0$ and that there be a distributed condensed-phase Arrhenius reaction with β large. The reaction may be exothermic, energetically neutral, or endothermic; in the latter case, Λ and $\hat{\alpha}$ are negative. Under these conditions, the condensed-phase reaction naturally does not control the deflagration rate, but equation (29) provides a relationship between m and T_i that is necessary for compatibility with the condensed-phase processes. Many polymeric binders of composite propellants experience endothermic gasification of this type. In fact, the basis of the analysis given here was first developed for endothermic linear pyrolysis of polymeric materials [44], [45].

A number of modifications and generalizations of the analysis presented here can be explored [14], [40], [45]. A reaction order of zero has been selected here for simplicity of development. By working with two simultaneous differential equations in the reaction zone, the formulation is easily extended to an arbitrary reaction order n. For $0 < n < 1$, a modified version of equation (29) is readily derived [45]. With the most natural assumption of unimolecular decomposition, we have $n = 1$, but for $n \geq 1$, a difficulty is encountered in that equations (15) and (16) allow Y_1 to approach unity only as $x \rightarrow +\infty$; just as in the gaseous flame (see Section 5.3.7), an infinite time is needed for complete conversion of reactant to product at these larger values of n. To place the interface at a finite location with $n \geq 1$, a value of Y_1 less than unity must be selected at the interface [45]. The selection may be made arbitrarily on the grounds of insensitivity of predictions to the value chosen within a reasonable range, or an additional physical phenomenon such as equilibrium vaporization, equilibrium solubility, or onset of dispersion may be introduced formally to fix the surface value of Y_1 [14]. Indeed, these physical phenomena may be relevant even for $0 \leq n < 1$, and equation (29) may be modified accordingly by taking $Y_1 < 1$ at the surface in carrying out the analysis. These refinements have been introduced mainly in conjunction with an assumption of surface adiabaticity [14]. In addition, adiabatic analyses are available for arbitrary $n \geq 0$ for situations in which the condensed phase extends to $x = +\infty$ [14]. Effects of nonadiabaticity associated with radiant flux incident on the surface of the solid have been analyzed by methods similar to that given here [40].

So long as the finite-rate chemistry occurs in a single reactive-diffusive zone extending to the surface of the condensed phase, the analysis need not be restricted to a one-step reaction. Known mechanisms of homogeneous polymer degradation may be taken into account—for example, [45]. The results continue to be expressible in formulas that resemble equation (29) but that usually are somewhat more complicated.

There are numerous possibilities for using equation (29) in conjunction with interface or gas-phase conditions to determine both m and T_i. For example, the condensed product of equation (14) conceivably might experience the rate process of equation (1) as soon as $Y_1 = 1$, so that equations (6) and (29) become two independent expressions for m and T_i. Alternatively and somewhat less unlikely, surface equilibrium may occur so that equation (12) determines T_i in equation (29); in this case, a gas-phase analysis is generally needed to find $p_{1,i}$. It appears that in most real homogeneous propellants, the products of the exothermic condensed-phase reactions are mainly gaseous, so that considerations of dispersion or possibly of gas-phase reactions are most relevant for determining T_i in equation (29), and equations (6), (11), or (12) are not directly useful.

The general similarity between equations (6) and (29) may be noted. Both exhibit an Arrhenius dependence of m on T_i, and measurements of the variation of m with T_i, therefore do not distinguish between the two equations, even though the processes represented fundamentally are quite different, one being an interface reaction and the other in reality a distributed condensed-phase reaction. The activation energy E for the surface reaction is seen to correspond to $E_1/2$, half of the activation energy for the bulk reaction, because of the square-root dependence of the deflagration rate on the homogeneous reaction rate. This square-root relationship clearly applies in endothermic linear pyrolysis as well. Thus in steady deflagration or linear pyrolysis, for many purposes the distributed condensed-phase reaction may be approximated as a surface reaction having half of the true activation energy, thereby achieving simplifications in the analysis. Fundamentally, equation (6) is seldom ever   correct for polymers, but since equation (29) often is, equation (6) may be employed, with suitably adjusted values of E and of the preexponential factor.

If the solid were inert and the interface gasification process were exothermic, releasing the energy q_i per unit mass of solid gasified, and if the system were adiabatic with negligible energy input from the gas, then $\int_{T_0}^{T_i} c_p \, dT = q_i$ would determine T_i, after which equation (6) would give m. This result, which constitutes the simplest complete model for solid-propellant deflagration, can be made to correspond to the adiabatic version of equation (29) by putting $E_1/2 = E$ and $q_i = q_s(1 - Y_{1,0})$ and equating the factors preceding the exponential. Gas-phase reaction rates may modify these results if heat conduction from the gas to the interface occurs.

7.5. DEFLAGRATION CONTROLLED BY GAS-PHASE REACTION RATES

If we overlook the processes occurring in the region $x \leq 0$ in Figure 7.1, then the illustrated burning mechanism of the solid closely resembles that of a premixed gaseous laminar flame. To proceed with an analysis, let us assume for illustrative purposes that the gaseous reaction is the one-step unimolecular process

$$\mathfrak{M}_1(g) \to \mathfrak{M}_2(g), \tag{30}$$

where the subscript 2 identifies the reaction product, and also adopt (in the gaseous region) the other basic simplifying assumptions employed in Section 5.3 (namely, low-speed flow and assumptions 1–11 of Chapter 5). Then the results of Section 5.3 can be applied to a deflagrating solid in which premixed gas-phase reactions influence the burning rate.

The purely gaseous deflagration extends from $x = -\infty$ to $x = +\infty$, with cold boundary conditions applied at $x = -\infty$; but the gaseous region of the burning solid extends only from $x = 0$ to $x = +\infty$, with cold boundary conditions applied at $x = 0$. This is the primary difference between a gaseous laminar flame and our present model of a burning solid. Any governing equations that can be derived for the gaseous flame without making reference to the cold boundary conditions will apply equally well to the burning solid in the region $x > 0$. If the normalized mass flux fraction of products is defined as

$$\epsilon \equiv 1 - (\epsilon_{2,\infty} - \epsilon_2)/\Delta\epsilon_2, \tag{31}$$

where ϵ_2 is the flux fraction of species 2, the subscript ∞ identifies conditions at $x = +\infty$, and $\Delta\epsilon_2$ is the total change in ϵ_2 in the gas, the normalized mass fraction of products is defined as

$$Y \equiv 1 - (Y_{2,\infty} - Y_2)/\Delta\epsilon_2, \tag{32}$$

and the dimensionless temperature is defined as

$$\tau \equiv 1 + c_p(T - T_\infty)/q, \tag{33}$$

where c_p is the (constant) specific heat at constant pressure for the gas and q is the total heat released in the gas per unit mass of the gas mixture, then, through a development similar to that given in Chapter 5, the reader may show that equations (5-24) and (5-42) can be derived without using the cold-boundary conditions. Thus the relations

$$Y = \tau \tag{34}$$

and

$$d\epsilon/d\tau = \Lambda\omega/(\tau - \epsilon), \tag{35}$$

where the dimensionless reaction-rate function $\omega(\tau)$ and the (constant) dimensionless mass-burning rate Λ are given by equations (5-43) and (5-45), respectively, remain valid in the gaseous region for the burning solid. Actually, in order to remove all reference to conditions at $x = -\infty$ in the purely gaseous problem, the quantity α' appearing in equation (5-43) must be defined as

$$\alpha' = q/(c_p T_\infty - q), \tag{36}$$

which is equivalent to equation (5-29) for the gaseous flame.

The boundary conditions for equation (35) at $x = +\infty$ for the burning solid are the same as those for the gaseous flame; namely, $\epsilon = 1$ and $\tau = 1$ [see equations (31) and (33)]. However, as we have already stated, the cold-boundary conditions differ in the two problems. For the burning solid, the definition of $\Delta\epsilon_2$ implies that

$$\Delta\epsilon_2 = \epsilon_{2,\infty} - \epsilon_{2,i},$$

where the subscript i identifies conditions in the gas at the interface ($x = 0$). Hence, equation (31) shows that $\epsilon = 0$ at $x = 0$. Equation (33) implies that $\tau = \tau_i$ at $x = 0$, where

$$\tau_i \equiv 1 + c_p(T_i - T_\infty)/q. \tag{37}$$

Therefore, the cold-boundary condition (at $x = 0$) for equation (35) in the burning solid is $\epsilon = 0$ and $\tau = \tau_i$.

It will be noted that these boundary conditions ($\epsilon = 0$ at $\tau = \tau_i$ and $\epsilon = 1$ at $\tau = 1$) are formally identical with the boundary conditions introduced in Section 5.3.2 in order to remedy the cold-boundary difficulty for the gaseous flame. In Chapter 5 the quantity τ_i was a dimensionless ignition temperature for which a pseudostationary value was chosen; for the present problem, on the other hand, τ_i is a well-defined dimensionless measure of the interface temperature. Thus the cold-boundary difficulty does not arise in the present problem; the presence of the condensed phase has removed it by cutting off the reaction at the interface.

A consequence of the equivalence of the two problems is that any of the procedures discussed in Section 5.3 for determining the laminar flame-speed eigenvalue Λ of equation (5-45) can be applied without modification for finding Λ as a function of τ_i, α', and β' [defined in equation (5-44)] in the present problem. In using these procedures, one should recognize that $\omega(\tau)$ is given by equation (5-43) only for $\tau > \tau_i$; $\omega \equiv 0$ for $\tau < \tau_i$. Results of the asymptotic analysis of Section 5.3.6 have been used in a number of investigations of solid-propellant deflagration [43], [46]–[49]. From the discussion associated with equation (5-75) (where now $\alpha = q/c_p T_\infty$ and $\beta = E_1 q/R^0 c_p T_\infty^2$), it is seen that Λ is then found to be independent of τ_i—that is, m is independent of T_i when the gas-phase properties and the overall energetics (T_∞) are fixed. In this respect, the gas-phase reaction controls m completely; other processes serve only to establish values for other quantities, such as T_i, when the asymptotic analysis that has been given is applicable.

With adiabatic combustion, departure from a complete control of m by the gas-phase reaction can occur only if the derivation of equation (5-75) becomes invalid. There are two ways in which this can happen; essentially, the value of m calculated on the basis of gas-phase control may become either too low or too high to be consistent with all aspects of the problem. If the gas-phase reaction is the only rate process—for example, if the condensed phase is inert and maintains interfacial equilibrium—then m may become arbitrarily small without encountering an inconsistency. However, if a finite-rate process occurs at the interface or in the condensed phase, then a difficulty arises if the value of m calculated with gas-phase control is decreased below a critical value. To see this, consider equation (6) or equation (29). As the value of m obtained from the gas-phase analysis decreases (for example, as a consequence of a decreased reaction rate in the gas), the interface temperature T_i, calculated from equation (6) or equation (29), also decreases. According to equation (37), this decreases τ_i. Eventually, at a sufficiently low value of m, the calculated value of T_i corresponds to $\tau_i = 0$. As this condition is approached, the gas-phase solution approaches one in which $dT/dx = 0$ at $x = 0$, and the reaction zone moves to an infinite distance from the interface. The interface thus becomes adiabatic, and the gas-phase processes are separated from the interface and condensed-phase processes.

At this limiting condition, the gas-phase deflagration becomes irrelevant to the solid-propellant deflagration. Either equation (6) or (29) can now be used to calculate the burning rate in conjunction with a condition of surface adiabaticity. If L denotes the net chemical enthalpy per unit mass needed for gasification of the condensed phase, then an energy balance for the condensed phase shows that this adiabaticity condition can be written as

$$\int_{T_0}^{T_i} c_p \, dT = -L, \tag{38}$$

which serves to determine T_i without reference to the gaseous phase. In principle, equation (38) can be applied irrespective of the sign of L and irrespective of whether the value of T_i obtained from it is greater than or less than T_0. For example, for the model that led to equation (29), if species 1 undergoes endothermic sublimation or vaporization after $Y_1 = 1$ with a heat of vaporization per unit mass of L_s, then $L = L_s - q_s(1 - Y_{1,0})$. For solid propellants, it is usual that $L < 0$ and $T_i > T_0$ when equation (38) applies.* An abrupt change from gas-phase control of m to interface or condensed-phase control occurs when surface adiabaticity is reached. In this sense, even if the conductive energy feedback from the gas is a small but positive fraction of the total amount of heat release that occurs prior to complete gasification, the gas-phase reaction controls the burning rate.

If the gas-phase reaction rate is reduced further, below a value corresponding to surface adiabaticity, then there no longer exists a gas-phase solution with an m that is consistent with condensed-phase and interface conditions. A "breakaway" phenomenon occurs in which the burning rate is controlled by condensed-phase or interface processes, and the gas-phase deflagration propagates more slowly than the solid-propellant deflagration. Under these conditions, after the gas-phase deflagration has moved away, the gas-phase chemistry occurs in a convective-reaction process with negligible molecular transport, anchored by the condensed-phase deflagration and analyzable by methods discussed in Section B.2.5, for example. Analyses have been given of this gas-phase reaction zone, which no longer influences the burning rate [14], [50], [51]; expressions for the distance from the interface to the region of major gas-phase reaction have been obtained by application of explosion theory.

The second way in which the derivation of equation (5-75) can fail is for the thickness of the reactive-diffusive zone in the gas to become comparable in size with the thickness of the convective-diffusive zone. This occurs if $(T_\infty - T_i)/T_\infty$ becomes of order β^{-1}, which would be favored by low overall

----

* This entire discussion has been subject to the restriction that L is not sufficiently large and positive to require, T_i in equation (38) to be less than absolute zero. If the condensed phase is sufficiently endothermic to violate this restriction, then equation (38) cannot be satisfied, surface adiabaticity cannot be approached, τ_i cannot approach zero, and solid-propellant deflagration with gasification cannot occur in the absence of some energy input from the gas phase. In this case the gas-phase reaction must always control the burning rate (at least when the burning involves complete gasification); if the surface equilibrium defined by equation (12) applies, then unusual shapes of curves of the burning rate as a function of pressure are obtained [8], [49].

activation energies for the gas-phase reaction. Since the presence of the condensed phase has removed the cold-boundary difficulty, β need not be large for a mathematically well-defined deflagration problem to exist. If β is not large, then there is a single gas-phase zone in which reaction, diffusion, and convection all are important. The methods of Sections 5.3.3 and 5.3.4 may be used to analyze the gas-phase deflagration in these cases; such applications of these methods have been made [52], [53]. If β is large, then the gas-phase reaction zone continues to involve a diffusive-reactive balance in the first approximation, but with equation (6) or equation (29) applicable, increasing m (by increasing the gas-phase reaction rate) eventually causes $(T_\infty - T_i)/T_\infty$ to become of order β^{-1}, whereupon the convective-diffusive zone in the gas disappears, and the interface with the condensed phase forms a boundary of the reactive-diffusive zone at a finite location in the stretched variable. The reaction zone in the gas then requires a modified analysis, which is readily developed, and a corresponding modification to equation (5-75) occurs. This same effect occurs with surface equilibrium [equation (12)] if increasing the pressure p is the means for increasing the gas-phase reaction rate (but not if p is held constant and the reaction rate is increased instead by increasing the preexponential constant in the rate expression). Analysis shows that when equation (5-75) fails for these reasons, the gas-phase reaction continues to control the burning rate, but m becomes bounded in the limit of an infinite gas-phase reaction rate.

Thus far our discussion pertains to conditions of overall adiabaticity. Nonadiabatic deflagration of solid propellants is of practical interest in many contexts. Enhancement or control of burning rates may be considered by external application of radiant energy fluxes [40]. In laboratory experiments on linear burning of strands of solid propellants, heat losses may be responsible for limits of deflagration.* In rocket motors these losses tend to be reduced by energy input from hot surroundings. Effects of energy losses may differ for different mechanisms of burning and for different types and locations of losses. To illustrate a few of the phenomena that may occur, let us consider radiant-energy loss from the interface between the condensed and gaseous phases, a process that often is significant in laboratory experiments with strand burners. For brevity, attention will be restricted to systems that are controlled by gas-phase reaction rates and for which equation (6) applies; discussions of other systems may be found in the literature [8], [49], [54].

If q_R denotes the radiant-energy flux from the surface, then under the assumption that this radiation is not absorbed elsewhere in the system, we find from an overall energy balance that

$$T_\infty = T_{\infty, a} - q_R/(mc_p), \tag{39}$$

---

* *Strands* are long cylinders with burning inhibited on all faces except one of those normal to the axis.

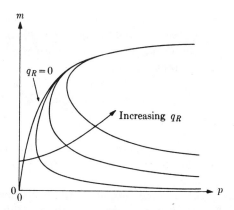

FIGURE 7.2. Schematic illustration of the pressure dependence of the mass burning rate when the gasification process is an unopposed rate process.

where $T_{\infty, a}$ is the value of T_∞ under adiabatic conditions ($q_R = 0$) for the same initial temperature T_0. Equation (39) can be derived formally from the general interface energy conservation condition given in equation (1-61) and the energy-conservation equation for the gas and the solid [for example, (5-9)]. In equation (39), q_R depends on the surface temperature T_i (typically q_R is proportional to T_i^4). Therefore, in equation (5-75) for m, T_∞ now depends on both T_i and m through equation (39), when $q_R \neq 0$. It may thus be seen in general that whenever equation (5-75) applies, heat losses affect the burning rate only through their influence on T_∞; the situation is somewhat more complicated if the gas-phase reaction zone extends to the surface of the condensed phase. Nevertheless, in both cases it is found that if $q_R > 0$, a solution no longer exists for m when the pressure p is less than a critical limiting value, and there are two solutions if p exceeds this value. This result, the derivation of which is somewhat complicated algebraically* even on the basis of equation (5-75), is illustrated schematically in Figure 7.2, which corresponds to a system in which the value of T_i in equation (38) is negative (that is, the equation cannot be satisfied) and the overall order (the pressure exponent) of the distributed gas-phase reaction is two. The absence of a solution at low pressures is an example of a prediction of a deflagration limit caused by heat loss (see Chapter 8). In the pressure range where two solutions for m exist, it is usually assumed that the lower burning rate represents an unstable solution and that the upper is correct physically (compare Sections

---

* The equation to be investigated may be shown to be roughly of the form

$$p = Am \exp\left(\left\{B - \left(\frac{C}{m}\right)\left[\ln\left(\frac{D}{m}\right)\right]^{-4}\right\}^{-1}\right),$$

where A, B, C, and D are positive constants; the increase in q_R is obtained by an increase in C.

3.4.4 and 8.2.3). Some further discussion of the function $m(p)$ will be given in the following section.

7.6. DISPERSION PHENOMENA AND OTHER INFLUENCES

Let us now consider combustion mechanisms that, strictly speaking, are not represented by the model illustrated in Figure 7.1. Specifically, at the interface in Figure 7.1, let us introduce the dispersed phase mentioned in Section 7.1. The consequent occurrence of two-phase flow adds substantial complication to the deflagration mechanism.

When dispersed-phase processes contribute in any essential way to the burning rate, the problem is simplest if the other phases may be treated as nonreactive and the reaction rate in the dispersed phase is strongly temperature-sensitive, since then the problem may be analyzed by asymptotic methods on the basis of purely dispersed-phase control. There are numerous possible models of this type, differing according to the equilibrium versus finite-rate character of the gasification step in the dispersed phase, the location of the heat release in the dispersed phase (for example, in the condensed or gaseous component or at their boundaries), geometric configuration of the dispersion (gaseous bubbles, liquid spheres, solid rods, and so on), and other attributes. Results are available for a few models that have dispersed-phase control [13]–[15], [55]. Polyvinylnitrate is a material that appears to burn in a process controlled by dispersed-phase combustion, and good agreement between theoretical predictions and experimental observations has been obtained [56].

If processes occurring in other zones are important as well, then the theoretical problem is more complicated because accurate conditions for the onset or termination of dispersion become necessary in developing analyses with reasonable accuracy. The new difficulties encountered are fundamental because the physics of these processes are not well understood. An example of a propellant in which conditions for onset of dispersion are relevant to the burning mechanism is nitrocellulose, for which it has been observed that the ratio of the energy feedback (energy transferred from the dispersed phase to the condensed phase) to the energy released in the condensed phase varies from about 0.3 at 5 atm to about 0.1 at 100 atm [18]. Dispersion might begin as a consequence of attainment of an equilibrium-like condition for transition to dispersion or in a rate process for subsurface generation of gases. For example, given the model of Section 7.4 for condensed-phase processes, species 1 of equation (14) might be presumed to initiate dispersion upon attainment of an equilibrium relationship of the form

$$Y_{1,i}/(1 - Y_{1,i}) = f(p)B_s e^{L_s/R^0 T_i}, \tag{40}$$

where L_s is a heat of solution, B_s is a solubility constant, and $f(p)$ is an increasing function of p. This function would be proportional to the partial pressure of species 1 in the gas if an equilibrium condition for absorption of this species applies, or it might be proportional to the total hydrostatic pressure if the phase transition is influenced by mechanical properties of a solid having a limiting "strength." If the transition to dispersion is a rate process, then rates of nucleation might be relevant, and a formula like equation (40) again may be derived, but with L_s representing an energy of nucleation and with B_s becoming proportional to m^2. At the high pressures encountered in the use of gun propellants (p on the order of 1000 atm), it is quite possible that the fundamental rate constants for the condensed-phase reactions become pressure dependent. Thus a wide variety of complications may arise.

Use of an equation like equation (40) as a downstream boundary condition for analysis of the condensed-phase reaction does not eliminate the necessity of analyzing the dispersed-phase processes. Equation (40) and results of a condensed-phase analysis provide two relationships among three unknowns, m, T_i, and $Y_{1,i}$. A third condition—for example, for specifying the energy feedback—is needed. Although the situation would be simplest if an adiabatic condition (zero temperature gradient) were achieved at the onset of dispersion, small amounts of energy feedback can be significant because of the strong dependences of the exponential factors on T_i. Thus a nonnegligible coupling between the reaction zones often remains, and the problem of developing a correct analysis is challenging. Much research remains to be done in analyzing the deflagration of most of the homogeneous propellants for which dispersed-phase processes are important.

When a given propellant formulation burns in the absence of externally applied perturbations, the value of m depends only on the two parameters T_0 and p. Temperature and pressure **sensitivity coefficients** are defined as $\pi_T = (\partial \ln m / \partial T_0)_p$ and $n = (\partial \ln m / \partial \ln p)_{T_0}$, respectively. A good theory of propellant deflagration, applicable for a range of values of T_0 and p, should agree with experiment concerning the values of m, π_T, and n. The identity $(\partial n / \partial T_0)_p = p(\partial \pi_T / \partial p)_{T_0}$ is a consequence of the definitions. Measurements of m, n, and π_T all are of use in helping to ascertain deflagration mechanisms. Various empirical relationships for m have been developed in the past. These include

$$m = cp^n/(\tilde{T_e} - T_0) \tag{41}$$

as well as relationships involving other functional dependences on p [8]. In equation (41), c, T_e, and n are three positive constants that serve to characterize m and its pressure and temperature sensitivities; T_e is called an explosion temperature and n is a pressure exponent of the burning rate, which must be less than unity if the propellant is to burn in a stable manner in a rocket chamber. Theoretical analyses often may provide agreement with experiment concerning c and n without predicting the temperature sensitivity correctly [40]; such analyses must be deemed incorrect in a fundamental sense. For homogeneous propellants, the factor $(T_e - T_0)$ in the denominator may often be raised to a positive power k different from unity without introducing disagreement with measurements, since measurements can be made only over a limited range of T_0 less than T_e. There are theories that predict positive values of k different from unity [40].

There are **plateau propellants** that exhibit an intermediate range of pressure over which m is practically independent of p and **mesa propellants** for which $m(p)$ achieves a maximum at a particular value of p then a minimum at a higher value. These effects may be produced in conventional double-base propellants by suitable addition of **burning-rate catalysts** (typically certain metal-organic salts) to the propellant formulation. It has been shown experimentally that these catalysts usually operate by modifying the interaction between the condensed-phase and dispersed-phase reaction zones [57], [58]. Thus dispersion phenomena are of importance to the deflagration of homogeneous propellants in a number of ways.

7.7. COMBUSTION OF HETEROGENEOUS PROPELLANTS

In principle, all the theoretical complexities that we have discussed for homogeneous propellants also may arise in attempts to analyze the deflagration of heterogeneous propellants.* Moreover, many other phenomena—for example, the occurrence of diffusion flames between fuel and oxidizer constituents—become relevant to the combustion process. To include all the potentially significant phenomena in a theoretical analysis becomes a formidable task that can be pursued only through the introduction of drastic idealizations for each of the phenomena considered. In recent years an appreciable amount of research has been devoted to the development of comprehensive models that take into account all effects that are known to be potentially relevant. The results are expressions for burning rates that contain several adjustable parameters, values for which can be selected within reason to obtain good agreement with measurements. These theories perform important functions beyond the correlation of existing data; for

---

* This is especially true if at least one of the propellant constituents can experience exothermic deflagration in the absence of the other. Most of the currently used heterogeneous propellants are of this type. Model propellants can be considered in which neither constituent can burn without the other. For these propellants gas-phase or liquid-phase heat release subsequent to mixing or heterogeneous heat release at points of contact between the two constituents must play an essential role in the deflagration mechanism. In general, gas-phase combustion appears to be of greater importance for heterogeneous propellants than for most homogeneous propellants.

example, they enable estimates to be made of burning rates of heterogeneous propellant formulations that differ somewhat from those that have been subjected to measurement, and they thereby indicate propellant modifications that may be tried for tailoring combustion characteristics in manners desired in applications. However, they tend not to lead to detailed understanding of specific aspects of the combustion processes. By focusing attention on limited aspects of the combustion mechanisms, better understanding of these aspects can be obtained (including unforeseen restrictions on complete models), and improved predictions of dependences of burning on certain parameters can be developed, without addressing questions of dependences on other parameters. Here we shall merely cite some literature and reviews of comprehensive models, and we shall discuss in a little more detail a few of the simplified models that address only certain specific phenomena. First, we shall consider composite propellants, primarily with ammonium perchlorate as oxidizer; then we shall make a few comments on the burning of black powder.

The origins of comprehensive models are usually traced to the work of Hermance [59], who included a gas-phase decomposition flame of ammonium perchlorate, fuel pyrolysis, and a heterogeneous reaction between fuel and oxidizer at their interface. Later, a gas-phase diffusion flame was added as well. The train of development may be followed in a sequence of publications [60]–[63]. The models contain various energy balances and different prescriptions for averaging over sizes and locations of oxidizer particles in the fuel matrix. The analytical formulations generally are expressed in terms of sets of simultaneous nonlinear algebraic equations that are solved by exercising routines for electronic computers. The criteria for successful models mainly have been taken to be agreement between theoretical and experimental dependences of burning rates on pressure for a variety of propellants involving different binders, different oxidizer particle size distributions, and so on. A topic of recent interest has been to account for effects of bimodal size distributions (for example, [63]). Although values of parameters in models have been selected successfully to produce good agreement with measured burning rates, the relationships between the values so obtained and independently determined values for the fundamental processes that they are intended to describe by and large remain uncertain. This is especially true for rate parameters such as overall activation energies. In fact, the more recent publications seldom ever even bother to list the values of the parameters that were employed. The principal trend of research on comprehensive models is toward additional complexity rather than toward independent evaluation of the appropriateness of the values of parameters employed. The intent is to develop codes that reproduce burning rates of propellants of practical interest rather than to clarify fundamentals of the chemical kinetics and of the physical processes that occur during deflagration.

A formidable problem encountered in modeling composite propellant deflagration concerns the statistics of oxidizer particle sizes and locations. Various methods (for example, [64]) have been proposed for averaging over these variables to obtain average burning rates from burning rates of individual particles, presumed known. One approach is to characterize the oxidizer particles as monopropellant spheres of a specified, constant radius with known, constant rates of normal surface regression, to characterize the fuel binder as a barrier having a burn-through time that is a known, increasing function of the shortest distance between the surfaces of burning and adjacent nonburning spheres and to characterize the configuration of the oxidizer spheres as being nearly a tightly packed hexagonal lattice with a probability density function assigned to a parameter that defines a small departure from symmetry [65]. By calculating an average time for propagation of burning along optimum paths, the average burning rate of the propellant can be expressed in terms of the regression rate of the oxidizer spheres, the average burn-through time of the binder, and geometric properties that include the particle size and statistical parameters of the particle configuration [65]. The model is beneficial in demonstrating how the packing configuration can influence even the pressure dependence of the average burning rate of the composite propellant. Thus, in terms of the mass burning rate of the oxidizer, m_O, (which depends upon p), a representative result for the mass burning rate of the propellant is [65]

$$m = m_O/(A + Bm_O), \tag{42}$$

where A and B depend on geometric parameters of the oxidizer distribution within the propellant and on the densities of the constituents and B, in addition, is proportional to the average burn-through time of the binder. The simplified model that gives equation (42) helps to explain potential influences of oxidizer particle size and particle loading on propellant burning rates.

The earliest simplified models of composite propellant combustion were models in which pyrolysis laws like equation (6) were applied separately to the fuel (F) and oxidizer (O) constituents [8], [28], [66]. Since the parameters A and E in expressions like $m = Ae^{-E/R^0T_i}$ differ for fuel and oxidizer, to achieve the same average values of m/ρ for each constituent, as needed for steady burning, it is necessary that the average surface temperatures T_i differ. This observation led to the two-temperature concept, in which the fuel and oxidizer components were assigned different surface temperatures that served to provide them with equal regression rates. By equating the two regression rates, a relationship between the two temperatures is obtained, and one additional relationship is then needed to determine completely the two temperatures and the regression rate of the propellant. This relationship has been taken to be an energy balance relating the energy feedback from a gas-phase flame to the energy needed for gasification of the propellant. If a pressure dependence of m is to be obtained in a simplified analysis, then the gas-phase flame must be premixed. It may be imagined that mixing of fuel and oxidizer gases is rapid and that the gas-phase flame is a premixed flame in the resulting gas mixture, or it may be considered that the mixing is relatively slow and unimportant and that the oxidizer gases support a decomposition flame that is responsible for the energy feedback. The former view might be better for sufficiently finely mixed propellant constituents and the latter for coarser formulations; the latter was employed in the original developments [28], [66]. Choices must be made in imposition of an energy balance—for example, in defining precisely what heats what—and there are consequent uncertainties and inaccuracies in the two-temperature models. However, the general conclusion that different constituents are likely to have different average surface temperatures is an important aspect that has found application in most of the newer modeling efforts.

Another class of simplified models focuses on diffusion flames between gases generated from fuel and oxidizer constituents. Summerfield [67] first discussed a model in which the rate of diffusion of gaseous fuel and oxidizer to a conical flame sheet above a fuel pocket controlled the overall burning rate of the propellant (see also [28]). This model has been augmented in various approximate ways to include premixed gas-phase flames as well [4], [8], [28], [67]–[69], thereby greatly improving agreement with measured pressure dependences of burning rates of propellants but also preventing detailed analyses of gas-phase flow fields from being accomplished with reasonable effort. To enable investigations of influences of diffusion flames to be pursued most simply, a sandwich model of propellant combustion has been conceived and subjected to a number of studies.

A **sandwich propellant** is a heterogeneous propellant composed of a series of parallel slabs of fuel and oxidizer, alternately placed side by side and bonded together. Burning occurs on a face that is perpendicular to the planes of the slabs. A heterogeneous propellant is approximated by an infinite number of slabs, but for purposes of clarifying processes occurring locally, investigators have often studied a small number of slabs—for example, only one slab of oxidizer and one of fuel or one of fuel between two of oxidizer. The advantage of the sandwich geometry is the simplification in analysis produced by the absence of variations in one of the directions. Many experimental studies have been made of the combustion of sandwich propellants (see [70] and references quoted therein). The concept was first discussed in the literature in [28]. An analysis of an infinite array of uniform fuel slabs alternating with uniform oxidizer slabs has been completed in a flamesheet approximation for diffusion flames that either extend from the lines of contact between solid fuel and oxidizer to infinity (under conditions of overall stoichiometry, [71]) or close over the fuel or oxidizer components (under nonstoichiometric conditions, [72]). The approximations in the gas-phase portion of the analysis are essentially the same as those discussed in Section 3.1 for the Burke-Schumann problem; the gas-phase equations are solved in a half-space with surface conditions describing interface species and energy conservation. Pyrolysis laws like equation (6) were applied to each propellant constituent, and the constituent most difficult to pyrolyze thereby determined the burning rate, while the pyrolysis parameters of the other constituent determined the difference in heights of the fuel and oxidizer surfaces in an approximation in which these surfaces were planar. The analysis predicts a decrease in burning rate with increasing slab thickness, in qualitative agreement with observed decreases in burning rates of heterogeneous propellants with increasing oxidizer particle sizes. However, there are many aspects in which the model did not attempt to achieve agreement with experiments performed with sandwich propellants, such as in details of the shape of the burning surface. In efforts to predict surface shapes better, theoretical analyses have been performed for a semi-infinite, endothermically pyrolyzing fuel slab adjacent to a semi-infinite, endothermically pyrolyzing oxidizer slab [73] and for this same configuration with the oxidizer supporting a gas-phase decomposition flame and with the diffusion flame neglected, the gaseous fuel being treated as inert [74]. These analyses have provided some new understanding but have not taken into account potential influences of finite-rate chemistry in the gas-phase diffusion flames—influences that have been suggested [75] to affect burning rates of heterogeneous propellants and that may be treated with help of the asymptotic methods discussed in Section 3.4.

Another type of sandwich propellant is the "sideways sandwich" suggested independently in [76] and [77]. In this model the propellant is burnt on a face parallel to the planes of the slabs. The model has no bearing on the occurrence of diffusion flames but instead is a vehicle for studying influences of heterogeneities within the solid on the combustion behavior. Heterogeneities in the solid's density, heat capacity, thermal conductivity, heat of gasification, heat of combustion, and rate parameters may be considered [77]. One-dimensional, time-dependent conservation equations describe the burning; there is a similarity to the analysis of oscillatory burning of homogeneous propellants (see Section 9.1.5). These heterogeneities result in local unsteadiness of the burning, and if mechanisms exist for communication of the unsteadiness to distant portions of the burning surface, then intrinsic oscillations of the propellant combustion may develop [77]. These oscillatory burning phenomena are discussed further in Chapter 9. Through nonlinear effects the heterogeneities themselves can modify the average burning rate [76], [77].

Black powder is a heterogeneous propellant of unusually complex structure in that three different solid phases are involved. It is a mixture of saltpeter, charcoal, and sulfur in weight percentages of about 75, 15, and 10. Although these percentages correspond to stoichiometry for the overall reaction

$$2KNO_3 + 3C + S \rightarrow K_2S + N_2 + 3CO_2, \tag{43}$$

the observed products of combustion are not entirely those of equation (43) but instead include a variety of species such as NO, NO_2, and K_2SO_4. Each of the three reactants in equation (43) is present in a separate phase in particulate form, with particle sizes between 1 and 100 μ. The particles are mixed and pressed into cakes that are ground into pellets having mean diameters typically between 300 μ and 3000 μ, and the pellets often are glazed with graphite before use. This ancient art produces a propellant that even today is better suited than any other for certain specialized applications (for example, for transmitting combustion from the initiator to the main propelling charge in gun propellants) because it provides a margin of safety by being relatively difficult to ignite while maintaining reliability since once ignited, it burns vigorously even at reduced pressures, spreading flames rapidly and in a reasonably reproducible manner through a somewhat loosely packed configuration of pellets.

The combustion of black powder has been subjected to scientific scrutiny since antiquity. Noble [78] and Vieille [79], for example, contributed to knowledge of its burning characteristics. Detailed experimental studies have been performed by use of more modern techniques [80]–[83]. Burning rates—that is, normal regression rates of surfaces of pellets or of tightly pressed powders—have been measured at reduced [84], elevated [85], and very high [86] pressures; results are shown in Figure 7.3, where the regression rate is $r \equiv m/\rho$, with ρ representing the density of the bulk propellant. In addition, rates of spread of flames among arrays of pellets

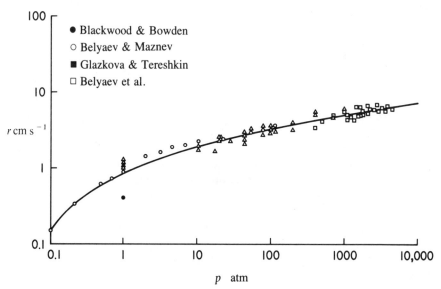

FIGURE 7.3. The normal regression rate of black powder as a function of pressure at 25°C [83].

have been measured [80] and found to be much larger than normal regression rates.

A theoretical model has been developed for the rate of normal regression of a burning surface of black powder [87]–[89]. The model involves a controlling two-phase (dispersed-phase) reaction zone in which carbon particles burn in a gaseous oxidant, mainly oxygen, produced from the decomposition of KNO_3. Despite contentions to the contrary [90], it appears that this model does not reasonably provide agreement with observed burning rates [91]. Melting of S (and possible of KNO_3) leading to mixing and exothermic reaction in a melt layer is known to be significant in ignition of black powder [80], and rates of heat release in a melt layer also may control the rate of normal regression [83]. No theories have been pursued yet for calculating the regression rate of black powder on the basis of control by condensed-phase reactions.

Flame spread among pellets of black powder appears to involve emission of fine droplets of hot salts containing K_2SO_4 from a burning liquid layer on the surface of a pellet and impingement of these particles on the surfaces of adjacent pellets, where they rapidly induce local ignitions [83]. At atmospheric pressure, particle velocities on the order of 5000 cm/s have been observed; these produce spread rates on the order of 60 cm/s. Two candidate mechanisms for theoretical description of the flame spread are the motion of hot gases through the gas-permeable propellant and a random walk of hot K_2SO_4 particles through the propellant matrix, with multiplication of the number of particles at each stop through a local surface ignition leading to renewed particle emissions [83]. There are theories for burning of gas-permeable propellants [92]–[95], which may be simplified and applied to flame spread in black powder to provide as an estimate of the spread velocity [83]

$$V = l(m/\rho_g)(\sigma/\epsilon), \tag{44}$$

where l is a representative length of the propellant charge, m is the normal mass burning rate of the propellant, ρ_g is the density of the product gas, σ is the surface-to-volume ratio (the total pellet surface area per unit chamber volume), and ϵ is the void fraction (the ratio of the volume not occupied by the pellets to the total chamber volume). The random-walk model is illustrated schematically in Figure 7.4, where v is the velocity of the hot salt particle and δ is the average distance between surfaces of adjacent blackpowder pellets; the four drawings illustrate the motions of salt particles at four successive time steps for a situation in which there is multiplication by a factor of three in each step. Although the theory of the process has not been developed, the resulting spread velocity appears to resemble that of equation (44) in some respects, with m/ρ_g replaced by v [83]. Further theoretical and experimental studies of the flame-spread processes could help to clarify the extent to which formulas like equation (44) may be applicable.

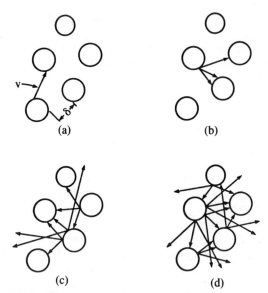

FIGURE 7.4.   Schematic model of random walk with multiplication [83].

7.8. EROSIVE BURNING

In all the models for the deflagration of solids discussed above, the average value of the component of velocity parallel to the surface of the solid has been assumed to be zero. However, practical designs for solid propellant rocket motors usually employ laterally burning charges such as that illustrated in Figure 7.5. Along the downstream surface of the propellant in these motors, the gas velocity parallel to the burning surface can attain high subsonic values. The consequent increased convective heat transfer to the downstream portion of the charge increases the burning rate in the downstream region. This phenomenon is termed **erosive burning**.

The two-dimensional nature of the process initially forced semi-empiricism upon theoretical descriptions of erosive burning. The first proposed empirical burning-rate formulas were of the form

$$m = m^0 + kM, \tag{45}$$

where m is the local mass burning rate, m^0 is the mass burning rate in the absence of erosion [as given by one of the above models], M is the local average mass flux of gas parallel to the surface (averaged over a cross section of the chamber—see Figure 7.5), and k is an *erosion constant*, which was taken to be independent of M [28]. More accurate experimental evidence indicates that contrary to equation (45), m is not exactly a linear function of M. In particular, a "threshold" value of M exists, below which M has

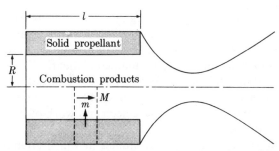

FIGURE 7.5. Schematic diagram of a laterally burning charge in a solid propellant rocket motor.

little or no effect on m [28]. A burning-rate formula that tends to reproduce this threshold phenomenon is [96]

$$m = m^0 + k_1(M^{4/5}/z^{1/5})e^{-k_2 m/M}, \qquad (46)$$

where z is the axial distance from the upstream point at which $M = 0$ and the (undetermined) constants k_1 and k_2 are independent of m^0, M, and z. The prediction that the last term in equation (46) does not depend on m^0 is in agreement with experiment [97], and the values of the constants k_1 and k_2 can be chosen in such a way that the function $m(M)$ agrees with results of reasonably accurate experiments [28], [98].

Equations (45) and (46) are only two of many formulas that have been used to describe erosive burning [8]. Most of the formulas that have been suggested are based on physical concepts of influences of crossflow on propellant burning. Among these concepts is the idea that high external velocities produce a turbulent boundary layer (see Chapter 12) on the propellant surface and thereby effectively increase the thermal diffusivity of the gas, which in turn increases the rate of heat transfer to the propellant and hence the burning rate [99]. The idea that turbulent convective heat transfer from the hot combustion products outside the boundary layer provides an additive contribution to the heat flux reaching the propellant surface and, therefore, to the burning rate underlies equation (46) [96]; the exponential factor in this equation is intended to account for reduction in this convective heat transfer through the "blowing" effect of mass transfer from the propellant surface, which becomes significant in upstream regions where m is of the same order as M. Reynolds numbers based on M and on R or z in Figure 7.5 ($0 \le z \le l$) typically vary from 0 to 10^8, and a number of different flow regimes may therefore be anticipated, ranging from negligible cross-flow near $z = 0$, to conditions of small crossflow in which boundary-layer ideas are inapplicable, to a regime having Reynolds numbers appropriate for laminar boundary layers (see Chapter 12), to a regime in which the Reynolds numbers are well above critical values for transition to turbulence, and turbulent boundary layers are anticipated. Most of the theoretical investigations emphasize the latter regimes rather than the conditions of small crossflow, which are below the threshold for erosion and within which "negative erosion" often is observed (a small *decrease* in m with increasing M over a range of values of M) [8]. Fundamental theories of negative erosion might be pursued through perturbation analyses of the deflagration theories discussed in previous sections, with the ratio of the crossflow velocity to the normal velocity treated as a small parameter; no work of this kind appears to have been undertaken.

There have been many newer studies in which attempts have been made to improve burning-rate formulas for erosive burning [8], [100]–[104] by refining or modifying the physical descriptions of boundary-layer flows with finite-rate chemistry. In addition, numerical integrations of the boundary-layer equations with finite-rate chemistry included have been performed for laminar and turbulent flows with erosive-burning applications in mind [105]–[109]; these investigations usually provide graphical results rather than burning-rate formulas. In boundary-layer studies, the external flow may be approximated as a steady, frictionless, adiabatic flow of an ideal gas with constant specific heats, in a constant-area channel with gas injection normal to the flow direction. This flow may be shown to be described by algebraic equations [8] such as

$$\frac{T}{T^0} = \frac{2}{1 + \psi} \frac{p}{p^0}, \qquad \frac{p}{p^0} = \frac{1 + \gamma\psi}{1 + \gamma}, \tag{47}$$

where γ is the ratio of specific heats, the superscript 0 identifies core conditions at $z = 0$ (where $M = 0$), and $\psi = [1 - (M/M^*)^2]^{1/2}$, in which $M^* \equiv \gamma p^0 / [2(\gamma^2 - 1)c_p T^0]^{1/2}$ represents the value of M at chocking conditions (a Mach number of unity). Equation (47) is used along with the mass balance

$$M = \int_0^z (P/A)m \, dz \tag{48}$$

(where the perimeter P and cross-sectional area A are given by $2\pi R$ and πR^2, respectively, for the circular cylindrical grain of Figure 7.5) in obtaining conditions outside the boundary layer. The main difficulties in analyses then arise in attempting to describe the boundary-layer flows. In one study [110], parabolic partial differential equations of turbulent reacting flow were integrated numerically across the entire chamber without breaking the flow into a core region and a boundary layer, a procedure that may be useful in accounting for rotationality in the core [111].

Although the most recent studies of erosive burning [102]–[110] have helped greatly to improve our understanding of the process, the theoretical problems remain difficult. Even in laminar boundary-layer flows, erosive burning of solid propellants presents one of the more challenging problems because simplifications achievable for nonpremixed combustion (Chapter

12) do not apply to the premixed solid-propellant system. In the turbulent boundary-layer flows, which are of greater practical interest, appreciable uncertainty remains concerning the proper partial differential equations that should be used to describe premixed combustion (see Chapter 10), and therefore the problem is fundamentally insoluble today. In view of these difficulties, the extent of progress that has been made recently in understanding erosive burning is remarkable. Nevertheless, much more research remains to be done on the subject, for example, in developing well-justified simplified approximations for laminar-flow situations and in ascertaining suitable descriptions of turbulent flows.

REFERENCES

1. G. P. Sutton, *Rocket Propulsion Elements*, New York: Wiley, 1949; 3d ed., 1963, Chapters 10–12, 310–387.
2. C. Huggett, "Combustion of Solid Propellants," in *Combustion Processes*, vol. II of *High Speed Aerodynamics and Jet Propulsion*, B. Lewis, R. N. Pease, and H. S. Taylor, eds. Princeton: Princeton University Press, 1956, 514–574.
3. M. Barrère, A. Jaumotte, B. Fraeijs de Veubeje, and J. Vandenkerckhove, *Rocket Propulsion*, New York: Elsevier Publishing Co., 1960, Chapters 4–6, 190–368.
4. S. S. Penner, *Chemical Rocket Propulsion and Combustion Research*, New York: Gordon and Breach, 1962, Chapter 4, 99–143.
5. V. M. Maltsev, M. I. Maltsev, and L. Y. Kashporov, *Basic Characteristics of Combustion*, Moscow: Chimia, 1977, Chapter 5, 269–317.
6. R. N. Wimpress, *Internal Ballistics of Solid-Fuel Rockets*, New York: McGraw-Hill, 1950.
7. N. N. Bakhman and A. F. Belyaev, *Combustion of Heterogeneous Condensed Systems*, Moscow: Nauka, 1967.
8. F. A. Williams, M. Barrère, and N. C. Huang, *Fundamental Aspects of Solid Propellant Rockets*, AGARDograph No. 116, Slough, England: Technivision Services, 1969.
9. A. G. Merzhanov and I. P. Borovinskaya, *CST* **10**, 195 (1975).
10. M. Barrère and H. Moutet, *Rech. aéro.* **9**, 31 (1956).
11. C. Guirao and F. A. Williams, *AIAA Journal* **9**, 1345 (1971).
12. M. Ben-Reuven, L. H. Caveny, R. J. Vichnevetsky, and M. Summerfield, *16yh Symp.* (1976), 1223–1233.
13. B. I. Khaikin and A. G. Merzhanov, *Dokl. Akad. Nauk SSSR* **173**, 1382 (1967).
14. A. G. Merzhanov, *C & F* **13**, 143 (1969).
15. B. I. Khaikin, A. K. Filonenko, S. I. Khudyaev, and T. M. Martem'yanova, *Fiz. Gor. Vzr.* **9**, 169 (1973).
16. S. S. Novikov, P. F. Pokhil, and Y. S. Ryazantsev, *Fiz. Gor. Vzr.* **4**, 469 (1968).
17. D. A. Frank-Kamenetskii, *Diffusion and Heat Transfer in Chemical Kinetics*, New York: Plenum Press, 1969, 317.
18. B. V. Novozhilov, *Nonstationary Combustion of Solid Rocket Fuels*, Moscow: Nauka, (1973).

19. R. E. Wilfong, S. S. Penner, and F. Daniels, *J. Phys. Chem.* **54**, 863 (1950).
20. R. D. Schultz and A. O. Dekker, *J. Chem. Phys.* **23**, 2133 (1955).
21. H. Spingler, *Z. phys. Chem.* **B52**, 90 (1942).
22. L. L. Bircumshaw and T. R. Phillips, *J. Chem. Soc.*, 4741 (1957).
23. A. K. Galwey and P. W. M. Jacobs, *J. Chem. Soc.*, 837 (1959), 5031 (1960).
24. S. H. Inami, W. A. Rosser, and H. Wise, *J. Phys. Chem.* **67**, 1077 (1963).
25. J. V. Davies, P. W. M. Jacobs, and A. Russell-Jones, *Trans. Faraday Soc.* **63**, 1737 (1967).
26. P. W. M. Jacobs and A. Russell-Jones, *J. Phys. Chem.* **72**, 202 (1968).
27. R. D. Schultz and A. O. Dekker, *5th Symp.* (1955), 260–267.
28. R. D. Schultz, L. Green, Jr., and S. S. Penner, *Combustion and Propulsion, Third AGARD Colloquium*, New York: Pergamon Press, 1958, 367–420.
29. M. K. Barsh et al, *Rev. Sci. Instr.* **29**, 392 (1958).
30. R. F. Chaiken, D. H. Sibbett, J. E. Sutherland, D. K. Van de Mark, and A. Wheeler, *J. Chem. Phys.* **37**, 2311 (1962).
31. J. Powling, *11th Symp.* (1967), 447–456.
32. P. W. M. Jacobs and G. S. Pearson, *C & F* **13**, 419 (1969).
33. C. Guirao and F. A. Williams, *J, Phys. Chem.* **73**, 4302 (1969).
34. W. Nachbar and F. A. Williams, *9th Symp.* (1963), 345–357.
35. R. H. Cantrell, *AIAA Journal* **1**, 1544 (1963).
36. W. H. Andersen, *AIAA Journal* **2**, 404 (1964).
37. R. H. Cantrell, *AIAA Journal* **2**, 406 (1964).
38. P. W. M. Jacobs and J. Powling, *C & F* **13**, 71 (1969).
39. F. A. Williams, *Int J. Heat Mass Transfer* **8**, 575 (1965).
40. M. M. Ibiricu and F. A. Williams, *C & F* **24**, 185 (1975).
41. G. Mohan and F. A. Williams, *AIAA Journal* **10**, 776 (1972).
42. T. Niioka and F. A. Williams, *17th Symp.* (1979), 1163–1171.
43. Y. B. Zel'dovich, *Zhur. Eksp. Teor. Fiz.* **12**, 498 (1942).
44. A. G. Merzhanov and F. I. Dubovitsky, *Dokl. Akad. Nauk SSSR* **129**, 153 (1959); **135**, 1439 (1960).
45. G. Lengellé, *AIAA Journal* **8**, 1989 (1970).
46. F. A. Williams, *AIAA Journal* **11**, 1328 (1973).
47. J. D. Buckmaster, A. K. Kapila, and G. S. S. Ludford, *Acta Astronautica* **3**, 593 (1976).
48. A. K. Kapila and G. S. S. Ludford, *Acta Astronautica* **4**, 279 (1977).
49. J. D. Buckmaster and G. S. S. Ludford, *Theory of Laminar Flames*, Cambridge: Cambridge University Press, 1982.
50. A. G. Merzhanov and A. K. Filonenko, *Dokl. Akad. Nauk SSSR* **152**, 143 (1963).
51. B. I. Khaikin, A. K. Filonenko, and S. I. Khudyaev, *Fiz. Gor. Vzr.* **4**, 591 (1968).
52. G. Rosen, *J. Chem. Phys.* **32**, 89 (1960).
53. W. E. Johnson and W. Nachbar, *8th Symp.* (1962), 678–689.
54. D. B. Spalding, *C & F* **4**, 59 (1960).
55. E. I. Maksimov, *Fiz Gor. Vzr.* **4**, 203 (1968).
56. E. I. Maksimov and A. G. Merzhanov, *Fiz. Gor. Vzr.* **2**, 47 (1966).
57. N. Kubota, T. J. Ohlemiller, L. H. Caveny, and M. Summerfield, *AIAA Journal* **12**, 1709 (1974).
58. N. Kubota, *17th Symp.* (1978), 1435–1441.
59. C. E. Hermance, *AIAA Journal* **4**, 1629 (1966).

60. M. W. Beckstead, R. L. Derr, and C. F. Price, *AIAA Journal* **8**, 2200 (1970); *13th Symp.* (1971), 1047–1056.
61. N. S. Cohen, *AIAA Journal* **18**, 277 (1980).
62. M. W. Beckstead, *18th Symp.* (1981), 175–185.
63. N. S. Cohen and L. D. Strand, *AIAA Journal* **20**, 1739 (1982).
64. R. L. Glick, *AIAA Journal* **12**, 384 (1974); **14**, 1631 (1976).
65. W. C. Strahle, *AIAA Journal* **16**, 843 (1978).
66. W. H. Andersen, K. W. Bills, E. Mishuck, G. Moe, and R. D. Schultz, *C & F* **3**, 301 (1959).
67. M. Summerfield, G. S. Sutherland, M. J. Webb, H. J. Tabak, and K. P. Hall, "Burning Mechanism of Ammonium Perchlorate Propellants," in *Solid Propellant Rocket Research*, vol. 1 of *Progress in Astronautics and Rocketry*, M. Summerfield ed., New York: Academic Press, 1960, 141–182.
68. M. Barrère and L. Nadaud, *Rech. aéro.* **17** (No. 98), 15 (1964).
69. J. A. Steinz, P. L. Stang, and M. Summerfield, "The Burning Mechanism of Ammonium Perchlorate-Based Composite Solid Propellants," AIAA Paper No. 68-658 (1968).
70. E. W. Price, J. C. Handley, R. R. Panyam, R. K. Sigman, and A. Ghosh, *AIAA Journal* **19**, 380 (1981).
71. W. Nachbar, "A Theoretical Study of the Burning of a solid Propellant Sandwich," in *Solid Propellant Rocket Research*, vol. 1 of *Progress in Astronautics and Rocketry*, M. Summerfield, ed., New York: Academic Press, 1960. 207–226.
72. W. Nachbar and G. B. Cline, Jr., "The Effects of Nonstoichiometric Composition and Particle Size on the Burning Rates of Composite Solid Propellants," *Tech. Note 3-64-62-1, Contract No. AF 49(638)-412*, Lockheed Missiles and Space Company, Palo Alto (1962); presented at Fifth AGARD Combustion and Propulsion Colloquium, Braunschweig (1962).
73. N. N. Bakhman and V. B. Librovich, *C & F* **15**, 143 (1970).
74. W. C. Strahle, *AIAA Journal* **13**, 640 (1975).
75. J. B. Fenn, *C & F* **12**, 201 (1968).
76. S. S. Novikov, P. F. Pokhil, Y. S. Ryantsev, and L. A. Sukanov, *Zhur. Prikl. Mekh. Tekhn. Fiz.*, no. 3, 128 (1968).
77. F. A. Williams and G. Lengellé, *Astronautica Acta* **14**, 97 (1969).
78. R. A. Noble and R. Abel, *Phil. Trans. Roy. Soc. London* **165A**, 49 (1875).
79. M. Vieille, *Memorial des Poudres et Salpetres* **6**, Paris: Gauthier-Vallars et Fils, 1893, Chapter II, 256–391.
80. J. D. Blackwood and F. P. Bowden, *Proc. Roy. Soc. London* **213A**, 285 (1952).
81. C. Campbell and G. Weingarten, *Trans. Faraday Soc.* **55**, 2221 (1959).
82. W. Hintze, *Explosivstoff* **2**, 41 (1968); *Nobel Hefte* **2**, 14 (1972).
83. F. A. Williams, *AIAA Journal* **14**, 637 (1976).
84. A. F. Belyaev and S. F. Maznev, *Dokl. Akad. Nauk SSSR* **131**, 887 (1960).
85. A. P. Glazkova and I. A. Tereshkin, *Zhur. Fiz. Khim.* **35**, 1622 (1961).
86. A. F. Belyaev, A. I. Korotkov, A. K. Parfenov, and A. A. Sulimov, *Zhur. Fiz. Khim*, **37**, 150 (1963).
87. N. N. Bakhman, *Dokl. Akad. Nauk SSSR* **129**, 824 (1959).
88. O. I. Leipunskii, *Zhur. Fiz. Khim.* **34**, 177 (1960).
89. B. V. Novozhilov, *Dokl. Akad. Nauk SSSR* **131**, 1400 (1960).
90. V. M. Zakharov and L. A. Klyachko, *Fiz. Gor. Vzr.* **8**, 15 (1972).

91. B. V. Novozhilov, *Zhur. Fiz. Khim.* **36**, 1803 (1962).
92. K. K. Kuo and M. Summerfield, *AIAA Journal* **12**, 49 (1974).
93. B. S. Ermolayev, A. A. Borisov, and B. A. Khasainov, *AIAA Journal* **13**, 1128 (1975).
94. P. S. Gough and F. J. Zwarts, *AIAA Journal* **17**, 17 (1979).
95. S. S. Gokhale and H. Krier, *Prog. Energy Combust. Sci.* **8**, 1 (1982).
96. J. M. Lenoir and G. Robillard, *6th Symp.* (1957), 663–667.
97. R. D. Geckler, *Selected Combustion Problems, Fundamentals and Aeronautical Applications*, AGARD, London: Butterworths Scientific Publications, 1954, 289–339.
98. T. Marklund and A. Lake, *ARS Journal*, **30**, 173 (1960).
99. J. Corner, *Trans. Faraday Soc.* **43**, 635 (1947).
100. J. A. Vandenkerckhove, *Jet Propulsion* **28**, 599 (1958).
101. P. J. Blatz, *8th Symp.* (1962), 745–752.
102. G. Lengellé, *AIAA Journal* **13**, 315 (1975).
103. M. King, *Journal of Spacecraft and Rockets* **15**, 139 (1978).
104. H. S. Mukunda, *CST* **18**, 105 (1978).
105. H. Tsuji, *9th Symp.* (1963), 384–393.
106. F. L. Schuyler and T. P. Torda, *AIAA Journal* **4**, 2171 (1966).
107. R. A. Beddini, *AIAA Journal* **16**, 898 (1978).
108. M. K. Razdan and K. K. Kuo, *AIAA Journal* **17**, 1225 (1979).
109. M. K. Razdan and K. K. Kuo, *AIAA Journal* **20**, 122 (1982).
110. R. A. Beddini, *AIAA Journal* **18**, 1345 (1980).
111. R. L. Glick, *AIAA Journal* **21**, 156 (1983).

CHAPTER 8

Ignition, Extinction, and Flammability Limits

A number of classical combustion phenomena are fundamentally more complex than the processes that have been considered thus far. These phenomena are related in one way or another to processes of ignition or extinction—processes that generally are multidimensional and time-dependent. Description of these phenomena by theories as rigorous as those presented in Chapters 2–6 is a formidable task that can be completed only by treating each problem individually. General principles are difficult to extract from detailed analyses of specific problems. Therefore, the presentation to be given here will be less precise, in an effort to expose general ideas. Analyses will be restricted to the simplest illustrative models. Although ignition, extinction, and flammability limits appear to be highly diverse phenomena, they share many common elements. Therefore, it is instructive to treat them from a relatively common viewpoint.

Ignition is the process whereby a material capable of reacting exothermically is brought to a state of rapid combustion. It has long been of practical interest from numerous viewpoints, ranging from the desire to prevent unwanted explosions in transportation and storage of combustibles to the need to initiate controlled combustion in furnaces, motors, and guns. Methods for achieving ignition include exposure of the combustible to a sufficiently hot surface, to an electrically heated wire, to a radiant energy source, to hot inert gases, to a small pilot flame, to an explosive charge, or to an electrical spark discharge. These methods have been listed here roughly in the order of increasing temperatures of the ignition sources. There tends to be an inverse relationship between the temperature or strength of the ignition source and the time of exposure required for ignition; strong sources ignite quickly, but weaker sources require longer exposures. In extreme cases of strong sources, detonation can be initiated directly (see Section 6.3.5); attention here is restricted to situations in which ignition results in a deflagration or a diffusion flame.

Extinction (quenching) of deflagrations or of diffusion flames may be achieved in various ways, including passing the flame through tubes of small diameter (a design principle of flame arrestors), removing an essential reactant from the system (an approach especially suitable for diffusion flames), adding sufficiently large quantities of a material (such as water or a chemical fire extinguishant) that slows combustion, or physically removing the flame from the reactant mixture by inducing high gas velocities (blowing the flame away, as accomplished, for example, by explosives in large production-well fires). Extinction phenomena are of interest in problems of safety and of fire suppression as well as in various aspects of control of combustion processes in industry and propulsion. General strategies for extinguishment may be listed as cooling, flame removal, reactant removal, and chemical inhibition; practical methods may be placed in one or more of these four categories. Since processes of flame extinction usually occur relatively rapidly if at all, more emphasis has been placed on ascertaining critical conditions for extinction than on extinction times.

Flammability limits are limits of composition or pressure beyond which a fuel-oxidizer mixture cannot be made to burn. They are of practical interest especially in connection with safety considerations because mixtures outside the limits of flammability can be handled without concern about ignition. For this reason, extensive tabulations of limits of flammability have been prepared [1], [2]. Meanings of these tabulations and their relationships to ignition and extinction phenomena will be considered here in Section 8.2.

Rates of chemical reactions always have a bearing on ignition, extinction, and flammability limits. There are many situations in which analyses of these phenomena reasonably may employ one-step, Arrhenius approximations to the rates. This fact enables common theories to be developed on the basis of *energy* considerations, which serve to correlate a number of different observed characteristics of ignition, quenching, and flammability limits. We shall focus our attention here on results explained by energy-conservation requirements and heat losses. In so doing, we exclude the consideration of special effects associated with finer details of chemical kinetics, such as radical diffusion or surface reactions.

Adoption of the energy-based viewpoint may be motivated by the fact that energetic aspects practically always are relevant to the phenomena to be studied. Seldom ever can energy conservation be neglected in addressing topics in combustion. Current uncertainties lie not at the level of asking whether energetic aspects or changes in chemical-kinetic mechanisms govern the phenomena, but rather involve questions of which of the two is of greater importance. In well-planned studies of influences of specific aspects of chemical-kinetic mechanisms, the energetic effects are taken into account; after their removal, the remaining influences may be attributed to particular aspects of chemical kinetics.

There are many situations in which chemical-kinetic details have been established as significant. For example, under appropriate conditions ignition of a fuel-air mixture by laser radiation is appreciably enhanced if the frequency of the light is suitable for dissociation of O_2; mechanisms of reaction of oxygen atoms are important in this ignition process. If bromine-containing species (for example, the chemical fire suppressant CF_3Br) are added to fuel-air flames, then extinction is promoted at least partially because the step $H + HBr \rightarrow H_2 + Br$ removes the very active H atom, replacing it with the less active Br atom, thereby reducing the overall rate of heat release. Although lists may be prepared of examples like these that illustrate the significance of specific chemical-kinetic mechanisms, we shall not discuss them further.

With the one-step, Arrhenius approximation to the chemistry, activation-energy asymptotics (Section 5.3.6) afford a useful vehicle for analyzing ignition and extinction phenomena. In Section B.2.5.4, corresponding ideas were introduced in studying the simplest problem related to ignition theory, the problem of homogeneous thermal explosions, and in Sections 3.4.3 and 3.4.4, activation-energy asymptotics were applied to the problem of diffusion-flame extinction. Here the method will be used first (Section 8.2) for analyzing premixed-flame propagation with homogeneously distributed heat losses and next (Section 8.3) in discussing a variety of ignition problems. It will be seen in Section 8.3 that the greater the degree of heterogeneity of the system, the more complicated the analyses tend to become. The considerations in Sections 8.1 and 8.2 are restricted to homogeneous mixtures of premixed combustibles. The analysis to be given in Section 8.2 will be seen to be relevant to premixed-flame extinction, to quenching distances (see Section 8.1), and to flammability limits. We begin in Section 8.1 with some rough physical ideas and observations that serve to define terms and to relate various ignition and extinction phenomena. References [3]–[5] are earlier books that present the material from viewpoints different from that of Sections 8.2 and 8.3 and that discuss the relevant experiments in greater detail.*

---

* More-recent reviews of flammability limits are given by L. A. Lovachev, V. S. Babkin, V. A. Bunev, A. V. V'yun, V. N. Krivulin, and A. N. Baratov, *C & F* **20**, 259 (1973) and by M. Hertzberg, "The Flammability Limits of Gases, Vapors and Dusts: Theory and Experiment," in *Fuel-Air Explosions*, J. H. S. Lee and C. M. Guirao eds., Waterloo, Canada: University of Waterloo Press, 1982, 3–48.

8.1. MINIMUM IGNITION ENERGIES AND QUENCHING DISTANCES

When a combustible mixture is ignited by heating a plane slab of the gas, it is found that the amount of energy (per unit surface area of the slab) added to the gas must exceed a definite minimum value for ignition to occur. Theoretically, ignition by means of a hot slab of gas represents one of the simplest conceivable ignition problems, since the plane, one-dimensional, time-dependent conservation equations describe the process. Analysis (for example, [6]) by numerical integration for slabs of various thickness, initially raised to the adiabatic flame temperature, shows (for instance, with one-step, Arrhenius kinetics) that for slabs thinner than a critical thickness, the temperature decays to the ambient temperature by heat conduction without producing ignition, while for thicker slabs, a propagating laminar flame develops. These results and those of various related experimental and theoretical studies (such as [4]–[8]) can be summarized by the following approximate rule. *Ignition will occur only if enough energy is added to the gas to heat a slab about as thick as a steadily propagating adiabatic laminar flame to the adiabatic flame temperature.*

This ignition criterion is immanently reasonable. It can also be shown to be roughly equivalent to a number of other reasonable physical statements—for example, that the rate of liberation of heat by chemical reactions inside the slab must approximately balance the rate of heat loss from the slab by thermal conduction. The rule also clearly represents an over-simplified approximation; discrepancies exceeding a factor of 2 or 3 will not be surprising. A formula expressing the rule is

$$H = (A\delta)\rho_0[c_p(T_\infty - T_0)], \qquad (1)$$

where H is **the minimum ignition energy**, A is the cross-sectional area of the slab, δ is the (adiabatic) laminar flame thickness, ρ_0 is the initial density of the mixture, c_p is the average specific heat at constant pressure, T_∞ is the adiabatic flame temperature, and T_0 is the initial temperature of the mixture. When use is made of equation (5-4) for δ, equation (1) reduces to the statement that the minimum ignition energy per unit area is

$$H/A = \lambda(T_\infty - T_0)/v_0, \qquad (2)$$

where λ is the average thermal conductivity of the gas and v_0 is the (adiabatic) laminar burning velocity.

A minimum ignition energy may be obtained explicitly from equation (2) by identifying a minimum cross-sectional area A for the slab. The minimum value of A has often been related to the results of flame-quenching experiments. Flames will not propagate through very narrow channels because of heat losses to the walls (and, possibly, other effects). Although experiments have been performed in a number of different configurations

[4], [5], [9]–[15], a popular procedure is to confine the combustible mixture between parallel flat plates and to define the **quenching distance** d as the minimum plate separation for which flame propagation can be achieved. It is reasonable to assume that multidimensional effects will be small, and therefore that the rule expressed by equation (1) will be valid, provided that the slab area A is greater than the area of a square with sides of length equal to the quenching distance (that is, $A \geq d^2$). When this condition is used in equation (2), the minimum ignition energy is found to be

$$H = d^2 \lambda (T_\infty - T_0)/v_0. \tag{3}$$

The approximate correlation between quenching distance and minimum ignition energy expressed by equation (3) is (roughly) substantiated experimentally [4].

Equation (3) will determine the minimum ignition energy in terms of fundamental properties of the system only if an expression for the quenching distance d can be found. For systems in which heat losses control the quenching process, d is determined by the factors discussed in Section 8.2. When the heat loss is assumed to occur through thermal conduction, the results of these heat loss studies agree with other heat loss theories [16], [17] and with experiment [4], [16], [17] in predicting that [see equation (30)]

$$d = a\delta, \tag{4}$$

where a(approximately 40) is a multiplicative factor that is not strongly dependent on the properties of the system, and the thickness δ of the (adiabatic) flame is given by equation (5-4). Equations (5-4) and (4) yield

$$d = a\lambda/c_p \rho_0 v_0, \tag{5}$$

whence equation (3) becomes

$$H = a^2 \lambda^3 (T_\infty - T_0)/c_p^2 \rho_0^2 v_0^3. \tag{6}$$

In view of equation (5-4), alternative forms of equation (6) are

$$H = a^2 \delta \lambda^2 (T_\infty - T_0)/c_p \rho_0 v_0^2 \tag{7}$$

and

$$H = a^2 \delta^3 \rho_0 c_p (T_\infty - T_0). \tag{8}$$

Equation (6) correctly predicts the approximate temperature and pressure dependences of the minimum ignition energy.

The results that have been given here for the minimum ignition energy are subject to various chemical-kinetic restrictions related to the nature of the ignition source. The chemistry involved in v_0, δ, and d evidently is that of a propagating deflagration. The same chemical mechanism therefore must be involved in H if correlations like those of equations (3) or (6)–(8) are to be obtained. If the ignition source selectively enhances particular steps in the kinetic mechanism, if it involves surface reactions, or if it produces temperatures much higher or much lower than the adiabatic flame temperature, then the basis of the correlations becomes questionable. Among the ignition procedures identified in the previous section (see also [18] for a classification), the hot-surface, electrical-wire, radiation-source, and hot-gas techniques often produce temperatures below the adiabatic flame temperature by amounts sufficient to raise doubts concerning the use of equations (6)–(8). From the viewpoint of chemical kinetics, pilot flames are ideally suited to the approach (because they provide temperatures and compositions in the range assumed), but experimentally it is difficult to introduce a pilot flame rapidly enough into a combustible mixture to avoid nonhomogeneities and associated heat losses that have not been taken into account here. Most of the experiments on which the correlations are based are spark-ignition experiments; although the initial temperatures produce ions and radicals and are much higher than the adiabatic flame temperature, this does not seem to exert a significant influence on the chemical kinetics of flame development. Sparks add measurable energy rapidly enough for their effects to be characterized in terms of an H.

The slab geometry that has been considered here is not representative of most spark-ignition arrangements; spherical or cylindrical ignition kernels are more realistic, although the actual shape of the region of elevated temperature may be quite complex (for example, toroidal regions of hot gas influenced by buoyancy have been observed*). Approximate analyses of spherical and cylindrical ignition sources [19]–[21] produce expressions that differ from equation (6) at most by a roughly constant multiplicative factor that can be absorbed into a^2. A considerable amount of attention has been devoted to studies of processes of spark ignition [4], [18]–[29], and many facts have been uncovered concerning the mechanism. For example, it is known that there is a critical electrode spacing giving a minimum value of H [4], [18], [22], [25], [28]; when the spacing is less than the quenching distance, the required energy for ignition may exceed the minimum ignition energy appreciably because of heat-conduction losses to the electrodes, but when the spark gap exceeds the quenching distance, the spark may heat a cylindrical region of length equal to the electrode spacing, thereby increasing the energy requirement as the spacing is increased. This last effect can be included approximately in equation (7) by replacing $a \delta$ there by the actual electrode spacing [20]. Simple formulas to account for increases in the ignition energy by convective gas flow past the electrodes [20], [23]–[27], by turbulence [20], [22], [24], [25], [28], and by the presence of liquid fuel droplets [27], [28] also have been devised.

Another factor influencing the minimum ignition energy for spark ignition is the spark duration [20], [23], [24], [26]. There is an optimum

---

* H. W. Emmons, personal communication.

duration for minimum energy, typically on the order of 50 μs. As the heating time is increased, the energy that must be supplied by the ignition source increases because of energy lost from the ignition region by conduction and radiation prior to ignition [30], but a significant amount of energy is also lost in shock waves, generated by the spark, that propagate away quickly without contributing to ignition [31]; this last loss increases as the spark duration decreases because of a consequent increase in shock strength. Thus the minimum value of the minimum spark-ignition energy, typically on the order of 1 mJ for stoichiometric gaseous hydrocarbon-air mixtures at atmospheric pressure, is attained at particular values of the electrode spacing and of the spark duration. In rare instances for spark or glow discharges of very long duration, and especially for heated surfaces, electrically heated wires, hot gases, or pilot flames, critical conditions for ignition are characterized better by an ignition temperature [18] (the source temperature needed to produce ignition) than by an ignition energy because losses through heat conduction and through igniter-induced fluid flow are substantial.

The material that has been presented in this section is of an approximate, physical nature and leaves open questions. For example, why is the value of a so large? These questions may be addressed by detailed analyses that begin with the conservation equations in differential form. Analyses of this type are developed in the following sections. An expression for a will be derived in Section 8.2, and ignition analyses will be considered in Section 8.3.

8.2. PREMIXED FLAMES WITH HEAT LOSSES

To address topics of flame quenching and of flammability limits more thoroughly, it is of interest to develop the theory of flame propagation with heat losses. In the introduction to this chapter, we indicated that although changes in chemical kinetics may influence limit phenomena, heat losses to the surroundings of the flame practically always are important. This fact was recognized in early work [4], [9], [32]-[34] and was used in predicting limit conditions [17], [35], [36]. To avoid the complexities of multidimensional analyses, the consequences of heat losses will be explored here within the framework of a one-dimensional, steady-flow model.

8.2.1. Methods of analysis

The earliest studies of heat-loss effects in premixed flames were based on analytical approximations to the solution of the equation for energy conservation [35]-[39]. Two such approximations that have been sufficiently popular to be presented in books are those of Spalding (see [40]) and of von Kármán (see [5]). Later work involved numerical integrations [41]-[43] and, more recently, activation-energy asymptotics [44]-[46].

Current computational capabilities enable two-dimensional formulation to be solved numerically [47]. Here we first address a model problem from the viewpoint of asymptotics.

The first objective of the following analysis will be to compute the burning velocity for a simplified model of a plane, one-dimensional, non-adiabatic flame. Under certain conditions involving nonvanishing heat losses, it will be found that no solution for the burning velocity exists, thus indicating the presence of flammability limits. Under other conditions, two burning velocities will be obtained, one of which presumably corresponds to an unstable flame.

We shall focus our attention on the steady-state temperature distribution, which is assumed to be determined by the equation

$$\lambda \, d^2T/dx^2 - mc_p \, dT/dx = -wq^0 + L, \tag{9}$$

where the flame is at rest in the adopted coordinate system, x is the co-ordinate normal to the flame, λ is a constant mean thermal conductivity, c_p is a constant average specific heat at constant pressure, m is the constant mass flow rate per unit area in the $+x$ direction, w is the chemical reaction rate (mass of fuel converted per unit volume per second), q^0 is the heat released in the chemical reaction per unit mass of fuel converted, and L is the heat loss per unit volume per second by radiation or conduction to the walls. Equation (9) follows from equation (1-43) and the one-dimensional form of equation (1-40) (with $\lambda = $ constant and $c_p = $ constant) provided that the phenomenological heat-loss term L is inserted; if multiplied by dx, it expresses a balance among heat conduction, thermal-enthalpy convection, heat production, and heat loss in an element of thickness dx. Since equation (9) is the only differential equation that will be used in the initial discussion, the approach may be described as a *thermal theory* of flame propagation.

In Section 5.3.6, activation-energy asymptotics have been applied to the adiabatic version of equation (9) for a particular rate function w; burning-rate formulas are given in Section 5.3.6 for this rate function and in Section 5.3.7 for others. Here it is convenient to presume that for $L = 0$, the burning rate is known on the basis of these results and to employ the known adiabatic mass burning rate m_a for the purpose of nondimensionalization. Thus, in analogy with equation (5-18), we introduce the nondimensional streamwise coordinate $\xi = m_a c_p x/\lambda$ and obtain the equation

$$d^2\tau/d\xi^2 - \mu \, d\tau/d\xi = -\Lambda_a \omega + \kappa, \tag{10}$$

where the burning-velocity ratio is

$$\mu = m/m_a, \tag{11}$$

the nondimensional temperature τ has been defined in terms of the initial temperature T_0 of the combustible mixture, and the initial mass fraction Y_{F0} of fuel as $\tau = c_p(T - T_0)/(q^0 Y_{F0})$, the nondimensional reaction rate $\lambda w/(Y_{F0} m_a^2 c_p)$ has been written as the product of a burning-rate eigenvalue

Λ_a of the adiabatic problem [defined by equation (5-45) for the adiabatic problem, for example] with a nondimensional function $\omega(\tau)$, and the non-dimensional heat-loss rate is

$$\kappa(\tau) = \lambda L/(q^0 Y_{F0} m_a^2 c_p). \tag{12}$$

The expansion parameter is the Zel'dovich number

$$\beta = E(T_{af} - T_0)/(R^0 T_{af}^2), \tag{13}$$

where $T_{af} = T_0 + q^0 Y_{F0}/c_p$ is the adiabatic flame temperature, R^0 is the universal gas constant, and E represents the overall activation energy. The limit $\beta \to \infty$ is considered with Λ_a times a suitable power of β held fixed [see, for example, the expression above equation (5-75)]. As exemplified by equation (5-66), the function $\omega(\tau)$ depends exponentially on β, and for large β we expect that there will be a narrow reaction zone centered about the position of maximum τ. Outside this reaction zone, the reaction-rate term in equation (10) will be exponentially small. Under the assumption that all the fuel is consumed in the reaction zone, we may readily derive the expression $mY_{F0} = \int_{-\infty}^{x_f} w \, dx$ from the fuel-conservation equation, where f identifies conditions just downstream from the reaction zone. Use of equations (10) and (11) in this expression gives

$$\mu(1 - \tau_f) = \int_{-\infty}^{\xi_f} \kappa \, d\xi - (d\tau/d\xi)_f \tag{14}$$

as a nondimensional representation of the influences of heat losses on the decrease in flame temperature.

The heat-loss term is analogous to that appearing in equation (B-52) and is assumed to be nonnegative and to vanish at the initial temperature ($\tau = 0$). The heat-loss distribution, $\kappa(\tau)$, is a critical factor in the structure of the nonadiabatic flame. It is found that if $\kappa(\tau)$ is of order unity or larger over a range of ξ of order unity, then flame-structure solutions do not exist with the present formulation; therefore, at first let us assume that $\kappa(\tau)$ is of order β^{-1} everywhere. In this case, upstream from the reaction zone in equation (10) there is a convective-diffusive balance with negligible loss in the first approximation, and the approximate solution

$$\tau = \tau_f e^{\mu\xi} \tag{15}$$

is obtained, where one constant has been selected to satisfy $\tau = 0$ at $\xi = -\infty$ and the other to place $\tau = \tau_f$ at $\xi = 0$, an arbitrary selection of the origin of the coordinate system. Downstream from the reaction zone only τ constant would be consistent with a convective-diffusive balance; with κ of order β^{-1}, a contraction of the coordinate is suggested by equation (10), such that $\zeta = \xi/\beta$ is treated as being of order unity, and in the first approximation the solution

$$\zeta = \mu \int_{\tau}^{\tau_f} (\beta\kappa)^{-1} \, d\tau \tag{16}$$

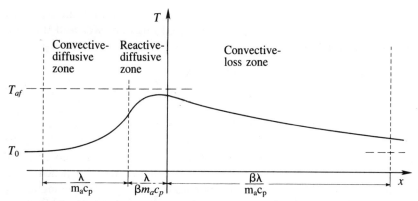

FIGURE 8.1. Schematic representation of the temperature distribution through a flame with heat losses.

is obtained. If, for example, L is proportional to $T - T_0$, so that from equation (12) κ can be written as $(c/\beta)\tau$ with c being a constant or order unity, then equation (16) explicitly gives $\tau = \tau_f e^{-c\xi/(\mu\beta)}$, an exponential decay of temperature over a length scale having ξ or order β. It thus becomes clear that in the first approximation, a convective-loss balance exists in equation (10) in the large downstream region, which therefore may be called a **convective-loss zone**. The effect of diffusion is a perturbation of first order in the convective-loss zone. The three-zone structure of the nonadiabatic flame is shown schematically in Figure 8.1.

The calculation of μ entails investigating the structure of the reaction zone, which is expected to exhibit a reactive-diffusive balance in the first approximation, with $\eta = \beta\xi$ being the relevant nondimensional space coordinate of order unity. Just as equation (5-71) was derived but with $y = \beta(\tau_f - \tau)/\tau_f$, we obtain from equation (10)

$$(dy/d\eta)^2 = (2\Lambda_a/\beta) \int_0^y \omega \, dy, \tag{17}$$

in the first approximation where a matching condition implied by equation (16), $dy/d\eta \to 0$ as $y \to 0$ to leading order, has been applied. A matching of this solution to equation (15) can be written as

$$\mu^2 = (2\Lambda_a/\beta) \int_0^\infty \omega \, dy. \tag{18}$$

The form of the reaction-rate function is germane to equation (18); the non-adiabatic form for a thermal theory possesses a greater degree of arbitrariness than does that for a complete theory with one-step, Arrhenius kinetics. To achieve agreement with the nonadiabatic version of the theory whose adiabatic reaction-rate function is given by equation (5-66), we may put

$$\omega = (\tau_f - \tau)\exp\{-\beta(1 - \tau)/[1 - \alpha(1 - \tau)]\}, \tag{19}$$

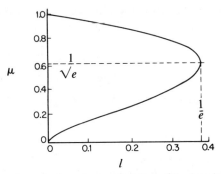

FIGURE 8.2. The burning-velocity ratio as a function of the heat-loss parameter for non-adiabatic flames.

where $\alpha = (T_{af} - T_0)/T_{af}$, and since equation (14) implies that $1 - \tau_f$ is of order β^{-1}, to the lowest order equation (18) then gives (with $\Lambda_a = \beta^2/2$)

$$\mu^2 = e^{-\beta(1-\tau_f)}. \tag{20}$$

Through investigation of complete theories with various other one-step Arrhenius rate functions (for example, two-reactant systems with reaction orders different from unity), it may be shown that equation (20) always follows from equation (18) to the present order of approximation.

Equation (14) may be used in equation (20) to provide an expression for μ in terms of a heat-loss parameter. From equation (16) it may be seen that $(d\tau/d\xi)_f = -\kappa(\tau_f)/\mu$, whence by use of equation (15) in the integral in equation (14) it may be shown that with the definition

$$l = \left[\kappa(\tau_f) + \int_0^{\tau_f} (\kappa/\tau)\, d\tau\right]\beta, \tag{21}$$

which represents a loss parameter of order unity, equation (20) may be written as

$$\mu^2 = e^{-l/\mu^2}. \tag{22}$$

The solution to equation (22) for l is $l = \mu^2 \ln(1/\mu^2)$, which enables l to be calculated explicitly as a function of μ. The result, plotted in the form of μ as a function of l, is shown in Figure 8.2. It is seen that $dl/d\mu = 0$ at $\mu = 1/\sqrt{e}$ —that is, at $l = 1/e$—and that there are no solutions for $l > 1/e$; for $l < 1/e$ there are two values of μ for each l, giving an upper and a lower burning velocity. These are the results that were identified at the beginning of this section. If the dimensional flame temperature, $T_f = T_0 + \tau_f q^0 Y_{F0}/c_p$, is introduced, then with L approximated as being linear in T, we find from equations (12) and (14) by use of equations (15) and (16) that

$$T_f = T_{af} - 2\lambda L(T_f)/(m^2 c_p^2), \tag{23}$$

where $L(T_f)$ is the loss function L evaluated at $T = T_f$, and the extinction condition is found from equations (12) and (21) to be

$$2L(T_f) = m_a^2 c_p^2 (T_{af} - T_0)/(\lambda \beta e) = m_a^2 R^0 c_p^2 T_{af}^2/(\lambda E e), \qquad (24)$$

where equation (13) has been employed in the last equality. To the order to which equation (20) has been derived, it may equally well be written as

$$\mu^2 = e^{-E/R^0 T_f}/e^{-E/R^0 T_{af}}, \qquad (25)$$

which shows directly that the reduction in the burning rate is due entirely to the influence of the reduced flame temperature on the Arrhenius factor in the reaction-rate expression. The double-valued solution arises through the influence of m on T_f in equation (23).

The development that has been presented here is not the most general asymptotic analysis. For example, by use of equation (5-18) it is readily possible to account for a variable thermal conductivity λ; the result is that λ is to be evaluated at T_f in the formulas that have been given. A generalization of greater fundamental significance would be to formulate the problem employing m instead of m_a in forming the nondimensional coordinate ξ and T_f instead of T_{af} in defining β; this permits order-unity departures of T_f/T_{af} from unity and exponentially small ratios m/m_a. Pursuit of this more general analysis for problems retaining the full species-conservation equations for one-step, Arrhenius processes reveals that solutions may be obtained when κ is of order unity over a range of ξ of order unity; the reaction-zone equations then differ from those typified by equation (5-70) and resemble those of the premixed-flame regime discussed in Section 3.4.4 [replace $\varphi^2 - \eta^2$ by $\varphi - \eta$ in equation (3-96)], there being a nonnegligible enthalpy gradient within the reaction zone. If β is sufficiently large, then the resulting extinction conditions reduce to those derived herein, and extinction occurs before the rate of heat loss becomes large enough to impose an appreciable enthalpy gradient within the thin reaction zone. It appears that generally β is large enough that the more general analysis, with its greater complexity, is not needed in applications having representative accuracies of 10%.

The rate functions considered in early analyses that did not employ asymptotics may be contrasted with equation (19). Spalding took $\omega = (1 - \tau)^n \tau^m$ with m and n constants, while von Kármán put ω constant over a finite range of ξ for which $\tau_i < \tau < 1$ and zero otherwise, where τ_i is a nondimensional ignition temperature. In these theories, large values of β may be approximated qualitatively by a large value of m or by τ_i near unity (for example, $\tau_i = 0.9$). The results thereby obtained are quite similar to those derived here. The asymptotic analysis possesses the advantage that the resulting formulas, such as equations (24) and (25), may be applied more directly with more readily available rate parameters (for example, E); it also reveals the structural aspects summarized in Figure 8.1.

8.2.2. The existence of two flame speeds

Figure 8.2, from equation (22), clearly shows the existence of steady-state solutions for two different flame speeds if $0 < l < 1/e$. Whether both solutions may be observed experimentally depends on the stability of the steady state. The stability is likely to be influenced by the experimental configuration and by parameters not appearing in equation (22). Static stability arguments have suggested that the upper branch is stable and the lower unstable [36], although by means of experiments with a flat-flame burner, in which the operator controls both the mass-flow rate and the rate of heat loss from a plane flame located above a cooled porous plate, a small portion of the lower branch in the vicinity of the maximum value of l apparently has been stabilized [36], [48], [49]. Analyses in which the flame is allowed to evolve only on a long time scale, having $tm_a^2 c_p/(\lambda\rho_0\beta)$ of order unity (where t denotes time and ρ_0 the initial density), also suggest stability only for the upper branch [46]. Experimentally, the upper branch appears stable, at least for sufficiently small values of l, and it is the upper branch that generally is ascribed physical meaning and along which the flame possesses its usual structure. At least for a one-reactant mixture with a Lewis number of unity, it seems quite likely that the entire upper branch is stable and that the practical extinction condition therefore can be identified with $l = 1/e$.

However, it should be emphasized here that the complete stability question is complex. Flame stability is the subject of the following chapter. For Lewis numbers different from unity, there are diffusive-thermal instabilities that are influenced by heat loss and that have led to various statements concerning the dependence of the stability in Figure 8.2 on the Lewis number [46]. We shall postpone any further discussion of these stability questions until the following chapter. It seems sufficient here to reemphasize that if special types of flame instabilities are not observed in an experiment, then the upper branch of Figure 8.2 may be expected to apply, with $l = 1/e$ providing a correct extinction criterion (within about 10% accuracy since corrections of order $1/\beta$ may be anticipated).

The foundations of the analysis producing Figure 8.2 fail on the lower branch when μ becomes sufficiently small. Use of the more general asymptotic approach identified in the preceding subsection can provide results that extend more accurately to smaller values of μ. In view of the established difficulty in obtaining stable flames on the lower branch, this refinement does not appear to warrant development.

8.2.3. Concentration limits of flammability

The dependence of the calculated flame speed on the initial reactant concentrations arises primarily through the influence of the concentrations on the adiabatic flame temperature T_{af}. If the reactant concentration is varied with the pressure and the initial temperature held fixed, then the quantity in equation (24) that varies most strongly is m_a^2, mainly because of its proportionality to $e^{-E/R^0 T_{af}}$. In a first approximation, the remaining factors may be considered constant in any given series of experiments, and a critical value of $e^{-E/R^0 T_{af}}$ may therefore be identified for extinctions from equation (24). The situation is illustrated schematically in Figure 8.3, where ϕ denotes the equivalence ratio for a fuel-oxidizer mixture (see Section 5.3.7). Rich and lean limits are seen to occur in Figure 8.3 where $e^{-E/R^0 T_{af}}$ achieves its critical value for extinction. The comparatively weak variation of T_{af} with ϕ causes a strong variation of $e^{-E/R^0 T_{af}}$, especially in the wings near the limits. Therefore, differences in values of parameters such as $L(T_f)$ between different experiments produce relatively small changes in the limiting values of ϕ. This enables flammability limits to be tabulated without extensive reference to the experimental configuration [2].

An accurate application of equation (24) to the calculation of flammability limits entails accounting for variations other than that of $e^{-E/R^0 T_{af}}$. Relevant parameters are exhibited explicitly in equation (24). That limits usually are somewhat wider for flames propagating upward in tubes than for flames propagating downward [1], [2] may be attributed to a larger value of $L(T_f)$ for downward propagation. The evaluation of $L(T_f)$ will be considered in Section 8.2.5, where questions will be addressed relating to the physics of losses and also to the accuracy with which equation (24) follows from equation (21). It has already been indicated in Section 5.3.7

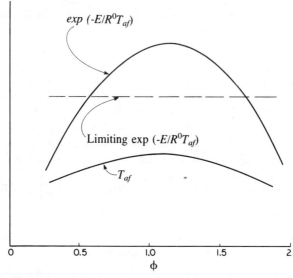

FIGURE 8.3. Illustration of the dependence of the adiabatic flame temperature and of the Arrhenius factor on the equivalence ratio, exhibiting extinction condition for defining flammability limits.

that parameters such as λ and E vary with ϕ; in calculating flammability limits, values that are appropriate in the vicinity of each limit separately should be employed.

8.2.4. Pressure limits of flammability

It is often observed experimentally that if the relative concentrations of the reactants and the initial temperature of the mixture are kept fixed but the pressure p is decreased, then a limiting pressure is reached, below which flame propagation cannot be achieved. The flammable range of ϕ usually narrows as the pressure is decreased, and below a critical pressure flame propagation does not occur for any value of ϕ. These observations are consistent with equation (24), in which the main pressure-dependent quantities are m_a^2 and $L(T_f)$. In terms of the overall order n (the pressure exponent) of the overall reaction rate, $m_a^2 \sim p^n$. It will be seen in the following section that $L(T_f) \sim p^m$, where $m \approx 0$ for conductive or convective losses and $0 \leq m \leq 1$ for radiative losses. Since $n > 1$ for practically all flames, the right-hand side of equation (24) decreases more rapidly than the left-hand side as p decreases, and the limiting equality may therefore be expected to be surpassed at a sufficiently low pressure.

8.2.5. Estimates of heat loss

Flammability limits can be truly independent of the experimental apparatus only if there is a critical point (for example, a critical value of ϕ) at which the exothermic chemistry is shut off completely, without heat loss. Although it is highly unlikely that this occurs, flammability limits may effectively be controlled by chemical kinetics if there is a value or a narrow range of ϕ at which the overall rate of heat release decreases abruptly, by one or two orders of magnitude, since then a very small heat loss may produce the same limit as a large loss. In cases of this kind (some of which may be envisaged from chemical-kinetic considerations but none of which have been established indisputably to occur for real combustible mixtures), the presence of heat loss is essential; however, the value of the rate of heat loss is irrelevant to the flammability limit over practical ranges of conditions. The behavior, in a sense, corresponds to a heat-loss theory with a very large value of E. An even more likely candidate for producing flammability limits that are independent of the apparatus over a very wide range of conditions is a solely radiative heat loss from an optically thin gas. In this case, by use of equation (E-50) in suitable versions of equations (1-15) and (1-28), equation (9) may be derived with

$$L = 4\sigma T^4 / l_p, \tag{26}$$

where σ is the Stefan-Boltzmann constant and l_p is the Planck mean absorption length. From the definitions of B_ν and of the absorptivity κ_ν given in

Section E.5.2, it may be seen from equation (E-50) that l_p is a local property of the gas mixture, not dependent on the experimental environment. Therefore, until the gas begins to become optically thick, which might not occur until sizes on the order of meters are reached, dimensions of the experimental apparatus do not influence any of the terms in equation (9), and predicted limits from radiant loss are apparatus-independent.

In equation (26), $1/l_p$ is proportional to the number density of emitting molecules, which, in turn, is proportional to pressure at constant temperature and relative composition (see Section E.5.2); when reabsorption is not negligible, the pressure dependence is weaker. These results were employed in the preceding subsection. In addition, $1/l_p$ varies strongly with position x in the flame because of the changes in the concentrations and degrees of excitation of the emitters. For this reason, the generality of equation (21), which admits a wide class of variations of L, is particularly helpful for use with radiant loss. An accurate application of equation (21) with $l = 1/e$ to calculate extinction with radiant loss can be made only if $\kappa(\tau)$ is known accurately. Although the needed details of flame-structure information usually are unavailable for real flames, a number of relevant general observations can be made.

First, we note that convergence of the integral in equation (21) requires $\kappa(\tau)$ to approach zero as τ approaches zero. For radiant loss often this has been achieved by putting L proportional to $T^4 - T_0^4$, with the last term viewed as representing absorption of radiation emitted by the surroundings. Typically, radiation at the ambient temperature is entirely negligible, and a better approach is to allow l_p in equation (26) to vary in such a way that $1/l_p$ goes to zero in the fresh mixture. In flames radiant losses often are negligible until a value of τ is reached at which radiatively active intermediate or product species are present; an exception to this rule may be the ozone decomposition flame, in which the reactant, ozone, is likely to radiate much more strongly than either the intermediate oxygen atom or the product oxygen molecule. If the radiation is dominated by chemiluminescence from an intermediate that exists only over a short range $\Delta\tau$ about the value $\tau = \tau_i$, then equation (21) becomes approximately $l = \beta\kappa(\tau_i)\Delta\tau_i/\tau_i$, where $\kappa(\tau_i)$ is obtained through equation (12) from an average value L in the region of emission. We may note that if $\Delta\tau_i$ is small (for example, of order $1/\beta$), then $\kappa(\tau_i)$ may be large (for instance, of order unity rather than of order $1/\beta$) while still enabling equation (21) with $l = 1/e$, to be used in calculating extinction. It might be expected that chemiluminescent emissions will often be strong but restricted to the main reaction zone; in this case, the asymptotic analysis requires modification in that a radiant loss term belongs in equation (17), and although the corresponding diffusion-flame problem has been investigated [50], extinction conditions for premixed flames are unavailable.*

---

* G. Joulin has completed the necessary analysis in recent work not yet published.

Estimates often indicate that although chemiluminescence may provide flames with their distinctive colors, the associated energy losses may be negligible in comparison with radiation from major stable products (such as from H_2O and CO_2 in hydrocarbon flames). If the concentration profiles of the products in the flame are similar to those of the normalized temperature and if the rate of radiant emission from the products is proportional to T^4, then we find from equations (12) and (26) that κ is proportional to $\tau(\tau + \alpha^{-1} - 1)^4$, whence l in equation (21) becomes proportional to $\tau_f(\tau_f + \alpha^{-1} - 1)^4 + [(\tau_f + \alpha^{-1} - 1)^5 - (\alpha^{-1} - 1)^5]/5$, which suggests that radiation emitted downstream from the reaction zone is about five times as important as that emitted upstream in contributing to extinction. This may be contrasted to the situation in which L is linear in τ, as represented by equation (24), wherein it was found that the upstream and downstream contributions were equal (their sum giving the factor of 2 in the equation). Since the downstream zone is convective in character, only the loss rate at its beginning can contribute to the reaction-zone behavior and to extinction [that is, only $\kappa(\tau_f)$ occurs in the first term in equation (21)]; in contrast, losses anywhere in the upstream convective-diffusive zone will influence l. It appears that for radiant losses from reaction products $\kappa(\tau_f)$ is often appreciably more important than the integral in equation (21), and that in this case equation (24) may therefore be used with the factor of 2 replaced by 1.2 or possibly unity.

Near rich limits of hydrocarbon flames, soot is sometimes produced in the flame. The carbonaceous particles—or any other solid particles—easily can be the most powerful radiators of energy from the flame. The function $\kappa(\tau)$ is difficult to compute for soot radiation for use in equation (21) because it depends on the histories of number densities and of size distributions of the particles produced; for example, an approximate formula for l_p for spherical particles of radius r_s with number density n_s, surface emissivity ϵ_s, and surface temperature T_s is $l_p = (T/T_s)^4/(\pi r_s^2 \epsilon_s n_s)$ [50]. These parameters depend on the chemical kinetics of soot production—a complicated subject. Currently it is uncertain whether any of the tabulated flammability limits are due mainly to radiant loss (since convective and diffusive phenomena will be seen below to represent more attractive alternatives), but if any of them are, then the rich limits of sooting hydrocarbon flames almost certainly can be attributed to radiant loss from soot.

Conductive or convective losses of energy from the flame necessarily depend on the geometry of the apparatus. Therefore, they often have been excluded as mechanisms that may be responsible for fundamental limits of flammability. However, in typical experiments these losses are larger than radiant losses, and we have indicated in Section 8.2.3 in discussing Figure 8.3 that near the limits, appreciable changes in the loss rate usually produce relatively small changes in the limits. Since precise values of flammability limits usually do, in fact, depend on the experimental configuration [1], convective losses are good candidates for explaining tabulated limits.

Special sets of circumstances may conspire to produce limits independent of apparatus dimensions even if convective or conductive losses are important. For example, in hydrogen-oxygen-nitrogen mixtures, hydrogen-lean flames propagating downward in open, vertical tubes experience a cellular instability (see Section 9.5.2) that leads to fingers of flames separated by a nonreactive, cool gas from which hydrogen has been removed by diffusion [51]. The sizes of the fingers are determined by the instability and are independent of the tube diameter. Conductive losses from the fingers to the cool gas are responsible for the lean flammability limit. Since the dimensions over which this conduction occurs are independent of the tube diameter, the lean limits do not depend on the size of the apparatus. On the other hand, for hydrogen-rich mixtures that do not experience the instability, the limits are observed to depend on the tube diameter [51], [52], as is normally expected for conductive losses. The observations thus are consistent with theoretical predictions [52], [53] that take into account both the instability and the conductive loss, and they show that diffusive-thermal effects (Section 9.5.2.3) can influence extinction phenomena. Unusual observations—for example, of lean methane-air flames that can propagate downward in a tube 2 cm in diameter but are extinguished in a tube 5 cm in diameter—may be explained on a similar basis [51].

Unlike the radiant loss from an optically thin flame, conductive or convective losses never can be consistent exactly with the plane-flame assumption that has been employed in our development. Loss analyses must consider non-one-dimensional heat transfer and should also take flame shapes into account if high accuracy is to be achieved. This is difficult to accomplish by methods other than numerical integration of partial differential equations. Therefore, extinction formulas that in principle can be used with an accuracy as great as that of equation (21) for radiant loss are unavailable for convective or conductive loss. The most convenient approach in accounting for convective or conductive losses appears to be to employ equation (24) with $L(T_f)$ estimated from an approximate analysis. The accuracy of the extinction prediction then depends mainly on the accuracy of the heat-loss estimate. Rough heat-loss estimates are readily obtained from overall balances.

The conductive heat loss per unit volume from a plane flame in a circular tube of diameter D can be estimated by the following simple reasoning. Consider an Eulerian element of gas of length dx in the tube whose walls are maintained at the temperature T_0. The energy per second conducted to the walls from this element is the product of the thermal conductivity λ, a mean temperature gradient $(T - T_0)/(D/2)$, and the wall area $\pi D\, dx$. The rate of heat loss per unit volume is then obtained through division by the volume of the element $dx\pi D^2/4$:

$$L = \frac{\lambda[(T - T_0)/(D/2)]\pi D\, dx}{dx\pi D^2/4} = \left(\frac{8\lambda}{D^2}\right)(T - T_0). \qquad (27)$$

This linear dependence of L upon T is consistent with the derivation of equation (24).

Influences of gas motion have been neglected in equation (27). These influences may depend strongly on the configuration of the apparatus in which the flame propagates. For flames propagating downward in open tubes, the velocity of the gas relative to the walls is approximately zero ahead of the flame and approximately the difference between burnt-gas and unburnt-gas velocities behind; for flames propagating upward, gas velocities in the direction of flame propagation may be influenced strongly by buoyancy and tend to be positive (upward) both before and after the flame passes. In contrast, in a forced flow that maintains the flame stationary in the laboratory frame, the gas velocity relative to the walls is directed from the cold to the hot side of the flame, at the burning velocity upstream and at the burnt-gas velocity downstream. Rates of heat transfer to the walls in these various configurations may be estimated by use of heat-transfer coefficients, which are obtained from heat-transfer correlations that depend on suitably defined Reynolds numbers (or perhaps Grashof numbers) [54]. The results may be described by replacing the factor 8 in equation (27) by a factor b that depends on a suitably defined Reynolds number, for example, as well as on the tube geometry and the type of flow. It is thereby found that

$$L(T_f) = (b\lambda/D^2)(T_f - T_0). \tag{28}$$

Typically, variations in b are relatively weak, and results quoted in [17] and [36], for example, indicate that $b \approx 15$ for circular tubes. Equation (28) thus shows that $L(T_f)$ is independent of pressure (as stated in Section 8.2.4) and that $L(T_f) \sim 1/D^2$.

The dependence $L(T_f) \sim 1/D^2$ implies that the left-hand side of equation (24) increases as the tube diameter D decreases. Since none of the other physical parameters in equation (24) depend on D, an extinction condition is reached when the tube diameter D becomes sufficiently small. This minimum tube diameter is called the **quenching diameter**, which is empirically approximately the same as (perhaps 25% or 50% greater than) the quenching distance d defined in Section 8.1. Quenching phenomena can thus be attributed to conductive heat losses.

An approximate formula for the quenching distance d can be obtained from equations (24) and (28). By substituting equation (28) into equation (24) and solving the resulting expression for D, we find

$$D = \sqrt{2e\beta b\lambda/(c_p m_a)}. \tag{29}$$

Equation (5-4) implies that equation (29) can be written as

$$D = \sqrt{2e\beta b}\,\delta, \tag{30}$$

where δ is the thickness of the adiabatic laminar flame. Neglecting the small difference between the quenching diameter and the quenching distance, we obtain equation (4) from equation (30), with $a = \sqrt{2e\beta b}$. Putting $\beta \approx 10$

and $b \approx 15$ in this formula gives $a \approx 30$, which is not very different from the value $a \approx 40$ quoted in Section 8.1. Thus we see that a is large as a consequence of two effects: First, the large nondimensional activation energy makes the flame highly sensitive to small rates of heat loss and capable of being extinguished by extraction of a small fraction of the heat released; second, the convective enhancement of heat loss combined with the geometrically favorable factor for conductive loss cause the loss rate to be about an order of magnitude greater than might be guessed in the absence of a loss model.

8.3. ACTIVATION-ENERGY ASYMPTOTICS IN IGNITION THEORY

The understanding of extinction phenomena, achieved in the preceding section through activation-energy asymptotics, has been matched and in some ways surpassed for ignition phenomena. In addition to specifications of critical conditions for ignition to occur, there are a number of results concerning ignition times that provide the history of the ignition process. Here we merely identify a few illustrations, introduce one example, and give reference to reviews for entrance to the literature.

If interest is focused on ascertaining critical ignition conditions, then analyses often can be performed on the basis of steady-state conservation equations (for example, Section 3.4.4). An example is the autoignition of a combustible within a container whose walls are maintained at a fixed temperature. Reviews of problems of this type are available (see the literature cited in Section B.2.5.4); typically it is found that multiple steady-state solutions exist and that the one with the smallest value of the maximum temperature represents a stable, slow reaction. Above a critical value of a parameter measuring the magnitude of the reaction rate, this slow-reaction solution no longer exists, and ignition occurs. The situation has been discussed in Section B.2.5.4 and an introduction from the viewpoint of asymptotics is available [46]. There is recent literature in which activation-energy asymptotics have been applied to the steady-state equations for various autoignition problems [55], [56].

Time-dependent conservation equations must be considered if the history of an ignition process is to be studied. There are autoignition problems for which progress is being made in describing ignition histories by asymptotic methods [57]–[59];* an introduction to these is available [46]. Problems of ignition by means of an externally applied stimulus have also been introduced from the viewpoint of activation-energy asymptotics [46];

---

* For an improved analysis see J. W. Dold, *Quart. J. Mech. Appl. Math.* **38**, 361 (1985).

an extensive review of research in this area that preceded activation-energy asymptotics is available [60], and a review of applications of asymptotic methods to these problems has been published [61]. Specific problems that have been addressed by asymptotics include the ignition of a reactive material that occupies a half-space and whose planar surface is subjected to a constant energy flux beginning at time zero [62], [63], ignition of the same material by a radiant energy flux with in-depth absorption of radiation [64], this same problem with heat loss at the surface to an inert material [65], [66], the endothermic gasification of a solid by a constant energy flux [67], [68], the ignition of a solid by processes involving exothermic heterogeneous reactions at its surface [69], [70], [71], the ignition of a solid by a radiant energy flux for situations in which the only exothermic reaction occurs in the gas phase—for example, after mixing of fuel vapors with an oxidizing gas [72]–[74]—and the ignition of a reactive material exposed at time zero to Newtonian heating [75] to an elevated surface temperature [76]. We shall briefly address the first of these problems as a vehicle for introducing the techniques.

Consideration of the history (to calculate quantities like ignition times) necessitates retention of time derivatives in the conservation equations. Just as in the previous section, to achieve the greatest simplicity we adopt a thermal theory, although in various applications that have been cited the full set of conservation equations has been considered. Let a reactive material occupy the region $x > 0$, and to avoid complications assume that the material remains at rest and has a constant density ρ, although coordinate transformations readily enable this assumption to be removed. Let the material, initially at temperature T_0, be exposed to a constant heat flux $q = -\lambda\, \partial T/\partial x$ at $x = 0$ for all time $t > 0$, where λ is the constant thermal conductivity of the material. The time-dependent equation for conservation of energy for the material, analogous to equation (9), is

$$\rho c_p\, \partial T/\partial t - \lambda\, \partial^2 T/\partial x^2 = w q^0, \tag{31}$$

where the remaining symbols and assumptions are the same as those of equation (9). Equation (31) may be derived formally from equations (1-11), (1-15), (1-21a) and (1-22) by introducing a list of explicit assumptions, but we shall not do that here. To achieve the simplest thermal theory, we shall neglect effects of reactant depletion on the reaction rate and write

$$w = \rho A e^{-E/R^0 T} \tag{32}$$

in equation (31), where A is a constant having the dimensions of a reciprocal time. Equation (31) expresses a balance of energy accumulation, heat conduction and chemical heat release, which may be contrasted to the balance appearing in equation (9). The problem that has been stated here is well-posed in a mathematical sense and, therefore, is soluble in principle for $T(x, t)$.

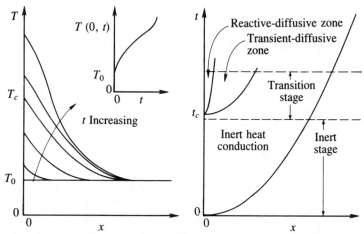

FIGURE 8.4. Illustration of the solution to the problem of ignition by a constant heat flux.

Ignition processes often are characterized by a gradual increase of temperature that is followed by a rapid increase over a very short time period. This behavior is exhibited in the present problem if a nondimensional measure of the activation energy E is large, as is true in the applications. Let t_c denote an ignition time, the time at which the rapid temperature increase occurs; a more precise definition of t_c arises in the course of the development. In the present problem, during most of the time that $t < t_c$, the material experiences only inert heat conduction because the heat-release term is exponentially small in the large parameter that measures E. The inert problem, with $w = 0$, has a known solution that can be derived by Laplace transforms, for example, and that can be written as

$$T/T_0 = 1 + 2\sqrt{\theta/\pi}e^{-z^2/(4\theta)} - z\,\mathrm{erfc}[z/(2\sqrt{\theta})] \equiv F(z, \theta), \qquad (33)$$

where the nondimensional time and space variables appropriate to the inert problem are

$$\theta = q^2 t/(\rho\lambda c_p T_0^2) \qquad (34)$$

and

$$z = qx/(\lambda T_0), \qquad (35)$$

respectively, and where erfc denotes the complementary error function, the properties of which are well known [77]. The solution given by equation (33) is illustrated schematically by the curves in Figure 8.4 for early t.

Let T_c denote $T(0, t_c)$ for the inert solution given by equation (33). Then

$$\epsilon = R^0 T_c^2/(ET_0\sqrt{\pi\theta_c}) \qquad (36)$$

is an appropriate small parameter of expansion, analogous to the ϵ of Section B.2.5.4 or to $1/\beta$ for the β of equation (13) or of Section 5.3.6. Here θ_c is the value of θ at $t = t_c$, as obtained from equation (34). The smallness of ϵ is attributable primarily to the necessary smallness of $R^0 T_c/E$; the inclusion of the additional factor $T_c/(T_0\sqrt{\pi\theta_c})$ in the small parameter serves to assure that ignition occurs prior to homogeneous thermal explosion if ϵ is to be small. Analogous factors appear in optimum selections of ϵ for other ignition problems; for example, for ignition by a constant surface temperature $T_w > T_0$, the definition $\epsilon = R^0 T_w^2/[E(T_w - T_0)]$ is best. A similar factor has appeared in β, as discussed in Section 5.3.6. The factor $T_c/(T_0\sqrt{\pi\theta_c})$ is irrelevant for large values of θ_c because equation (33) shows that this factor then approaches $2/\pi$, a quantity of order unity. The only reason for not putting $\pi/2$ as a factor in equation (36) as well is to achieve some simplification in the final formulas; this cannot be seen until the analysis is completed. It is sufficient, if less elegant, simply to select $\epsilon = R^0 T_0/E$ and to assume that θ_c and T_c/T_0 are of order unity; the additional restrictions then appear in the course of the analysis.

A first approximation to the ignition time may be obtained by equating the heat-release term in equation (31) to either of the other two terms, as calculated from the inert solution, evaluated at $x = 0$. In view of equations (32) and (33), this gives

$$\mathrm{D_{II}} \equiv (q^0 \rho \lambda T_0 A/q^2)e^{-E/R^0 T_c} = 1/\sqrt{\pi\theta_c}, \tag{37}$$

where $\mathrm{D_{II}}$ is a Damköhler number, the ratio of a reaction rate to a heat-conduction rate. The motivation for writing equation (37) is that the chemistry will remain insignificant everywhere until the maximum temperature, that at $x = 0$, becomes large enough for the rate of heat release to equal the inert rate of increase of thermal enthalpy, and thereafter—because of the Arrhenius factor—the chemistry will become dominant and cause a rapid temperature increase, identified with ignition. To obtain an ignition time t_c from equation (37), use is made of equation (34) and of the solution

$$T_c = T_0(1 + 2\sqrt{\theta_c/\pi}) \tag{38}$$

given by equation (33); the result is a transcendental equation for t_c. A more primitive approximation, obtainable by similar reasoning simply from a nondimensionalization of equation (31), is $\mathrm{D_{II}} = 1$ [in place of the final equality in equation (37)], which may be solved explicitly to give

$$t_c = \left(\frac{\pi\rho\lambda c_p T_0^2}{4q^2}\right)\left\{\left(\frac{E}{R^0 T_0}\right)\left[\ln\left(\frac{q^0 \rho \lambda T_0 A}{q^2}\right)\right]^{-1} - 1\right\}^2. \tag{39}$$

In the rough approximation $\mathrm{D_{II}} = 1$, the temperature

$$T_c = \left(\frac{E}{R^0}\right)\left[\ln\left(\frac{q^0 \lambda T_0 A}{q^2}\right)\right]^{-1} \tag{40}$$

plays the role of an ignition temperature, in that the reaction is negligible when the temperature is below this value and dominant when above. According to equation (40), the ignition temperature depends on the externally applied heat flux but is independent of the heating time; equation (37) gives an ignition temperature that decreases slowly as the ignition time increases. Although $D_{II} = 1$ is easy to use, it is less accurate than equation (37), which in turn is appreciably less accurate than the result of a more thorough asymptotic analysis.

To proceed with an improved theory, we may investigate the process of transition to ignition by stretching the time variable about $t = t_c$. To accomplish this, let

$$s = (\theta - \theta_c)/(\epsilon\pi\theta_c) + c, \tag{41}$$

where c is a constant translation to be selected so that an event best identifiable with ignition occurs at a value of s of order unity. Since it is convenient to work with the difference between the true temperature and that given by the inert solution, the definition

$$\psi = [(T/T_0) - F(z, \theta)]/(\epsilon\sqrt{\pi\theta_c}) \tag{42}$$

is introduced, where F is defined by equation (33). To consider a small perturbation about the inert solution with a nontrivial expansion of the Arrhenius factor, ψ is to be treated as being of order unity. Substitution into the differential equation then eventually reveals that space stretching also is needed if space derivatives are to appear at the lowest order in ϵ. It is found that the variables $X = z/(\epsilon\sqrt{\pi\theta_c})$ and $Z = z/\sqrt{\epsilon\pi\theta_c}$ both occur, with X of order unity in an inner zone and Z of order unity in an outer zone [or, more precisely, in an intermediate zone, since equation (33) shows that T differs appreciably from T_0 when $z/\sqrt{\pi\theta_c}$ is of order unity]. To construct an expansion for small ϵ in the stretched variables, the expansion of equation (33) about $z = 0$ and $\theta = \theta_c$, namely, $F(z, \theta) = F(0, \theta_c) + (\theta - \theta_c)/\sqrt{\pi\theta_c} - z + \cdots$, is needed. Use of this expansion to achieve a linearization within the exponential enables the combination of equations (31) and (32) to be written in the variables s and X as

$$\partial\psi/\partial s - (\partial^2\psi/\partial X^2)/\epsilon = (1/\sqrt{\epsilon})e^{\psi + s - X} + \cdots, \tag{43}$$

with $\partial\psi/\partial X = 0$ as the heat-flux boundary condition at $X = 0$. Here the selection

$$c = \ln(D_{II}\sqrt{\epsilon\pi\theta_c}) \tag{44}$$

has been introduced as a first approximation in equation (41) to achieve an appropriate scaling. In equation (43) and also in the equation in the (Z, s) coordinates, matching to the inert solution requires that $\psi \to 0$ as $s \to -\infty$ (and also as $Z \to \infty$). Solutions may be sought as power series in $\sqrt{\epsilon}$, and a two-term expansion of the inner solution may be developed and matched with the leading order of the intermediate solution to produce the problem

$$\left. \begin{array}{ll} \partial\psi/\partial s = \partial^2\psi/\partial Z^2, & \psi(Z, -\infty) = \psi(\infty, s) = 0, \\ \partial\psi/\partial Z = -e^{\psi+s} & \text{at } Z = 0. \end{array} \right\} \tag{45}$$

Investigation of the properties of equation (45) provides the improved formula for the ignition time.

Equation (45) describes a transient heat-conduction problem having a surface heat flux that increases exponentially with the surface temperature. This incremental heat flux is caused by the chemical reaction that, according to equation (43), occurs in a narrow zone at the surface that may be described as a reactive-diffusive zone because the time derivative is seen to be of higher order in ϵ there. The exponential nonlinearity in the boundary condition in equation (45) is known to cause $\psi(0, s)$ to approach infinity at a finite value of s; at larger values of s, the equation cannot be meaningful everywhere. The infinity of ψ may be described as a thermal runaway that develops in the stretched temperature variable. It seems natural to identify this runaway as a unique ignition event and to equate the time at which it appears to the ignition time. A numerical integration of equation (45) has shown that $\psi(0, s) = \infty$ occurs at $s = -0.431$. Since we originally defined $\theta = \theta_c$ at ignition, equation (41) shows that $c = -0.431$, and equation (44) therefore yields

$$D_{II} = 0.65/\sqrt{\epsilon\pi\theta_c} \tag{46}$$

as an improved equation defining the ignition time. Equation (46) differs from equation (37) by the factor $0.65/\sqrt{\epsilon}$, with ϵ given by equation (36). Carrying the asymptotic expansion to higher order would produce $1 + \mathcal{O}(\sqrt{\epsilon})$ as a correction factor for equation (46).

The view of the ignition process that arises from this analysis is illustrated in Figure 8.4. There is an initial stage of inert heat conduction, followed by a shorter stage of transition to an ignited state. The depth of the heated layer and the elevation of surface temperature increase in proportion to $\sqrt{t}$ during the inert stage under the influence of the surface heat flux. There are two additional zones during the transition stage, a thin reactive-diffusive zone at the surface, where the chemical heat release causes an accelerating increase in temperature, and a broader transient-diffusive zone where the additional heat flux caused by the heat release is felt. The surface temperature as a function of time exhibits an inflection and then a runaway during the transition stage, but these departures from the inert solution are small, of order ϵ.

Of course, the divergence at a finite t is not contained in the original problem defined by equations (31) and (32). It is a consequence of the linearization within the exponential produced by the expansion in ϵ and has been encountered in a simpler context associated with thermal explosions in Section B.2.5.4. In fact, near t_c the true surface temperature increases at a rapidly accelerating rate over time scales small compared with those having s of order unity. For the problem as formulated, the acceleration begins to diminish only when surface temperatures on the order of E/R^0 are reached, temperatures much higher than any of practical interest. In reality, the acceleration lessens at much lower temperatures, on the order of an adiabatic flame temperature, as a consequence of the onset of reactant depletion. Subsequently, a flame develops and propagates into the material. At very long times the process returns to one of inert heat conduction under the influence of the surface flux, now in a material which, in effect, initially is approximately at the adiabatic flame temperature. These subsequent stages can also be described by activation-energy asymptotics [63].

Ignition problems typically exhibit a distinguished stage of transition to ignition for small ϵ, and this stage must be analyzed to obtain expressions giving ignition times accurately. The nature of the mathematical problem encountered in the transition stage varies with the type of ignition problem considered. It may involve partial differential equations as in equation (45), ordinary differential equations in space or time, or integral equations that cannot be reduced to differential equations [76]. The nonlinear driving term of chemical heat release, which is characteristic of these problems, may occur in a differential equation or in a boundary condition [such as equation (45)] and may act over different time and space scales. Thus a great variety of mathematical problems is encountered in asymptotic analyses of ignition.

Fundamentally, there always exists a degree of arbitrariness in defining an ignition time. The definition based on thermal runaway that has been introduced here is a mathematical definition embedded within activation-energy asymptotics. Other definitions more closely tied to experiment include the time at which a selected, sufficiently high temperature is achieved at a specified position in the reacting system, the time at which a specified rate of increase of temperature occurs, the time at which luminous radiant output from the system is first observed, the time at which radiation in a selected wavelength band associated with reaction intermediates or products is detected, and a go, no-go definition. With this last definition, the ignition stimulus is applied for a fixed period of time and then removed instantaneously, and the subsequent history of the system is observed. Often the maximum temperature decreases after the ignition source is turned off and then begins to increase rapidly a short time later, soon satisfying an ignition criterion according to one of the other definitions given above. The shortest exposure time that results in satisfaction of an ignition criterion within a reasonable time after shut-off—for example, within a period of one exposure time—is then taken to represent the ignition time.

This go, no-go definition has been employed for our example problem in an analysis by methods in which the original partial differential equation, obtained from equations (31) and (32), was integrated numerically [78], and the resulting ignition time was found to correlate well with equation (46). This agreement implies that for the problem considered, the ignition time is practically independent of the definition employed. An independence of this type often arises in ignition problems because ϵ is relatively small in ignition. The low temperatures often associated with ignition events can cause $\epsilon \approx 10^{-2}$, in comparison with 10^{-1} for typical flame and extinction processes; thus activation-energy asymptotics tend to be more accurate for ignition. However, there are many ignition problems for which the value of the ignition time depends strongly on the definition adopted. This occurs both experimentally and theoretically and is attributable to the process being more complex than that analyzed here; for example, more than one chemical process may be involved, or the system may be close to a condition of transition from one ignition mechanism to another. Although methods of the type illustrated here may be employed in analyzing these more complex processes, the theoretical problems become challenging.

REFERENCES

1. H. F. Coward and G. W. Jones, "Limits of Flammability of Gases and Vapors," *Bulletin 503*, U.S. Bureau of Mines, U.S. Government Printing Office, Washington, D.C. (1952).
2. M. G. Zabetakis, "Flammability Characteristics of Combustible Gases and Vapors," *Bulletin 627*, U.S. Bureau of Mines, U.S. Government Printing Office, Washington, D.C. (1965).
3. W. Jost, *Explosion and Combustion Processes in Gases*, New York: McGraw-Hill, 1946.
4. B. Lewis and G. von Elbe, *Combustion, Flames and Explosions of Gases*, 1st ed., New York: Academic Press, 1951.
5. B. P. Mullins and S. S. Penner, *Explosions, Detonations, Flammability and Ignition*, New York: Pergamon Press, 1959.
6. D. B. Spalding, *Aircraft Engineering* **25**, 264 (1953); *Proc. Roy. Soc. London* **245A**, 352 (1958).
7. G. Rosen, *J. Chem. Phys.* **30**, 298 (1959); **31**, 273 (1959).
8. S. S. Penner and F. A. Williams, *Jet Propulsion* **27**, 544 (1957).
9. A. A. Putnam and R. A. Jensen, *3rd Symp.* (1949), 89–98.
10. R. E. Cullen, *Trans. Am. Soc. Mech. Eng.* **75**, 43 (1953).
11. D. M. Simon, F. E. Belles, and A. E. Spakowski, *4th Symp.* (1953), 126–138.
12. A. L. Berlad, *J, Phys. Chem.* **58**, 1023 (1954).
13. A. L. Berlad and A. E. Potter, Jr., *5th Symp.* (1955), 728–735.
14. A. E. Potter, Jr., *Progress in Combustion Science and Technology*, vol. 1, New York: Pergamon Press, 1960, 145–181.
15. D R. Ballal and A. H. Lefebvre, *16th Symp.* (1977), 1689–1698.
16. A. E. Potter, Jr., and A. L. Berlad, *6th Symp.* (1957), 27–36; NACA Rept. No. 1264 (1956).

17. E. Mayer, *C & F* **1**, 438 (1957).
18. F. E. Belles and C. C. Swett, "Ignition and Flammability of Hydrocarbon Fuels," Chapter III of *Basic Considerations in the Combustion of Hydrocarbon Fuels with Air*, NACA Rept. No. 1300 (1959), 277–318.
19. J. B. Fenn, *Ind. Eng. Chem.* **43**, 2865 (1951).
20. C. C. Swett, Jr., *6th Symp.* (1957), 523–532; NACA Res. Memo. No. E54F29a (1954).
21. W. H. Avery and H. L. Olsen, *Selected Combustion Problems II*, AGARD, London: Butterworths Scientific Publications, 1956, 145–147.
22. G. G. de Soete, *13th Symp.* (1971), 735–743.
23. D. R. Ballal and A. H. Lefebvre, *C & F* **24**, 99 (1975).
24. D. R. Ballal and A. H. Lefebvre, *15th Symp.* (1975), 1473–1481.
25. D. R. Ballal and A. H. Lefebvre, *Proc. Roy. Soc. London* **357A**, 163 (1977).
26. M. Kono, K. Iinuma, S. Kumagai, and T. Sakai, *CST* **19**, 13 (1978).
27. D. R. Ballal and A. H. Lefebvre, *C & F* **35**, 155 (1979).
28. D. R. Ballal and A. H. Lefebvre, *18th Symp.* (1981), 1737–1746.
29. R. Maly, *18th Symp.* (1981), 1747–1754.
30. C. H. Yang, *C & F* **6**, 215 (1962).
31. H. G. Adelman, *18th Symp.* (1981), 1333–1342.
32. R. Friedman, *3rd Symp.* (1949), 110–120.
33. Th. von Kármán and Millán, *4th Symp.* (1953), 173–178.
34. A. A. Putnam and L. R. Smith, *4th Symp.* (1953), 708–714.
35. Y. B. Zel'dovich, *Zhur. Eksp. Teor. Fiz.* **11**, 159 (1941).
36. D. B. Spalding, *Proc. Roy. Soc. London* **240A**, 83 (1957).
37. T. N. Chen and T. Y. Toong, *C & F* **4**, 313 (1960).
38. A. L. Berlad and C. H. Yang, *C & F* **4**, 325 (1960).
39. J. Adler, *C & F* **7**, 39 (1963).
40. I. Glassman, *Combustion*, New York: Academic Press, 1977.
41. C. H. Yang, *C & F* **5**, 163 (1961).
42. J. Adler and D. B. Spalding, *Proc. Roy. Soc. London* **261A**, 53 (1961).
43. M. Gerstein and W. B. Stine, *14th Symp.* (1973), 1109–1118.
44. G. Joulin and P. Clavin, *Acta Astronautica* **3**, 223 (1976).
45. J. Buckmaster, *C & F* **26**, 151 (1976).
46. J. D. Buckmaster and G. S. S. Ludford, *Theory of Laminar Flames*, Cambridge: Cambridge University Press, 1982.
47. S. L. Aly and C. E. Hermance, *C & F* **40**, 173 (1981).
48. J. P. Botha and D. B. Spalding, *Proc. Roy. Soc. London* **225A**, 71 (1954).
49. D. B. Spalding and V. S. Yumlu, *C & F* **3**, 553 (1959).
50. S. H. Sohrab, A. Liñán, and F. A. Williams, *CST* **27**, 143 (1982).
51. B. Bregeon, A. S. Gordon, and F. A. Williams, *C & F* **33**, 33 (1978).
52. T. Mitani and F. A. Williams, *C & F* **39**, 169 (1980); *Archivum Combustionis* **1**, 61 (1981).
53. T. Mitani, *CST* **23**, 93 (1980).
54. F. A. Williams, *Prog. Energy Combust. Sci.* **8**, 333 (1982).
55. A. K. Kapila and B. J. Matkowsky, *SIAM J. Appl. Math.* **36**, 373 (1979); **39**, 391 (1980).
56. A. K. Kapila, B. J. Matkowsky, and J. Vega, *SIAM J. Appl. Math.* **38**, 382 (1980).
57. D. R. Kassoy and J. Poland, *SIAM J. Appl. Math.* **39**, 412 (1980); **41**, 231 (1981).

58. A. K. Kapila, *SIAM J. Appl. Math.* **39**, 21 (1980); **41**, 29 (1981).
59. J. Poland and D. R. Kassoy, *C & F* **50**, 259 (1983).
60. A. G. Merzhanov and A. E. Averson, *C & F* **16**, 89 (1971).
61. F. A. Williams, "Asymptotic Methods in Ignition Theory," *Memoria del VII Congreso de la Academia Nacional de Ingenieria*, Oaxaca, Mexico (1981), 224–227.
62. A. Liñán and F. A. Williams, *CST* **3**, 91 (1971).
63. A. K. Kapila, *International Journal of Engineering Science*, **19**, 495 (1981).
64. A. Liñán and F. A. Williams, *C & F* **18**, 85 (1972).
65. W. B. Bush and F. A. Williams, *Acta Astranautica* **2**, 445 (1975).
66. W. B. Bush and F. A. Williams, *C & F* **27**, 321 (1976).
67. M. Kindelán and F. A. Williams, *CST* **10**, 1 (1975).
68. T. Niioka and F. A. Williams, *17th Symp.* (1979), 1163–1171.
69. A. Liñán and A. Crespo, *CST* **6**, 223 (1972).
70. M. Kindelán and A. Liñán, *Acta Astronautica* **5**, 1199 (1978).
71. T. Niioka, *CST* **18**, 207 (1978).
72. M. Kindelán and F. A. Williams, *Acta Astronautica* **2**, 955 (1975).
73. M. Kindelán and F. A. Williams, *CST* **16**, 47 (1977).
74. T. Niioka, *18th Symp.* (1981), 1807–1813.
75. T. Niioka and F. A. Williams, *C & F* **29**, 43 (1977).
76. A. Liñán and F. A. Williams, *SIAM J. Appl. Math*, **36**, 587 (1979); **40**, 261 (1981).
77. M. Abramowitz and I. A. Stegun, *Handbook of Mathematical Functions*, New York: Dover Publications, 1965.
78. H. H. Bradley, Jr., *CST* **2**, 11 (1970).

CHAPTER 9

Combustion Instabilities

Instabilities arise in combustion processes in many different ways; a thorough classification is difficult to present because so many different phenomena may be involved. In one approach [1], a classification is based on the components of a system (such as a motor or an industrial boiler) that participate in the instability in an essential fashion. Three major categories are identified: **intrinsic instabilities**, which may develop irrespective of whether the combustion occurs within a combustion chamber, **chamber instabilities**, which are specifically associated with the occurrence of combustion within a chamber, and **system instabilities**, which involve an interaction of processes occurring within a combustion chamber with processes operative in at least one other part of the system. Within each of the three major categories are several subcategories selected according to the nature of the physical processes that participate in the instability. Thus intrinsic instabilities may involve chemical-kinetic instabilities, diffusive-thermal instabilities, or hydrodynamic instabilities, for example. Chamber instabilities may be caused by acoustic instabilities, shock instabilities, or fluid-dynamic instabilities within chambers, and system instabilities may be associated with feed-system interactions or exhaust-system interactions, for example, and have been assigned different specific names in different contexts.

Here we shall try to mention, at least in passing, most of the known instabilities that have been observed and studied; more detailed presentation will be reserved for a few of the types that have been investigated more thoroughly. The arrangement of sections does not directly follow the categorization just given. We begin by considering a particular chamber instability for which a substantial literature exists, acoustic instability of solid-propellant rockets. Next, an intrinsic instability of solid fuels, which builds on the material presented in Section 7.4, will be addressed. Section 9.3 concerns chamber instabilities and system instabilities of liquid-propellant rocket engines, another subject that has received extensive analysis. System instabilities in other combustion devices are discussed in Section 9.4. The final section addresses intrinsic instabilities of premixed flames, a fundamental topic of central relevance to both laminar and turbulent flame propagation.

Among the areas not covered here is that of intrinsic instabilities associated with chemical-kinetic mechanisms, as exhibited in cool-flame phenomena, for example; these subjects are touched briefly in Section B.2.5.3. Intrinsic instabilities of detonations were considered in Section 6.3.1 and will not be revisited. Certain aspects of intrinsic instabilities of diffusion flames were mentioned briefly in Section 3.4.4; diffusion flames appear to exhibit fewer intrinsic instabilities than premixed flames, although under appropriate experimental conditions their effects can be observed, as indicated at the end of Section 9.5.2. Certain chamber instabilities that are not related to acoustic instabilities (such as Coanda effects—oscillatory attachment of flows to different walls) will not be discussed here, but reviews are available [1].

9.1. ACOUSTIC INSTABILITIES IN SOLID-PROPELLANT ROCKET MOTORS

9.1.1. Oscillation modes

At an early stage in the development of solid-propellant rocket motors, unexpected irregular histories of the chamber pressure as a function of time occasionally were observed in place of the smooth history for which the motor was designed [2]. In extreme cases the "secondary peaks" of pressure became large enough to cause structural failure of the motor. The irregularities often were accompanied by regular pressure oscillations having frequencies comparable with the natural vibrational frequencies of sound waves in the rocket chamber [3] and can be caused by amplification of acoustic waves [4], [5], [6]. Instabilities that result from sound-wave amplification are called **acoustic instabilities**. Extensive reviews of acoustic instabilities in solid-propellant rockets have been published (for example, [7]).

Acoustic waves in chambers possess various modes of oscillation. To be specific we may consider cylindrical chambers of the type illustrated in Figure 7.5, with burning assumed to occur only at the inner surface of the propellant charge and to go to completion in a distance small compared with all characteristic chamber dimensions. To develop acoustic equations for the product gas in the chamber, we first consider small departures from a uniform, quiescent state and neglect body forces and all transport fluxes. The equations for conservation of mass, momentum, and energy are then equations (4-45), (4-46) and (4-47), respectively, but by making suitable use of equation (4-49) we need not retain equation (4-47). Equations (4-45) and (4-46) can be written as

$$\partial \rho / \partial t + \mathbf{V} \cdot (\rho \mathbf{v}) = 0 \tag{1}$$

and

$$\partial \mathbf{v} / \partial t + (\mathbf{V}p)/\rho = 0 \tag{2}$$

if $\mathbf{v} \cdot \mathbf{V}\mathbf{v}$ is neglected in equation (4-46) on the assumptions that the mean velocity is negligibly small and that velocity fluctuations are small enough for terms quadratic in them to be negligible. From equation (4-49) we see that if the flow is either chemically frozen or in chemical equilibrium, then the entropy does not vary with time along a particle path, and if the entropy is uniform at some time, it remains constant. We assume that this condition applies and write the sound speed as

$$a = \sqrt{dp/d\rho}, \tag{3}$$

where the derivative is evaluated at constant entropy and at either frozen or equilibrium conditions (compare Section 4.3.4) in the uniform state. Then for acoustic perturbations, $dp = a^2 d\rho$, which implies that with $p = \bar{p}(1 + p')$ and $\rho = \bar{\rho}(1 + \rho')$ (where $\bar{p}$ and $\bar{\rho}$ are constant and where p' and ρ' are small compared with unity),

$$p' = \gamma \rho', \tag{4}$$

where $\gamma \equiv a^2 \bar{\rho}/\bar{p}$, which may be shown to reduce to the ratio of specific heats for an ideal gas with constant heat capacities. By taking the time derivative of equation (1) and the divergence of equation (2), $\partial(\mathbf{V} \cdot \mathbf{v})/\partial t$ may be eliminated with the aid of the linearization and use may be made of equation (4) to obtain

$$\partial^2 p' / \partial t^2 - a^2 \mathbf{V}^2 p' = 0, \tag{5}$$

which is the wave equation of acoustics.

If p' is known, then the velocity field may be obtained from the time integral of the equation

$$\partial \mathbf{v} / \partial t = - (\bar{p}/\bar{\rho}) \mathbf{V} p', \tag{6}$$

which follows from equation (2). Alternatively, a velocity potential φ may be introduced such that

$$\mathbf{v} = \mathbf{V}\varphi, \qquad p' = - (\bar{\rho}/\bar{p}) \, \partial \varphi / \partial t, \tag{7}$$

consistent with equation (6); φ also satisfies the wave equation. The derivation of equation (5) may be shown to require that $|\mathbf{v}/a|$ be small.

Properties of the solution to equation (5) are well understood (see, for example, [8]). The method of separation of variables, for example, shows that solutions representing oscillatory wave propagation in a cylindrical chamber are conveniently expressed as a sum of terms, each of which is given by the product of a Bessel function of the radial coordinate r, a trigonometric function of the axial coordinate z, a trigonometric function of the azimuthal angle φ, and a trigonometric function of time. The precise form of the oscillatory solution depends, of course, on the boundary conditions which must be applied at the ends of the chamber and at the surface of the propellant. The simplest boundary conditions are for rigid walls, namely, that the normal component of velocity must be zero [and therefore that $(\nabla p')_{\text{normal}} = 0$; see equation (6)] at the propellant surface ($r = R$) and at the upstream ($z = 0$) and downstream ($z = l$) ends of the chamber. The general oscillatory solution to equation (5), subject to these conditions, is [5]

$$p' = \sum_{k=0}^{\infty} \sum_{m=0}^{\infty} \sum_{n=0}^{\infty} J_m(\alpha_{mk} r/R)\cos(n\pi z/l)$$

$$\times \left[b_{kmn} \cos(m\varphi + \omega_{kmn} t - \beta_{kmn}) + c_{kmn} \cos(m\varphi - \omega_{kmn} t - \gamma_{kmn}) \right], \quad (8)$$

where k, m, and n are integers, J_m is the Bessel function of the first kind of order m, α_{mk} is the kth root of the equation $dJ_m(\alpha)/d\alpha = 0$, b_{kmn} and c_{kmn} are amplitude constants, β_{kmn} and γ_{kmn} are constant phase angles, and the (circular) frequencies ω_{kmn} are determined by

$$\omega_{kmn} = a\sqrt{(\alpha_{mk}/R)^2 + (n\pi/l)^2}. \quad (9)$$

Terms with $m = k = 0$ ($n \neq 0$) describe longitudinal modes, those with $m = n = 0$ ($k \neq 0$) correspond to radial modes, and those with $k = n = 0$ ($m \neq 0$) represent tangential modes. From equation (9) we see that the lowest (nonzero) frequency corresponds to the fundamental axial mode ($n = 1$, $k = m = 0$, $\alpha_{mk} = 0$, $\omega = a\pi/l$) when l/R is sufficiently large and to a tangential mode ($k = n = 0$, $\omega = a\alpha_{m0}/R$) when l/R is sufficiently small. Frequencies corresponding to tangential modes with $m = 1, 2, \ldots, 5$ have been observed [3].

Oscillations with only one frequency are **monochromatic waves**. Thus each normal mode of oscillation [each term in equation (8)] defines a monochromatic wave. There are special shapes of chambers for which more than one mode may have the same frequency; this is called **degeneracy** and admits an infinite variety of monochromatic wave forms (for example, tangential modes in cylindrical chambers). Most of the normal modes describe standing waves, waves having **nodal points** for the velocity (points where the velocity is always zero) and for the amplitude of the pressure oscillations. Thus, according to equation (8), longitudinal modes have pressure nodes at $nz/l = \frac{1}{2}, \frac{3}{2}, \ldots$, and they have velocity nodes at $nz/l = 0, 1, 2, \ldots$, as may be found by use of equation (6). For these modes, **antinodes** (points at which the amplitudes achieve maxima) occur for pressure at $nz/l = 0, 1, 2, \ldots$ and for velocity at $nz/l = \frac{1}{2}, \frac{3}{2}, \ldots$. From this example, we see that pressure nodes and velocity antinodes coincide, as do velocity nodes and pressure antinodes. Between the nodal and antinodal positions, the pressure and velocity oscillations are out of phase by an amount $\pi/2$, in that if one is proportional to $\cos(\omega_{00n}t - \beta_{00n})$, the other is proportional to $\sin(\omega_{00n}t - \beta_{00n})$. Transverse modes (those having $n = 0$) represent standing waves if $b_{km0} = \pm c_{km0}$. If $m \neq 0$ and either $b_{km0} = 0$ or $c_{km0} = 0$, then the transverse mode describes a traveling wave that propagates in the $\pm\varphi$ direction with angular frequency ω_{km0} (that is, the amplitude remains constant along $\varphi = \mp \omega_{km0}t/m$). The component of velocity in the direction of propagation of a traveling wave is in phase with the pressure amplitude, as may be verified by use of equation (6). In some cases, evidence favoring traveling tangential waves in rocket motors has been obtained [3].

Departures from the assumption of small-amplitude oscillations about a uniform, quiescent state can modify the modes of oscillation. Also, the boundary conditions influence the solutions [for example, if there were an open isobaric boundary, then $p' = 0$ would be appropriate there, and equation (8) would be changed]. Nevertheless, the results that have been given form good first approximations for use in approaches to the analysis of acoustic instabilities. Frequencies and spatial dependences of amplitudes are less strongly influenced by flow nonuniformities that are the magnitudes of the amplitudes of the various modes. Nothing in what has been presented so far provides a basis for calculating the constants b_{kmn}, c_{kmn}, β_{kmn}, and γ_{kmn}. Identification of the energy in an acoustic field followed by development of a balance equation for the acoustic energy is a useful approach to the estimation of magnitudes of amplitudes. This enables questions of the occurrence of acoustic instability to be addressed by considering mechanisms of damping and amplification of acoustic waves.

9.1.2. Conservation of acoustic energy

To see how damping and amplification affect amplitudes of acoustic vibrations, let us first define the acoustic energy. The sum of the internal energy and kinetic energy per unit mass at any position and time in a gas is $u + v^2/2$, where $v = |\mathbf{v}|$. Multiplying this quantity by the density and using equation (1-10), which may be written as $u = h - p/\rho$, where h is the enthalpy per unit mass, we obtain an expression for the energy per unit volume. Since entropy changes are of higher order, we may treat the thermodynamic quantities ρ and h as unique functions of p in an expansion through second order about the uniform state. With the definition $p = \bar{p}(1 + p')$ and with $\bar{\rho}$ and $\bar{h}$ representing the values in the uniform state, we then obtain— from a Taylor expansion through second order—

$$\bar{\rho}\bar{h} + \bar{p}p'd(\rho h)/dp + (\bar{p}p')^2 \left[d^2(\rho h)/dp^2\right]/2 - \bar{p} - \bar{p}p' + \rho v^2/2$$

for the energy per unit volume. Here $\bar{\rho}\bar{h} - \bar{p}$ represents the energy of the uniform state, and the rest of the expression is therefore to be associated with the energy of the acoustic field. Since $dh = dp/\rho$ under the present conditions according to the definition of h and to equation (4-49) and the equation preceding it, $d(\rho h)/dp = 1 + h/a^2$ in the expression for the energy of the acoustic field, where use has been made of equation (3). Differentiating this result with use of equation (3) gives $d^2(\rho h)/dp^2 = 1/(\rho a^2) + h d^2\rho/dp^2$. By employing these results in the preceding expression, we find that the portion of the energy per unit volume associated with the acoustic field may be written as

$$\bar{p}p'\bar{h}/a^2 + (\bar{p}p')^2[\bar{h}d^2\rho/dp^2]/2 + (\bar{p}p')^2/(2\bar{\rho}a^2) + \bar{\rho}v^2/2,$$

to second order. Since, to second order with $\rho = \bar{\rho}(1 + \rho')$, we have $\bar{\rho}\rho' = \bar{p}p' \, d\rho/dp + (\bar{p}p')^2(d^2\rho/dp^2)/2$, we see from equation (3) that a shorter expression for the above energy is

$$\bar{\rho}\rho'\bar{h} + (\bar{p}p')^2/(2\bar{\rho}a^2) + \bar{\rho}v^2/2.$$

If this expression were integrated over a volume containing a fixed mass, then the integral of the first term here would vanish because the volume integral of $\bar{\rho}\rho'$ is zero; the first term thus is associated with energy changes that are due to mass changes, and it is not counted as an energy density of the sound field. The local, instantaneous energy per unit volume in the acoustic field is therefore defined as

$$e = \bar{p}p'^2/(2\gamma) + \bar{\rho}v^2/2, \tag{10}$$

with γ defined after equation (4).

A similar derivation of equation (10) may be found in [9]. Equation (10) shows that since p' and $\mathbf{v}$ are first-order quantities, the energy e is a quantity of second order, although expressions for p' and $\mathbf{v}$ valid to first order may be employed in the equation to evaluate e. A conservation equation for e may be developed from equations (1), (4), (6), and (10). The partial derivative of equation (10) with respect to time is

$$\partial e/\partial t = (\bar{p}/\gamma) \, p'\partial p'/\partial t + \bar{\rho}\mathbf{v} \cdot \partial \mathbf{v}/\partial t,$$

in which $\partial p'/\partial t = \gamma \partial \rho'/\partial t$ may be employed by use of equation (4). The substitution of equation (6) here then gives $\partial e/\partial t = \bar{p}p' \, \partial \rho'/\partial t - \bar{p}\mathbf{v} \cdot \nabla p'$. Since the linearized form of equation (1) is $\partial \rho'/\partial t = -\nabla \cdot \mathbf{v}$, we see that

$$\partial e/\partial t + \bar{p}\nabla \cdot (p'\mathbf{v}) = 0, \tag{11}$$

which is the simplest form of the desired conservation equation. Equation (11) shows that $\bar{p}p'\mathbf{v}$ may be interpreted as the local, instantaneous flux of acoustic energy.

The derivation of equation (11) has neglected the mean flow of gas in the chamber, spatial variations of mean thermodynamic properties, and homogeneous dissipative effects such as the viscous terms in the equation for momentum conservation. To take these effects into account by starting with the full set of conservation equations is a complicated problem that has been addressed partially in the literature from various viewpoints. Especially challenging is the task of including flow nonhomogeneities that are expected to lead to refraction of sound waves and to be of significance in real rocket chambers. A number of studies have been directed toward investigating the interaction between the sound and the flow [10]–[16], usually on the basis of expansions for small values of the Mach number of the mean flow. Although the Mach number is not small in a choked nozzle, boundary conditions may be applied at a nozzle entrance plane having a low Mach number [4], [17], with use made of special analyses [18]–[21] of acoustic waves in the nozzle flow for ascertaining the appropriate boundary conditions. To avoid a lengthy development of equations, we may include effects of mean flow and of homogeneous dissipation in an ad hoc, phenomenological manner by adding to equation (11) the divergence of the rate of convection of acoustic energy by the local mean velocity $\bar{\mathbf{v}}$ and the local instantaneous rate of dissipation of acoustic energy per unit volume Φ. The modified conservation equation then becomes

$$\partial e/\partial t + \bar{p}\mathbf{V} \cdot (p'\mathbf{v}) + \mathbf{V} \cdot (e\bar{\mathbf{v}}) + \Phi = 0, \qquad (12)$$

in which $\bar{\mathbf{v}}$ is to be obtained from steady-flow analyses with acoustics neglected and expressions for Φ may be derived by considering specific mechanisms of dissipation.

A time-average value of e may be defined by integrating over a suitable time period. If the acoustic field is monochromatic, then the integration may be performed over one period of oscillation. If it is a superposition of monochromatic waves, then the integration is extended over a time long compared with the longest oscillation period. Let angular brackets identify the time average; the average acoustic energy is $\langle e \rangle$. With p' represented as a superposition of monochromatic waves, both $\langle e \rangle$ and $\langle p'\mathbf{v} \rangle$ also may be written as a sum over monochromatic waves because the average of

$$\cos(\omega_1 t - \omega_2 t - \beta)$$

is zero for $\omega_1 \neq \omega_2$; that is, all cross terms average to zero. If there is degeneracy for a particular ω, then a further decomposition of p' into a sum over degenerate modes can be made, but wave interference may prevent $\langle e \rangle$ from being expressed as a single sum over degenerate modes; the simple additivity of $\langle e \rangle$ is lost, and double sums may be needed for $\langle e \rangle$ and $\langle p'\mathbf{v} \rangle$. A conservation equation for $\langle e \rangle$ may be obtained from the time average of equation (12) if the time over which $\langle e \rangle$ varies appreciably is long compared with the averaging time. The Q of an acoustic cavity is essentially the ratio of the time over which $\langle e \rangle$ varies to the longest oscillation period, and the averaging procedure is useful only for high Q, a situation that usually is encountered in practice. With the restriction to high Q, the average of equation (12) is

$$\partial\langle e\rangle/\partial t + \bar{p}\mathbf{V}\cdot(\langle p'\mathbf{v}\rangle) + \mathbf{V}\cdot(\langle e\rangle\bar{\mathbf{v}}) + \langle\Phi\rangle = 0. \tag{13}$$

When balance equations for acoustic energy are used, they almost invariably refer to time averages of the type indicated here.

Space averages of e also may be defined by integrating over a fixed volume $\mathscr{V}$ of the chamber. Use of the divergence theorem in equation (12) shows that

$$\frac{d}{dt}\int_{\mathscr{V}} e\, d\mathscr{V} = -\int_{\mathscr{A}} \bar{p}p'\mathbf{v}\cdot\mathbf{n}\, d\mathscr{A} - \int_{\mathscr{A}} e\bar{\mathbf{v}}\cdot\mathbf{n}\, d\mathscr{A} - \int_{\mathscr{V}} \Phi\, d\mathscr{V}, \tag{14}$$

where $\mathbf{n}$ is the unit outward-pointing normal vector at the boundary and the area integrals are carried over the entire boundary. The first term on the right-hand side of equation (14) is the rate at which the boundaries do work on the acoustic field. Equation (14) shows that the homogeneous dissipation, the net convection of acoustic energy across the boundary, and the boundary work all influence the average e in the chamber. The volume average also may be applied to equation (13) to obtain an equation like equation (14) for $\langle e\rangle$.

Let the volume average of $\langle e\rangle$ be $\bar{e} = \int_{\mathscr{V}}\langle e\rangle\, d\mathscr{V}/\mathscr{V}$, and define an amplification rate α by

$$\alpha = \left[-\int_{\mathscr{A}} \bar{p}\langle p'\mathbf{v}\rangle\cdot\mathbf{n}\, d\mathscr{A} - \int_{\mathscr{A}} \langle e\rangle\bar{\mathbf{v}}\cdot\mathbf{n}\, d\mathscr{A} - \int_{\mathscr{V}} \langle\Phi\rangle\, d\mathscr{V}\right]\Big/(2\bar{e}\mathscr{V}). \tag{15}$$

Then, if α remains constant, the balance like equation (14) for $\bar{e}$ gives

$$\bar{e} = \bar{e}_0 e^{2\alpha t}, \tag{16}$$

where $\bar{e}_0$ is the value of $\bar{e}$ at $t = 0$. Depending on the sign of α, the acoustic energy in the chamber either increases or decreases exponentially with time. Acoustic instability therefore may be considered to occur when $\alpha > 0$. The problem of evaluating acoustic instability then reduces to the problem of calculating α. Of the three contributions to α that appear in equation (15), we shall be concerned mainly with the first and the third; the convection term vanishes if the acoustic energy convected into the chamber is convected out at the same rate.

9.1.3. The acoustic admittance

A clearer interpretation of the boundary work may be obtained by neglecting convection and homogeneous dissipation and by focusing attention on a monochromatic wave field of frequency ω. A boundary at which a rigid-wall condition ($\bar{\mathbf{v}}\cdot\mathbf{n} = 0$) or an isobaric condition ($p' = 0$) is exactly applicable clearly has $\langle p'\mathbf{v}\rangle\cdot\mathbf{n} = 0$ and therefore no boundary work. For monochromatic waves, the boundary work depends on the phase difference between p' and $\mathbf{v} \cdot \mathbf{n}$. Suppose that at the boundary point of interest

$$p' = P' \cos(\omega t - \beta), \qquad \mathbf{v} \cdot \mathbf{n} = V \cos(\omega t - \gamma), \tag{17}$$

where the constant P' and V are the positive amplitudes of p' and $\mathbf{v} \cdot \mathbf{n}$, and β and γ, respectively, are their constant phase angles. The average rate of work done per unit area on the acoustic field at the boundary point then is

$$- \bar{p}\langle p'\mathbf{v}\rangle \cdot \mathbf{n} = - \bar{p}P'V \int_0^{2\pi/\omega} \cos(\omega t - \beta) \cos(\omega t - \gamma)\, dt(\omega/2\pi)$$

$$= - (\bar{p}P'V \cos \delta)/2, \tag{18}$$

where $\delta \equiv \beta - \gamma$ is the phase difference between pressure and outward-normal velocity oscillations. This is negative, corresponding to energy removal from the vibrations at the boundary, if $-\pi/2 < \delta < \pi/2$ and positive, corresponding to energy addition into the vibrations at the boundary, if $-\pi < \delta < -\pi/2$ or $\pi/2 < \delta \leq \pi$. On the average, an increase in pressure is associated with a positive outward velocity in the former case and with a negative outward velocity (a positive inward velocity) in the latter.

Complex representations of acoustic waves are convenient for many purposes. Equation (17) may be written as

$$p' = P'e^{i(\omega t - \beta)}, \qquad \mathbf{v} \cdot \mathbf{n} = Ve^{i(\omega t - \gamma)} \tag{19}$$

with the understanding that the real parts of these expressions represent the physical variables. With this notation, boundary conditions near velocity nodes are expressed most easily in terms of the complex acoustic **admittance** $\mathscr{Y}$ of the boundary, which is defined as the ratio of the complex outward velocity perturbation to the complex pressure perturbation,

$$\mathscr{Y} \equiv \mathbf{v} \cdot \mathbf{n}/(\bar{p}p'), \tag{20}$$

where $\mathbf{v} \cdot \mathbf{n}$ and p' are given by equation (19). We see from equations (19) and (20) that

$$\mathscr{Y} = [V/(\bar{p}P')]\, e^{i\delta}, \tag{21}$$

the real part of which has the same sign as $\cos \delta$. The average rate of energy input per unit area to the acoustic field is found from equations (18) and (21) to be expressible as

$$- \bar{p}\langle p'\mathbf{v}\rangle \cdot \mathbf{n} = - [(\bar{p}P')^2/2]\mathrm{Re}\{\mathscr{Y}\}, \tag{22}$$

where $\mathrm{Re}\{\ \}$ denotes the real part. The admittance and its reciprocal, the *normal specific acoustic impedance*, play many roles in analyses of monochromatic wave fields [22]. It can be shown, for example, that a traveling wave incident on the boundary is reflected with a diminished amplitude if $\mathrm{Re}\{\mathscr{Y}\} > 0$ and with an increased amplitude if $\mathrm{Re}\{\mathscr{Y}\} < 0$.

An appropriate nondimensional measure of the admittance is

$$y = \mathscr{Y}\bar{p}/a, \tag{23}$$

in which $\bar{p}/a$ is the characteristic impedance of the internal gaseous medium. From equation (20) we see that $y = 0$ at a rigid-wall boundary and $y = \infty$ at an isobaric boundary; thus, both large and small values of the magnitude $|y|$ may be expected to be found at various positions within a chamber. At solid boundaries like the propellant surface in rocket applications, $|y| \ll 1$, and $y = 0$ may be introduced as a first approximation at these boundaries in analyzing the wave field. This enables us to obtain a simplified expression for $\bar{e}$ for monochromatic waves, thereby facilitating amplification-rate estimates.

If the complex representations

$$p' = \mathrm{Re}\{Pe^{i\omega t}\}, \qquad \mathbf{v} = \mathrm{Re}\{Ve^{i\omega t}\} \tag{24}$$

are introduced, where the complex amplitudes P and V are functions of position, then substitution into equations (5) and (6) yields the Helmholtz equation for P,

$$\nabla^2 P + (\omega^2/a^2)P = 0, \tag{25}$$

and the relationship

$$i\omega\mathbf{V} = -(\bar{p}/\bar{\rho})\nabla P \tag{26}$$

for $\mathbf{V}$. The identity $\nabla \cdot (P\nabla P) = \nabla P \cdot \nabla P - P^2\omega^2/a^2$, obtained by use of equation (25) implies that if either $P = 0$ or $\mathbf{V} \cdot \mathbf{n} = 0$ at all boundaries, then

$$\int_{\mathscr{V}} (\bar{p}P^2/\gamma)\, d\mathscr{V} + \int_{\mathscr{V}} (\bar{\rho}\mathbf{V} \cdot \mathbf{V})\, d\mathscr{V} = 0, \tag{27}$$

where equation (26) and the definition of γ following equation (4) have been employed. By use of equation (24) and the result given in equation (27) in the definition of e that appears in equation (10), it may be shown that the $\bar{e}$ defined above equation (15) is expressible as

$$\bar{e} = \bar{p}\int_{\mathscr{V}} |P|^2\, d\mathscr{V}/(2\gamma\mathscr{V}) = \bar{\rho}\int_{\mathscr{V}} |\mathbf{V} \cdot \mathbf{V}|\, d\mathscr{V}/(2\mathscr{V}), \tag{28}$$

where $|P|^2$ represents the square of the magnitude of the complex pressure amplitude. Since $|P| = P'$ at the boundary according to equations (19) and (24), the amplitude measure appearing in equation (28) is the same as that in equation (22). Equations (22) and (28) may be used in equation (15) to show that for conditions under which $\bar{\mathbf{v}} = 0$ and $\Phi = 0$,

$$\alpha = -(\gamma a/2)\int_{\mathscr{A}} |P|^2\, \mathrm{Re}\{y\}\, d\mathscr{A}/\int_{\mathscr{V}} |P|^2\, d\mathscr{V}, \tag{29}$$

where the definition in equation (23) has been introduced.

The contribution of the boundary work or of radiation of acoustic energy through the boundary to the rate of amplification of the acoustic field within the chamber can be estimated more readily from equation (29) than from equation (15). Natural vibrational frequencies of sound in the chamber [see equation (9), for example] are on the order of $\omega \approx a\mathscr{A}/\mathscr{V}$, since $\mathscr{V}/\mathscr{A}$ is a characteristic chamber dimension. Therefore, equation (29) shows that

$$\alpha \approx -\omega \, \overline{\text{Re}\{y\}}, \tag{30}$$

where $\overline{\text{Re}\{y\}}$ is obtained from a suitable average of the nondimensional admittance over the boundary area. Since the derivation of equations (15) and (16) implicitly required $|\alpha| \ll \omega$, we see from equation (30) that

$$|\overline{\text{Re}\{y\}}| \ll 1$$

is needed for the approach adopted to be valid; if $|y|$ were of order unity, then the boundary would be well matched with the acoustic cavity, and all the acoustic energy could be transmitted through the boundary in a time on the order of an oscillation period. Studies of boundary effects on acoustic instability in rockets therefore involve expansions for $|y| \ll 1$. The approximate constancy of α as oscillations grow or decrease in magnitude arises from the cancellation that results from the presence of $|P|^2$ in both the numerator and the denominator of equation (29). In equation (28) it may be considered that $\bar{e}$ varies with time only because the pressure amplitude $|P|$ is proportional to $e^{\alpha t}$ at every point in the chamber, a dependence consistent with equation (16). If amplitudes of oscillation become sufficiently large, then nonlinearities begin to become important, and departures from the exponential dependence on t develop.

In addition to the growth or decay associated with $\text{Re}\{y\}$, there are modifications to the natural frequencies of vibration associated with the imaginary part, $\text{Im}\{y\}$. We shall not discuss these modifications, which are small when $|y|$ is small. Although attention here has been focused on monochromatic waves, the fact that energy densities averaged over a cycle are additive may be considered to justify application of the present results separately to each frequency for acoustic fields composed of a superposition of monochromatic waves. Therefore, the main task in analyzing influences of boundary work on amplification or damping of acoustic energy may be viewed as that of finding the real part of the acoustic admittance at the boundary.

9.1.4 Damping mechanisms

9.1.4.1. Relative importance

In the linear range, results for the α of equation (15) may be obtained separately for the various mechanisms of damping or amplification, and the values of α for each may be added to assess the stability. Damping mechanisms provide negative contributions to α. There are many mechanisms of acoustic damping in chambers. Losses through the choked end nozzle often are most important for acoustic modes that have longitudinal components. Solid propellants often have fine condensed particles in their combustion products, and these particles can provide appreciable damping. The chamber walls and the propellant itself can participate in the oscillations, especially for modes having radial components, and viscoelastic damping within an internally burning propellant grain then may become significant. Viscous and (to a lesser extent) heat-transfer processes at the walls can provide nonnegligible contributions to damping for chambers with relatively large wall areas. Chemical relaxation processes in the gas may produce small contributions to damping, but homogeneous gas-phase dissipation through viscosity and heat conduction generally is entirely negligible except at high frequencies, where it may be responsible for the general absence of very high modes. Here we shall briefly consider theories for each of these effects, beginning first with processes of boundary damping. A somewhat more thorough presentation is given in [7], and there is a useful summary in [12]. Many additional damping mechanisms may be envisioned under various circumstances [7]; if problems of acoustic instability are encountered, then special damping devices (related, for example, to the acoustic liners [23]–[26] or baffles [27] that have been used for sound attenuation in liquid-propellant rocket motors and in other equipment) may be employed.* We shall not address these topics further.

9.1.4.2. Nozzle damping.

The propagation of longitudinal acoustic waves in choked nozzles has been analyzed on the basis of the one-dimensional, time-dependent forms of equations (4-45) and (4-46) by introducing linearizations of the previously indicated type (for example, $p = \bar{p}(1 + p')$) for the streamwise velocity v as well—that is, $v = \bar{v}(1 + v')$—and by allowing the mean quantities $\bar{p}$, $\bar{\rho}$, and $\bar{v}$ to vary with the streamwise distance z through the nozzle, in a manner presumed known from a quasi-one-dimensional, steady-flow nozzle analysis [20]. The perturbation equations

$$\partial \rho'/\partial t + \bar{v}(\partial \rho'/\partial z + \partial v'/\partial z) = 0 \tag{31}$$

and

$$\partial v'/\partial t + \bar{v}\,\partial v'/\partial z = (p' - \rho' - 2v')\,d\bar{v}/dz - [\bar{p}/(\bar{\rho}\bar{v})]\,\partial p'/\partial z \tag{32}$$

---

* See Section 9.3 for definitions of these damping devices.

are obtained, in which equations (4) and (24) (with $\mathbf{v}$ replaced by v') may be employed to derive the pair of first-order ordinary differential equations

$$\gamma \bar{v}\, dV/dz + \bar{v}\, dP/dz + i\omega P = 0 \quad (33)$$

and

$$\gamma \bar{v}\, dV/dz + (a^2/\bar{v})\, dP/dz + \gamma(2 d\bar{v}/dz + i\omega)V - (\gamma - 1)(d\bar{v}/dz)P = 0. \quad (34)$$

The nozzle throat is a singularity of this system because $\bar{v} = a$ there, and the condition that waves not be transmitted upstream from the supersonic section of the nozzle translates into the requirement that the solution in the subsonic part remains regular at the throat. Equations (33) and (34) may be expanded about the throat, the constant multiplying of the singular terms may be set equal to zero, the arbitrary constant in the regular part of the expansion of the solution may be assigned any convenient value, and a numerical integration may then be performed in the direction of decreasing z, with

$$y = (\bar{v}/a)(V/P) \quad (35)$$

evaluated at each point to obtain the local nondimensional admittance, defined in equation (23), that would apply if the nozzle-chamber interface were to occur at that point. Methods of numerical integration are available not only for these quasi-one-dimensional equations but also for two-dimensional equations that allow for modes having components that are not longitudinal [28]; three-dimensional nozzle calculations have also been performed [29].

If $\bar{v}$ is a linear function of z, then equations (33) and (34) may be reduced to the hypergeometric equation [20]. Subtracting equation (33) from equation (34) yields an expression for V that can be substituted into equation (33) to provide a second-order differential equation for P. With the subscript t identifying throat conditions, it may be shown that when $\bar{v}$ is linear in z, then

$$x \equiv (\bar{v}/a_t)^2 = (\gamma + 1)M^2/[2 + (\gamma - 1)M^2],$$

where $M = \bar{v}/a$ is the local Mach number, and the differential equation

$$x(1 - x)\, d^2P/dx^2 - 2[1 + ik/(\gamma + 1)]x\, dP/dx - [ik(1 + ik/2)/(\gamma + 1)]P = 0$$

may be derived, where

$$k = \omega(z_t - z_l)/(a_t - \bar{v}_l), \quad (36)$$

in which the subscript l identifies conditions at the nozzle entrance. In terms of the hypergeometric function F, the solution to this differential equation that remains regular at $x = 1$ is $F(a, b; c; 1 - x)$,* where $c = 1 +$

---

* The notation here is the same as that of M. Abramowitz and I. A. Stegun, *Handbook of Mathematical Functions*, New York: Dover, 1965

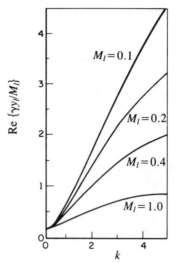

FIGURE 9.1. The real part of the nondimensional nozzle admittance $\gamma y_l/M_l$ as a function of the nondimensional frequency k for various values of the nozzle-entrance Mach number M_l with $\gamma = 1.2$ for longitudinal modes in a nozzle having a velocity linear with distance [20].

$a + b = 2[1 + ik/(\gamma + 1)]$ and $ab = ik(1 + ik/2)/(\gamma + 1)$. This solution can be used along with the expression

$$V = [(\gamma - 1 + ik)P - (\gamma + 1)(1 - x)\,dP/dx]/[(2 + ik)\gamma]$$

to calculate y_l from equation (35). The result is shown in Figure 9.1 for $\gamma = 1.2$.

Expansions of y_l for the small values of M_l that are of greatest practical interest may be obtained from the expansion of F for small value of x [20]. Also, expansions for large and small values of k may be obtained more generally from equations (33) and (34), without the restriction to a linear variation of $\bar{v}$ with z. It is found from the solution for $\bar{v}$ linear in z that $\text{Re}\{y_l\} > 0$, that $\text{Re}\{y_l\}$ approaches $1/\gamma$ as k approaches infinity, and that $\text{Re}\{y_l\}$ approaches $M_l(\gamma - 1)/(2\gamma)$ as k approaches zero. The last of these results is of greatest practical interest because [as seen from equation (36) for small M_l] k is the frequency made nondimensional by the ratio of the length of the subsonic portion of the nozzle to the sound speed at the throat and is therefore roughly on the order of the ratio of the subsonic nozzle length to the chamber length, which generally is small. It is seen from this result that for small M_l, $\text{Re}\{y_l\}$ is small, as assumed in Section 9.1.3. On the other hand, for long nozzles or for values of M_l approaching unity, $\text{Re}\{y_l\}$ begins to become of order unity, and nozzle losses become large, invalidating the approach of Section 9.1.3 and changing the wave form within the chamber substantially. There exists a degree of experimental verification of the predictions [30], [31]. Transverse modes are predicted to be much less strongly influenced by the end nozzle than are modes with longitudinal components

[21] and might even be amplified slightly by the nozzle interaction. There is a body of literature on acoustic radiation from subsonic orifices [10]–[12], [32]–[34]; for a small subsonic vent of area $\mathscr{A}$ radiating energy from a chamber mode having its acoustic velocity perpendicular to the vent axis, the rate of energy loss by a pistonlike mechanism is roughly $\bar{p}a|P|^2\mathscr{A}^2/l^2$, where l is the wavelength of the mode and the amplitude $|P|$ is evaluated at the vent location.

9.1.4.3. Wall damping

Damping by wall friction may be addressed on the basis of the theory of oscillatory boundary layers [35], [36]. Under conditions in real motors, the oscillatory boundary layer at the wall is much thinner than the mean-flow boundary layer [37], and the tangential acoustic velocities just outside the oscillatory boundary layer are (T_w/T_c) times those outside the mean-flow boundary layer, where T_w is the wall temperature and T_c the gas temperature in the chamber. Dissipation in the oscillatory boundary layer may be analyzed by considering a flat element of the surface exposed to the complex amplitude $V(T_w/T_c)$ of velocity oscillation parallel to the surface, where equation (24) has been employed for the acoustic field in the chamber, with the coordinate system locally aligned in the direction of velocity oscillation. Neglecting locally the effects of density changes and the spatial variations of acoustic amplitudes, we write the combination of equations (1-2) and (1-5) for the velocity v parallel to the wall as

$$\rho_w \, \partial v/\partial t = \partial(\mu_w \, \partial v/\partial y)/\partial y, \tag{37}$$

where a noninertial coordinate system has been employed in which the wall oscillates and the gas just outside the oscillatory boundary layer is at rest, so that the pressure-gradient term will be cancelled by a noninertial effective body-force term in momentum conservation. Here the subscript w indicates that the density ρ and viscosity μ are to be evaluated in the gas at the wall, and y is the coordinate normal to the wall, whence $v = (T_w/T_c) \, \text{Re}\{Ve^{i\omega t}\}$ at $y = 0$ and $v = 0$ at $y = \infty$ are boundary conditions. The solution to equation (37) is readily found to be

$$v = (T_w/T_c)\text{Re}\{V \exp[i\omega t - (1 + i)\sqrt{\omega\rho_w/2\mu_w}\,y]\}; \tag{38}$$

the time-average rate of energy dissipation per unit wall area can then be shown to be

$$\int_0^\infty \frac{1}{2}\mu_w \left|\left(\frac{T_w}{T_c}\right)V \exp\left[-(1 + i)\sqrt{\frac{\omega\rho_w}{2\mu_w}}\,y\right]\right|^2 dy = \frac{1}{2}\left(\frac{T_w}{T_c}\right)^2 |V|^2 \sqrt{\frac{\omega\rho_w\mu_w}{2}}. \tag{39}$$

This outline of a theory for wall dissipation shows first, from equation (38), that the thickness of the oscillatory boundary layer is on the order of

$\sqrt{2\mu_w/(\rho_w\omega)}$, which practically always is less than 1 mm. The dissipative energy flux is seen from equation (39) to increase with the frequency in proportion to $\sqrt{\omega}$. In the same way that acoustic velocity oscillations produce an oscillatory velocity boundary layer, acoustic temperature oscillations give rise to an oscillatory temperature boundary layer at the wall, so long as the solid temperature does not exactly follow the temperature oscillations of the gas (which is usual because of the greater thermal capacity of the solid). The amplitude of acoustic temperature oscillations is proportional to $|P|(\gamma - 1)/\gamma$ rather than to $|V|$. If the wall temperature remains constant, then an analysis of the temperature oscillations, analogous to that indicated above for the velocity oscillations, shows that the ratio of the rate of dissipation of acoustic energy by the oscillatory temperature boundary layer to that by the oscillatory velocity boundary layer is approximately $C(a/\gamma)^2(|P|^2/|V|^2)$, where $C = (\gamma - 1)(T_c/T_w)\sqrt{\lambda_w/\mu_w c_{pw}}$, in which λ and c_p are the thermal conductivity and the specific heat at constant pressure for the gas. Since the constant C usually is less than unity, wall dissipation by velocity oscillations usually is greater than that by temperature oscillations.

9.1.4.4. Homogeneous damping

The homogeneous rate of dissipation of acoustic energy per unit volume by viscosity in the gas is of the order of $\mu|V|^2/(a/\omega)^2$, since a/ω is the wavelength that determines the velocity gradient associated with the sound field. From this estimate and equation (39) we see that the ratio of the homogeneous rate of viscous dissipation to the rate of viscous dissipation at the walls is roughly $\sqrt{[\mu/(\rho a)](\omega/a)}\,(w/a)(\mathcal{V}/\mathcal{A})$. Here $\mu/(\rho a)$ is on the order of a molecular mean free path [see, for example, equations (E-27) and (E-34)], while for lower modes, a/ω is of the order of a characteristic chamber dimension, as is $\mathcal{V}/\mathcal{A}$. Therefore, for viscosity, in the lower modes the ratio of homogeneous damping to wall damping is on the order of the square root of the ratio of a mean free path to a chamber dimension, which clearly is a small number under all conditions of interest. The proportionality of this ratio to $\omega^{3/2}$ indicates that high modes can be relatively strongly damped by homogeneous viscous dissipation. Similar estimates can be made (with similar conclusions) for homogeneous dissipation by heat conduction between temperature peaks and valleys of the sound waves.

9.1.4.5. Solid vibrations

Homogeneous dissipation within a viscoelastic solid propellant can produce significant damping if the propellant participates in the oscillations. Usually the solid does not participate appreciably, but under special conditions it does. To see what those special conditions are we may first consider undamped oscillations in a one-dimensional, two-medium system having gas in the region $0 < z < fl$ and propellant in the region $fl < z < l$, where l is the length of the cavity and f is the fraction of its volume occupied by gas [38]. Although the wave equation for the solid involves two sound speeds [7], for simplicity we may apply equation (5) separately to each medium, with the a for the solid representing its dilational sound speed, and we enforce zero-displacement conditions ($\partial p'/\partial z = 0$) at $z = 0$ and at $z = l$ as well as equal-displacement and equal-pressure conditions for solid and gas at their interface, $z = fl$. Letting $R = (\rho_s a_s)/(\rho a)$ be the ratio of impedance of the solid to that of the gas and $S = a_s/a$ be the ratio of the sound speed of the solid to that of the gas, we may readily derive the equation

$$R \sin\left(\frac{\omega l f}{a}\right) = A \sin\left[\frac{\omega l(1 - f)}{aS}\right] \tag{40}$$

for the natural vibrational frequencies, where

$$A = \cos\left(\frac{\omega l f}{a}\right)\bigg/\cos\left[\frac{\omega l(1 - f)}{aS}\right] \tag{41}$$

is the ratio of the pressure amplitude of the oscillation in the solid to that in the gas. Typically, R is on the order of 10^3 and S is on the order of unity; solutions to equation (40) for $R = 10$ and $S = \frac{1}{2}$ are shown by the solid lines in Figure 9.2, where the relatively small value of R was selected to enable the relationships among the curves to be seen more easily. Also shown in Figure 9.2 are frequencies for gas modes that correspond to vibrations of the gas only (with zero-displacement conditions applied at the propellant surface) and frequencies of solid modes for solid vibrations with free-boundary conditions applied to the solid at both of its surfaces.

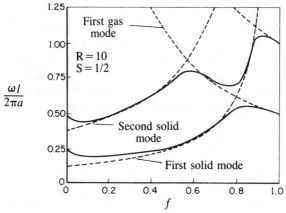

FIGURE 9.2. Natural vibrational frequencies as functions of the fraction f occupied by gas for a one-dimensional, two-medium system [38].

For large values of R, equation (40) shows that if A is of order unity, then $\omega l/f \approx n\pi a$, where n is an integer. This condition defines gas modes, for which we then see [from equation (41)] that the pressure amplitudes are of the same order of magnitude in the gas and solid, while [for example, from equation (6)] the velocity amplitudes are much larger in the gas. With R large, if it is not true that $\omega l/f \approx n\pi a$, then equation (40) requires A to be large, which [from equation (41)] is seen to imply that $\omega l(1 - f) \approx (m + \frac{1}{2})$ $\pi a S$ with m an integer. This condition defines solid modes having a free-boundary condition at the gas interface, with roughly equal velocity amplitudes in the two media but a much larger pressure amplitude in the solid. From these results, we see through equation (10) that for solid modes the acoustic energy per unit volume, e, is much larger in the solid than in the gas (that per unit mass, e/ρ, is comparable), while for gas modes the acoustic energy per unit mass, e/ρ, is much larger in the gas than in the solid (that per unit volume is comparable). For values of f such that oscillation frequencies nearly coincide for a gas mode and a solid mode—namely, $n(1 - f) \approx$ $(m + \frac{1}{2})fS$—more careful investigation of equations (40) and (41) is needed, and it is found that A is of an intermediate order of magnitude and that the frequency behavior seen in Figure 9.2 in the vicinity of the mode-crossing points (intersections of dotted lines) occurs. As f is varied continuously, sharp transitions between gas modes and solid modes occur when R is large, but the frequency eigenvalues never cross. Resonance between gas and solid oscillations occurs at mode crossings, while between crossings the large impedance difference of the two media effectively isolates one from the other.

Solid modes are not observed in acoustic instability of solid-propellant rockets, partly because of viscoelastic damping but mainly because their free-boundary (constant-pressure) interface condition removes the major source of amplification (see Section 9.1.5). Therefore, we should focus our attention on gas modes, and we see that the oscillations of the solid are likely to be negligible, except at the mode-crossing conditions of resonance. At resonance, appreciable acoustic energy from the gas is transmitted into the solid and may be dissipated there viscoelastically. Since the gas fraction f in a rocket chamber increases with time during burning, the resonance may be responsible for an intermittent disappearance of instability; this may be true even if viscoelastic damping is negligible because the transfer of acoustic energy into the solid reduces that in the gas. A number of analyses of influences of solid vibrations [38]–[41] contribute to the understanding of their unique attributes.

9.1.4.6. Relaxation damping

Homogeneous dissipation by chemical or molecular relaxation processes in the gas may be addressed on the basis of a formulation like that given in Section 4.3.4. Wave equations arise in which a relaxation time τ

and both equilibrium and frozen sound speeds appear; a one-dimensional version was written in equation (4-113). Instead of looking for solutions precisely harmonic in time, as was done in Section 4.3.4.4, we may seek solutions harmonic in space, as needed to satisfy boundary conditions for normal modes. For example, by substituting $v = \text{Re}\{Ve^{ikx+st}\}$ into equation (4-113), where V, k, and s are constants, k being a wave number, we obtain

$$\tau s^3 + bs^2 + \tau\omega^2 s + \omega^2 = 0,$$

where $b \equiv a_{f0}^2/a_{e0}^2$ and $\omega^2 \equiv a_{f0}^2 k^2$. Here the frequency ω is real by definition and $b - 1$ is small but positive. Because of this we may set $s = i\omega + \hat{\alpha}$ and assume that $|\hat{\alpha}| \ll |\omega|$, evaluating s^2 and s^3 by expansion to first order in $|\hat{\alpha}/\omega|$. The solution to the resulting equation for $\hat{\alpha}$ is

$$\hat{\alpha} = -(\omega/2)(b - 1)(\omega\tau + ib)/[b^2 + (\omega\tau)^2],$$

which exhibits both wave damping and a small decrease in frequency. The contribution to the previously defined amplification constant $\alpha = \text{Re}\{\hat{\alpha}\}$, to the order of approximation to which this result applies, is

$$\alpha = -(\omega/2)(a_{f0}^2/a_{e0}^2 - 1)(\omega\tau)/[1 + (\omega\tau)^2]. \tag{42}$$

More thorough derivations of formulas equivalent to equation (42) may be found in the literature [42], [43]. The result here shows that the dimensionless decay rate, $-\alpha/\omega$, vanishes as $\omega\tau$ approaches zero or infinity and attains a maximum value of $(a_{f0}^2/a_{e0}^2 - 1)/4$ at $\omega\tau = 1$. Thus the dissipation rate is greatest when the reciprocal of the relaxation time equals the frequency of the acoustic oscillations and is negligible for long or short relaxation times (that is, for frozen or equilibrium behavior, respectively; compare Section 4.3.4.4). Most vibrational relaxations are too rapid to contribute significantly to α, and most chemical times are either too short or too long; values of a_{f0}/a_{e0} for the few that are not typically produce $\alpha \approx -10^{-2}\omega$, which is not a large damping effect.

9.1.4.7. Particle damping

As indicated in Section 9.1.4.1, particle damping can be of greater significance. The theory of sound attenuation by spherical inert particles in the gas is well developed [44]–[49]. A number of different regimes may occur [7], but that of greatest practical interest has

$$r_s \ll \sqrt{\mu/(\rho\omega)} \ll l,$$

with the Stokes drag law applicable. Here r_s is the particle radius, $\sqrt{\mu/(\rho\omega)}$ is the previously identified thickness of an oscillatory boundary layer on a flat surface, and l is the wavelength of the acoustic field. Gas-particle flow with Stokes drag was considered in Section 4.2.6. In the notation of that section, generalized to three-dimensional flow by representing the velocities as vectors, it may be seen from equation (4-40) that the force per unit volume exerted on the gas by the particles is $n_s 6\pi r_s \mu \, (\mathbf{v}_s - \mathbf{v}_g)$. The dissipation rate Φ of equation (12) is the dot product of the negative of this force with gas velocity $\mathbf{v}_g$, that is, the negative of the rate of doing work on the gas by the particles, per unit volume. Thus, from equation (15), the consequent contribution to the amplification rate is

$$\alpha = - \int_{\mathscr{V}} \langle n_s(\tfrac{4}{3}\pi r_s^3 \rho_s)(\mathbf{v}_g - \mathbf{v}_s) \cdot \mathbf{v}_g/\tau \rangle \, d\mathscr{V}/(2\bar{e}\mathscr{V}), \qquad (43)$$

where the response time

$$\tau = 2r_s^2 \rho_s/(9\mu), \qquad (44)$$

identified in Section 4.2.6, has been introduced. The mass of particles per unit volume appears in equation (43) and can be expressed in terms of the gas density ρ_g and the mass of gas per unit mass of the system, Z, introduced in Section 4.2.6, by the formula

$$n_s(\tfrac{4}{3}\pi r_s^3 \rho_s) = \rho_g(1 - Z)/Z. \qquad (45)$$

Evaluation of the average in equation (43) necessitates investigation of the oscillatory velocity fields. If equation (24) is employed for the gas velocity $\mathbf{v}_g$ and a similar representation, $\mathbf{v}_s = \mathrm{Re}\{\mathbf{V}_s e^{i\omega t}\}$, is introduced for the particle velocity, then by use of the equation of motion for a particle,

$$\tfrac{4}{3}\pi r_s^3 \rho_s \, d\mathbf{v}_s/dt = 6\pi r_s \mu(\mathbf{v}_g - \mathbf{v}_s), \qquad (46)$$

[which is readily inferred from equation (4-40) for example] it is found that

$$\mathbf{V}_s = \mathbf{V}(1 - i\omega\tau)/[1 + (\omega\tau)^2]. \qquad (47)$$

By substitution of equations (28), (45), and (47) into equation (43), we can show that

$$\alpha = - \int \frac{(1 - Z)(\omega\tau)^2 |\mathbf{V} \cdot \mathbf{V}|}{(Z\tau)[1 + (\omega\tau)^2]} \, d\mathscr{V} \bigg/ \left[2 \int |\mathbf{V} \cdot \mathbf{V}| \, d\mathscr{V} \right], \qquad (48)$$

which reduces to equation (42) with

$$a_{e0} = \sqrt{Z} \, a_{f0} \qquad (49)$$

if Z and τ are uniform throughout the chamber. Thus equilibrium and frozen sound speeds can be defined for particle-gas mixtures in a manner entirely analogous to that for chemical relaxation, as becomes clear from an alternative analysis [47]. Under frozen conditions, which occur for large particles [large τ in equation (44)], the particles remain stationary, while under equilibrium conditions (for small particles giving small τ) the particles move with the gas. The optimum particle radius for maximum damping, corresponding to $\omega\tau = 1$, is found from equation (44) to be $\sqrt{9\mu/(2\rho_s\omega)}$, which typically lies in the range of 10^{-3} cm to 10^{-4} cm. The value of α at this maximum is $-\omega(1 - Z)/(4Z)$, which may achieve a value on the order of $-10^{-1}\,\omega$, since values of $1 - Z$ as large as 0.3 are not uncommon for certain metalized propellants. The linearity implies that if there is a particle size distribution, with $G(r_s)\,dr_s$ being the number of particles per unit volume in the radius range dr_s about r_s, then

$$\alpha = -(\omega/2)(\rho_s/\rho_g) \int_0^\infty \{(\tfrac{4}{3})\pi r_s^3(\omega\tau)/[1 + (\omega\tau)^2]\}G(r_s)\,dr_s, \qquad (50)$$

in which τ is given by equation (44).

In addition to the velocity-lag effect analyzed here, there can be temperature-lag effects on particle damping associated with heat transfer to the particles. For a discussion of these effects and for a more detailed presentation of the velocity-lag analysis, see [7]. Predictions of particle damping have been found to agree well with measurement [50], as may be seen from Figure 9.3. The experiments shown here involved a distribution of particle sizes, and equation (50) was used for the theoretical curve in the figure, with an additional term representing the heat-transfer contribution added; in the figure, $\bar\alpha = -(\alpha/\omega)3\rho_g/[2\pi\rho_s \int_0^\infty r_s^3 G(r_s)\,dr_s]$, and $\bar\tau$ is defined as τ in equation (44) with r_s replaced by $\int_0^\infty r_s^3 G(r_s)\,dr_s/\int_0^\infty r_s^2 G(r_s)\,dr_s$.

In Section 9.1.4, many different approaches to the analysis of various types of damping mechanisms were indicated. The variety of formulations exposed may help to show alternative paths to damping calculations.

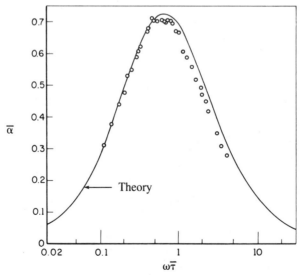

FIGURE 9.3. Comparison of theory and experiment concerning particle attenuation, showing dimensionless damping coefficient as a function of dimensionless particle radius [50].

9.1.5. Amplification mechanisms

9.1.5.1. Relative importance

Among the mechanisms that may lead to amplification of acoustic oscillations are combustion at the burning surface, finite-rate chemistry in exothermic reactions that occur homogeneously throughout the chamber, and conversion of mean-flow energy into acoustic energy through aerodynamic noise-generation processes of various types. In solid-propellant rockets, the first of these sources generally is the most important. The second would be more relevant to liquid-propellant rockets, for example. The third usually is negligible, although it may become significant under special circumstances; much research has been done on this fluid-mechanical topic in the field of aeroacoustics, and reviews are available (for example, [25]). Our attention here will be focused on the first mechanism.

9.1.5.2. Amplification criteria

The mechanism of sound amplification by heat release was understood by Rayleigh [51]. Qualitatively, if we erroneously neglect any velocity changes just to get a quick result, then equation (10) is $e = \bar{p}p'^2/(2\gamma)$ and equation (4-54) is $Dp/Dt = a^2\dot{Q}/(c_p T)$, where $\dot{Q}$ denotes the rate of heat release per unit volume; multiplying the second equation by p'/γ and averaging over a cycle, by use of the first equation, we obtain

$$d\langle e\rangle/dt = \langle p'\dot{Q}\rangle(\gamma - 1)/\gamma, \tag{51}$$

which in fact turns out to be correct for heat addition to an inviscid, constant-property, ideal gas experiencing small-amplitude acoustic oscillations about a uniform, quiescent state. Rayleigh recognized, as implied by equation (51), that amplification results if, on the average, heat addition occurs in phase with the pressure increases during the oscillation; this is the Rayleigh criterion for sound generation by periodic addition and abstraction of heat [52]. It explains qualitatively many observations of singing and vibrating flames (and vibrations in tubes containing heated elements as well), described beautifully by Tyndall [53]* and dating back at least to an observation made by Dr. B. Higgins in 1777. Of course, the amplification at a burning propellant surface is considerably more complex than the phenomena described by equation (51), since molecular transport processes are involved in an essential way for example (as was seen for the steady-state combustion in Chapter 7). However, the pistonlike mechanism implied by equation (51) when the heat release is localized and the result is expressed in terms of an acoustic admittance [as in equation (29)] is tied closely to the amplification by propellant combustion.

---

* I am indebted to S. Temkin for this reference.

In analyzing solid-propellant combustion in Chapter 7, we employed a coordinate x, normal to the propellant surface, which ranged from $-\infty$ deep in the solid to $+\infty$ far into the gas. However, in comparison with chamber dimensions, the distances over which changes occurred were small, on the order of heat-conduction lengths for deflagrations, typically a few millimeters or less. Therefore, from the viewpoint of acoustic instability, it is appropriate to approximate the combustion zone as an interface and to employ admittances to describe its influence on the sound field. The pressure oscillations of the acoustic field are imposed on the combustion zone, where the dimensions and velocities are small enough that the time-varying pressure is independent of x to a good approximation [54]. The combustion-zone structure must be analyzed subject to these pressure oscillations, and from the solutions the admittance $\mathscr{Y}$, defined in equation (20) (with $\mathbf{v}$ now representing the difference of the velocity from its mean value) is evaluated at $x = -\infty$ and at $x = +\infty$ for use in calculating acoustic amplification. The difference of these two values arises in the acoustic analysis [39], but at $x = -\infty$, $\mathbf{v} \cdot \mathbf{n}$ is on the order of the ratio of the gas density to the solid density times that at $x = +\infty$, and since this ratio is small, it is sufficient to evaluate $\mathscr{Y}$ at $x = +\infty$ for use in amplification expressions such as equation (29).* Since $\mathbf{n}$ is the unit normal pointing outward from the chamber in equation (20), $\mathbf{v} \cdot \mathbf{n}$ is the negative of the oscillating part of the gas velocity in the $+x$ direction at $x = +\infty$.

Although it is the velocity that appears most directly in the admittance that is responsible for the amplification, analysis of the combustion zone usually is performed more conveniently in terms of the mass flux m. Putting $m = \bar{m}(1 + m')$ and assuming that $|m'| \ll 1$, we may express y of equation (23) as

$$y = - [\bar{m}/(\bar{\rho}a)](\mu - r), \tag{52}$$

where μ is the ratio of the complex amplitude of m' to the complex amplitude of p', r is the ratio of the complex amplitude of ρ' to the complex amplitude

---

* For the solid modes discussed in Section 9.1.4.5 and identified in Figure 9.2, the nondimensional acoustic impedance, seen by the solid at its interface, is small (since the interface is nearly an isobaric pressure node) and is therefore more appropriate than the nondimensional admittance y (its reciprocal) for use in equation (22). Thus the right-hand side of equation (22) is best expressed as $- V^2/2$ times the real part of the impedance. The difference of the impedance at $x = -\infty$ and that at $x = +\infty$ appears, but the latter is negligible because of the density ratio. The combustion-zone analysis may be performed in a noninertial frame that moves with the acoustic velocity. In this frame, an apparent body force per unit volume of order $\rho \omega V$ appears in the equation for momentum conservation. In terms of the thickness δ of the combustion zone, for a given energy of the acoustic field, the ratio of the rate of doing work on the field per unit area at the pressure-node condition of the solid mode to that at the velocity-node condition of the gas mode is on the order of $\delta \rho \omega a (v_g \rho_g/\rho_s)/(p v_g)$, which is roughly $(\delta \omega/a)(\rho_g/\rho_s)$, the value of which seldom is as large as 10^{-4}. Therefore, the rate of amplification for solid modes is much smaller than that for gas modes; effectively, solid modes are not excited.

of p', the velocity amplitude has been eliminated by use of the expansion $m/\rho = (\bar{m}/\bar{\rho}) (1 + m' - \rho' + \cdots)$, and all quantities are to be evaluated in the gas at $x = +\infty$.

Equation (4) implies that $r = 1/\gamma$ in equation (52). Although this value assuredly is correct formally, a complication has been identified on the basis of studies of the combustion-zone structure [55], [56]. Models based on the hypothesis of a quasisteady gas-phase flame predict a nonisentropic relationship between p and ρ in the gas at the downstream edge of the gas-phase reaction zone; the density and temperature oscillations are found to be related to the pressure oscillations by conditions that are not isentropic. Since the combustion has been completed, the periodically generated non-isentropic component is transported with fixed form at the mean gas velocity, say v_∞, as implied by equation (4-49), for example. The consequent spatial and temporal oscillations of entropy have been termed **entropy waves**, although this is somewhat of a misnomer, since equation (4-49) is not the wave equation. These "waves" decay through nonquasisteady heat conduction between their temperature peaks and valleys; this dissipation is much more rapid than the corresponding dissipation for acoustic waves because the wavelength, of order v_∞/ω, is shorter (typically 10^{-1} cm to 1 cm). By balancing the rate of heat conduction per unit area with the rate of change of thermal enthalpy per unit area in a half-wave, a decay time may be estimated as $(v_\infty/\omega)^2$ divided by a thermal diffusivity, and when this is multiplied by v_∞ to obtain a characteristic decay length, representative values (with $v_\infty \approx 10^2$ cm/s and $\omega \approx 10^3$/s) are found to lie between 1 cm and 10 cm, which in extreme cases may reach a significant fraction of the chamber dimension.* Thus the combustion response may create a region large in comparison with the thickness of the steady-state combustion zone, within which the wave field is modified by nonisentropic effects. It is strictly proper to employ equation (52) in an acoustic boundary condition for amplification calculations only outside the nonisentropic zone, and $r = 1/\gamma$ holds there. However, small changes in μ that may occur between this location and the downstream edge of the reaction zone usually are not addressed in analyses; often it is most convenient to evaluate μ just after quasisteady combustion. When this is done, r may be evaluated in equation (52) from the quasisteady combustion-zone analysis, in the hope that use of the value so obtained may account approximately for dissipation in the nonisentropic zone. In the isothermal limit, $p' = \rho'$ and $r = 1$; it is expected that $1/\gamma \leq \mathrm{Re}\{r\} \leq 1$, and therefore influences of uncertainties in the value of r are not large. We shall employ $r = 1/\gamma$ and view dissipation in the nonisentropic zone to be counted as an additional damping contribution.

---

* Since the ratio of the decay length to the chamber dimension theoretically increases in proportion to the chamber dimension, the effect is predicted to be greatest in large chambers.

By use of equation (52) in equation (29), for example, we see that a necessary condition for amplification by the propellant combustion is

$$\text{Re}\{\mu\} > 1/\gamma. \tag{53}$$

The ratio μ often is termed the **response function**, or the **burning-rate response**, and combustion-zone analyses typically provide magnitudes of order unity for it. Therefore, equation (52) indicates that y is of the same order of magnitude as the Mach number of mean flow of the burnt gas. This observation enables the order of magnitude of the growth or decay rate to be estimated directly from equation (30). More accurate computations necessitate calculating $\text{Re}\{\mu\}$. The many different possibilities for steady-state structures of the combustion zone indicated in Chapter 7 imply that many different analyses can be relevant to the calculation of μ. Here we shall outline only two and comment on other approaches.

9.1.5.3. Time-lag theories

Many early analyses of time-dependent solid-propellant combustion employed the concept of a combustion **time lag** [4], [17], [57]–[60]. A given element of reactant material requires a certain amount of time to burn, and if a small pressure pulse is applied to a steadily burning propellant, then the regression rate will take a certain amount of time to reach its new steady level. As the simplest idealization of these observations, assume that a pressure change applied at time t has no effect on m until time $t + \tau_l$, at which time m instantaneously assumes its steady-state value, at all x, appropriate to the new pressure. The interval τ_l is the time lag, assumed here to be a known constant. If n is the pressure sensitivity [appearing, for example, in equation (7-41)], then in $m(t) = \bar{m}[1 + m'(t)]$ and $p(t) = \bar{p}[1 + p'(t)]$, we have $m'(t) = np'(t - \tau_l)$, so that

$$\mu = ne^{-i\omega\tau_l}. \tag{54}$$

Thus the criterion in equation (53) becomes $n\cos(\omega\tau_l) > 1/\gamma$, which requires that $n > 1/\gamma$ and that $\omega\tau$ be in an appropriate range for amplification to occur; typically $\omega\tau_l$ is small, whence amplification is predicted at low frequencies, $\omega < (1/\tau_l)\cos^{-1}[1/(n\gamma)]$. Among the improvements to equation (54) is the introduction [17], [59] of a sensitive part of τ_l that varies with pressure and gives results more versatile than equation (54).

A critique and more thorough review of the time-lag ideas has been published [7]. The principal value of time-lag concepts lies in the wide range of problems to which they can be applied with relative ease. Their principal deficiency lies in the difficulty of relating τ_l to the fundamental processes occurring. If the one-dimensional, time-dependent conservation equations are linearized about any one of the steady-state solutions of Chapter 7 and μ is calculated from a perturbation analysis, then it is found to be difficult to express the results in terms of distributions of time lags. Time-lag approaches are useful for a first look at instability problems (including liquid-propellant rocket instability), but more detailed analyses of the flow processes are better when they can be performed.

9.1.5.4. Combustion response

Although many different processes may be involved in solid-pro-pellant combustion, estimates suggest that characteristic times for all of them may be short compared with the characteristic time for heat conduction in the solid. This latter characteristic time is

$$\tau_s = \rho_s \lambda_s / (c_{ps} \bar{m}^2), \tag{55}$$

where the subscript s identifies properties of the solid. Representative values of τ_s lie in the range 10^{-3} s to 10^{-4} s, which is comparable with typical values of $1/\omega$. Thus it is attractive to assume that all processes remain quasisteady except the solid-phase heat conduction. This idea dates back to the work of Zel'dovich [61], [62], who considered time-dependent burning within the context of an approximation of constant temperature at the solid-gas interface. Oscillatory burning has been analyzed under this assumption [63], and a more complete model has been explored, in which use is made of the deflagration analysis of Section 7.5 [64]. It is of interest to present an analysis of the last of these models because it is relatively simple, and yet the results of many of the more recently published theories bear a close resemblance to its results.

The combustion mechanism addressed involves inert heat conduction in the solid, surface gasification by an Arrhenius process and a gas-phase deflagration having a high nondimensional activation energy. With the density, specific heat, and thermal conductivity of the solid assumed constant, the equation for energy conservation in the solid becomes

$$\rho_s c_{ps} \partial T / \partial t + m c_{ps} \partial T / \partial x = \lambda_s \partial^2 T / \partial x^2. \tag{56}$$

The coordinate system is selected so that the solid-gas interface, where conditions are identified by the subscript i, is maintained at $x = 0$. By mass conservation, m is independent of x everywhere, but it varies with t in this coordinate system. Boundary conditions for equation (56) are $T = T_0$, the initial temperature of the propellant, at $x = -\infty$ and $T = T_i$ at $x = 0$. The interface Arrhenius law, given by equation (7-6) but also interpretable in terms of a distributed solid-phase reaction in a thin zone (as indicated at the end of Section 7.4), is written here in the form

$$m = B_s \exp[-E_s / (R^0 T_i)], \tag{57}$$

where the constants B_s and E_s denote the preexponential factor and the overall activation energy for the surface process. Under the assumption of quasisteady, adiabatic, gas-phase combustion, the interface energy conservation condition readily (for example, [65]) may be shown to be expressible as

$$-\lambda_s(\partial T/\partial x)_i = m[L_s - q + c_p(T_\infty - T_i)], \qquad (58)$$

where L_s is the energy per unit mass required for gasification, q is the (constant) energy per unit mass released in the gas-phase reaction, c_p is the (constant) specific heat at constant pressure for the gas, and T_∞ is the flame temperature (which varies with time). An overall energy balance may be applied to steady-state conditions (identified by an overbar) to give $q = c_{ps}(\bar{T}_i - T_0) + \bar{L}_s + c_p(\bar{T}_\infty - \bar{T}_i)$, and thermodynamics require L_s to vary with T_i if $c_{ps} \neq c_p$—that is, $L_s = \bar{L}_s + (c_p - c_{ps})(T_i - \bar{T}_i)$. Analysis of the quasisteady gas-phase flame by activation-energy asymptotics [65] shows that also

$$m = Cp^n \exp[-E_1/(2R^0 T_\infty)], \qquad (59)$$

where C and n are positive constants (n being half of the pressure exponent for the homogeneous gas-phase reaction) and where E_1 is the overall activation energy for the gas-phase combustion [see Section 7.5 and equation (5.75)]. Equations (56)–(59) provide a well-posed problem for addressing influences of harmonic pressure oscillations; the pressure appears here only in equation (59).

The steady-state solution to equation (56) is easily seen to be

$$\bar{T} = T_0 + (\bar{T}_i - T_0)\exp(x\bar{m}c_{ps}/\lambda_s). \qquad (60)$$

By putting $T = \bar{T} + \text{Re}\{\Delta T e^{i\omega t}\}$ and linearizing equation (56) about $\bar{T}$ (under the assumption that $|\Delta T| \ll \bar{T}$), we may readily derive the perturbation equation

$$i\omega\tau_s\Delta T + d\Delta T/d\xi + \mu P\, d\bar{T}/d\xi = d^2\Delta T/d\xi^2, \qquad (61)$$

where τ_s is defined in equation (55), P in equation (24), and $\xi = x\bar{m}c_{ps}/\lambda_s$. A particular solution to equation (61) is $-(i\omega\tau_s)^{-1}\mu P(\bar{T}_i - T_0)e^\xi$, and the solution to the homogeneous equation is a constant times $e^{s\xi}$, where

$$s = (1 + \sqrt{1 + 4i\omega\tau_s})/2, \qquad (62)$$

the positive sign of the square root having been selected to satisfy $\Delta T = 0$ at $\xi = -\infty$. From the general solution, represented as the sum of the particular solution and a solution to the homogeneous equation, the relationship between the heat-flux perturbation and the temperature perturbation at the interface,

$$(d\Delta T/d\xi)_i = s(\Delta T)_i + (s - 1)\mu P(\bar{T}_i - T_0)/(i\omega\tau_s), \qquad (63)$$

may be derived. The perturbations of equations (57), (58), and (59) give

$$\mu P = (\Delta T)_i E_s/(R^0 \bar{T}_i^2) = nP + (\Delta T)_\infty E_1/(2R^0 \bar{T}_\infty^2) \qquad (64)$$

and

$$(d\Delta T/d\xi)_i = \mu P(\overline{T}_i - T_0) - (c_p/c_{ps})(\Delta T)_\infty + (\Delta T)_i, \tag{65}$$

where the overall adiabaticity and the L_s variation have been employed with equation (58) in obtaining the last expression. Use of equations (62)–(65) to remove $i\omega\tau_s$, $(d\Delta T/d\xi)_i$, $(\Delta T)_i$ and $(\Delta T)_\infty$ from the system then yields

$$\mu = nBs/[s^2 - (1 + A - B)s + A], \tag{66}$$

where

$$A = (\overline{T}_i - T_0)E_s/(R^0\overline{T}_i^2) \tag{67}$$

and

$$B = 2E_s c_p \overline{T}_\infty^2/(E_1 c_{ps} \overline{T}_i^2). \tag{68}$$

Equations (62) and (66) provide μ as a function of $\omega\tau_s$ with n, A, and B appearing as the relevant steady-state combustion parameters.

An idea of the predictions concerning amplification may be obtained from expansions for small and large values of $\omega\tau_s$. From equations (62) and (66) we find that

$$\text{Re}\{\mu\} = n\{1 + (\omega\tau_s)^2[(B + 1)(2A - 1) - A^2]/B^2 + \cdots\}$$

for $\omega\tau_s$ small and $\text{Re}\{\mu\} = nB/\sqrt{8\omega\tau_s} + \cdots$ for $\omega\tau_s$ large. These results demonstrate the quasisteady response for $\omega\tau_s = 0$, amplification that decreases in proportion to $1/\sqrt{\omega\tau_s}$ as $\omega\tau_s$ increases for $\omega\tau_s$ large and a behavior for small $\omega\tau_s$ showing a decreasing response with increasing $\omega\tau_s$ if A is small or large but an increasing response for intermediate values of A, values between $(B + 1) - \sqrt{(B + 1)^2 - (B + 1)}$ (for $A > \frac{1}{2}$) and $(B + 1) + \sqrt{(B + 1)^2 - (B + 1)}$. Typically, $\text{Re}\{\mu\}$ exhibits one broad maximum, in the vicinity of $\omega\tau_s \approx 1$, the height of which depends on the values of A and B and exceeds n for a range of values of A and B. The enhancement is caused by the transient buildup of energy in the solid-phase heat-conduction zone and is typical of predictions of more-complicated theories. In general, large values of A and of A/B lead to larger peak responses through the greater sensitivity provided by the high activation energies. The present model also exhibits an intrinsic instability if A or A/B is too large, as will be discussed in Section 9.2. A few representative theoretical response curves are shown in Figure 9.4, where the envelope curves represent the maximum achieved for any value of A and the envelope asymptote is related to the intrinsic instability, which occurs for $A \gtrsim 6$ in this case.

Many analyses of the general type outlined here, for calculating the response function from models that do not rely on time lags, have appeared in the literature (for example, [66]–[74]); reviews are available [7], [12], [75], [76]. Some of the models admit a certain amount of nonquasisteady

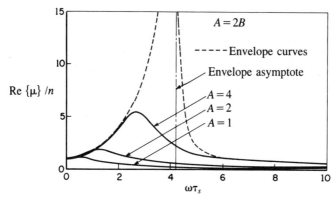

FIGURE 9.4. Representative burning-rate response curves obtained from equation (66), showing a nondimensional measure of the amplification rate as a function of the nondimensional frequency for various values of the nondimensional activation energy for gasification, A, with $A = 2B$.

behavior in the gas phase, which can be important at high frequencies. The equations allowing fully nonquasisteady combustion in the gas phase are complicated [54] and have not been explored completely. These models all pertain to homogeneous propellants, for which it was seen in Chapter 7 that the combustion mechanisms may be complex, particularly with respect to condensed-phase and dispersed-phase processes. A useful philosophy is to attempt to develop general descriptions that are independent of specific details of combustion models [76].* In fact, most of the mechanisms described in Chapter 7 have not been investigated in the context of oscillatory response, even on the basis of quasisteady gas-phase approximations. Nevertheless, it seems likely that the main elements of the response function for homogeneous propellants are reasonably represented by the analyses that have been performed. This belief is based on the observations that gas-phase combustion is usually at least of small importance for transmitting energy in propellant burning, while condensed-phase (or possibly dispersed-phase) processes have longer response times, which allow them to exhibit transient energy accumulation. Thus even if the model that has been analyzed here is inaccurate in detail, it seems likely to retain the qualitatively essential physics for describing response to pressure oscillations. Although experimental measurements of the response function are difficult to perform (except by inference from measured growth rates of pressure oscillations in chambers), the few results that have been reported for homogeneous propellants tend to agree qualitatively with predictions [77]–[80].

---

* It can be shown that by reinterpreting the parameters n, A, and B in terms of sensitivities of m to p, T_i, and T_∞, results for any model that maintains quasisteady gas-phase processes can be expressed in terms of the results that have been derived here.

9.1.5.5. Heterogeneity effects

For heterogeneous propellants, the current situation is much less satisfactory. The complexity of the combustion process was discussed in Section
7.7. To employ a result like equation (66) directly is questionable, although
attempts have been made to evaluate parameters like A and B of equations
(67) and (68) from complicated combustion models for use in response-
function calculations [81], [82]. Relatively few theories have been addressed
specifically to the acoustic response of heterogeneous propellants [82].
Applications of time-lag concepts to account for various aspects of heterogeneity have been made [60], [83], a simplified model—including transient
variations in stoichiometry—has been developed [84], and the sideways
sandwich model, described in Section 7.7, has been explored for calculating
the acoustic response [85]. There are reviews of the early studies [7] and of
more recent work [82].

A peculiar aspect of the response of heterogeneous propellants is
associated with the periodic uncovering of different propellant constituents
during normal regression [85]. If the uncovering time is short compared
with an acoustic period, then an average is perceived by the acoustic field;
if the uncovering time is long, then low-frequency, nonacoustic oscillations
may be superposed (at least locally) on the acoustic oscillations. Uncovering
times often can be comparable with acoustic periods, thereby leading to the
possibility of significant interaction between heterogeneity and acoustics.
Description of the interaction necessitates augmentation of response-
function concepts and entails considering a synchronization time needed
for the establishment of a coherent phase of the heterogeneity over the propellant surface by action of the acoustic oscillations [85]. When the propellant surface responds coherently, an additional term arises in equation
(14) that is proportional to the product of the pressure amplitude and the
heterogeneity amplitude, the latter measuring the magnitude of a heterogeneity of the propellant and being independent of the acoustic amplitude.
An equation of the form

$$d\sqrt{\bar{e}}/dt = \alpha\sqrt{\bar{e}} + \beta \qquad (69)$$

therefore is obtained, wher α and β are constants [85]. The solutions to
equation (69) differ from equation (16) and show that the heterogeneity
interactions can lead to linear rather than exponential growths of acoustic
amplitudes with time at low amplitudes ($\alpha > 0$, $\beta > 0$), to amplitude thres-
holds that must be exceeded for growth to occur ($\alpha > 0$, $\beta < 0$) and to limiting
amplitudes that are approached at long times ($\alpha < 0$, $\beta > 0$). Preferred-
frequency phenomena, in which instabilities are strongest in specific, narrow,
frequency ranges, have been observed experimentally for composite propellants [86] and may be attributable to the heterogeneity of the solid.
Response curves like Figure 9.4 may be found to possess more than one

maximum. Some aspects of the results qualitatively resemble those anticipated from nonlinear effects.

9.1.6. Nonlinear effects

Nonlinear phenomena, usually associated with high amplitudes of
the acoustic field, can introduce many interesting effects into acoustic
instability [76]. Here we shall discuss only three topics involving nonlinearity:
the response of the combustion zone to transverse velocity oscillations
(conventionally termed **velocity coupling**), changes in the mean burning
rate of the propellant in the presence of an acoustic field, and instabilities
that involve the propagation of steep-fronted waves (identified in the introduction as shock instabilities).

Erosive burning was addressed in Section 7.8 where it was indicated
that the burning rate m depends on the gas velocity parallel to the propellant
surface. In addition to the mean-flow erosion, there is acoustic erosion from
most of the modes in a chamber experiencing acoustic instability. It has long
been recognized that acoustic erosion can affect the instability [7], [12],
[87]–[93], and the idea of introducing a velocity-coupled response function,
on par with the pressure-coupled response function μ, arose early. Although
there are special conditions under which this makes sense, the influences
of velocity coupling usually are nonlinear, even at low acoustic amplitudes,
and they require special consideration.

If the acoustic velocity in the gas adjacent to the burning surface is
small compared with the local mean erosion velocity, then it may be expected
to produce small-amplitude oscillations of m that locally may be related
to the amplitude of the acoustic velocity oscillation through an admittance-
like expression similar to equation (52). Of course, a corresponding response
to pressure oscillations also is to be anticipated. The oscillations of m can
excite acoustic oscillations if there are pressure oscillations at the point in
question to enable work to be performed on the acoustic field. The combustion-zone analyses needed for calculating the response would be complicated
because of the complexity of erosive burning, indicated in Section 7.8. An
analysis has been published [94] of a related problem, that of combustion
of a solid fuel with a gaseous oxidizer in a laminar boundary layer. However,
little work has been done on the corresponding solid-propellant problem,
which is more difficult because of the essential role played by finite-rate
chemistry. Even for these relatively simple problems, a difficulty in applying
results of the analyses to the calculation of growth rates of acoustic fields
arises from the fact that the velocity oscillations associated with oscillations
of m are normal to the velocity oscillations of the driving field. Thus a mode
transverse to the driving mode is generated, and the new mode is not likely
to be a normal mode of the chamber. The process of transfer of normal
oscillations to tangential oscillations must be studied to ascertain the efficiency with which the response can produce amplification.

   All other types of velocity response are nonlinear in one way or another. For example, if there is no mean erosion and if frequencies are low enough for the propellant to respond to the acoustic erosion in an entirely quasisteady fashion, then equation (7-45) or equation (7-46), for example, shows that the fluctuating response m' is a function of $|v|$, where v is the acoustic velocity tangent to the surface. With v proportional to $\cos \omega t$, this causes m' to be a function of $|\cos \omega t| = \{[1 + \cos(2\omega t)]/2\}^{1/2}$, the Fourier expansion of which exhibits contributions from frequencies 2ω, 4ω, 6ω, ... but none from frequency ω. Thus the response generates harmonics but does not affect oscillations at the driving frequency. This same effect would be found in the presence of mean erosion if the acoustic velocity were normal to the mean erosive velocity. More generally, with a nonzero mean erosive velocity and a component of the acoustic velocity parallel to it, contributions at frequency ω, as well as harmonics, are found. The amplitudes of the harmonics in m' increase as $|v|$ increases. It is theoretically possible that, as the intensity of the acoustic field increases, the erosive response extracts energy from the fundamental and feeds it into harmonics, eventually establishing a steady condition in which the amplitudes of the fundamental and of all harmonics are independent of time, even though the acoustic field itself remains entirely in a range of linearity. This type of process has been described for heater-excited acoustics oscillations in tubes [95]. It provides a mechanism for reaching limiting amplitudes that is an alternative to nonlinearities in the bulk of the chamber, in boundary damping or in the pressure-coupled response.

   Another aspect of nonlinearity is a change in the mean burning rate, which is of importance in producing "secondary peaks" of the mean pressure. It is easy to write formulas for this change if the response is assumed to be entirely quasisteady. For example, if equation (7-41) is taken to apply under conditions of oscillatory pressure, then with $m = \bar{m}_0(1 + m')$ and $p = \bar{p}(1 + p')$, where $\bar{m}_0$ is the value of $\bar{m}$ at $p' = 0$ (so that $\bar{m}' \neq 0$ at finite amplitude), we find through expansion to second order (by use of the average $\bar{p}' = 0$) that

$$\bar{m}' = \overline{p'^2} n(n - 1)/2 + \cdots, \tag{70}$$

which exhibits a decrease in the mean burning rate with increasing pressure amplitude for $0 < n < 1$. By contrast, erosive effects (in the absence of an erosion threshold or of negative erosion) produce an increase in the mean burning rate with increasing velocity amplitude. For example, if the erosive velocity is $v = v_0(1 + v')$ and v' is sinusoidal in time (so that its maximum is $2\sqrt{\overline{v'^2}}$), then through adoption of equation (7-45) with $M = M_0(1 + v')$, it may be shown by evaluating $\int_0^{2\pi} (|1 + 2\sqrt{\overline{v'^2}} \sin \theta| - 1) \, d\theta/2\pi$ that $\bar{m}' = 0$ for $2\sqrt{\overline{v'^2}} < 1$ and

$$\bar{m}' = (kM_0/\bar{m}_0)(2/\pi)\{\sqrt{4v'^2 - 1} - \cos^{-1}[1/(2\sqrt{\overline{v'^2}})]\} \tag{71}$$

for $2\sqrt{\overline{v'^2}} > 1$, where the inverse cosine lies between 0 and $\pi/2$. In this case, a threshold for onset of an acoustic-erosive contribution to $\overline{m}$ occurs at the beginning of rectification, that is, when the maximum acoustic erosive velocity equals the mean erosive velocity. At a fixed value of the dimensional amplitude, $2v_0\sqrt{\overline{v'^2}}$, the largest $\overline{m}'$ in equation (71) occurs at $v_0 = 0$ (as expected, since this corresponds to a zero threshold) and is given by

$$4kM_0\sqrt{\overline{v'^2}}/(\pi\overline{m}_0).$$

For a few propellants, experimental evidence exists showing that a decrease in $\overline{m}'$ is produced by pressure oscillations and an increase by erosive velocity oscillations [90], in qualitative agreement with the predictions of equations (70) and (71). However, the same results are not expected to apply at frequencies, high enough that departure from quasisteady response occurs. A few analyses of nonquasisteady changes in the mean burning rate have been completed for various models of pressure-coupled response [76] and suggest that as the frequency increases, $\overline{m}'$ decreases from the value given in equation (70). Therefore, the increases in $\overline{m}$ that are of greatest practical concern for producing increases in $\overline{p}$ in choked chambers in all probability are usually associated with effects of acoustic erosion.

As acoustic amplitudes increase, the wave forms begin to depart from those of the normal modes. The departures remain small if nonlinearities develop in damping processes at sufficiently low acoustic amplitudes and serve to limit the amplitudes to low levels. Theoretical methods for analyzing nonlinear growth and limiting amplitudes in this low-amplitude regime have been investigated [96]-[98]. There are systems in which the amplitudes become large and the acoustic waves begin to exhibit a steep-fronted character like shock waves. In addition, there are motors that are linearly stable but nonlinearly unstable to finite-amplitude, shocklike disturbances. Particular examples involve axial modes in cylindrical motors with internally burning tubular grains [99], [100]. A theoretical approach has been developed for analyzing these shock instabilities [101], [102]. The approach is based on methods that have been applied to related problems in nonreacting flows [103]-[108].

In the theoretical analysis of shock instability, shock waves that are not too strong are presumed to propagate axially back and forth in a cylindrical chamber, bouncing off a planar combustion zone at one end and a short choked nozzle at the other [101], [102]. The one-dimensional, time-dependent conservation equations for an inviscid ideal gas with constant heat capacities are expanded about a uniform state having constant pressure $\overline{p}$ and constant velocity $\overline{v}$ in the axial (z) direction. Since nonlinear effects are addressed, the expansion is carried to second order in a small parameter ϵ that measures the shock strength; discontinuities are permitted across the normal shock, but the shock remains isentropic to this order of approximation. Boundary conditions at the propellant surface ($z = 0$) and at the nozzle entrance ($z = l$) may be expressed in terms of real admittances, Y_0 and Y_l, respectively, by expansion to second order:

$$\mp (v - \bar{v})_{0, l} = Y_{0,l}^{(1)}(p - \bar{p})_{0, l} + Y_{0,l}^{(2)}(p - \bar{p})_{0, l}^2 + \cdots. \tag{72}$$

The small parameter of expansion for the shock strength is

$$\epsilon = (\bar{p}/\bar{v})(-Y_0^{(1)} - Y_l^{(1)}); \tag{73}$$

here $\epsilon > 0$, so that more mechanical energy is fed into the wave pattern at the combustion zone than is extracted at the nozzle. The analysis makes use of characteristics (Section 4.3.3) and may be performed in different ways [101], [102]. For example characteristics may be introduced as independent variables and space and time may be treated as dependent variables. Solutions cyclic in time are sought, and, as is usual in perturbation procedures of this type, it is necessary to carry the analysis to second order if the complete solution at first order is required. The results relate the pressure and velocity fields and the shock strength to the admittances. For example, for small Mach numbers $\overline{M}$ of mean flow in the chamber, it is found that the pressure increase across the shock is

$$\Delta p = \epsilon \bar{p} \overline{M}[4\gamma^2/(\gamma + 1)]. \tag{74}$$

Representative pressure and velocity fields are illustrated in Figure 9.5, where the sawtooth shapes of the waves may be seen.

The solutions obtained describe the long-time behavior of slightly nonlinear waves. It is known that nonlinearity causes waves to steepen with time, eventually approaching shocks. In rocket-chamber experiments, the steep fronts may develop from sinusoidal waves or may be established directly by an imposed pressure pulse. The limiting amplitude, given by equation (74), may be thought of as being set by a balance between the net work input at the boundaries and a kind of dissipation in the shock. Thus in the first approximation the shock strength depends on the boundary admittances [see equations (73) and (74)]; only $Y^{(1)}$ appears for small values of $\overline{M}$, but both $Y^{(1)}$ and $Y^{(2)}$ occur at larger $\overline{M}$ [7], [101]. Calculation of these admittances entails analysis of the combustion zone and of the nozzle flow subject to shocklike perturbations. Few of the needed admittance analyses have been completed; constancy of the Mach number at the nozzle entrance has been employed [101], [102], and a time-lag approach to the description of the combustion zone has been adopted [102]. It appears that, in general, the admittances could be functions of the pressure history over a cycle at the boundary, thereby introducing additional complications into the formulation. For axial modes in tubular grains, the driving mechanism must involve a continuous non-one-dimensional energy input from the response of the propellant combustion to the passage of the shock over its surface and probably exhibits a significant erosive component.

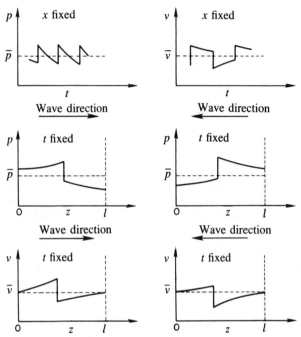

FIGURE 9.5. Theoretically calculated nonsinusoidal wave shapes of pressure and velocity fields for shock instability [101].

If ϵ of equation (73) is not small, then cyclic solutions cannot be obtained by the perturbation approach that has been outlined here. In general, methods of numerical integration are needed in describing large-amplitude behaviors accurately. Accurate use of numerical methods for acoustic instability is difficult to achieve because so many physical processes with poorly known parameters usually are involved. There are intrinsic instabilities in burning solids that involve fewer processes and that are therefore better suited to investigation by numerical approaches. Nonlinear perturbation methods, somewhat analogous to the method just described, also can be applied to these intrinsic instabilities, as will be indicated in the following section.

9.2. INHERENT OSCILLATIONS OF BURNING SOLIDS

As an especially simple example of an intrinsic instability, let us first consider the planar, adiabatic, gasless combustion of a solid, mentioned at the beginning of Section 7.1 and discussed in the middle of Section 7.4. The statement of energy conservation in the solid may be taken to be equation (56) with a heat-release term, say w_q, added to the right-hand side. Although w_q properly depends on the reactant concentration, a temperature-explicit

formulation may be achieved by adopting an expression like equation
(7-18), with $w_q = q_s w_1$ for $x < 0$ in the notation of Section 7.4. Here $x = 0$
has been selected as the position at which the reactant has been consumed
completely, and $w_q = 0$ for $x > 0$. The burning rate m is expressed in terms
of w_1 by equation (7-20). Use of these equations along with $T = T_0 =$
constant at $x = -\infty$, $dT/dx = 0$ at $x = +\infty$ and a suitable initial profile
for $T(x)$ provides a well-posed problem for calculating $T(x, t)$ and $m(t)$, so
long as attention is restricted to times small compared with the time for
homogeneous thermal explosion of the reactant at temperature T_0.* As can be
inferred from the time-dependent generalization of equation (7-15), the
formulation in employing equation (7-20) excludes a transient behavior
within the reaction zone; if time-dependent versions of equations (7-15) and
(7-16) describe the original problem, then this can be shown to restrict the
formulation to conditions such that changes with time are not too rapid.

It is attractive to seek solutions by use of activation-energy asymptotics. If there is a narrow reaction zone in the vicinity of $x = 0$, then by
stretching the coordinate about $x = 0$ and excluding variations on a very
short time scale, the augmented version of equation (56) becomes
$\lambda_s \partial^2 T/\partial x^2 = -q_s w_1$ to the first approximation in the reaction zone. By
use of equation (7-20), the integral of this equation across the reaction zone
is seen to be expressible as

$$(\partial T/\partial x)_{x=0-} - (\partial T/\partial x)_{x=0+} = m q_s(1 - Y_{1,0})/\lambda_s, \qquad (75)$$

where the subscript $x = 0-$ identifies conditions "at minus infinity" in the
stretched variable, which—through matching—correspond to conditions
just on the negative side of $x = 0$ for the outer solution in the unstretched
x variable. Multiplying the differential equation by $\partial T/\partial x$ and integrating
over the same range yields

$$[(\partial T/\partial x)_{x=0-}]^2 - [(\partial T/\partial x)_{x=0+}]^2 = (2q_s/\lambda_s) \int_{-\infty}^{0} w_1(\partial T/\partial x)\, dx, \qquad (76)$$

in which the asymptotic expansion of the integral may be evaluated relatively
conveniently for temperature-explicit rates. It is consistent within the formal
context of the asymptotics to restrict attention to problems in which
$(\partial T/\partial x)_{x=0+}$ is of higher order in the nondimensional activation energy
(the Zel'dovich number). Then, with use of equation (7-18), equations (75)
and (76) give, in the first approximation,

$$\frac{\lambda_s}{q_s(1 - Y_{1,0})}\left(\frac{\partial T}{\partial x}\right)_{x=0-} = m = \left[\frac{2R^0 \rho_s B_1 T_i^{\alpha_1+2} \lambda_s e^{-E_1/R^0 T_i}}{E q_s(1 - Y_{1,0})^2}\right]^{1/2}, \qquad (77)$$

---

* On a thermal-explosion time scale, imposition of $T = $ constant at $x = -\infty$ is inconsistent with the differential equation.

the last equality of which may also be inferred from equation (7-29). Here T_i denotes the value of T at $x = 0$.

Equation (77) and $(\partial T/\partial x)_{x=0+} = 0$ provide first approximations to the results obtained by a more formal derivation of jump conditions across the thin reaction zone. Equation (56) applies on each side of the reaction zone in this three-zone problem. To the lowest order in the Zel'dovich number, T remains constant for $x > 0$. The problem for $x < 0$ becomes identical to that described by equations (56)–(58), with the replacements $E_s = E_1/2$,

$$B_s = \{2R^0\rho_s B_1 T_i^{\alpha_1+2}\lambda_s/[E_1 q_s(1 - Y_{1,0})^2]\}^{1/2}$$

and $L_s - q + c_p(T_\infty - T_i) = -q_s(1 - Y_{1,0}) = $ constant. The resulting form of equation (58) corresponds to a surface that is adiabatic on the side $x = 0+$ and at which a constant energy per unit mass is released. Studies of the stability of the steady state therefore may be pursued as a special case of the stability analysis of the problem defined by equations (56)–(59). Specifically, to achieve surface adiabaticity within the context of a linear expansion about equation (60), we may put $n = 0$ in equation (59) and $B = c_p/c_{ps}$ in equation (68), obtaining $(\Delta T)_\infty = (\Delta T)_i$ from equation (64); to enforce a constant surface heat release per unit mass, we must then put $c_p = c_{ps}$ in equation (65). Thus for describing small perturbations about the steady state in gasless combustion, the formulation of Section 9.1.5.4 may be employed with $n = 0$ and $B = 1$, and the sole remaining parameter is A [defined in equation (67)], which has been assumed to be large in the course of the activation-energy asymptotics.

A stability analysis of the steady state of equations (56)–(59) may be developed in a straightforward manner [64]. A relatively simple technique is to look for solutions of the form $T = \bar{T} + \text{Re}\{\Delta T e^{(\alpha+i\omega)t}\}$, where α is the growth rate and ω the frequency. Intrinsic instability to planar disturbances occurs if solutions having $\alpha > 0$ exist. From the derivation leading to equation (66), it may then readily be seen that time-dependent solutions in the absence of external pressure perturbations may occur if

$$s^2 - (1 + A - B)s + A = 0, \tag{78}$$

where now

$$s = [1 + \sqrt{1 + 4(\alpha + i\omega)\tau_s}]/2. \tag{79}$$

Equation (78) may be written as $(A - B)\sqrt{1 + 4(\alpha + i\omega)\tau_s} = A + B + 2(\alpha + i\omega)\tau_s$, the square of which yields

$$AB + (\alpha^2 - \omega^2 + 2i\alpha\omega)\tau_s^2 = [(A - B)^2 - (A + B)](\alpha + i\omega)\tau_s,$$

from which it is seen that for $\omega \neq 0$, we have $\alpha\tau_s = \frac{1}{2}[(A - B)^2 - (A + B)]$ and $\omega = \sqrt{AB/\tau_s^2 - \alpha^2}$. This solution applies when $(A - B)^2 + 1 \leq 2(A + B)$; otherwise, $\omega = 0$ and

$$\alpha\tau_s = \frac{1}{2}[(A - B)^2 - (A + B) - (A - B)\sqrt{(A - B)^2 + 1 - 2(A + B)}],$$

which is positive (implying instability) throughout its region of validity. A stability boundary in the AB plane may be defined by setting the first expression for α equal to zero; that is, the combustion is intrinsically unstable unless $(A - B)^2 < A + B$. A stability diagram in the parameter space is shown in Figure 9.6.

In applying these results to the model of Section 9.1.5.4, it may be noted from equations (67) and (68) that A and A/B represent nondimensional activation energies for the interface and gas-phase processes, respectively. Figure 9.6 therefore shows that intrinsic instability occurs (even in the absence of pressure perturbations) when the nondimensional activation energies are sufficiently large. If $A < 1$ or if $A/B < 1$, then according to the criterion obtained, the combustion is intrinsically stable, but, for example, if $A > 1$, then there is a critical value of A/B above which instability occurs. Since equation (59) relies on an asymptotic expansion for large values of A/B, we may conclude that in a strict interpretation of the range of validity of the analysis, instability is to be anticipated whenever $A > 1$. In fact, the cold-boundary difficulty is not present in this model, a generalization of equation (59) may be developed for finite values of A/B (still within the context of a quasisteady gas phase), and the resulting stability boundary will not be expected to differ greatly from that shown in Figure 9.6 even for relatively small values of A/B; with suitable redefinitions of A and B, it may be shown that Figure 9.6 remains exactly correct. Thus Figure 9.6 is expected to provide at least a qualitatively reasonable approximation to the stability boundary for all values of A and A/B.

To apply the results to the gasless combustion problem introduced at the beginning of this section, we let $B = 1$ and find that intrinsic instability

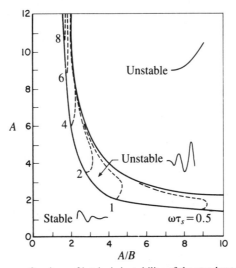

FIGURE 9.6. Diagram of regions of intrinsic instability of the steady-state combustion given by equation (60).

is predicted to occur if $A > 3$. As formulated, this gasless combustion problem possesses the cold-boundary difficulty, and the results are therefore restricted to large values of A. Since 3 may not be a sufficiently large number, apparently we should conclude that the steady-state solution to the adiabatic version of the problem of Section 7.4 is intrinsically unstable. The possibility of the occurrence of instabilities such as this in burning solids was known to Zel'dovich [61], and stability analyses for various models are available [76], [109], [110]. The mechanism of the instability encountered here can be understood physically.

Under the present conditions of negligible diffusion, flame propagation in the solid is associated with an excess enthalpy per unit area given by $\int_{-\infty}^{0} \rho_s c_{ps}(T - T_0)\,dx$ just ahead of the reaction sheet. This excess provides a local reservoir of heated reactant in which a flame may propagate at an increased velocity. If $\alpha_s \equiv \lambda_s/(\rho_s c_{ps})$ denotes the thermal diffusivity of the solid, then for the steady-state solution, the thickness of the heated layer of reactant is on the order of $x_1 = \alpha_s/v_1$, where v_1 is the steady-state flame speed. From the second equality in equation (77), we see that

$$v_1 = C \exp[-E_1/(2R^0 T_{i1})],$$

where

$$C \equiv \{2R^0 B_1 T_i^{\alpha_1 + 2}\lambda_s)/[\rho_s E_1 q_s(1 - Y_{1,0})^2]\}^{1/2} \approx \text{constant},$$

and where $T_{i1} = T_0 + q_s(1 - Y_{1,0})/c_{ps}$, the adiabatic flame temperature. If a flame now begins to propagate adiabatically in the heated reactant, then its velocity will be roughly $v_2 = C \exp[-E_1/(2R^0 T_{i2})]$, where

$$T_{i2} = T_{i1} + q_s(1 - Y_{1,0})/c_{ps} = T_0 + 2q_s(1 - Y_{1,0})/c_{ps},$$

a value considerably greater than the adiabatic flame temperature of the unheated reactant. Approximate conditions for the existence of this hot flame are reasonably well satisfied because its thickness, $x_2 = \alpha_s/v_2$, is much less than the thickness x_1 of the heated layer in which it propagates, as a consequence of the fact that v_1/v_2 is small for large Zel'dovich numbers. The hot flame may propagate until it consumes all the heated reactant; this occurs over a distance on the order of x_1 and, therefore, lasts for a time of about $t_2 = x_1/v_2$. When the hot flame encounters cold reactant, it is halted, and thereafter practically no propagation of the reaction front can occur until a new layer having thickness of order x_1 is heated. The transient heat-conduction time required for this heating is roughly $t_1 = x_1^2/\alpha_s$. After the heating is completed, an adiabatic flame may again be established, and the hot flame again may be initiated at its reaction front.

The mechanism of the instability that has been described here suggests that intrinsic instability of gasless combustion evolves to an inherently pulsating mode of propagation rather than to a sustained explosion or extinction. This conclusion has been substantiated theoretically by more detailed analyses [109], [111], and pulsating propagation has been observed experimentally for solids [112]–[114]. These results verify that during the rapid portions of pulsating propagation, flame temperatures in excess of the steady-state adiabatic flame temperature occur. Estimates of characteristics of pulsating propagation may be obtained from the model that has been described for the mechanism. The ratio of the duration of a pulse to the time between pulses is roughly

$$\frac{t_2}{t_1} = \frac{v_1}{v_2} = \exp\left[- \left(\frac{E_1}{2R^0}\right)\left(\frac{1}{T_{i1}} - \frac{1}{T_{i2}}\right)\right] \approx \exp\left[- \frac{E_1 q_s(1 - Y_{1,0})}{2R^0 c_{ps} T_{i1}^2}\right],$$

which is small when the Zel'dovich number is large. The average velocity of propagation is

$$x_1/(t_1 + t_2) = v_1/[1 + (t_2/t_1)],$$

which is less than the velocity v_1 of the steady flame. These results are in qualitative agreement with predictions of accurate theories. It can be imagined that for sufficiently large Zel'dovich numbers, additional, more-rapid flames might develop and follow the first hot flame; a tendency for this to occur has been found in numerical integrations [109] which show, at the larger values of the activation energy, an irregular series of pulses, with each series separated by quiet periods.

We have seen that there are interpretive difficulties associated with the use of activation-energy asymptotics for calculating limits of intrinsic instability in gasless combustion. Real condensed-phase processes might not be strictly Arrhenius; for example, there might be a critical temperature exceeding T_0, perhaps corresponding to a phase transition, below which the reaction effectively ceases. The cold-boundary difficulty then disappears, and a well-posed deflagration problem can then be defined without introducing asymptotic concepts. For a problem of this kind, it is to be expected that there will be a critical value of a parameter that measures the temperature sensitivity of the heat-release rate, such that the steady-state, one-dimensional combustion is stable for values of the parameter below the critical value, while the deflagration is inherently oscillatory at larger values. The critical value would represent a **bifurcation point**, a value at which a splitting into more than one possible solution occurs (for example, a steady solution and an oscillatory solution) and at which the character of the solution to the problem may change. If the reaction is of the Arrhenius form for temperatures above the switch-on temperature, then A is a suitable parameter for measuring the temperature sensitivity of the rate. Although stability analyses based on activation-energy asymptotics may not provide accurate values of A at the bifurcation point, they may nevertheless yield stability limits that are roughly correct. Numerical integrations [109] seem to indicate that for gasless combustion, $A \gtrsim 6$ typically may be a better criterion for intrinsic planar instability than the criterion $A > 3$, obtained above from the asymptotic analysis. Stability analyses in which equations (75) and (76) are employed with a δ-function heat-release rate but without $(\partial T/\partial x)_{x=0+} = 0$, so that inert heat conduction occurs for $x > 0$ as well, give $A > 2 + \sqrt{5}$ as the criterion for instability [111]; the resulting bifurcation value of A, approximately 4.24, is closer to the value inferred from the numerical integrations than is that obtained from the more formally correct asymptotic analysis.

If attention is restricted to the vicinity of the bifurcation point, then a nonlinear perturbation analysis can be developed for describing analytically the nature of the pulsating mode [111]. In effect, the difference between A and its bifurcation value is treated as a small parameter, say ϵ, and oscillatory solutions for temperature profiles are calculated as perturbations about the steady solution in the form of a power series in $\sqrt{\epsilon}$. The departure of the oscillation frequency from its value at bifurcation is expressed in the same type of series. The methods of analysis possess a qualitative similarity to those of the shock-instability analysis discussed at the end of the previous section. The results exhibit the same general behavior that was found from the numerical integrations [109] for conditions near bifurcation.

For solid propellants, the pulsating instability may be eliminated through stabilizing influences provided by energy input from the gas-phase flame. Even if the solid propellant burns by a mechanism predominantly involving condensed-phase reactions of large nondimensional activation energy, relatively small rates of heat feedback to the surface through processes that are not too strongly dependent on temperature may produce stability of the steady deflagration. At sufficiently low pressures, the gas-phase flame may be absent, and under these conditions pulsating deflagration of a homogeneous solid propellant is anticipated. This phenomenon has been observed at low pressures in strand-burner and rocket-motor tests of certain propellants; it has often been called *chuffing*, particularly in the limit in which the pulse duration is short compared with the time between pulses. Although the theoretical explanations that have been proposed for chuffing [115]–[117] appeal to concepts of homogeneous thermal explosions and usually introduce more than one chemical step, a mechanism like that just described may be relevant.

Heterogeneous solid propellants possess additional mechanisms with potentials for producing oscillatory burning. For example, certain metalized composite propellants have been observed to burn in the laboratory with identifiable ranges of frequencies of oscillation [118]; in this case, the mechanism may involve chemical interactions between the metal and the oxidizer. It was indicated in Section 9.1.5.5 that heterogeneities introduce at least local periodicities and that, in the presence of a mechanism for synchronizing the phases of the oscillations over the surface of the propellant, sustained coherent oscillations of the combustion will occur. A review is available that includes discussion of influences of preferred-frequency combustion in heterogeneous propellants [119].

The ideas of oscillatory burning that have been considered thus far in the present section have been restricted to one-dimensional, time-dependent phenomena. Spinning propagation of deflagrations in cylindrical samples of certain homogeneous solid combustibles has been observed experimentally [114]. A model for spinning propagation has been proposed [120], and a linear stability analysis has been worked out [121], admitting nonplanar perturbations in the model of [111] described previously. Axial propagation in a cylinder of radius a is considered, with an adiabaticity condition applied at the wall, $r = a$. An algebraic expression of the type of equation (78) is obtained; an additional nondimensional parameter appears—namely, $k_{mn} = \varphi_{mn} v_0 \tau_s / a$, where v_0 is the flame speed of the unperturbed wave and φ_{mn} is the mth root of the equation $dJ_n(\varphi)/d\varphi = 0$ for the Bessel function of order n. Since k_{mn} represents a nondimensional wave number of the disturbance in the transverse direction, the equation analogous to equation (78) becomes a dispersion relation, giving the complex frequency as a function of the wave number. The stability limit ($\alpha = 0$) may then be obtained in the form of a curve of k_{mn} as a function of A. The results show that the minimum value of A along this limit curve occurs at $k_{mn} = 1/2$ and has $A = 4$, which is below the value $2 + \sqrt{5}$ that was obtained for the planar instability. This result suggests that there may be materials for which $4 < A < 2 + \sqrt{5}$ that are incapable of experiencing pulsating combustion under any conditions but that can exhibit spinning combustion if the diameter of the cylinder is sufficiently large.* Since the smallest nonzero value of φ_{mn} is $\varphi_{11} \approx 1.84$, the smallest cylinder radius for the occurrence of this spinning propagation is roughly $a \approx 3.7 \, v_0 \tau_s$ if $A \approx 4$. In general, the permissible number of spin cells increases as the diameter increases. There is a geometrical analogy here to spinning detonations (Section 6.3.1), although the physical mechanisms involved of course are very different.

The inherent oscillations that have been addressed here in fact correspond to a limiting case of the diffusive-thermal intrinsic instabilities of flames, which will be introduced in Section 9.5. At large values of the Lewis number the character of the diffusive-thermal instability is that which has been described here. It will be found in Section 9.5 that additional types of instability behavior may occur if diffusion of chemical species is not negligible.

---

* F. J. Higuera and A. Liñán, "Stability of Solid-Propellant Combustion to Non-Planar Perturbations" in *Dynamics of Flames and Reactive Systems*, vol. 95 of *Progress in Astronautics and Aeronautics*, J. R. Bowen, N. Manson, A. K. Oppenheim and R. I. Soloukhin, eds., New York: American Institute of Aeronautics and Astronautics, 1984, 248–258, have shown that a similar phenomenon occurs for the model that led to equation (78). When transverse disturbances are admitted, the stability boundary in Figure 9.6 is changed to $\sqrt{A} = \sqrt{B} + 1/\sqrt{2}$, which approaches the boundary shown in Figure 9.6 for large values of A but gives $A = 1/2 (< 1)$ at large values of A/B.

9.3. OSCILLATORY BURNING IN LIQUID-PROPELLANT ROCKET MOTORS

Liquid-propellant rocket motors are no less susceptible to oscillatory burning than are solid-propellant motors. Rarely has a development program for a liquid-propellant rocket been completed without encountering a problem of combustion instability. The two major types of combustion instabilities for liquid-propellant motors are acoustic instabilities (with frequencies greater than roughly 10^3 cycles per second) and instabilities of lower frequencies that typically involve interactions between chamber processes and flow in the propellant feed lines. The former type has been called *screaming* and the latter *chugging*, although various other descriptive terms also have been applied. Screaming often is the more severe of the two in that, by locally enhancing rates of heat transfer, it can rapidly burn holes in propellant injectors or chamber walls. A thorough presentation of the subject of combustion instability in liquid-propellant rockets has been published in a volume of more than 600 pages [122]. Here we shall merely give a brief discussion first of screaming, then of chugging, and finally of instabilities at intermediate frequencies.

Acoustic instabilities in liquid-propellant rocket motors share many features in common with those of solid-propellant motors. Therefore, most of the materials in Section 9.1 is relevant—especially the discussions of oscillation modes and of damping mechanisms, although viscoelastic damping and solid-particle damping usually are absent for liquid propellants. Because of their fewer natural damping mechanisms, liquid-propellant motors more often need instability "fixes" like the baffles and liners mentioned at the end of Section 9.1.4.1. The former are solid surfaces extending into the flow to interfere with establishment of low-frequency acoustic modes and to enhance attenuation, and the latter involve perforated sheets on the chamber walls, designed according to the principles of the Helmholtz resonator (described in the following section) to increase acoustic damping by dissipation in the oscillatory flow through the orifices. The main difference between the two types of motors lies in the character of the amplification mechanisms. Combustion is distributed more or less homogeneously throughout the chamber of a liquid-propellant motor and generally involves droplet combustion, at least to some extent. Diffusion flames also may be established elsewhere than around droplets, for example, between fuel and oxidizer streams, and responses of these diffusion flames to acoustic oscillations may contribute to amplification. Conditions might occur under which mixing is rapid and combustion slower so that homogeneous gas-phase chemical rate processes provide some amplification. Equation (51) is useful for calculating contributions of homogeneous processes to amplification.

Analysis of acoustic amplification by combustion processes in liquid-propellant rockets is more diffcult than that for solid-propellant motors because the direct use of admittances through approximations of planar burning surfaces seldom can be justified. Therefore, time-lag approximations of the type discussed in Section 9.1.5.3 have been applied more often for liquid-propellant motors. Time-lag hypotheses have become relatively sophisticated for acoustic instabilities in liquid-propellant rockets [20], [122]. Numerical integrations of nonlinear differential equations, with chemical kinetics and diffusion processes treated in overall quasisteady approximations, also have been applied [122], [123]. A few analyses have been pursued concerning time-dependent vaporization and diffusion-flame combustion in which linearizations about known quasisteady solutions were developed with the objective of ultimately being able to calculate amplitude and phase relationships for use in an amplification expression like equation (51) [94], [122], [124]–[127].* The three specific diffusion-flame problems analyzed were the stagnation-point boundary layer on a vaporizing fuel [126], the laminar boundary layer on a flat plate of vaporizing fuel [94], [125], and the laminar jet diffusion flame [124], [127]. Numerical estimates suggest that the last of these three problems, interpreted, for example, as a model for combustion in the wake region of a vaporizing droplet, possesses the greatest potential relevance to acoustic instability in liquid-propellant rockets [129]. However, because of the complexity of the problem, the application to instability calculations has not been carried beyond the point of drawing rough, general inferences about aspects of scaling [122]. Further research on problems like these could improve our fundamental understanding of amplification mechanisms for combustion instabilities in liquid-propellant motors.

Chugging is a system instability having frequencies less than roughly 10^2 cycles per second and involving oscillations in the propellant feed lines and injection systems as well as in the combustion chamber. These frequencies are generally sufficiently low compared with the natural acoustic vibrational frequencies of the chamber that the pressure can be considered to be spatially uniform throughout the chamber. Consequently, the oscillations depend in a less-detailed manner on the nature of the processes occurring in the chamber, and ordinary linear differential equations with time lags are well suited for describing the instability. These equations may involve a number of time lags, which correspond to delays in the feed system, in the injection system, and in various processes occurring in the combustion chamber. Crocco [20]

---

* Amplification expressions in general are not as simple as equation (51) for two-phase flows in rocket motors [128]. Equilibrium and frozen sound speeds with respect to droplet behavior may be identified (compare Sections 9.1.4.6 and 4.2.6), and a mass-source term may arise in equation (51) in the development of a conservation equation for the acoustic energy in the gas. From the viewpoint of equation (14), the quantity Φ must include amplification effects that arise from both energy sources and mass sources. Acoustic theories may be derived from the conservation equations of Chapter 11 for the purpose of analyzing acoustic amplification in spray combustion.

has made a considerable amount of theoretical and experimental progress in deducing properties of the time lags of the conversion processes in the combustion chamber. Theoretical analyses of low-frequency oscillations involve determining the response of one section of the rocket system to vibrations in another section, delineating regions of self-excited instabilities in the entire rocket system, and developing feedback servomechanism techniques for stabilizing the system. Tsien [130] has given clear illustrations of these analyses. A number of reviews are available [20], [122], [128], [129], [131]–[134].

There are mechanisms for oscillations at intermediate frequencies, higher than chugging but lower than screaming [122]. For example, in propellant motors, the injection rates of fuel and oxidizer may differ in their sensitivities to chamber pressure, so that a change in the pressure at the injector produces a change in the injected fuel-oxidizer mixture ratio. The mixture-ratio variations are convected through the chamber and, upon combustion, result in variations in the temperature of the burnt gas which, in turn, may modify the nozzle flow, producing pressure perturbations that propagate back upstream acoustically to influence the pressure at the injector face. It is easy to imagine that a sustained oscillation in a longitudinal mode may be established in this manner, and since half of the process is acoustic and the other half convective (which is slower), the frequency will be less than an acoustic frequency of the chamber. A similar mechanism can readily be envisioned for an end-burning solid-propellant motor; if the combustion is quasisteady and the consequent nonisentropic variations in burnt-gas temperature with pressure, discussed in Section 9.1.5.2, do not decay before reaching the nozzle, then a sustained oscillation can develop, without variations in the mixture ratio. A closely related mechanism also can be involved in *rumble* instabilities in ramjet combustors.* A more detailed discussion of a phenomenon of this type was given in Section 6.3.1, from which it may be seen that one-dimensional instabilities of the planar detonation structure involve qualitatively the same type of process. The term *entropy wave*, which often has been applied to instabilities of this type in motors, fails to distinguish mechanisms involving mixture-ratio variations from those that do not, and the term is therefore insufficiently descriptive. Convective-acoustic instabilities of this general kind also could involve the convection of vortices with axes transverse to the flow direction or of swirls having longitudinal axes, whose generation may be enhanced by pressure waves in suitable geometrical configurations. Thus vorticity may play a role in intermediate-frequency oscillations under appropriate conditions.

---

* G. E. Abouseif, J. A. Keklak, and T. Y. Toong, *CST* **36**, 83 (1984).

9.4. SYSTEM INSTABILITIES IN COMBUSTION EQUIPMENT

Chugging in liquid-propellant rockets is only one of many examples of system instabilities in combustion devices [135], [136]. As may be seen from the contribution of Putnam to [137], a considerable amount of research has been performed on mechanisms of these instabilities. Interactions of processes occurring in intakes and exhausts with those occurring in the combustion chamber typically are involved, and it may or may not be necessary to consider acoustic wave propagation in one or more of these components in theoretical analyses [138]–[142]. Here we shall not address problems involving acoustic wave propagation; we shall restrict our attention to bulk modes, in which spatial variations of the pressure in the combustion chamber are negligible.

The simplest example of an oscillation in a bulk mode is the classical Helmholtz resonator. Consider a gas-filled chamber of volume $\mathscr{V}$, into which is inserted a tube of length l and of cross-sectional area $\mathscr{A}$ that is open to the atmosphere. Assume that the atmospheric pressure remains constant at the outer end of the tube at a value $\bar{p}$, and consider slowly varying processes for which the pressure p at the inner end of the tube is equal to the spatially uniform pressure within the chamber. With friction neglected, a force balance on the gas in the tube is $\mathscr{A}(p - \bar{p}) = (\bar{\rho}\mathscr{A}l)\,dv/dt$, where $\bar{\rho}$ is the average gas density in the tube and v is the average outward velocity of gas in the tube. Conservation of mass for the gas of density ρ in the chamber is $\mathscr{V}\,d\rho/dt = \bar{\rho}\mathscr{A}v$. Applying equation (3) to the gas in the chamber, we have $d\rho/dt = (dp/dt)/a^2$. By substituting this relation into the one preceding it, differentiating, and then substituting the result into the first expression, we find that the pressure history is described by

$$(l\mathscr{V}/\mathscr{A})\,d^2p/dt^2 = a^2(p - \bar{p}). \tag{80}$$

Equation (80) possesses solutions that are oscillatory in time, with frequency

$$\omega_H = a/\sqrt{l\,\mathscr{V}/\mathscr{A}}. \tag{81}$$

The derivation of equation (81) is invalid unless the resulting Helmholtz frequency ω_H is small compared with the lowest natural frequency of acoustic waves in the chamber, since otherwise an acoustic field with a nonuniform p would have to be considered; a related restriction limits the length l of the exit tube. If l is comparable with a characteristic length of the chamber, then a representative acoustic frequency is a/l, and the ratio $\omega_H/(a/l)$ is the square root of the ratio of the volume of the tube to that of the chamber, which is small for small tubes in large chambers—the configuration envisaged in the Helmholtz model.

Since frictional damping always is present but has been neglected in the derivation of equation (80), a source of excitation, such as a suitably phased rate of heat release by combustion in the chamber [see equation (51)], must be available if sustained oscillations are to occur in the Helmholtz mode; growth of oscillations would need an excitation source even if there were no damping. Equation (81), which is directly useful in design of the acoustic liners defined in the previous section, also often provides a reasonable approximation for frequencies of bulk modes in combustion devices even when the mean velocity of flow out of the chamber exceeds the oscillation velocity.

An example of bulk modes in solid-propellant rockets is afforded by the low-frequency, or L^*, instability [7]. A characteristic length of importance in rocket design is the ratio of the gas volume in the chamber to the throat area of the nozzle; this ratio often is denoted by L^*, and its ratio to a characteristic exhaust velocity provides an estimate of the residence time of a fluid element in the gas phase inside the chamber. A mass balance for the gas inside a rocket chamber with a choked nozzle is

$$Km = pC_D + d(\rho L^*)/dt, \tag{82}$$

where m is mass burning rate per unit area for the propellant, K is the ratio of the burning area to the throat area, and C_D is a discharge coefficient for the nozzle, defined as the ratio of the mass flow rate through the nozzle to the product of the throat area with the chamber pressure. Since in rocket designs the fractional rate of increase of L^* is small compared with the rate of development of the instability, by subtracting the steady-state mass balance from equation (82), and by employing equation (3), we find that

$$\omega_L^{-1}\, dp/dt = (\bar{p}/\bar{m})(m - \bar{m}) - (p - \bar{p}), \tag{83}$$

where a characteristic frequency for the instability is

$$\omega_L = a^2 C_D/L^*. \tag{84}$$

Since equation (84) shows that ω_L decreases as L^* increases, it is reasonable to neglect spatial variations of the pressure within the chamber, which are produced by propagation of acoustic waves, if L^* is sufficiently large.

The response of the burning rate to pressure variations is needed for use in equation (83). If the mechanism of Section 9.1.5.4 is considered, then either by use of Laplace transforms or (more simply) by seeking solutions proportional to $e^{(\alpha + i\omega)t}$, it is readily found from equation (83) that

$$1 + (\alpha + i\omega)/\omega_L = nBs/[s^2 - (1 + A - B)s + A], \tag{85}$$

where A and B are given by equations (67) and (68) and s by equation (79) [compare equation (66)]. Instability occurs when $\alpha > 0$ in the solution to equation (85). The condition for neutral stability—that is, the stability limit—is defined by putting $\alpha = 0$ in equation (85). In view of equation (79), it is seen that $\omega\tau_s$ is a suitable nondimensional frequency, and therefore the neutral stability condition provides a relationship among the dimensionless parameters n, A, B, and $\omega_L \tau_s$. Thus it may be expected that for a given propellant (that is, for given values of n, A, and B), there is a critical value of $\omega_L \tau_s$ for the occurrence of instability. In view of equation (55), we may therefore write

$$(a^2 C_D/L^*)\rho_s \lambda_s/(c_{ps}\overline{m}^2) = f(n, A, B) \tag{86}$$

as an equation for the stability limit, where the function f is to be obtained from equation (85). It is found that L^* instability occurs when $\omega_L \tau_s$ is above this critical value—that is, when $\omega_L^{-1} \, dp/dt$ becomes sufficiently small compared with $p - \bar{p}$ in equation (83)—which corresponds to a chamber with a low capacity, a relatively short gas-phase residence time. Since $\overline{m}$ is proportional to $\bar{p}^n$ according to equation (59), we see from equation (86) that for a given propellant $L^*\bar{p}^{2n}$ assumes a critical value at the stability limit, and the instability appears when this product is less than its critical value. This prediction is in accord with experimental observations on L^* instability [143]. The two characteristic times involved in the mechanism are the gas-phase residence time in the chamber and the condensed-phase, heat-conduction time in the burning solid propellant.

There have been many studies of L^* instability that support various inferences of the analysis outlined above [117], [143]–[147]. Theoretical investigations also have addressed questions of influences of spatial gas-phase temperature variations [148] and of solid-phase heterogeneity [149] on the instability. If $L^*\bar{p}^{2n}$ is progressively decreased below the critical value for instability, then it is observed that the nearly sinusoidal low-frequency instability progressively evolves into a nonlinear phenomenon having the character of chuffing, discussed in Section 9.2. The mechanism involved in this chuffing process need not necessarily be the same as that of the corresponding intrinsic instability since an essential role may be played by the fact that the ratio of the chamber residence time to the solid heat-conduction time is small. In other words, a propellant that does not exhibit the non-sinusoidal oscillations while burning in the open may experience them in a suitable chamber. Therefore, the term *chuffing* as used in practice may apply to different fundamental phenomena.

9.5. HYDRODYNAMIC AND DIFFUSIVE INSTABILITIES IN PREMIXED FLAMES

9.5.1. Formulation through asymptotic methods

We now restrict our attention to premixed gaseous combustibles and address the question of the intrinsic stability of their deflagrative processes. The topic is a large one that has undergone extensive advancement in recent years. Here we shall only be able to touch on some of the highlights. The reader must consult reviews for more thorough presentations [137], [150]–[153].

Most of the early analyses of the stability of planar deflagrations treated only one-dimensional perturbations [154]–[159]. Some of the results indicated unconditional stability in adiabatic propagation for the classes of flames investigated [154], [158], [159], while others identified regions of instability, the limits of which depend on the thermodynamic, transport and chemical-kinetic properties of the system [155]–[157].* At the time that these investigations were performed, an objective was to equate stability limits with flammability limits of the combustible mixture. Attempts to correlate flammability limits with stability limits have been unsuccessful, and the analysis of Section 9.2 suggests that an instability is likely to evolve to an oscillatory propagation rather than to extinction. Although an oscillatory mode might turn out to be more susceptible to extinction by heat loss, limits of flammability under perfectly adiabatic conditions are not anticipated to exist; fundamentally, a direct correspondence between stability and flammability limits is not to be expected. The results of Section 9.2 suggest that even for gaseous reactants, oscillatory propagation might occur if the thermal diffusivity is sufficiently greater than the diffusivity of the limiting reactant (the stoichiometrically deficient one whose concentration vanishes at equilibrium). Asymptotic analyses have demonstrated this to be true [160], [161], and numerical integrations with real kinetics have encountered the same phenomenon under certain conditions [162]. Upstream heat loss to the flame holder of a flat-flame burner can broaden the range of conditions over which the oscillations occur; analyses of pulsations of these "burner-stabilized" flames have been performed by both numerical and asymptotic methods [163]–[169]. In particular, a nonlinear ordinary differential equation with a time delay† may be derived by asymptotic methods for describing fully nonlinear pulsations [168]; a sawtooth-type history of the reaction-zone location is a typical result [164], [168].

Here we shall be concerned with the stability of planar deflagrations to non-one-dimensional perturbations. We shall see that a richer variety

---

* Additional studies with findings of unconditional stability are those of J. M. Richardson, *4th Symp.* (1953), 182–189, who addressed nonplanar disturbances as well, and of G. I. Barenblatt and Y. B. Zel'dovich, *Prikl. Mat. Mekh.* **21**, 856 (1957) and Y. B. Zel'dovich and G. I. Barenblatt, *C & F* **3**, 61 (1959).

† The equation is of a type that has been encountered in other contexts and may be written in nondimensional variables as $adx/dt = e^{-x(t-1)} - 1$, where $x(t-1)$ signifies the function $x(t)$ evaluated at $t-1$. The constant a, of order unity, is $e^{\xi}/(\beta\xi)$, where ξ represents the nondimensional distance of equation (5-18) between the flame holder and the steady-state reaction-sheet position and β is the Zel'dovich number appearing in equation (5-66). The delay is the time for a temperature perturbation to travel from the reaction sheet to the flame holder and back. The steady-state solution $x = 0$ is unstable if $a < 2/\pi$; the oscillations tend to be sinusoidal if a is near $2/\pi$ and sawtooth at small values of a. For more recent, continuing work on this problem see, for example, A. C. McIntosh and J. F. Clarke, *CST* **38**, 161 (1984).

of behaviors is encountered. Attention will be restricted to a one-step, two-reactant process of large nondimensional activation energy. Departures from one-step kinetics can produce significant influences on flame stability even if the normal flame speed can be described well by a one-step approximation. For example, it may readily be imagined that in the presence of a nonplanar disturbance, reaction intermediaries may diffuse laterally from one part of the flame to another and thereby affect the response of the flame to the disturbance. Since research on this topic has not yet progressed to a point at which clear general conclusions can be drawn, we shall not discuss influences of reaction mechanisms on stability. However, one review article is available that summarizes current knowledge on the subject [153].

At large values of the Zel'dovich number, the chemical reaction is confined to a thin sheet in the flow. For all purposes except the analysis of the sheet structure, the sheet may be treated as a surface—for example, $G(\mathbf{x}, t) = 0$—which in terms of the Cartesian coordinates (x, y, z) may be written locally as $x = F(y, z, t)$ if the x coordinate is not parallel to the sheet in its local orientation. When analyses are pursued in outer-scale variables, it is convenient to work in a coordinate system that moves with the sheet. For the undisturbed flow, let the x coordinate be normal to the planar flame, with the unburnt gas extending to $x = -\infty$ and the burnt gas to $x = +\infty$. Employ the steady, adiabatic, laminar flame speed, measured in the burnt gas, $v_\infty \equiv \rho_0 v_0 / \rho_\infty$ with v_0 given by equation (5-78), and the thermal diffusivity evaluated at the adiabatic flame temperature, $\alpha_\infty \equiv \lambda_\infty / (\rho_\infty c_{p,\infty})$, for defining nondimensional space and time variables as $\xi = x v_\infty / \alpha_\infty - f$, $\eta = y v_\infty / \alpha_\infty$, $\zeta = z v_\infty / \alpha_\infty$ and $\tau = t v_\infty^2 / \alpha_\infty$, where $f(y, z, t) = F(y, z, t) v_\infty / \alpha_\infty$. Under the assumption that the reaction sheet is smooth and nowhere parallel to the x axis, equation (1-1), through a transformation of variables, can be written as

$$\partial R / \partial \tau + \partial (Ru) / \partial \xi + \mathbf{V}_\perp \cdot (R \mathbf{v}_\perp) = 0, \tag{87}$$

where $R = \rho / \rho_\infty$, $\mathbf{V}_\perp$ is a two-dimensional gradient having dimensionless components $(\partial / \partial \eta, \partial / \partial \zeta)$, $\mathbf{v}_\perp$ is a transverse velocity vector composed of the velocity components in the y and z directions made nondimensional through division by v_∞, and the longitudinal mass flux relative to the reaction sheet is

$$\rho u v_\infty = \rho (v_1 - \partial F / \partial t - \mathbf{v} \cdot \nabla F), \tag{88}$$

in which v_1 is the x component of the velocity vector $\mathbf{v}$, and ∇F has no longitudinal component since by definition F is not a function of x.

The coordinate transformation also enables a nondimensional version of the equation for momentum conservation to be written from equations (1-2) and (1-5). For this purpose, Prandtl numbers based on the first and a modified second coefficient of viscosity, respectively, may be defined as $\mathrm{Pr}_1 = \mu_\infty c_{p,\infty} / \lambda_\infty$ and $\mathrm{Pr}_2 = (\frac{1}{3}\mu_\infty + \kappa_\infty) c_{p,\infty} / \lambda_\infty$. To allow for variable viscosities, the nondimensional functions $\mu_1 = \mu / \mu_\infty$ and $\mu_2 = (\frac{1}{3}\mu + \kappa)/(\frac{1}{3}\mu_\infty + \kappa_\infty)$ may be introduced. An appropriate nondimensional measure of the pressure is $P = (p - p_\infty)/(\rho_\infty v_\infty^2)$, where p_∞ is the downstream pressure of the steady, adiabatic deflagration, and a nondimensional body force per unit mass may be expressed as $\mathbf{F} = \sum_{i=1}^N Y_i \mathbf{f}_i \alpha_\infty/v_\infty^3$, the longitudinal ($x$) and transverse components of which will be written as $F_\parallel$ and $\mathbf{F}_\perp$. The longitudinal and transverse components of momentum conservation then become, respectively,

$$R\left(\frac{\partial}{\partial\tau} + u\frac{\partial}{\partial\xi} + \mathbf{v}_\perp\cdot\mathbf{V}_\perp\right)\left(u + \frac{\partial f}{\partial\tau} + \mathbf{v}_\perp\cdot\mathbf{V}_\perp f\right)$$

$$= -\frac{\partial P}{\partial\xi} + RF_\parallel + \mu_1\mathrm{Pr}_1\Delta\left(u + \frac{\partial f}{\partial\tau} + \mathbf{v}_\perp\cdot\mathbf{V}_\perp f\right)$$

$$+ \mathrm{Pr}_1 G_\parallel + \mathrm{Pr}_2\frac{\partial}{\partial\xi}\left[\mu_2\left(\frac{\partial u}{\partial\xi} + \mathbf{V}_\perp\cdot\mathbf{v}_\perp\right)\right] \tag{89}$$

and

$$R\left(\frac{\partial\mathbf{v}_\perp}{\partial\tau} + u\frac{\partial\mathbf{v}_\perp}{\partial\xi} + \mathbf{v}_\perp\cdot\mathbf{V}_\perp\mathbf{v}_\perp\right)$$

$$= -\mathbf{V}_\perp P + (\mathbf{V}_\perp f)\frac{\partial P}{\partial\xi} + RF_\perp + \mu_1\mathrm{Pr}_1\Delta\mathbf{v}_\perp + \mathrm{Pr}_1 G_\perp$$

$$+ \mathrm{Pr}_2\left[\mathbf{V}_\perp - (\mathbf{V}_\perp f)\frac{\partial}{\partial\xi}\right]\left[\mu_2\left(\frac{\partial u}{\partial\xi} + \mathbf{V}_\perp\cdot\mathbf{v}_\perp\right)\right], \tag{90}$$

where the Laplacian operator in the transformed coordinate system is

$$\Delta = (1 + |\mathbf{V}_\perp f|^2)\frac{\partial^2}{\partial\xi^2} + \mathbf{V}_\perp\cdot\mathbf{V}_\perp - 2(\mathbf{V}_\perp f)\cdot\mathbf{V}_\perp\frac{\partial}{\partial\xi} - (\mathbf{V}_\perp\cdot\mathbf{V}_\perp f)\frac{\partial}{\partial\xi}, \tag{91}$$

and $G_\parallel$ and $G_\perp$ are quantities that are proportional to derivatives of μ_1. Most of the analyses that have been performed treat μ and κ as constant, so that $\mu_1 = 1, \mu_2 = 1, G_\parallel = 0$ and $G_\perp = 0$, and it is often assumed that $\kappa = 0$, so that $\mathrm{Pr}_2 = \frac{1}{3}\mathrm{Pr}_1$. Eventually, we shall adopt these assumptions; the more-general forms exhibited here are of interest for discussing influences of variable transport properties and for pursuing further investigations.

The equation for conservation of energy will be written under the assumptions that the Mach number is small and that the work done by body forces is negligible, both of which are accurate approximations for the deflagration problems considered. In addition, we shall neglect the Dufour effect (and later the Soret effect), so that by use of equation (1-6) in equation (3-72), in which $h = \sum_{i=1}^N h_i Y_i$, we obtain

$$\rho\frac{\partial h}{\partial t} + \rho\mathbf{v}\cdot\nabla h = \frac{\partial p}{\partial t} + \nabla\cdot(\lambda\nabla T) - \nabla\cdot\left(\rho\sum_{i=1}^N h_i Y_i \mathbf{V}_i\right) - \nabla\cdot\mathbf{q}_R. \tag{92}$$

It is not obvious that the cross-transport effects are unimportant in deflagration stability because it will be found that changes of diffusion coefficients or Lewis numbers by amounts on the order of 10% can be significant, and Soret effects typically may reach 10% of those of the ordinary diffusion processes. Analysis has shown that there are conditions under which the Soret effect may even change the character of the instability [170], but for usual flames, including mixtures containing hydrogen (for which the effect might be anticipated to be greatest), its only influence appears to be a relatively small, quantitative modification to stability boundaries [153]. Therefore, in demonstrating the results of interest here, we can take advantage of the appreciable simplification afforded by neglect of cross-transport effects. In addition, to circumvent complexities associated with the multicomponent diffusion equation, we shall neglect body-force and pressure-gradient diffusion (likely good approximations) and shall employ the diffusion equation that would apply if the species $i = N$ were present in great excess, namely, $Y_i V_i = -D_i \nabla Y_i$, $i = 1, \ldots, N - 1$, where D_i is the binary diffusion coefficient for the species pair i and N [compare equation (E-25)]. With the nondimensionalizations $\theta = T/T_\infty$, $H_i = h_i/(c_{p,\infty} T_\infty)$, $\lambda_1 = \lambda/\lambda_\infty$, $\delta_i = \rho D_i/(\rho_\infty D_{i,\infty})$, and $L = \lambda_\infty (\nabla \cdot \mathbf{q}_R)/[(\rho_\infty v_\infty c_{p,\infty})^2 T_\infty]$, in the transformed coordinates, equation (92) then becomes

$$R\left(\frac{\partial}{\partial \tau} + u\frac{\partial}{\partial \xi} + \mathbf{v}_\perp \cdot \nabla_\perp\right)\left(\sum_{i=1}^{N} H_i Y_i\right)$$

$$= \lambda_1 \Delta\theta + \sum_{i=1}^{N-1} \delta_i(H_i - H_N)\Delta Y_i/\text{Le}_i$$

$$+ \frac{\partial \lambda_1}{\partial \xi}\left[(1 + |\nabla_\perp f|^2)\frac{\partial\theta}{\partial\xi} - (\nabla_\perp f)\cdot(\nabla_\perp\theta)\right]$$

$$+ (\nabla_\perp \lambda_1)\cdot\left[\nabla_\perp\theta - (\nabla_\perp f)\frac{\partial\theta}{\partial\xi}\right]$$

$$+ \sum_{i=1}^{N-1}\frac{1}{\text{Le}_i}\left\{\left[(1 + |\nabla_\perp f|^2)\frac{\partial Y_i}{\partial\xi} - (\nabla_\perp f)\cdot(\nabla_\perp Y_i)\right]\frac{\partial}{\partial\xi}\left[\delta_i(H_i - H_N)\right]\right.$$

$$\left. + \left[\nabla_\perp Y_i - (\nabla_\perp f)\frac{\partial Y_i}{\partial\xi}\right]\cdot\nabla_\perp\left[\delta_i(H_i - H_N)\right]\right\} - L, \tag{93}$$

where the Lewis numbers are $\text{Le}_i = \lambda_\infty/(\rho_\infty c_{p,\infty} D_{i,\infty})$. The term $\partial p/\partial t$ has disappeared because it is of the order of the square of the Mach number if changes in P are assumed to be of order unity. The somewhat lengthy form of equation (93) is a consequence of including variable transport coefficients as well as variable heat capacities, which appear through the H_i with h_i given by equation (1-11), in which it will be convenient to let the reference temperature T^0 be T_∞.

The equations for conservation of chemical species contain the reaction-rate terms that will be nonnegligible only in a narrow region about $\xi = 0$. For the one-step, two-reactant chemistry to be considered, we let $i = 1, 2$ identify the reactants and consider $i = 1$ to be the limiting or deficient reactant, consistent with equation (5-78). If specific heats vary, $\lambda_1 \neq 1$ or $\delta_i \neq 1$, then equation (5-78) is not quite right, but a revised asymptotic analysis of the steady, adiabatic flame would give the correct formula. For near-stoichiometric conditions, an interesting phenomenon may occur in the perturbation about the steady, planar solution; instead of $i = 1$, the species $i = 2$ may become the limiting reactant at various times and positions along the flame sheet, and diffusion flames emanating from the stoichiometric points on the flame sheet may then be present in the downstream region. An analysis of this effect has been developed [171], but we shall not discuss it here and shall assume that attention is restricted to conditions under which it does not occur. For the reaction $v'_{1,1}\mathfrak{M}_1 + v'_{2,1}\mathfrak{M}_2 \to$ products, equation (1-8), modified as suggested by equation (B-16), implies that the ratio $w_i/(W_i v''_{i,1})$ for any product is equal to $-w_1/(W_1 v'_{1,1}) = -w_2/(W_2 v'_{2,1}) = (\rho_\infty v_\infty^2/\alpha_\infty)w/(W_1 v'_{1,1})$, where a dimensionless reaction-rate function is

$$w = \frac{\alpha_\infty W_1 v'_{1,1} B_1 T^{\alpha_1}}{\rho_\infty v_\infty^2} e^{-E_1/(R^0 T)} \left(\frac{\rho Y_1}{W_1}\right)^{n_1}\left(\frac{\rho Y_2}{W_2}\right)^{n_2}. \tag{94}$$

With the preceding assumptions and in the transformed coordinates, equation (1-4) then becomes

$$R\left(\frac{\partial Y_i}{\partial \tau} + u\frac{\partial Y_i}{\partial \xi} + \mathbf{v}_\perp \cdot \mathbf{\nabla}_\perp Y_i\right)$$

$$= \frac{\delta_i}{\text{Le}_i}\Delta Y_i + v_i w$$

$$+ \frac{1}{\text{Le}_i}\left\{\frac{\partial \delta_i}{\partial \xi}\left[(1 + |\mathbf{\nabla}_\perp f|^2)\frac{\partial Y_i}{\partial \xi} - (\mathbf{\nabla}_\perp f)\cdot(\mathbf{\nabla}_\perp Y_i)\right]\right.$$

$$\left. + (\mathbf{\nabla}_\perp \delta_i)\cdot\left[\mathbf{\nabla}_\perp Y_i - (\mathbf{\nabla}_\perp f)\frac{\partial Y_i}{\partial \xi}\right]\right\}, \qquad i = 1, \ldots, N - 1, \tag{95}$$

where $v_1 = -1$, $v_2 = -W_2 v'_{2,1}/(W_1 v'_{1,1})$, $v_i = W_i v''_{i,1}/(W_1 v'_{1,1})$ for reaction products and $v_i = 0$ for nonreactive species. Equations (87), (89), (90), (93), and (95) are suitable conservation equations for describing the evolutions of u, P, $\mathbf{v}_\perp$, θ, and Y_i, with R related to these quantities through the equation of state, $R\theta \sum_{i=1}^N (Y_i/W_i) = \sum_{i=1}^N (Y_{i,\infty}/W_i)$.

Appropriate integrations of the differential equations across the reaction sheet serve to eliminate the rate term w from the system, thereby providing a set of reaction-free equations with reaction-sheet jump conditions for analyzing the dynamics of the flame. The reaction-zone equations to be integrated may be developed formally by stretching the ξ coordinate about $\xi = 0$ by a factor β (that is, treating $\beta\xi$ as being of order unity) and then expanding the resulting equations in the small parameter β^{-1}. The relevant Zel'dovich number is

$$\beta = E_1 Q/(R^0 T_\infty), \tag{96}$$

where the nondimensional heat released per unit mass of mixture has been defined as $Q = Y_{1,0}(h_1^0 - \sum_{i=2}^{N} v_i h_i^0)/(c_{p,\infty} T_\infty)$, in which $Y_{1,0}$ is the initial mass fraction of the deficient reactant and the h_i^0 are the standard enthalpies of formation that appear in equation (1-11) and that are to be evaluated at the reference temperature T_∞. Since the transverse coordinates and the time variable are not stretched, all the jump conditions except those of equation (95) are obtained relatively easily, in the formulation that has been given here, simply by integrating the equations across the reaction sheet. In this way it is found from equation (87) that Ru is continuous, from suitable combinations of equations (89) and (90) that $P - (\text{Pr}_1 + \text{Pr}_2)\partial u/\partial \xi$ and $(\mathbf{V}_\perp f)[\partial u/\partial \xi + (\mathbf{V}_\perp f) \cdot \partial \mathbf{v}_\perp/\partial \xi] + \partial \mathbf{v}_\perp/\partial \xi$ are continuous, and from equation (93) that $\partial\theta/\partial\xi + \sum_{i=1}^{N-1} (H_{i,\infty} - H_{N,\infty})(\partial Y_i/\partial\xi)/\text{Le}_i$ is continuous. These results require further that u, $\mathbf{v}_\perp$, R, θ, and Y_i be continuous in the first approximation and rely on the assumptions that μ_1, μ_2, λ_1, and δ_i are continuous. The conditions obtained from equations (89), (90), and (93) in effect involve only the longitudinal components of the stress tensor and of the heat-flux vector. The first of the conditions quoted from equations (89) and (90) expresses a pressure discontinuity that balances the discontinuity in the viscous stress tensor, and the second states that the streamwise gradients of the components of velocity tangent to the sheet are continuous.

When equation (95) is integrated across the sheet by use of equation (91) it is found that the discontinuity in $\partial Y_i/\partial\xi$ is

$$-(1 + |\mathbf{V}_\perp f|^2)^{-1} \text{Le}_i v_i \int w \, d\xi,$$

but this result is not sufficient because the integral is not readily expressed in terms of other variables. It does imply that (since $v_1 = -1$)$(\partial Y_i/\partial\xi)[\text{Le}_1/(\text{Le}_i v_i)] + \partial Y_1/\partial\xi$ is continuous [as may be derived directly from equation (95) by dividing by v_i and adding to the equation for Y_1 to eliminate w prior to integration], and this result may be used in the jump condition obtained from energy conservation to show that

$$\partial\theta/\partial\xi - \left[\sum_{i=1}^{N-1} v_i(H_{i,\infty} - H_{N,\infty})\right](\partial Y_1/\partial\xi)/\text{Le}_1$$

is continuous, which (since $\sum_{i=1}^{N} v_i = 0$ by mass conservation in the chemical reaction) may be expressed in terms of the nondimensional heat release that appears in equation (96) as the requirement that $\partial\theta/\partial\xi + (\partial Y_1/\partial\xi)Q/(Y_{1,0}\text{Le}_1)$ be continuous. However, a more thorough asymptotic analysis of the reaction-zone structure is needed for deriving a jump condition for $\partial Y_1/\partial \xi$.

The necessary application of activation-energy asymptotics parallels that given in Sections 9.2, 5.3.6, and 8.2.1 and has been developed for both one-reactant [172], [173] and two-reactant [174], [175] systems. For most problems (for example, for stability analyses), gradients of the total enthalpy and of the mixture ratio of the reactants are negligible in the first approximation within the reaction sheet, so that in the scaled variables appropriate to the reaction zone, both $\theta + Y_1 Q/(Y_{1,0} \text{Le}_1)$ and $Y_1 + Y_2 \text{Le}_1/(v_2 \text{Le}_2)$ are constant. Evaluation of these constants is aided by the further result that $Y_1 = 0$ (and hence $\partial Y_1/\partial \xi = 0$) downstream from the reaction zone, at least in the first approximations, so that for the reaction-zone analysis, the expressions

$$\theta = \theta_f - Y_1 Q/(Y_{1,0} \text{Le}_1), \qquad Y_2 = Y_{2,f} - Y_1 v_2 \text{Le}_2/\text{Le}_1 \qquad (97)$$

may be employed, with θ_f and $Y_{2,f}$ constants representing the local, instantaneous values just downstream from the reaction sheet. Because of the strong dependence of the reaction rate on temperature, in the formulation the order of magnitude of $|1 - \theta_f|$ cannot be greater than β^{-1}; also, to achieve the greatest flexibility in the analysis, $Y_{2,f}$ may be treated as being of order β^{-1}. The differential equation describing the reaction-zone structure in the first approximation, obtained from equations (94) and (95) (for $i = 1$) through appropriate scalings and stretching, is

$$\frac{1}{\text{Le}_1} (1 + |\nabla_\perp f|^2) \frac{\partial^2 Y_1}{\partial \xi^2} = \frac{\beta^{n_1 + n_2 + 1} Y_1^{n_1} Y_2^{n_2} \exp[(\beta/Q)(1 - 1/\theta)]}{2G(n_1, n_2, a) Y_{1,0}^{n_1 + n_2 - 1} \text{Le}_1^{n_1} \text{Le}_2^{n_2}(-v_2)^{n_2}}, \qquad (98)$$

in which equation (5-78) has been employed as a definition of $G(n_1, n_2, a)$. Introduction of equation (97) into equation (98) renders the right-hand side a function of Y_1, thereby enabling the equation to be integrated to provide, at the leading order in β^{-1}, the matching condition

$$\left(\frac{\partial Y_1}{\partial \xi} \right)_{\xi=0-} = -\left\{ \frac{G\left(n_1, n_2, \dfrac{\beta Y_{2,f}}{-v_2 \text{Le}_2 Y_{1,0}}\right) \exp\left[\dfrac{\beta(\theta_f - 1)}{Q} \right]}{(1 + |\nabla_\perp f|^2) G(n_1, n_2, a)} \right\}^{1/2}, \qquad (99)$$

where the function G is defined below equation (5-78).

Equation (99) relates the concentration gradient of the limiting reactant just upstream from the reaction sheet to the temperature and the concentration of the reactant in excess at the reaction sheet. Therefore, it contains all the essential information associated with the reaction-zone structure. The continuity conditions derived previously enable other gradients to be obtained from this result. For example, since $\theta - 1$ is of order β^{-1} downstream, we have $(\partial\theta/\partial\xi)_{\xi=0-} = -(\partial Y_1/\partial\xi)_{\xi=0-} Q/(Y_{1,0} \text{Le}_1)$ to the lowest order. The corresponding one-reactant problem is readily recoverable by putting $n_2 = 0$ [so that the functions G in equation (99) cancel] and then letting $v_2 \to 0$. It may be noted that if a becomes large so that $Y_{2,f}$ becomes of order unity, then $\beta Y_{2,f}/(-v_2 Le_2 Y_{1,0} a) - 1$ becomes small, and it is found from the definition of G that the ratio of G's in equation (99) approaches unity, which implies that a mixture with a sufficient excess of species 2 behaves like a one-reactant mixture. With equation (99) and the jump conditions that have now been obtained, equations (87), (89), (90), (93), and (95) with $w = 0$ comprise a reaction-free description of the flame dynamics. Stability analyses are conveniently pursued by perturbing these equations about their steady, planar solutions.

A significant contribution to the complexity of the equations that have been given here is the fact that the coordinate system adopted is non-inertial and nonorthogonal. These complications are offset by the simplification achieved by placing the reaction sheet at a fixed location, $\xi = 0$. It may be possible to obtain an orthogonal system of coordinates by generating the coordinates on the basis of the instantaneous flame shape. This would not be likely to introduce any appreciable simplification into the local analysis of the flame structure and dynamics, but it could help in describing large-scale flame motions and in allowing for situations in which the reaction sheet becomes parallel to the x axis at various points. These geometric aspects of global flame dynamics have not been investigated.

Restrictions on the formulation pertain to wavelengths and frequencies of the disturbances. For example, the wavelength must be large compared with the thickness of the reaction sheet and the frequency low compared with the transit time through the sheet. These restrictions are satisfied well in a vast majority of problems. Disturbances with wavelengths on the order of the flame thickness (ξ of order unity) are permitted here. If wavelengths are large compared with the entire flame thickness and frequencies are small compared with the transit time through the flame, then a further integration across the flame can be performed to derive a single partial differential equation for $f(y, z, t)$. Results of this type have been developed within the approximation that the extent of thermal expansion across the flame is small [151], [176] and also for more general situations [177]–[180]. These results aid in simplifying analyses of flame dynamics and contribute to the understanding of factors influencing flame motions (see Section 10.3.3).

9.5.2. Cellular flames

There are three basic distinct types of phenomena that may be responsible for intrinsic instabilities of premixed flames with one-step chemistry: **body-force effects, hydrodynamic effects** and **diffusive-thermal effects**. Cellular flames—flames that spontaneously take on a nonplanar shape—often have structures affected most strongly by diffusive-thermal

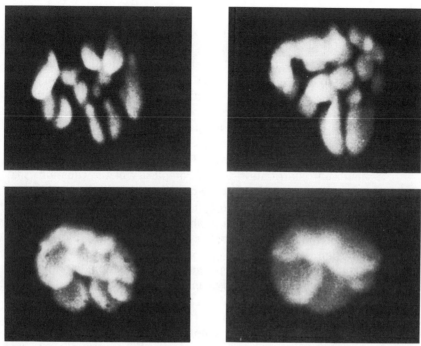

FIGURE 9.7. Cellular premixed hydrogen-oxygen flames, propagating downward in an open tube 5 cm in diameter, with a mole fraction of nitrogen of 0.8 and a molar ratio of hydrogen to oxygen of 1.03 (upper left), 1.57 (upper right), 2.01 (lower left), and 2.34 (lower right); visibility provided by addition of CF_3Br in molar percentage of about 0.2 (photographs taken by B. Bregeon).

phenomena; for example, the cellular flames shown in Figure 9.7 are largely controlled by a particular diffusive-thermal process. However, body-force and hydrodynamic phenomena also play interesting roles. Therefore, it is desirable to discuss all three types of instabilities here. We shall first address body-force effects, then hydrodynamic effects, and finally diffusive-thermal effects; the ordering, in a sense, is from large-scale to small-scale phenomena.

9.5.2.1. Body-force instabilities

It is well known* that a less-dense fluid beneath a more-dense fluid is buoyantly unstable [181]. Therefore, flames propagating upward may be expected to experience an instability that mathematically is a consequence of the term **F** in equations (89) and (90). If the entire flame is viewed as a discontinuity moving at velocity v_0 with respect to the fresh gas and v_∞ with respect to the burnt mixture, then through dimensional reasoning, a characteristic transverse length of the instability in the presence of a gravitational acceleration g may be guessed to be of order $v_0 v_\infty/g$ and a characteristic

---

* See pages 371, 372, and 458–462 of [45].

growth time of order $\sqrt{v_0 v_\infty}/g$. As might be expected physically, this last estimate suggests that the severity of the instability increases as the flame speed decreases. In earth gravity ($g \approx 10^3$ cm/s^2), typical wavelengths are estimated to be on the order of a few centimeters and growth times in tens of milliseconds.

Experimentally, it has been observed that for initially quiescent mixtures ignited at the closed base of a vertical tube open at the top in the laboratory, an elongated flame having a roughly parabolic shape with its nose pointing upward rapidly develops, propagates upward, and achieves an approximately constant rise velocity of the nose (if the mixture is sufficiently weak to avoid the buildup of pressure that initiates transition toward detonation). Consistent with the preceding estimates, the single-lobe shape that is usually observed suggests that the wavelength of the buoyant instability is comparable with or larger than typical tube diameters. For weak combustible mixtures of different fuels in tubes of various diameters D, the measured rise velocities have been shown [182] to agree well with the value $v_b \approx 0.3\sqrt{gD}$, obtained theoretically [183] for the velocity at which an air column rises into a vertical tube filled with water, sealed at the top and open to the atmosphere at the bottom. Thus, so long as $v_0 \ll v_b$, the flame motion in these upward-propagation experiments is affected primarily by buoyancy; the flame shape is quite similar to the shape of the air column. Since the nose of the flame is curved and finds itself propagating into a nonuniform flow, the streamlines of which diverge in approaching the flame, the flame structure at the nose will differ from that of a normal flame (see Section 10.3.2).

Buoyant instabilities have been observed for flames propagating in larger vented enclosures [184]-[186]. The same mechanism is operative for steady flames of flat-flame burners with the fuel flow directed downward. This configuration is similar to that in which the buoyantly convective Rayleigh instability involving formation of Benard cells often has been studied for nonreactive fluids, but in the flame experiments stabilization effects for nonplanar disturbances are provided by interactions with the burner and by the divergence of the streamlines. Nevertheless, instabilities have been seen in this configuration for diffusion flames on sufficiently large burners (greater than about 10-cm diameter) [187], and similar phenomena would be expected for premixed flames under appropriate conditions.

Acceleration of a flame sheet produces effects of the same type as those of body forces. In a noninertial coordinate system that moves with the flame (or in which the flame moves at a constant velocity), an effective body force appears as a consequence of flame acceleration. This may be seen from equation (89) by bringing $R\partial^2 f/\partial\tau^2$ to the right-hand side of the equation; the effective nondimensional body force per unit volume becomes $R(F_{\parallel} - \partial^2 f/\partial\tau^2)$. Thus a flame acceleration in the direction of its motion (a positive rate of increase of $-\partial f/\partial\tau$) is equivalent to a body force directed from the fresh mixture to the burnt gas. A body-force instability is therefore associated with flame acceleration. By the preceding reasoning, the wavelength and growth time of the instability may be estimated as $v_0 v_\infty / |\partial^2 F/\partial t^2|$ and $\sqrt{v_0 v_\infty}/|\partial^2 F/\partial t^2|$. These quantities decrease as the acceleration increases.

For decelerating flames, flames propagating downward, or burner-stabilized flames with the flow upward, the body-force effects are stabilizing. Because of other mechanisms of instability, to be discussed later, the ease with which stable laminar deflagrations are observed in the laboratory may be attributable largely to the stabilizing influence of buoyancy. Normal buoyancy effects are relatively far-field phenomena in that within the thickness of the thin flame (including the preheat zone as well as the reaction zone), the buoyancy terms in the conservation equations are comparatively small. Therefore, buoyancy may stabilize a deflagration without appreciably modifying the flame structure. Indeed, analyses of flame structure generally neglect buoyancy, with good justification.* Nevertheless, if buoyancy is responsible for observed flame stability, then it must have a significant influence on conditions for the occurrence of cellular flames. Therefore, it appears to be necessary to include buoyancy in analyses of stability boundaries. This has been done with the effect of flame curvature on the burning velocity included in a phenomenological way [137], as well as on the basis of the conservation equations of the previous section [178].

There are special situations in which buoyancy can be responsible for instabilities in configurations that might appear to be buoyantly stable. For example, a relatively slow flame propagating downward in an initially quiescent, weak combustible mixture contained in a chamber open at the top experiences conductive cooling of the burnt gas at the walls of the chamber. This cooling increases the density of the burnt gas near the walls, in comparison with that of the combustion products in the center of the chamber, and buoyancy forces may then promote downward motion of the gases near the wall. A pattern of buoyant convection then begins to be established, with the flame near the wall moving well ahead of the flame in the central part of the chamber. Phenomena of this general type have been observed experimentally [188], [189]; a net rate of buoyant rise of downward-burning combustible gas in the center of a large chamber or in the open is a common phenomenon. Thus body forces may produce a variety of observable effects on premixed-flame propagation.

9.5.2.2. Hydrodynamic instabilities

A startling discovery in flame theory was made independently by Darrieus[†] and Landau [190]; a stability analysis that *neglects* body forces and that treats the entire flame simply as a discontinuity in density that

---

* Body forces relevant to flame structure may be electrical forces acting on ionized species for flames in strong electric fields; we do not discuss influences of electric fields.

† Because of conditions in France during the second world war, the work of G. Darrieus never was published, but he presented a paper entitled "Propagation d'un front de flamme" at two conferences in France, La Technique Moderne (1938) and Congrès de Méchanique Appliquée, Paris (1945).

propagates normal to itself at a constant speed v_0 with respect to the more dense gas demonstrates that since thermal expansion across the flame causes $\rho_\infty < \rho_0$ (that is, $v_\infty > v_0$), the planar deflagration is unconditionally unstable. This result, largely contradictory to common experience in the laboratory, nevertheless is both correct and readily comprehensible (see below) and certainly requires consideration in connection with cellular-flame phenomena [137]. Clear expositions of the character of this hydrodynamic instability of flame fronts may be found in [9] and [153].

The hydrodynamic instability can be derived directly from equations (87), (89) and (90). The first step is to obtain jump conditions across the flame by integrating the equations with respect to ξ through the entire flame; with time derivatives and transverse gradients neglected within the flame, equation (87) shows that Ru is constant throughout the flame, the integral of equation (89) gives $Ru(u_+ - u_-) = P_- - P_+$ (where the subscripts $+$ and $-$ identify conditions just downstream and upstream from the flame, respectively), and integration of equation (90) yields $Ru(\mathbf{v}_{\perp+} - \mathbf{v}_{\perp-}) = \mathbf{V}_\perp f(P_+ - P_-)$, which—when combined with the preceding relationship—produces the condition of continuity of the velocity components tangent to the flame, $\mathbf{v}_{\perp+} + u_+ \mathbf{V}_\perp f = \mathbf{v}_{\perp-} + u_- \mathbf{V}_\perp f$. Since the flame speed is considered to be unmodified by the instability, $u_+ = 1$ and $u_- = 1/R_0$ (which is the ratio of densities of burnt and unburnt gas). The second step is to linearize the differential equations for the constant-density flows on each side of the flame about the steady, planar solution, which is given by $u = U$ (a constant defined to be unity for $\xi > 0$ and $1/R_0$ for $\xi < 0$), $P = 0$ for $\xi > 0$ and $1 - 1/R_0$ for $\xi < 0$, and $\mathbf{v}_\perp = 0$; the linearization gives

$$\partial u/\partial \xi + \mathbf{V}_\perp \cdot \mathbf{v}_\perp = 0, \quad R(\partial u/\partial \tau + \partial^2 f/\partial \tau^2 + U \partial u/\partial \xi) = -\partial P/\partial \xi,$$

and

$$R(\partial \mathbf{v}_\perp/\partial \tau + U \partial \mathbf{v}_\perp/\partial \xi) = -\mathbf{V}_\perp P$$

(with $R = 1$ for $\xi > 0$ and $R = R_0 = $ constant for $\xi < 0$). The third step is to seek solutions to these linear equations with constant coefficients for perturbations that are sinusoidal in the transverse direction (for example, proportional to $\mathrm{Re}\{e^{i\mathbf{k}\cdot\mathbf{x}_\perp}\}$, where the components of $\mathbf{x}_\perp$ are η and ζ), which satisfy the jump conditions at $\xi = 0$ and vanish as $\xi \to \pm\infty$ [thus requiring, in view of equation (88), that u approach $U - \partial f/\partial \tau$ as $\xi \to \pm\infty$]. A dispersion relation may be sought by letting the time dependence of the perturbations be proportional to $e^{\sigma\tau}$, where σ is the nondimensional growth rate; if $\mathrm{Re}\{\sigma\} > 0$, the parts of the perturbations that vanish at infinity are then found to be proportional to $e^{k\xi}$ for $\xi < 0$ and to a linear combination of $e^{-k\xi}$ and $e^{-\sigma\xi}$ for $\xi > 0$, where $k = |\mathbf{k}|$. A quadratic equation for σ is obtained, the acceptable solution to which is

$$\sigma = k[\sqrt{1 + (R_0^2 - 1)/R_0} - 1]/(1 + R_0). \tag{100}$$

Since $R_0 > 1$, equation (100) gives $\sigma > 0$ and hence implies instability.

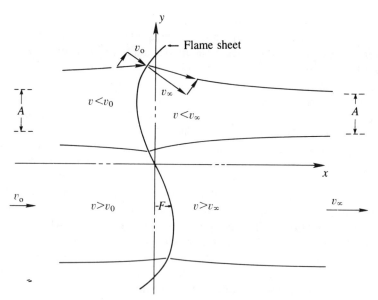

FIGURE 9.8. Schematic illustration of the mechanism of hydrodynamic instability.

In dimensional variables, equation (100) states that the growth rate of the disturbance is proportional to the product of the transverse wave number with the geometric-mean flame speed $\sqrt{v_\infty v_0}$; the factor of proportionality, which depends only on the density ratio, may be written as

$$(\sqrt{v_\infty^2 + v_\infty v_0 - v_0^2} - \sqrt{v_\infty v_0})/(v_\infty + v_0),$$

which vanishes if $v_\infty = v_0$, is positive when $v_\infty > v_0$, approaches $(v_\infty - v_0)/(2v_0)$ at small values of $(v_\infty - v_0)/v_0$, and approaches unity as $v_\infty/v_0 \to \infty$. Since the flame speed is the only dimensional parameter that appears in this stability problem, it is clear on dimensional grounds that the growth rate must be proportional to the product of the wave number and flame speed. That the constant of proportionality is positive may be traced to the fact that the flame-shape perturbations are convected with the fluid. Referring to Figure 9.8, we see that continuity of the tangential component of velocity across a tilted flame element required the streamlines to be deflected toward the local normal to the sheet in passage through the flame. This causes a stream tube of initial area A (which must be the same far upstream and far downstream, since the perturbations vanish there) to be larger at the flame than far away if the sheet is convex toward the unburnt gas, as illustrated. In incompressible flow, the area increase causes a velocity decrease, as indicated.* Since a flame-sheet perturbation with an upstream flame displace-

---

* Compressibility and acoustics are absent from the equations of the previous section; they play negligible roles in the problems addressed because the Mach numbers involved are very small.

ment ($F < 0$) geometrically must have a curvature convex toward the unburnt gas, it encounters a decreased fluid velocity and therefore (propagating at its normal speed) will tend to move even farther upstream. Conversely, a downstream flame displacement results in stream-tube contraction and, therefore, a tendency toward further displacement downstream. A formula that contributes to this explanation is equation (88), which shows that since the perturbation of u vanishes at the flame, $\partial F/\partial t$ equals the perturbation of v_1. The instability is of a purely diverging, nonoscillatory character (σ is real).

From equation (100) we see that the growth rate is inversely proportional to the wavelength l of the disturbance; small wrinkles grow more rapidly than large wrinkles. As the wavelength decreases and approaches the thickness δ of the deflagration, it is no longer permissible to treat the flame as a discontinuity. Even for wavelengths on the order of 10 to 100 times the deflagration thickness, diffusive and thermal effects within the flame may cause the flame speed to depend appreciably on the flame shape and on the local flow field [137]. Discussion of these diffusive-thermal phenomena is postponed to the following subsection, but it may be anticipated that they often introduce stabilizing influences, thereby causing the growth rate to achieve a maximum value at a particular wavelength appreciably greater than the flame thickness and to decrease thereafter with decreasing wavelength, possibly becoming negative at a critical wavelength, $l = l_c$. If a laboratory flame is contained within an apparatus whose walls exclude disturbances of wavelength greater than l_c, then the hydrodynamic instability will not be observed. Unfortunately, there are difficulties with this popular resolution of the contradiction between theory and experiment concerning hydrodynamic instability; stable flames can be observed in apparatuses with characteristic dimension D too large (for example, $D/\delta \gtrsim 100$ [153]) to enforce a condition $l < l_c$ for any reasonable value of l_c. For these flames, the resolution must lie in geometric or nonlinear hydrodynamic aspects of flame propagation and in effects of buoyancy.

For stationary, planar flames in a vertically upward flow, buoyancy must be a significant contributor to stability. It can be shown [137], [178] that if the hydrodynamic stability problem is addressed as before [190] but with the body-force terms in equations (89) and (90) retained in the differential equations, then equation (100) becomes

$$\sigma = \frac{k}{1 + R_0} \left\{ \sqrt{1 + \left(\frac{R_0^2 - 1}{R_0}\right)\left[1 - \left(\frac{\alpha_\infty/v_\infty}{k}\right)\left(\frac{g}{v_0 v_\infty}\right)\right]} - 1 \right\}, \quad (101)$$

in which the new term is the ratio of a dimensional wavelength characteristic of the transverse disturbance, $2\pi(\alpha_\infty/v_\infty)/k \equiv l$, to the body-force wavelength, $2\pi v_0 v_\infty/g \equiv l_b$. Gravity thus introduces stability for disturbances of large wavelengths, $l > l_b$. The range of wavelengths for instability now becomes finite, $l_c < l < l_b$; if $l_b < l_c$, then the hydrodynamic instability is eliminated entirely. It is likely that this last condition applies to experiments in which large planar flames are stabilized in a uniform flow [153]. Since l_b is proportional to $v_0 v_\infty$, for a given value of g the stabilization effect increases as the flame speed decreases; the effect often can be substantial because typical values of l_b for horizontal flame sheets in earth gravity are on the order of centimeters. Even for nonhorizontal flames such as bunsen-burner flames, the stabilizing influence of buoyancy may not be negligible.

In most configurations flame-induced flow nonuniformities complicate matters. In the absence of disturbances like acoustic perturbations (which can flatten flames under appropriate circumstances [191]), flames propagating downward in tubes and channels tend to become convex toward the unburnt gas. The flow streamlines then assume the pattern illustrated in the upper part of Figure 9.8, modified by allowing the area A far downstream to be less than that far upstream, since flow separation occurs at the wall as the flame discontinuity passes, leaving a stagnant layer adjacent to the wall downstream from the flame sheet [192]. The velocity-jump conditions across the flame sheet readily show that all curved flames that are not everywhere normal to the incoming flow produce vorticity; downward-propagating flames in tubes or channels with quiescent or irrotational flow upstream therefore have rotational flow downstream, but the rotationality per se does not appear to be a significant aspect of stability. Buoyancy again may help greatly to stabilize the downward-propagating flame against hydrodynamic instability. However, an additional factor of significance that may be dominant, for example, for flame propagation in channels in the absence of gravity is a ray-optics aspect of the development of hydrodynamic instability [193]–[195]. For an expanding spherical flame in an initially quiescent medium the increase of the area of the flame sheet with time is known to delay the onset of hydrodynamic instability [196], [197]; for propagation in channels this same effect is present because of the curvature of the flame, and moreover, except for the ray propagating along the centerline of the channel, the time available for growth of the instability is limited by the intersection of the ray with the wall. Thus steadily propagating, curved flames in channels or tubes in the absence of buoyancy may represent a stable configuration achieved by the long-time, nonlinear evolution of hydrodynamic instability.

In the early investigations of hydrodynamic instability [9], [190], it was presumed that the instability evolves to a chaotic state characteristic of turbulence. Thus self-turbulization of premixed flames was attributed to hydrodynamic instability (analogous, in a sense, to the development of turbulence in shear flows). This viewpoint must be revised if the instability evolves to stable nonplanar structures, as suggested above. From numerical experiments with equations describing the self-evolution of flame surfaces in the limit of small values of the density change across the flame, it has been inferred [152], [198]–[200] that the hydrodynamic instability evolves to flames with steady, corrugated shapes that have sharp ridges pointed toward the burnt gas (while the diffusive-thermal instability evolves to a somewhat similar but unsteady shape that exhibits a chaotic behavior). Thus evidence is growing against the idea that in the absence of other physical phenomena, the hydrodynamic instability of deflagrations leads to turbulence. However, the question certainly is not completely resolved. It is observed experimentally that as large (up to 10-m radius) spherical flame sheets expand, they develop pebbled surface structures (with average sizes on the order of 10 cm) that appear to be chaotic in some respects and, on the average, propagate at about twice the normal burning velocity [152]. Although it would seem that the hydrodynamic instability should be the dominant phenomenon in these large flames, diffusive phenomena may play a role beyond their obvious relevance to the cusp structure at the ridges of the sheet.

9.5.2.3. Diffusive-thermal instabilities

The principal features of the flame shown in Figure 9.7 cannot be understood on the basis of the instabilities discussed thus far because the flame speed and the density change vary little from one photograph to another, but the flame structure clearly varies greatly. The changes in flame structure are associated with variations of the effective diffusivity of the limiting reactant [201], [202]. Significant diffusive influences in cellular flames were established in early work [137]; the first reported observations of these effects appear to be those for tent flames with periodic ridges on bunsen-type burners, the so-called polyhedral flames [203]. The effects were first addressed theoretically in a semiphenomenological manner [204] by allowing the flame speed to depend on the radius of curvature of the flame sheet (the reciprocal of the mean curvature, here denoted by $\mathscr{R}$ and taken to be positive for convexity toward the burnt gas and negative for convexity toward the fresh mixture). The flame speed with respect to the unburnt mixture is written as a constant times $1 + \mathscr{L}/\mathscr{R}$, where the Markstein length, $\mathscr{L}$, is a constant intended to describe influences of flame-structure modifications on the flame speed; the formulation may be viewed as one that retains only the first term in an expansion for small flame curvatures. Stability analyses like those discussed earlier can then be pursued for any value of $\mathscr{L}$, without further inquiry into the flame-structure phenomena that determine $\mathscr{L}$. In particular since $1/\mathscr{R} = -\nabla^2 F$ for small deflections of the flame sheet, a Fourier decomposition will introduce a correction term proportional to k^2 in a formula like equation (100) for σ, and the new term will be most important for large values of k (small wavelengths). Of course, a complete analysis entails studying the influences of the perturbations on the internal structure of the flame—for example, with the objective of calculating $\mathscr{L}$ from the fundamental chemical and transport properties of the system.

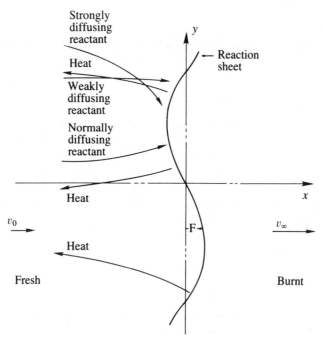

FIGURE 9.9   Schematic illustration of the mechanism of diffusive-thermal instability.

Extensive research on the theory of these flame-structure modifications has been pursued [170]–[180], [205]–[225], and we now have a reasonably good general understanding of the influence on flame stability of the diffusive and thermal processes that occur within the flame.

The essential physical ideas apparently were first known to Zel'dovich.* The instability mechanisms can be explained by reference to Figure 9.9, where the arrows indicate directions of net fluxes of heat and species. A key element in the explanation is the fact that for large values of the Zel'dovich number, the burning velocity depends mainly on the local flame temperature. This intuitively clear result is not entirely obvious from the formulation of Section 9.5.1 but may be motivated from it by employing $Ru\, \partial Y_1/\partial\xi = \partial^2 Y_1/\partial\xi^2$ as a rough approximation to equation (95) in the preheat (convective-diffusive) zone for $i = 1$; if this equation is integrated with Ru constant and with the boundary conditions that $Y_1 = Y_{1,0}$ at $\xi = -\infty$ and $Y_1 = 0$ at $\xi = 0$, then application of equation (99) to the solution [which is $Y_1 = Y_{1,0}(1 - e^{Ru\xi})$] shows that Ru is proportional to $\exp[\beta(\theta_f - 1)/(2Q)]$, that is, the burning velocity increases with increasing

---

* They were expressed in his early book, *Theory of Combustion and Gas Detonation*, Moscow: Akad. Nauk SSSR, 1944.

flame temperature in the expected Arrhenius fashion. In neglecting non-planar influences within the preheat zone, this reasoning is restricted to disturbances with transverse wavelengths that are large compared with the flame thickness. It is evident that a bulge in the reaction sheet extending toward the unburnt gas serves as a local sink for reactants and a local source for heat. Increasing the rate of diffusion of the limiting reactant to the sheet increases the rate of heat release there and hence tends to increase the flame temperature T_f, while increasing the rate of heat conduction from the sheet tends to reduce T_f. At the bulge, the local gradients will be steepened in comparison with those of the normal flame, and the reactant flux and conductive flux will both, therefore, be increased. If the thermal diffusivity equals the diffusivity of the limiting reactant ($\text{Le}_1 = 1$), then these increases are balanced in such a way that T_f is unchanged. However, if the thermal diffusivity is greater ($\text{Le}_1 > 1$, weakly diffusing reactants), then T_f decreases at the bulge; if it is less ($\text{Le}_1 < 1$, strongly diffusing reactant), then T_f increases (as a consequence of the preferential diffusion of the reactant with respect to heat). In this last situation, the local increase in flame speed, associated with the increase in T_f, causes the bulge to tend to become larger, thereby giving rise to the diffusive-thermal instability, which thus is seen to occur for $\text{Le}_1 < 1$.

This reasoning suggests that the diffusive and thermal effects within the flame tend to stabilize the hydrodynamic instability (that is, produce a critical wavelength $l_c > 0$ such that the flame is stable to disturbances with transverse wavelengths $l < l_c$) only if $\text{Le}_1 > 1$.* However, there is an additional, fundamentally simpler physical mechanism that is purely thermal in character and that tends to provide a stabilizing influence for $\text{Le}_1 = 1$. When $\text{Le}_i = 1$ and $\delta_i = \lambda_1$ for all species i, it can be shown rigorously from equation (93) with $L = 0$ that $\sum_{i=1}^{N} H_i Y_i$ remains constant and from equation (95) that $Y_i + v_i Y_1$ remains constant. From this, wherever $Y_1 = 0$, the temperature remains constant and equal to the adiabatic flame temperature. In Figure 9.9 the temperature is therefore constant everywhere on the reaction sheet and to the right thereof for this adiabatic system with equal diffusivities. In the lower part of the figure, where the hot gas tends to surround the cold combustible, the heating rate is greater than that for the plane flame, and the local propagation speed therefore tends to be increased. Conversely, heat is lost to the side from the bulge in the upper part of the figure, so that the propagation speed is reduced there (see Section 8.2). These effects cause the wrinkled reaction sheet to tend to become planar and therefore constitute a stabilizing influence. In the absence of hydrodynamic effects and any other perturbations, the equation for evolution of the reaction sheet with $\lambda_1 = 1$ and with all perturbation lengths large compared with the flame thickness

---

* It must, however, be less than the critical value for pulsating instability, as indicated at the beginning of Section 9.5.1.

can be shown to be $1 + \partial f/\partial \tau = \mathbf{V}_\perp \cdot (\mathbf{V}_\perp f)$, a two-dimensional, time-dependent heat-conduction equation for $f + \tau$ in this case. The stable solution to this equation, approached as $\tau \to \infty$, is the plane-flame solution $\partial f/\partial \tau = -1$. Since the Lewis-number effects described in the previous paragraph involve changes in T_f that make themselves felt through the Arrhenius factor, they may be expected to be large compared with the smoothing effect discussed here if β is large and if Le_1 is not too close to unity.

In two-reactant systems with $Le_1 \neq Le_2$, there is an additional purely diffusive effect, often termed **preferential diffusion** of reactants, that can be understood even more readily than either of the two effects indicated above. The more strongly diffusing reactant can reach the sink at the upstream-pointing bulge more readily than its weakly diffusing partner. Therefore, the wrinkles induce local mixture-ratio variations. Upstream bulges are relatively enriched in the strongly diffusing reactant and downstream bulges are consequently enriched in the weakly diffusing reactant. These variations make themselves felt on the dynamics of the reaction sheet mainly in their (thermal) effect on the flame speed through T_f; therefore, in near-stoichiometric mixtures their influences are comparable in magnitude to those of the diffusive-thermal interactions. If the limiting component is relatively strongly diffusing, then the mixture-ratio variations aggravate instability; the mixture-ratio variation of a weakly diffusing limiting reactant is a stabilizing effect. At stoichiometric conditions for a mixture that has its maximum normal burning velocity at stoichiometry, upstream bulges can tend to be suppressed and downstream bulges can tend to be amplified by the mixture-ratio variations.

Contrary to the suggestion made in Section 9.5.2.2, the diffusive and thermal phenomena discussed here evidently do not always produce stabilizing influences for nonplanar disturbances. When they do not, there is nothing to stabilize the hydrodynamic instability at small wavelengths, and planar flames in a uniform flow then cannot exist. Our reasoning suggests that there is a critical value of the Lewis number of the limiting component, say Le_c, such that diffusive-thermal instability occurs for $Le_1 < Le_c$, and planar flames will not be observable in mixtures that satisfy this criterion. Even for $Le_1 > Le_c$, hydrodynamic instability may prevent the plane flame from occurring unless walls or buoyancy provide sufficiently strong stabilizing influences. The consequences of the instability in unstable situations are understood only partially. For $Le_1 < Le_c$, nonplanar diffusive-thermal phenomena will be enhanced as the instability develops, and it may be reasoned [201] the flame will evolve to relatively stable, nonplanar cellular shapes, like those seen at the top of Figure 9.7 (perhaps moving somewhat chaotically). For $Le_1 > Le_c$, the steepening gradients associated with development of the hydrodynamic instability accentuate magnitudes of nonplanar diffusive-thermal effects, which may therefore exert a largely controlling influence on the final cellular shape (a shape somewhat like that seen at the bottom right of Figure 9.7). Although the curvature at the nose regions of large cells may be thought to be too small for diffusive-thermal phenomena to be relevant there, the associated divergence of the streamlines as they approach the sheet provides a strain configuration known [221] to be strongly stabilizing to ripples for diffusive-thermal processes. Therefore, even if the original source of the instability is hydrodynamic, the shapes of the cellular flames that develop may be dominated largely by diffusive-thermal effects.

For estimating when diffusive-thermal instabilities may occur, it is important to know the value of the critical Lewis number, Le_c. Although our reasoning suggested that Le_c is just slightly below unity, we ignored the finite thickness of the reaction sheet (an effect of relative order $1/\beta$) and did not specifically address influences, within the flame, of the gas expansion associated with the heat release. That the latter influences are also stabilizing may be understood by realizing that the decrease in density with increasing temperature increases transverse distances through transverse convection and therefore reduces transverse gradients of temperature and of concentrations locally in the vicinity of the reaction sheet; the diffusive-thermal variations in T_f thereby are reduced (in effect, the reaction sheet can be considered to be moved to a less accessible position, with a partially insulating, low-density layer of fluid placed in front of it), and the smoothing effect associated with constancy of T_f is relatively enhanced. Theoretical analysis of a one-reactant, adiabatic system with constant specific heats, for small reaction-sheet curvatures and large Zel'dovich numbers, has shown [177] that if the transport properties are constant ($\mu_1 = \mu_2 = \lambda_1 = \delta_i = 1$), then

$$Le_c = 1 - \frac{1}{\beta}(2R_0 \ln R_0) \bigg/ \int_0^{R_0-1} x^{-1}\ln(1 + x)\,dx, \qquad (102)$$

in which the factor multiplying $1/\beta$ approaches $2[1 + 3(R_0 - 1)/4]$, as $R_0 \to 1$, equals $(48/\pi^2)\ln 2 \approx 3.37$ at $R_0 = 2$ and diverges as $4R_0/\ln R_0$ for $R_0 \to \infty$. Since the density ratio R_0 typically has a value of about 5, with the estimate $10 \lesssim \beta \lesssim 20$, the corresponding value of Le_c certainly does not exceed 0.5, a value that may be compared with the limit 0.9 (which is obtained at $\beta = 20$ for $R_0 = 1$) to demonstrate the large influence of gas expansion. In fact, the influence probably is somewhat greater than would be implied by equation (102) because the decrease in diffusivity with decreasing temperature helps to make the reaction sheet less accessible to reactants, as has been shown in a further investigation of variable-property effects [223]. Thus it seems safe to conclude that $Le_1 > Le_c$ for most real combustible mixtures and, therefore, that the observed instabilities, which lead to cellular-flame structures affected mainly by diffusive-thermal phenomena, are in fact first initiated by the hydrodynamic instability.

By viewing the coefficient of $1/\beta$ in equation (102) as a quantity of order unity (even though it may be somewhat large numerically), we see that the difference between Le_c and unity formally is of order $1/\beta$. For the investigation of diffusive-thermal effects, this ordering is dictated by the fact that $\exp[\beta(\theta_f - 1)/Q]$ appears in equation (99). Changes in the flame temperature of an order no larger than $1/\beta$ times the adiabatic flame temperature are required if the response of the reaction sheet is not to become exponentially large. There are nonplanar disturbances that, through equation (93), induce relative diffusion of the reactant and heat to an extent sufficient to produce fractional changes of order unity in the flame temperature if Le_1 differs from unity by an amount of order unity. Therefore, theoretical analyses of wrinkled flames based on activation-energy asymptotics usually treat departures of Le_i from unity as being of order $1/\beta$ [170], [171], [173], [175]-[180], [198]-[200], [217]-[225]. The known diffusive-thermal instabilities all appear for $|Le_i - 1|$ of order $1/\beta$, and in real gas mixtures this ordering usually is reasonable numerically.

The influence of viscosity within the flame on the hydrodynamic instability has been a subject of uncertainty since the early studies of instabilities [137]. Through activation-energy asymptotics, within the context of an expansion for long wavelengths of disturbances, it has now been shown [223] that Pr_2 does not appear in the dispersion relation and that, through Pr_1, shear viscosity introduces a very small stabilizing effect if the coefficient of viscosity increases as the fluid passes through the flame. If the coefficient of viscosity is constant, then there is no effect [178], [219], the viscosity-induced modifications in the convective-diffusive zone being cancelled by those in the reactive-diffusive zone. Physically, the viscosity modifies the gas expansion only slightly. The asymptotic analyses [178], [179], [223] also provide expressions for the length $\mathscr{L}$ (introduced at the beginning of the present subsection), showing that gas-expansion effects increase $\mathscr{L} V_\infty/\alpha_\infty$ appreciably, while variable-property effects decrease it somewhat; $\mathscr{L}$ is greater than predicted by constant-density, constant-property analyses.

With few exceptions [177]-[180], [223]-[225], recent analyses of diffusive-thermal phenomena in wrinkled flames have employed approximations [208] of nearly constant density and constant transport coefficients, thereby excluding the gas-expansion effects discussed above. Although results obtained with these approximations are quantitatively inaccurate, the approach greatly simplifies the analysis and thereby enables qualitative diffusive-thermal features shared by real flames to be studied without being obscured by the complexity of variations in density and in other properties. In particular, with this approximation it becomes feasible to admit disturbances with wavelengths less than the thickness of the preheat zone (but still large compared with the thickness of the reactive-diffusive zone). In this approach it is usual to set $\mathbf{v} = 0$; equations (87)-(90) are no longer needed, and equations (93) and (95) are simplified somewhat. It is possible to include effects of heat losses by retaining L in equation (93). Dispersion relations have been developed through activation-energy asymptotics for one-reactant flames, both adiabatic [213] and nonadiabatic [173] ($L \neq 0$), as well as for adiabatic two-reactant flames [175]. The non-adiabatic result is [173]

$$1 + 4(\sigma + k^2) = \left(\frac{Le_1 - 1}{2/\beta}\right)\left[\frac{\sigma\Gamma(\sigma,k)}{2(\sigma + k^2)} - 1\right] + [\ln(1/\mu)]\Gamma(\sigma, k), \quad (103)$$

where $\Gamma(\sigma, k) \equiv 1 + \sqrt{1 + 4(\sigma + k^2)}$, in which the real part of the radical is to be taken to be positive. Here the heat loss has been assumed to be linear in the temperature increment [Newtonian, L proportional to $\theta - \theta_0$ in equation (93) or κ proportional to τ in equation (8-10)], and μ, the ratio of the burning velocity of the planar flame with heat loss to that under adiabatic conditions, is defined by equation (8-11) so that in terms of the loss parameter l, defined in equation (8-21), we have [from equation (8-22)], $\ln(1/\mu) = l/(2\mu^2)$. For two-reactant flames [175], equation (103) may be applied, with Le_1 replaced by the effective Lewis number $Le_1 + (Le_2 - Le_1) \partial G(n_1, n_2, a)/\partial n_2$, in which G and a are defined below equation (5-78).

The implications of equation (103) are illustrated in Figure 9.10, where various stability boundaries are shown by solid lines in a plane of

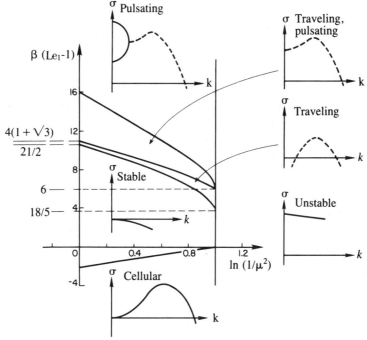

FIGURE 9.10. Regions of diffusive-thermal instability in a Lewis-number, heat-loss plane, with dispersion relations illustrated by insets.

the measure $\beta(\text{Le}_1 - 1)$ of the departure from unity of the effective Lewis number of the limiting reactant as a function of the nondimensional measure $l/\mu^2 = \ln(1/\mu^2)$ of the rate of heat loss. Although these results depend quantitatively on the approximations, such as Newtonian loss, they are likely to remain qualitatively correct under more general conditions. As the heat loss increases, the region of diffusive-thermal stability is seen in Figure 9.10 to narrow from $-2 < \beta(\text{Le}_1 - 1) < 21/2$ under adiabatic conditions ($\mu = 1$) to $0 < \beta(\text{Le}_1 - 1) < 18/5$ at the flammability limit ($\mu = 1/\sqrt{e}$); consistent with our earlier reasoning, this region includes $\text{Le}_1 = 1$. The steady-state solutions at values of μ below $1/\sqrt{e}$ [that is, $\ln(1/\mu^2) > 1$ in Figure 9.10, corresponding to the lower branch of the burning-velocity curve in Figure 8.2] are found to be unstable for all values of Le_1. The stable and unstable behaviors are illustrated by the insets in Figure 9.10, which are sketches of the dispersion relation, the solution to equation (103) for σ as a function of k. The critical value Le_c of Le_1 below which the cellular instability occurs is seen to increase from $1 - 2/\beta$ under adiabatic conditions to unity at the flammability limit; the former value, of course, agrees with that given by equation (102) for $R_0 = 1$. In the range of cellular instability, the dispersion relation exhibits a maximum growth rate at a preferred wavelength and damping at sufficiently small wavelengths (that is, at values of k above that at which σ passes through zero); this stabilization at short wavelengths, producing a critical wavelength $l_c > 0$ for stability even in the range of diffusive-thermal instability, could not be obtained in our earlier reasoning, which was based on the restriction that the wavelength be large compared with the flame thickness.

Three additional types of diffusive-thermal instability appear in Figure. 9.10 for sufficiently large values of the Lewis number. One is the planar pulsating instability, indicated at the beginning of Section 9.5.1 and discussed in Section 9.2; this occurs at the largest values of the Lewis number. The other two are instabilities that involve waves traveling along the reaction sheet. For traveling waves, k is positive and σ is complex, giving f proportional to $\exp[\tau \, \text{Re}\{\sigma\}]\text{Re}\{e^{i[\mathbf{k} \cdot \mathbf{x}_\perp + \tau \, \text{Im}\{\sigma\}]}\}$. In the dispersion relations sketched in Figure 9.10, the dashed curves represent $\text{Re}\{\sigma\}$ for conditions under which σ is complex. There is a narrow range of parameters in which instability is experienced ($\text{Re}\{\sigma\} > 0$) only in the traveling modes and another range in which instability is experienced in both traveling and pulsating modes (that is, the range of k over which $\text{Re}\{\sigma\} > 0$ extends to $k = 0$). The character of these traveling-wave instabilities has been investigated theoretically [220]. It has been found experimentally [153], [226]–[229] that in many circumstances oscillations are capable of being transmitted directly along the flame surface. Because of the constant-density approximation that underlies equation (103), it is not yet clear whether observed traveling waves correspond to the traveling diffusive-thermal instabilities illustrated in Figure 9.10.

It is seen in Figure 9.10 that the stable range of parameters is narrowed near the flammability limit, and the cellular and traveling-wave instabilities may be expected to be observable more readily. In particular, over most of the range of heat loss, the traveling-wave instabilities appear to occur at values of Le_1 greater than those of real gas mixtures, but near the limit these values decrease into an accessible range. Near the flammability limit, there are special effects that can widen somewhat the range of propagation of cellular flames [222]. Research on near-limit behavior is continuing.

The cells seen in Figure 9.7 arise largely through preferential diffusion of hydrogen to the bright zones, leaving a mixture between the cells that is too lean to burn. The cell sizes can be correlated on the basis of the wavelength for the maximum rate of amplification of the cellular instability, seen at the bottom of Figure 9.10 [202], [230], [231]. The increase in size with increasing hydrogen-oxygen ratio is associated with the increase in this wavelength, produced by the increase in the effective Lewis number identified earlier. There must be inaccuracies in the correlation because of the constant-density approximation and because of the evidently nonlinear character of the structures in Figure 9.7. Nonlinear analyses of flame shapes of cellular flames are needed.*

The diffusive-thermal instabilities that have been considered here pertain entirely to premixed flames. Diffusive-thermal instabilities rarely occur for diffusion flames, but they can be observed under certain circumstances. In a diffusion flame between air and hydrogen-nitrogen mixtures, in the forward stagnation region of a porous cylinder, patterns have been found [232] (at sufficiently large dilutions of the stream of hydrogen coming from the cylinder) in which the combustion occurs in periodically positioned stripes wrapped around the cylinder, with no observable reaction between the stripes. The mechanism of this instability must be preferential diffusion of hydrogen, through nitrogen and combustion products, to the combustion zones, with the consequence that there is insufficient hydrogen to support combustion in between. A similar mechanism could apply to earlier observations of instability in coflowing hydrogen-air diffusion flames [233]. Neither data on the stripe size and spacing nor theoretical analyses of the instability have yet been published.

REFERENCES

1. M. Barrère and F. A. Williams, *12th Symp.* (1969), 169–181.
2. R. N. Wimpress, *Internal Ballistics of Solid-Fuel Rockets*, New York: McGraw-Hill, 1950.
3. R. P. Smith and D. F. Sprenger, *4th Symp.* (1953), 893–906.

---

* A step in this direction has been taken by T. Mitani, *CST* **36**, 235 (1984).

4. H. Grad, *Comm. Pure Appl. Math.* **2**, 79 (1949).

5. R. D. Geckler, *5th Symp.* (1955), 29–40.

6. L. Green, Jr., *Jet Propulsion* **26**, 655 (1956).

7. F. A. Williams, M. Barrère, and N. C. Huang, *Fundamental Aspects of Solid Propellant Rockets*, AGARDograph No. 116, Slough, England: Technivision Services, 1969.

8. P. M. Morse and H. Feshback, *Methods of Theoretical Physics*, vol. 1, New York: McGraw-Hill, 1953.

9. L. D. Landau and E. M. Lifshitz, *Fluid Mechanics*, London: Pergamon Press, 1959.

10. F. T. McClure, R. W. Hart, and R. H. Cantrell, *AIAA Journal* **1**, 586 (1963).

11. R. H. Cantrell and R. W. Hart, *J. Acoust. Soc. Am.* **36**, 697 (1964).

12. R. W. Hart and F. T. McClure, *10th Symp.* (1965), 1047–1065.

13. F. E. C. Culick, *Astronautica Acta* **12**, 113 (1966).

14. F. E. C. Culick, *CST* **7**, 165 (1973).

15. F. E. C. Culick, *CST* **10**, 109 (1975).

16. W. K. Van Moorhem, *AIAA Journal* **20**, 1420 (1982).

17. S. I. Cheng, *Jet Propulsion* **24**, 27, 102 (1954).

18. H. S. Tsien, *J. Am. Rocket Soc.* **22**, 139 (1952).

19. L. Crocco, *Aerotecnica* **33**, 46 (1953).

20. L. Crocco and S. I. Cheng, *Theory of Combustion Instability in Liquid Propellant Rocket Motors*, AGARDograph No. 8, London: Butterworths Scientific Publication, 1956.

21. F. E. C. Culick, *AIAA Journal* **1**, 1097 (1963).

22. P. M. Morse, *Vibration and Sound*, New York: McGraw-Hill, 1948.

23. M. R. Baer, C. E. Mitchell and W. R. Espander, *AIAA Journal* **12**, 475 (1974).

24. C. E. Mitchell and M. R. Baer, *AIAA Journal* **13**, 1107 (1975).

25. J. E. Ffowcs Williams, *Annual Review of Fluid Mechanics* **9**, 447 (1977).

26. W. Koch and W. Möhring, *AIAA Journal* **21**, 200 (1983).

27. M. R. Baer and C. E. Mitchell, *AIAA Journal* **15**, 135 (1977).

28. R. J. Astley and W. Eversman, *Journal of Sound and Vibration* **74**, 103 (1981).

29. L. Crocco and W. A. Sirignano, *Behavior of Supercritical Nozzles under Three-Dimensional Oscillatory Conditions*, AGARDograph No. 117, Paris: The Advisory Group for Aerospace Research and Development, NATO, 1967.

30. L. Crocco, R. Monti, and J. Grey, *ARS Journal* **31**, 771 (1961).

31. F. G. Buffum, Jr., G. L. Dehority, R. O. Slates, and E. W. Price, *AIAA Journal* **5**, 272 (1967).

32. R. F. Lambert, *J. Acoust. Soc. Am.* **25**, 1068 (1953).

33. R. B. Lindsay, *Mechanical Radiation*, New York: McGraw-Hill, 1960.

34. R. W. Hart and R. H. Cantrell, *J. Acoust. Soc. Am.* **35**, 18 (1963).

35. K. Stewartson, "The Theory of Unsteady Laminar Boundary Layers," in *Advances in Applied Mechanics*, vol. VI, New York: Academic Press, 1960, 1–37.

36. K. Stewartson, *The Theory of Laminar Boundary Layers in Compressible Fluids*, Oxford: Clarendon Press, 1964, Chapter 6.

37. R. H. Cantrell, F. T. McClure, and R. W. Hart, *J. Acoust. Soc. Am.* **35**, 500 (1963).

38. F. T. McClure, R. W. Hart, and J. F. Bird, "Solid Propellant Rocket Motors as Acoustic Oscillators," in *Solid Propellant Rocket Research*, vol. 1 of *Progress in*

Astronautics and Rocketry, M. Summerfield, ed., New York: Academic Press, 1960, 295–358.

39. F. T. McClure, R. W. Hart, and J. F. Bird, *J. Appl. Phys.* **31**, 884 (1960).
40. J. F. Bird, *J. Acoust. Soc. Am.* **32**, 1413 (1960).
41. O. J. Deters, *ARS Journal* **32**, 378 (1962).
42. V. Blackman, *J. Fluid Mech.* **1**, 61 (1956).
43. J. F. Clarke, *J. Fluid Mech.* **7**, 577 (1960).
44. C. J. T. Sewell, *Phil. Trans. Roy. Soc. London* **210A**, 239 (1910).
45. H. Lamb, *Hydrodynamics*, 6th ed. 1932, New York: Dover, 1945, 645–663.
46. P. S. Epstein and R. R. Carhart, *J. Acoust. Soc. Am.* **25**, 553 (1953).
47. G. F. Carrier, *J. Fluid Mech.* **4**, 376 (1958).
48. R. A. Dobbins and S. Temkin, *AIAA Journal* **2**, 1106 (1964).
49. R. A. Dobbins and S. Temkin, *AIAA Journal* **5**, 2182 (1967).
50. S. Temkin and R. A. Dobbins, *J. Acoust. Soc. Am.* **40**, 1016 (1966).
51. Lord Rayleigh, *Nature* **18**, 319 (1878).
52. Lord Rayleigh, *The Theory of Sound*, New York: Dover, 1945, 226–235.
53. J. Tyndall, *Sound*, New York: D. Appleton and Co., 1867.
54. F. A. Williams, *J. Appl. Phys.* **33**, 3153 (1962).
55. R. W. Hart and R. H. Cantrell, *AIAA Journal* **1**, 398 (1963).
56. H. Krier, J. S. T'ien, W. A. Sirignano, and M. Summerfield, *AIAA Journal* **6**, 278 (1968).
57. L. Green, Jr., *Jet Propulsion* **28**, 386 (1958).
58. W. Nachbar and L. Green, Jr., *J. Aero. Sci.* **26**, 518 (1959).
59. S. I. Cheng, *8th Symp.* (1962), 81–96.
60. M. Barrère and J. J. Bernard, *8th Symp.* (1962), 886–894.
61. Y. B. Zel'dovich, *Zhur. Eksp. Teor. Fiz.* **12**, 498 (1942).
62. Y. B. Zel'dovich, *Zhur. Prikl. Mekh. Tekhn. Fiz.*, no. 3, 126 (1964).
63. A. G. Smith, "A Theory of Oscillatory Burning of Solid Propellants Assuming a Constant Surface Temperature," in *Solid Propellant Rocket Research*, vol. 1 of *Progress in Astronautics and Rocketry*, M. Summerfield, ed., New York: Academic Press, 1960, 375–392.
64. M. R. Denison and E. Baum, *ARS Journal* **31**, 1112 (1961).
65. F. A. Williams, *AIAA Journal* **11**, 1328 (1973).
66. R. W. Hart and F. T. McClure, *J. Chem. Phys.* **30**, 1501 (1959).
67. R. H. Cantrell, R. W. Hart and F. T. McClure, *AIAA Journal* **2**, 1100 (1964).
68. S. S. Novikov and Y. S. Ryazantsev, *Zhur. Prikl. Mekh. Tekhn. Fiz.*, no. 6, 77 (1964).
69. R. H. Cantrell, F. T. McClure and R. W. Hart, *AIAA Journal* **3**, 418 (1965).
70. S. S. Novikov and Y. S. Ryazantsev, *Zhur. Prikl. Mekh. Tekhn. Fiz.*, no. 2, 57 (1966).
71. J. C. Friedly and E. E. Petersen, *AIAA Journal* **4**, 1604 (1966).
72. M. Imber, *AIAA Journal* **4**, 1610 (1966).
73. F. E. C. Culick, *Astronautica Acta* **13**, 221 (1967).
74. J. S. T'ien, *CST* **5**, 47 (1972).
75. F. E. C. Culick, *AIAA Journal* **6**, 2241 (1968).
76. B. V. Novozhilov, *Nonstationary Combustion of Solid Rocket Fuels*, Moscow: Nauka, 1973.
77. M. D. Horton and E. W. Price, *9th Symp.* (1963), 303–310.

78. R. Strittmater, L. Watermeier, and S. Pfaff, *9th Symp.* (1963), 311–315.
79. R. S. Brown, F. E. C. Culick, and B. T. Zinn, "Experimental Methods for Combustion Admittance Measurements", in *Experimental Diagnostics in Combustion of Solids*, vol. 63 of *Progress in Astronautics and Aeronautics*, T. L. Boggs and B. T. Zinn, eds., New York: American Institute of Aeronautics and Astronautics, 1978, 191–200.
80. L. H. Caveny, K. L. Collins, and S. W. Cheng, *AIAA Journal* **19**, 913 (1981).
81. R. N. Kumar and F. E. C. Culick, *CST* **15**, 179 (1977).
82. N. S. Cohen, *AIAA Journal* **19**, 907 (1981).
83. S. I. Cheng, "Combustion Instability in Solid Rockets Using Propellants with Reactive Additives," in *Solid Propellant Rocket Research*, vol. 1 of *Progress in Astronautics and Rocketry*, M. Summerfield, ed., New York: Academic Press, 1960, 393–422.
84. W. A. Wood, *9th Symp.* (1963), 335–344.
85. F. A. Williams and G. Lengellé, *Astronautica Acta* **14**, 97 (1969).
86. J. L. Eisel, M. D. Horton, E. W. Price, and D. W. Rice, *AIAA Journal* **2**, 1319 (1964).
87. R. W. Hart and F. T. McClure, "The Influence of Erosive Burning on Acoustic Instability in Solid Propellant Rocket Motors," in *Solid Propellant Rocket Research*, vol. 1 of *Progress in Astronautics and Rocketry*, M. Summerfield, ed., New York: Academic Press, 1960, 423–451.
88. F. T. McClure, J. F. Bird, and R. W. Hart, *ARS Journal* **32**, 374 (1962).
89. R. W. Hart, J. F. Bird, R. H. Cantrell, and F. T. McClure, *AIAA Journal* **2**, 1270 (1964).
90. J. E. Crump and E. W. Price, *AIAA Journal* **2**, 1274 (1964).
91. E. W. Price, *10th Symp.* (1965), 1067–1082.
92. E. W. Price, *12th Symp.* (1969), 101–113.
93. E. W. Price, *AIAA Journal* **17**, 799 (1979).
94. F. A. Williams, *AIAA Journal* **3**, 2112 (1965).
95. G. F. Carrier, *Quart. Appl. Math.* **12**, 383 (1955).
96. F. E. C. Culick, *CST* **2**, 179 (1970).
97. F. E. C. Culick, *CST* **3**, 1 (1971).
98. M. S. Padmanabhan, B. T. Zinn, and E. A. Powell, *CST* **20**, 179 (1979).
99. L. A. Dickinson, *ARS Journal* **32**, 643 (1962).
100. W. G. Brownlee, *AIAA Journal* **2**, 275 (1964).
101. W. A. Sirignano and L. Crocco, *AIAA Journal* **2**, 1285 (1964).
102. C. E. Mitchell, L. Crocco and W. A. Sirignano, *CST* **1**, 35 (1969).
103. M. J. Lighthill, *Phil. Mag.* **40**, 1179 (1949).
104. G. B. Whitham, *Comm. Pure Appl. Math.* **5**, 338 (1952).
105. C. C. Lin, *Journal of Math. and Phys.* **33**, 117 (1954).
106. B. T. Chu, *Phys. Fluids* **6**, 1625 (1963).
107. B. T. Chu, *Phys. Fluids* **6**, 1638 (1963).
108. W. Chester, *J. Fluid Mech.* **18**, 1 (1964).
109. K. G. Shkadinskii, B. I. Khaikin, and A. G. Merzhanov, *Fiz. Gor. Vzr.* **7**, 19 (1971).
110. Y. B. Zel'dovich, O. I. Leipunskii, and V. B. Librovich, *Theory of Non-steady Powder Combustion*, Moscow: Nauka, 1975.
111. B. J. Matkowsky and G. I. Sivashinsky, *SIAM J. Appl. Math.* **35**, 465 (1978).

112. E. I. Maksimov, A. G. Merzhanov, and V. M. Shkiro, *Fiz. Gor. Vzr.* **1**, 24 (1965).
113. A. G. Merzhanov, A. K. Filonenko, and I. P. Borovinskaya, *Dokl. Akad. Nauk SSSR* **208**, 892 (1973).
114. A. G. Merzhanov and I. P. Borovinskaya, *CST* **10**, 195 (1975).
115. J. D. Huffington, *Trans. Faraday Soc.* **50**, 942 (1954).
116. D. M. Clemmow and J. D. Huffington, *Trans. Faraday Soc.* **52**, 385 (1956).
117. R. A. Yount and T. A. Angelus, *AIAA Journal* **2**, 1307 (1964).
118. Y. H. Inami and H. Shanfield, *AIAA Journal* **2**, 1314 (1964).
119. E. W. Price, "Review of the Combustion Instability Characteristics of Solid Propellants," in *Advances in Tactical Rocket Propulsion*, S. S. Penner, ed., AGARD Conference Proceedings no. 1, Maidenhead, England: Technivision Services, 1968, 139–194.
120. T. P. Ivleva, A. G. Merzhanov, and K. G. Shkadinskii, *Dokl. Akad. Nauk SSSR* **239**, 255 (1978).
121. G. I. Sivashinsky, *SIAM J. Appl. Math.* **40**, 432 (1981).
122. D. T. Harrje and F. H. Reardon, eds., *Liquid Propellant Rocket Combustion Instability*, NASA SP-194, U.S. Government Printing Office, Washington, D.C. (1972).
123. R. J. Priem, *9th Symp.* (1963), 982–992.
124. W. C. Strahle, *AIAA Journal* **3**, 957 (1965).
125. W. C. Strahle, *AIAA Journal* **3**, 1195 (1965).
126. W. C. Strahle, *10th Symp.* (1965), 1315–1325.
127. W. C. Strahle, *11th Symp.* (1967), 747–754.
128. L. Crocco, *10th Symp.* (1965), 1101–1128.
129. L. Crocco, *12th Symp.* (1969), 85–99.
130. H. S. Tsien, *Engineering Cybernetics*, New York: McGraw-Hill, 1954, 94–110.
131. C. C. Ross and P. P. Datner, *Selected Combustion Problems, Fundamentals and Aeronautical Applications*, AGARD, London: Butterworths Scientific Publications, 1954, 352–376.
132. S. S. Penner and P. P. Datner, *5th Symp.* (1955), 11–28.
133. M. Barrère and J. Corbeau, "Les Instabilités de Combustion dans les Fusées à Propergol Liquide," presented at Fifth AGARD Combustion and Propulsion Colloquium, Braunschweig (1962).
134. R. J. Fontaine, R. S. Levine, and L. P. Combs, "Secondary Nondestructive Instability in Medium Size Liquid Fuel Rocket Engines," in *Advances in Tactical Propulsion*, S. S. Penner, ed., AGARD Conference Proceedings no. 1, Maidenhead, England: Technivision Services, 1968, 383–419.
135. A. A. Putnam, *Combustion Driven Oscillations in Industry*, New York: Elsevier, 1971.
136. A. A. Putnam and D. J. Brown, "Combustion Noise: Problems and Potentials," in *Combustion Technology: Some Modern Developments*, H. B. Palmer and J. M. Beér, eds., New York: Academic Press, 1974, 127–162.
137. G. H. Markstein, ed., *Non-Steady Flame Propagation*, New York: Macmillan, 1964.
138. M. W. Thring, *7th Symp.* (1959), 659–663.
139. F. Mauss, E. Perthuis and B. Salé, *10th Symp.* (1965), 1241–1249.
140. P. H. Kydd, *12th Symp.* (1969), 183–192.
141. J. M. Pariel and L. de Saint Martin, *12th* Symp. (1969), 193–201.

142. C. J. Lawn, *19th Symp.* (1982), 237–244.

143. R. Sehgal and L. Strand, *AIAA Journal* **2**, 696 (1964).

144. R. Akiba and M. Tanno, "Low Frequency Instability in Solid Rocket Motors," in *Proceedings of the First Symposium (International) on Rockets and Astronautics*, Yokendo, Bunkyo-Ku, Tokyo (1959), 74–82.

145. M. W. Beckstead, N. W. Ryan, and A. D. Baer, *AIAA Journal* **4**, 1622 (1966).

146. M. W. Beckstead and E. W. Price, *AIAA Journal* **5**, 1989 (1967).

147. M. W. Beckstead, H. B. Mathes, E. W. Price, and F. E. C. Culick, *12th Symp.* (1969), 203–211.

148. J. S. T'ien, W. A. Sirignano and M. Summerfield, *AIAA Journal* **8**, 120 (1970).

149. C. K. Law and F. A. Williams, *CST* **6**, 335 (1973).

150. Y. B. Zel'dovich, G. I. Barenblatt, V. B. Librovich and G. M. Makhviladze, *The Mathematical Theory of Combustion and Explosion*, Moscow: Nauka, 1980, Chapter 6.

151. J. D. Buckmaster and G. S. S. Ludford, *Theory of Laminar Flames*, Cambridge: Cambridge University Press, 1982, Chapter 11.

152. G. I. Sivashinsky, *Annual Review of Fluid Mechanics* **15**, 179 (1983).

153. P. Clavin, *Prog. Energy Combust. Sci.* (1985), to appear

154. D. Layzer, *J. Chem. Phys.* **22**, 222, 229 (1954).

155. J. B. Rosen, *J. Chem. Phys.* **22**, 733, 743 (1954).

156. J. B. Rosen, *6th Symp.* (1957), 236–241.

157. J. F. Wehner and J. B. Rosen, *C & F* **1**, 339 (1957).

158. J. Menkes, *Proc. Roy. Soc. London* **253A**, 380 (1959).

159. J. Adler, *Appl. Sci. Research* **A11**, 65 (1962).

160. B. J. Matkowsky and D. O. Olagunju, *SIAM J. Appl. Math.* **39**, 290 (1980).

161. S. B. Margolis and B. J. Matkowsky, *CST* **27**, 193 (1982).

162. V. I. Golovichev, A. M. Grishin, V. M. Agranat, and V. N. Bertsun, *Dokl. Akad. Nauk SSSR* **241**, 305 (1978).

163. J. F. Clarke and A. C. McIntosh, *Proc. Roy. Soc. London*, **372A**, 367 (1979).

164. S. B. Margolis, *CST* **22**, 143 (1980); *18th Symp.* (1981), 679–693.

165. A. C. McIntosh and J. F. Clarke, "The Time Response of a Premixed Flame to Changes in Mixture Strength," in *Combustion in Reactive Systems*, vol. 76 of *Progress in Astronautics and Aeronautics*, J. R. Bowen, N. Manson, A. K. Oppenheim, and R. I. Soloukhin, eds., New York: American Institute of Aeronautics and Astronautics, 1981, 443–462.

166. B. J. Matkowsky and D. O. Olagunju, *SIAM J. Appl. Math.* **40**, 551 (1981).

167. G. Joulin, *CST* **27**, 83 (1981).

168. G. Joulin, *C & F* **46**, 271 (1982).

169. J. Buckmaster, *SIAM J. Appl. Math.* **43**, 1335 (1983).

170. P. L. García-Ybarra and P. Clavin, "Cross-Transport Effects in Nonadiabatic Premixed Flames," in *Combustion in Reactive Systems*, vol. 76 of *Progress in Astronautics and Aeronautics*, J. R. Bowen, N. Manson, A. K. Oppenheim and R. I. Soloukhin, eds., New York: American Institute of Aeronautics and Astronautics, 1981, 463–481.

171. G. I. Sivashinsky, *SIAM J. Appl. Math.* **39**, 67 (1980).

172. G. I. Sivashinsky, *J. Chem. Phys.* **62**, 638 (1975).

173. G. Joulin and P. Clavin, *C & F* **35**, 139 (1979).

174. G. I. Sivashinsky, *ASME Trans. Ser. C; Journal of Heat Transfer* **96**, 530 (1974).

175. G. Joulin and T. Mitani, *C & F* **40**, 235 (1981).
176. G. I. Sivashinsky, *Acta Astronautica* **4**, 1177 (1977).
177. P. Clavin and F. A. Williams, *J. Fluid Mech.* **116**, 251 (1982).
178. P. Pelcé and P. Clavin, *J. Fluid Mech.* **124**, 219 (1982).
179. P. Clavin and G. Joulin, *Journal de Physique-Lettres* **44**, L-1 (1983).
180. M. Matalon and B. J. Matkowsky, *J. Fluid Mech.* **124**, 239 (1983).
181. G. I. Taylor, *Proc. Roy. Soc. London* **201A**, 192 (1950).
182. A. Levy, *Proc. Roy. Soc. London* **283A**, 134 (1965).
183. R. M. Davies and G. I. Taylor, *Proc. Roy. Soc. London* **200A**, 385 (1950).
184. R. G. Zalosh, "Gas Explosion Research at Factory Mutual," in *Fuel-Air Explosions*, J. H. S. Lee and C. M. Guirao, eds., Waterloo, Canada: University of Waterloo Press, 1982, 771–785.
185. D. M. Solberg, J. A. Pappas, and E. Skramstad, *18th Symp.* (1981), 1607–1614.
186. D. M. Solberg, "Gas Explosion Research Related to Safety of Ships and Off-shore Platforms," in *Fuel-Air Explosions*, J. H. S. Lee and C. M. Guirao, eds., Waterloo Canada: University of Waterloo Press, 1982, 787–819.
187. L. Orloff and J. deRis, *13th Symp.* (1971), 979–992.
188. V. N. Krivulin, L. A. Lovachev, E. A. Kudryavtsev, and A. N. Baratov, *Fiz. Gor. Vzr.* **11**, 890 (1975).
189. L. A. Lovachev, *C & F* **27**, 125 (1976).
190. L. D. Landau, *Zhur. Eksp. Teor. Fiz.* **14**, 240 (1944).
191. D. J. Parks, N. J. Alvares, and D. G. Beason, *Fire Safety Journal* **2**, 237 (1979).
192. M. S. Uberoi, *Phys. Fluids* **2**, 72 (1959).
193. Y. B. Zel'dovich, *Zhur. Prikl. Mekh. Tekhn. Fiz.*, no. 1, 102 (1966).
194. Y. B. Zel'dovich, A. G. Istratov, N. I. Kidin, and V. B. Librovich, *CST* **24**, 1 (1980).
195. Y. B. Zel'dovich, *C & F* **40**, 225 (1981).
196. A. I. Rozlovskii and Y. B. Zel'dovich, *Dokl. Akad. Nauk SSSR* **57**, 365 (1954).
197. A. G. Istratov and V. B. Librovich, *Astronautica Acta* **14**, 433 (1969).
198. D. M. Michelson and G. I. Sivashinsky, *Acta Astronautica* **4**, 1207 (1977).
199. G. I. Sivashinsky, *Acta Astronautica* **6**, 569 (1979).
200. D. M. Michelson and G. I. Sivashinsky, *C & F* **48**, 211 (1982).
201. B. Bregeon, A. S. Gordon, and F. A. Williams, *C & F* **33**, 33 (1978).
202. T. Mitani and F. A. Williams, *C & F* **39**, 169 (1980).
203. A. Smithells and H. Ingle, *J. Chem. Soc.* **61**, 204, 217 (1882).
204. G. H. Markstein, *J. Aero. Sci.* **18**, 199 (1951).
205. H. Einbinder, *J. Chem. Phys.* **21**, 480 (1953).
206. W. Eckhaus, *J. Fluid Mech.* **10**, 80 (1961).
207. B. T. Chu and J. Y. Parlange, *J. de Méchanique* **1**, 293 (1962).
208. G. I. Barenblatt, Y. B. Zel'dovich, and A. G. Istratov, *Zhur. Prikl. Mekh. Tekhn. Fiz.* no. 4, 21 (1962).
209. A. G. Istratov and V. B. Librovich, *Prikl. Mat. Mekh.* **30**, 451 (1966).
210. J. Y. Parlange, *J. Chem. Phys.* **48**, 1843 (1968).
211. A. P. Aldushin and B. I. Khaikin, *Fiz. Gor. Vzr.* **11**, 128 (1975).
212. G. I. Sivashinsky, *Acta Astronautica* **3**, 889 (1976).
213. G. I. Sivashinsky, *CST* **15**, 137 (1977).
214. J. Buckmaster, *C & F* **28**, 225 (1977).
215. A. P. Aldushin and S. G. Kasparov. *Dokl. Akad. Nauk SSSR* **244**, 67 (1979).

216. B. J. Matkowsky and G. I. Sivashinsky, *SIAM J. Appl. Math.* **37**, 669 (1979).
217. B. J. Matkowsky, L. J. Putnick, and G. I. Sivashinsky, *SIAM J. Appl. Math.* **38**, 489 (1980).
218. G. I. Sivashinsky and B. J. Matkowsky, *SIAM J. Appl. Math.* **40**, 255 (1981).
219. M. L. Frankel and G. I. Sivashinsky, *CST* **29**, 207 (1982).
220. B. J. Matkowsky and D. O. Olagunju, *SIAM J. Appl. Math.* **42**, 486, 1138 (1982).
221. G. I. Sivashinsky, C. K. Law, and G. Joulin, *CST* **28**, 155 (1982).
222. G. Joulin and G. I. Sivashinsky, *CST* **31**, 75 (1983).
223. P. Clavin and P. L. García-Ybarra, *J. de Méchanique Théorique et Appliquée* **2**, 245 (1983).
224. P. Clavin and C. Nicoli, *C & F* **60**, 1 (1985).
225. M. Matalon and B. J. Matkowsky, *SIAM J. Appl. Math.* **44**, 327 (1984).
226. G. H. Markstein, *3rd Symp.* (1949), 162–167.
227. G. H. Markstein, *Jet Propulsion* **26**, 490 (1956).
228. R. E. Petersen and H. W. Emmons, *Phys. Fluids* **4**, 456 (1961).
229. T. Maxworthy, *C & F* **6**, 233 (1962).
230. T. Mitani, *CST* **23**, 93 (1980).
231. T. Mitani and F. A. Williams, *Archivum Combustionis* **1**, 61 (1981).
232. S. Ishizuka and H. Tsuji, *18th Symp.* (1981), 695–703.
233. M. R. Dongworth and A. Melvin, *CST* **14**, 177 (1976).

CHAPTER 10

Theory of Turbulent Flames

Although Mallard and le Chatelier [1] recognized that turbulence affected burning velocities, scientific investigations of turbulent flames began in 1940 with Damköhler's classical theoretical and experimental study [2] of premixed flames. Damköhler's work was soon extended by Shchelkin [3] and others. However, the theoretical understanding of premixed turbulent flames contributed by subsequent works prior to 1965 was not much greater than that originally developed by Damköhler. Accounts of a number of these early turbulent-flame studies have appeared in textbooks [4]-[6]. Theoretical analyses of turbulent diffusion flames began in the same decade [7]-[11], and reviews of much of the early work on this subject as well are available [4], [6], [12].

In the first edition of this book the comment was made that, in a sense, the topic of premixed turbulent flames did not belong in a book on combustion theory because the theoretical basis of the subject was practically undeveloped. This statement is no longer true. In recent years remarkable progress has been made in the theory of turbulent flames, and a number of new approaches to the subject are available for both premixed and nonpremixed systems. Moreover, a few of these approaches are sufficiently firmly based to enable analyses to be pursued with high confidence in limiting cases. Therefore, it is now entirely appropriate to discuss theories of turbulent combustion here. In fact, the subject is so broad that we shall be able to cover only a portion of it and must refer to recent reviews [13]-[40] for more extensive presentations. In particular, [27] is a 250-page book devoted to the subject.

The most fundamental question in turbulent combustion concerns the applicability of a continuum description. It is conventional to assume that the full Navier-Stokes equations (that is, the Navier-Stokes equations augmented by chemical-kinetic laws) govern turbulent combustion; the most convincing derivation of this result for gases is based on kinetic theory (Appendixes D and E). Molecular chaos is a key hypothesis underlying the derivation in that it enables the collision contribution $\delta f_i/\delta t$ in equation (D-2) to be expressed in terms of integrals involving only the one-particle velocity distribution function $f_i(\mathbf{x}, \mathbf{v}, t)$, $i = 1, \ldots, N$. Without molecular chaos, equation (D-2) is not closed because a two-particle distribution function, representing the joint probability that a molecule is in one volume element of phase space, while at the same time another molecule is in another volume element, appears in an integral expression for $\delta f_i/\delta t$. The Boltzmann equation must be supplemented by an equation for the two-particle distribution in the absence of molecular chaos. Certain studies in kinetic theory in which this has been accomplished [41], [42] through introduction of a weakened chaos hypothesis have revealed long-range correlations that persist on macroscopic time scales and thereby modify the full Navier-Stokes equations, especially the chemical reaction-rate expression. There are particular problems in turbulent combustion, such as the homogeneous thermal explosion of turbulent hydrogen-oxygen mixtures, in which these kinetic-theory effects have been found to be significant, for example, in reducing explosion times appreciably [43]-[45]. Therefore, it cannot be claimed that theories of turbulent combustion should necessarily be developed from the Navier-Stokes equations; a more fundamental level of description may be needed to address certain phenomena. However, a Navier-Stokes description appears to suffice for most phenomena. The following presentation is based entirely on this continuum view. There are a number of excellent background books on Navier-Stokes turbulence in nonreactive fluids [46]-[52].

Navier-Stokes turbulence, characterized by fluctuations of velocities and state variables in space and time, occurs when the inertial forces are sufficiently large in comparison with the viscous forces—that is, at sufficiently large values of a representative Reynolds number Re (see Section 3.1.7). Turbulence develops from fluid-mechanical instabilities of flows [53] when Re is large enough to prevent viscous damping from suppressing fluctuations. The Navier-Stokes equations, combined with a proper set of initial and boundary conditions, are deterministic in that they are believed to possess a unique solution. However, at high Re, the solution is strongly sensitive throughout most of the flow field to the initial and boundary conditions, and in reality the conditions cannot be specified accurately enough to obtain the deterministic solution, either theoretically or experimentally. It then becomes reasonable to introduce statistical methods and to seek only probabilistic aspects of the solutions, that is, to treat turbulence as a random process.

Basics of the probabilistic description of turbulence will be presented in the following section. Conceptually, we may consider the initial and boundary conditions to be given only in a probabilistic sense. Although the statistical properties of the flow might be expected to depend on those assigned to the initial and boundary conditions, this input information is modified so greatly by the many fluid instabilities that the statistical aspects of the turbulence are found to be virtually independent of those of initial and boundary fluctuations. In this sense, turbulence possesses a "strange attractor" that draws its statistics away from input statistics and toward statistics that are characteristic of the flow system [54]. It seems clear that even if a totally deterministic problem were specified, the solution in most of the flow would be pseudorandom in that it would pass tests for randomness. The solution to the deterministic problem is said to exhibit chaos in that, for example, small changes in boundary or initial conditions lead to large and irregular changes in the solution away from the boundaries. Studies of strange attractors are just beginning and have not yet found specific applications in turbulent combustion.

After presenting some background information on theories of turbulent combustion in the following section, we shall discuss theoretical descriptions of turbulent diffusion flames (Section 10.2) and of turbulent premixed flames (Section 10.3). Although there are situations in which turbulence can tend to wash out the distinction between premixed and non-premixed systems, the best approaches to analyses of these two limits remain distinct today, and thus it is convenient to treat them separately.

10.1. PROBABILISTIC DESCRIPTIONS

10.1.1. Probability-density functionals

With turbulent combustion viewed as a random (or stochastic) process, mathematical bases are available for addressing the subject. A number of textbooks provide introductions to stochastic processes (for example, [55]). In turbulence, any stochastic variable, such as a component of velocity, temperature, or the concentration of a chemical species, which we might call v, is a function of the continuous variables of space $\mathbf{x}$ and time t and is, therefore, a stochastic function. A complete statistical description of a stochastic function would be provided by a *probability-density functional*, $\mathscr{P}[v(\mathbf{x}, t)]$, defined by stating that the probability of finding the function in a small range $\delta v(\mathbf{x}, t)$ about a particular function $v(\mathbf{x}, t)$ is $\mathscr{P}[v(\mathbf{x}, t)]\delta v(\mathbf{x}, t).$*

---

* Although it is essential conceptually to make a distinction between a stochastic function and a particular function $v(\mathbf{x}, t)$ that appears as the argument of $\mathscr{P}$, we choose not to do so notationally for the sake of simplicity. Symbols like v always will refer to arguments of probabilities unless there is an indication to the contrary.

Here, $\delta v(\mathbf{x}, t)$ is a volume element in the function space, and from the fundamental definition of measure or probability, $\int \cdots \int \mathscr{P}[v(\mathbf{x}, t)]\delta v(\mathbf{x}, t) = 1$, where $\int \cdots \int$ effectively involves a continuously infinite number of real-variable integrations extending over the entire function space. For incompressible Navier-Stokes turbulence, it can be shown [56] from the Navier-Stokes equations that if $\mathbf{v}$ represents the velocity vector, then $\mathscr{P}[\mathbf{v}(\mathbf{x}, t)]$ satisfies a functional differential equation that is linear and that in principle contains all the statistical information concerning the velocity field. For reacting flows, there is a functional differential equation, which has the same properties, for $\mathscr{P}[v(\mathbf{x}, t)]$ if v is interpreted as a vector in $N + 4$ dimensions whose components represent values of velocity, density, temperature, and all chemical compositions in an N-component mixture [27]. Although this fact and the ability to write the functional differential equation symbolically are comforting, they are of little help because useful methods of solution are unavailable.

To make progress in analyzing turbulent reacting flows, we must seek a description at a less-detailed level than that of the probability-density functional. One line of research is to attempt to solve the time-dependent Navier-Stokes equations for the reacting flow numerically;* although some interesting results have been obtained in this way for relatively simple problems (for example, [57] and [58]) by working with simplified equations such as the Euler equations, the current capabilities of electronic computers in general are much too limited to enable this approach to be pursued. Seeking less-detailed levels of description invariably leads to formulations for which, in one sense or another, there are more unknowns than equations, since any simplification must delete some potentially relevant information contained in the probability-density functional. This difficulty is the infamous "closure" problem of turbulence, which exhibits itself in many different forms, depending on the approach adopted.

An avenue that has received exploration is the development of equations for evolution of probability-density functions. If, for example, attention is restricted entirely to particular, fixed values of $\mathbf{x}$ and t, then the variable whose value may be represented by v becomes a random variable instead of a random function, and its statistics are described by a probability-density function. The *probability-density function* for v may be denoted by $P(v)$, where $P(v)\,dv$ is the probability that the random variable lies in the range dv about the value v. By definition $P(v) \geq 0$, and $\int_{-\infty}^{\infty} P(v)\,dv = 1$. One approach to obtaining an equation of evolution for $P(v)$ is to introduce the ensemble average of a fine-grained density, as described by O'Brien in [27], for example; another is formally to perform suitable integrations in

---

* Depending on the problem, this would be done either once for a sufficiently long period of time or many times, so that an ensemble of solutions is obtained, from which the statistical properties can be calculated in accordance with a frequency interpretation of the stochastic process.

the functional differential equation for $\mathscr{P}$. Irrespective of the method, an expression is obtained for $\partial P/\partial t$ that involves averages fundamentally not representable solely in terms of P. Closure hypotheses then are introduced to relate the expression for $\partial P/\partial t$ to P, with the result that its $\mathbf{x}$ derivatives (for example, ∇P)—and often its v derivatives as well (for example, $\partial^2 P/\partial v^2$)—appear in a closed equation for P [30], [37]. Although we still must contend with a formulation involving a large number of independent variables (particularly if v is a vector), methods are available for solving the types of evolution equations that have been obtained for P, with present-day computers, for a number of problems of interest.

Although closure is achieved at a relatively fundamental level in these formulations involving equations for evolution of probability-density functions, uncertainties nevertheless remain concerning the closure hypotheses, not with respect to chemical source terms (which are handled with relative ease in the approach) but rather with respect to influences of convection and diffusion. For example, the equations often involve a diffusion approximation in v-space, the fundamental justification for which can be challenged. Because of the complexities in solving the evolution equation, this type of approach seems best adapted to comparatively simple flows, such as in systems with high degrees of spatial uniformity. We shall not here develop the methods based on equations for evolution of probability-density functions.

10.1.2. Averages

The oldest approach to theoretical analyses of turbulent flows works only with averages and does not introduce probabilities at all. Methods of this type often are called **moment methods** because averages are moments of probability-density functions (as discussed later). If an average of any of the Navier-Stokes equations for reacting flows is taken, then an equation is obtained for the average value of a flow variable, say $\bar{v}$. However, because of the nonlinearity of the conservation equations, an equation for $\bar{v}$ contains averages of products of flow variables, and since averages of products in general are not equal to products of averages, there always are more unknowns than equations in any selected set of averaged Navier-Stokes equations. Therefore, with moment methods the closure problem is manifest in the need to relate higher moments (averages of products of larger numbers of flow variables) to lower moments (averages of products with fewer variables). Enough closure approximations, which are also termed **turbulence modeling**, are needed so that the number of unknowns becomes the same as the number of equations.

This type of closure for constant-density, nonreacting turbulent flows has a long and rather successful history of development and is described in many textbooks (for example, [48]). From the inertial terms in momentum conservation arise averages of products of velocity components; certain forms of these are called **Reynolds stresses** and typically are assumed to be proportional to gradients of average velocities, with constants of proportionality called **turbulent viscosities** or **turbulent diffusivities**, depending on the units. Similarly, the convective terms in equations for energy and species conservation produce averages of the product of a velocity component with temperature or a species concentration, **Reynolds transport**, and these may correspondingly be modeled as proportional to the gradient of the average temperature or of the average species concentration, with the coefficients again being turbulent diffusivities, now for energy or for chemical species. This sketchy description will become defined better later, for example in the vicinity of equation (38).

In use of moment methods in reacting flows, modeling is needed for the chemical source terms as well (in contrast to methods based on equations for evolution of probability-density functions) because of their nonlinearity. As a consequence of this, the chemistry can introduce severe difficulties in moment methods for reacting flows [17], [28]. Even if the difficulties in the chemical source terms are overcome, problems that stem from the chemistry remain in modeling the Reynolds transport [28], [40]; in a rough physical sense, the chemistry interacts with the transport modeling through processes in which fluid elements are transported by turbulence to a location, react there, and then are transported back. A consequence of this interaction is the occurrence of **counter-gradient diffusion**; the signs of the turbulent diffusivities must change within the flow if a moment-method formulation is adopted.* Counter-gradient diffusion now is a well-established, real phenomenon for certain turbulent reacting flows [59]–[63]. Since it cannot be handled well by moment methods, these methods are ill-suited for many turbulent combustion problems.†

Because of these difficulties with moment methods for reacting flows, we shall not present them here. A number of reviews are available [22], [25], [27], [32]. There are classes of turbulent combustion problems for which moment methods are reasonably well justified [40]. Since the computational difficulties in use of moment methods tend to be less severe than those for many other techniques (for example, techniques involving evolution equations for probability-density functions), they currently are being applied to turbulent combustion in relatively complex geometrical configurations [22], [31], [32]. Many of the aspects of moment methods play important roles in other approaches, notably in those for turbulent diffusion flames (Section 10.2). We shall develop those aspects later, as they are needed.

---

* In normal, or *gradient*, diffusion, the transport-flux and gradient vectors have opposite directions; in counter-gradient diffusion, they have the same direction. When both gradient and counter-gradient diffusion occur in different regions of the flow, the sign of the diffusivity changes somewhere.

† Related problems can arise in the modeling needed for methods based on evolution of probability-density functions.

Nevertheless, definitions of averages per se require further discussion here because they arise in one way or another in all approaches, and there are a number of different ways to define averages.

In statistical methods, the most fundamental average is an ensemble average. The stochastic process is imagined to be reproduced many times; the collection of reproductions is the ensemble, and the ensemble average is the sum of the values at the x and t of interest in each member of the ensemble, divided by the total number of members of the ensemble. Ensemble averages are relatively well adapted to application in reciprocating-engine combustion, for example, where each cycle may be viewed as a separate member of the ensemble.* There are conditions under which ensemble averages may be replaced by space or time averages that may be obtained more simply; ergodic theorems establish the equivalence of ensemble and time averages under suitable conditions of statistical *stationarity* of the stochastic process. Although there are different definitions of stationarity, we may consider a process to be stationary in a strict sense if all of its statistical properties are independent of the selection of the origin of time. During steady operation, the turbulent flow through a continuous combustor such as a gas turbine or a ramjet appears to be stationary on physical grounds. In these flows, averages may be viewed as time averages, and simplifications in statistical descriptions arise because of the stationarity.

There are other flows, called *homogeneous*, in which ensemble averages may be viewed instead as space averages. **Homogeneous turbulence** [47] is turbulence whose statistical properties are independent of the selection of the origin of x; there are corresponding simplifications in its statistical description. Turbulent combustion in a well-stirred reactor may be approximated as if it possessed homogeneity. There are turbulent flows in which homogeneity applies only in particular coordinate directions. For example, a turbulent flame positioned normal to the average flow, downstream from a uniform grid in a suitably designed combustion tunnel, is nearly homogeneous in the two transverse directions parallel to the grid, in that the statistical properties are invariant under translations in these directions, and ensemble averages are equivalent to averages over either of the coordinates with respect to which the process is homogeneous. Homogeneity must not be confused with *isotropy*, which refers to invariance of statistical properties under rotation of the coordinate system. The statistics of streamwise and transverse velocity fluctuations must be the same in isotropic turbulence, for example; isotropy occurs only approximately even without combustion, and a flame can be a source for generation of further departures from isotropy. The experimental evaluation of some averages that depend on coordinate directions can be simplified somewhat by use of isotropy if it applies.

---

* There are difficulties here associated with *cycle-to-cycle variation* [64], which is considered to be different from turbulence.

Irrespective of whether space, time, or ensemble averages are employed, questions arise as to what variables are to be selected for use (for example, for which averages are to be calculated) in analyses of turbulent combustion. Considering equation (1-1) we may guess that ρ and $\mathbf{v}$ are reasonable variables, and we may therefore choose to work with the averages $\bar{\rho}$ and $\bar{\mathbf{v}}$. The ensemble average of equation (1-1) (with overbars signifying averages) then is

$$\partial \bar{\rho}/\partial t + \mathbf{V} \cdot (\overline{\rho \mathbf{v}} + \overline{\rho' \mathbf{v}'}) = 0, \tag{1}$$

where the local, instantaneous fluctuations are $\rho' = \rho - \bar{\rho}$ and $\mathbf{v}' = \mathbf{v} - \bar{\mathbf{v}}$, which have $\bar{\rho}' = 0$ and $\bar{\mathbf{v}}' = 0$ by definition. The average of a product of fluctuations appears in equation (1). A simplification can be achieved in variable-density flow by selecting the product $\rho \mathbf{v}$ as a fundamental variable to be employed instead of $\mathbf{v}$. With ρ and $\rho \mathbf{v}$ as variables, a product of fluctuations does not arise in the average of the continuity equation; the divergence term is then simply $\mathbf{V} \cdot (\overline{\rho \mathbf{v}})$. A formulation in which this modification was effected has been developed by Favre [65].

Favre defines the mass-weighted average of a variable v as $\tilde{v} = (\overline{\rho v})/\bar{\rho}$ and denotes the departure of v from $\tilde{v}$ as $v'' = v - \tilde{v}$. Then the average $\overline{v''}$ is not zero, but $\overline{\rho v''} = 0$ by definition. In this notation, equation (1) becomes

$$\partial \bar{\rho}/\partial t + \mathbf{V} \cdot (\bar{\rho} \tilde{\mathbf{v}}) = 0. \tag{2}$$

The simplification of equation (1) represented by equation (2) is relatively unimpressive, but in the average of the equation for momentum conservation, an appreciably greater compactness is obtained in the inertial terms; the dyadic $\overline{\rho \mathbf{v} \mathbf{v}} - \bar{\rho} \tilde{\mathbf{v}} \tilde{\mathbf{v}}$ is found to be $\overline{\rho \mathbf{v}'' \mathbf{v}''} = \bar{\rho}(\widetilde{\mathbf{v}'' \mathbf{v}''})$, which involves four terms if the more conventional average is employed. The simplifications that arise in convective and inertial terms are offset somewhat by complications produced in molecular transport terms, but in many uses of averaged equations, modeling is introduced for effects of molecular transport irrespective of the averaging selected. Therefore, notable economies generally are achieved through introduction of these mass-weighted averages.

The desirability of working with Favre averages is a topic of current controversy. The controversy cannot be resolved on the basis of ease of measurement because some instruments provide $\bar{v}$ most directly, while others provide $\tilde{v}$. It is unclear whether the successful constant-density modeling of Reynolds stresses is extendable more readily to $\overline{\rho \mathbf{v}' \mathbf{v}'}$ or to $\overline{\rho \mathbf{v}'' \mathbf{v}''}$. In thorough studies there is unlikely to be a fundamental reason for making a choice because the identity $\bar{\rho} \tilde{v} = \bar{\rho} \bar{v} + (\overline{\rho' v'})$ demonstrates that one average is recoverable from the other once $\bar{\rho}$ and $\overline{\rho' v'}$ are obtained, as they must be for completeness. The economies associated with the introduction of Favre averaging therefore appear to constitute the determining factor in favor of its adoption. Probability-density functions appropriate for use in connection with mass-weighted averages have been defined by Bilger (see [27]), as indicated at the end of the following section.

10.1.3. Properties of probability-density functions

The basic definition of a probability-density function is given in Section 10.1.1. From $P(v)$, the *average* of the random variable is readily calculated as

$$\bar{v} = \int_{-\infty}^{\infty} vP(v)\, dv. \tag{3}$$

This is also the first moment of $P(v)$ about zero because the nth moment about zero is defined as

$$\overline{v^n} = \int_{-\infty}^{\infty} v^n P(v)\, dv. \tag{4}$$

The mean-square value, $\overline{v^2}$, is the second moment about zero. The *variance* is the second moment about the mean, also termed the *second central moment*, and conventionally is denoted by

$$\overline{v'^2} = \sigma^2 = \int_{-\infty}^{\infty} (v - \bar{v})^2 P(v)\, dv = \overline{v^2} - \bar{v}^2, \tag{5}$$

where the last equality follows directly from the definitions by expanding $(v - \bar{v})^2$. Either the variance or its positive root, σ, the *standard deviation*, measures the width of the distribution of $P(v)$; here σ represents the root-mean-square intensity of the fluctuations about the mean (that is, average) value. The third and fourth central moments may be written as

$$s\sigma^3 = \int_{-\infty}^{\infty} (v - \bar{v})^3 P(v)\, dv \tag{6}$$

and

$$\kappa\sigma^4 = \int_{-\infty}^{\infty} (v - \bar{v})^4 P(v)\, dv, \tag{7}$$

where s and κ are called the *skewness* and *kurtosis* (or flatness factor), respectively. Knowledge of the values of $\bar{v}$, σ, s, and κ would provide a significant amount of information about $P(v)$; for example, s is a measure of asymmetry of $P(v)$, with positive s implying a more gradual tailing off toward positive values of v than toward negative values, and κ is a measure of how kinky $P(v)$ is, with larger values of κ corresponding to a shape more like a sharp, rectangular mesa than like a gently curved hill. However, in general, $P(v)$ cannot be reconstructed completely even from a knowledge of the values of all of its moments because this is possible in principle only if $P(v)$ obeys various continuity and convergence requirements that many probability-density functions of practical interest do not satisfy.

The average of any single-valued function of v, say $g(v)$, is

$$\bar{g} = \int_{-\infty}^{\infty} g(v)P(v)\,dv, \tag{8}$$

which may be evaluated directly by integration. If $g(v)$ possesses a power-series expansion that converges wherever $P(v) \neq 0$, then $\bar{g}$ may be evaluated in terms of the moments of $P(v)$. Given $g(v)$ and $P(v)$, a probability-density function for a new random variable with its value represented by g, say $Q(g)$, may be constructed according to the formula

$$Q(g) = \sum_{i=1}^{n} \left[P(v_i) \middle/ \left| \frac{dg}{dv} \right|_{v=v_i} \right], \tag{9}$$

where v_i $(i = 1, \ldots, n)$ are the values of v for which $g(v)$ takes on the value of g. Equation (9) may be understood best by first considering functions $g(v)$ that are monotonic, so that there is only one value of v for any given g (that is, $n = 1$), and the equation becomes $Q(g)|dg| = P(v)|dv|$, stating that all members of the ensemble for which the original random variable lies in the range dv about v also have the new random variable lying in the range dg about g, where $dg = |dg/dv|\,dv$; equation (9) extends this result to non-monotonic functions by adding the contributions from the members of the ensemble having the original random variable in the range $dg/|dg/dv|$ about each v_i for which $g(v_i) = g$. If g remains constant over a finite range of v—for example, $g = g_0$ for $v_1 < v < v_2$—then equation (9) simply produces infinity at $g = g_0$, but through consideration of the fraction of members of the ensemble for which $g = g_0$, it becomes clear that in the vicinity of $g = g_0$, there is a delta-function contribution to Q given by

$$Q(g) = \left[\int_{v_1}^{v_2} P(v)\,dv \right] \delta(g - g_0);$$

in turbulent combustion, delta functions often arise in probability-density functions for temperature and species concentrations. Considerations like those indicated here enable $Q(g)$ to be generated for any single-valued function $g(v)$. If $g(v)$ is multiple-valued, then neither $\bar{g}$ nor $Q(g)$ can be obtained even in principle from $P(v)$ because it lacks information concerning the relative frequencies with which each of the different possible values of g occurs in the ensemble; in equilibrium diffusion flames, the mixture fraction usually is a double-valued function of temperature (see Section 3.4.2), and therefore in the absence of additional information, a probability-density function for it cannot be generated from one for temperature (which fails to distinguish rich from lean mixtures).

Knowledge of $P(v)$ for one random variable at fixed $\mathbf{x}$ and t offers very limited information concerning a stochastic process because no aspects of relationships obeyed by the random variable at adjacent values of $\mathbf{x}$ and t are contained therein. To increase the amount of statistical information, joint probability-density functions may be introduced. If v_1 is a value of a stochastic function at a particular position and time $(\mathbf{x}_1, t_1)$, and v_2 is a value of the same stochastic function at a different position and time $(\mathbf{x}_2, t_2)$, then a relevant *joint probability-density function* (for two random variables) may be denoted by $P(v_1, v_2)$, where $P(v_1, v_2)\, dv_1\, dv_2$ is defined as the probability that the first random variable lies in the range dv_1 about v_1 and the second in the range dv_2 about v_2, viewed as the fraction of members of the ensemble in which both of the random variables are in their respective ranges. Here $P(v_1, v_2)$ is a two-point, two-time joint probability-density function of one stochastic function. If $\mathbf{x}_1 = \mathbf{x}_2$ but $t_1 \neq t_2$, then the adjectives one-point, two-time would be applied to $P(v_1, v_2)$, while if $t_1 = t_2$ but $\mathbf{x}_1 \neq \mathbf{x}_2$, the terminology two-point, one-time would be used. Two-point probability-density functions play important roles in describing stochastic processes whose underlying conservation equations involve derivatives with respect to $\mathbf{x}$, for example. Even for $\mathbf{x}_1 = \mathbf{x}_2$ and $t_1 = t_2$, joint probability-density functions (now one-point, one-time) are of interest in turbulent combustion; here v_1 and v_2 may be values of two different components of velocity or of one velocity component and temperature, for example, as needed in evaluating Reynolds stresses and Reynolds transports. If v_1 and v_2 relate to the same stochastic function, then the further qualifier *auto* often is used; if different, *cross*. In a similar manner, joint probability-density functions may be defined for any number of random variables. Thus $P(v_1, v_2, \ldots, v_n)$ may be introduced, and allowing n to approach infinity suitably in principle could result in a complete statistical description of a stochastic process. The definition requirements $P(v_1, v_2, \ldots, v_n) \geq 0$ and

$$\int_{-\infty}^{\infty} \int_{-\infty}^{\infty} \cdots \int_{-\infty}^{\infty} P(v_1, v_2, \ldots, v_n)\, dv_1\, dv_2 \ldots dv_n = 1 \tag{10}$$

always apply. In practice it is seldom convenient to consider $n > 3$.

It is always possible to recover $P(v_1)$ from $P(v_1, v_2)$ because

$$\int_{-\infty}^{\infty} P(v_1, v_2)\, dv_2 = P(v_1) \tag{11}$$

from the underlying definitions. In equation (11) $P(v_1)$ would be called a *marginal probability-density function*, having been generated on a margin (the v_1 axis) from the projection of a joint function. Often it is desirable to consider not all members of the ensemble, but only those in which one random variable (for instance, the second) takes on a specified value or lies in a specified range; restricting attention to these members is establishing a condition on the probability, and the resulting probability-density functions are called *conditioned probability-density functions*. It is conventional to place a vertical bar inside the parentheses in the notation for P and to write the conditioning restriction after the bar. If the conditioning is that the second random variable have the value v_2, then the notation $P(v_1|v_2)$ is used for the conditioned probability-density function. Thus $P(v_1|v_2)\,dv_1$ is the probability that the first random variable lies in the range dv_1 about v_1 while the second takes on the value v_2. The meanings of the various probabilities show that

$$P(v_1, v_2) = P(v_1|v_2)P(v_2). \tag{12}$$

A result of the definitions, the meaning of which should be clear in spite of the somewhat imprecise notation, is **Bayes's theorem**, $P(v_1|v_2)P(v_2) = P(v_2|v_1)P(v_1)$, in which all four P's are different functions.

Conditioning often is needed in turbulent combustion. For example, often there is interest in flows (for example, jets and wakes) that are partly turbulent and partly nonturbulent (in that the flow is partly irrotational), and the conditioning may be that rotational fluid be present. These flows are said to exhibit **intermittency** of the turbulence; an intermittency function may be defined as a stochastic function having a value $I(\mathbf{x}, t) = 0$ in irrotational flow and $I(\mathbf{x}, t) = 1$ in rotational flow, and conditioning on $I = 1$ may be desirable for a number of purposes. We may decompose a probability-density function as

$$P(v) = (1 - \bar{I})P_0(v) + \bar{I}P_1(v), \tag{13}$$

where $P_0(v)$ is the probability-density function in the irrotational fluid and $P_1(v)$ is that for the rotational fluid. In turbulent diffusion flames, conditioning on the requirement that fuel be present is another example of conditioning that sometimes can be useful. Conditioned averages and conditioned moments can be defined in a straightforward manner since conditioned probability-density functions obey all the properties of unconditioned probability-density functions.

A random variable with a value denoted by v_1 is said to be statistically independent from a random variable with a value denoted by v_2 if $P(v_1|v_2)$ is independent of v_2 for all v_2—that is, if the conditioning does not affect the probability-density function. In this case $P(v_1|v_2) = P(v_1)$—that is, the conditioned probability-density function is the same as the marginal probability-density function. From equation (12) it then follows that

$$P(v_1, v_2) = P(v_1)P(v_2) \tag{14}$$

for statistically independent random variables. Statistical independence is a strong property that is difficult to justify. For example, the value of the temperature at a point in a turbulent flame may be expected to be related statistically to the value of the fuel concentration there, and the value of the velocity at a point will be related to the value of the velocity at a nearby point.

A much weaker property, implied by statistical independence but not implying it, is vanishing of the *covariance*, the second cross central moment, sometimes called the *cross correlation*,

$$R_{12} = \int_{-\infty}^{\infty} \int_{-\infty}^{\infty} (v_1 - \bar{v}_1)(v_2 - \bar{v}_2)P(v_1, v_2) \, dv_1 \, dv_2. \tag{15}$$

A nondimensional measure of the covariance is the *correlation coefficient*,

$$c_{12} = R_{12}/\sqrt{\overline{v_1'^2} \, \overline{v_2'^2}}, \tag{16}$$

where the $\overline{v_i'^2}$ for $i = 1, 2$ represent mean-square fluctuations (variances), defined in equation (5). Use of equations (3) and (14) in equation (15) readily shows that $R_{12} = 0$ if the two random variables are statistically independent. If $R_{12} = 0$, then the random variables are said to be uncorrelated.

The joint probability-density function for the density of the gas and another random variable may be denoted by $P(\rho, v)$. A mass-weighted (or Favre) probability-density function for the second random variable may then be defined as

$$\tilde{P}(v) = \int_{0}^{\infty} \rho P(\rho, v) \, d\rho/\bar{\rho}. \tag{17}$$

By definition, $\tilde{P}(v)$ possesses the necessary nonnegative and normalization properties of probability-density functions. It is especially useful in connection with the Favre-averaged formulation of the conservation equations, since corresponding averages are obtained from $\tilde{P}(v)$ by the usual rules for averaging. Thus, for the $\tilde{v}$ and v'' defined above equation (2),

$$\left. \begin{aligned} \tilde{v} &= \int_{-\infty}^{\infty} v\tilde{P}(v) \, dv, \\ \widetilde{v''^2} &= \int_{-\infty}^{\infty} (v - \tilde{v})^2 \, \tilde{P}(v) \, dv, \end{aligned} \right\} \tag{18}$$

the first of which is similar to equation (3) and the second of which is like equation (5).

10.1.4. Fourier decompositions

In analyses of turbulent combustion, it is helpful for various purposes to view the processes in frequency or wave-number spaces instead of in time or physical space. Fourier transforms provide the desired decompositions. Transforms with respect to different variables arise. In working with probability-density functions for velocities, we may transform these functions with respect to velocity components to obtain "characteristic functions" from which the probability-density functions can be recovered [55]. Transforms with respect to spatial coordinates or time are encountered more often. The former is especially useful for coordinates in which the stochastic process is homogeneous, and the latter if the process is statistically stationary. The main advantage of introducing transforms in these situations is that differentiation becomes multiplication by the transform variable, thereby effecting simplifications in the analysis. The quantities transformed may be the velocity components or state variables themselves or various averages of them. Particularly for velocity and state variables, the formalism of Fourier-Stieltjes transforms may be introduced to allow for anticipated nonanalytic behavior [47], or else this refinement can be bypassed by accepting ill-behaved transforms (having delta functions in them). Here we shall consider only transforms of certain averages, namely, correlations.

The space-time *velocity correlation tensor* is defined as

$$\mathbf{R}(\mathbf{x}_1, t_1; \mathbf{x}_2, t_2) = \overline{\mathbf{v}'(\mathbf{x}_1, t_1)\mathbf{v}'(\mathbf{x}_2, t_2)}. \tag{19}$$

Its components could be evaluated from equation (15) if suitable probability-density functions were known. With $\mathbf{x} \equiv \mathbf{x}_2 - \mathbf{x}_1$ and $\tau \equiv t_2 - t_1$, we may define a four-dimensional Fourier transform as

$$\Phi(\mathbf{k}, \omega; \mathbf{x}_1, t_1) = \int_{-\infty}^{\infty} \int_{-\infty}^{\infty} \int_{-\infty}^{\infty} \int_{-\infty}^{\infty} e^{i\omega\tau} e^{i\mathbf{k}\cdot\mathbf{x}} \mathbf{R}(\mathbf{x}_1, t_1; \mathbf{x}_2, t_2)\, d\mathbf{x}\, d\tau, \tag{20}$$

where $d\mathbf{x}$ is a volume element in $\mathbf{x}$ space. In equation (20), which assumes that $\mathbf{R}$ is defined for all real values of $\mathbf{x}$ and t (for example, by putting $\mathbf{R} = 0$ for values outside the bounds of the flow of interest), the transform variable ω is the frequency, and the transform variable $\mathbf{k}$ is the wave-number vector. For a scalar field such as the temperature T, corresponding quantities are the autocorrelation,

$$R(\mathbf{x}_1, t_1; \mathbf{x}_2, t_2) = \overline{v'(\mathbf{x}_1, t_1)v'(\mathbf{x}_2, t_2)} \tag{21}$$

with $v' = T'$ (the temperature fluctuation) and the transform

$$\Phi(\mathbf{k}, \omega; \mathbf{x}_1, t_1) = \int_{-\infty}^{\infty} \int_{-\infty}^{\infty} \int_{-\infty}^{\infty} \int_{-\infty}^{\infty} e^{i\omega\tau} e^{i\mathbf{k}\cdot\mathbf{x}} R(\mathbf{x}_1, t_1; \mathbf{x}_2, t_2)\, d\mathbf{x}\, d\tau. \tag{22}$$

Cross correlations such as between temperature T and a mass fraction Y_i—namely, $\overline{T'(\mathbf{x}_1, t_1)Y_i'(\mathbf{x}_2, t_2)}$—may be treated similarly. Correlations are readily recovered from their transforms by the inversion formula; for example, with $\varphi(\omega) = \int_{-\infty}^{\infty} e^{i\omega\tau} f(\tau)\, d\tau$, we have $f(\tau) = \int_{-\infty}^{\infty} e^{-i\omega\tau}\varphi(\omega)\, d\omega/2\pi$, which would be applied four times for the space-time transforms.

Equations (20) and (22) in their full forms are seldom useful. For statistically stationary turbulence, $\mathbf{R}(\mathbf{x}_1, t_1; \mathbf{x}_2, t_2)$ and $R(\mathbf{x}_1, t_1; \mathbf{x}_2, t_2)$

depend on t_1 and t_2 only through the difference $\tau = t_2 - t_1$, and their transforms with respect to τ alone depend only on ω and the two positions; these functions for $\mathbf{x}_2 = \mathbf{x}_1$, denoted by $\boldsymbol{\Phi}(\omega; \mathbf{x}_1)$ and $\Phi(\omega; \mathbf{x}_1)$, are transforms (or spectra) of time correlations at a particular position and can be especially useful. Similarly, if there is homogeneity in some coordinate directions, then transforms with respect to those coordinates may be useful.

Ideas about the dynamics of turbulence have been developed by reference to fully homogeneous turbulence, for which the three-dimensional transforms with $t_2 = t_1$, denoted here by $\boldsymbol{\Phi}(\mathbf{k}; t_1)$ and $\Phi(\mathbf{k}; t_1)$ and called *spectra* of space correlations, play central roles. Half of the trace of $\boldsymbol{\Phi}(\mathbf{k}; t_1)$, to be denoted here by $E(\mathbf{k}; t_1)$, is related to the average turbulent kinetic energy per unit mass—say $q(\mathbf{x}_1, t_1)$—since, from equation (19), its inverse transform is

$$\frac{1}{2}\overline{\mathbf{v}'(\mathbf{x}_1, t_1) \cdot \mathbf{v}'(\mathbf{x}_1 + \mathbf{x}, t_1)} = \frac{1}{(2\pi)^3} \int_{-\infty}^{\infty} \int_{-\infty}^{\infty} \int_{-\infty}^{\infty} e^{-i\mathbf{k}\cdot\mathbf{x}} E(\mathbf{k}; t_1) \, d\mathbf{k}, \quad (23)$$

which is independent of $\mathbf{x}_1$ for homogeneous turbulence; putting $\mathbf{x} = 0$ in equation (23) gives $q(\mathbf{x}_1, t_1)$ at time t_1 as an integral over wave-number space.

For isotropic turbulence, the space correlation obtained from equation (19) by setting $t_2 = t_1$ depends only on $r \equiv |\mathbf{x}_2 - \mathbf{x}_1|$ at fixed values of $\mathbf{x}_1$ and t_1, and its spatial transform therefore depends only on $k \equiv |\mathbf{k}|$, that is,

$$\boldsymbol{\Phi}(k; \mathbf{x}_1, t_1) = \int_0^{\infty} (kr)^{-1} \sin(kr) \mathbf{R}(\mathbf{x}_1, t_1; \mathbf{x}_2, t_1) 4\pi r^2 \, dr; \quad (24)$$

in this case, equation (23) shows that the average kinetic energy per unit mass can be expressed as an integral over a scalar wave number, namely,

$$q(\mathbf{x}_1, t_1) = \int_0^{\infty} e(k; \mathbf{x}_1, t_1) \, dk, \quad (25)$$

where the *energy spectrum*, $e(k; \mathbf{x}_1, t_1)$, is $(k/2\pi)^2$ times the trace of $\boldsymbol{\Phi}(k; \mathbf{x}_1, t_1)$ and becomes $4\pi k^2 E(\mathbf{k}; t_1)/(2\pi)^3$ in equation (23).

A spectral decomposition for intensities of fluctuations of scalar fields may be obtained from equation (21) in a similar manner, for example,

$$\overline{v'^2}(\mathbf{x}_1, t_1) = \int_0^{\infty} [4\pi k^2 \Phi(k; \mathbf{x}_1, t_1)/(2\pi)^3] \, dk. \quad (26)$$

If $1/k$ is interpreted as a turbulent eddy dimension, then the integrand in equation (25) or in equation (26) may be viewed as the contribution (per unit range of wave number) to the turbulent kinetic energy or to the fluctuation intensity from eddies of size $1/k$. Spectral decompositions in frequency for turbulent kinetic energy and for intensities of scalar fluctuations may be introduced in an analogous way for statistically stationary turbulence.

10.1.5. Scales of turbulence

The dependence of $e(k; \mathbf{x}_1, t_1)$ on k affords a convenient focus for discussing many properties of turbulent flows as well as various scales of turbulence. Although the approximation of isotropy is not exactly correct even for the simplest nonreacting turbulent flows in the laboratory and, moreover, may be poorer for many turbulent reacting flows, the simplifications afforded by speaking in terms of a scalar k offset the greater precision associated with retaining $\mathbf{k}$ as a vector. For nonisotropic turbulence, we may consider our remarks to apply to an average over directions. The general physical concepts of turbulent motion, thereby derived, have been verified reasonably well. Isotropy tends to be achieved more closely at small scales than at larger scales; initially our discussion will mainly concern the dynamics at smaller scales. Here we shall write $e(k)$, suppressing the dependence on $\mathbf{x}_1$ and t_1, and we shall occasionally appeal implicitly to results for homogeneous turbulence because the ideas have been developed most thoroughly in that case.

Equation (25) suggests that a continuum of scales of turbulence contributes to q. The contribution from each scale can be obtained experimentally by measuring $e(k)$. In view of equations (19) and (24), this involves measuring velocity fluctuations simultaneously at two different locations for the full range of distances between these locations. This is an ambitious task that has been completed successfully for a number of turbulent flows. The classical experiment concerns the flow in a constant-area wind tunnel downstream from turbulence-producing grids. The measurement task can be simplified appreciably in this experiment if it can be assumed that the turbulence is convected with negligible change past the fixed measurement station at the constant average velocity $\bar{\mathbf{v}}$. If this is acceptable, then $\mathbf{v}'(\mathbf{x}_1, t_1) = \mathbf{v}'(\mathbf{x}_2, t_2)$ for $\mathbf{x}_2 = \mathbf{x}_1 + \bar{\mathbf{v}}(t_2 - t_1)$, so that with $\mathbf{x}_2 - \mathbf{x}_1$ in the direction of $\bar{\mathbf{v}}$, $\mathbf{v}'(\mathbf{x}_1, t_1)\mathbf{v}'(\mathbf{x}_2, t_1) = \mathbf{v}'(\mathbf{x}_1, t_1)\mathbf{v}'(\mathbf{x}_1, t_1 - |\mathbf{x}_2 - \mathbf{x}_1|/|\bar{\mathbf{v}}|)$, and $e(k)$ can be obtained from a measurement of the time history of the velocity fluctuation at a single point. An approximation of this type, which has been called a *Taylor hypothesis*, is reasonably good in many nonreacting flows for finding $e(k)$ at all but the largest scales. Taylor hypotheses are much more questionable for many turbulent flames because of rapid changes of fluctuations with position, produced by the influence of combustion on the turbulence through the effects of heat release. Measurements of $e(k)$ show that it vanishes at small and large values of k and reaches a maximum at an intermediate value. The maximum usually occurs where $1/k$ is on the order of the integral scale of the turbulence.

Integral scales (also called *correlation lengths*) may be defined explicitly in terms of correlation coefficients. Consider any one component of equation (19) for $t_2 = t_1$, $\mathbf{x}_2 \neq \mathbf{x}_1$, let $r \equiv |\mathbf{x}_2 - \mathbf{x}_1|$, and allow v_1 and v_2 in equation (15) to represent the selected components of $\mathbf{v}(\mathbf{x}_1, t_1)$ and $\mathbf{v}(\mathbf{x}_2, t_1)$, respectively, for equation (19). Then, in terms of the c_{12} of equation (16), the integral scale becomes

$$l_{12} = \int_0^\infty c_{12}\, dr, \tag{27}$$

in which the integration is performed with the point $\mathbf{x}_1$ fixed and with $\mathbf{x}_2$ varying from $\mathbf{x}_1$ to a boundary of the fluid (indicated by ∞ in the integral) in such a way that the unit vector from $\mathbf{x}_1$ toward $\mathbf{x}_2$ maintains a fixed direction. In general, l_{12} depends on the direction of the unit vector, on $\mathbf{x}_1$ and t_1, and on the component of equation (19) selected. For homogeneous, isotropic turbulence at fixed t_1, there are only two different values of l_{12}, one for diagonal elements of $\mathbf{R}$ and the other for off-diagonal elements. Under more general circumstances, a unique *integral scale* l at $(\mathbf{x}_1, t_1)$ may be defined as an average over the diagonal elements only and over all directions. Integral scales may be defined for scalar fluctuations in a similar manner by considering equation (21) instead of equation (19); the question of selection of a component of a tensor then does not arise. For other purposes integral time scales also are introduced; for example, if we were to consider equation (21) with $\mathbf{x}_2 = \mathbf{x}_1$ and let v_1 and v_2 in equation (15) refer to the variables at times t_1 and t_2, respectively, then the *integral time scale* or *correlation time* would be

$$t_{12} = \int_0^\infty c_{12}\, d\tau \tag{28}$$

with the c_{12} of equation (16) and with $\tau = t_2 - t_1$. If a Taylor hypothesis applies, then a relationship between integral space and time scales can be developed. From its definition we see that the integral scale l physically measures the distance over which the fluctuations occur in a relatively fixed phase—that is, exhibit a correlation.

Since $e(k)$ peaks near $k = 1/l$, the largest concentration of turbulent kinetic energy occurs at scales in the vicinity of l. The eddies with sizes in this range have been called the *energy-containing eddies*. For values of k below those corresponding to the energy-containing eddies, $e(k)$ falls rapidly toward zero with decreasing k; the size of the apparatus imposes an upper limit on the maximum eddy size achievable (the minimum k having $e \neq 0$). For values of k above $1/l$, there may be significant additional structure in the function $e(k)$ if the *turbulence Reynolds number*, R_l, is sufficiently large. The definition of R_l is like that of Re of Section 3.1.7, but with v and a there replaced by $\sqrt{2q}$ and l, respectively [where the q is that of equation (25)]. Thus the integral scale and the root-mean-square velocity fluctuation are employed in defining the turbulence Reynolds number;

$$R_l = \bar{\rho}\sqrt{2q}\, l/\bar{\mu}. \tag{29}$$

Since R_l is based on properties of the turbulence, it is associated more closely with regimes and dynamics of turbulent motion than is Re. Both scale (l) and intensity* ($\sqrt{2q}$) of turbulence must be known to characterize it properly. Turbulence dies rapidly if $R_l \ll 1$. For $R_l \approx 1$, the peak of $e(k)$ is followed with increasing k by a tailing off that is roughly symmetrical with the rise. For $R_l \gg 1$, a condition often encountered in practice, measurement and physical reasoning agree in ascribing a distinctive shape to $e(k)$.

The viscous terms are the most highly differentiated terms in the equation for momentum conservation and therefore are largest relative to other terms at the smallest scales (largest k's). When $R_l \gg 1$, there is an intermediate range of k between $1/l$ (where turbulence is generated from mean-flow instabilities) and $1/l_k$ (defined below, where turbulence is dissipated by viscosity), over which only the inertial terms in the equation for momentum conservation are significant. The dependence of $e(k)$ on k in this *inertial subrange* or *Kolmogorov subrange* was deduced by Kolmogorov through dimensional analysis. A spectral transfer of energy, which is a cascade of energy from small to large wave numbers, is considered to occur through interactions between components of different wave numbers, produced by the nonlinearity of the inertial terms. The interaction becomes evident if the convolution theorem is applied to the Fourier transform of the inertial terms in the equation for momentum conservation in homogeneous turbulence [47]; the contribution at any given wave number then involves an integral over all other wave numbers, thereby causing spectral mixing. During the cascade, only k and the average rate of spectral transfer of energy can affect $e(k)$. The average rate of spectral transfer must be the same as the average rate of dissipation of turbulent kinetic energy and dimensionally may be estimated as $\epsilon = (2q)^{3/2}/l$. The only quantity with the dimensions of $e(k)$ obtainable from ϵ and k is a dimensionless constant times $\epsilon^{2/3} k^{-5/3}$. Therefore, in the inertial subrange, $e(k)$ decreases with increasing k in proportion to $k^{-5/3}$. This deduction agrees with experimental observations for $R_l \gg 1$.

The inertial subrange ends when k becomes large enough for viscosity to be important. The characteristic length at which this occurs, the *Kolmogorov length*, l_k, is obtained through dimensional analysis as the length that may be constructed from ϵ and a representative kinematic viscosity, $v \equiv \bar{\mu}/\bar{\rho}$, namely, $l_k = (v^3/\epsilon)^{1/4}$. From equation (29) and the definitions, we see that

$$l_k = l/R_l^{3/4}, \tag{30}$$

which shows explicitly that $l_k < l$ when $R_l > 1$. For $k \gg 1/l_k$, the viscous

---

* Often the intensity in R_l is defined in terms of just one component of velocity so that with isotropy $\frac{1}{3}\sqrt{2q}$ appears instead. Related constant factors that also occur in other definitions such as equation (31) are not introduced herein.

terms are much larger than the inertial terms in the equation for momentum conservation, and the equation becomes linear, whence it is straightforward to demonstrate for homogeneous, constant-density turbulence that $e(k)$ becomes proportional to e^{-2vk^2t} [47]; at any given time t, this represents an exponential decrease of $e(k)$ with increasing k^2, a much more rapid drop-off than the power-law behavior in the inertial subrange. Therefore, in a first approximation for $R_l \gg 1$, it may be assumed that l_k represents a cutoff length below which no turbulent eddies exist. Experiments confirm that the cascade is terminated abruptly by viscous dissipation approximately at $k = 1/l_k$.

We have now seen that for $R_l \gg 1$, there are at least two special length scales, l and l_k. A third, the *Taylor scale*, arises in estimates of the magnitude of the average rate of viscous dissipation. If through suitable substitutions, multiplications and averaging in the equation for momentum conservation an (unclosed) equation for $\partial q/\partial t$ is derived, then this equation is found to contain terms that may be interpreted as convection by the mean flow, turbulent diffusion, production of turbulent kinetic energy, and dissipation of turbulent kinetic energy. The dissipation term is v times the mean-square value of the magnitude of the vorticity vector ($\mathbf{V} \times \mathbf{v}$) and can be written as a nondimensional constant times $v(2q)/l_t^2$, where l_t is the Taylor length. By equating this quantity to ϵ, we find through use of equation (29) that

$$l_t = l/\sqrt{R_l}. \tag{31}$$

Comparison of equations (30) and (31) shows that for $R_l \gg 1$, $l_k \ll l_t \ll l$; that is, l_t is of a magnitude intermediate between that of l and l_k. Nothing special occurs in the energy spectrum at $k = 1/l_t$. Time scales related to the various lengths may be defined through division by various velocities, and Reynolds numbers based on l_t or l_k instead of l may be introduced if desired (and often are).

The cascade concepts may be applied to intensities of scalar fields as well as to the turbulent kinetic energy. Passive scalars (those that do not influence the velocity field) in the absence of chemical reactions can experience spectral transfers as a consequence of the convective terms in their conservation equations, and an inertial-convective subrange can exist in which the integrand of equation (26) exhibits a power-law dependence on k analogous to that of $e(k)$ [66]. The average rate of scalar dissipation,

$$\bar{\chi} = \overline{2D(\mathbf{V}v') \cdot (\mathbf{V}v')}, \tag{32}$$

for the variable in equation (26), plays a role analogous to that of ϵ in the analyses. Here D is a representative coefficient of diffusion for the scalar field. Dimensional analysis again provides the power of k in the inertial-convective subrange. There have been studies of influences of chemical reactions on the spectra of passive scalars [16]. Nonlinearities of the chemical production terms can modify cascade aspects of the spectra. A conserved scalar is one that possesses no chemical source term (see Section 3.4.2); it may be inferred that if a scalar is not conserved, then there are hazards associated with attempts to apply the usual cascade concepts to it. Partly for this reason, the formulation in the following section focuses on conserved scalars, or mixture fractions.*

Most of the modern research on turbulence dynamics is addressing aspects that differ from those of the cascade concepts. This research emphasizes coherent structures that occur in turbulent flows [67]. The volume in which [29] appears is concerned with this subject. The focus is placed on frequency-halving processes instead of the frequency-doubling effects operative in the cascade. For example, the increase in width of a turbulent shear layer with increasing distance downstream may be associated largely with vortex-pairing processes that create larger structures from smaller structures, a counter-cascade phenomenon that meshes well with the ideas of [54]. Relationships are being identified to the onset of turbulence and to the development of spottiness of turbulence at large Reynolds numbers through vortex-stretching mechanisms that tend to concentrate fine structures in localized regions of space. These effects are complementary (not contradictory) to the cascade dynamics because they occur mostly at scales on the order of l or larger. They may be responsible for an increase in l with increasing time as the turbulence evolves, a process that must occur, for example, if shear layers are to grow in width (as, of course, they do). Much of the early research on turbulent flows [48] acknowledges scale-increase phenomena.

There are approaches to analyses of turbulent combustion that, although not deductively based on the Navier-Stokes equations, nevertheless appeal to concepts of coherent structures [68], [69]. We shall not have space here to present these approaches and must refer instead to reviews [18], [27], [40]. These methods share some aspects in common with "age" theories of stirred reactors [19], theories that we also shall forego discussing for the sake of brevity. Instead, we shall consider a promising approach to the theoretical analysis of turbulent diffusion flames.

10.2. TURBULENT DIFFUSION FLAMES

10.2.1. Objectives of Analyses

We saw in the last section that disparate length scales and time scales exist for turbulent flows. Various time scales also are associated with the

---

* If Prandtl or Schmidt numbers differ appreciably from unity, as in liquids or ionized gases, then a number of additional interesting aspects of cascade phenomena arise, and additional turbulence scales with distinguished attributes may be defined. We do not discuss these topics but instead refer the reader to C. H. Gibson, *Phys. Fluids* 11, 2305 (1968).

chemistry. It seems clear that, depending on values of ratios of turbulence times to chemical times, a number of different regimes of turbulent combustion may occur. The proper nondimensional parameters that delineate the boundaries of the different regimes are not well known at present. Nevertheless, theoretical analyses are best directed toward specific regimes rather than being aimed at including all problems within a single framework (which is not likely to be possible).

For initially nonpremixed reactants, two limiting cases may be visualized, namely, the limit in which the chemistry is rapid compared with the fluid mechanics and the limit in which it is slow. In the slow-chemistry limit, extensive turbulent mixing may occur prior to chemical reaction, and situations approaching those in well-stirred reactors (see Section 4.1) may develop. There are particular slow-chemistry problems for which the previously identified moment methods and age methods are well suited. These methods are not appropriate for fast-chemistry problems. The primary combustion reactions in ordinary turbulent diffusion flames encountered in the laboratory and in industry appear to lie closer to the fast-chemistry limit. Methods for analyzing turbulent diffusion flames with fast chemistry have been developed recently [15], [20], [27]. These methods, which involve approximations of probability-density functions using moments, will be discussed in this section.

There are many reasons for interest in turbulent diffusion flames. Some of the specific questions asked are: What is the flame length or flame height? What is the average local volumetric rate of heat release? What are the critical conditions for a turbulent-jet diffusion flame to be lifted off the fuel duct? What are the radiant energy fluxes emitted by the flame? How complete is the combustion? What is the rate of production of oxides of nitrogen? What is the rate of production and liberation of smoke and of unburnt hydrocarbons? What are the properties of noise emission from the flame? A number of these questions will be addressed here. Some that will not nevertheless have been considered to some extent in the literature. For example, questions of noise have been treated by Strahle and others [35], [70], [71], and influences of radiation have been studied by Tamanini [72]. Four physical phenomena that can introduce appreciable uncertainties into turbulent-combustion analyses are radiation [72], buoyancy [72], [73], high Mach numbers [74], and multiphase flow [75]; we shall not address these complexities. Instead, throughout most of the discussions we shall consider gaseous fuels at low Mach numbers with negligible buoyancy and radiant interactions, and we shall focus our attention mainly on two-stream problems, that is, those in which the various inlet streams, no matter how complex geometrically, are of two distinct types, one fuel and the other oxidizer, each having a constant (but different) chemical composition and temperature (see Section 3.4.2). We shall see that good methods are available for analyzing problems of this type.

In the face of the difficulties imposed by the chemistry in turbulent-combustion problems, a rational philosophy is to attempt to reduce the problem to one involving a nonreacting turbulent flow. In one way or another, this philosophy underlies the methods to be discussed in the remaining sections of this chapter. The unique aspects associated with the chemistry will be emphasized; after achieving a reduction to a nonreacting problem, discussion mainly will concern differences from constant-density, Navier-Stokes turbulence. Because of the many unresolved questions in Navier-Stokes turbulence, this limited objective seems reasonable here. We cannot expect to know more about reacting turbulent flows than we know about nonreacting turbulent flows; for example, closure problems for nonreacting turbulence remain. Predictive methods and experimental results for non-reacting flows may readily be applied if a suitable reduction of the turbulent-combustion problem is achieved, and experiments generally are easier without combustion.

10.2.2. Use of coupling functions

Coupling functions play a central role in reducing problems of turbulent diffusion flames to problems of nonreacting turbulent flows. In Section 1.3 and in Chapter 3, we emphasized that coupling functions are helpful for analyses of nonpremixed combustion. From the analysis of Section 3.4.2, it may be deduced that their utility extends to turbulent flows. Mixture fractions, which are conserved scalars (Section 10.1.5), were defined and identified as normalized coupling functions in Section 3.4.2. The presentation here will be phrased mainly in terms of the mixture fraction Z of equation (3-70).

It is true in general that the (thermal plus chemical) enthalpy and all atom (or element) mass fractions are conserved scalars. If reaction rates are sufficiently rapid for chemical equilibrium to be maintained everywhere, then given local, instantaneous values of the pressure, enthalpy, and atom mass fractions, we may use the equations of chemical equilibrium to calculate local, instantaneous values for all the state variables. In this respect, whenever full chemical equilibrium applies in turbulent combustion, the problem may be expressed as one involving a nonreacting turbulent flow of a fluid with complicated equations of state. Under these conditions, to work with equations involving reaction rates would only serve to introduce potential sources of error; it is better to work with the conserved scalars, for which turbulence modeling is much more firmly based. Influences of the complicated equations of state on appropriate modeling procedures have not been explored very much; the greatly reduced densities and increased molecular diffusivities that tend to prevail near points where the mixture is stoichiometric may affect the best-suitable closure approximations and methods of analysis. However, these complications are likely to be small in comparison with uncertainties introduced by finite-rate chemistry. Current practice, which may be reasonably accurate, is to apply available modeling of constant-density turbulence with Favre averaging employed to include density variations.

The method of approaching the nonreacting problem must be selected to enable calculations to be made of the quantities of interest for the turbulent flow in chemical equilibrium. The average temperature and the root-mean-square temperature fluctuation are examples of quantities that often are of interest. Moment methods for the conserved scalars in the nonreacting problem do not provide this information because the nonlinear dependence of T on Z [for example, equation (3-84)] prevents moments of T from being obtained directly from those of Z. In general, any desired one-point, one-time information concerning temperature and species concentrations may be obtained from the corresponding joint probability-density function for all conserved scalars (pressure, enthalpy, and atom mass fractions). This fact focuses interest on methods that involve probability-density functions. Use of an equation for evolution of probability-density functions suffers from uncertainties in closure approximations and usually entails performing extensive computations. An alternative, found satisfactory for many problems, is to select a general class of shapes for the probability-density functions of the conserved scalars, parameterizing the class with a small number of parameters, and to calculate the evolution of the parameters by moment methods applied to the conserved scalars. This last procedure is the one to be discussed here.

Shapes of probability-density functions of conserved scalars have been studied extensively by Bilger and are discussed by him in [27]. A variety of different shapes have been found depending on the type of flow and the position in the flow field, as illustrated for $P(Z)$ in Figure 10.1. Flows that possess intermittency exhibit delta functions in $P(Z)$ at $Z = 0$ or at $Z = 1$, represented in Figure 10.1 by vertical arrows whose heights measure the integral of $P(Z)$. It may be seen from Figure 10.1 that there are strong tendencies for delta functions to appear in mixing layers, while stirred reactors tend to have probability-density functions that are more-nearly Gaussian.

All the shapes in Figure 10.1 may be approximated well by a re-normalized, truncated Gaussian function with delta functions added at the ends, namely,

$$P(Z) = \alpha\delta(Z) + \beta\delta(1 - Z) + \gamma e^{-(Z-\mu)^2/(2\sigma^2)}, \qquad (33)$$

where α, β, σ, and μ are nonnegative constants and γ is a nonnegative constant for $0 < Z < 1$ and zero otherwise [compare equation (13)]. The normalization of $P(Z)$ gives (for $0 < Z < 1$)

$$\gamma = (1 - \alpha - \beta)\bigg/\bigg\{\sqrt{\pi/2}\,\sigma\bigg[\text{erf}\bigg(\frac{1-\mu}{\sqrt{2}\sigma}\bigg) + \text{erf}\bigg(\frac{\mu}{\sqrt{2}\sigma}\bigg)\bigg]\bigg\}, \qquad (34)$$

where erf denotes the error function.

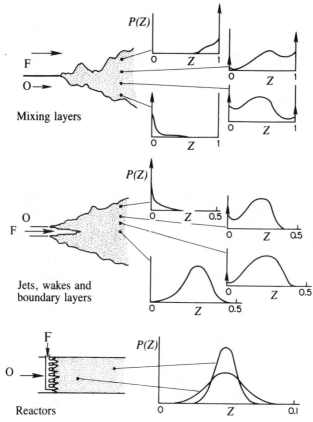

FIGURE 10.1. Illustrations of shapes of probability-density functions for the mixture fraction in many turbulent flows (adapted from the contribution of R. W. Bilger to [27]).

Equation (33) is versatile because it possesses four independent parameters. This is advantageous for fitting a wide variety of shapes but disadvantageous if the evolution of $P(Z)$ is to be calculated from theories for the evolution of moments of Z because equations for evolution of four moments are then needed, so that the analysis becomes somewhat complicated. A parameterization involving only two parameters would enable $P(Z)$ to be calculated from moment equations for evolution of the average and of the mean-square fluctuation of Z. For many flows this can be achieved from equation (33) by putting either $\alpha = 0$ or $\beta = 0$ and by imposing a relationship among σ, μ, and the remaining intermittency—for example, perhaps $\sigma = \mu/2$ for jets; for reactors, $\alpha = \beta = 0$ provides the two-parameter, clipped-Gaussian distribution. An alternative is to employ a poorer approximation for $P(Z)$ from the outset, such as the beta-function distribution, $P(Z) = 0$ for $Z < 0$ or $Z > 1$ and

$$P(Z) = Z^{\alpha - 1}(1 - Z)^{\beta - 1}\Gamma(\alpha + \beta)/[\Gamma(\alpha)\Gamma(\beta)] \qquad (35)$$

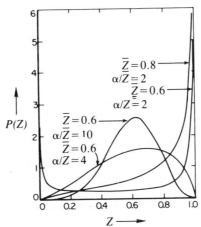

FIGURE 10.2. Various shapes of beta-function densities for the mixture fraction (from N. Peters).

for $0 \le Z \le 1$, where α and β are nonnegative constants and Γ denotes the gamma function. From Figure 10.2 it is seen that equation (35) [which has $\overline{Z} = \alpha/(\alpha + \beta)$ and $\overline{Z'^2} = \overline{Z}(1 - \overline{Z})/(1 + \alpha + \beta)$] simulates delta functions by infinities at $Z = 0$ and at $Z = 1$ for certain values of α and β. Equation (35) appears to give adequate accuracy for many purposes [76].

To illustrate how an analysis may be completed when a two-parameter representation of $P(Z)$ is adopted, let us assume that Z obeys equation (3-71) and that ρ and D are unique functions of Z. Let us consider statistically stationary flows and work with Favre averages, seeking equations for the calculation of $\tilde{Z}$ and $\widetilde{Z''^2}$. According to the discussions of equations (33) and (35), we need $\overline{Z}$ and $\overline{Z'^2}$ instead, but with $\rho(Z)$ known, it is straightforward to generate one pair from the other. For the joint probability-density function, we have $P(\rho, Z) = P(Z)\delta[\rho - \rho(Z)]$, so in equation (17), $\tilde{P}(Z) = \rho(Z)P(Z)/\bar{\rho}$, which may be used in the relations in (18) to provide expressions for $\tilde{Z}$ and $\widetilde{Z''^2}$ that involve only α and β if, for example, equation (35) is adopted. Since $\overline{Z}$ and $\overline{Z'^2}$ also have been expressed in terms of α and β, the correspondence between the two pairs of variables is established. The integrals needed must be evaluated numerically, even for relatively simple functions $\rho(Z)$. This can be avoided if the representation in equation (33) or equation (35) is taken initially to apply to $\tilde{P}(Z)$ instead of to $P(Z)$ since then, for example, equation (18) shows that the expressions following equation (35) involve $\tilde{Z}$ and $\widetilde{Z''^2}$ instead of $\overline{Z}$ and $\overline{Z'^2}$. With $\rho(Z)$ known, with $P(Z) = \bar{\rho}\tilde{P}(Z)/\rho(Z)$, and with $\bar{\rho} = 1/\int_0^1 [\tilde{P}(Z)/\rho(Z)]\,dZ$, there are no special difficulties in evaluating averages by using $\tilde{P}(Z)$ instead of $P(Z)$. At present, there is no experimental basis for believing that equation (33) or equation (35) would apply better to one of these probability-density functions than to the other.

For statistically stationary flows, the average of equation (3-71) is

$$\nabla \cdot (\bar{\rho}\tilde{\mathbf{v}}\tilde{Z}) = \nabla \cdot (\overline{\rho D \nabla Z} - \overline{\rho \mathbf{v}'' Z''}). \tag{36}$$

Multiplication of equation (3-71) by Z'' and averaging can be shown to produce

$$\begin{aligned}
\bar{\rho}\tilde{\mathbf{v}} \cdot \nabla \widetilde{Z''^2} = &-2\overline{\rho \mathbf{v}'' Z''} \cdot \nabla \tilde{Z} - \nabla \cdot (\overline{\rho \mathbf{v}'' Z''^2}) - 2\overline{\rho D \nabla Z'' \cdot \nabla Z''} \\
&+ \nabla \cdot \overline{[\rho D \nabla (Z''^2)]} + 2\overline{Z'' \nabla \cdot (\rho D \nabla \tilde{Z})},
\end{aligned} \tag{37}$$

while equation (2) reduces to $\nabla \cdot (\bar{\rho}\tilde{\mathbf{v}}) = 0$. In equations (36) and (37), $\overline{\rho \mathbf{v}'' Z''}$ represents the Reynolds transport; the first and second terms on the right-hand side of equation (37) are interpreted as the production and diffusion of fluctuations of Z, respectively, and the third (the first involving D) describes dissipation. The simplest approach having justification for high Reynolds numbers is to neglect $\overline{\rho D \nabla Z}$ in equation (36) and the last two terms in equation (37) on the basis that turbulent transport dominates molecular transport and that $|\overline{\nabla Z''}| \gg |\nabla \tilde{Z}|$; to employ a gradient approximation for turbulent transport,

$$\overline{\rho \mathbf{v}'' Z''} = -\bar{\rho} D_T \nabla \tilde{Z}, \quad \overline{\rho \mathbf{v}'' Z''^2} = -\bar{\rho} D_T \nabla \widetilde{Z''^2}, \tag{38}$$

where D_T is a turbulent diffusion coefficient; and to relate the dissipation to a length scale for the turbulence, namely, for an instantaneous scalar dissipation defined as $\hat{\chi} = 2D\nabla Z'' \cdot \nabla Z''$ [compare equation (32)], employ $\tilde{\chi} = CD_R \widetilde{Z''^2}/l_Z^2$, where C is a constant, D_R a reference value of D, and l_Z a Taylor length for Z. Fixed values are ascribed to the constants C and D_R, while D_T and l_Z either are estimated or are calculated from auxiliary differential equations such as an equation for evolution of a dissipative length scale of turbulence [27]. A correspondingly averaged and modeled equation for momentum conservation is needed for obtaining $\tilde{\mathbf{v}}$, which appears in equations (36) and (37). Although many modeling approximations are seen to be involved in achieving closure, the methods are relatively well substantiated because of the absence of the chemical source term. In particular, no firm reason has been found to question the gradient-diffusion approximation in equation (38).

After $P(Z)$ or $\tilde{P}(Z)$ is obtained at each point in the flow by the methods that have now been described, it is usually possible to calculate any desired averages involving temperature and species concentrations at each point. As an illustration, let us assume that equations (3-82), (3-83) and (3-84) hold true. These relationships are shown schematically in Figure 10.3. A convenient approach is to use equation (9) to generate probability-density functions for Y_F, Y_O, and T from $P(Z)$. If the shape of $P(Z)$ is the second one shown for the jet in Figure 10.1, then the probability-density functions obtained are those illustrated in Figure 10.4. For the fuel, the contribution to $P(Z)$ for $0 \le Z \le Z_c$ collapses into a delta function at $Y_F = 0$, as is readily

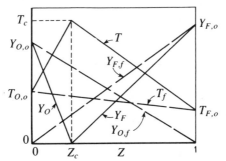

FIGURE 10.3. Illustration of the dependences of the temperature and of the fuel and oxidizer mass fractions on the mixture fraction in the flame-sheet approximation, as given by equations (3-80)–(3-85).

inferred from Figure 10.3. For the oxidizer, two delta functions appear, one from that at $Z = 0$ in $P(Z)$ and the other from the fact that the contribution to $P(Z)$ for $Z_c \leq Z \leq 1$ collapses into a δ-function at $Y_O = 0$. For the temperature, since Z is a double-value function of T, there are two additive contributions to $P(T)$ in the range $T_{0,0} < T < T_c$; each of these is indicated by a dashed line in Figure 10.4, and the summation in equation (9) contains two terms instead of just one. From these results, we see that the shapes of the probability-density functions for reactant concentrations and temperature are considerably more complex than those of the mixture fraction and may not be characterized readily by simple formulas. Nevertheless, averages such as $\bar{T}$ and $\overline{T'^2}$ may be obtained in a straightforward manner—for example, from equation (8). With general equilibrium chemistry, the calculations require use of electronic computers but are still straightforward.

Distributions like those in Figure 10.4, for example, indicate that $\bar{Y}_F$ or $\bar{T}$ differs from $Y_F(\bar{Z})$ or $T(\bar{Z})$, respectively. If mixing were complete in the sense that all probability-density functions were delta functions and fluctuations vanished, then differences like $\bar{T} - T(\bar{Z})$ would be zero. That this situation is not achieved in turbulent diffusion flames has been described qualitatively by the term *unmixedness* [7]. Although different quantitative definitions of unmixedness have been employed by different authors, in one way or another they all are measures of quantities such as $\bar{Y}_F - Y_F(\bar{Z})$ or $\bar{T} - T(\bar{Z})$. The unmixedness is readily calculable from $P(Z)$, given any specific definition (see Bilger's contribution to [27]).

The *visible flame length* of turbulent-jet diffusion flames is a quantity that has been measured, beginning with the earliest studies in the field [7]. It is logical to identify the flame position as the location of the stoichiometric surface and thereby to use Z directly for calculating flame lengths. Thus in the flame-sheet approximation, the instantaneous flame length at any given time would be given by the maximum value of the axial distance from the jet exit to the point at which $Z = Z_c$, at that time. The average flame length

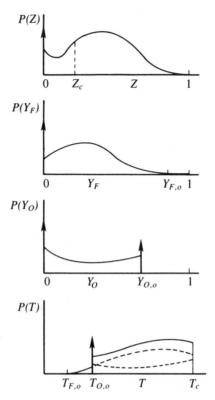

FIGURE 10.4. Illustration of probability-density functions for the mixture fraction, fuel mass fraction, oxidizer mass fraction, and temperature for a jet-type diffusion flame in the flame-sheet approximation.

is the average value of this quantity and can be estimated directly from the profile of $\bar{Z}$ (or $\tilde{Z}$ if the Favre average is wanted) along the center line. Thus setting $\bar{Z} = Z_c$ on the center line gives an average flame length, without need for calculating probability-density functions. The functional dependences of flame lengths, obtained from $\tilde{Z}$ when equation (38) is used in equation (36) with the usual modeling for D_T, agree with those given in Section 3.1.7 for the turbulent flow regime.

When closer scrutiny is given to the visible flame length, a number of questions arise. If the flame length is a random variable with a probability-density function denoted by $P(h)$, then it will have an average value $\bar{h} = \int_0^\infty h P(h)\, dh$ and a mean-square fluctuation $\overline{h'^2} = \int_0^\infty (h - \bar{h})^2 P(h)\, dh$. Since $P(h)\, dh$ is the probability that the flame length lies in the range dh about h, it is quite different from $P(Z)$, which always refers to a given point; in principle, there is no way to calculate $P(h)$ solely from knowledge of the functions $P(Z)$ at all points in the flow. The problem encountered here is a type of zero-crossing problem of stochastic processes [55], and its solution would involve considering spatial correlations for Z of the type defined in equation (21) with $t_2 = t_1$; little is known about two-point, joint probability-density functions for Z. Thus, even if the flame length is identified with $Z = Z_c$, interesting statistical aspects of it lie beyond current calculational abilities. Moreover, there are fundamental uncertainties in relating visible flame lengths to the criterion that $Z = Z_c$. In hydrocarbon flames, blue emissions of radiation typically occur near $Z = Z_c$, but often the blue is obscured by yellow emissions from soot in fuel-rich regions that provide the dominant contribution to the radiant energy flux. The yellow observed need not correlate so well with $Z = Z_c$. Results of measurements of visible flame lengths depend to a degree on the method of measurement, for example (with some photographic methods) on the exposure time of a photographic plate. In view of these various uncertainties, accuracies better than 10% probably should not be sought in comparisons of measured and calculated visible flame lengths.*

The results developed here rely on equation (3-71) and on the assumption that the functions $\rho(Z)$ and $D(Z)$ exist; the results for T and Y_i rely further on the assumption that these quantities are unique functions of Z. In principle, the procedures can be generalized by working with all the conserved scalars (pressure, enthalpy, and atom mass fractions) and by introducing and approximating joint probability-density functions for these. However, difficulties would be encountered, especially through equation (3-72) for the enthalpy h, which contains terms for which turbulence modeling is not well developed. All the assumptions introduced in Section 3.4.2, up to equation (3-74), are needed for developing a manageable formulation. In particular, negligible buoyancy, low Mach numbers, and negligible radiant energy loss are required if the approach is to apply for temperature fields. Also, a negligible effect of $\partial p/\partial t$ on the enthalpy fluctuations is needed, thereby possibly ruling out transient processes such as those occurring in piston engines. The two-stream configuration, negligible fluctuations of state variables in the feed streams, and adiabatic walls (if they are impermeable to chemical species) are further formal restrictions needed in relating h and Y_i to Z on a space-time resolved basis. These last requirements appear to be relatively mild points of detail in that the extent of departures from them, encountered in most diffusion flames, are not expected to be sufficiently great to impair the accuracies of the results noticeably. Some thought is being given to methods for extension to three-stream problems, which arise in some applications.

---

* Flame lengths are employed in calculations of radiant energy emissions from flames [see papers coauthored by C. L. Tien in the *16th* (page 1481) and *18th* (page 1159) *Symp.*]. Because of the complexity of the problem, good fundamental approaches to the analysis of rates of radiant energy emission from turbulent flames are unavailable.

The most important restriction on the method that has been presented is chemical equilibrium, and the second most important is equal diffusivities. How critical each of these is in diffusion flames is a topic to which research recently has been devoted. In sufficiently fuel-rich portions of hydrocarbon-air diffusion flames, the chemical-equilibrium approximation is not good (see Figure 3.8 and the discussion in Section 3.4.1), but empirical approaches apparently still can be employed to relate nonequilibrium concentrations uniquely to Z with reasonable accuracy for main species [77]. In addition, the extent to which the burning locally proceeds to CO or to CO_2 may vary with the fuel, local stoichiometry, and characteristic flow times; methods to account for this are being developed [78], [79]. The theoretical methods that have been applied in studying the validity of the two major approximations are expansions for Lewis numbers near unity [80] and expansions in reaction-rate parameters for near-equilibrium flows [27], [28], [81]. The results of the research tend to support a rather broad range of applicability for the predictions obtained by the approach that has been described [27]. However, continuing research is needed on the limitations of the technique.*

10.2.3. Production of trace species

Oxides of nitrogen and soot-related species are examples of chemical components present in very low concentrations in turbulent diffusion flames; they are trace species. Trace species that maintain chemical equilibrium pose no special problems in that they may be handled directly by the methods of the preceding subsection. However, the trace species of interest often are far from equilibrium, as soot-related species always are and oxides of nitrogen almost always are. The fact that the concentrations of these species are low means that they affect the thermochemistry to a negligible extent and that finite-rate effects for them can therefore be analyzed more easily than those for major species. Methods of analysis have been developed in the literature [15], [27], [28], [82], [83]. Here we shall indicate how calculations of interest may be performed.

The local average rates of production of nonequilibrium trace species generally are of the greatest concern. Their local average concentrations are of lesser interest. This is fortunate because the production rates often can be obtained without introducing a conservation equation for the concentration of the species of interest. It is difficult to analyze the conservation equations for species experiencing finite-rate chemistry in turbulent flows, irrespective of whether the species are present in trace amounts. We shall address only production rates and shall begin with a general formulation, following [28], before discussing specific applications.

Let there be T trace species out of equilibrium, identified by subscript $i = 1, \ldots, T$, and assume (quite reasonably) that their chemistry involves at

---

* For a recent application see M. K. Razdan and J. G. Stevens, $C \& F$ **59**, 289 (1985).

most binary collisions between trace species. The equilibrium diffusion flame is to be analyzed first, using the methods of the preceding subsection and neglecting the trace species. This gives the temperature and the concentrations of all other species in terms of Z. The mass rates of production w_i, equation (1-8), of the nonequilibrium trace species may then be expressed in terms of Z and the mass fractions Y_i of the trace species in the general form

$$w_i = w_{i0}(Z) + \sum_{j=1}^{T} w_{ij}(Z)Y_j + \sum_{j=1}^{T} \sum_{k=1}^{T} w_{ijk}(Z)Y_j Y_k, \quad i = 1, \ldots, T, \quad (39)$$

in view of the assumed chemistry. Here the first term arises from steps in which none of the reactants are nonequilibrium trace species, the second from steps involving one such trace species as a reactant, and the third from steps involving two such species. For purposes of characterizing the chemistry, the reasonable approximation of constant pressure has been employed here. In view of equation (39), the local average production rate evidently is

$$\bar{w}_i = \int_0^1 w_{i0}(Z)P(Z)\, dZ + \sum_{j=1}^{T} \int_0^1 \int_0^1 Y_j w_{ij}(Z)P(Z, Y_j)\, dY_j\, dZ$$

$$+ \sum_{j=1}^{T} \sum_{k=1}^{T} \int_0^1 \int_0^1 \int_0^1 Y_j Y_k w_{ijk}(Z)P(Z, Y_j, Y_k)\, dY_k\, dY_j\, dZ, \quad i = 1, \ldots, T,$$

$$(40)$$

where $P(Z, Y_j)$ and $P(Z, Y_j, Y_k)$ are joint probability-density functions. Thus $\bar{w}_i$ may be evaluated directly from knowledge of the appropriate probability-density functions.

Given the methods of the preceding subsection, calculation of the first term of equation (40) by numerical integration poses no difficulty in principle. The second and third terms can be evaluated only if the appropriate joint probability-density functions can be obtained. This would entail either measurement or theoretical consideration of the conservation equations for the nonequilibrium trace species. These are challenging problems; typical interactions through chemical kinetics are strong enough, for example, to cause hypotheses of statistical independence for $P(Z, Y_j)$ or for $P(Z, Y_j, Y_k)$ to be questionable. Therefore, well-justified approaches for obtaining $\bar{w}_i$ are available today only if the second and third terms are negligible, that is, only if the nonequilibrium trace species participate to a negligible extent in the net production of the species of interest. This is often true, as the following examples will illustrate. Therefore, useful methods are available for obtaining the average rates of production in some interesting cases.

First consider production of NO by the Zel'dovich mechanism given at the end of Section B.2.5 [15], [83]. The molar rate of production of NO is $\omega_{NO} = 2k_f c_{N_2} c_O$, where k_f is the specific reaction-rate constant for the first step. If equilibrium is maintained for O_2 dissociation, then the concentration of O is related to that of O_2 by $c_O^2 = K_c c_{O_2}$, where K_c is the equilibrium constant for concentrations in the dissociation reaction. Both k and K_c depend strongly on the temperature T, since [83]

$$k_f = 7 \times 10^{13} \exp(-37{,}750/T)\, \text{cm}^3/\text{mol} \cdot \text{s}$$

and $\sqrt{K_c} = 4.1 \exp(-29{,}150/T)(\text{mol}/\text{cm}^3)^{1/2}$, with T in degrees Kelvin. Thus

$$\omega_{NO} = 5.7 \times 10^{14} c_{N_2} \sqrt{c_{O_2}}\, e^{-66{,}900/T}. \tag{41}$$

Since N_2 and O_2 are major species, the first of which is practically inert and the second of which usually nearly maintains chemical equilibrium, the equilibrium chemistry for the diffusion flame gives $\omega_{NO}(Z)$ from equation (41). Then, as in equation (40),

$$\bar{\omega}_{NO} = \int_0^1 \omega_{NO}(Z)P(Z)\, dZ. \tag{42}$$

Equation (42) provides an estimate of the average rate of production of NO.

Because of the strong dependence of ω_{NO} on T in equation (41), evaluation of $\bar{\omega}_{NO}$ from equation (42) by numerical integration requires very fine step sizes. However, the integral may readily be evaluated as an asymptotic expansion for large values of T_a/T_c, where $T_a = 66{,}900\text{K}$. If a flame-sheet approximation for the diffusion-flame structure is adopted, then $c_{O_2} = 0$ for $Z \geq Z_c$, so the integral in equation (42) extends only from 0 to Z_c; near $Z = Z_c$, c_{O_2} is linear in $Z_c - Z$, so that $c_{O_2} = c(Z_c - Z)$, where c is constant. On the oxygen side of the flame sheet, T is also linear in $Z_c - Z$, say $T = T_c[1 - b(Z_c - Z)]$, where b is constant. Employing the expansion $[1 - b(Z_c - Z)]^{-1} \approx 1 + b(Z_c - Z)$ in the argument of the exponential, we obtain the following asymptotic expansion from equations (41) and (42):

$$\bar{\omega}_{NO} = 5.7 \times 10^{14} e^{-T_a/T_c} \frac{c_{N_2,c}\sqrt{c}\, P(Z_c)}{[(T_a/T_c)b]^{3/2}} \int_0^{\infty} \sqrt{\xi}\, e^{-\xi}\, d\xi, \tag{43}$$

where the substitution $\xi = (T_a/T_c)b(Z_c - Z)$ has been made. The integral in equation (43) is a gamma function with value $\sqrt{\pi}/2$. Equation (43) shows that the average production rate is closely proportional to the value of the probability-density function for Z at the stoichiometric point Z_c. Experimentally, the production rate has a maximum on the fuel-rich side of the average stoichiometric surface in a turbulent-jet diffusion flame [82]; this is readily explained if for the profiles illustrated in Figure 10.1, $P(Z_c)$ is greater at a fuel-rich point than at a point on the average stoichiometric surface.

The calculation outlined here is inaccurate in that equilibrium dissociation of oxygen is not attained in most diffusion flames. It is known that superequilibrium concentrations of oxygen atoms occur; these may lead to appreciably increased values of $\bar{\omega}_{NO}$. An approach to take this into account without changing the formulation is to employ an empirical correlation for $c_O(Z)$, much like that indicated at the end of Section 10.2.2 for nonequilibrium species in fuel-rich regions. It might be thought that to achieve better accuracy, O atoms could be treated in a framework similar to that of NO, since they also are nonequilibrium trace species. However, this approach encounters difficulties because reactions such as $O + H_2 \rightarrow OH + H$, involving O atoms and major species, are important, so that a joint probability-density function may have to be considered in equation (40); moreover, accounting for three-body recombinations, such as $O + OH + M \rightarrow HO_2 + M$, brings the three-variable (trivariate) joint probability-density functions into equation (40). The ultimate utility of the general formulation for these more complicated processes is unclear because the approach that has been presented is relatively unexplored.

Equation (42) cannot be used if NO concentrations approach their equilibrium values, since the net production rate then depends on the concentration of NO, thereby bringing bivariate probability-density functions into equation (40). Also, if reactions involving nitrogen in fuel molecules are important, then much more involved considerations of chemical kinetics are needed. Processes of soot production similarly introduce complicated chemical kinetics. However, it may be possible to characterize these complex processes in terms of a small number of rate processes, with rates dependent on concentrations of major species and temperature, in such a way that a function $w_i(Z)$ can be identified for soot production. Rates of soot-particle production in turbulent diffusion flames would then readily be calculable, but in regions where soot-particle growth or burnup is important as well, it would appear that at least a bivariate probability-density function should be considered in attempting to calculate the net rate of change of soot concentration.

10.2.4. Average rates of heat release

In view of the results just given, we might guess that local average rates of chemical heat release in turbulent diffusion flames are calculable in some fashion from $P(Z)$. However, this is not possible even in principle for equilibrium flows. As equilibrium is approached, the chemical production terms in the equations for species conservation become indeterminate, involving differences of large numbers that cancel (for example, see Section B.2.5.2). A more circuitous route is therefore needed to find the average rate of heat release [15], [20], [27], [28]. The necessary expressions will be developed here.

In view of equation (1-1), equation (1-4) can be written as

$$w_i = \partial(\rho Y_i)/\partial t + \nabla \cdot [\rho(\mathbf{v} + \mathbf{V}_i)Y_i], \quad i = 1, \ldots, N. \tag{44}$$

Near conditions of chemical equilibrium, averages involving w_i are best computed from the right-hand side of equation (44). Under assumptions 1, 2, and 3 of Section 1.3 and the approximation of equal binary diffusion coefficients, equation (44) becomes

$$w_i = \partial(\rho Y_i)/\partial t + \nabla \cdot (\rho \mathbf{v} Y_i) - \nabla \cdot (\rho D \nabla Y_i), \quad i = 1, \ldots, N, \tag{45}$$

an acceptable approximate equation with which we shall work here. In equilibrium, the Y_i in equation (45) may be viewed as known functions of Z. With this assumption, equation (45) may be written as

$$w_i = \frac{dY_i}{dZ}\left[\frac{\partial(\rho Z)}{\partial t} + \nabla \cdot (\rho \mathbf{v} Z) - \nabla \cdot (\rho D \nabla Z)\right]$$

$$- \rho D |\nabla Z|^2 \frac{d^2 Y_i}{dZ^2}, \quad i = 1, \ldots, N. \tag{46}$$

In view of equation (1-1) and equation (3-71), the quantity in the square brackets in equation (46) vanishes. Therefore, in a turbulent flow under the given approximations, the local, instantaneous rate of production of species i per unit volume is*

$$w_i = -\rho D(d^2 Y_i/dZ^2)(\nabla Z) \cdot (\nabla Z), \quad i = 1, \ldots, N. \tag{47}$$

Multiplication by heats of formation and summation over i provides a similar expression for the local, instantaneous rate of heat release.

Although $d^2 Y_i/dZ^2$ is a function of Z in equation (47), the dot product $(\nabla Z) \cdot (\nabla Z)$ is not. Therefore, knowledge of $P(Z)$ is insufficient for obtaining the average $\bar{w}_i$ from equation (47); a joint probability-density function for Z and for $\chi \equiv 2D\nabla Z \cdot \nabla Z$ must be known. The nonnegative quantity χ is closely related to the instantaneous scalar dissipation $\hat{\chi}$, defined after equation (38), and it will be called by the same name here. For high turbulence Reynolds numbers, these two dissipations almost always are practically equal because gradients of fluctuations then are much larger than gradients of averages. The occurrence of χ in equation (47) underscores the importance of a joint probability-density function, $P(\chi, Z)$, for the mixture fraction Z and its dissipation rate χ. In terms of $P(\chi, Z)$, the average of equation (47) is

$$\bar{w}_i = -\tfrac{1}{2}\int_0^1 \int_0^\infty \rho(d^2 Y_i/dZ^2)\chi P(\chi, Z)\, d\chi\, dZ, \quad i = 1, \ldots, N. \tag{48}$$

By introducing the heat of formation per unit mass for each species i at temperature $T(Z)$, say h_i^0, we may express the local, instantaneous rate of heat release per unit volume as $\dot{q} = -\sum_{i=1}^N h_i^0 w_i$, so that the average rate of heat release per unit volume becomes

$$\bar{q} = \tfrac{1}{2}\int_0^1 \int_0^\infty \rho\left[\sum_{i=1}^N h_i^0(d^2 Y_i/dZ^2)\right]\chi P(\chi, Z)\, d\chi\, dZ. \tag{49}$$

These formulas show that knowledge of the average production rates of near-equilibrium species and of heat relies on knowledge of the joint probability-density function for Z and χ. Alternatively, since D is a function

---

* This result was first derived explicitly by R. W. Bilger, *CST* **13**, 155 (1976); see also [20].

of Z (through its dependence on T and Y_i), the joint probability-density function for Z and the magnitude of its gradient could have been introduced [15]. Quite generally, the latter interpretation is more readily understandable, as can be seen most simply by considering a flame-sheet picture; the magnitude of the gradient is a measure of the rate of diffusion of reactants into the reaction sheet. The relationship to scalar dissipation is fortuitous, in a sense, since the reaction does not directly involve dissipation. It merely happens that since dissipation of the conserved scalar field and reaction in the flame sheet both involve molecular diffusion, their rates can be expressed in similar forms.

Thermochemical calculations enable $d^2 Y_i/dZ^2$ at equilibrium to be obtained in general cases. If equations (3-82), (3-83), and (3-84) apply, then $d^2 Y_i/dZ^2$ are delta functions located at the flame sheet, $Z = Z_c$ in Figure 10.3. In this approximation, $P(\chi, Z)$ need not be known for all values of Z to obtain $\bar{w}_i$ and $\bar{q}$; it is sufficient to know $P(\chi, Z)|_{Z=Z_c} \equiv P(\chi, Z_c)$, where the notation, although imprecise, should be clear enough. Since, according to equation (12), $P(\chi, Z) = P(\chi|Z)P(Z)$ we see that $P(\chi, Z_c) = P(\chi|Z_c)P(Z_c)$ in the present notation; that is, in addition to $P(Z)$ at $Z = Z_c$, we must know the conditioned-average rate of dissipation, $\bar{\chi}_c \equiv \int_0^\infty \chi P(\chi|Z_c)\, d\chi$, to obtain $\bar{w}_i$ or $\bar{q}$ from equation (48) in the flame-sheet approximation. Since knowledge of the joint or conditioned functions is practically absent [27], statistical independence is often hypothesized or else it is merely assumed, less restrictively, that the conditioned-mean dissipation equals the unconditioned mean. A small amount of data is available on unconditioned-average rates of scalar dissipation in turbulent flows (see discussions in [83]–[86]), and additional measurements are being made. These results allow estimates of $\bar{\chi}_c$ to be made, even though accurate calculations are beyond current capabilities.

After $\bar{\chi}_c$ is obtained, the local average rates of fuel consumption and of heat release may be calculated, in the flame-sheet approximation, from the formulas

$$\bar{w}_F = -\tfrac{1}{2}\rho_c \bar{\chi}_c P(Z_c) Y_{F,0}/(1 - Z_c) \tag{50}$$

and

$$\bar{q} = \tfrac{1}{2}\rho_c \bar{\chi}_c P(Z_c)[(q^0 Y_{F,0})/(W_F \nu_F)]/(1 - Z_c), \tag{51}$$

which are derived from equations (48) and (49) by using equations (3-82) and (3-84) to obtain the delta functions. For example, we see from equation (3-82) that $d^2 Y_F/dZ^2 = [Y_{F,0}/(1 - Z_c)]\delta(Z - Z_c)$. In equation (51), $q^0/(W_F \nu_F)$ is the heat released per unit mass of fuel consumed, as in Section 3.4.2, and Z_c is given by equation (3-81). The density ρ_c is the function $\rho(Z)$ evaluated at $Z = Z_c$—that is, the density at the relevant adiabatic flame temperature. Subject to the conditioning, the density remains constant so that there are no differences between Favre averaging and ordinary averaging. If statistical independence had been assumed first and a Favre average introduced through $\tilde{\chi} = \int_0^\infty \chi \tilde{P}(\chi)\, d\chi$ with $\tilde{P}(\chi)$ defined as in equation (17), then $\bar{\rho}\tilde{\chi}$ would appear in equations (50) and (51) in place of $\rho_c\bar{\chi}_c$, where $\bar{\rho}$ is the (unconditioned) average density at the point in question. Equations (50) and (51) indicate that the influence of the fluid mechanics emerges in the proportionality of the local, average rates to the product $\bar{\chi}_c P(Z_c)$; in various flows $\bar{\chi}_c$ may range perhaps from $10^{-1}s^{-1}$ to $10^4 s^{-1}$, while $P(Z_c)$ generally is somewhere between 10^{-3} and 10.

Equation (47) may be treated as the first term in a perturbation about equilibrium flow [27], [81]. Given a scheme for the chemical kinetics, equation (1-8) may be employed to generate an expansion of w_i in the departures of T and of Y_i from their chemical-equilibrium dependences on Z. A general equation for the departures from equilibrium may then be derived from equation (45) [27], [81], and suitable modeling in this equation then provides the first correction to $\bar{w}_i$. The procedure, never simple, is least complicated if the departure from chemical equilibrium can be characterized fully in terms of a single variable. More research remains to be done on the subject.

10.2.5. Effects of strain on flame sheets

The view of turbulent diffusion flames that emerges from the preceding considerations is that in many situations they can be described as being composed of wrinkled, moving, laminar sheets of reaction. Effects of finite-rate chemistry involving trace species (for example, leading to yellow radiant emissions from soot) may be distributed more broadly throughout the flow, but the major heat release often occurs in narrow regions about stoichiometric surfaces. Structures of the reaction sheets were analyzed in Section 3.4. The results of Section 3.4 may be applied fully to turbulent diffusion flames that consist of ensembles of wrinkled, laminar, diffusion-flame sheets, or flamelets. This enables many conclusions to be drawn concerning the structure and dynamics of turbulent diffusion flames.

The analysis of Sections 3.4.3 and 3.4.4 may be applied directly to flamelets in turbulent flows. The result in equation (3-98), with the definition in equation (3-97), shows that if $|\nabla Z|^2$ becomes too large at any given position and time at a point on the flame sheet in the turbulent flow, then extinction occurs at that point. According to equation (47), an increase in $|\nabla Z|^2$ produces an increase in the magnitude of the local, instantaneous, diffusion-controlled reaction rate. Within the context of activation-energy asymptotics, this increase terminates abruptly with extinction at a critical maximum value, beyond which the chemistry is not fast enough to keep pace with the rate at which the reactants are diffusing into the reaction zone. Since $|\nabla Z|^2$ is proportional to the local, instantaneous rate of dissipation of the conserved scalar, which in turn is proportional to the square of a strain rate (a velocity gradient) in the turbulent flow, sufficiently large strain rates cause local flamelet extinctions.

Criteria for local flamelet extinction are readily stated from equation (3-98). These criteria provide restrictions that must be satisfied if the turbulent flame is to be composed of an ensemble of flamelets. What happens at higher strain rates is not known today. Local extinctions must be followed by a certain degree of nonreactive molecular mixing of fuel and oxidizer. If conditions are favorable, then the mixed reactants may ignite and burn in a premixed fashion. We shall see in the following section that for the premixed combustion to occur in thin fronts, the strain rate again cannot be too large. Perhaps premixed flamelets never can develop subsequent to extinction of diffusion flamelets. The overall residence times available in the flow may be too short for the mixed reactants to burn at all. Another alternative is that combustion and heat release of the molecularly mixed reactants occur in broad zones in a roughly homogeneous manner, like that of a stirred reactor. This mode of burning would be favored in the presence of generally high overall levels of temperature because of the strong dependence of typical reaction rates on temperature; since the rates would then accelerate as heat is released, processes akin to homogeneous thermal or branched-chain explosions may then be envisioned to occur. Bases for specifying conditions for relevance of any of these processes currently are unavailable. There are many studies that could help to improve our understanding of processes that may occur subsequent to flamelet extinction. For example, time-dependent analyses of the dynamics of extinction and analyses of the evolution of a hole in a continuous, planar, diffusion-flame sheet would provide helpful information. Can conditions be identified for holes to grow or contract? Can rates of growth or contraction be calculated? These and other questions must be explored if understanding of turbulent diffusion-flame structures at high strain rates is to be obtained.

A working hypothesis that seems reasonable for many turbulent diffusion flames—notably for those in open environments, with cold reactants that release large amounts of energy over narrow ranges of stoichiometry at rates strongly dependent on temperature—is that combustion is not reestablished after flamelet extinction. This hypothesis can be employed to investigate liftoff and blowoff of turbulent-jet diffusion flames [86], [87], [88]. In the turbulent range, as the velocity of the fuel issuing from the jet is increased, the average flame height changes little. At sufficiently low fuel velocities, the base of the flame extends to the mouth of the duct, where it is stabilized. When a critical exit velocity, the liftoff velocity, is exceeded, the flame is abruptly detached from the duct and acquires a new configuration of stabilization in which combustion begins a number of duct diameters downstream. The axial distance from the duct exit to the plane at which the flame begins, the liftoff height of the lifted diffusion flame, increases as the exit velocity is increased further. When the exit velocity exceeds a second critical value, the blowoff velocity, the flame can no longer be stabilized in the mixing region, and combustion ceases (blowoff occurs). Liftoff heights and blowoff velocities are of both fundamental and applied interest, and a number of measurements of them may be found in the literature [89]–[91].

The occurrence of liftoff may be understood in terms of critical strain rates for flamelet extinction [86]. As the exit velocity is increased, the average local strain rate increases, thus causing an increase in $\bar{\chi}_c$. An increasing fraction of flamelets then encounters extinction conditions, $\delta \leq \delta_E$, with δ obtained from equation (3-97) and δ_E from equation (3-98). When too many of the flamelets are extinguished, a network of diffusion flamelets no longer is connected to the burner, and liftoff must occur. In a first, rough approximation, this may be assumed to happen when $\bar{\chi}_c \geq \chi_E$, where χ_E is the value of $2D|\nabla Z|^2$ obtained from equation (3-97) by putting $\delta = \delta_E$. Since this value can be calculated from thermochemical and chemical-kinetic parameters alone, the liftoff criterion expresses a comparison between a fluid-mechanical quantity ($\bar{\chi}_c$) and a chemical property of the system. Thus we see that $\bar{\chi}_c$ is relevant not only to average rates of heat release [equation (51)] but also to liftoff phenomena.

After liftoff occurs, the lower temperatures at the duct exit produce lower molecular diffusivities—that is, higher Reynolds numbers—so that $\bar{\chi}_c$ is increased appreciably in the vicinity of the exit. Farther downstream, $\bar{\chi}_c$ is reduced as the jet spreads. Beyond some distance from the duct exit, the condition $\bar{\chi}_c \leq \chi_E$ again is established, and the turbulent diffusion flame may exist. The liftoff height then may be determined as the height at which $\bar{\chi}_c = \chi_E$ [86], [88]. This height increases as the jet velocity increases because at any fixed height, $\bar{\chi}_c$ increases with jet velocity. Although there are significant uncertainties in estimating $\bar{\chi}_c$ for turbulent jets, reasonable agreements between measured and estimated liftoff heights have been obtained.

This discussion represents only a broad outline of the ideas. There are a number of questions of detail, for example, concerning the radial position in the jet at which a condition like $\bar{\chi}_c = \chi_E$ should be applied. A position might best be selected at which the concentration of stoichiometric surfaces is greatest; this lies off the axis of symmetry, at least for low liftoff heights. At sufficiently high jet velocities, very few stoichiometric surfaces will occur anywhere with $\bar{\chi}_c \leq \chi_E$, and blowoff must occur. Thus the viewpoint is consistent qualitatively with observations.

An alternative way to calculate liftoff heights is to set the local average-flow velocity in the jet equal to a premixed turbulent flame speed [89]–[91]. This is done on the basis of the assumption that thorough molecular-scale mixing occurs in the cold jet upstream from the point of flame stabilization. The universal validity of this assumption is questionable and in need of further study [86]. If it is a reasonable approximation, then associated with the radial mixture-ratio profile of the nonreactive jet at each axial location is a flame-speed profile (see the following section) that, when balanced against the average velocity profile through identification of an axial distance that results in a tangency condition, gives the flame-base position for each exit velocity. Further research may help to establish ranges of validity for the two alternative views.

10.3. TURBULENT PREMIXED FLAMES

10.3.1. Objectives of analysis

Unlike diffusion flames, premixed laminar flames are appropriately characterized in terms of a laminar flame speed v_0 [for example, equation (5-2)] that can be expressed very roughly as $v_0 = \sqrt{D_L/\tau_c}$, where D_L is a laminar (molecular) diffusivity and τ_c is a representative chemical conversion time in the hot part of the laminar flame. A useful nondimensional measure of the intensity of turbulence for premixed turbulent combustion is the ratio of the root-mean-square velocity fluctuation to the laminar flame speed, namely, $\sqrt{2q}/v_0$. A laminar-flame quantity against which turbulence scales can be measured is the flame thickness, $\delta = D_L/v_0$ [equation (5-4)]. Regimes of premixed turbulent combustion can be considered in a parameter plane whose coordinates are the nondimensional intensity and the ratio of a scale of turbulence to δ. Figure 10.5 is an illustration of the parameter plane, with the Kolmogorov scale [equation (30)] selected for the turbulence length. A plot equivalent to Figure 10.5 but with a different selection of coordinates was given by Bray in [27]; turbulence Reynolds numbers are employed there, instead of l_k/δ, for the horizontal scale, and lines of constant values of l_k/δ then appear in the figure. The specific selection of coordinates is a somewhat arbitrary matter of individual preference, since the same information can be illustrated irrespective of the selection. Although use of l_k/δ exhibits the influence of this scale ratio more clearly, the presence of τ_c in δ causes the chemistry to affect both coordinates in Figure 10.5, while the Reynolds numbers are properties of the turbulence alone [39], [40].

For the purposes of Figure 10.5, all molecular diffusivities have been set equal, so D_L is the same as the kinematic viscosity, $v = \bar{\mu}/\bar{\rho}$, which appears in equation (29). By use of equations (29) and (30), it is then readily found that the turbulence Reynolds number based on the integral scale is $R_l = [(\sqrt{2q}/v_0)(l_k/\delta)]^4$, a relationship that enables lines of constant values of R_l to be plotted, as shown. Turbulence Reynolds numbers quoted in the literature are often based on the Taylor scale, equation (31), instead of the integral scale; these are directly related to R_l and are denoted by R_t in Figure 10.5. In addition to the ratio (l_k/δ) of the smallest turbulence scale to the laminar-flame thickness, the ratio of the largest scale (the integral scale) to the flame thickness, l/δ, is a relevant parameter. Lines of constant values of l/δ, generated from equation (30), also are shown in Figure 10.5.

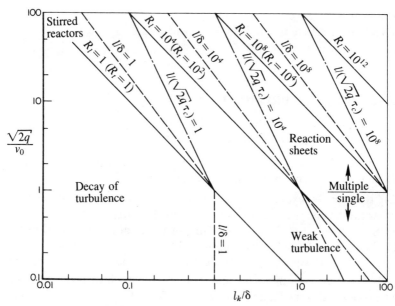

FIGURE 10.5. Parameter plane of nondimensional intensity and nondimensional Kolmogorov scale of turbulence for premixed turbulent combustion, showing regimes of combustion and lines for constant values of turbulence Reynolds numbers, nondimensional integral scale, and a Damköhler number.

Ratios of fluid-dynamical or transport times to chemical times are Damköhler numbers (see Sections 3.4 and 5.4) that are of relevance in identifying regimes of reacting flows. One definition of a Damköhler number for turbulent reacting flows is the ratio of a large-eddy time, $l/\sqrt{2q}$, to τ_c; lines of constant values of this quantity are shown in Figure 10.5. Another definition would be the ratio of a Kolmogorov-eddy time to τ_c. The characteristic time for a Kolmogorov eddy, $(\nu/\epsilon)^{1/2} = [D_L l/(2q)^{3/2}]^{1/2}$, is the shortest time scale associated with the dynamics of the turbulence; the Damköhler number based on this time can be shown to be $(l_k/\delta)^2$, whence vertical lines in Figure 10.5 correspond to constant values of this Damköhler number. Since chemical times can be defined for nonpremixed combustion as well, by working with Damköhler numbers the regimes indicated in Figure 10.5 may be interpreted for turbulent diffusion flames, although the coordinates have been selected to apply more directly to premixed turbulent combustion.

At turbulence Reynolds numbers below unity, special distinctions between different scales of turbulence disappear, and turbulence decays rapidly. Therefore, there is relatively little interest in this portion of Figure 10.5. For $R_l < 1$ and to the left of the vertical line labeled $l/\delta = 1$, that is, in the part of the diagram identified as "decay of turbulence," a representative decay time for turbulence is less than the transit time through a laminar flame. In the low-intensity portion of this regime, the influence of turbulence on the combustion is negligible. Although this regime at sufficiently high intensity (for example, $\sqrt{2q}/v_0 > 1$) may correspond to combustion in certain stirred reactors where turbulence is maintained by stirring, most stirred reactors have $R_l > 1$. We shall no longer concern ourselves with the regime of decay of turbulence.

Two important limiting regimes are identifiable in Figure 10.5, the regime of stirred reactors and the regime of reaction sheets. In the former the turbulent mixing is rapid compared with the chemistry, thereby causing the combustion to occur in a distributed reaction zone, and in the latter the chemistry in hot regions is rapid compared with the turbulent processes, so that the combustion tends to occur in thin sheets wrinkled by the turbulence. This general classification is applicable for both nonpremixed and premixed combustion; here we are concerned with premixed systems. Ascertaining the character of premixed combustion in the stirred-reactor regime is a difficult problem that will not be addressed here, although a number of comments on this regime will be made in Section 10.3.5. The main objective throughout the rest of this section is to investigate premixed turbulent combustion in the regime of reaction sheets. This regime may be inferred from Figure 10.5 to encompass a wide range of conditions of practical interest.

Although it is clear that the two limiting regimes exist, the precise location of the boundary separating them is not well understood. Damköhler [2] first recognized the two regimes and suggested that $l/\delta = 1$ may be an appropriate boundary. From Figure 10.5 it is seen that this would allow the reaction-sheet regime to extend farther than would a criterion based on a large-eddy Damköhler number of unity. Kovasznay [92] offered the alternative criterion that $v_0/\delta = \sqrt{2q}/l_t$, which by use of equation (31) can be shown to be equivalent to $(l_k/\delta)^2 = 1$, thereby giving an even narrower reaction-sheet regime in Figure 10.5 (the right half of the figure). Since δ/v_0 is a residence time in a laminar flame, this last criterion can be interpreted as a requirement that the product of the residence time with the representative strain rate ($\sqrt{2q}/l_t$) of the turbulent field be unity [93]. It seems clear that $\delta \ll l_k$ is at least a sufficient condition for occurrence of the reaction-sheet regime, and $\delta \gg l$ a sufficient criterion for the regime of distributed reaction. It is possible that if $l_k \ll \delta \ll l$, then an intermediate regime occurs with distinguished characteristics of its own that may involve elements of both limiting cases. In fact, there may be more than one distinguished intermediate regime—for example, if turbulent-flame structures were to depend on whether $\delta < l_t$ or $\delta > l_t$. Furthermore, within either of the two limiting regimes, dominant attributes of the turbulent combustion may vary with parameters; for example, in the reaction-sheet regime highly convoluted, folded, and multiply connected reaction sheets may occur at high turbulence intensities and weakly wrinkled, simply connected sheets in the portion marked "weak turbulence." Thus there are a number of different mechanisms of premixed turbulent combustion.

   In the regime of reaction sheets, premixed combustion involves propagation of wrinkled laminar flames. The problem of describing burning in this regime is difficult when the turbulence intensity is high. Interesting general ideas on the subject may be found in the literature [94], [95], [96]. The turbulent dispersion of fluid elements becomes sufficiently large to cause different portions of wrinkled flame sheets to approach each other. Laminar propagation of the sheets may then cause the reaction fronts to collide and cut off pockets of unburnt gas that decrease in size as the flame sheets surrounding them propagate inward [94]; a corresponding mechanism for forming pockets of burnt gas does not exist. The combustion might be viewed as a balance between the generation and consumption of area of the wrinkled laminar flames [95]. An approximate theory has been developed in which the turbulent dispersion is offset by the wrinkled laminar flame-front advance [96]. Under these high-intensity conditions, turbulence exerts the main influence on the motion of reaction regions, but the influence of propagation of the laminar flame fronts is never negligible [96]. Because of the complexity of the problem, we shall not concern ourselves here with this high-intensity behavior.

   The limit beyond which it is necessary to consider effects of the type just mentioned has been suggested to be approximately $\sqrt{2q}/v_0 = 1$ [96], as indicated by a boundary between multiple wrinkled sheets and single wrinkled sheets on the right-hand side of Figure 10.5. The boundary shown represents the best guess currently available, but its position is uncertain. Unless l_k/δ is sufficiently large, diffusive-thermal effects (Section 9.5.2.3) cause the boundary location to depend on this ratio and therefore to depart from a horizontal line in Figure 10.5. Even at very large values of l_k/δ, hydrodynamic instability (Section 9.5.2.2) may cause the level of a horizontal boundary to vary with heat release.* At large values of l_k/δ, the laminar flame is of negligible thickness and is, therefore, characterized solely by v_0 but at a fixed value of $\sqrt{2q}/v_0$, the variation of the turbulence Reynolds number with l_k/δ may cause the tendency toward multiply connected sheets to vary with this parameter; the value of $\sqrt{2q}/v_0$ at the boundary might be expected to decrease slightly with increasing l_k/δ because of the increasing range of turbulence scales. Without addressing the difficult question of the boundary of validity, the following two sections will be concerned entirely with single reaction sheets. We shall investigate the dynamics of wrinkled flames in turbulent flows.

---

   * The nonlinear interaction between turbulence and the hydrodynamic instability is an important, difficult, and unresolved problem of turbulent-flame theory for large-scale turbulence in the reaction-sheet regime.

At sufficiently low intensities of turbulence, expansions for small values of $\sqrt{2q}/v_0$ can be relevant [15]. This kind of approach could be applicable even for $R_l < 1$ and thus might describe turbulent flames for which l_k/δ is not large. However, there is little interest in this low-intensity regime, and suitable analyses for it have not been developed. The approach to be adopted here instead is an expansion for large values of l_k/δ. Since explicit restrictions on turbulence intensities do not appear, the results may apply at intensities of practical interest. The aim is not to analyze the dynamics of the turbulence but rather to investigate the dynamics of the flame in the turbulent flow and the influence of the flame on any given turbulence.

The turbulent-combustion questions of practical concern in premixed systems include those indicated in Section 10.2.1. In addition, in premixed turbulent combustion the average linear rate of propagation of the region of reaction normal to itself, the turbulent flame speed, is of central interest. The parts of Figure 10.5 in which turbulent flame speeds exist have not been properly established theoretically, and for practical configurations of flame propagation there have been discussions of turbulent-flame accelerations instead of speeds. However, there is a large body of experiments, summarized in [33], in which turbulent flame speeds have been observed and their values measured. Moreover, the theories developed for the various regimes have been directed largely to the calculation of turbulent flame speeds and have succeeded in obtaining expressions for them. The principal focus of our presentation will be the turbulent flame speed, which we shall assume to exist in the regime analyzed and to be a property of the gas mixture and of the turbulence. Before proceeding to consider turbulent flames, we shall indicate briefly the modifications that the strain in turbulent flows can produce in premixed laminar flames.

10.3.2. Effects of strain on laminar flames

Laminar flames in turbulent flows are subjected to strain and develop curvature as consequences of the velocity fluctuations. These influences modify the internal structure of the flame and thereby affect its response to the turbulence. The resulting changes are expected to be of negligible consequence at sufficiently large values of l_k/δ in Figure 10.5, but as turbulence scales approach laminar-flame thicknesses, they become important. Therefore, at least in part of the reaction-sheet regime, consideration of these effects is warranted. The effects of curvature were discussed in Section 9.5.2.3. Here we shall focus our attention mainly on influences of strain.

Given any velocity field $\mathbf{v}$ for a fluid, the motion in the vicinity of a point on the reaction sheet can be resolved into a uniform translation with velocity $\mathbf{v}$, a rigid-body rotation with angular velocity $\frac{1}{2}\nabla \times \mathbf{v}$ and a pure straining motion [97]. The first two of these motions have no effect on the internal structure of a locally planar reaction sheet. The third is described by the (symmetric) rate-of-strain tensor $\mathbf{e} = \frac{1}{2}[(\nabla\mathbf{v}) + (\nabla\mathbf{v})^T]$. There are many ways to decompose the effect of a general $\mathbf{e}$ into a superposition of simpler flows. Insofar as the influence of $\mathbf{e}$ on the local internal structure of a locally planar sheet is concerned, stagnation-point flow is the relevant element of the decomposition [93]. Two parameters characterize a general, three-dimensional, stagnation-point flow [98], [99]. One is the normal element of the rate-of-strain tensor, $b \equiv \mathbf{n} \cdot \mathbf{e} \cdot \mathbf{n}$, where $\mathbf{n}$ is a unit vector locally normal to the sheet, and the other is the ratio, c, of the principal axes of the intersection of the rate-of-strain quadratic with the plane of the sheet [$c = (\partial v/\partial y)/(\partial w/\partial z)$ if y and z are coordinates in the directions of these principal axes and v and w are velocity components in directions y and z]. Although the second of these parameters affects details of the sheet structure (for example, distinguishing between two-dimensional and axisymmetric stagnation points), the first exhibits the major qualitative influence. Thus analyses of flat flames in counterflowing streams (the premixed version of the flow shown in Figure 3.7) for $-\infty < b < \infty$ with at least one value of c are needed for investigating the influence of strain on laminar flames (just as the analyses of Section 3.4 were needed in Section 10.2.5 for diffusion flames).

The first analysis of a premixed flame in counterflowing streams of reactants and products is that of Klimov [93], who addressed adiabatic systems with Lewis numbers of unity and one-step chemistry with strongly temperature-dependent rates of heat release, deriving a number of important aspects of the flame behavior. The relevant differential equations are readily obtained from those of Section 9.5.1 by letting f, R, θ, and Y_i be independent of y and z (for example, $\nabla_\perp f = 0$). From equation (9-95), for example, it may be inferred that the principal difference in structure from the unstrained flame arises from the variation of the streamwise component of the mass flux Ru with the normal distance ξ through the flame. The source of this variation is understood most simply by considering a two-dimensional, constant-density, stagnation-point flow; the rate of change of the transverse component of velocity with transverse distance then is $\mathbf{t} \cdot \mathbf{e} \cdot \mathbf{t}$, where $\mathbf{t}$ is a unit vector tangent to the sheet in the plane of the flow. By continuity $\mathbf{t} \cdot \mathbf{e} \cdot \mathbf{t} = -b$, so that, for example, a decrease in the normal mass flux with distance through the flame ($b < 0$) is reflected in a net transverse outflow ($\mathbf{t} \cdot \mathbf{e} \cdot \mathbf{t} > 0$) of fluid from a volume element. This type of process was called **flame stretch** by Karlovitz [100]. Thus studies of influences of strain on flames are flame-stretch studies.

A general definition of flame stretch for planar flames is the time derivative of the logarithm of an area of the flame sheet [15], [93], the boundary of the area being considered to move with the local transverse component of the fluid velocity at the sheet. This definition is applied to an infinitesimal element of surface area at each point on the flame sheet to provide the distribution of stretch over the sheet. Thus at any given point on the sheet, the flame stretch becomes $-\mathbf{n} \cdot \nabla \times (\mathbf{v} \times \mathbf{n}) = -(\mathbf{n} \cdot \nabla)(\mathbf{v} \cdot \mathbf{n}) + \nabla \cdot \mathbf{v}$ (since $\mathbf{n}$ is a constant vector here), which is $-b$ when $\mathbf{e}$ is evaluated (conveniently) for the constant-density flow of reactants just ahead of the flame or of products just behind. With the nondimensionalization of Section 9.5.1, a suitable nondimensional flame stretch becomes $\kappa = -b\alpha_\infty/v_\infty^2$, which is roughly $-b$ times a residence time (δ/v_0) in the flame and is also the reciprocal of an appropriate Damköhler number. The nondimensional stretch κ appears in the nondimensional differential equations for the flame structure.

The influence of stretch on flame structure can be seen qualitatively without going through a formal analysis of the equations of Section 9.5.1. For illustrative purposes, it is sufficient first to put $R = 1, \delta_1 = 1$, and $Le_1 = 1$ in equation (9-95) for $i = 1$, thereby obtaining (with $\nabla_\perp f = 0$ and Y_1 dependent only on ξ and τ)

$$\frac{\partial Y_1}{\partial \tau} + (u_s - \kappa\xi)\frac{\partial Y_1}{\partial \xi} = \frac{\partial^2 Y_1}{\partial \xi^2} - w, \tag{52}$$

subject to $Y_1 = Y_{1,0}$ at $\xi = -\infty$ and $Y_1 = 0$ at $\xi = \infty$. Here use has been made of equation (9-87) in the constant-density case ($R = 1$) with b locally constant to write u as a linear function of ξ, $u = u_s - \kappa\xi$, where the integration constant u_s may be written as $u_s = \kappa\xi_s$ if ξ_s is the value of ξ at the stagnation point (where $u = 0$) in flame-fixed coordinates. The spatial variation of the mass flux, produced by stretch, appears explicitly in equation (52) as $\kappa\xi$.

The nature of the solution to equation (52) depends on the sign of κ. Positive stretch corresponds to $\kappa > 0$ and negative stretch, flame-sheet compression, corresponds to $\kappa < 0$. For positive stretch, according to the sign of $u_s - \kappa\xi$ in equation (52), material flows toward the flame from distances far upstream or far downstream, and therefore the strain tends convectively to thin the flame sheet and to steepen the temperature and concentration gradients within it. Conversely, for compression ($\kappa < 0$), material tends to flow away from the flame and thereby smooths and thickens the profiles. In fact, a steady-state solution to equation (52) does not exist for $\kappa < 0$. This may be understood by observing that with $\kappa < 0$, $u = u_s - \kappa\xi$ always becomes negative at sufficiently large negative values of ξ, (that is, $-\xi > -u_s/\kappa$), so that in coordinates moving with the reaction sheet, convection is in the upstream direction, the direction of flame propagation, at all distances ahead of the flame beyond a critical distance; since a convective-diffusive balance of thermal energy is needed in the steady flame structure ahead of the reactive-diffusive zone and since conduction of heat is in the upstream direction, convection of thermal energy must be directed downstream if the steady balance is to be established. When κ is negative but sufficiently small, the region of essential unsteadiness lies sufficiently far ahead of the reaction sheet for an approximate steady-state solution to be defined through most of the flame [93]. For $\kappa \geq 0$, there always exists a steady-state solution of some kind. Most studies of the influences of flame stretch are restricted to positive stretch ($\kappa > 0$) and work with steady-state conservation equations [for example, equation (52) with $\partial Y_1/\partial \tau = 0$].

Many further analyses of structures of stretched laminar flames have now been published [101]–[113]. With one exception [113], they essentially apply activation-energy asymptotics for one-step kinetics; multistep kinetics can have a significant influence on the responses of flames to strain [113], through enhanced diffusive losses of reaction intermediaries from the reaction zone. Most of the analyses employ a constant-density approximation, but a few [106], [110]–[112] have taken density variations into account. As a consequence of these studies, we now have a good understanding of many aspects of the influences of strain.

Unless κ is small, the definition of burning velocity becomes ambiguous because the mass flux then varies appreciably with distance through the flame. The most natural definition of burning velocity would employ the mass flux at the reactive-diffusive zone because it is at this plane that the heat release occurs. Since the reactive-diffusive zone lies on the downstream end of the flame, it is understandable that an increase in κ tends to decrease the burning velocity according to this definition; for example, at constant density $\partial u/\partial \xi$ is negative when κ is positive, so velocities in the downstream part of the flame are less than those upstream. At sufficiently large values of κ, in flame-fixed coordinates the stagnation point may lie within the flame, so that the velocity is negative at the reactive-diffusive zone. This gives negative burning velocities at large κ [93], a result that is in no way paradoxical, since reactants still can enter the reaction zone by diffusion against the unfavorable convection in the convective-diffusive zone. These results suggest that the rate of heat release or of reactant consumption per unit flame area for most purposes will be a better measure of the response of the flame to strain than will the burning velocity. For one-step reactions with Lewis numbers of unity in adiabatic systems, the rate of heat release per unit area is decreased by strain [93]; it is unchanged until the reaction zone begins to exhibit a reactive-convective-diffusive balance (see Section 3.4) and thereafter decreases with increasing κ, approaching zero as κ approaches infinity [106]. Of course, as a consequence of the increase in area of a flame-sheet element with time under stretch, the rate of heat release per unit *initial* flame area increases with time [15], [93].

Nonadiabaticity and Lewis numbers differing from unity modify the rate of heat release per unit area. Let us rule out distributed heat loss and consider nonadiabaticity associated with the temperature of the product stream at infinity, T_∞, differing from the adiabatic flame temperature, T_{af}. If the product stream is hotter (a superadiabatic condition), then by enhancing heat conduction (through reducing distances over which heat conduction occurs) and by bringing the reactive-diffusive zone closer to the product side of the stagnation point, an increase in κ results in an increase in the flame temperature at the reactive-diffusive zone and thereby increases the reaction rate and the rate of heat release per unit area of the flame [110]. Through this same mechanism, an increase in κ results in a decrease in the heat-release rate if the temperature of the product stream is below the adiabatic flame temperature (a subadiabatic condition). In this last situation, if the product stream is cold enough, then an abrupt transition in the flame structure occurs when κ is increased past a critical value [110] that depends on the degree of subadiabaticity through a nondimensional parameter, which can be taken to be

$$H \equiv \beta(T_{af} - T_\infty)/T_{af} = (T_{af} - T_\infty)(T_{af} - T_0)E_1/(R^0 T_{af}^3) \qquad (53)$$

when the specific heat is constant; here E_1 is the overall activation energy and T_0 is the initial temperature of the cold reactants. For $H \lesssim 3$ the reaction rate varies continuously with κ, but at larger values of H, abrupt transitions occur.

The behavior obtained is illustrated in Figure 10.6, where μ is the ratio of the rate of heat release per unit area of the strained flame to that of the adiabatic unstrained flame with the same reactants and where, in the

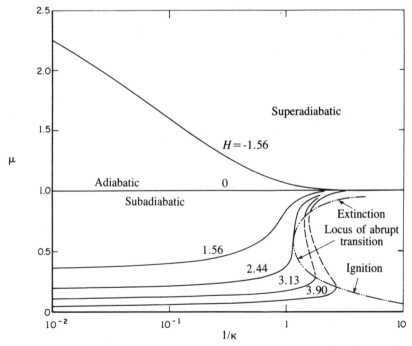

FIGURE 10.6. The ratio of the rate of heat release of a strained flame to that of an unstrained flame as a function of the reciprocal of the nondimensional strain rate for various values of the nonadiabaticity parameter H; calculations were performed for a one-reactant, first-order reaction in an ideal gas with constant specific heat and Lewis number of unity, with thermal conductivity proportional to temperature and an adiabatic flame temperature four times the initial temperature [110].

definition of κ, the strain rate $(-b)$ was evaluated in the constant-density product flow at the adiabatic flame temperature. In Figure 10.6 we see that at the larger values of H there are **S**-shaped curves giving abrupt transitions that occur from conditions of relatively rapid rates of heat release to conditions of relatively slow rates with increasing strain rates. For an experiment in which κ is caused to increase gradually, the transition would be identified as an abrupt extinction event. Thus at values of H above a limiting value of order unity, there is a critical nondimensional strain rate κ for extinction, which is a function of H. Since κ is the reciprocal of a suitably defined Damköhler number, this result implies that there is a critical Damköhler number (dependent on H) below which extinction occurs. At the extinction point, the temperature in the reaction zone has decreased to a value at which the rate of heat release no longer is sufficient for an appreciable amount of reaction to occur in the residence time available in the reaction zone.

The abruptness of the transition in Figure 10.6 at larger H is a consequence of the strong dependence of the reaction rate on temperature. At a fixed extent of nonadiabaticity $[(T_{af} - T_\infty)/T_{af}$ fixed$]$, a decrease in the Zel'dovich number β in equation (53) produces a decrease in H and, therefore, a tendency away from abruptness. For all values of H, it is found that μ approaches zero gradually as κ approaches infinity [110] (subsequent to the abrupt transition, if it occurs), as has been indicated above for adiabatic flames; this gradual approach occurs at values of κ that are roughly of order β^2, and it is not indicated in Figure 10.6, which has been obtained from an analysis that treats κ as being of order unity. For values of β that are small enough (or for sufficiently small extents of subadiabaticity at any given β, as well as for all superadiabatic conditions) only the gradual variations occur. There are many experimental conditions under which the abrupt transition should be observable. If κ is gradually decreased (at a fixed value of H that is large enough) for a flow in which the rate of heat release is very small, then a point is reached at which an abrupt transition to a state of rapid reaction occurs. This transition would be interpreted experimentally as ignition*; loci for both ignition and extinction are indicated in Figure 10.6. When the critical value of κ is plotted as a function of H, the graph exhibits a cusplike shape, with the point of the cusp occurring at a maximum value of κ where the ignition and extinction loci meet.

Analogous behavior is attributable to Lewis numbers differing from unity in adiabatic systems [108], [111]. In terms of the Lewis number Le_1 of the reactant in a one-reactant system, the relevant parameter that corresponds to H is

$$J \equiv \beta(Le_1 - 1)/Le_1 = (Le_1 - 1)(T_{af} - T_0)E_1/(R^0 T_{af}^2 Le_1), \qquad (54)$$

---

* In reality, ignition often occurs in a regime of distributed reaction, which differs from the regime calculated in Figure 10.6 [110].

and a multiple-valued dependence of μ upon κ occurs if $J \gtrsim 12$. The source of the variation of μ with κ can be understood on the basis of the influence of Le_1 on the variation of the flame temperature. For the unstrained adiabatic flame, the flame temperature is T_{af}, independent of Le_1. By increasing gradients, the strain increases the rate of reactant diffusion into the reaction sheet and also the rate of conductive heat loss from the reaction sheet to the gas upstream. The former effect tends to increase the flame temperature because of the heat release that occurs upon reactant consumption, while the latter tends to decrease the flame temperature. If $\mathrm{Le}_1 = 1$, then these two effects counterbalance each other so that the flame temperature remains at T_{af}, and the rate of heat release per unit area is independent of the strain rate. If the thermal diffusivity is less than the diffusion coefficient of the reactant ($\mathrm{Le}_1 < 1, J < 0$), then the effect of reactant diffusion is the greater of the two, and the flame temperature increases (so that μ increases) as κ increases; in this case the variations are gradual, analogous to the super-adiabatic case of Figure 10.6.* If the thermal diffusivity exceeds the diffusion coefficient of the reactant ($\mathrm{Le}_1 > 1, J > 0$), then the heat-conduction effect is the greater, the flame temperature and μ decrease as κ increases, and there is the possibility of an abrupt transition to a condition of slow reaction (an extinction transition) like that described earlier for subadiabatic systems. † The characteristics of the abrupt transitions and their interpretations in terms of extinction and ignition parallel those discussed in connection with Figure 10.6. A schematic illustration of the portion of the curve of μ as a function of κ that exhibits abrupt extinction is shown in the upper part of Figure 10.7. The condition $J \gtrsim 12$, needed for abrupt extinctions to occur in adiabatic systems, is encountered for relatively few reactant combinations, since it requires reactants with rather high molecular weights or unusually large overall activation energies. Small degrees of subadiabaticity can appreciably enhance the tendency toward the existence of abrupt extinctions [111] with $\mathrm{Le}_1 \neq 1$.

The discussion thus far has concerned counterflowing streams of reactants and products in an infinite domain. Related configurations include the impingement of a reactant stream on a flat wall and counterflowing streams of reactants. Although less directly relevant to turbulent combustion, these configurations recently have been subjects of laboratory measurements in laminar flows [114]–[121] and therefore afford good possibilities for comparisons between theory and experiment. The symmetric problem of counterflowing reactants, in which two identical reactant streams are directed toward each other and identical flames are established in each

---

* In fact, physically explicable abrupt transition may occur at higher strain rates even in this case [111], but such transitions require values of parameters that are unrealistic.

† The condition $\mathrm{Le}_1 > 1$ is the opposite from that ($\mathrm{Le}_1 < 1$) identified in Section 9.5.2.3 as conducive to the formation of cellular flames.

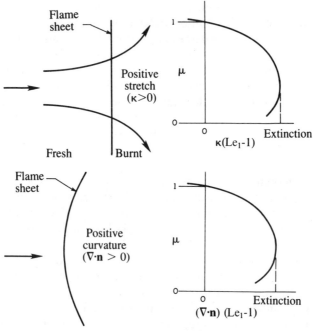

FIGURE 10.7. Schematic illustrations of influence of flame stretch and of flame curvature, through Lewis-number effects, on rates of heat release per unit area for adiabatic, one-reactant flames with one-step chemistry of large overall activation energy.

stream, is equivalent by symmetry to a one-stream problem in which one of the reactant streams is directed normally onto an infinite adiabatic flat plate at which the no-slip condition is replaced by a zero-stress condition; a number of theoretical analyses of this problem have been published (for example, [101], [102], [103], [107], [112], [118]),* and nonadiabatic problems also have been studied [112], [118]. The results of these analyses are qualitatively similar to those just discussed, but the multiple-valued dependence of μ upon κ is now found to occur for $J \geq 4$ in the symmetric problem; by removing the possibility of thermal adjustments occurring in the product stream, the introduction of symmetry strengthens the dependence of the flame temperature on the Lewis number and enables abrupt extinctions to be encountered for many real reactant mixtures that have $Le_1 > 1$. There are clear experimental confirmations of this qualitative prediction [114], [120]. As κ is increased for reactants with $Le_1 > 1$ to a sufficient extent, the two flames move closer together but experience abrupt extinction at a critical value of κ before they reach the stagnation point; for reactants with $Le_1 < 1$, the two flames tend to merge at the plane of symmetry prior to abrupt extinction.

---

* Also J. Sato and H. Tsuji, *CST* **33**, 193 (1983).

The theoretical results that have been discussed here allow κ to be of order unity, a condition that sometimes has been referred to as one of *strong strain*, although the term *moderate strain* seems better, with *strong strain* reserved for large values of κ. Small values of κ are identified as conditions of *weak strain*; under these conditions the reaction sheet remains far to the reactant side of the stagnation point, and by integrating across the convective-diffusive zone, a formulation in terms of the location of the reaction sheet can be derived [102], like that discussed at the end of Section 9.5.1. By combining the approximations of weak strain and weak curvature, convenient approaches to analyses of wrinkled flames in turbulent flows can be obtained [38].

10.3.3. Theory of wrinkled laminar flames

At the beginning of Section 10.3.2 we indicated that turbulence induces both strain and curvature in laminar flames. The influence of curvature on the flame structure can be understood physically by the same type of reasoning that was just employed. The essential physical ideas were given in Section 9.5.2.3—for example, for a one-reactant, adiabatic system, by enhancing rates of upstream transport, a convexity toward the unburnt gas decreases the flame temperature and thus the heat-release rate if $Le_1 > 1$. Again, when J of equation (54) is large enough, abrupt extinction transitions occur, as illustrated in the lower part of Figure 10.7, where $\mathbf{n}$ represents a unit vector normal to the reaction sheet and directed from the burnt gas toward the fresh mixture, so that $\mathbf{V} \cdot \mathbf{n}$ is the sum of the reciprocals of the two principal radii of curvature of the sheet, each taken to be positive when the sheet is concave toward the burnt gas; that is, $\mathbf{V} \cdot \mathbf{n} = -1/\mathscr{R}$ for the radius $\mathscr{R}$ defined at the beginning of Section 9.5.2.3.

At least for weak strain and weak curvature, the influences of these strain and curvature phenomena on the flame structure can be characterized in terms of a single quantity, an effective curvature of the flame with respect to the flow or the total stretch of the flame surface produced by the flow with respect to the moving, curved flame. By evaluating b from the rate-of-strain tensor in the products just behind the flame, the last of these quantities may be expressed nondimensionally as

$$\kappa = (\alpha_\infty/v_\infty^2)(-b + v_\infty \mathbf{V} \cdot \mathbf{n})$$

$$= \mathbf{V}_\perp \cdot \mathbf{v}_\perp + [\mathbf{V}_\perp^2 f + (\partial/\partial\tau + \mathbf{v}_\perp \cdot \mathbf{V}_\perp)\sqrt{1 + |\mathbf{V}_\perp f|^2}]/\sqrt{1 + |\mathbf{V}_\perp f|^2}, \quad (55)$$

where the special coordinate system and the nondimensional notation of Section 9.5.1 have been introduced in the last equality, in which the transverse velocity $\mathbf{v}_\perp$ is that on the hot side of the flame. The derivations of these general results involve various aspects of geometry and kinematics, which recently have been clarified completely [104], [122], [123]. There are many equivalent forms in which equation (55) can be written; for example, the expressions may be referred to quantities ahead of the flame instead of behind or (at the expense of small additional complications) to quantities evaluated at an arbitrarily selected location within the flame. The first form shown in equation (55) has an especially simple interpretation, since the curvature term (involving $\mathbf{V} \cdot \mathbf{n}$) is the stretch that the sheet would experience if it were to move in a quiescent medium at the normal burning velocity v_∞; it is merely necessary to add this stretch associated with the curvature (through the relative motion) to the previously identified stretch associated with the flow in order to obtain the total stretch. If $|\mathbf{V}_\perp f| \ll 1$, then the second form shown in equation (55) reduces to $\mathbf{V}_\perp \cdot \mathbf{v}_\perp + \mathbf{V}_\perp^2 f$, which is readily understood as a stretch contribution plus a curvature contribution, since the nondimensional form of $\mathbf{V} \cdot \mathbf{n}$ becomes $\mathbf{V}_\perp^2 f$ (because the transverse components of $\mathbf{n}$ simplify to $\mathbf{V}_\perp f$, while the normal component is nearly the constant unity).

Although strain and curvature effects can be combined as in equation (55), it cannot be concluded that they are of equal importance for wrinkled laminar flames in turbulent flows. If it is assumed that the flame shape is affected mainly by the large eddies, then in terms of the flame thickness δ and the integral scale l, the nondimensional curvature is of order δ/l. This may be compared with the corresponding relevant nondimensional strain rate [93] $(\sqrt{2q}/l_t)(\delta/v_0)$; their ratio, $(l/l_t)(\sqrt{2q}/v_0)$, increases with increasing turbulence intensity. At a sufficiently large turbulence Reynolds number R_l, the strain effect is found to be large in comparison with the curvature effect, as was reasoned in the early study of Klimov [93]. Thus, unless flame instabilities induce curvatures large compared with those produced by the turbulence, the strain effects are anticipated to predominate in most turbulent flames. Nevertheless, the curvature effects are the larger at sufficiently low turbulence intensities and often may be included as easily as the strain effects in analyses of wrinkled flames in turbulent flows.

A fruitful approach to studying wrinkled flames in turbulence has been developed by Clavin and his co-workers [38], who begin with the formulation of Section 9.5.1 and subsequently introduce formal expansions for small values of δ/l to address flames in large-scale turbulence. The analyses sometimes involve Fourier decompositions in transverse spatial coordinates and in time for turbulence that is statistically stationary and homogeneous in transverse directions. Irrespective of whether the Fourier decompositions are introduced, the expansions are applied locally, in effect separately for each Fourier component; therefore, strictly speaking, they require δ/l_k to be small. The flame-structure developments correspond to expansions for weak curvature and weak strain; the κ of equation (55) locally must be small. By use of the estimates in the preceding paragraph and the observation that δv_0 is very roughly of order v, the weak-strain restriction may be shown by use of equations (30) and (31) to require that δ/l_k be small, irrespective of the value of R_l. Thus the theories are, fundamentally, expansions for small ratios of the laminar-flame thickness to the Kolmogorov scale of the turbulence. Subject to this restriction, there are no further basic limitations on the turbulence intensity; $\sqrt{2q}$ may be large compared with v_0.

The formulation of Section 9.5.1 has served to remove the chemistry from the field equations, replacing it by suitable jump conditions across the reaction sheet. The expansion for small δ/l_k subsequently serves to separate the problem further into near-field and far-field problems. The domains of the near-field problems extend over a characteristic distance of order δ on each side of the reaction sheet. The domains of the far-field problems extend upstream and downstream from those of the near-field problems over characteristic distances of orders from l_k to l. Thus the near-field problems pertain to the entire wrinkled flame, and the far-field problems pertain to the regions of hydrodynamic adjustment on each side of the flame in essentially constant-density turbulent flow. Either matched asymptotic expansions or multiple-scale techniques are employed to connect the near-field and far-field problems. The near-field analysis has been completed for a one-reactant system with allowance made for a constant Lewis number differing from unity (by an amount of order $1/\beta$) for ideal gases with constant specific heats and constant thermal conductivities and coefficients of viscosity [122], [124], [125]; the results have been extended to ideal gases with constant specific heats and constant Lewis and Prandtl numbers but thermal conductivities that vary with temperature [126]. The far-field analysis has been completed only in a linear approximation that requires $\sqrt{2q}/v_0$ to be small [38].

The near-field analysis provides expressions for the wrinkled-flame motion and the reaction-sheet temperature in terms of the flame shape and the gas velocities at the edge of the wrinkled flame. To the first order in the small parameter δ/l_k, the equation for the flame motion may be written, in the nondimensional notation of Section 9.5.1, as

$$\frac{\partial f}{\partial \tau} + \mathbf{v}_\perp \cdot \nabla_\perp f = v_\parallel - \sqrt{1 + |\nabla_\perp f|^2}\left\{1 - \kappa\left[I_1 + \frac{1}{2}\beta(\mathrm{Le}_1 - 1)I_2\right]\right\}, \quad (56)$$

where $v_\parallel = v_1/v_\infty$ [with v_1 representing the dimensional longitudinal velocity component appearing in equation (9-88)],

$$I_1 = \frac{R_0}{R_0 - 1}\int_{1/R_0}^1\left(\frac{\lambda_1}{\theta}\right)d\theta \quad (57)$$

and

$$I_2 = \frac{R_0}{R_0 - 1}\int_{1/R_0}^1\left(\frac{\lambda_1}{\theta}\right)\ln\left(\frac{R_0 - 1}{\theta R_0 - 1}\right)d\theta. \quad (58)$$

Here κ is given by equation (55), and just as in equation (55), the nondimensional turbulent velocity components $\mathbf{v}_\perp$ and $v_\parallel$ are to be evaluated locally in the burnt gas just downstream from the wrinkled flame. If it is desired to measure b and the turbulence velocities in the fresh mixture just ahead of the flame, then $\kappa = (\alpha_0/v_0^2)(-b + v_0 \mathbf{V} \cdot \mathbf{n})$ is an appropriate nondimensional total stretch; with v_0 and α_0 used instead of v_∞ and α_∞ in the nondimensionalizations of Section 9.5.1, equation (56) is again obtained in the new variables, but with I_2 now replaced therein by I_2/R_0 and with λ_1 now representing λ/λ_0 in equations (57) and (58). With these new definitions, the nondimensional flame temperature θ_f of Section 9.5.1 is found in the first approximation to be

$$\theta_f = 1 - \kappa(\mathrm{Le}_1 - 1)I_2(R_0 - 1)/R_0^2, \tag{59}$$

which agrees with preceding reasoning in demonstrating that positive stretch reduces θ_f if $\mathrm{Le}_1 > 1$.

The meaning of the results typified by equation (56) can be explained by addressing the various terms sequentially. If the turbulence velocities are large compared with the flame speeds, then equation (56) reduces to $\partial f/\partial \tau + \mathbf{v}_\perp \cdot \mathbf{V}_\perp f = v_\parallel$, which defines the motion of a sheet attached to fluid elements. In the coordinates adopted, the velocity with which the sheet moves normal to itself with respect to the fluid locally is given by

$$v_\infty(v_\parallel - \partial f/\partial \tau - \mathbf{v}_\perp \cdot \mathbf{V}_\perp f)/\sqrt{1 + |\mathbf{V}_\perp f|^2},$$

which is seen from equation (56) to be the burning velocity v_∞ if $\kappa = 0$. The term involving κ in equation (56) is a correction to the burning velocity v_∞ resulting from the structural modifications to the wrinkled flame produced by stretch and curvature. The corresponding correction to the burning velocity v_0 is somewhat different; the normal velocity of the sheet with respect to the unburnt gas is given by $v_0\{1 - \kappa[I_1 + \frac{1}{2}\beta(\mathrm{Le}_1 - 1)I_2/R_0]\}$, with the revised definitions of κ and λ_1 indicated above. The Markstein length $\mathscr{L}$, defined at the beginning of Section 9.5.2.3, is seen by comparison to be $(\alpha_0/v_0)[I_1 + \frac{1}{2}\beta(\mathrm{Le}_1 - 1)I_2/R_0]$ according to the present results.* Since κ is of order δ/l_k here, the burning-velocity corrections decrease as l_k increases, and for a number of purposes the term involving κ in equation (56) can be ignored.

With κ omitted and the turbulence velocities viewed as given, equation (56) is a partial differential equation of first order for the evolution of the flame sheet. From equation (55) it is seen that if κ is retained, then the differential equation involves second spatial derivatives. If the analysis were

---

* Numerical evaluations then give $\mathscr{L} > 0$ for real combustible gas mixtures (with the possible exception of fuel-lean mixtures having hydrogen as the fuel) [38], [122], [126]. On the other hand, the quantity $I_1 + \frac{1}{2}\beta(\mathrm{Le}_1 - 1)I_2$, relevant to the burning velocity with respect to the burnt gas, is negative for a number of mixtures that have $\mathrm{Le}_1 < 1$ [38]. These differences arise from changes in the normal mass flux in crossing the flame and have a number of experimental implications [38], [122], for example, in explaining observed velocity changes of outwardly propagating spherical flames [38], [127].

pursued to higher order in δ/l_k, then fourth spatial derivatives also would appear, as has been found [128], [129], [130] in evolution equations derived on the basis of an approximation of nearly constant density. These higher derivatives are not likely to be important in flows with small values of δ/l_k. What is important is that the turbulence velocities just upstream or just downstream from the wrinkled flame are not known. Far-field analyses must be pursued to solve the propagation problem completely. Results of the near-field analysis—specifically equation (56), the corresponding equation involving velocities at the unburnt edge of the flame, and jump conditions for velocities and pressure across the flame [131]—are needed for approaching the far-field analysis.

In the far field, effects of hydrodynamic instability (Section 9.5.2.2), make themselves felt. In an approximation of nearly constant-density flow, integrations over the far field can be performed to obtain an evolution equation for the flame sheet that includes hydrodynamic effects [128], [129], [130], but this approximation is not physically realistic for real flames. In some early analyses that treated the wrinkled flames as discontinuities, the hydrodynamic instability was simply ignored, thereby producing erroneous results. Ascertaining the influence of the turbulence on the hydrodynamic instability is a difficult problem that is only beginning to be addressed. It might be guessed that if the turbulence velocities are sufficiently large, then they may overpower the hydrodynamic instability; even lower turbulence velocities may have suitable phase relationships to negate hydrodynamic instability. Since gravity (in a favorable orientation) and diffusive-thermal effects (included here through κ) together may eliminate hydrodynamic instability, analyses of problems of turbulent flame propagation in which hydrodynamic instability does not occur may be pursued. Of course, the evolution equation may describe wrinkled-flame motions of increasing amplitudes in the presence of hydrodynamic instability; for most real flames, the diffusive-thermal effects stabilize the short wavelengths of the hydrodynamic instability—wavelengths whose growth would invalidate the approximation of small δ/l_k. Far-field nonlinearities that develop at large amplitudes make the far-field analysis difficult in the presence of hydrodynamic instability. The nonlinearities are present even in the absence of the instability unless $\sqrt{2q/v_0}$ is small. The results to be discussed in the following section do not include the far-field effects.

In the reaction-sheet regime, the structure of the turbulent flame is determined by the dynamics of wrinkled laminar flames. Thus the thickness of the turbulent flame (if it is large compared with that of the laminar flame) is controlled by the distance to which fluctuations in the laminar-flame position may extend. Statistical aspects of distributions of temperature and of species concentrations in the turbulent flame can be expressed entirely in terms of statistics of the laminar-flame position (through f), orientation (through $\nabla_\perp f/|\nabla_\perp f|$), and structure (through κ). The simplest example is that of a turbulent flame propagating in the $-x$ direction and composed of wrinkled laminar flames for which κ is negligibly small and whose local normal vectors deviate only slightly from the x direction; in dimensional variables, if $Q(x)$ denotes the distribution of a property through a laminar flame having its reactive-diffusive zone at $x = 0$, then for this turbulent flame the average of Q is $\bar{Q}(x) = \int_{-\infty}^{\infty} Q(x - F)P(F)\,dF$, where $P(F)$ is the probability-density function for the (dimensional) laminar-flame position F. Thus the difficult problem of calculating $P(F)$ from statistics of the approaching turbulent flow must be addressed if the average profile $\bar{Q}(x)$ for the turbulent flame is to be predicted. Qualitatively, $\bar{Q}(x)$ will be a more gently varying function of x than $Q(x)$. Profiles of average fluctuations through the turbulent flame—for example, the mean-square temperature fluctuation $\overline{T'^2}(x)$—typically achieve large values, comparable with those associated with the difference between burnt-gas and unburnt-gas values, throughout most of the turbulent flame and decrease gradually to zero on both the upstream and downstream sides. Structures of this type make gradient modeling unphysical for streamwise turbulent transport (Reynolds transport) of reactants or thermal energy. An explicit analysis for a simplified model has shown that if an attempt is made to force gradient modeling into the problem then the turbulent diffusion coefficient must change sign within the interior of the turbulent flame [59].

Since the thickness of the turbulent flame is controlled by excursions of the wrinkled laminar flame, it is affected by turbulent motions of fluid elements (turbulent dispersion or turbulent diffusion [51]), by laminar-flame propagation with respect to fluid elements, and by far-field hydrodynamic adjustments that are associated with the density change across the laminar flame. Let us consider a statistically stationary turbulent flow in an Eulerian coordinate system in which the turbulent-flame position is fixed, the streamwise coordinate being identified as that normal to the turbulent flame. In such a flow, the first of the three contributions to the turbulent-flame thickness is roughly the root-mean-square value of the streamwise turbulent displacement of fluid elements in the Eulerian frame, the streamwise Eulerian displacement [59] (the time integral of the streamwise component of velocity at a fixed point in the Eulerian frame). Even though the velocity field is statistically stationary, the streamwise displacement experimentally is not [38], [125]; its Lagrangian counterpart (which, in fact, embodies a fuller representation of the first of the three contributions to the flame thickness) describes the manifestly nonstationary process of turbulent dispersion. Thus if only the first of the three contributions is operative, then the turbulent-flame thickness increases with time (in proportion to $\sqrt{t}$ at sufficiently large t). An offsetting effect from the second contribution, laminar-flame propagation, is responsible for producing the time-independent thickness of the turbulent flame that is observed experimentally, whence the observed flame thickness must depend additionally on processes of laminar-flame propagation. In weak turbulence of large scale, modifications to the burning velocity by diffusive-thermal effects, through the term involving κ in equation (56), are important at low frequencies in causing the turbulent-flame thickness to remain constant [38].

The jump conditions for velocities across a laminar flame approximated as a discontinuity enable turbulence intensities upstream and downstream from a wrinkled flame to be related [125]. If the angles between the local normals to the flame sheet and the mean direction of propagation remain small, then velocity deflections by tilted elements of the sheet enhance velocity fluctuations in transverse directions, inducing anisotropy in an initially isotropic turbulence. The relative intensities (normalized with respect to the local mean flow velocity) are not increased by the flame (in fact, the longitudinal component of the relative intensity is decreased appreciably, a kind of dilatation effect), but the absolute intensities in transverse directions are increased significantly, so that there is an overall increase in the total absolute intensity, a type of **flame-generated turbulence**. Findings of *self-turbulization* by flames propagating in nonturbulent fluids [130] may be considered to represent another kind of flame-generated turbulence. In more-complex configurations, turbulent flames propagating on the average at an angle to a mean flow generate nonuniform profiles of mean velocity (through density changes) that may generate increased turbulence intensities.

10.3.4 Turbulent flame speeds

Expressions for turbulent burning velocities may be obtained from evolution equations of the type given in equation (56) by taking suitable averages. Techniques of this type will be indicated here prior to discussion of alternative methods. We shall make use of the formula like equation (56) that applies to properties ahead of the flame, as discussed above equation (59), because turbulence characteristics usually are specified for the reactant mixture instead of for the burnt gas. If κ is negligibly small, so that the laminar burning velocity may be considered to be unaffected by the turbulence, then a general approach may be employed to write an expression for the turbulent burning velocity v_T in terms of the constant laminar flame speed v_0 and the shape of the wrinkled flame sheet. This fundamental fact was first observed by Damköhler [2]. As illustrated in Figure 10.8, since all of the fluid flowing through the turbulent flame also passes through the wrinkled laminar flame, it is necessary, by consideration of an overall mass balance, that on the average $\rho_0 v_T \mathscr{A} = \rho_0 v_0 \mathscr{A}_f$, where $\mathscr{A}$ and $\mathscr{A}_f$ are the cross-sectional area of the turbulent flow and the average area of the wrinkled flame sheet, respectively. Thus

$$v_T = v_0(\mathscr{A}_f/\mathscr{A}). \tag{60}$$

There is a general way to write the area ratio $\mathscr{A}_f/\mathscr{A}$ of equation (60) in terms of the function $G(\mathbf{x}, t)$, defined at the beginning of Section 9.5.1 for describing the shape of the reaction sheet. Let $\mathbf{e}$ be a constant unit vector

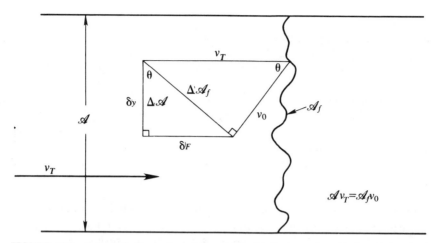

FIGURE 10.8. Schematic illustration of the relationship between the turbulent flame speed and the wrinkled flame area for a premixed turbulent flame consisting of a wrinkled laminar flame.

pointing in the direction of propagation of the turbulent flame. Then at any given point on the reaction sheet (where $G = 0$), the ratio of an area element of the sheet to the projection of that area element on the direction of propagation is $|\nabla G|/|\mathbf{e} \cdot \nabla G|$. Therefore, with an overbar denoting a suitable average, we have

$$v_T = v_0 \overline{(|\nabla G|/|\mathbf{e} \cdot \nabla G|)}_{G=0} \tag{61}$$

from equation (60). The type of average intended in equation (61) is fundamentally an ensemble average but may be understood in more detail by considering turbulent flames that are statistically homogeneous in transverse directions. In this case the definition

$$\overline{\left(\frac{|\nabla G|}{|\mathbf{e} \cdot \nabla G|}\right)}_{G=0} = \lim_{\mathscr{A} \to \infty} \left\{ \frac{1}{\mathscr{A}} \iint \left(\frac{|\nabla G|}{|\mathbf{e} \cdot \nabla G|}\right)_{G=0} d\mathscr{A} \right\} \tag{62}$$

may be employed, expressing the average as an average over the cross-sectional area of the flow (*not* as an average over surface area of the wrinkled flame). Alternatively, if a coordinate system can be identified in which the flow is statistically stationary, then the average can be represented as a time average for a tube having a fixed infinitesimal cross-sectional area element. In all cases, the position in the direction of turbulent-flame propagation at which the integrand is evaluated varies to maintain $G = 0$ (reaction-sheet presence). The definition given here is entirely unrestricted with respect to the shape of the reaction sheet; even multiply connected sheets are admissable. If the sheet is multiply connected or tends to fold back on itself, then there is more than one position in the direction of flame propagation at which

$G = 0$, and the *sum* of all such positions must be included in the integral in equation (62) or in the time or ensemble average.

If the wrinkled flame does not tend to fold back on itself—that is, if the flame sheet nowhere becomes parallel to the vector **e**—then the location of the reaction sheet everywhere may be described by putting $G(\mathbf{x}, t) = x - F(y, z, t)$, where F is defined at the beginning of Section 9.5.1, and x is a coordinate in the propagation direction **e** of the turbulent flame. With this selection we find that $|\mathbf{e} \cdot \nabla G| = 1$ and $|\nabla G| = \sqrt{1 + |\nabla_\perp f|^2}$, so that equation (61) reduces to

$$v_T = v_0 \sqrt{1 + |\nabla_\perp f|^2}. \tag{63}$$

This formula is readily interpreted by reference to Figure 10.8. The tangent of the angle θ between the local normal to the flame sheet and the average stream direction is $\delta F/\delta y$, whence the average area ratio, $\overline{\mathscr{A}_f/\mathscr{A}} = \overline{1/\cos \theta} = \overline{\sqrt{1 + \tan^2 \theta}}$, is the average that appears in equation (63).

Writing equation (60), equation (61), or equation (63) does not solve the problem of calculating v_T; it is necessary to relate the average area ratio to properties of the incoming turbulence. Under certain restrictions this can be done [59], [122], [125]. Consider a turbulent flame, such as that illustrated in Figure 10.8, in a coordinate system in which the turbulence processes are statistically stationary, and assume that homogeneity prevails in transverse directions (y and z). This configuration approximates a flame normal to the flow in grid turbulence that has been generated in a wind tunnel. Introduce the approximations that lead to equation (56), and write that equation in terms of the velocity field just ahead of the flame (so that $v_\parallel = v_1/v_0$, for example). In the adopted coordinate system, $v_1 = v_T + v'$, where v' is the fluctuation of the longitudinal component of velocity in this Eulerian frame just ahead of the flame. Since κ is small for sufficiently small gradients, in a first approximation we shall neglect the κ term in equation (56). The average of that equation then may be shown [38], [122] to give equation (63).* If the transverse gradients of f are small, then the lowest approximation to equation (56) is readily seen to be [59], [125]

$$\partial f/\partial \tau = v'/v_0, \tag{64}$$

the time integral of which relates f to the upstream turbulence. Thus if the streamwise Eulerian displacement of fluid elements by the turbulence is defined as $a = \int v' \, dt$, equation (63) reduces in this approximation to

$$v_T = v_0 \sqrt{1 + (\partial a/\partial y)^2 + (\partial a/\partial z)^2}. \tag{65}$$

---

* The derivation employs $\overline{v'} = 0$ (by definition), $\partial \overline{f}/\partial \tau = 0$ (from stationarity), and $\nabla_\perp \cdot (\overline{\nabla_\perp f}) = 0$ (from homogeneity), among other results.

Equation (65) indicates that for the flow under consideration the average area ratio $\mathscr{A}_f/\mathscr{A}$ involves an average of the transverse gradient of the streamwise Eulerian displacement. This is understandable in that the Eulerian displacement defines the extent of motion of the flame sheet under the current approximations, and the transverse gradient of this provides the differences in the extent of longitudinal motion at adjacent transverse locations of points on the sheet. These differences are responsible for the increased area of the flame sheet. It may be noted that although (as indicated in Section 10.3.3) the Eulerian displacement is not likely to be statistically stationary, its transverse gradients are, so that v_T is predicted to be independent of time. Although the derivation given here is restricted to small values of $\partial a/\partial y$ and $\partial a/\partial z$, equation (65) may continue to be approximately true for larger values. The averages appearing in equation (61) or in equation (63) are more difficult to relate to properties of the upstream turbulence for conditions of higher amplitudes, and the necessary correspondences have not yet been developed.

The statistical properties of $\partial a/\partial y$ and $\partial a/\partial z$, needed for evaluating the average in equation (65), are often unavailable. Further approximations may be introduced to express these quantities in terms of more readily available turbulence properties. If the upstream turbulence is isotropic, then statistics of transverse derivatives of longitudinal velocity fluctuations are the same as those of longitudinal derivatives of transverse velocity fluctuations. With a Taylor hypothesis, the time integral of the longitudinal derivative of a fluctuating quantity is approximately $1/v_T$ times the fluctuating quantity. Hence, in isotropic turbulence, the statistics of $\partial a/\partial y$ are approximately the same as those of a transverse velocity fluctuation divided by v_T, which in turn (again for isotropy) are the same as those of the nondimensional longitudinal fluctuation v'/v_T. Thus equation (65) may be approximated as

$$v_T = v_0\sqrt{\overline{1 + 2(v'/v_T)^2}}\,, \tag{66}$$

which may be evaluated if a probability-density function $P(v'/v_T)$ is known.

An idea of the implications of equation (66) may be obtained by bringing the average inside the square root. For isotropic turbulence, $\overline{v'^2} = 2q/3$; an attempt may be made to take into account departures from isotropy (and possibly inaccuracy of the Taylor hypothesis as well) by inserting an additional factor C, the value of which is unknown, into this relationship. Then equation (66) becomes $v_T = v_0\sqrt{1 + 2C(2q/v_T^2)/3}$, which (after suitable algebraic manipulations that require solving a quadratic equation) gives

$$\frac{v_T}{v_0} = \left\{\frac{1}{2}\left[1 + \sqrt{1 + \frac{8}{3}C\left(\frac{2q}{v_0^2}\right)}\,\right]\right\}^{1/2}. \tag{67}$$

Equation (67) predicts that

$$\frac{v_T}{v_0} = 1 + \left(\frac{C}{3}\right)\left(\frac{2q}{v_0^2}\right) \tag{68}$$

for small values of $\sqrt{2q}/v_0$, that v_T/v_0 depends roughly linearly on $\sqrt{2q}/v_0$ at intermediate values of this relative intensity, and that a square-root dependence,

$$\frac{v_T}{v_0} = \left(\frac{2C}{3}\right)^{1/4}\left(\frac{\sqrt{2q}}{v_0}\right)^{1/2}, \tag{69}$$

is approached at high intensities. This last result, although qualitatively consistent with certain experimental data, is not well justified because equations (64), (65), (66), and (67) all are of questionable validity if $\sqrt{2q}/v_0$ is too large.

It is important to keep in mind that in equations (64)–(69), v', a, and q refer to properties that the turbulence has just ahead of the wrinkled flame, after it has passed through the far-field upstream region of hydrodynamic adjustment. Studies relating these properties to those of the turbulence upstream from the far-field region are in progress [38]. Modifying factors must appear, multiplying the gradients in equation (65) or the velocity fluctuation in equation (66), if these quantities are to be interpreted as corresponding to the turbulence far upstream. The modifications must depend on the density ratio across the flame because there is no modification in the limit of zero density change. The modifications also depend on parameters associated with the physical processes that remove the hydrodynamic instability, such as body forces and diffusive-thermal phenomena [38]; stationarity is unachievable, at least in a linear approximation, in the presence of hydrodynamic instability. In a very rough sense, the modifications might be viewed as included in the parameter C, so that equation (67) may then be applied with q referred to the turbulence far upstream. If this approach is taken, then experimental results for velocity spectra [38] at small values of $\sqrt{2q}/v_0$ appear to indicate that C typically is roughly on the order of 0.1—that is, the influence of the hydrodynamic adjustment is appreciable. With this modified procedure, the value of C may depend upon $\sqrt{2q}/v_0$, since comparisons for high-intensity turbulence [39] suggest that $C > 10$ is needed to achieve agreement with certain data.

At small near-field values of $\sqrt{2q}/v_0$, equation (65) reduces to

$$v_T = v_0\left[1 + \frac{1}{2}\overline{\left(\frac{\partial a}{\partial y}\right)^2} + \frac{1}{2}\overline{\left(\frac{\partial a}{\partial z}\right)^2}\right]. \tag{70}$$

For stable flames at small κ, there are no uncertainties in equation (70); it cannot be wrong. Approximations of the type that have been discussed can be introduced in equation (70) to derive equation (68), in which q may refer to the turbulence far upstream if C accounts for the hydrodynamic adjustment. Although there are a number of uncertainties in obtaining this C, for stable flames at small κ, none of them affect the fundamental prediction that $v_T - v_0$ is proportional to $(\sqrt{2q}/v_0)^2$ at low far-field intensities. This quadratic dependence is consistent with the low-intensity limit of the formula $(v_T/v_0)^2 = 1 + C(2q/v_0^2)$, deduced through simple physical reasoning by Shchelkin [3] for large-scale turbulence, although he later revised his thinking to obtain a different result [94]. It also was derived by Tucker [132] (who ignored the possibility of hydrodynamic instability) as well as in an analysis of a model problem involving a low-intensity expansion [133]. Although the quadratic dependence must be correct under the conditions stated, it is not obtained in more recently published theories, perhaps partially because there is no clear experimental confirmation of it through direct measurements of v_T. Experiments on flames in turbulence having scales sufficiently large and intensities sufficiently small are difficult to perform with the necessary accuracy [38]; equation (70) does not apply at the intensity conditions of major practical interest.

At higher near-field intensities such that $\sqrt{2q}/v_0$ is of order unity, the transverse gradients of f no longer are small*, although the curvature (appearing in κ) still is for stable wrinkled flames at small δ/l_k. Equation (61), and possibly equation (63), remain good under these conditions, but equation (64) no longer is accurate because it is seen from equation (56) that transverse gradients now appear through the flame-sheet tilt and transverse convection. Equation (70) becomes bad, and equation (65) can represent only a rough approximation. Flame-speed formulas of the form

$$v_T = v_0 + C\sqrt{2q} \tag{71}$$

often have been proposed, sometimes with a constant multiplying v_0 as well [2], [5], [94], [134], [135]—for example, from postulates of particular shapes of flame wrinkles, sometimes augmented by considerations of influences of flame-generated turbulence [94], [134], [136]. Expressions like equation (71) are in rough agreement with approximations derived from equation (67) and with a number of measurements of v_T over this intermediate range of intensities [39].

At large near-field values of $\sqrt{2q}/v_0$, equation (61) may remain valid for sufficiently small δ/l_k, but equations (63) and (64) do not, and the utility of equations (65) and (67) becomes quite questionable. As indicated near the end of Section 10.3.1, alternative viewpoints become more useful in this high-intensity flame-sheet regime. An approximate result that has been derived for this regime is [96]

$$v_T/v_0 = C(\sqrt{2q}/v_0)^{0.7}, \tag{72}$$

---

* If the hydrodynamic adjustment reduces intensities appreciably, as has been found experimentally in particular stable configurations [38], then equation (70) in near-field variables may apply at reasonably high far-field intensities.

which with $C = 2.4$ is in acceptable agreement with a reasonable body of data.

It may be noted, as seen explicitly in equation (66), that v_T is predicted to be independent of turbulence scale in the large-scale regimes under consideration; it may depend on ratios of transverse and longitudinal scales, for example, but not on the magnitude of any one scale. According to Shchelkin [3], Zel'dovich first deduced this result from dimensional reasoning in unpublished work. There are data that tend to support this prediction [137]. The turbulent-flame thickness increases with scale, in proportion to the scale at large l_k/δ, but the first scale-dependent correction to the turbulent flame speed is proportional to a representative magnitude of κ^2 [122], [124], [125], which varies as $(\delta/l_k)^2$. When the intensities are low enough to permit equation (63) to be used in the first approximation, it is found that the second approximation, which includes scale effects, is

$$v_T = v_0[\sqrt{1 + |\nabla_\perp f|^2} - (\overline{\kappa^2 \sqrt{1 + |\nabla_\perp f|^2}})F(R_0, \beta, \mathrm{Le}_1)], \qquad (73)$$

as might be inferred from the average of equation (56), extended to include a correction term of order κ^2 inside the braces. Thus influences of turbulence scales on v_T arise through second-order terms involving modifications of structures of the wrinkled laminar flames. To date, the function $F(R_0, \beta, \mathrm{Le}_1)$ has been calculated only in the limit of negligible density change, in which case it was found to be [124], [125]

$$F(1, \beta, \mathrm{Le}_1) = \beta(\mathrm{Le}_1 - 1)[1 + \beta(\mathrm{Le}_1 - 1)/8]. \qquad (74)$$

This formula indicates that modifications of the wrinkled-flame structure by the turbulence tend to decrease v_T if $\mathrm{Le}_1 > 1$ and to increase it for cellularly stable flames with $\mathrm{Le}_1 < 1$. It seems likely [125] that through changes in factors like I_1 and I_2 of equations (57) and (58) with R_0, $F(R_0, \beta, \mathrm{Le}_1)$ will decrease in magnitude with increasing R_0 and typically produce only small effects on v_T for flames of practical interest. No experimental tests of equation (73) have been made.

There are many experimental results on turbulent burning velocities [33]. No one has obtained predictions that agree with all of the data (see, for example, the discussion by Bray in [27]). Burning-velocity formulas are best tested against measurements that fall within the regime to which they were designed to apply. For example, equation (67) is restricted to stable flames in large-scale turbulence. Measurements, mainly under large-scale conditions, have been made for stoichiometric hydrogen-air flames at reduced pressures, in which the flames were observed to be remarkably stable [138]. The results for v_T exhibit independence of scale and, within the accuracy of the data, consistency with equation (67) concerning the dependence on intensity, including the small rate of increase of v_T with increasing $\sqrt{2q}$ at small values of $\sqrt{2q}/v_0$ and the qualitative tendency to approach a square-root dependence on $\sqrt{2q}/v_0$ at large values. Since thorough comparisons have not been made, this agreement is only suggestive. Experimentally, there are differences in burning velocities of open and enclosed turbulent flames at the same values of scales and intensities far upstream. This may reflect influences of confinement on the hydrodynamic adjustments occurring in the far field—influences viewed as affecting the value of C in equation (67) if q there refers to conditions far upstream.

There are data that graphically illustrate the occurrence of different regimes of propagation. In Figure 10.9, measured turbulent burning velocities [137] are plotted as a function of a turbulence scale at various values of the turbulence intensity far upstream, for propane-air flames, with the values $v_0 = 45$ cm/s and $\delta = 0.005$ cm used for nondimensionalization. We see that as l/δ becomes sufficiently large, v_T tends to become independent of scale but increases with increasing intensity, as theory predicts. The initial departures from independence of l as l/δ decreases from large values may be consistent with equation (73). In the regime of small scales and low intensities (open symbols), there is a tendency for v_T to increase strongly with increasing scale at fixed intensity (and also to increase with increasing intensity at fixed scale); these results are consistent with an embryonic analysis [15] based on a formal expansion in $\sqrt{2q}/v_0$ for low intensity that does not involve a scale restriction such as a presupposition of a structure involving wrinkled flames. The data shown at the smaller scales and higher intensities may approach another regime, to be discussed in the following subsection. Thus Figure 10.9 appears to illustrate three different regimes of turbulent flame propagation.

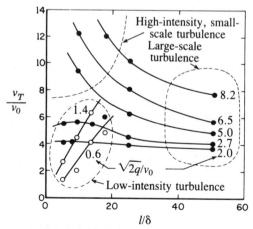

FIGURE 10.9. The ratio of the turbulent burning velocity to the laminar burning velocity as a function of the ratio of the integral scale to the laminar-flame thickness, for various turbulence intensities, for stoichiometric propane-air flames at atmospheric pressure, and initially at room temperature [137].

10.3.5. Flames in turbulence of high intensity or small scale

At sufficiently large Kolmogorov scales, indefinitely increasing the turbulence intensity (moving upward along a vertical line on the right of Figure 10.5) will continually increase the number of flame sheets per unit cross-sectional area of the turbulent flame brush but does not lead to a violation of equation (61). In fact, sufficiently large turbulence intensities even may tend to suppress influences of hydrodynamic and diffusive-thermal instabilities for unstable flames, enabling equation (61) to describe a steadily propagating turbulent flame composed of laminar flamelets that would propagate unsteadily if the turbulence were not present. Of course, the accurate reduction of equation (61) to a formula that can be used in flame-speed calculations continues to offer interesting challenges for high-intensity turbulence.

At smaller values of l_k/δ, influences of modifications to flamelet structure by the turbulence become important. For systems in which the planar laminar flame is unstable but in laminar flow evolves to a stationary structure of finite amplitude under influences of hydrodynamic and diffusive-thermal effects (cellular flames, Section 9.5.2), studies of perturbations about the finite-amplitude structure may help to reveal characteristics of the turbulent flame, at least at low turbulence intensities. Since no investigations of this type have been performed, thorough knowledge of structures of turbulent flames in cellularly unstable regimes is unavailable. The theoretical analyses that are needed appear to be difficult but tractable, at least if attention is restricted to limited regions of parameter spaces like the one appearing in Figure 9.10. Irrespective of whether the planar laminar flame is stable or unstable, at the smaller values of l_k/δ, the modifications of the structures of the laminar flames by the turbulence may become great enough to lead to local extinctions, as was indicated in Section 10.3.2 (see Figure 10.7). Since the extinctions typically occur at values of κ on the order of unity (Figure 10.6), following the reasoning of Section 10.3.3 we find that they become important in the turbulent flame when $(\sqrt{2q}/l_t)(\delta/v_0) = \delta^2/l_k^2$ is of order unity [139]. Nonadiabaticity was inferred in Section 10.3.2 to be an important element in the extinctions, but enhanced diffusive losses of radicals from the reaction zone (an effect not included in one-step kinetic models) also may be quite important and lessen the importance of nonadiabaticity. Perturbation analyses may be performed near the limit of extinction to investigate the development of holes in reaction sheets in turbulent flames.

When disruptions of flame sheets become sufficiently extensive, there is appreciable nonreactive mixing of reactants and products at molecular scales. The extent of disruption increases as l_k/δ decreases; certainly if l/δ becomes small compared with unity, then the turbulent flame no longer can be composed of wrinkled laminar flames. The true structures of turbulent flames in the limit of small values of l/δ are unknown.

Damköhler [2] introduced the superficially reasonable hypothesis that in the limit $l \ll \delta$, the turbulence effectively modifies transport coefficients within the flame, enabling equation (5-2) to be used to estimate the turbulent burning velocity by replacing λ/c_p therein by $(\lambda/c_p)_T$, the laminar value plus a turbulent contribution that is readily obtained from turbulent-mixing studies [46, 48, 50]. This suggestion remains attractive [133] until it is realized that the strong dependence of the rate of heat release on temperature is an essential aspect of premixed flame propagation, so that whenever $(\lambda/c_p)_T$ differs measurably from the laminar value, the reaction-rate factor in equation (5-2) is modified greatly [17]. Thus at small values of l/δ in turbulence of sufficiently low intensity it may clearly [15] be inferred that the relationship

$$v_T \approx (1/\rho_0)\sqrt{(\lambda/c_p)w_T} \qquad (75)$$

is a better approximation than that proposed originally, where w_T is an effective reaction rate that includes perturbation effects of turbulence. With this approximation, equation (5-1) shows that the ratio of the thickness δ_T of the turbulent flame to that of the laminar flame is $\delta_T/\delta = \sqrt{w/w_T}$, where w measures the reaction rate in the laminar flow. If the experimental observation that $\delta_T > \delta$ applies, then $w_T < w$, and we find from equation (75) that $v_T < v_0$, contrary to most experiments; however, there are no good experiments at low intensities with $l \ll \delta$, and the limiting case represented by equation (75) is without practical interest.

It has been suggested [140] that for $l \ll \delta$ at turbulence intensities of interest, the convection of reaction regions by turbulence leads to a distributed reaction zone, with heat release occurring more or less homogeneously throughout the turbulent flame brush and with local fluctuations in temperature and composition being small. If this were true, then the moment methods discussed at the beginning of Section 10.1.2 would be appropriate for calculations of turbulent-flame structure and propagation, but the cold-boundary difficulty (Section 5.3.2) would arise in a manner that cannot be circumvented, and turbulent burning velocities would not exist in the simplest configurations. The high-intensity, small-scale regime in Figure 10.9, the upper, left-hand portion of Figure 10.5, is of appreciable practical importance but has not received proper theoretical elucidation. The possibility of studying it by perturbation methods remains distant because structures about which perturbations might be developed are unknown.

An alternative view [36] is that if l/δ becomes too small, then extinctions of laminar flamelets are reflected in extinction of the turbulent flame. According to this idea, there is a region to the upper left in Figure 10.5 in which turbulent flame propagation cannot occur. It seems physically that phenomena of this type may pertain to confined turbulent flows in reactors of small volume, where they would reflect influences of turbulence properties on blowout of the turbulent combustion in the reactor, and they could have a bearing on blowoff of burner-stabilized turbulent flames, but for open turbulent flames in which residence times are permitted to approach infinity, the turbulent reactant mixture *must* be able to find a way to burn. Flamelet extinction followed by reactant-product mixing would lead later to more nearly homogeneous reactions, akin to ignition or thermal explosion. Nevertheless, abrupt transitions in modes of propagation of turbulent flames cannot be excluded, although there is no clear experimental demonstration of such transitions.

Simplified models continue to be sought for mechanisms of propagation of premixed turbulent flames at values of l_k/δ that are too small for the wrinkled-flame structure to apply [23], [33], [141]. Although the models cannot be right, the complexity of the problem makes it difficult to ascertain the most significant ways in which they are wrong. Current knowledge of turbulent structures of nonreacting turbulent flows must play an important part in model development [33]. Among the processes that estimates show to have a significant effect are the different responses of the (more dense) reactants and the (less dense) products to pressure changes [35], [61], [142], tending to lead to enhanced rates of motion of products along randomly oriented vortex cores of eddies. It seems that in the absence of a well-founded, general theoretical framework for analysis, simplified models will continue to be speculative.

There are confined flows for which turbulent flame spread can be calculated reasonably well at high levels of turbulence intensity, even if $l \ll \delta$, in spite of the severe uncertainties that have been emphasized above. These are flows that are dominated by the fluid mechanics, in the sense that effects of the flame, other than the fact that it releases heat, are of secondary importance. A typical example is a rod-stabilized premixed turbulent flame in a high-speed ramjet combustor or in a turbojet afterburner. The burning velocity in the continuous flow is the average velocity of the reactant stream times the sine of the angle of flame spread with respect to the stream. The angle of spread, and therefore the burning velocity, is fairly insensitive to chemical-kinetic factors; it is controlled largely by rates of turbulent mixing in high-intensity turbulent flows having local turbulence properties dependent on mean profiles of flow variables.* The average flow may be described by averages of equations for mass and momentum conservation, along with averages of equations for species and energy conservation—for example, the average of equation (45) for the fuel. Moment methods are found to be useful in that, because of the dominance of the fluid mechanics, turbulent-diffusivity approximations like equation (38) are found to work well enough even for nonconserved quantities like Y_i. To be able to calculate the spread

---

* This configuration-dependent interaction would introduce complications in attempts to apply the more fundamental theories that have been discussed.

angle it is necessary only to obtain an expression for the average reaction rate—for example, $\overline{w}_F$—appearing in the average of equation (45) for the fuel, $i = F$. In turbulence of high intensity, $\overline{w}_F$ is to be expressed in terms of parameters relevant to turbulent mixing, independent of molecular-level reaction rates.

Spalding developed a successful *eddy-breakup model* for describing rates of flame spread in high-intensity flows of this type [143]–[146]. There are a number of different versions of the model; one employs

$$\overline{w}_F = -\rho C_F \sqrt{\overline{Y_F'^2}}\, \epsilon/q, \tag{76}$$

in terms of the root-mean-square concentration fluctuation, the dissipation rate ϵ, and intensity q (defined in Section 10.1.5). Here C_F is an eddy-breakup constant that may be ascribed a value that produces good agreement between calculated and measured spread rates and profiles of mean fuel concentration ($10^{-1} < C_F < 10^2$, depending on the formulation, but typically C_F is of order unity). In performing a calculation with equation (76), ϵ/q may be estimated from mean-flow properties of the turbulence or calculated from additional moment-method differential equations, for example, equations for the turbulent kinetic energy and for a length scale of turbulence; an estimate in terms of $\overline{Y}_F$ or a moment-method conservation equation for mean-square concentration fluctuations may be used for obtaining the square-root factor in equation (76), for example. The idea behind placing ϵ/q in equation (76) is that the rate of reaction is controlled by the rate of turbulent dissipation (eddy-breakup). The degree of success achieved in using expressions like equation (76) [27], [143]–[146] is an indication of the extent to which the average rate of reaction is independent of the chemical kinetics at high turbulence intensities in the configurations of confinement considered. Thus there are practical problems that can be addressed with reasonable success in spite of the current state of uncertainty concerning fundamentals of premixed flames in high-intensity, small-scale turbulence.

REFERENCES

1. E. Mallard and H. L. leChatelier, *Ann. Mines* **4**, 343 (1883).
2. G. Damköhler, *Z. Elektrochem.* **46**, 601 (1940); English translation, NACA Tech. Memo. No. 1112 (1947).
3. K. I. Shchelkin, *Zhur. Tekhn. Fiz.* **13**, 520 (1943); English translation, NACA Tech. Memo. No. 1110 (1947).
4. B. Lewis and G. von Elbe, *Combustion, Flames and Explosions of Gases*, 1st ed., New York: Academic Press, 1951, 480–507, 550–557.
5. B. Karlovitz, "Combustion Waves in Turbulent Gases," in *Combustion Processes*, vol. II of *High Speed Aerodynamics and Jet Propulsion*, B. Lewis, R. N. Pease, and H. S. Taylor, eds., Princeton: Princeton University Press, 1956, 312–364.

6. S. S. Penner, *Chemistry Problems in Jet Propulsion*, New York: Pergamon Press, 1957, 293–295.

7. W. R. Hawthorne, D. S. Weddell, and H. C. Hottle, *3rd Symp.* (1949), 266–288.

8. K. Wohl, C. Gazley, and N. Kapp, *3rd Symp.* (1949), 288–300.

9. S. Yagi and K. Saji, *4th Symp.* (1953), 771–781.

10. M. W. Thring and M. P. Newby, *4th Symp.* (1953), 789–796.

11. I. Sawai, M. Kunugi, and H. Jinno, *4th Symp.* (1953), 806–814.

12. K. Wohl and C. W. Shipman, "Diffusion Flames," in *Combustion Processes*, vol. II of *High Speed Aerodynamics and Jet Propulsion*, B. Lewis, R. N. Pease, and H. S. Taylor, eds., Princeton: Princeton University Press, 1956, 381–404.

13. G. E. Andrews, D. Bradley, and S. B. Lwakabamba, *C & F* **24**, 284 (1975).

14. S. N. B. Murthy, ed., *Turbulent Mixing in Nonreactive and Reactive Flows*, New York: Plenum Press, 1975.

15. F. A. Williams, "A Review of Some Theoretical Considerations of Turbulent Flame Structure," in *Analytical and Numerical Methods for Investigation of Flow Fields with Chemical Reactions, Especially Related to Combustion*, M Barrère, ed., AGARD Conference Proceedings No. 164, AGARD, Paris, 1975, II1-1 to II1-25.

16. J. C. Hill, *Annual Review of Fluid Mechanics* **8**, 135 (1976).

17. P. A. Libby and F. A. Williams, *Annual Review of Fluid Mechanics* **8**, 351 (1976).

18. F. V. Bracco, ed., "Special Issue on Turbulent Reactive Flows," *CST* **13**, (1976), 1–275.

19. D. T. Pratt, *Prog. Energy Combust. Sci.* **1**, 73 (1976).

20. R. W. Bilger, *Prog. Energy Combust. Sci.* **1**, 87 (1976).

21. R. J. Tabaczynski, *Prog. Energy Combust. Sci.* **2**. 143 (1976).

22. R. A. Edelman and P. T. Harsha, *Prog. Energy Combust. Sci.* **4**, 1 (1978).

23. J. Chomiak, *Prog. Energy Combust. Sci.* **5**, 207 (1979).

24. K. N. C. Bray, *17th Symp.* (1979), 223–233.

25. W. P. Jones, "Models for Turbulent Flows with Variable Density and Combustion," in *Prediction Methods for Turbulent Flows*, W. Kollmann, ed., New York: Hemisphere Publishing Corp., 1980, 379–421.

26. J. N. Mattavi and C. A. Amann, eds., *Combustion Modeling in Reciprocating Engines*, New York: Plenum Press, 1980.

27. P. A. Libby and F. A. Williams, eds., *Turbulent Reacting Flows*, Berlin: Springer-Verlag, 1980.

28. P. A. Libby and F. A. Williams, *AIAA Journal* **19**, 261 (1981).

29. N. Peters and F. A. Williams, "Coherent Structures in Turbulent Combustion," in *The Role of Coherent Structures in Modelling Turbulence and Mixing*, J. Jimenez, ed., Berlin: Springer-Verlag, 1981, 364–393

30. E. E. O'Brien, *AIAA Journal* **19**, 366 (1981).

31. T. D. Butler, L. D. Cloutman, J. K. Dukowicz, and J. D. Ramshaw, *Prog. Energy Combust. Sci.* **7**, 293 (1981).

32. W. P. Jones and J. H. Whitelaw, *C & F* **48**, 1 (1982).

33. R. G. Abdel-Gayed and D. Bradley, *Phil. Trans. Roy. Soc. London* **301A**, 1 (1981).

34. H. Eickhoff, *Prog. Energy Combust. Sci.* **8**, 159 (1982).

35. W. C. Strahle, *19th Symp.* (1982), 337–347.

36. R. G. Abdel-Gayed and D. Bradley, "The Influence of Turbulence Upon the Rate of Turbulent Burning," in *Fuel-Air Explosions*, J. H. S. Lee and C. M. Guirao, eds., Waterloo, Canada: University of Waterloo Press, 1982, 51–68.

37. S. B. Pope, *Prog. Energy Combust. Sci.* **11**, 119 (1985).
38. P. Clavin, *Prog. Energy Combust. Sci.* **11**, 1 (1985).
39. J. Abraham, F. A. Williams and F. V. Bracco, "A Discussion of Turbulent Flame Structure in Premixed Charges," SAE Paper 850345, Society of Automotive Engineers, Detroit, 1985.
40. F. A. Williams, "Turbulent Combustion," in *The Mathematics of Combustion*, J. D. Buckmaster, ed., vol. 2 of *Frontiers in Applied Mathematics*, Philadelphia: Society for Industrial and Applied Mathematics, 97–131, 1985.
41. S. Tsugé, *Phys. Fluids* **17**, 22 (1974).
42. S. Tsugé and K. Sagara, *Phys. Fluids* **19**, 1478 (1976).
43. S. Tsugé and K. Sagara, *CST* **18**, 179 (1978).
44. K. Sagara, *CST* **21**, 191 (1980).
45. K. Sagara and S. Tsugé, *Phys. Fluids* **25**, 1970 (1982).
46. S. Goldstein, *Modern Developments in Fluid Dynamics*, vol. I, Chapter V, Oxford: Oxford University Press, 1938,
47. G. K. Batchelor, *The Theory of Homogeneous Turbulence*, Cambridge: Cambridge University Press, 1953.
48. H. Schlichting, *Boundary-Layer Theory*, 4th ed., New York: McGraw-Hill, 1960, Chapters XVIII and XIX.
49. A. T. Townsend, *The Structure of Turbulent Shear Flow*, Cambridge: Cambridge University Press, 1956.
50. J. O. Hinze, *Turbulence*, New York: McGraw-Hill, 1959.
51. H. Tennekes and J. L. Lumley, *A First Course in Turbulence*, Boston: MIT Press, 1972.
52. A. S. Monin and A. M. Yaglom, *Statistical Fluid Mechanics*, Boston: MIT Press, vol. 1, 1971, vol. 2, 1975.
53. C. C. Lin, *The Theory of Hydrodynamic Stability*, Cambridge: Cambridge University Press, 1955.
54. M. J. Feigenbaum, "Universal Behavior in Nonlinear Systems," in *Los Alamos Science*, Los Alamos, New Mexico, Summer (1980), 4–27.
55. A. Papoulis, *Probability, Random Variables and Stochastic Processes*, New York: McGraw-Hill, 1965.
56. E. Hopf, *J. Rat. Mech. Anal.* **1**, 87 (1952).
57. A. F. Ghoniem, A. F. Chorin, and A. K. Oppenheim. *18th Symp.* (1981), 1375–1383.
58. R. G. Rehm and H. R. Baum. *CST* **40**, 55 (1984).
59. P. Clavin and F. A. Williams, *J. Fluid Mech.* **90**, 589 (1979).
60. J. B. Moss, *CST* **22**, 119 (1980).
61. P. A. Libby and K. N. C. Bray, *AIAA Journal* **19**, 205 (1981).
62. T. Yanagi and Y. Mimura, *18th Symp.* (1981), 1031–1039.
63. I. G. Shepherd, J. B. Moss, and K. N. C. Bray, *19th Symp.* (1982), 423–431.
64. D. R. Lancaster, *Society of Automotive Engineers Transactions* **85**, 651 (1976).
65. A. Favre, "Statistical Equations of Turbulent Gases," in *Problems of Hydrodynamics and Continuum Mechanics*, M. A. Lavrent'ev, ed., Society for Industrial and Applied Mathematics, Philadelphia, 1969, 231–266.
66. C. H. Gibson, *Phys. Fluids* **11**, 2316 (1968).
67. A. Roshko, *AIAA Journal* **14**, 1349 (1976).
68. D. B. Spalding, *CST* **13**, 3 (1976).
69. A. S. C. Ma, D. B. Spalding, and R. L. T. Sun, *19th Symp.* (1982), 393–402.

70. W. C. Strahle, *Prog. Energy Combust. Sci.* **4**, 157 (1958).
71. K. Ramohalli, "Some Fundamental Acoustic Observations in Combusting Turbulent Jets," in *Combustion in Reactive Systems*, vol. 76 of *Progress in Astronautics and Aeronautics*, J. R. Bowen, N. Manson, A. K. Oppenheim, and R. I. Soloukhin, eds., New York: American Institute of Aeronautics and Astronautics, 1981, 295–313.
72. F. Tamanini, *17th Symp.* (1979), 377–387.
73. S. M. Jeng, L. D. Chen, and G. M. Faeth, *19th Symp.* (1982), 349–358.
74. J. S. Evans and C. J. Schexnayder, *AIAA Journal* **18**, 188 (1980).
75. P. J. Smith, T. H. Fletcher, and L. D. Smoot, *18th Symp.* (1981), 1285–1293.
76. J. Janicka and W. Kollman, *17th Symp.* (1979), 421–430.
77. R. W. Bilger, *C & F* **30**, 277 (1977).
78. S. K. Liew, K. N. C. Bray, and J. B. Moss, *CST* **27**, 69 (1981).
79. R. W. Bilger and S. H. Stårner, *C & F* **51**, 155 (1983).
80. R. W. Bilger, *AIAA Journal* **20**, 962 (1982).
81. R. W. Bilger, *CST* **22**, 251 (1980).
82. R. W. Bilger and R. E. Beck, *15th Symp.* (1975), 541–552.
83. N. Peters, *CST* **19**, 39 (1978).
84. R. A. Antonia, B. R. Satyaprakash, and A. K. M. F. Hussain, *Phys. Fluids* **23**, 695 (1980).
85. F. C. Lockwood and H. A. Moneib, *CST* **22**, 63 (1980).
86. N. Peters and F. A. Williams, *AIAA Journal* **21**, 423 (1983).
87. N. Peters, *CST* **30**, 1 (1983).
88. J. Janicka and N. Peters, *19th Symp.* (1982), 367–374.
89. T. A. Brzustowski, *Physico-Chemical Hydrodynamics*, **1**, 27 (1980).
90. G. T. Kalghatgi, *CST* **26**, 233 (1981).
91. R. Günther, K. Horch, and B. Lenze, "The Stabilization Mechanism of Free Jet Diffusion Flames," *Colloque International Berthelot-Vieille-Mallard-leChatelier, First Specialists Meeting (International) of The Combustion Institute*, Pittsburgh: The Combustion Institute, 1981, 117–122.
92. L. S. G. Kovasznay, *Jet Propulsion* **26**, 485 (1956).
93. A. M. Klimov, *Zhur. Prikl. Mekh. Tekhn. Fiz.*, No. 3, 49 (1963).
94. K. I. Shchelkin and Y. K. Troshin, *Gasdynamics of Combustion*, Baltimore: Mono Book Corp., 1965.
95. V. R. Kuznetsov, *Fiz. Gor. Vzr.* **11**, 574 (1974).
96. A. M. Klimov, *Dokl. Akad. Nauk SSSR* **221**, 56 (1975); "Premixed Turbulent Flames—Interplay of Hydrodynamic and Chemical Phenomena," in *Flames, Lasers and Reactive Systems*, vol. 88 of *Progress in Astronautics and Aeronautics*, J. R. Bowen, N. Manson, A. K. Oppenheim and R. I. Soloukhin, eds., New York: American Institute of Aeronautics and Astronautics, 1983, 133–146.
97. G. K. Batchelor, *An Introduction to Fluid Mechanics*, Cambridge: Cambridge University Press, 1967, 79–84.
98. L. Howarth, *Phil. Mag.* **42**, 335, 1433 (1951).
99. P. A. Libby, *AIAA Journal*, **5**, 507 (1967).
100. B. Karlovitz, D. W. Denniston, Jr., D. H. Knapschaefer, and F. E. Wells, *4th Symp.* (1953), 613–620.
101. V. M. Gremyachkin and A. G. Istratov, "On a Steady Flame in a Stream with a Velocity Gradient," *Gorenie i Vzriv.*, Moscow: Nauka, 1972, 305–308.

102. G. I. Sivashinsky, in *Acta Astronautica* **3**, 889 (1976).
103. J. D. Buckmaster, *17th Symp.* (1979), 835–842.
104. J. D. Buckmaster, *Acta Astronautica* **6**, 741 (1979).
105. Y. B. Zel'dovich, G. I. Barenblatt, V. B. Librovich, and G. M. Makhviladze, *The Mathematical Theory of Combustion and Explosion*, Moscow: Nauka, 1980, 272–277.
106. P. A. Libby and F. A. Williams, *C & F* **44**, 287 (1982).
107. P. A. Durbin, *J. Fluid Mech.* **121**, 141 (1982).
108. J. D. Buckmaster and D. Mikolaitis, *C & F* **47**, 191 (1982).
109. J. D. Buckmaster and G. S. S. Ludford, *Theory of Laminar Flames*, Cambridge: Cambridge University Press, 1982, Chapter 10.
110. P. A. Libby and F. A. Williams, *CST* **31**, 1 (1983).
111. P. A. Libby, A. Liñán, and F. A. Williams, *CST* **34**, 257 (1983).
112. P. A. Libby and F. A. Williams, *CST* **37**, 221 (1984).
113. K. Seshadri and N. Peters, *CST* **33**, 35 (1983).
114. H. Tsuji and I. Yamaoka, "An Experimental Study of Extinction of Near-Limit Flames in a Stagnation Flow," *Colloque International Berthelot-Vieille-Mallard-le Chatelier, First Specialists Meeting (International) of The Combustion Institute*, Pittsburgh: The Combustion Institute, 1981, 111–116.
115. C. K. Law, S. Ishizuka, and M. Mizomoto, *18th Symp.* (1981), 1791–1798.
116. S. Ishizuka, K. Miyasaka, and C. K. Law, *C & F* **45**, 293 (1982).
117. S. Ishizuka and C. K. Law, *19th Symp.* (1982), 327–335.
118. H. Daneshyar, J. M. C. Mendes-Lopes, and G. S. S. Ludford, *19th Symp.* (1982), 413–421.
119. H. Tsuji and I. Yamaoka, *19th Symp.* (1982), 1533–1540.
120. J. Sato, *19th Symp.* (1982), 1541–1548.
121. S. Sohrab, Z. Y. Ye, and C. K. Law, "An Experimental Study of Flame Interactions," Western States Section, Pasadena, Calif.: The Combustion Institute, 1983; *20th Symp.* 1957–1965 (1985).
122. P. Clavin and G. Joulin, *Journal de Physique-Lettres* **44**, L-1 (1983).
123. M. Matalon, *CST* **31**, 169 (1983).
124. P. Clavin and F. A. Williams, "Effects of Lewis Number on Propagation of Wrinkled Flames in Turbulent Flow," in *Combustion in Reactive Systems*, vol. 76 of *Progress in Astronautics and Aeronautics*, J. R. Bowen, N. Manson, A. K. Oppenheim, and R. I. Soloukhin, eds., New York: American Institute of Aeronautics and Astronautics, 1981, 403–411.
125. P. Clavin and F. A. Williams, *J. Fluid Mech.* **116**, 251 (1982).
126. P. Clavin and P. L. García-Ybarra, *J. de Méchanique Theorique et Appliqué* **2**, 245 (1983).
127. M. L. Frankel and G. I. Sivashinsky, *CST* **31**, 131 (1983).
128. G. I. Sivashinsky, *Acta Astronautica*, **4**, 1177 (1977).
129. J. D. Buckmaster and G. S. S. Ludford, *Theory of Laminar Flames*, Cambridge: Cambridge University Press, 1982, Chapter 11.
130. G. I. Sivashinsky, *Annual Review of Fluid Mechanics* **15**, 179 (1983).
131. P. Pelcé and P. Clavin, *J. Fluid Mech.* **124**, 219 (1982).
132. M. Tucker, NACA Rept. No. 1277 (1956).
133. F. A. Williams, *J. Fluid Mech.* **40**, 401 (1970).
134. B. Karlovitz, D. W. Denniston, Jr., and F. E. Wells, *J. Chem. Phys.* **19**, 541 (1951).
135. K. Wohl, L. Shore, H. von Rosenberg, and C. W. Weil, *4th Symp.* (1953), 620–635.

136. A. C. Scurlock and J. H. Grover, *4th Symp.* (1953), 645–658.

137. D. R. Ballal and A. H. Lefebvrc, *Proc. Roy. Soc. London* **344A**, 217 (1975).

138. D. R. Ballal, *Proc. Roy. Soc. London* **367A**, 485 (1979).

139. F. A. Williams, *C & F* **26**, 269 (1976).

140. M. Summerfield, S. H. Reiter, V. Kebely, and R. W. Mascolo, *Jet Propulsion* **25**, 377 (1955).

141. D. R. Ballal, *Proc. Roy. Soc. London* **367A**, 353; **368A**, 267, 283, 295 (1979).

142. J. Chomiak, *16th Symp.* (1977), 1655–1673.

143. D. B. Spalding, *7th Symp.* (1959), 595–603.

144. D. B. Spalding, *11th Symp.* (1967), 807–815.

145. D. B. Spalding, *13th Symp.* (1971), 649–657.

146. D. B. Spalding, *16th Symp.* (1977), 1657–1663.

CHAPTER 11

Spray Combustion

Many practical, important devices, ranging from home oil heaters to chemical rocket motors, involve the burning of liquid (or solid) particles in a gas. The term **spray** will here be defined to include all such systems in which there are so many particles that only a statistical description of their behavior is feasible. Thus in addition to being relevant to processes occurring in diesel engines or gas-turbine combustors, for example, which are commonly considered to experience spray combustion, the material to be discussed here will also apply to pulverized-coal burners, for example, which do not involve sprays, according to conventional usage of the term. This degree of generality is achievable only because attention is restricted to fundamental principles and to idealized illustrative examples, without probing too far into specific details of the physical processes that occur. Although it would be more precise to call the subject of the present chapter "two-phase combustion for which a condensed phase is dispersed in a gas phase," the abbreviated terminology adopted should be clear enough. An understanding of the mechanism of spray combustion requires a knowledge of (1) the burning mechanism of individual particles, (2) the statistical methods for describing groups of particles, and (3) the manner in which the groups of particles modify the behavior of the gas in flow systems. The first of these problems was treated in Chapter 3; the second and third are considered here.

Spray combustion is a complicated subject because it involves many different processes. A typical sequence of events would be the injection and atomization of liquid fuel, mixing of droplets with oxidizing gas, heat transfer to droplets producing evaporation of liquid, mixing of fuel vapor with gas—possibly followed by gas-phase ignition processes and certainly by either non-premixed or premixed gas-phase combustion—all accompanied by various additional aspects of finite-rate chemistry, such as fuel pyrolysis, production of oxides of nitrogen, and extinction phenomena. In focusing attention on a general framework for analysis, we shall not have to address many of these complications. An experimentally oriented description of spray combustion has been published [1], and reviews of current problems in the field may be found in the literature [2–6].

Of major interest concerning these problems are influences of turbulence in spray combustion [5]. The turbulent flows that are present in the vast majority of applications cause a number of types of complexities that we are ill-equipped to handle for two-phase systems (as we saw in Section 10.2.1). For nonpremixed combustion in two-phase systems that can reasonably be treated as a single fluid through the introduction of approximations of full dynamic (no-slip), chemical and interphase equilibria, termed a *locally homogeneous flow model* by Faeth [5], the methods of Section 10.2 can be introduced reasonably successfully [5], but for most sprays these approximations are poor. Because of the absence of suitable theoretical methods that are well founded, we shall not discuss the effects of turbulence in spray combustion here. Instead, attention will be restricted to formulations of conservation equations and to laminar examples. If desired, the conservation equations to be developed can be considered to describe the underlying dynamics on which turbulence theories may be erected—a highly ambitious task.

In practical applications the turbulence is especially important to the mixing of injected liquid fuel and sprays with the ambient oxidizing gas. Systems often are mixing-limited in that overall rates of heat release, for example, are controlled mainly by the mixing and atomization steps rather than by combustion. Two-phase mixing and atomization processes themselves pose many fundamental problems [4–8], which certainly are not simplified in the presence of hot environments; we shall offer only a few comments related to these subjects at the end of this chapter. Spray-combustion processes themselves, however, remain important in a number of ways, even in situations that mainly are mixing-limited, for example, in relationship to determinations of combustion efficiencies that may be controlled by the burnout of the larger droplets. The material to be presented here centers on the combustion processes, and the examples pertain to systems possessing a strong degree of spatial homogeneity.

In this chapter, the equation governing the statistical counting procedure for sprays is first derived (Section 11.1) and is applied (Section 11.2) to a very simplified model of rocket-chamber combustion in order to obtain an estimate of the combustion efficiency. This illustrative example and others given by Shapiro and Erickson [9] indicate that some progress can be made on the basis of statistics (item 2 in the first paragraph), without considering the effect of the particles on the gas (item 3). Since the interaction of the spray and the gas is, however, of importance for most spray-combustion problems, the conservation equations for a gaseous fluid containing a dilute spray are presented next (Section 11.3). Solutions to these equations are given for an improved model of rocket combustion (Section 11.5) and for the hetero-geneous laminar burning velocity of a spray (Section 11.6). Some of the more recently developed concepts in spray combustion will be indicated in Section 11.7.

11.1. SPRAY STATISTICS

11.1.1. Particle size and shape

It will be assumed here for simplicity that one parameter r (the radius in the case of a spherical liquid droplet) is sufficient to specify the size and shape of a particle. For solid particles (or liquid droplets), this assumption will be valid in spray combustion when either the particles are geometrically similar or their shape is of no consequence in the combustion process. Liquid droplets will obey this hypothesis in particular if they are spherical, which will not be true unless (1) they collide with each other so seldom that collision-induced oscillations are viscously damped to a negligible amplitude for most droplets, and (2) their velocity relative to the gas is sufficiently low. An alternative parameter to the radius is the mass of the droplet [10]; the choice between this, the droplet volume, or the radius of a sphere of equal volume is a matter of individual preference.

For liquid droplets, requirement (1) typically means that the spray must be dilute (that is, the ratio of the volume occupied by the condensed phase to the volume occupied by the gas must be small) because collisions tend to be frequent when the volume of particles per unit volume of space becomes too large. Since the mass density of the particles greatly exceeds that of the gas in many sprays and the stoichiometry of most hydrocarbon-oxidizer systems is such that the mass of the fuel is considerably less than that of the gaseous oxidizer in stoichiometric mixtures, the hypothesis of a dilute spray often is valid in hydrocarbon spray combustion.

It has been shown [11] that the degree of deformation and the amplitude of oscillation of a liquid droplet depend on the ratio of the dynamic force to the surface-tension force, which is given by the Weber number, $We = 2r\rho_g|\mathbf{v} - \mathbf{u}|^2/S$. Here S is the surface tension of the liquid, ρ_g is the gas density, $\mathbf{v}$ represents the droplet velocity, and $\mathbf{u}$ is the velocity of the gas. When $We \ll 10$, droplets are nearly spherical; as We increases, the droplets deform and eventually break up at $We \approx 10$. Hence, for systems involving liquid droplets, the present formulation should be applied only for dilute sprays with low Weber numbers. There has been a considerable amount of more recent research on droplet breakup that provides better understanding and improved criteria for the occurrence of breakup in various flows [12–22].

11.1.2. The distribution function

A statistical description of the spray may be given by the distribution function (or density function)

$$f_j(r, \mathbf{x}, \mathbf{v}, t) \, dr \, d\mathbf{x} \, d\mathbf{v},$$

which is the probable number of particles of chemical composition j in the radius range dr about r located in the spatial range $d\mathbf{x}$ about $\mathbf{x}$ with velocities in the range $d\mathbf{v}$ about $\mathbf{v}$ at time t. Here $d\mathbf{x}$ and $d\mathbf{v}$ are abbreviations for the three-dimensional elements of physical space and velocity space, respectively. The variables r, $\mathbf{x}$, and $\mathbf{v}$ (and sometimes others, such as the droplet temperature) must appear in the distribution function because, in a spray, conditions are not known well enough to permit one to specify the exact size, position, or velocity of each particle. The fact that a phenomenologically more complete statistical description than that contained in f is usually not required tends to be borne out in applications.

11.1.3. The spray equation

An equation describing the time rate of change of the distribution function f_j may be derived phenomenologically by using reasoning analogous to that employed in the kinetic theory of gases (Appendix D). The force per unit mass on a particle of kind j at $(r, \mathbf{x}, \mathbf{v}, t)$ is denoted by $\mathbf{F}_j = (d\mathbf{v}/dt)_j$, and the rate of change of the size r of a particle of kind j at $(r, \mathbf{x}, \mathbf{v}, t)$ is defined as $R_j = (dr/dt)_j$. Both R_j and $\mathbf{F}_j$ are allowed to depend on r, $\mathbf{x}$, $\mathbf{v}$, t, and the local properties of the gas. The rate of increase of f_j with time through particle formation or destruction by processes such as nucleation or liquid breakup will be denoted by $\hat{Q}_j$. Finally, Γ_j will represent the rate of change of the distribution function (for particles of kind j) caused by collisions with other particles. These collisions must occur sufficiently seldom that the aerodynamic contributions to Γ_j are separable from those in $\mathbf{F}_j$. Adding the changes in f_j resulting from particle growth, the motion of particles into and out of the spatial element $d\mathbf{x}$ by virtue of their velocity $\mathbf{v}$, the change of the number of particles in the velocity element $d\mathbf{v}$ because of the acceleration $\mathbf{F}_j$, sources of droplets $\hat{Q}_j$, and collisions Γ_j, we find that

$$\frac{\partial f_j}{\partial t} = -\frac{\partial}{\partial r}(R_j f_j) - \nabla_x \cdot (\mathbf{v} f_j) - \nabla_v \cdot (\mathbf{F}_j f_j)$$

$$+ \hat{Q}_j + \Gamma_j, \quad j = 1, \ldots, M. \tag{1}$$

Here the subscripts on the gradient operators distinguish derivatives with respect to spatial and velocity coordinates, and M is the total number of different kinds of particles (classified according to their chemical composition). The details of the procedure for deriving equation (1) (considering the volume element $dr\,d\mathbf{x}\,d\mathbf{v}$ and counting what enters and leaves) are so familiar in fluid dynamics and in kinetic theory that they need not be repeated here. Equation (1) will be called the *spray equation*.

The terms $\hat{Q}_j$ and Γ_j in equation (1) probably are of dominant importance in regions where atomization or coalescence occurs, Bases are available [23], [24], [25] for addressing questions related to the evaluation of the term Γ_j. In many combustors, the intensity of burning is comparatively low in the neighborhood of the atomizer, and the main part of the combustion occurs in regions where particle interactions and sources are of no more than secondary importance. We shall focus our attention on the burning process and shall therefore neglect $\hat{Q}_j$ and Γ_j. In the illustrative examples consideration will also be restricted to steady processes, whence $\partial f_j/\partial t = 0$. Unsteady spray combustion is of considerable significance in relation to such phenomena as diesel-engine combustion or combustion instability in liquid-propellant rocket motors (see Section 9.3). The form of the spray equation to be used in the present chapter becomes

$$\frac{\partial}{\partial r}(R_j f_j) + \nabla_x \cdot (\mathbf{v} f_j) + \nabla_v \cdot (\mathbf{F}_j f_j) = 0, \quad j = 1, \ldots, M. \tag{2}$$

11.2. SIMPLIFIED MODEL OF COMBUSTION IN A LIQUID-PROPELLANT ROCKET MOTOR

11.2.1. The model

The objective of the analysis given in this section is to illustrate the use of equation (2) by considering the problem of determining the combustion efficiency of a variable-area, quasi-one-dimensional rocket chamber such as that illustrated in Figure 11.1, in which M different kinds of liquid droplets are present. In order to avoid considering the behavior of the gas, we must assume that the material burns to completion as soon as it evaporates. The amount of heat released will then be proportional to the mass evaporated, thus making it possible to relate the combustion efficiency to the mass of the spray present. Even in this case, the equations contain parameters, such as R_j, which depend on the local gas properties. However, estimates of these parameters are often obtainable without solving for the gas flow, so that, while the theory is essentially incomplete, it is not entirely useless.

Probert [26] was the first to give an analysis of the following type; his work has been extended and refined by others [23], [24], [27], [28].

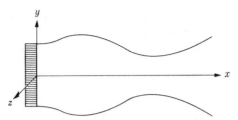

FIGURE 11.1.   Schematic illustration of variable-area rocket chamber.

11.2.2. Simplified spray equation

The velocity dependence of the distribution function, which is not of primary interest here, may be eliminated from the spray equation by integrating equation (2) over all velocity space. Since $f_j \to 0$ very rapidly (at least exponentially) as $|\mathbf{v}| \to \infty$ for all physically reasonable flows, the divergence theorem shows that the integral of the last term in equation (2) is zero, whence

$$\frac{\partial}{\partial r}(\bar{R}_j G_j) + \nabla_x \cdot (\bar{\mathbf{v}}_j G_j) = 0, \quad j = 1, \dots, M, \tag{3}$$

where the number of droplets per unit volume per unit range of radius is

$$G_j \equiv \int f_j \, d\mathbf{v}, \quad j = 1, \dots, M, \tag{4}$$

and the bar denotes an average over all velocities, that is,

$$\bar{R}_j = \int R_j f_j \, d\mathbf{v}/G_j, \quad j = 1, \dots, M, \tag{5}$$

and

$$\bar{\mathbf{v}}_j = \int \mathbf{v} f_j \, d\mathbf{v}/G_j, \quad j = 1, \dots, M. \tag{6}$$

If the local cross-sectional area of the chamber is denoted by $A(x)$, then, since the mean flow is in the x direction (parallel to the motor axis) and the velocities $\bar{\mathbf{v}}_j$ are parallel to the chamber wall at the wall, integrating equation (3) over the two spatial coordinates normal to the x axis yields

$$\frac{\partial}{\partial r}(\bar{R}_j G_j) + \frac{1}{A}\frac{\partial}{\partial x}(A\bar{v}_j G_j) = 0, \quad j = 1, \dots, M. \tag{7}$$

Here $\bar{v}_j$ is the x component of $\bar{\mathbf{v}}_j$, and the quantities $\bar{R}_j$, $\bar{v}_j$, and G_j have been assumed to be essentially independent of the spatial coordinates normal to x (alternatively, these quantities may be interpreted as averages over the cross section).

From the results of Chapters 3 (see also [24]), it may be shown that for nearly all droplet vaporization mechanisms, the dependence of R_j on droplet size may be expressed approximately by the equation

$$R_j = -\chi_j/r^{k_j}, \quad j = 1, \ldots, M, \tag{8}$$

where $0 \le k_j \le 1$ and $\chi_j (\ge 0)$ is independent of r. When molecular processes control the vaporization, $k_j = 1$ for vaporization without combustion and for a burning fuel droplet in an oxidizing atmosphere (see Section 3.3.4). According to Section 3.5, $0 \le k_j \le 1$ for a burning monopropellant droplet. When radiative transfer controls the vaporization, $k_j = 0$. Convective processes produce some modifications to equation (8) (see Section 3.3.6). Here we shall adopt equation (8) and shall assume that k_j is a constant.

It will also be assumed that, although the velocity distributions may differ for particles of different sizes, $\bar{v}_j$ is independent of r. This will be true, for example, when the droplets are so small that they all travel with the velocity of the gas, or when the effect of the gas is negligible but droplets of all sizes are injected with the same average velocity.

The approximations given above enable us to integrate equation (7).

11.2.3 Solution of the spray equation

By substituting equation (8) into equation (7), we find that

$$-\chi_j \, \partial\psi_j/\partial r + \bar{v}_j r^{k_j} \, \partial\psi_j/\partial x = 0, \quad j = 1, \ldots, M, \tag{9}$$

where

$$\psi_j \equiv A\bar{v}_j G_j/r^{k_j}, \quad j = 1, \ldots, M. \tag{10}$$

Upon transformation to the new independent variables

$$\xi_j \equiv \left[r^{k_j+1} - (k_j + 1) \int_0^x (\chi_j/\bar{v}_j)\, dx \right]^{1/(k_j+1)}, \quad j = 1, \ldots, M, \tag{11}$$

and

$$\eta_j \equiv \left[r^{k_j+1} + (k_j + 1) \int_0^x (\chi_j/\bar{v}_j)\, dx \right]^{1/(k_j+1)}, \quad j = 1, \ldots, M, \tag{12}$$

equation (9) becomes

$$\partial\psi_j/\partial\xi_j = 0, \quad j = 1, \ldots, M. \tag{13}$$

The solution to equation (13) is $\psi_j(\xi_j, \eta_j) = \psi_j(\eta_j)$. Letting the subscript 0 identify conditions at the injector ($x = 0$), we may write this solution as

$$\psi_j = \psi_{j,0}(\eta_j), \quad j = 1, \ldots, M, \tag{14}$$

since $\eta_j = r$ at $x = 0$ according to equation (12). From equation (10) it is then seen that the expression

$$G_j = [(A_0 \bar{v}_{j,0})/(A\bar{v}_j)](r/\eta_j)^{k_j} G_{j,0}(\eta_j), \quad j = 1, \ldots, M, \tag{15}$$

determines the size distribution $G_j(r, x)$ at any position x in terms of the distribution $G_{j,0}(r)$ at $x = 0$.

The final step in the analysis is to obtain the combustion efficiency for a chamber of length x from the size distribution at position x. Let Q_j denote the heat released per unit mass of material evaporated from a droplet of kind j, and let $\rho_{l,j}$ represent the density of the liquid in droplets of kind j. The mass of the spray of kind j per unit volume of space is therefore $\int_0^\infty \frac{4}{3}\pi r^3 \rho_{l,j} G_j \, dr$, and the corresponding mass flow rate (mass per second) is $A\bar{v}_j$ times this. Hence, the total amount of heat released per second by spray j between the injector and position x is

$$Q_j\left(A_0 \bar{v}_{j,0} \int_0^\infty \tfrac{4}{3}\pi r^3 \rho_{l,j} G_{j,0} \, dr - A\bar{v}_j \int_0^\infty \tfrac{4}{3}\pi r^3 \rho_{l,j} G_j \, dr\right).$$

Since the maximum possible heat-release rate (obtained by burning all the droplets) is $\sum_{j=1}^M \dot{Q}_j$, where

$$\dot{Q}_j \equiv Q_j A_0 \bar{v}_{j,0} \int_0^\infty \tfrac{4}{3}\pi r^3 \rho_{l,j} G_{j,0} \, dr \tag{16}$$

is the maximum heat-release rate obtainable by burning all droplets of kind j, the ratio of the actual heat-release rate in a motor of length x to the maximum heat-release rate (the combustion efficiency) is

$$\eta_c = \frac{\displaystyle\sum_{j=1}^M Q_j\left(A_0 \bar{v}_{j,0} \int_0^\infty \tfrac{4}{3}\pi r^3 \rho_{l,j} G_{j,0} \, dr - A\bar{v}_j \int_0^\infty \tfrac{4}{3}\pi r^3 \rho_{l,j} G_j \, dr\right)}{\displaystyle\sum_{j=1}^M Q_j A_0 \bar{v}_{j,0} \int_0^\infty \tfrac{4}{3}\pi r^3 \rho_{l,j} G_{j,0} \, dr}. \tag{17}$$

From equation (15) it follows that

$$\eta_c = 1 - \frac{\displaystyle\sum_{j=1}^M (Q_j \rho_{l,j} \bar{v}_{j,0}) \int_0^\infty r^3 (r/\eta_j)^{k_j} G_{j,0}(\eta_j) \, dr}{\displaystyle\sum_{j=1}^M (Q_j \rho_{l,j} \bar{v}_{j,0}) \int_0^\infty r^3 G_{j,0}(r) \, dr}, \tag{18}$$

which, coupled with equation (12), determines the combustion efficiency for an arbitrary initial droplet size distribution.

11.2.4. Droplet size distributions

In order to obtain numerical results for η_c, we must know $G_{j,0}(r)$. A functional form for $G_{j,0}(r)$, which agrees well with observed size distributions for real sprays [4], [28–33] and simplifies many theoretical computations, is

$$G_{j,0} = b_j r^{t_j} \exp(-a_j r^{s_j}), \quad j = 1, \ldots, M, \tag{19}$$

where a_j, b_j, s_j, and t_j are independent of r. The form given in equation (19), which could be termed the *generalized Rosin–Rammler distribution*, was first proposed by Tanasawa [27]. The special case $t_j = 2$ is the Nukiyama–Tanasawa distribution [33], $t_j = s_j - 4$ is the Rosin–Rammler distribution [34], and $s_j = 1$ is the chi-square distribution [35]. Other distributions that have been proposed, such as the log-normal distribution [36], complicate analytical work. Since four adjustable parameters appear in equation (19), measurements of greater accuracy than normally possible are required to demonstrate deviations from this equation.

Actually, two of the parameters in equation (19) are determined by the total number of droplets per unit volume and by the average droplet radius; the other two govern the shape of the distribution about the mean (for example, the standard deviation). The total number of droplets of kind j per unit volume (irrespective of their size) will be denoted by

$$n_j = \int_0^\infty G_j \, dr, \quad j = 1, \ldots, M, \tag{20}$$

and angular brackets will be used to identify number-weighted averages with respect to the size distribution $G_j(r)$—for example,

$$\langle r \rangle_j = \int_0^\infty r G_j \, dr / n_j, \quad j = 1, \ldots, M. \tag{21}$$

If s_j and t_j are retained for describing the shape of the distribution, then it is reasonable to replace a_j and b_j by the physically more significant variables $\langle r \rangle_{j,0}$ and $n_{j,0}$. The substitution of equation (19) into equations (20) and (21) yields two equations which may be solved for a_j and b_j. When these results are substituted into equation (19), it is found that

$$G_{j,0} = \frac{n_{j,0} s_j [\Gamma\{(t_j + 2)/s_j\}]^{t_j+1}}{\langle r \rangle_{j,0} [\Gamma\{(t_j + 1)/s_j\}]^{t_j+2}} \left(\frac{r}{\langle r \rangle_{j,0}} \right)^{t_j}$$
$$\times \exp\left\{ -\left[\frac{r}{\langle r \rangle_{j,0}} \frac{\Gamma\{(t_j + 2)/s_j\}}{\Gamma\{(t_j + 1)/s_j\}} \right]^{s_j} \right\}, \quad j = 1, \ldots, M, \tag{22}$$

where Γ denotes the gamma function. There are special cases ($-4 < t_j \le -1$) for which equation (19) is useful even if n_j [as given by equation (20)] is infinite; in these cases, use of equation (22) necessitates introducing limit concepts, and mass-weighted rather than number-weighted averages may be employed to write an alternative expression for $G_{j,0}$ that does not involve divergences.

For initial droplet-size distributions given by equation (22), the combustion efficiency and other spray properties may be expressed in terms of the function

$$H(a, b, c, z) \equiv \frac{e^{-z}}{\Gamma(ab + c)} \int_0^\infty [(y + z)^a - z^a]^b (y + z)^{c-1} e^{-y} \, dy. \tag{23}$$

When b is a nonnegative integer, $H(a, b, c, z)$ is easily evaluated in terms of the incomplete gamma function,

$$\Gamma(c, z) \equiv \int_0^z y^{c-1} e^{-y} \, dy,$$

which has been tabulated [35], [37]. Specifically, when $b = 0$, it is seen by transforming from the variable y to $(y + z)$ in the integral in equation (23) that

$$H(a, 0, c, z) = 1 - [\Gamma(c, z)/\Gamma(c)]. \tag{24}$$

From Probert's numerical results [26], $H[2(1 - c)/3, \frac{3}{2}, c, z]$ may be evaluated for various values of c and z; this case will be seen to determine the combustion efficiency when $k_j = 1$ and $G_{j,0}$ is of the original Rosin–Rammler form. Tanasawa and Tesima [28] numerically evaluated functions that are somewhat similar to (but not exactly equivalent to) $H(2, \frac{3}{2}, c, z)$ for various values of c and z; this may be seen to correspond to the Nukiyama–Tanasawa form for $G_{j,0}$ with $k_j = 1$. Tables of $H(a, b, c, z)$ for other values of a and b apparently are not yet available.

From equation (23) it is possible to obtain expansions of $H(a, b, c, z)$ that are valid for either large or small values of z, the parameter which will be seen to correspond to the chamber length. For small values of z, an expansion of equation (23) in powers of z yields

$$H(a, b, c, z) = 1 - bz^a[\Gamma(ab + c - a)/\Gamma(ab + c)] + \cdots \tag{25}$$

for all physically important values of a and c when $b > 0$. In the special case $b = 0$, an expansion of equation (24) in powers of z gives (for $c > 0$)

$$H(a, 0, c, z) = 1 - z^c/\Gamma(c + 1) + \cdots \tag{26}$$

near $z = 0$. Expanding the integral in equation (23) in powers of $1/z$ shows that

$$H(a, b, c, z) = a^b z^{ab+c-b-1} e^{-z}[\Gamma(b + 1)/\Gamma(ab + c)] + \cdots \tag{27}$$

for large values of z. Equations (25)–(27) can be useful for determining the behavior of the spray in limiting ranges of x.

11.2.5. The combustion efficiency and other spray properties

Equations for the representative spray properties n_j and $\langle r \rangle_j$ may be derived for systems in which $G_{j,0}$ is given by equation (22). If equation (15) is substituted into equation (20), use is made of equations (12) and (22), and the variable of integration is changed from r to

$$
y \equiv \left[\frac{\Gamma\{(t_j + 2)/s_j\}}{\Gamma\{(t_j + 1)/s_j\}}\right]^{s_j} \left\{\left[\left(\frac{r}{\langle r \rangle_{j,0}}\right)^{k_j+1} + \frac{(k_j + 1)}{\langle r \rangle_{j,0}^{k_j+1}} \int_0^x \frac{\chi_j}{\bar{v}_j} \, dx\right]^{s_j/(k_j+1)}\right.
$$
$$
\left. - \left[\frac{(k_j + 1)}{\langle r \rangle_{j,0}^{k_j+1}} \int_0^x \frac{\chi_j}{\bar{v}_j} \, dx\right]^{s_j/(k_j+1)}\right\}, \quad j = 1, \ldots, M, \tag{28}
$$

then it is found from equation (23) that

$$n_j = [(A_0 \bar{v}_{j,0})/(A\bar{v}_j)]n_{j,0} H(1, 0, \frac{t_j + 1}{s_j}, z_j), \quad j = 1, \dots, M, \quad (29)$$

where

$$z_j \equiv \left[\frac{\Gamma\{(t_j + 2)/s_j\}}{\Gamma\{(t_j + 1)/s_j\}} \right]^{s_j} \left[\frac{(k_j + 1)}{\langle r \rangle_{j,0}^{k_j+1}} \int_0^x \frac{\chi_j}{\bar{v}_j} dx \right]^{s_j/(k_j+1)}, \quad j = 1, \dots, M. \quad (30)$$

Carrying out a similar operation with equation (21) yields

$$\langle r \rangle_j = \langle r \rangle_{j,0} \frac{H\left(\frac{k_j + 1}{s_j}, \frac{1}{k_j + 1}, \frac{t_j + 1}{s_j}, z_j \right)}{H\left(1, 0, \frac{t_j + 1}{s_j}, z_j \right)}, \quad j = 1, \dots, M, \quad (31)$$

where use has been made of equation (29). Expressions for other average quantities in terms of the H function may easily be obtained by analogous procedures.

The number-flux ratio $n_j A \bar{v}_j / n_{j,0} A_0 \bar{v}_{j,0}$, computed from equation (29) by use of equation (24), is presented in Figure 11.2, which shows how the number of droplets decreases as the spray travels downstream (increasing z_j) because of the disappearance of the smallest burning droplets ($r \rightarrow 0$). From the definition of z_j [equation (30)], it is seen that large values of χ_j and small values of $\langle r \rangle_{j,0}$, k_j, and $\bar{v}_j$ lead to rapid rates of decrease of the number of droplets. Although Figure 11.2 might appear to imply that large values of s_j and small values of t_j favor more rapid rates of decrease of n_j, this conclusion is not always valid because s_j and t_j also appear in z_j [see equation (30)]. It may, in fact, be shown from equation (27) that far downstream, large values of both s_j and t_j (that is, more nearly uniform size distributions) tend to produce smaller values of $n_j A \bar{v}_j / n_{j,0} A_0 \bar{v}_{j,0}$ at a given position x. The reason for this result is that the small number of very large droplets require a very long time to burn.

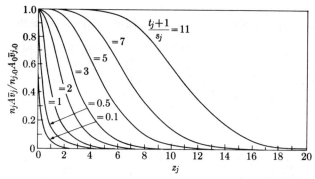

FIGURE 11.2. The ratio of the number flux of droplets of kind j to its value at $x = 0$ as a function of the dimensionless distance z_j for various values of $(t_j + 1)/s_j$.

The property of the spray that is of greatest practical importance is the combustion efficiency, which is given by equation (18). When $G_{j,0}$ is of the generalized Rosin–Rammler form [equation (22)], by using equations (12) and (23) in equation (18), we find that

$$\eta_c = 1 - \left\{ \sum_{j=1}^{M} \left[\dot{Q}_j H \left(\frac{k_j + 1}{s_j}, \frac{3}{k_j + 1}, \frac{t_j + 1}{s_j}, z_j \right) \right] \middle/ \sum_{j=1}^{M} \dot{Q}_j \right\}, \quad (32)$$

where $\dot{Q}_j$ is given by equation (16).

It is of interest to consider a bipropellant rocket motor ($M = 2$) in which liquid fuel ($F, j = 1$) and liquid oxidizer ($O, j = 2$) undergo the overall reaction $v_F F + v_O O \to$ products. Since the present model prevents us from considering gaseous reactions, the fuel and oxidizer must evaporate in stoichiometric proportions—that is, by the same kind of reasoning that was used in Section 11.2.3.

$$\left[A_0 \bar{v}_{1,0} \int_0^\infty \tfrac{4}{3}\pi r^3 \rho_{l,1} G_{1,0} \, dr - A\bar{v}_1 \int_0^\infty \tfrac{4}{3}\pi r^3 \rho_{l,1} G_1 \, dr \right] \middle/ v_F W_F$$

$$= \left[A_0 \bar{v}_{2,0} \int_0^\infty \tfrac{4}{3}\pi r^3 \rho_{l,2} G_{2,0} \, dr - A\bar{v}_2 \int_0^\infty \tfrac{4}{3}\pi r^3 \rho_{l,2} G_2 \, dr \right] \middle/ v_O W_O,$$

where W_F and W_O are the molecular weights of the fuel and oxidizer, respectively. A consequence of this relationship is that some degree of arbitrariness remains in the definition of Q_j. A consistent definition is $Q_1 = 0, Q_2 = Q_0$ for rich or stoichiometric mixtures, and $Q_1 = Q_F, Q_2 = 0$ for lean mixtures, where $Q_F = Q_0(v_O W_O)/(v_F W_F)$, and Q_F and Q_O are the heats of reaction (for *liquid* reactants) per unit mass of fuel consumed and oxidizer consumed, respectively. The difference between lean and rich mixtures arises from the fact that when burning is completed ($\eta_c \to 1$), there are still some oxidizer droplets present in the lean case, but the opposite is true for rich mixtures. For rich or stoichiometric mixtures, $\dot{Q}_1 = 0$ from the above definition, and equation (32) becomes

$$\eta_c = 1 - H \left(\frac{k_2 + 1}{s_2}, \frac{3}{k_2 + 1}, \frac{t_2 + 1}{s_2}, z_2 \right). \quad (33)$$

For lean mixtures, the subscript 2 is replaced by the subscript 1 in equation (33);

$$H \left(\frac{k_1 + 1}{s_1}, \frac{3}{k_1 + 1}, \frac{t_1 + 1}{s_1}, z_1 \right)$$

$$= H \left(\frac{k_2 + 1}{s_2}, \frac{3}{k_2 + 1}, \frac{t_2 + 1}{s_2}, z_2 \right) \qquad \text{for stoichiometric mixtures.}$$

When $M = 1$ or, as shown in the previous paragraph, for fuel-oxidizer systems with $M = 2$, η_c is completely determined by $H[(k_j + 1)/s_j, 3/(k_j + 1),$ $(t_j + 1)/s_j, z_j]$ for droplets of one kind j. Expressions for η_c that are valid for

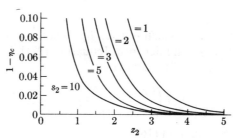

FIGURE 11.3. The asymptotic dependence of the combustion efficiency on the dimensionless chamber length z_2 for various values of the exponent s_2 in the ordinary Rosin-Rammler distribution ($t_2 = s_2 - 4$) with $k_2 = 1$.

small and large values of x may therefore be obtained from equations (25) and (27). In particular, when η_c is given by equation (33), equation (27) shows that for large values of z,

$$
\eta_c \approx 1 - \left(\frac{k_2 + 1}{s_2}\right)^{3/(k_2 + 1)}
$$

$$
\times \left[\frac{\Gamma\{(k_2 + 4)/(k_2 + 1)\}}{\Gamma\{(t_2 + 4)/s_2\}}\right] z_2^{(t_2 + 4)/s_2 - (k_2 + 4)/(k_2 + 1)} e^{-z_2}, \qquad (34)
$$

where z_2 is given by equation (30). The combustion efficiency predicted by equation (34), which holds for the values of x that are of greatest practical interest because it corresponds to η_c near unity, is plotted in Figure 11.3 for various values of s_2 with $k_2 = 1$ and with $t_2 = s_2 - 4$ (ordinary Rosin–Rammler distribution). In the representative case $s_2 = 4$, it may be seen that η_c reaches 99% at $z_2 = 2.5$, which [from equation (30)] corresponds to a physical distance x determined by $\int_0^x (\chi_2/\bar{v}_2) dx / \langle r \rangle_{2, 0}^2 = 3.3$.

From equation (34), as well as from the expansion of η_c for small values of x, it is found that large values of χ_j, s_j, and t_j and small values of $\langle r \rangle_{j, 0}$ and $\bar{v}_j$ lead to higher combustion efficiencies. All these effects, except possibly the dependence on s_j and t_j, are expected a priori on physical grounds. Since large values of s_j and t_j correspond to uniform droplet size distributions, injectors producing monodisperse sprays should give a higher combustion efficiency than those producing a nonuniform droplet-size distribution with the same average droplet radius.

11.3. THE CONSERVATION EQUATIONS FOR DILUTE SPRAYS

11.3.1. Motivation

A natural question that arises in the preceding analysis is: What really determines the parameters χ_j and $\bar{v}_j$, which were treated as known quantities? Both of these quantities depend on the interaction of the spray with the gas and must be evaluated by reasonable guesses if the preceding analysis is to be useful without amplification. In the present section, we shall derive equations that govern χ_j, $\bar{v}_j$, and all other parameters that depend on the local properties of the gas. The equations to be derived here are applicable to many two-phase flows in which the particles modify the behavior of the gas in a nonequilibrium fashion.

Consideration will be restricted to dilute sprays, so that the statistical fluctuations in the flow, which are induced by the random motion of individual particles, may be neglected. Therefore, our objective is to obtain the hydrodynamic equations for the (local) *average* properties of the gas. These equations will be derived by phenomenological reasoning and will be shown to be equivalent to the ordinary equations of fluid dynamics, with suitably added source terms accounting for the average effect of the spray. For the sake of generality, allowance will be made for M different kinds of droplets and N different chemical species in the gas.

11.3.2. Overall continuity

In vaporizing or subliming, particles will add

$$-\sum_{j=1}^{M} \iint \rho_{l,j} 4\pi r^2 R_j f_j \, dr \, d\mathbf{v}$$

grams per unit volume per second to the gas. Consequently, the overall continuity equation for the gas is

$$\partial \rho_f / \partial t + \nabla_x \cdot (\rho_f \mathbf{u}) = - \sum_{j=1}^{M} \iint \rho_{l,j} 4\pi r^2 R_j f_j \, dr \, d\mathbf{v}, \qquad (35)$$

where ρ_f, which will be called the *fluid density*, is the mass of gas per unit volume of physical space. A detailed derivation of equation (35)—obtained, for example, by accounting for the gas entering and leaving a small volume element fixed in space and for the gas produced in that volume element by vaporization—will not be given here because it is quite familiar in fluid dynamics. Similar steps will be omitted in the derivations of the other conservation equations.

The usual density of the gas ρ_g, that is, the mass of gas per unit volume of space available to the gas, is not precisely equal to ρ_f because of the volume occupied by the particles. This difference already has been encountered and discussed in Section 4.2.6. Since the fraction of the total volume of physical space occupied by the particles is $\sum_{j=1}^{M} \iint \frac{4}{3}\pi r^3 f_j \, dr \, d\mathbf{v}$, the fluid and gas densities are related by the expression

$$\rho_f / \rho_g = 1 - \sum_{j=1}^{M} \iint \frac{4}{3}\pi r^3 f_j \, dr \, d\mathbf{v}. \qquad (36)$$

According to the definition of a dilute spray, the fractional volume occupied by the particles must be small. Hence equation (36) implies that one of our basic assumptions would be violated if ρ_f differed greatly from ρ_g.

11.3.3. Species conservation

In spray combustion, cases in which chemical reactions occur in the immediate neighborhood of individual particles are of particular interest. In such cases, the chemical species that flow from the particle to the main part of the gas are not the same as those that evaporate from the surface of the particle. In order to obtain a meaningful species conservation equation, it is therefore convenient to define the *surface layer* of a particle as the region (surrounding the particle) in which the fluid properties differ appreciably from their local average values. This definition is clearly somewhat arbitrary, but the development of the conservation equations will remove some of the ambiguity. Theories or experiments on single particles, using the local average gas properties for boundary conditions at infinity, are supposed to determine the flow within the surface layer. The concept is thus analogous to a boundary-layer approximation (Section 12.1).

The radial mass-flux fraction of chemical species k, averaged over all directions for nonspherically symmetrical surface layers, flowing to the gas from the outer edge of the surface layer of a droplet of kind j, will be denoted by $\Omega_{k,j}$. As will be seen in the subsequent applications, these quantities usually are determined by the overall stoichiometry of the surface-layer reaction and by the composition of the particle. Since $-\sum_{j=1}^{M} \iint \rho_{l,j} 4\pi r^2 R_j \Omega_{k,j} f_j \, dr \, d\mathbf{v}$ is then the mass of species k per unit volume per second added to the gas by the spray, the equation for the conservation of mass of component k in the gas becomes

$$\partial(\rho_f Y_k)/\partial t + \nabla_x \cdot [\rho_f(\mathbf{u} + \mathbf{U}_k)Y_k] = \omega_k - \sum_{j=1}^{M} \iint \rho_{l,j} 4\pi r^2 R_j \Omega_{k,j} f_j \, dr \, d\mathbf{v},$$

$$k = 1, \ldots, N, \quad (37)$$

where Y_k is the mass fraction of species k in the gas, $\mathbf{U}_k$ is the diffusion velocity of species k, and ω_k is the mass rate of production of species k by homogeneous chemical reactions in the gas.

11.3.4. Momentum conservation

The force per unit mass exerted on a particle of kind j by the surrounding gas will be denoted by $\mathbf{F}_j$. Among the effects that may contribute to $\mathbf{F}_j$ are (1) skin friction and separation drag, (2) gravity and other body forces,

(3) rotation of the particle with respect to the gas, and (4) pressure gradients in the gas. It can be shown [24] that the first of these effects is usually the largest in sprays. For effect (1), we may express $\mathbf{F}_j$ in terms of a drag coefficient C_D through the equation

$$\mathbf{F}_j = \tfrac{3}{8}(\rho_g/\rho_{l,j})(1/r)|\mathbf{u} - \mathbf{v}|(\mathbf{u} - \mathbf{v})C_D. \tag{38}$$

The value of C_D depends mainly on the Reynolds number

$$\text{Re} \equiv \rho_g|\mathbf{v} - \mathbf{u}|2r/\mu_g, \tag{39}$$

where μ_g is the viscosity of the gas. In most sprays the particles are so small that the flow about them is laminar and the classical Stokes [38] or Oseen [39] formulas for C_D are approximately valid. An experimental result for nonburning solid and liquid particles in sprays is [40]

$$C_D = 27/\text{Re}^{0.84}. \tag{40}$$

A number of reviews giving expressions for C_D are available (for example, [41] and [42]); for thick sprays (sprays that are not dilute), there is a significant dependence of C_D on the fraction of the volume occupied by the liquid [25].

In terms of the force exerted on individual particles, the average force per unit volume exerted on the spray by the gas is

$$\sum_{j=1}^{M} \int\!\!\int \rho_{l,j}\tfrac{4}{3}\pi r^3 \mathbf{F}_j f_j \, dr \, d\mathbf{v}.$$

The equation for the conservation of momentum of the gas therefore becomes

$$\rho_f \partial\mathbf{u}/\partial t + \rho_f \mathbf{u}\cdot\nabla_x\mathbf{u} = -\nabla_x\cdot\mathbf{P} + \rho_f \sum_{k=1}^{N} Y_k\mathbf{f}_k - \sum_{j=1}^{M} \int\!\!\int \rho_{l,j}\tfrac{4}{3}\pi r^3 \mathbf{F}_j f_j \, dr \, d\mathbf{v}$$

$$- \sum_{j=1}^{M} \int\!\!\int \rho_{l,j} 4\pi r^2 R_j(\mathbf{v} - \mathbf{u})f_j \, dr \, d\mathbf{v}, \tag{41}$$

in which the last term accounts for the momentum carried to the gas by the material that vaporizes from the particles. In equation (41), $\mathbf{P}$ denotes the total pressure tensor, which is related to the hydrostatic pressure p and to the viscous stress tensor $\mathbf{T}$ through the expression

$$\mathbf{P} = p\mathbf{U} + \mathbf{T}, \tag{42}$$

where $\mathbf{U}$ is the unit tensor; $\mathbf{f}_k$ is the external body force per unit mass acting on species k in the gas.

11.3.5. Energy conservation

The equation for the conservation of energy is found, by similar physical reasoning, to be

$$
\frac{\partial}{\partial t}\left[\rho_f\left(h_f + \frac{u^2}{2}\right)\right] + \nabla_x \cdot \left[\rho_f \mathbf{u}\left(h_f + \frac{u^2}{2}\right)\right]
$$

$$
= - \nabla_x \cdot \mathbf{q} - \nabla_x \cdot (\mathbf{T} \cdot \mathbf{u}) + \frac{\partial p}{\partial t} + \rho_f \sum_{k=1}^{N} Y_k(\mathbf{u} + \mathbf{U}_k) \cdot \mathbf{f}_k
$$

$$
- \sum_{j=1}^{M} \iint \rho_{l,j} \tfrac{4}{3}\pi r^3 (\mathbf{F}_j \cdot \mathbf{v}) f_j \, dr \, d\mathbf{v}
$$

$$
- \sum_{j=1}^{M} \iint \rho_{l,j} 4\pi r^2 R_j \left(h_j + \frac{v^2}{2}\right) f_j \, dr \, d\mathbf{v}, \tag{43}
$$

where h_f is the total enthalpy per unit mass of the gas, $\mathbf{q}$ is the heat-flux vector, and h_j, which is determined by the processes occurring in the surface layer, is defined as the total enthalpy flowing to the gas from the outer edge of the surface layer of a particle of type j per unit mass of material evaporating. In equation (43), the last two terms account for the work done on the gas by the spray and the energy added to the gas by the material vaporizing.

11.3.6. Comments on formulations

For the applications discussed in the following sections, the preceding conservation equations are supplemented by the ideal gas equation of state, in which ρ_g (not the fluid density ρ_f) enters. Since f_j appears in each of the conservation equations, it is apparent that they are coupled to the spray equation, which therefore must also be included to obtain a complete set of integrodifferential equations describing spray combustion.

The formulation that has been given here is not the only approach to the description of two-phase flows with nonequilibrium processes. Many different viewpoints have been pursued; textbooks are available on the subject [43], [44], and a reasonably thorough review recently has been published [45]. Combustion seldom has been considered in this extensive literature. Most of the work that has addressed combustion problems has not allowed for a continuous droplet distribution function but instead has employed a finite number of different, discrete droplet sizes in seeking computer solution sets of conservation equations [5]. The present formulation admits discrete sizes as special cases (through the introduction of delta functions in f_j) but also enables influences of continuous distributions to be investigated. A formulation of the present type recently has been extended to encompass thick sprays [25]. Some other formulations of problems of multiphase reacting flows have been mentioned in Sections 7.6 and 7.7.

Computational capabilities have now progressed to a point at which it is no longer unreasonable to consider trying to solve the equations that have been given here numerically for particular flows; steps have been taken in this direction [46]–[49], producing results that can be compared with experiment, but further development of numerical methods is needed. Another worthwhile avenue is to introduce different types of simplifying approximations that enable analytical solutions to be obtained. Comparisons of different simplified solutions with experimental results can shed light on the relative importance of various processes. The examples given next illustrate the development of approximate analytical solutions for two different problems.

11.4. SIMPLIFIED CONSERVATION EQUATIONS

11.4.1. Assumptions

In the remainder of the chapter, attention will be restricted to steady-state, one-dimensional, constant-area flows in which all particles travel at the same velocity and have the same chemical composition. Hence, $\partial/\partial t = 0$, $\nabla_x \to \mathbf{e}_x \, \partial/\partial x$, $\nabla_v \to \mathbf{e}_x \, \partial/\partial v$, $M = 1$ (and the subscript j will usually be omitted), and

$$f(r, \mathbf{x}, \mathbf{v}) = G(r, x)\delta(\mathbf{v} - \bar{\mathbf{v}}), \tag{44}$$

where x is the flow direction, $\bar{\mathbf{v}} = \bar{v}\mathbf{e}_x$ is the velocity of all particles, and $\mathbf{e}_x$ is a unit vector in the x direction. In equation (44), which serves to define the size distribution function G, δ is the three-dimensional delta function. For the purpose of illustrating analytical techniques, the appropriate simplified forms of the preceding governing equations are derived here. A reduction somewhat similar to ours, but assuming that $\bar{\mathbf{v}} = \mathbf{u}$ and allowing for area variations and for M different kinds of particles, is given in [50].

A few additional minor assumptions which are usually valid will appear in the following development.

11.4.2. Overall continuity

With the preceding assumptions, the integral over $\mathbf{v}$ in equation (35) can be performed, showing that

$$d(\rho_f u)/dx = -\int_0^\infty \rho_l 4\pi r^2 \, RG \, dr, \tag{45}$$

where u is the x component of $\mathbf{u}$ and R equals R_j evaluated at $\mathbf{v} = \bar{\mathbf{v}}$. The mass of the condensed phase per unit spatial volume is

$$\rho_s \equiv \int_0^\infty \rho_l \tfrac{4}{3}\pi r^3 \, G \, dr, \tag{46}$$

which will be called the *spray density* to distinguish it from the density of the condensed phase ρ_l. If the spray equation [equation (2)] is multiplied by $\rho_l \frac{4}{3}\pi r^3$ and is integrated over $\mathbf{v}$ and r, then it is seen that

$$d(\rho_s \bar{v})/dx = \int_0^\infty \rho_l 4\pi r^2 \, RG \, dr. \tag{47}$$

In obtaining equation (47), use is made of the fact that r, $\mathbf{x}$, and $\mathbf{v}$ are all independent variables in order to bring $\rho_l \frac{4}{3}\pi r^3$ inside the $\mathbf{x}$ and $\mathbf{v}$ derivatives. The divergence theorem and the fact that $f = 0$ when $\mathbf{v} \neq \bar{\mathbf{v}}$ are then used to show that the last term in equation (2) vanishes when integrated over $\mathbf{v}$. The relationship $\int f \, d\mathbf{v} = G$, which follows from equation (44), is next employed in the other two terms. The first term is then integrated by parts (with respect to r), after which use is made of the fact that $\frac{4}{3}\pi r^3 \rho_l RG$ approaches zero as $r \to 0$ and as $r \to \infty$. These are the kinds of elementary operations that are used repeatedly in the analysis, and discussion of them will be omitted in the future.

Equations (45) and (47) show that the overall continuity equation may be written in the form

$$d(\rho_f u + \rho_s \bar{v})/dx = 0, \tag{48}$$

the integral of which is

$$\rho_f u + \rho_s \bar{v} \equiv m = \text{constant}, \tag{49}$$

where m is the total mass-flow rate per unit area. Equation (49) is precisely what one would obtain from a phenomenological mass balance for all material (gas and condensed phase) flowing across a unit area.

11.4.3. Species conservation

Under the previous assumptions and the additional hypothesis that $\Omega_{k,j} \equiv \Omega_k$ is independent of r (that is, all particles have the same Ω_k), equation (37) becomes

$$\frac{d}{dx}[\rho_f(u + U_k)Y_k] = \omega_k - \Omega_k \frac{d}{dx}(\rho_s \bar{v}), \quad k = 1, \ldots, N, \tag{50}$$

where U_k is the x component of $\mathbf{U}_k$. Here use has been made of equation (47).

11.4.4. Momentum conservation

If equation (2) is multiplied by $\rho_l \frac{4}{3}\pi r^3 \mathbf{v}$ and is integrated over $\mathbf{v}$ and r, then (after a few integrations by parts) it is found that

$$\frac{d}{dx}\left(\bar{v}^2 \int_0^\infty \rho_l \frac{4}{3}\pi r^3 G \, dr\right) - \bar{v} \int_0^\infty \rho_l 4\pi r^2 RG \, dr - \int_0^\infty \rho_l \frac{4}{3}\pi r^3 FG \, dr = 0,$$

where F is the x component of $\mathbf{F}_j$. By making use of equations (46) and (47), this expression may be written in the form

$$\rho_s \bar{v}\, d\bar{v}/dx = \int_0^\infty \rho_l \tfrac{4}{3}\pi r^3 FG\, dr, \tag{51}$$

which is a momentum equation for the spray. Under the present assumptions, the integral over $\mathbf{v}$ in equation (41) is easily evaluated. Substituting equations (42) and (51) into equation (41) and neglecting body forces $\mathbf{f}_k$ then yields

$$\frac{d}{dx}(\rho_f u^2 + \rho_s \bar{v}^2) = -\frac{dp}{dx} - \frac{d\tau_{xx}}{dx}, \tag{52}$$

where τ_{xx} is the xx component of the shear stress tensor $\mathbf{T}$ and use has been made of equations (47) and (48). Equation (52), which may be written in the integrated form

$$\rho_f u^2 + \rho_s \bar{v}^2 + p + \tau_{xx} \equiv P = \text{constant}, \tag{53}$$

may also be derived from an overall momentum balance for gas plus condensed material.

11.4.5. Energy conservation

Applying the present assumptions to equation (43), we find that

$$\frac{d}{dx}\left[\rho_f u\left(h_f + \frac{u^2}{2}\right)\right] = -\frac{dq_x}{dx} - \frac{d}{dx}(\tau_{xx}u) - \frac{d}{dx}\left[\rho_s \bar{v}\left(\frac{\bar{v}^2}{2}\right)\right]$$
$$- \int_0^\infty \rho_l 4\pi r^2 R h_j G\, dr, \tag{54}$$

where q_x is the x component of $\mathbf{q}$. Here use has been made of equations (47) and (51).

It will be assumed (as is often the case) that there is no production of total (thermal plus chemical) enthalpy in the surface layers of the particles. The quantity h_j may then be evaluated at the *inner* edge of the surface layer (that is, at the particle surface) and thereby may be related to the total enthalpy per unit mass of the condensed phase, h_l. The rate of increase of total enthalpy of a particle of kind j is, in general, given by

$$\frac{d}{dt}(\rho_{l,j}\tfrac{4}{3}\pi r^3 h_{l,j}) = \rho_{l,j} 4\pi r^2 R_j h_{l,j}$$
$$+ \rho_{l,j}\tfrac{4}{3}\pi r^3\left(R_j \frac{\partial h_{l,j}}{\partial r} + \mathbf{v}\cdot\nabla_x h_{l,j} + \mathbf{F}_j\cdot\nabla_v h_{l,j}\right), \tag{55}$$

since $d/dt = R_j \partial/\partial r + \mathbf{v}\cdot\nabla_x + \mathbf{F}_j\cdot\nabla_v$. The preceding hypothesis implies that

$$\rho_{l,j} 4\pi r^2 R_j h_j = \frac{d}{dt}(\rho_{l,j}\tfrac{4}{3}\pi r^3 h_{l,j}). \tag{56}$$

Combining equations (55) and (56), multiplying by f_j, and integrating over r and $\mathbf{v}$ yields, for the present case,

$$\int_0^\infty \rho_l 4\pi r^2 R h_j G \, dr = \frac{d}{dx}\left(\int_0^\infty \rho_l \tfrac{4}{3}\pi r^3 h_l \bar{v} G \, dr\right), \tag{57}$$

where use has been made of the relation

$$\nabla_x \cdot \iint \rho_{l,j} \tfrac{4}{3}\pi r^3 h_{l,j} \mathbf{v} f_j \, dr \, d\mathbf{v} = \iint \rho_{l,j} 4\pi r^2 R_j h_{l,j} f_j \, dr \, d\mathbf{v}$$
$$+ \iint \rho_{l,j} \tfrac{4}{3}\pi r^3 \left(R_j \frac{\partial h_{l,j}}{\partial r} + \mathbf{v}\cdot\nabla_x h_{l,j} + \mathbf{F}_j\cdot\nabla_v h_{l,j}\right) f_j \, dr \, d\mathbf{v},$$

which is obtained by multiplying equation (2) by $\rho_{l,j}\tfrac{4}{3}\pi r^3 h_{l,j}$ and integrating over $\mathbf{v}$ and r.

We shall consider only systems for which h_l is independent of r. Substituting equation (57) into equation (54) and using equation (46) then gives

$$\frac{d}{dx}\left[\rho_f u\left(h_f + \frac{u^2}{2}\right) + \rho_s \bar{v}\left(h_l + \frac{\bar{v}^2}{2}\right)\right] = \frac{dq_x}{dx} - \frac{d}{dx}(\tau_{xx} u). \tag{58}$$

An energy equation like equation (58) also would be obtained from a phenomenological energy balance applied to the gas and condensed material simultaneously. A first integral of equation (58) is

$$\rho_f u\left(h_f + \frac{u^2}{2}\right) + \rho_s \bar{v}\left(h_l + \frac{\bar{v}^2}{2}\right) + q_x + \tau_{xx} u \equiv \mathcal{H} = \text{constant}. \tag{59}$$

The reader will note that all the conservation equations except the species equation have been integrated once as a consequence of our simplifying assumptions. The final equations of the present section are used in the following applications.

11.5. EXTENDED MODEL OF COMBUSTION IN A LIQUID-PROPELLANT ROCKET MOTOR

11.5.1. The model

In the model of rocket combustion given in Section 11.2, expressions for the parameters χ_j and $\bar{v}_j$ remained undeveloped. An analysis given by Spalding [51], [52] determines these parameters by addressing gas flow. When the interaction of the particles with the gas is considered, the equations become so complicated that exact analytical results are obtainable only in very simple cases. Consequently, in addition to the approximations enumerated in the preceding section (for example, constant area), we shall assume here that all particles are of the same size (monodisperse sprays); that is,

$$G(r, x) = n(x)\delta(r - \langle r \rangle), \tag{60}$$

where n is the number of particles per unit volume, and δ is the one-dimensional delta function. Both $\langle r \rangle_0$ and $\bar{v}_0$, the droplet radius and the velocity of droplets at the injector ($x = 0$), are assumed to be known, and the gas velocity u is assumed to go to zero at $x = 0$. As in Section 11.2, the material will be assumed to react as soon as it vaporizes, and also the fluid density ρ_f will be taken to be constant. Under these conditions, the only gas conservation equation that is really required is the overall continuity equation.

Although analyses in which some of these assumptions are removed, for example, by accounting for finite gas-phase reaction rates [53], for non-uniform particle-size distributions [54], [55], [56] and for droplet breakup [57], [58], are more realistic in that correlations with observed rocket-motor performance sometimes can be obtained, they involve numerical integrations which may tend to obscure the essential ideas. Simple analytical results have, however, been developed for a model that accounts in an approximate way for nonuniform size distributions [59]. Comprehensive reviews of related studies may be found in [58] and [60]. The major drawback to all of these analyses is the one-dimensional flow approximation, which excludes from consideration the three-dimensional flows generally observed in real engines.

Since the droplets are all of the same size in our present model, they will all disappear simultaneously at a *finite* distance from the injector, contrary to the results of Section 11.2. Therefore, instead of focusing attention on the combustion efficiency, we shall attempt to determine the minimum chamber length required for complete combustion ($\eta_c = 100\%$), which will be denoted by x^*.

11.5.2. The spray equation

By substituting equations (60) and (44) into equation (2) and by integrating over r and v, we find that the spray equation assumes the very simple form

$$d(n\bar{v})/dx = 0, \tag{61}$$

the integral of which,

$$n\bar{v} = n_0 \bar{v}_0, \tag{62}$$

simply states that the total number of particles crossing any plane normal to the x axis per second is constant.

11.5.3. Droplet vaporization rate

Substituting equation (60) into equations (46) and (47) yields

$$\rho_s = \rho_l \tfrac{4}{3}\pi\langle r\rangle^3 n \tag{63}$$

and

$$d(\rho_s \bar{v})/dx = \rho_l 4\pi\langle r\rangle^2 Rn. \tag{64}$$

When equations (62) and (63) are used in equation (64), we find that

$$\bar{v}\,d\langle r\rangle/dx = R, \tag{65}$$

which, of course, must be valid in the present simple case because of the definition of R.

As previously indicated (Section 11.2.2), the expression for R depends on the vaporization mechanism. An attractive formula for liquid rockets employing volatile oxidizers is that of a burning fuel droplet in an infinite quiescent oxidizing atmosphere [equation (3-58)], namely,

$$R = -\frac{\lambda}{c_p \rho_l}\frac{1}{\langle r\rangle}\ln\!\left(1 + \frac{T - T_l}{L/c_p} + \frac{Q_o}{L}\,Y_o\right), \tag{66}$$

where λ, c_p, T, and Y_O are the thermal conductivity, specific heat as constant pressure, temperature, and mass fraction of oxidizer for the gas; T_l is the temperature of the liquid; L is the heat of vaporization per unit mass, and Q_O is the heat reaction (for *gaseous* fuel) per unit mass of oxidizer consumed. It is reasonable to assume that when the material reacts as soon as it vaporizes, the pressure, temperature, and composition of the gas remain constant. Hence, only $\langle r\rangle$ varies in equation (66), and therefore equation (66) may be written as

$$R = R_0(\langle r\rangle_0/\langle r\rangle), \tag{67}$$

where the constant R_0 is the value of R at $x = 0$. Since various other vaporization mechanics also lead to equation (67), this equation has a certain degree of generality, but it must be modified if convective corrections to R become important. Spalding's extension of this analysis [52] to account for convective effects shows that in a typical case, the present formulas overestimate x^* by a factor of about 2 by neglecting the increased vaporization rate and increased drag.

11.5.4. Droplet drag

By substituting equation (60) into equation (51), it is found that

$$\rho_s \bar{v}\,d\bar{v}/dx = \rho_l \tfrac{4}{3}\pi\langle r\rangle^3 Fn, \tag{68}$$

which reduces to the expected result

$$\bar{v}\,d\bar{v}/dx = F \tag{69}$$

when use is made of equation (63). Adoption of the Stokes drag law gives [39]

$$C_D = 24/\text{Re} \tag{70}$$

and

$$F = \tfrac{9}{2}(\mu_g/\rho_l)(1/\langle r \rangle^2)(u - \bar{v}) \tag{71}$$

from equations (38) and (39).

11.5.5. Continuity

The assumption that the gas properties are constant (Section 11.5.3) requires that ρ_g be constant. From equations (36) and (60) it is seen that

$$\rho_f/\rho_g = 1 - \tfrac{4}{3}\pi\langle r \rangle^3 n. \tag{72}$$

Since the product $n\langle r \rangle^3$ is, in general, not constant, the previously stated assumption $\rho_f = \text{constant}$ is valid only for dilute sprays, $\tfrac{4}{3}\pi\langle r \rangle^3 n \ll 1$. When attention is restricted to dilute sprays, equation (49) (the continuity equation) provides an expression for the gas velocity u. Since $u = 0$ at $x = 0$, $m = \rho_{s,0}\bar{v}_0$. Hence, equation (49) becomes

$$\rho_f u = m[1 - (\rho_s\bar{v})/(\rho_{s,0}\bar{v}_0)]. \tag{73}$$

Using equations (62) and (63) in equation (73), we obtain

$$u = (m/\rho_f)(1 - \langle r \rangle^3/\langle r \rangle_0^3). \tag{74}$$

11.5.6. Solution to the problem

The dimensionless droplet radius may be defined as

$$\sigma = \langle r \rangle/\langle r \rangle_0, \tag{75}$$

and a dimensionless droplet velocity is

$$\varphi = \bar{v}/\bar{v}_0. \tag{76}$$

Substituting equation (67) into equation (65) then gives

$$\varphi \, d\sigma/dx = [R_0/(\bar{v}_0 \langle r \rangle_0)]/\sigma, \tag{77}$$

and substituting equations (71) and (74) into equation (69) yields

$$\varphi \frac{d\varphi}{dx} = \left(\frac{9\mu_g}{2\rho_l\langle r \rangle_0^2\bar{v}_0}\right)\frac{1}{\sigma^2}\left[\left(\frac{m}{\rho_f\bar{v}_0}\right)(1 - \sigma^3) - \varphi\right]. \tag{78}$$

Since all quantities in equations (77) and (78) except σ, φ, and x are constants, these equations comprise two first-order ordinary differential equations describing the flow; the boundary conditions are $\sigma = \varphi = 1$ at $x = 0$.

Dividing equation (78) by equation (77) produces the linear differential equation

$$d\varphi/d\sigma = -(\beta_2/\sigma)[(1 - \sigma^3)/\beta_1 - \varphi] \tag{79}$$

with the boundary condition $\varphi = 1$ at $\sigma = 1$, where

$$\beta_1 \equiv \bar{v}_0 \rho_f/m \tag{80}$$

and

$$\beta_2 \equiv -(9\mu_g)/(2\rho_l\langle r\rangle_0 R_0) \tag{81}$$

are the only two dimensionless constants entering the problem. From equation (80) we see that β_1 is a dimensionless measure of the injection velocity. Equation (81), in which the minus sign is included to make $\beta_2 > 0$ since $R_0 < 0$, implies that β_2 is a dimensionless measure of the droplet drag. Once equation (79) is solved, the properties may be related to the distance x through the integral of equation (77), which is

$$x = (-\langle r\rangle_0 \bar{v}_0/R_0) \int_\sigma^1 (\varphi\sigma)\, d\sigma. \tag{82}$$

An analytical solution to equation (79) is easily obtained, for example, by transforming from φ to $\varphi\sigma^{-\beta_2}$ as the dependent variable. The solution is

$$\varphi = \frac{1}{\beta_1}\left[\left(\beta_1 - \frac{3}{3 - \beta_2}\right)\sigma^{\beta_2} + 1 + \left(\frac{\beta_2}{3 - \beta_2}\right)\sigma^3\right]. \tag{83}$$

By substituting this result into equation (82), we find that

$$x = \left(-\frac{\langle r\rangle_0 \bar{v}_0}{R_0}\right)\frac{1}{\beta_1}\left[\left(\beta_1 - \frac{3}{3 - \beta_2}\right)\left(\frac{1}{2 + \beta_2}\right)(1 - \sigma^{2 + \beta_2})\right.$$
$$\left. + \tfrac{1}{2}(1 - \sigma^2) + \frac{\beta_2}{5(3 - \beta_2)}(1 - \sigma^5)\right]. \tag{84}$$

It may be seen from equation (83) that when the drag is zero ($\beta_2 = 0$), $\varphi = 1$; that is, the droplet velocity remains constant, equal to its injection velocity as expected. In the opposite limiting case of infinite drag ($\beta_2 = \infty$), equation (83) shows that $\varphi = (1 - \sigma^3)/\beta_1$, which, according to equations (74) and (80), implies that $\bar{v} = u$; the droplets immediately decelerate to the gas velocity and the gas and droplet velocities remain equal thereafter. In intermediate cases, the droplets first decelerate until the gas velocity exceeds the droplet velocity, at which time they begin to accelerate, finally reaching the gas velocity again when $\langle r\rangle \to 0$. Profiles for the typical case $\beta_2 = 1$, $\beta_1 = 0.5$ are shown in Figure 11.4.

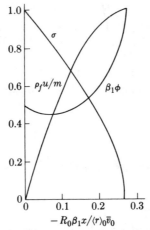

FIGURE 11.4. Profiles of droplet velocity, gas velocity, and droplet radius in a rocket motor with the drag parameter $\beta_2 = 1$ and the dimensionless injection velocity $\beta_1 = 0.5$ (from [52]).

11.5.7. The chamber length for complete combustion

The droplets disappear when $\sigma \to 0$. By setting $\sigma = 0$ in equation (84), the chamber length required for complete combustion is found to be

$$x^* = (-\langle r \rangle_0 \bar{v}_0 / R_0)(1 + 3\beta_2/10\beta_1)/(2 + \beta_2). \tag{85}$$

Equation (85), which is plotted in Figure 11.5, reasonably predicts that in order to obtain a short minimum chamber length with a given mass flow rate m, the droplet radius $\langle r \rangle_0$ should be small and the vaporization rate R_0 should be large. Furthermore, Figure 11.5 and equation (85) show that for

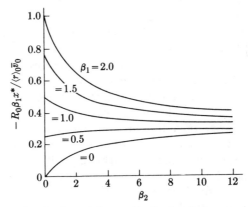

FIGURE 11.5. The dimensionless minimum chamber length for complete combustion as a function of the drag parameter β_2 for various values of the dimensionless injection velocity β_1 (from [52]).

fixed m and ρ_f, decreasing β_1 decreases x^*; that is, a low injection velocity favors a short chamber length. This last effect is mitigated by accounting for nonuniform droplet size distributions [59]. These trends are consistent with those predicted in Section 11.2.

The present analysis illustrates techniques for taking particle-gas interactions into account simply, by introducing an explicitly stated sequence of assumptions.

11.6. DEFLAGRATIONS IN SPRAYS

11.6.1. Description

A mixture of air and a combustible spray can support a propagating laminar flame in the same types of experiments that are described in Section 5.1.1. The laminar burning velocity of the spray can therefore be determined by measuring the cone angle of the flame that it produces in a steady-flow Bunsen-burner type of apparatus [61]. Flames of this general type are spray deflagrations or heterogeneous flames (heterogeneous in the sense that the initial combustible mixture is not homogeneous). Many experiments have been performed on spray deflagrations [61]–[69]. Similar phenomena occur for suspensions of solid fuel particles in oxidizing gases, such as pulverized coal in air [70]; reviews are available on deflagrations in these systems [71], [72]. In addition to deflagrations, detonations can occur in two-phase systems with dispersed fuels; these were discussed in Section 6.4.

Two distinct modes of spray deflagration may be identified. If the droplets are small enough and sufficiently volatile, they may vaporize completely and mix well with the oxidizer in the convective-diffusive zone of the deflagration before they reach the reactive-diffusive zone. In this limit, the structure of the reactive-diffusive zone is unmodified, and the analysis of Section 5.3.6, for example, may be employed for calculating the burning velocity. The flame speed is less than that for the corresponding gaseous fuel only because of the energetics associated with the heat needed for vaporization of the fuel, as a theoretical analysis of this regime has shown [73]. In the opposite limit of large particles or nonvolatile fuels, particle combustion is germane to the deflagration structure. Deflagrations of this type represent combustion in premixed systems (since the deflagration concept fundamentally pertains to premixed combustibles) with a nonpremixed substructure (since particle burning is a nonpremixed process). The heterogeneity becomes an essential aspect of the flame-propagation process, and the burning velocity may be called the *heterogeneous burning velocity* of the spray. This limit of heterogeneous flame propagation is the limit that will be considered in the present section.

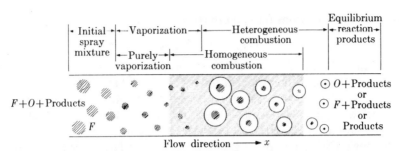

FIGURE 11.6.   Rough schematic diagram of model of spray deflagration.

The deflagration process is illustrated pictorially in Figure 11.6. Transport of radiation can be an important mode for heating the fuel under appropriate conditions [71], [72], especially for solid combustibles that radiate strongly as they burn. Radiation will be neglected in the analysis to be outlined, as it is directed toward liquid sprays, which often radiate negligibly. The motion of sufficiently large particles relative to the gas under the influence of gravity can be important [65], [74]; the occurrence of this phenomenon is excluded here by neglecting body forces. Since consideration is restricted to hetereogeneous burning, the homogeneous combustion illustrated in Figure 11.6 will be absent, and gaseous fuel will not be found in the initial mixture.

It is straightforward to identify approximately a boundary between homogeneous and heterogeneous propagation. For homogeneous propagation, enough vaporization must occur during the transit time through the convective-diffusive zone to support a gaseous deflagration. From equation (5-4) this time is roughly $t_r \approx \delta/v_0 \approx \lambda/(c_p \rho_0 v_0^2)$. From equations (3-33) and (3-65), the vaporization time of a droplet of initial radius r_0 is roughly $t_v \approx 4r_0^2/K \approx r_0^2 \rho_l c_p/[2\lambda \ln(1 + B)]$ [since ρ_d of equation (3-65) is ρ_l here]. Assuming that homogeneous burning occurs if $t_v \lesssim t_r$, we then find the criterion for homogeneous burning to be $r_0^2 \lesssim 2\delta^2(\rho_0/\rho_l)\ln(1 + B)$, which with $\ln(1 + B) \approx 5, \rho_0/\rho_l \approx 10^{-3}$, and $\delta \approx 0.02$ cm gives $r_0 \lesssim 20 \mu$, in reasonable agreement with corresponding experimental observations [61], [62] of about 10μ for the droplet radius in a monodispersive spray at the transition from homogeneous to heterogeneous burning. Thus in typical experiments, homogeneous propagation occurs for $r_0 \lesssim 10 \mu$ and heterogeneous propagation at larger initial radii.

In the regime of heterogeneous propagation, the burning velocity may be estimated from equation (5-2) if w therein is approximated from results for droplet burning. If $(1 + v^{-1})$ denotes the mass of the mixture reacted per unit mass of fuel burnt (where v is the stoichiometric fuel-gas ratio), then an estimate from equation (47), for example, is

$$w \approx (1 + v^{-1}) \int_0^\infty \rho_l 4\pi r^2 (-R)G \, dr,$$

which—with equations (60) and (66)—gives roughly

$$w \approx 4\pi(1 + v^{-1})n_0\langle r\rangle_0(\lambda/c_p)\ln(1 + B),$$

where the subscript 0 identifies initial conditions. Thus with ρ representing the total mass of the mixture per unit volume, we have from equation (5-2) $v_0 \approx [\lambda/(\rho c_p)][4\pi(1 + v^{-1})n_0\langle r\rangle_0 \ln(1 + B)]^{1/2}$. At a fixed overall fuel-oxidizer ratio, $n_0\langle r\rangle_0^3$ remains constant at constant pressure, so that v_0 is seen to vary inversely with $\langle r\rangle_0$ in this regime, and from equation (5-4), the flame thickness is then proportional to $\langle r\rangle_0$. For large values of $\langle r\rangle_0$, the burning velocity is less than that of the homogeneous flame, but as $\langle r\rangle_0$ is decreased, v_0 may increase to a value more than twice that of the homogeneous flame prior to the transition to homogeneous burning. At fixed $\langle r\rangle_0$, to keep the overall fuel-oxidizer ratio constant n_0 must increase in proportion to the pressure, so that $v_0 \sim 1/\sqrt{p}$, as would be obtained for a first-order gaseous reaction. In the rest of this section, we shall indicate how a more systematic formulation of the spray-deflagration problem can be developed to reach these same qualitative conclusions.

In particular, the one-step chemical process $v_F F + v_O O \rightarrow$ products will be investigated, where v_F and v_O are the stoichiometric coefficients for fuel F and oxidizer O appearing as reactants. It will be convenient to adopt a coordinate system in which the combustion wave is at rest, the combustible mixture approaches from $x = -\infty$ and equilibrium reaction products move away toward $x = +\infty$; conditions become uniform as $x \rightarrow \pm\infty$ in Figure 11.6. All the conservation equations derived in Section 11.4 will be needed here, and all the simplifications in Section 11.4 are assumed to be valid. Since the initial relative velocity of the droplets and the gas is zero and the velocity gradients may not be too large, all droplets will be assumed to travel at the same velocity as the gas ($\bar{v} = u$). Estimates of the droplet acceleration using equation (71) indicate that this additional approximation is valid in the present problem if the droplets are not too large. Other assumptions will be stated in the course of the illustrative analysis.

11.6.2. Overall continuity and the spray equation

Since $\bar{v} = u$, equation (49) becomes

$$\rho u = m = \text{constant}, \tag{86}$$

where $\rho \equiv \rho_f + \rho_s$ is the total mass (of gas plus liquid) per unit volume.

It is convenient to define the mass fraction of gas as

$$Z = \rho_f/\rho, \tag{87}$$

whence equation (45) reduces to

$$dZ/dx = -\int_0^\infty \rho_l 4\pi r^2 RG \, dr/m. \tag{88}$$

Since the spray is composed of fuel droplets and since $\bar{v} = u$, in equation (88) R will be given by equation (66), which may be written in the form of equation (8) with $k = 1$ and with

$$\chi = \frac{\lambda}{c_p \rho_l} \ln\left(1 + \frac{T - T_l}{L/c_p} + \frac{Q_o}{L} Y_o\right). \tag{89}$$

Since $M = 1$, the subscript j has been omitted.

Under our present assumptions, the spray equation simplifies to equation (7) with the additional conditions $M = 1$, $\bar{v} = u$, and $A = $ constant. Since the solution to equation (7) given in Section 11.2 will also be valid in the present problem, equation (15) is applicable here and reduces to

$$G = (u_0/u)(r/\eta)G_0(\eta), \tag{90}$$

where, according to equation (12),

$$\eta = \left[r^2 + 2\int_{-\infty}^{x} (\chi/u)\, dx\right]^{1/2}. \tag{91}$$

The subscript 0 is used to identify the known conditions far upstream (at $x = -\infty$) instead of conditions at $x = 0$ as in Section 11.2.

Equation (90) may be substituted into equation (88), giving

$$dZ/dx = (4\pi\rho_l u_0/m)(\chi/u)\int_0^{\infty} (r^2/\eta)G_0(\eta)\, dr. \tag{92}$$

Since the initial drop-size distribution is known, the integral in equation (92) is a known function of $\int_{-\infty}^{x} (\chi/u)\, dx$, [see equation (91)]. Equation (92) is therefore an integrodifferential equation for determining the rate of disappearance of the liquid.

11.6.3. Species conservation

Since homogeneous reactions are neglected ($\omega_k = 0$), equation (50) becomes

$$\frac{d}{dx}\left[Z\left(1 + \frac{U_k}{u}\right)Y_k\right] = \Omega_k \frac{dZ}{dx}, \quad k = 1, \ldots, N, \tag{93}$$

where use has been made of equations (86) and (87). Thermal diffusion will be neglected, and the binary diffusion coefficients of all pairs of gaseous species will be assumed equal. Making use of the result that the pressure is constant (Section 11.6.4) we then find that Fick's law diffusion is valid here, that is,

$$Y_k U_k = -D\, dY_k/dx, \quad k = 1, \ldots, N, \tag{94}$$

where D is the binary diffusion coefficient. Equation (94) may be substituted into equation (93), and (since Ω_k is a constant) one integration may readily be performed. Evaluating the constant of integration at $x = -\infty$, where $dY_k/dx = 0$, we obtain

$$(\rho_f D/m)(dY_k/dx) = Z(Y_k - \Omega_k) - Z_0(Y_{k,0} - \Omega_k), \quad k = 1, \ldots, N, \quad (95)$$

where use has been made of equations (86) and (87).

From the overall stoichiometry of the reaction in the surface layer and from the definition of Ω_k (Section 11.3.3), it follows that for the fuel $\Omega_F = 0$ (since all of the fuel is consumed in the droplet surface layer) and for the oxidizer $\Omega_O = -(v_O W_O)/(v_F W_F)$, where W_k is the molecular weight of chemical species k. The values of Ω_k for reaction products may similarly be expressed in terms of their molecular weights and stoichiometric coefficients.

Since no fuel escapes from the surface layers and $Y_{F,0} = 0$, no fuel will be present in the gas, that is, $Y_F = 0$ is the solution of equation (95) with $k = F$. In fact, it is consistent to neglect homogeneous reactions only when this is true. Hence, equation (95) will be used only for oxidizer and products.

11.6.4. Momentum and energy conservation

Equation (53) for momentum conservation becomes

$$\rho u^2 + p + \tau_{xx} = P = \text{constant}. \quad (96)$$

The fact that the Mach number is low implies that $\rho u^2 \ll p$. The velocity gradients are sufficiently small that the approximation $\tau_{xx} = 0$ is acceptable. Hence equation (96) simply states that $p = \text{constant}$ is a good assumption.

The condition of low Mach number may also be written as $u^2/2 \ll h_f$. Since $\tau_{xx} = 0$, equation (59) for energy conservation then reduces to

$$Z h_f + (1 - Z)h_l + q_x/m = \mathscr{H}/m = \text{constant}, \quad (97)$$

where equations (86) and (87) have been employed.

The enthalpy of the gas is

$$h_f = \sum_{k=1}^{N} Y_k h_k, \quad (98)$$

where

$$h_k = h_k^0 + c_p(T - T^0), \quad k = 1, \ldots, N, \quad (99)$$

in which h_k^0 represents the standard enthalpy of formation per unit mass of gaseous species k at the standard temperature T^0, and the specific heat at constant pressure for the gas, c_p, has been assumed to be constant (independent of temperature) and to be identical for all gaseous species k. Under similar assumptions, the enthalpy of the liquid becomes

$$h_l = h_l^0 + c_{p,l}(T_l - T^0), \quad (100)$$

where h_l^0 is the standard enthalpy of formation per unit mass of liquid fuel and $c_{p,l}$ is the specific heat at constant pressure of the liquid. The standard heat of formation of gaseous fuel, h_F^0, and the standard heat of vaporization of the fuel, L^0, are related to h_l^0 by the expression

$$h_l^0 = h_F^0 - L^0, \tag{101}$$

and L^0 is related to the heat of vaporization at the actual liquid temperature through the relationship

$$L = L^0 - (c_{p,l} - c_p)(T_l - T^0). \tag{102}$$

The x component of the heat-flux vector is

$$q_x = -\lambda \, dT/dx + \sum_{k=1}^{N} h_k \rho_f Y_k U_k, \tag{103}$$

where λ is the thermal conductivity of the gas.

By substituting equations (94) and (98)–(103) into equation (97), it is found that

$$-\frac{\lambda}{m}\frac{dT}{dx} - \left(\frac{\rho_f D}{m}\right)\sum_{k=1}^{N} h_k^0 \frac{dY_k}{dx} + \sum_{k=1}^{N} h_k^0 Z Y_k - Z(h_F^0 - L) + c_{p,l}(T_l - T_{l,0})$$

$$+ Zc_p(T - T_l) = \sum_{k=1}^{N} h_k^0 Z_0 Y_{k,0} - Z_0(h_F^0 - L_0) + Z_0 c_p(T_0 - T_{l,0}), \tag{104}$$

in which the constant $\mathcal{K}$ has been evaluated at $x = -\infty$, where $dT/dx = dY_k/dx = 0$. Equation (95) may be utilized in equation (104) to show that

$$(\lambda/m)(dT/dx) = Z(c_p T - Q') - Z_0(c_p T_0 - Q_0') + c_{p,l}(T_l - T_{l,0}), \tag{105}$$

where

$$Q' \equiv Q_F - L + c_p T_l, \tag{106}$$

in which

$$Q_F \equiv h_F^0 - \sum_{k=1}^{N} h_k^0 \Omega_k = -\Omega_0 Q_0 \tag{107}$$

is the standard (gaseous) heat of reaction per unit mass of fuel. Equation (105) is the form of the energy equation that is most useful here.

It is necessary to introduce an equation of state for the gas, which will be taken to be that of an ideal gas [equation (1-9)],

$$p = \rho_g R^0 T / \overline{W}, \tag{108}$$

where R^0 is the universal gas constant and

$$\overline{W} = \left[\sum_{k=1}^{N} (Y_k/W_k)\right]^{-1} \tag{109}$$

is the average molecular weight of the gas mixture. Since $\rho_g \approx \rho_f$ and since constancy of p follows from the momentum equation, using equation (87) in equation (108) yields

$$Z\rho T/\overline{W} = p/R^0 = \text{constant.} \tag{110}$$

11.6.5. The mathematical problem and boundary conditions

A complete set of equations has not yet been obtained because, in general, the temperature of the liquid varies in a droplet and is determined by the initial conditions and by the heat-transfer equation in the droplet. However, after a flame has developed in the droplet surface layer, T_l remains constant within a few degrees of the boiling temperature for most fuels (see Section 3.3.4). Initially T_l must equal T_0 for thermal equilibrium to exist. Therefore, if T_0 is sufficiently close to the boiling temperature, it is valid to assume that $T_l = T_0 = \text{constant}$. An estimate (using the energy equation) indicates that this is an accurate approximation when the energy required to raise the liquid temperature from its initial value to the boiling point is much less than the heat of combustion, which is often true. It will be assumed here that $T_l = T_0$, in which case $L = \text{constant}$ and Q' becomes a constant and plays the role of an effective heterogeneous heat of reaction. Analyses have been performed in which this assumption is not introduced [75], [76].

The unknown dependent variables of the problem become ρ, u, Z, T, and Y_k. Equations (86) and (110) provide two algebraic equations relating these variables; the other $N + 2$ variables are determined by the $N + 1$ differential equations given by equations (95) and (105) and by the integro-differential equation given in equation (92). The initial values ρ_0, Z_0, T_0, and $Y_{k,0}$ are controlled by the experimenter. Attention will be restricted to lean mixtures, whence $Z \to 1$ as $x \to \infty$ for chemical equilibrium to exist at the hot boundary. Since all the differential equations are of the first order, the additional boundary condition suggests that solutions will exist only for particular values of a parameter, which physically is expected to be the burning velocity u_0.

The number of differential equations can be decreased if the Lewis number is unity. Equations (95) and (105) are so similar in form that when $\lambda/c_p = \rho_f D$, it is possible to solve for Y_k as a function of T. In a manner similar to that in which equation (5-24) was derived, it may be shown [73] that

$$Y_k = \Omega_k + (Y_{k,0} - \Omega_k)[(Q' - c_p T)/(Q' - c_p T_0)], \quad k = 1, \ldots, N, \tag{111}$$

which determines all mass fractions in terms of the temperature. It is interesting that the simplification associated with the approximation $\text{Le} = 1$ for gaseous-flame problems also extends to sprays.

Further simplifications arise through equation (92) for certain size distributions $G_0(r)$; for uniform size distributions, equation (92) becomes an ordinary differential equation. When

$$G_0(r) = n_0 \delta(r - \langle r \rangle_0), \tag{112}$$

it can be shown [50] that equation (92) reduces to

$$dZ/dx = (4\pi \rho_l u_0 n_0 \langle r \rangle_0/m)(\chi/u)[(1 - Z)/(1 - Z_0)]^{1/3}. \tag{113}$$

Equation (113) may be seen to be valid by utilizing equations (8), (62), (63), (86), and (87) in equation (64).

The two remaining differential equations are equations (105) and (113); the other dependent variables are determined by the algebraic expressions given in equations (86), (110), and (111). Since the spatial coordinate x appears in none of these equations except as d/dx in equations (105) and (113), x is easily eliminated from the problem to obtain a single differential equation,

$$\frac{dT}{dZ} = \frac{Z(c_p T - Q') - Z_0(c_p T_0 - Q')}{\left(\dfrac{4\pi\lambda^2 n_0 \langle r \rangle_0}{c_p m^2}\right)\left(\dfrac{Z_0}{Z}\right)\left(\dfrac{T_0}{T}\right)\left(\dfrac{\overline{W}}{\overline{W}_0}\right)\left(\dfrac{1 - Z}{1 - Z_0}\right)^{1/3} \ln\left(1 + \dfrac{T - T_0}{L/c_p} + \dfrac{Q_0}{L} Y_0\right)}, \tag{114}$$

the derivation of which involves use of equations (86), (89), and (110). In equation (114), $\overline{W}$ is related to Y_k by equation (109), and Y_k is related to T through equation (111). The boundary conditions for equation (114) are $T = T_0$ at $Z = Z_0$ and

$$T = Z_0 T_0 + (1 - Z_0)Q'/c_p = T_0 + (1 - Z_0)(Q_F - L)/c_p \equiv T_\infty$$

at $Z = 1$. The last condition is a consequence of equation (105) because, as $Z \to 1$, $x \to \infty$ and $dT/dx \to 0$; this result simply expresses the phenomenological overall energy balance for the adiabatic system. The mass-burning rate m is expected to be an eigenvalue of equation (114).

11.6.6. Solution to the problem

Normalized variables and nondimensional parameters may be identified from equation (114) [73], and an iterative method of solution may be devised [50] and applied [73] to obtain an approximate formula for $m = \rho_0 u_0 = \rho_0 v_0$. This formula exhibits the attributes of the burning velocity for the heterogeneous regime that were discussed in Section 11.6.1. Thus a more formal development leads to the qualitative results that have been inferred from physical reasoning. It is instructive to have seen a representative sequence of approximations involved in a formal development. The iterative method [50] has been applied successfully [77] with fewer approximations, through use of techniques for numerical integration, to derive results that extend beyond the heterogeneous regime. Results of numerical integrations of time-dependent partial differential equations for discrete sets of droplet sizes also have been reported [78].

A relevant question in any formulation of the problem concerns the criterion for initiation of heterogeneous burning. This question presents noticeable difficulties in the formulation that has been given because when the droplets are assumed to begin at their vaporization temperature, diffusion flames already may exist around each droplet; the cold-boundary difficulty (Sections 2.1.2 and 5.3.2) then arises in a severe form because the rate of droplet burning is not sufficiently strongly dependent on the gas temperature for a pseudostationary burning rate m to exist for the deflagration. In the analysis [73], an ignition temperature was introduced to make the problem well-posed; to begin the iterative solution, an expansion of equation (114) about the hot boundary was employed so that in the first iteration (but not in subsequent iterations) m was independent of the ignition temperature. If the droplet temperature initially is below the vaporization temperature, then a natural procedure is to introduce a two-zone structure, a heat-up zone in which the droplet temperature increases to the vaporization temperature without combustion and a droplet-burning zone that is assumed to begin when the vaporization temperature is reached. Although the cold-boundary difficulty then disappears (in a manner analogous to that of the solid-pro-pellant deflagration problem of Section 7-5), this procedure still fails to consider what physics really are involved in initiation of the droplet burning. Quasisteady droplet ignition theory (Section 3.4.4) has been applied as an initiation criterion [78], but this procedure is analogous to use of a thermal-explosion analysis for laminar-flame propagation and is therefore unlikely to describe properly the dominant initiation influences of streamwise molecular transport of species and energy in the deflagration. There is need for further study of physically correct criteria for initiation of heterogeneous burning in spray deflagrations.

11.7. SPRAY PENETRATION AND CLOUD COMBUSTION

In the analysis considered in the preceding section, it was assumed that a uniform spray had been established initially and that once ignited, each droplet burned with an envelope flame around it. These conditions have been achieved reasonably well in the laboratory for various fuel-lean sprays [65]. However, in practical systems the sprays are not uniform, the manner in which the spray penetrates the oxidizing gas is important, and a cloud-burning mode of combustion (in which diffusion flames surround groups of droplets, see the last paragraph of Section 3.3.6) may occur [2], [79]. These realities motivate studies of spray penetration and cloud combustion.

Literature on theories of collective effects of droplet interactions was cited in Section 3.3.6. An approximate criterion for the occurrence of a cloud-burning mode with diffusion flames surrounding droplet clouds is readily stated. The overall fuel-gas ratio ϕv (where v is the stoichiometric mass ratio and ϕ is the equivalence ratio) is $\phi v \approx \frac{4}{3}\pi\langle r\rangle^3 n\rho_l/\rho_f$. From equation (3-63), for an isolated droplet the ratio of the flame radius to the droplet radius is $r_f/r_l = (1 + B)/\ln(1 + v)$. Since the average droplet spacing is approximately $1/n^{1/3}$, the flame radius exceeds the spacing if $\langle r\rangle n^{1/3} \gtrsim \ln(1 + v)/\ln(1 + B)$, which from the previous result can be written as

$$\phi v \gtrsim \tfrac{4}{3}\pi(\rho_l/\rho_f)[\ln(1 + v)/\ln(1 + B)]^3. \tag{115}$$

For typical hydrocarbon sprays in air, $v \approx \frac{1}{20}$, and by using $B \approx 10$ and $\rho_l/\rho_f \approx 10^3$, we find that equation (115) becomes $\phi \gtrsim 0.7$. Although there are a number of inaccuracies here, for example, associated with the use of equation (3-63), it may be concluded that stoichiometric hydrocarbon sprays in air may be expected to experience at least some cloud-burning effects. Analyses of spray deflagrations with droplet-group burning have not been pursued.

Spray penetration is a complicated process that typically involves liquid jet penetration, atomization, and droplet-gas interactions. Steady-flow, cross-stream penetration (for example, for supersonic combustion [7], [80]–[82]) as well as transient penetration (for example, for diesel combustion [4], [8], [49]) are of interest. Rough estimates of penetration distances can be made if the mechanism of penetration is known. For example, since the lifetime of a vaporizing droplet of initial radius r_l with an evaporation constant K is $4\,r_l^2/K$, the evaporation-controlled penetration distance (the distance moved prior to completion of vaporization) of a droplet is $4vr_l^2/K$ if v is the average droplet velocity. From equations (38) and (69) it may be estimated that a momentum-controlled penetration distance (the distance required for a droplet of fixed radius r_l to slow to the gas velocity) is about $(8/3)(\rho_l/\rho_g)(r_l/C_D)$, which is less than the evaporation-controlled distance (and therefore more likely to be controlling) if $(2/3)(\rho_l/\rho_g)(K/C_D) < vr_l$. Rough calculations of this type can help in identifying useful approximation that can be employed in more thorough theories.

REFERENCES

1. A. F. Williams, *Combustion of Sprays of Liquid Fuels*, London: Elek Science, 1976.
2. N. A. Chigier, *Prog. Energy Combust. Sci.* **2**, 97 (1976).
3. A. Williams, *Prog. Energy Combust. Sci.* **2**, 167 (1976).
4. M. M. Elkotb, *Prog. Energy Combust. Sci.* **8**, 61 (1982).

5. G. M. Faeth, *Prog. Energy Combust. Sci.* **9**, 1 (1983).

6. W. A. Sirignano, *Prog. Energy Combust. Sci.* **9**, 21 (1985).

7. F. A. Williams, *Astronautica Acta* **15**, 547 (1970).

8. R. D. Reitz and F. V. Bracco, *Phys. Fluids* **25**, 1730 (1982).

9. A. H. Shapiro and A. J. Erickson, *Trans. Am. Soc. Mech. Eng.* **79**, 775 (1957).

10. F. A. Williams, "General Description of Combustion and Flow Processes," in *Liquid Propellant Rocket Combustion Instability*, D. T. Harrje and F. H. Reardon, eds., NASA SP-194, Washington, D.C.: U.S. Government Printing Office, 1972, 37–45.

11. J. O. Hinze, *Appl. Sci. Research* **1**, 263, 273 (1948).

12. W. R. Lane, *Ind. Eng. Chem.* **43**, 1312 (1951).

13. O. G. Engle, *J. Research, National Bureau of Standards* **60**, 245 (1958).

14. A. R. Hanson, E. G. Domich, and H. S. Adams, *Phys. Fluids* **6**, 1070 (1963).

15. A. A. Ranger and J. A. Nicholls, *AIAA Journal* **7**, 285 (1969).

16. A. A. Ranger and J. A. Nicholls, *Int. J. Heat Mass Transfer* **15**, 1203 (1972).

17. E. Y. Harper, G. W. Grube, and I. D. Chang, *J. Fluid Mech.* **52**, 565 (1972).

18. P. G. Simpkins and E. L. Bales, *J. Fluid Mech.* **55**, 692 (1972).

19. G. D. Waldman, W. G. Reinecke, and D. C. Glen, *AIAA Journal* **10**, 1200 (1972).

20. D. Barthès-Biesel and A. Acrivos, *J. Fluid Mech.* **61**, 1 (1973).

21. G. E. Fox and E. K. Dabora, *14th Symp.* (1973), 1365–1373.

22. B. D. Fishburn, *Acta Astronautica* **1**, 1267 (1974).

23. F. A. Williams, *Phys. Fluids* **1**, 541 (1958).

24. F. A. Williams, *8th Symp.* (1962), 50–69.

25. P. J. O'Rourke and F. V. Bracco, "Modeling of Drop Interactions in Thick Sprays and Comparison with Experiment," in *Proc. Inst. Mech. Engrs., Conference on Stratified Charge Engines*, London (1980), 103–116.

26. R. P. Probert, *Phil Mag.* **37**, 94 (1946).

27. Y. Tanasawa, *Tech. Rept. Tohoku Univ.* **18**, 195 (1954).

28. Y. Tanasawa and T. Tesima, *Bulletin JSME* **1**, 36 (1958).

29. Y. Tanasawa and K. Kobayasi, *Tech. Rept. Tohoku Univ.* **19**, 135 (1955).

30. Y. Tanasawa and S. Toyoda, *Tech. Rept. Tohoku Univ.* **20**, 27 (1955).

31. R. D. Ingebo and H. H. Foster, NACA Tech. Note No. 4087 (1957).

32. R. D. Ingebo, NACA Tech. Note No. 4222 (1958).

33. S. Nukiyama and Y. Tanasawa, *Trans. Soc. Mech. Eng. Japan* **5**, 62 (1939).

34. P. Rosin and E. Rammler, *Zeit. des Vereins Deutscher Ing.* **71**, 1 (1927); *J. Inst. Fuel* **7**, 29 (1933).

35. M. Abramowitz and I. A. Stegun, *Handbook of Mathematical Functions*, New York: Dover, 1965.

36. R. A. Mugele and H. D. Evans, *Ind. Eng. Chem.* **43**, 1317 (1951).

37. E. Jahnke and F. Emde, *Tables of Functions*, New York: Dover, 1945, 22–23.

38. H. Lamb, *Hydrodynamics*, Cambridge: Cambridge University Press, 1924, Sections 337 and 338.

39. S. Goldstein, *Modern Developments in Fluid Dynamics*, Oxford: Oxford University Press, 1950, vol. II, Chapter XI, Section 1.

40. R. D. Ingebo, NACA Tech. Note No. 3762 (1956).

41. C. T. Crowe, J. A. Nicholls, and R. B. Morrison, *9th Symp.* (1963), 395–405.

42. G. M. Faeth, *Prog. Energy Combust. Sci.* **3**, 191 (1977).

43. S. L. Soo, *Fluid Dynamics of Multiphase Systems*, Waltham, Mass.: Blaisdell, 1967.

44. G. B. Wallis, *One-Dimensional Two-Phase Flow*, New York: McGraw-Hill, 1969.

45. D. A. Drew, *Annual Review of Fluid Mechanics* **15**, 261 (1983).
46. F. V. Bracco, *CST* **8**, 69 (1973).
47. C. K. Westbrook, *16th Symp.* (1977), 1517–1526.
48. H. C. Gupta and F. V. Bracco, *AIAA Journal* **16**, 1053 (1978).
49. L. Martinelli, R. D. Reitz, and F. V. Bracco, "Comparison of Computed and Measured Dense Spray Jets," in *Dynamics of Flames and Reactive Systems*, vol. 95 of *Progress in Astronautics and Aeronautics*, J. R. Bowen, N. Manson, A. K. Oppenheim and R. I. Soloukhin, eds., New York: American Institute of Aeronautics and Astronautics, 1984, 484–512.
50. F. A. Williams, *C & F* **3**, 215 (1959).
51. D. B. Spalding, *Aero. Quart.* **10**, 1 (1959).
52. D. B. Spalding, "A One-Dimensional Theory of Liquid Fuel Rocket Combustion," *A.R.C. Tech. Rept. No. 20-175, Current Paper No. 445* (1959).
53. J. Adler, "A One-Dimensional Theory of Liquid Fuel Rocket Combustion. Part II. The influence of Chemical Reaction," *A.R.C. Tech. Rept. No. 20-189, Current Paper No. 446* (1959).
54. F. A. Williams, S. S. Penner, G. Gill, and E. F. Eckel, *C & F* **3**, 355 (1959).
55. R. J. Priem and M. F. Heidmann, *ARS Journal* **29**, 836 (1959).
56. F. V. Bracco, *AIAA Journal* **12**, 1534 (1974).
57. S. Z. Burstein, S. S. Hammer, and V. D. Agosta, "Spray Combustion Model with Droplet Breakup: Analytical and Experimental Results," in *Detonation and Two-Phase Flow*, vol. 6 of *Progress in Astronautics and Rocketry*, S. S. Penner and F. A. Williams, eds., New York: Academic Press, 1962, 243–267.
58. S. Lambiris, L. P. Combs, and R. S. Levine, "Stable Combustion Processes in Liquid Propellant Rocket Engines," Fifth AGARD Combustion and Propulsion Colloquium, Braunschweig, April 1962.
59. F. A. Williams, "A Simplified Model of Liquid Rocket Combustion," *Tech. Rept. No. 32, Contract No. DA-04-495-Ord-1634*, Pasadena, Calif.: California Institute of Technology, 1960.
60. D. T. Harrje and F. H. Reardon, eds., *Liquid Propellant Rocket Combustion Instability*, NASA SP-194, Washington, D.C.: U.S. Government Printing Office, 1972, chapter 2.
61. J. H. Burgoyne and L. Cohen, *Proc. Roy. Soc. London* **225A**, 375 (1954).
62. J. H. Burgoyne, D. M. Newitt, and A. Thomas, *The Engineer* **198**, 165 (1954).
63. J. A. Bolt and T. A. Boyle, *Trans. Am. Soc. Mech. Eng.*, **78**, 609 (1956).
64. J. H. Burgoyne, *Chem. Eng. Progress* **53**, 121 (1957).
65. Y. Mizutani and M. Ogasawara, *Int. J. Heat Mass Transfer* **6**, 921 (1965).
66. Y. Mizutani and T. Nishimoto, *CST* **6**, 1 (1972).
67. Y. Mizutani and A. Nakajima, *C & F* **20**, 343, 351 (1973).
68. C. E. Polymeropoulos and S. Das, *C & F* **25**, 247 (1975).
69. S. Hayashi and S. Kumagai, *15th Symp.* (1975), 445–452; *CST* **15**, 169 (1977).
70. L. D. Smoot, M. D. Horton, and G. A. Williams, *16th Symp.* (1977), 375–387.
71. L. D. Smoot and M. D. Horton, *Prog. Energy Combust. Sci.* **3**, 235 (1977).
72. L. D. Smoot and D. T. Pratt, *Pulverized Coal Combustion and Gasification*, New York: Plenum Press, 1979.
73. F. A. Williams, "Monodisperse Spray Deflagration," in *Liquid Rockets and Propellants*, vol. 2 of *Progress in Astronautics and Rocketry*, L. E. Bollinger, M. Goldsmith, and A. W. Lemmon, Jr., eds., New York: Academic Press, 1960, 229–264.

74. F. V. Bracco, *AIChE Symp. Series* **138**, 70 (1974).
75. Y. Mizutani, *CST* **6**, 11 (1972).
76. P. B. Patil, M. Sichel, and J. A. Nicholls, *CST* **18**, 21 (1978).
77. C. E. Polymeropoulos, *CST* **9**, 197 (1974).
78. B. Seth, S. K. Aggarwal, and W. A. Sirignano, *C & F* **39**, 149 (1980).
79. Y. Onuma and M. Ogasawara, *15th Symp.* (1975), 453–465.
80. I. Catton, D. E. Hill, and R. P. McRae, *AIAA Journal* **6**, 2084 (1968).
81. A. Sherman and J. A. Schetz, *AIAA Journal* **9**, 666 (1971).
82. S. I. Baranovsky and J. A. Schetz, *AIAA Journal* **18**, 625 (1980).

CHAPTER 12

Flame Attachment and Flame Spread

Combustion processes in which flames interact with solid or liquid surfaces are relatively complicated, yet are of importance under many conditions of interest. Thus, for example, in a study of flame quenching, von Kármán and Millán [1] considered the interaction of a premixed flame with a cool wall. Related processes arise in flame holding by bodies in combustion chambers [2] and in fire spread through condensed fuels [3]. Simplifications in problems of this type occur if the boundary-layer approximation [4] is applicable. Phenomena including hypersonic flight in air, flame stabilization by plates and rods in premixed flowing gases, thermal protection of reentry vehicles by ablative heat shields, and the burning of a solid or liquid fuel surface in an oxidizer stream involve chemical reactions in boundary layers. Origins of theoretical investigations of these subjects may be traced in early reviews [5], [6], [7], the last of which is a detailed presentation of work accomplished prior to 1965. Numerous problems involving chemically reacting boundary layers now have been analyzed. There is an extensive literature on boundary-layer theory [4], [8]–[11] and on its application to high-speed flows [12], [13].

Because of the importance of boundary-layer theory to problems related to flame attachment and to flame spread, we shall begin with a presentation of this subject as applied to reacting flows. It will be seen that formulations based on coupling functions (Section 1.3) are particularly useful for reacting boundary layers. A general theoretical framework will be developed (Section 12.1.2), within which a wide variety of reacting boundary-layer problems may be addressed. To illustrate the application of the theory, the burning of a flat plate of fuel (maintained at a specified temperature) in an oxidizing stream, will be considered in some detail (Section 12.2). Because of the initially unmixed character of the flow, the coupling-function formulation is most helpful in this problem. Problems having a premixed character are more complicated in that finite-rate chemistry must be considered and local departures from boundary-layer flow often arise.

In the final two sections, mechanisms of flame stabilization and processes of flame spread will be considered. In the first of these sections, we shall see that the onset of combustion can cause the boundary-layer approximation to fail. Physical aspects of various types of flame stabilization will be reviewed. The discussion of flame spread also will focus on the many different types of physical processes that may be involved. The presentation should serve to emphasize approximate unifying concepts as well as various currently outstanding unknowns.

12.1. THE BOUNDARY-LAYER APPROXIMATION FOR LAMINAR FLOWS WITH CHEMICAL REACTIONS

12.1.1. Derivation of simplified governing equations

Let us consider a steady flow for which a Cartesian coordinate system (x, y, z) can be established such that $+x$ is the principal flow direction and all flow properties are independent of z. In this two-dimensional (x, y) flow, it will further be assumed that except in a layer extending parallel to the principal flow direction, all flow properties vary so slowly that transport effects are negligibly small. For convenience, the viscous, diffusive, and heat-conducting layer will be placed in the vicinity of the plane $y = 0$ (which, for example, may represent a stationary flat plate, or may divide two parallel

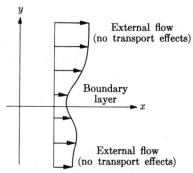

FIGURE 12.1. Schematic illustration of the profile of a representative flow property in a boundary layer.

streams), as illustrated in Figure 12.1. We shall assume that the flow field far from the plane $y = 0$ (which will be termed the external flow field) is known and shall focus our attention on the flow in the vicinity of $y = 0$.

The primary approximation of boundary-layer theory is the assumption that the layer is thin, in the sense that the characteristic distance δ in the y direction over which properties change appreciably (the boundary-layer thickness) is small compared with the characteristic length l over which properties change appreciably in the x direction. This approximation is discussed in detail for nonreacting flows in [4] (for example). In order to present the corresponding approximation for reacting systems, we shall initially adopt all the assumptions of the coupling-function formulation given in Section 1.3. In our present (two-dimensional, steady) problem, the continuity equation then becomes [see equation (1-34)]

$$\partial(\rho u)/\partial x + \partial(\rho v)/\partial y = 0. \tag{1}$$

The equations for the conservation of momentum in the x and y directions become [see equation (1-2) and (1-5) with $\partial/\partial z = 0$, $\partial/\partial t = 0$, $\mathbf{f}_i = 0$, $\kappa = 0$, and $p = $ constant]

$$\underbrace{u\frac{\partial u}{\partial x} + v\frac{\partial u}{\partial y}}_{u^2/l}$$

$$= \frac{1}{\rho}\left\{ -\underbrace{\frac{\partial}{\partial x}\left[\frac{2}{3}\mu\left(\frac{\partial u}{\partial x} + \frac{\partial v}{\partial y}\right)\right] + \frac{\partial}{\partial x}\left(2\mu\frac{\partial u}{\partial x}\right)}_{(\rho l u/\mu)^{-1}(u^2/l)} + \underbrace{\frac{\partial}{\partial y}\left[\mu\left(\frac{\partial v}{\partial x} + \frac{\partial u}{\partial y}\right)\right]}_{(\rho l u/\mu)^{-1}(u^2 l/\delta^2)}\right\} \tag{2}$$

and

$$\underbrace{u\frac{\partial v}{\partial x} + v\frac{\partial v}{\partial y}}_{u^2\delta/l^2}$$

$$= \frac{1}{\rho}\left\{ -\underbrace{\frac{\partial}{\partial y}\left[\frac{2}{3}\mu\left(\frac{\partial u}{\partial x} + \frac{\partial v}{\partial y}\right)\right] + \frac{\partial}{\partial y}\left(2\mu\frac{\partial v}{\partial y}\right)}_{(\rho l u/\mu)^{-1}(u^2/\delta)} + \underbrace{\frac{\partial}{\partial x}\left[\mu\left(\frac{\partial u}{\partial y} + \frac{\partial v}{\partial x}\right)\right]}_{(\rho l u/\mu)^{-1}(u^2\delta/l^2)}\right\}, \tag{3}$$

respectively, and the energy and species conservation equations may be expressed in the form [see equations (1-44) and (1-45)]

$$L(\alpha) = \underbrace{\frac{\partial}{\partial x}(\rho u\alpha) + \frac{\partial}{\partial y}(\rho v\alpha)}_{(\rho u\alpha/l)} - \underbrace{\frac{\partial}{\partial x}\left(\rho D\frac{\partial \alpha}{\partial x}\right)}_{(lu/D)^{-1}(\rho u\alpha/l)} - \underbrace{\frac{\partial}{\partial y}\left(\rho D\frac{\partial \alpha}{\partial y}\right)}_{(lu/D)^{-1}(\rho u\alpha/\delta^2)} = \omega, \tag{4}$$

with ω given by equation (1-43) and α given by equation (1-46) or (1-47). Here u and v are the components of the velocity $\mathbf{v}$ in the x and y directions, respectively.

The boundary-layer approximation states that the orders of magnitude of the gradients are

$$\partial u/\partial x \sim u/l \tag{5}$$

and

$$\partial u/\partial y \sim u/\delta, \tag{6}$$

with $\delta/l \ll 1$. In view of equation (5), equation (1) implies that

$$\partial v/\partial y \sim u/l, \tag{7}$$

the integral of which shows that

$$v \sim u\delta/l. \tag{8}$$

Equations (5)–(8) may easily be used to obtain the estimates of the orders of magnitude of the terms in equations (2)–(4) indicated beneath these equations. Since $\delta/l \ll 1$, the indicated orders of magnitude imply that the last term in equation (2) is large compared with the other terms on the right-hand side, the last term in equation (3) is small compared with the other terms on the right-hand side, and the third term in $L(\alpha)$ [equation (4)] is small compared with the fourth (last) term. Equation (2) then shows that the dominant (last) viscous term will be of the same order as the convective terms appearing on the left-hand side only if the Reynolds number,

$$\mathrm{Re} \equiv \rho l u/\mu, \tag{9}$$

is sufficiently large, namely,

$$\mathrm{Re} \sim (l/\delta)^2. \tag{10}$$

With this assumption, all remaining terms in equation (3) become small (of order δ/l) compared with the remaining terms in equation (2), and the transverse momentum equation [equation (3)] can be omitted. Equations (2) and (4) reduce to

$$u\frac{\partial u}{\partial x} + v\frac{\partial u}{\partial y} = \frac{1}{\rho}\frac{\partial}{\partial y}\left(\mu\frac{\partial u}{\partial y}\right) \tag{11}$$

and

$$L(\alpha) = \frac{\partial}{\partial x}(\rho u\alpha) + \frac{\partial}{\partial y}(\rho v\alpha) - \frac{\partial}{\partial y}\left(\rho D\frac{\partial\alpha}{\partial y}\right) = \omega, \tag{12}$$

respectively. The diffusive term in equation (12) is of the same order as the convective terms when $lu/D \sim (l/\delta)^2$, that is, when the Schmidt number is $\mathrm{Sc} \equiv \mu/(\rho D) \sim 1$ [see equations (9) and (10)], as is generally true for gases (see Appendix E).

In some problems (as emphasized in Chapter 3), known boundary conditions enable us to obtain the desired properties of the solution (such as overall burning rates) by considering only the coupling function β (defined in Section 1.3) instead of α. In such cases, equation (1-49) indicates that the nonhomogeneous rate term ω in equation (12) does not have to be considered. In view of equation (1), the operator $L(\alpha)$ given in equation (12) can be written as

$$L(\alpha) = \rho u \frac{\partial \alpha}{\partial x} + \rho v \frac{\partial \alpha}{\partial y} - \frac{\partial}{\partial y}\left(\rho D \frac{\partial \alpha}{\partial y}\right). \tag{13}$$

Equation (13) serves to emphasize the similarity in the forms of equations (11) and (12). In particular, if the Schmidt number is unity, equations (1-49) and (11) can both be written as

$$L(\beta) = \rho u \frac{\partial \beta}{\partial x} + \rho v \frac{\partial \beta}{\partial y} - \frac{\partial}{\partial y}\left(\mu \frac{\partial \beta}{\partial y}\right) = 0, \tag{14}$$

with $\beta = \beta_T \equiv \alpha_T - \alpha_1$, $\beta = \beta_i \equiv \alpha_i - \alpha_1$, and $\beta = \beta_u \equiv u$. Thus, the differential equation for the coupling functions for temperature and composition is identical with the differential equation for the x component of velocity in boundary-layer flows when Sc $= 1$. Since the coupling-function theory is based on the premise that the Lewis number Le is unity, the simplification indicated in equation (14) holds only if the Prandtl number, Pr $=$ Sc/Le, is unity as well.

12.1.2. Generalizations

The special geometry considered in Section 12.1.1, as well as the assumptions involved in the coupling-function theory, restrict the range of applicability of equations (1), (11), and (12). In this section we shall indicate how equations (1), (11), and (12) must be modified when various assumptions are removed. For this purpose, we shall consider the time-dependent, axisymmetric boundary layer on a body of radius $R(x)$. Here x will represent the streamwise coordinate, measured along the surface of the body, and y will denote the coordinate normal to the boundary layer (normal to x) with $y = 0$ at the body surface. Since separation of the boundary layer from the surface of the body is not considered, the curvature, $-d^2R/dx^2$, cannot be too large. Allowance will be made for a spatially uniform, accelerating or decelerating motion of the coordinate system in the $-x$ direction at a velocity $U(t)$, measured in an arbitrary inertial frame. If $U(t)$ is not a constant, then $R(x)$ must be practically constant if the formulation to be developed is to be applicable. Accelerations in the y direction through body forces such as gravity or through motion, deformation, or strong curvature of the body are assumed to be negligible. The formulation will exclude consideration of axisymmetric deformable bodies; of cross-flow effects that may arise, for example, from body spin and that produce three-dimensional boundary layers; of three-dimensional or axisymmetric wakes, jets, and stagnation points; and of strong normal accelerations that may give rise, for example, to buoyantly produced boundary layers on horizontal heated plates in gravity fields. Although different but related formulations will be needed for the problems just enumerated, the approach to be given encompasses a wide variety of time-dependent, axisymmetric, and planar boundary-layer flows. Since the methods employed in deriving these generalized boundary-layer equations are just like those introduced in the preceding subsection, a comparatively abbreviated development will be presented here.

Assuming that $R(x)$ is large compared with the boundary-layer thickness δ, we may express equation (1-1) as

$$\frac{\partial(R\rho)}{\partial t} + \frac{\partial(R\rho u)}{\partial x} + \frac{\partial(R\rho v)}{\partial y} = 0. \tag{15}$$

The operator $L(\alpha)$ is now defined as

$$L(\alpha) = R\rho\frac{\partial\alpha}{\partial t} + R\rho u\frac{\partial\alpha}{\partial x} + R\rho v\frac{\partial\alpha}{\partial y} - \frac{\partial}{\partial y}\left(R\mu\frac{\partial\alpha}{\partial y}\right). \tag{16}$$

By use of equation (1-5) we may then write the x component of equation (1-2) in the boundary-layer approximation as

$$L(u) = -R\frac{\partial p}{\partial x} + R\rho\frac{dU}{dt} + R\rho f_x, \tag{17}$$

where f_x is the x component of $\sum_{i=1}^{N} Y_i\mathbf{f}_i$, the body force per unit mass. The term involving dU/dt in equation (17) arises from a transformation to the noninertial frame of reference that is fixed to the body; dU/dt thus represents an apparent body force per unit mass. Of the two coefficients of viscosity that appear in equation (1-5), only the shear viscosity survives in equation (17) because terms involving bulk viscosity are at most of order δ/l compared with the terms retained. The y component of equation (1-2), with use made of equation (1-5), shows that the fractional change in pressure across the boundary layer is at most of order $(\delta/l)^2$ and that, in equation (17), p may therefore be treated as being a function only of x and t; the external pressure gradient is impressed on the boundary layer.

The appropriate form of equation (1-3), energy conservation, is derived most easily from the alternative form given in equation (3-72). By use of equations (1-5), (1-6) and (1-11), we then find that

$$L\left(h + \frac{1}{2}u^2\right) = R\frac{\partial p}{\partial t} + R\rho u\left(\frac{dU}{dt} + f_x\right) - \frac{\partial(Rq_R)}{\partial y}$$

$$- \frac{\partial}{\partial y}\left[R(\mu c_p - \lambda)\frac{\partial T}{\partial y} + R\sum_{i=1}^{N} h_i\left(\rho Y_i V_i + \mu\frac{\partial Y_i}{\partial y}\right)\right], \tag{18}$$

where q_R denotes the y component of the radiant heat-flux vector $\mathbf{q}_R$ and V_i represents the y component of the diffusion velocity $\mathbf{V}_i$. The Dufour effect in equation (1-6), which is usually negligible, has been omitted here for the sake of brevity but could readily be added if desired. The reasoning showing that only the y components of $\mathbf{q}_R$ and of $\mathbf{V}_i$ survive in the boundary-layer approximation parallels that which led to the last term on the left-hand side of equation (12) and employs the general diffusion equation, equation (1-7). Diffusion velocities in equation (3-72) contribute negligibly to the work term, $R\rho u(dU/dt + f_x)$, in equation (18) because $|V_i| \ll u$ according to the boundary-layer estimates.

The generalized boundary-layer form of the equation for species conservation, equation (1-4), is

$$L(Y_i) = Rw_i - \frac{\partial}{\partial y}\left[R\left(\rho Y_i V_i + \mu \frac{\partial Y_i}{\partial y}\right)\right], \quad i = 1, \ldots, N. \tag{19}$$

To find V_i in equations (18) and (19) we must use the boundary-layer approximation to the y component of equation (1-7). Since the body-force and pressure-gradient terms are negligible in the boundary-layer approximation for the y component, the diffusion equation becomes

$$\frac{\partial X_i}{\partial y} = \sum_{j=1}^{N} \frac{X_i X_j}{D_{ij}}(V_j - V_i) + \frac{\kappa_i}{T}\frac{\partial T}{\partial y}, \quad i = 1, \ldots, N, \tag{20}$$

where

$$\kappa_i \equiv \sum_{j=1}^{N} \left(\frac{X_i X_j}{\rho D_{ij}}\right)\left(\frac{D_{T,j}}{Y_j} - \frac{D_{T,i}}{Y_i}\right), \quad i = 1, \ldots, N, \tag{21}$$

which accounts for thermal diffusion (the Soret effect). Equations (15)–(20) comprise $2N + 3$ equations in the $2N + 3$ unknowns v, u, T, Y_i, and V_i. Equations (1-9), (1-11) with $h = \sum_{i=1}^{N} Y_i h_i$, and (1-12) relate ρ, h and X_i, which appear in these equations, to the dependent variables identified here.

In equations (15)–(19), the function $R(x)$ has variably been placed inside or outside of t and y derivatives to make it clear that the Mangler transformation [14], $R\rho \to \rho'$, $R\mu \to \mu'$, $R\lambda \to \lambda'$, $Rp \to p'$, $Rq_R \to q_R'$, $Rw_i \to w_i'$, and $R\,\partial p/\partial x \to (\partial p/\partial x)'$, reduces the axisymmetric equations to equations for planar boundary layers in the new variables. The indicated transformation concerning the pressure gradient means that the axisymmetric flow for a given pressure gradient corresponds to a two-dimensional flow under a different pressure gradient. If the subscript ∞ identifies external-flow variables just outside the boundary layer, then from equations (15), (16), and (17) we find that

$$\partial(R\rho_\infty)/\partial t + \partial(R\rho_\infty u_\infty)/\partial x = 0 \tag{22}$$

and

$$\rho_\infty\,\partial u_\infty/\partial t + \rho_\infty u_\infty\,\partial u_\infty/\partial x = -\partial p/\partial x + \rho_\infty(dU/dt + f_x), \tag{23}$$

which together relate $\partial p/\partial x$ to u_∞ and its derivatives in the external stream; thus a modification of the pressure gradient implies a modification of the external velocity and, therefore, of the boundary conditions for equation (17).

Following the Mangler transformation, the Howarth-Dorodnitzyn transformation [15], [16]—in which the y coordinate is replaced by

$$z = \int_0^{y} R\rho \, dy = \int_0^{y} \rho' \, dy, \tag{24}$$

a mass coordinate—serves to reduce the boundary-layer equations for variable-density flow to corresponding equations for constant-density flow, with $\rho' v + u \int_0^{y} (\partial \rho'/\partial x)_{y,t} \, dy$ playing the part of the transverse velocity, $\rho' \mu'$, the kinematic viscosity, $\rho' \lambda'/c_p$, the thermal diffusivity, and w_i'/ρ', the reaction-rate function in the equivalent incompressible flow having a density of unity. For time-dependent boundary layers, the Moore-Stewartson transformation [17], [18] combines the Howarth-Dorodnitzyn transformation with the introduction of a stream function, ψ, to achieve further simplifications in the formulation. Specifically, the substitutions

$$u = \left(\frac{\partial \psi}{\partial z}\right)_{x,t}, \qquad R\rho v = -\left(\frac{\partial \psi}{\partial x}\right)_{z,t} - u\left(\frac{\partial z}{\partial x}\right)_{y,t} - \left(\frac{\partial z}{\partial t}\right)_{x,y} + V(x, t) \tag{25}$$

are made in the boundary-layer equations, where for clarity and conciseness a notation like that used in thermodynamics has been adopted, in which the variables held constant in differentiations are indicated by subscripts. Substitution verifies that equation (25) automatically satisfies equation (15), so that the continuity equation no longer needs to be included in the governing set when equation (25) is employed with an arbitrary function $V(x, t)$.

Many boundary-layer problems possess similarity solutions—solutions in which appropriately identified dependent variables may be expressed as functions of a single independent variable, η, the nondimensional similarity variable, through functions defined by ordinary differential equations in η. To facilitate derivations of similarity solutions, after employing the Moore-Stewartson transformation, we shall further transform the equations to a general similarity variable and introduce a nondimensional stream function $F(x, \eta, t)$, according to

$$\eta = z/g(x, t), \qquad F = \psi/[g(x, t)u_\infty(x, t)], \tag{26}$$

where $g(x, t)$ is an arbitrary function to be specified later for achieving desired simplifications in particular problems. This will enable us to present a formulation suitable for analysis of many boundary-layer problems.

The complete transformation of equations (15)–(20) to be introduced here is one from the independent variables (x, y, t) to the independent variables (x, η, t), where $\eta = [\int_0^{y} R\rho \, dy]/g$, so that

$$y = g \int_0^{\eta} (R\rho)^{-1} \, d\eta, \tag{27}$$

according to equations (24) and (26). The dependent variable F of equation (26) replaces u, where

$$u = u_\infty F' \tag{28}$$

from equations (25) and (26). Here and below, the prime denotes a partial derivative with respect to η, with x and t held fixed; the thermodynamic notation no longer will be employed since x, η, and t will be the independent variables everywhere. From the expression for $R\rho v$ in equation (25) it can be shown that

$$R\rho v = -\frac{\partial}{\partial x}(gu_\infty F) + R\rho u_\infty F'\frac{\partial}{\partial x}\left[g\int_0^\eta (R\rho)^{-1}\,d\eta\right] - \eta\frac{\partial g}{\partial t}$$

$$+ R\rho\frac{\partial}{\partial t}\left[g\int_0^\eta (R\rho)^{-1}\,d\eta\right] + V, \tag{29}$$

which serves to determine v and thus replaces equation (15). To allow for variations in the product $\rho\mu$, it is conventional to define a nondimensional function $C(x, \eta, t)$ as

$$C = (\rho\mu)/(\rho_\infty\mu_\infty). \tag{30}$$

For brevity, it is also useful to introduce four nondimensional functions of x and t, defined as

$$\left.\begin{aligned}G_1 &= \frac{g}{R^2\rho_\infty\mu_\infty}\frac{\partial(gu_\infty)}{\partial x}, & G_2 &= \frac{g}{R^2\rho_\infty\mu_\infty}\frac{\partial g}{\partial t}, \\[2mm] G_3 &= \frac{g^2}{R^2\rho_\infty\mu_\infty}\frac{\partial u_\infty}{\partial x}, & G_4 &= \frac{g^2}{R^2\rho_\infty\mu_\infty u_\infty}\frac{\partial u_\infty}{\partial t},\end{aligned}\right\} \tag{31}$$

and a length function, $l(x, t)$, given by

$$l = (g^2 u_\infty)/(R^2\rho_\infty\mu_\infty). \tag{32}$$

A nondimensional operator replacing $L(\alpha)$ is then found from equation (16) with $\eta_0 \equiv V/(\partial g/\partial t)$, to be

$$\mathscr{L}(\alpha) \equiv \frac{-g^2}{R^3\rho\rho_\infty\mu_\infty}L(\alpha) = (C\alpha')' + G_1 F\alpha' + (\eta - \eta_0)\alpha^1$$

$$+ l\left[\left(\frac{\partial F}{\partial x}\right)\alpha' - F'\left(\frac{\partial\alpha}{\partial x}\right) - \frac{1}{u_\infty}\left(\frac{\partial\alpha}{\partial t}\right)\right]. \tag{33}$$

The remaining boundary-layer equations, equations (17)–(20), may then readily be expressed in the new variables in terms of the operator $\mathscr{L}(\alpha)$.

The conservation of streamwise momentum, equation (17), is the only equation that involves the functions G_3 and G_4; it may be written as

$$\mathscr{L}(F') = G_3 F'^2 + G_4 F' + \frac{l}{u_\infty^2}\left[\frac{1}{\rho}\left(\frac{\partial p}{\partial x}\right) - \left(\frac{dU}{dt} + f_x\right)\right]. \tag{34}$$

Energy conservation, equation (18), may be expressed as

$$\mathscr{L}\left(h + \frac{1}{2}u^2\right) = -\frac{l}{u_\infty}\left[\frac{1}{\rho}\frac{\partial p}{\partial t} + u_\infty F'\left(\frac{dU}{dt} + f_x\right) - \frac{1}{\rho}\frac{\partial q_R}{\partial y}\right]$$
$$+ \left\{C\left[\left(1 - \frac{\lambda}{\mu c_p}\right)c_p T' + \sum_{i=1}^{N} h_i\left(\frac{gY_i V_i}{R\mu} + Y_i'\right)\right]\right\}'. \quad (35)$$

Species conservation, equation (19), becomes

$$\mathscr{L}(Y_i) = -\frac{l}{u_\infty}\left[\frac{w_i}{\rho}\right] + \left\{C\left(\frac{gY_i V_i}{R\mu} + Y_i'\right)\right\}', \quad i = 1, \ldots, N. \quad (36)$$

The diffusion velocities in equations (35) and (36) are to be obtained from equation (20) in the new variables:

$$X_i' = \sum_{j=1}^{N} \frac{gX_i X_j}{R\rho D_{ij}}(V_j - V_i) + \frac{\kappa_i}{T}T', \quad i = 1, \ldots, N. \quad (37)$$

It is seen that $R\mu/g$ arises naturally as a velocity for nondimensionalizing the diffusion velocities in equations (35) and (36) and that if this non-dimensionalization is introduced, then Schmidt numbers, $\mu/(\rho D_{ij})$, appear in equation (37); the Prandtl number, $\mu c_p/\lambda$, appears in equation (35). If the Prandtl and Schmidt numbers are unity and $\kappa_i = 0$, then the terms inside the braces in equations (35) and (36) vanish. Similarity solutions may exist even if these terms do not vanish. The forcing terms in equations (34), (35) and (36) (those multiplied by l and appearing inside brackets) or nonsimilar boundary or initial conditions are the primary causes of departures from similarity.

Various different types of boundary-layer flows may be addressed by selecting g so that one of the four functions defined in equation (31) is a specified, nonzero constant. In many problems, a number of the functions in equation (31) vanish; for steady flows we put $G_2 = G_4 = \eta_0 = 0$, while for time-dependent boundary layers that do not have an x dependence $G_1 = G_3 = 0$. Steady flows with a constant, nonzero external velocity u_∞ have $G_2 = G_3 = G_4 = 0$ [and $R\rho_\infty$ constant, from equation (22)] and may be described by setting $G_1 = 1$, so that (when the singularity in η is placed at $x = 0$)

$$g = \sqrt{2\int_0^x (R^2\rho_\infty \mu_\infty/u_\infty)\,dx}; \quad (38)$$

the similarity form of equation (34) without forcing terms, for example, then becomes

$$(CF'')' + FF'' = 0, \quad (39)$$

which describes self-similar flows in mixing layers or over bodies with a vanishing external pressure gradient.

Planar boundary layers properly correspond to the limit $R \to \infty$ but also may be obtained more simply by just putting $R = 1$, that is, by ignoring R throughout. Steady, planar jet flows may be described by a similarity solution $F(\eta)$ by putting $G_1 = -G_3 = 1$ and interpreting the variable u_∞, appearing in equation (26) and in subsequent relations, as the velocity at the plane of symmetry.* Steady, planar wakes require a more general assumption regarding F, namely, $F = f_1(x)f(\eta) - f_2(x)\eta$. Steady, planar viscous flows in converging or diverging channels have nonvanishing pressure gradients and variable centerline velocities u_∞; equations for the corresponding self-similar solutions $F(\eta)$ may be obtained by letting $G_3 = \pm 1$ with $G_1 = G_2 = G_4 = 0$. By selecting appropriate constant values of G_1 and G_3 with $G_2 = G_4 = 0$, all the Falkner-Skan boundary-layer flows with pressure gradients [9] are recovered. Unsteady self-similar flows over infinite plates (with $\partial/\partial x = 0$) for motions impulsively started from rest at $t = 0$ and having u_∞ constant thereafter are described by setting $G_2 = 1$ with $G_1 = G_3 = G_4 = 0$ to develop an ordinary differential equation for $F(\eta)$. There are self-similar solutions to time-dependent problems for which u_∞ grows or decreases exponentially with time, obtained by putting $G_4 = \pm 1$ with $G_1 = G_2 = G_3 = 0$. Time-dependent flows of more general characters may be considered by selecting different constant values for G_2 and G_4, and certain flows with both x and t dependences correspond to nonzero constant values of three or all four of the functions defined in equation (31). The formulation is a convenient starting point for investigating small departures from self-similar flows encountered for example, by imposing a slow time dependence on steady boundary layers or a weak x dependence on unsteady self-similar boundary layers. A representative application to combustion in oscillatory boundary layers on semi-infinite flat plates may be found in [19].

12.2. COMBUSTION OF A FUEL PLATE IN AN OXIDIZING STREAM

12.2.1. Definition of the problem

As an example of the use of the preceding equations, let us consider the problem illustrated in Figure 12.2. A stationary semi-infinite flat plate consisting of solid or liquid fuel is placed parallel to a gaseous oxidizing stream with a uniform approach velocity u_∞. Gaseous fuel emerges from the plate by sublimation or vaporization and reacts with the oxidizer in the boundary layer. The products of the reaction are diffused and convected away. Our primary objective is to determine the burning rate of the fuel,

---

* Equations (22) and (23) then do not apply to this u_∞; the external velocity, the pressure gradient, and body forces vanish.

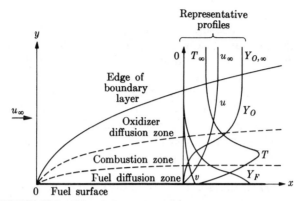

FIGURE 12.2. Schematic diagram of the combustion of a fuel plate.

$(\rho v)_0$ (mass per unit area per second leaving the surface). The similarity solution to this problem was developed by Emmons [20].

In our analysis, following [20], we shall adopt all the assumptions of Section 12.1.1 and shall employ certain of the transformations of Section 12.1.2. Equation (1-42) now represents the overall combustion reaction. Species 1 will be taken to be a principal oxidizing component which will be identified by the subscript O. Just as in the droplet-burning problem of Section 3.3, the temperature-oxidizer coupling function β_T will be found to be useful for obtaining the burning rate.

12.2.2. Boundary conditions

Boundary conditions at $y = \infty$ (where properties are identified by the subscript ∞) are $\alpha_O = \alpha_{O,\infty} = $ constant (a measure of the known oxidizer concentration in the free stream), $\alpha_T = \alpha_{T,\infty} = $ constant (a measure of the known temperature of the free stream), and $u = u_\infty = $ constant (the known velocity of the free stream). The mass fraction of the fuel, Y_F, must approach zero as $y \to \infty$.

The boundary conditions that will be adopted at $y = 0$ (where conditions will be identified by the subscript 0) are $\alpha_O = 0$, $\alpha_T = \alpha_{T,0} = $ constant (a measure of the sublimation or boiling temperature), and $u = 0$. The first of these conditions would follow from a flame-sheet hypothesis and is accurate for many liquid fuels (see Section 3.3.4 or [21]). The validity of the second condition is discussed in Section 3.3.4; this condition will be most accurate for a volatile liquid fuel. The third condition is rigorously true only for a solid fuel but is an excellent approximation for liquid fuels if longitudinal flow of the liquid is suppressed (for example, by extruding the liquid through a porous solid material).*

---

* When $u_0 \neq 0$, a two-phase boundary layer with liquid flow must be considered. A number of analyses of problems of this type have been reported [22]-[28].

The final boundary condition to be employed at $y = 0$ is the requirement that enough heat must be conducted into the surface to vaporize the fuel leaving the surface and to supply the heat lost from the surface (by conduction to the interior of the condensed phase and/or by radiation to the surroundings). This condition, which is physically evident, can be derived formally from the equations of Section 1.4 [primarily, equation (1-61)]. One can also show that the other interface conditions given in Section 1.4 are trivially satisfied and need not be considered in determining the burning rate. A formula expressing the interface energy conservation condition stated above is

$$(\lambda \, \partial T/\partial y)_0 = (\rho v)_0 (L + Q),$$

where λ is the thermal conductivity of the gas, L is the heat of gasification per unit mass for the fuel, and Q is the heat lost from the surface per unit mass of fuel gasified. For convenience, we shall set the standard reference temperature equal to the surface temperature ($T^0 \equiv T_0$), whence equation (1-46) and our assumption that $\mathrm{Pr} = 1$ imply that the preceding expression can be written in the form

$$\mu_0 (\partial \alpha_T/\partial y)_0 = (\rho v)_0 q, \tag{40}$$

where

$$q \equiv (L + Q) \bigg/ \left. \sum_{i=1}^{N} h_i^0 W_i (v_i' - v_i'') \right. \tag{41}$$

is the ratio of the heat conducted into the surface to the heat liberated in the reaction. Since our hypothesis $T_0 = $ constant usually implies that L and Q are constant, we shall assume that q is independent of x ($q = $ constant).

The boundary conditions for α_T and α_O developed above can be translated into boundary conditions on the coupling function $\beta_T = \alpha_T - \alpha_O$; namely,

$$\beta_{T,\infty} = \alpha_{T,\infty} - \alpha_{O,\infty} = \text{constant} \tag{42}$$

at $y = \infty$,

$$\beta_{T,0} = \alpha_{T,0} = \text{constant} \tag{43}$$

at $y = 0$, and

$$\mu_0 (\partial \beta_T/\partial y)_0 = (\rho v)_0 q, \tag{44}$$

at $y = 0$. Equation (44) relies on the assumption (valid when $\alpha_{O,0} = 0$ is a reasonable approximation) that $(\partial \alpha_O/\partial y)_0$ is negligible in comparison with the value of $(\partial \alpha_T/\partial y)_0$ given in equation (40). The boundary conditions on $u(= \beta_u)$ can be restated as

$$\beta_{u,\infty} = u_\infty = \text{constant} \tag{45}$$

at $y = \infty$ and

$$\beta_{u,0} = 0 = \text{constant} \tag{46}$$

at $y = 0$. Equations (42)–(46) comprise a sufficient number of boundary conditions to determine the burning rate from equations (1), (11), and (14) (for β_T) alone; that is, we need not determine explicit composition or temperature profiles, and equation (12) can be ignored.

12.2.3. Solution

Since β_T and β_u both obey the same linear homogeneous equation, equation (14), a particular (but not general) solution for β_T in terms of β_u is $\beta_T = A\beta_u + B$, where A and B are constants.* Equations (42)–(46) show that a solution of this type does, in fact, satisfy the appropriate boundary conditions. This solution is

$$\beta_T = \beta_{T,0} + (\beta_u - \beta_{u,0})(\beta_{T,\infty} - \beta_{T,0})/(\beta_{u,\infty} - \beta_{u,0}),$$

which can be written as

$$\beta_T = \alpha_{T,0} + (u/u_\infty)(\alpha_{T,\infty} - \alpha_{T,0} - \alpha_{0,\infty}) \tag{47}$$

[see equations (42), (43), (45), and (46)].

Since equation (47) expresses β_T in terms of u, only equations (1) and (11) remain to be solved. The boundary conditions for these equations are [see equations (44)–(46)],

$$u = u_\infty = \text{constant} \tag{48}$$

at $y = \infty$,

$$u = 0 \tag{49}$$

at $y = 0$, and

$$\mu_0(\partial u/\partial y)_0 = (\rho v)_0 u_\infty q(\alpha_{T,\infty} - \alpha_{T,0} - \alpha_{0,\infty})^{-1} \tag{50}$$

at $y = 0$. Equation (50) is obtained by substituting equation (47) into equation (44). The problem described by equations (1) and (11) with these boundary conditions is formally identical to the problem of a nonreacting boundary layer on a flat plate with gas blowing through the plate at a given velocity v_0. Equation (50) serves to determine the blowing rate (or burning rate) from the solution to the blowing problem.

The nonreacting problem defined here possesses a similarity solution. To find this solution, we may employ the transformations of Section 12.1.2

---

* Aerodynamicists often refer to equations of this type as *Crocco relations* because L. Crocco, in an early paper, derived and employed a linear expression of this type relating the enthalpy to the velocity in nonreacting flows.

with $G_1 = 1$ and $G_2 = G_3 = G_4$ in equation (31). Since the problem is planar, R is ignored throughout. Equation (38) may be expressed as $g = \sqrt{2\xi/u_\infty}$, where

$$\xi = \int_0^x \rho_\infty \mu_\infty \, dx. \tag{51}$$

Since the forcing terms in equation (34) all vanish in this problem, we obtain equation (39), in which for simplicity we shall introduce the further assumption that $C = 1$—that is, $\rho\mu = \rho_\infty \mu_\infty$ [see equation (30)]. This assumption (that $\rho\mu$ does not vary across the boundary layer) often is reasonable for gases; if changes in the average molecular weight are negligible, then—because of the constancy of the pressure—the ideal-gas law implies that $\rho \sim 1/T$, in which case constancy of $\rho\mu$ corresponds to $\mu \sim T$, a dependence close to the kinetic-theory predictions discussed in Appendix E. With $C = 1$, equation (39) is the Blasius equation [4], $F''' + FF'' = 0$, and in view of equation (28), the boundary conditions implied by equations (48) and (49) are $F'(\infty) = 1$ and $F'(0) = 0$. Use may be made of the present formula for g, $C = 1$, $F'(0) = 0$, and equations (27) and (29) to ascertain the boundary condition implied by equation (50); the calculation results in

$$F''(0) = -F(0)q(\alpha_{T,\infty} - \alpha_{T,0} - \alpha_{O,\infty})^{-1}, \tag{52}$$

which states that $F''(0)/[-F(0)]$ is a positive quantity whose value is known from the boundary conditions and the energetics. We thus obtain a non-linear, ordinary differential equation of the third order with three boundary conditions, suitable for numerical solution. Only one parameter occurs in the problem: $F''(0)/[-F(0)]$. Emmons [20] has computed extensive numerical solutions for the boundary conditions given here and has also presented a number of approximate analytical formulas that correlate his numerical results.

With $F(\eta)$ known, we may readily calculate u from equation (28) and β_T from equation (47). Obtaining profiles in physical coordinates involves first evaluating $\rho(\eta)$—for example, from β_T and the ideal-gas law by introduction of a flame-sheet approximation—and then performing the integration in equation (27) for y. Profiles of the normal mass flux, ρv, are given by equation (29), which—for the present problem—reduces to

$$\rho v = \rho_\infty \mu_\infty (2\xi/u_\infty)^{-1/2} \left(F'\rho \int_0^\eta \rho^{-1} \, d\eta - F \right). \tag{53}$$

The mass burning rate at the fuel surface is readily seen from equations (51) and (53) to be

$$\rho_0 v_0 = \rho_\infty u_\infty \left[(2u_\infty/\mu_\infty^2) \int_0^x \rho_\infty \mu_\infty \, dx \right]^{-1/2} [-F(0)]. \tag{54}$$

If we choose to specify $-F(0)$ instead of $F''(0)/[-F(0)]$, then the problem becomes entirely equivalent to one of a Blasius boundary layer with a prescribed, self-similar distribution of blowing along the surface of the flat plate.

12.2.4. The burning rate

Equation (54) gives the local burning rate in physical variables. The ratio $(\rho_0 v_0)/(\rho_\infty u_\infty)$ is a local dimensionless burning rate or mass-transfer coefficient. An average dimensionless burning rate for a plate of length l can be defined as

$$C_B \equiv (\mu_\infty/u_\infty) \int_0^{\xi_l} (v_0/\mu_0)\, d\xi \bigg/ \int_0^{\xi_l} d\xi = [(2\mu_\infty^2)/(u_\infty \xi_l)]^{1/2}[-F(0)], \quad (55)$$

where μ_∞ is evaluated at $x = l$, $\xi_l \equiv \int_0^l (\rho\mu)\, dx$, and equation (54) has been employed in obtaining the last equality. If the product $\rho\mu$ is independent of x (as well as of y), then C_B reduces to the conventional average mass-transfer coefficient; that is, equation (55) becomes

$$C_B \equiv \int_0^l (\rho_0 v_0)\, dx/(\rho_\infty u_\infty l) = (2/\mathrm{Re}_\infty)^{1/2}[-F(0)], \quad (56)$$

where $\mathrm{Re}_\infty \equiv \rho_\infty u_\infty l/\mu_\infty$ is the Reynolds number based on properties at infinity for a plate of length l.

A graph of $C_B(u_\infty \xi_l/\mu_\infty^2)^{1/2}$ as a function of the dimensionless parameter $(\alpha_{T,\infty} - \alpha_{T,0} - \alpha_{0,\infty})/q$, plotted from the numerical results of Emmons [20], is shown in Figure 12.3. According to equation (55), the ordinate on this graph is simply $-\sqrt{2}F(0)$, which is

$$\left(\frac{\rho_0 v_0}{\rho_\infty u_\infty}\right)\left[\frac{4u_\infty}{\mu_\infty^2}\int_0^x (\rho\mu)\, dx\right]^{1/2}$$

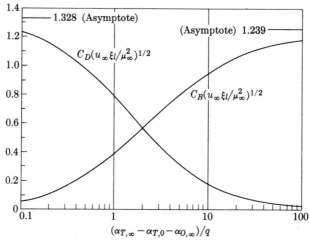

FIGURE 12.3. The dimensionless burning rate and the dimensionless drag on the plate for the combustion of a fuel plate.

according to equation (54), and $C_B \operatorname{Re}_\infty^{1/2}$ if equation (56) is valid. From the definitions of q [equation (41)] and of the α's, it can be seen that the abscissa in Figure 12.3 is the ratio of the difference between the total (thermal plus chemical) enthalpy per unit mass of the gas in the free stream ($y = \infty$) and the total (thermal plus chemical) enthalpy per unit mass of the gas at the surface ($y = 0$), to the energy required at the surface to gasify a unit mass of the condensed fuel. This ratio may be interpreted as a driving force for mass transfer and is often assigned the symbol B (see Section 3.3.6). Since C_B represents a (dimensionless) mass flux of fuel entering the gas, Figure 12.3 can be interpreted as a graph of the dimensionless flux versus the dimensionless driving force for the present problem. Generalized formulations of this concept constitute the basis of mass-transfer theory (see [29], for example).

12.2.5. The force on the plate

Also of some interest in our present problem is the effect of the mass transfer upon the viscous forces exerted on the plate. The local shear stress at the surface of the plate is $\mu_0 (\partial u/\partial y)_0$ [see, for example, equation (1-5)]. According to equations (27), (28), (38), and (51), this is given by

$$\mu_0(\partial u/\partial y)_0 = (\rho\mu)u_\infty(u_\infty/2\xi)^{1/2}F''(0).$$

The conventional dimensionless measure of the local shear stress (the friction coefficient) is

$$\frac{\mu_0(\partial u/\partial y)_0}{\rho_\infty u_\infty^2/2} = \left(\frac{2\mu_\infty^2}{u_\infty \xi}\right)^{1/2} F''(0), \tag{57}$$

where the equality is a consequence of the preceding relationship. An average dimensionless friction force on a plate of length l can be defined as

$$C_D \equiv (\rho\mu) \int_0^{\xi_l} [\mu_0(\partial u/\partial y)_0/(\rho\mu)]\, d\xi \bigg/ \left[(\rho_\infty u_\infty^2/2) \int_0^{\xi_l} d\xi\right], \tag{58}$$

whence equation (57) yields

$$C_D = 2(2\mu_\infty^2/u_\infty \xi_l)^{1/2}F''(0). \tag{59}$$

When $\rho\mu$ is independent of x, C_D becomes the ordinary drag coefficient for the plate and equations (58) and (59) reduce to

$$C_D \equiv \int_0^l \mu_0(\partial u/\partial y)_0\, dx/[\rho_\infty u_\infty^2 l/2] = 2(2/\operatorname{Re}_\infty)^{1/2}F''(0), \tag{60}$$

where Re_∞ appears in equation (56).

In Figure 12.3, $C_D(u_\infty \xi_l/\mu_\infty^2)^{1/2}$ is plotted as a function of

$$(\alpha_{T,\infty} - \alpha_{T,0} - \alpha_{0,\infty})/q$$

from the results of Emmons [20]. According to equation (59), this curve is equivalent to a graph of $2\sqrt{2}F''(0)$, which is $2\,(u_\infty \xi/\mu_\infty^2)^{1/2}$ times the local friction coefficient, in view of equation (57). When equation (60) is valid, $C_D(u_\infty \xi_l/\mu_\infty^2)^{1/2}$ reduces to $C_D \operatorname{Re}_\infty^{1/2}$. The graph shows clearly that the mass transfer reduces the viscous drag on the plate. This phenomenon is well known in mass-transfer theory [29], [30]; finite rates of transfer of mass away from a body always tend to reduce the viscous drag on the body (but usually increase the pressure drag).

12.2.6. Related studies

In many of the investigations related to the problem that has been analyzed here, the principal objective has been to predict the rate of heat transfer to the condensed material. One motivation has been to develop methods for calculating the rate of heat transfer to bodies traveling at hypersonic speeds through the atmosphere [12], [13].

The earliest analyses were concerned with the flow of dissociating gases over solid bodies of various shapes with no mass addition to the boundary layer [31]–[40]. Two limiting cases considered were (1) that in which all reaction rates are fast enough to maintain chemical equilibrium throughout the boundary layer and (2) that in the reaction rates are negligibly small everywhere in the gas. Two subcases of case (2) were identified, namely (a) that in which there are no reactions at the walls of the body (a noncatalytic wall) so that the chemical composition remains uniform everywhere and (b) that in which the walls are perfectly catalytic so that chemical equilibrium conditions apply at the walls. Since finite-rate chemistry need not be included in the formulation in cases (1), (2a), and (2b), approaches analogous to that which have been illustrated here for non-premixed combustibles may be employed. If none of these limiting cases apply, then more-complicated analyses are needed because finite rates of chemical reactions must then be considered either at the surface (see Section B.4) or in the gas, if not in both locations, and the problem becomes one of a premixed flow with finite-rate chemical kinetics. Various premixed configurations will be considered in the following section; a review of some of the analyses and of results obtained is available [7].

Later analyses dealt with boundary layers in which mass addition is employed as a method of reducing the rate of heat transfer to the body, either by injection of gases or by ablation of the solid material itself. Systems in-volving the injection of dissociating materials, the sublimation of inert materials, surface combustion of the solid, the injection of combustible fuels, and melting and vaporization of the solid have been studied. Reviews of some of this work have been published [7]. Analyses of combustion in natural-convection boundary layers are relevant to fire problems; reference to some studies of this type will be made in Section 12.4.

The investigations mentioned thus far are concerned exclusively with laminar boundary layers. Although studies of turbulent boundary layers in-volving mass transfer have resulted in useful correlations, analyses of chemical reactions between nonpremixed combustibles in turbulent boundary layers continue to pose interesting questions, as may be inferred from Section 10.2. For sufficiently rapid or sufficiently slow rates of chemical reactions the heat-transfer and mass-transfer problems are reducible to corresponding problems involving nonreacting turbulent boundary-layer flows. For intermediate rates, good approaches with firm fundamental bases generally are unavailable. Among the motivations for investigations in this area have been interests in predicting burning rates of hybrid (solid fuel-gaseous oxidizer) rocket motors [41], [42], [43], in calculating structures of supersonic turbulent fuel-oxidizer mixing regions within supersonic-combustion ramjet engines [44], [45] and in finding flame shapes and rates of heat and mass transfer in turbulent wall fires [46]. Studies of reactions between *premixed* combustibles in turbulent boundary layers motivated, for example, by erosive-burning problems in solid-propellant rocket motors, are fundamentally even more challenging (see Section 7.8).

12.3. MECHANISMS OF FLAME STABILIZATION

Let us now consider continuous flows of premixed combustible gases and address the question of conditions necessary to retain a flame in the system [2]. This question is of practical significance for many power-production devices. To achieve high power densities, gas velocities in combustors exceed flame velocities, and so means must be found to stabilize flames against **blowout**, a condition at which the flames are transported through the exit of the burner so that combustion ceases. There are two main classes of stabilization techniques, stabilization by fluid streams and stabilization by solid elements. Although other stabilization methods may be envisioned, such as continuous or intermittent deposition of radiant or electrical energy, in the vast majority of practical continuous-flow systems, stabilization is obtained by techniques that fall within one of the two main classes. Stabilization by solid elements will be discussed first; then stabilization by fluid streams will be considered.*

Boundary-layer theory finds wide application in analyses of flame stabilization. The simplest configuration is the stagnation-point boundary layer, studies of which provide information concerning stabilization of a flame ahead of a bluff body placed in the flow; some analyses and ideas

---

* Although flows in combustors usually are turbulent, analyses of flame stabilization are often based on equations of laminar flow. This may not be as bad as it seems because in the regions of the flow where stabilization occurs, distributed reactions may be dominant, since reaction sheets may not have had time to develop; an approximation to the turbulent flow might then be obtained from the laminar solutions by replacing laminar diffusivities by turbulent diffusivities in the results. Improved approximations may be sought by moment methods (Section 10.1.2). Turbulent-flow theories are not discussed here, but some comments on results for turbulent flows are made. A review of theories for stabilization in turbulent flows is available [2].

relevant to this configuration were reviewed in Section 10.3.2, from which it may be inferred that unless the strain or losses to the body cause extinguishment, the reduction in the flow velocity enables the flame to exist. Another simple configuration representative of stabilization by solid elements is the boundary layer on a semi-infinite heated flat plate with its surface oriented parallel to the incoming flow of the cool combustible. By raising the temperature of the adjacent gas, the heated plate decreases the reaction time and thus permits combustion to begin near the surface and to propagate into the flow as an oblique flame. The plates considered in the theoretical analyses [47]–[55] could be viewed as approximating heated walls of combustors. Because of the parabolic character of the boundary-layer equations, the problem bears similarities to some of the one-dimensional, time-dependent ignition problems discussed in Section 8.3, although greater complexity usually is encountered, and ellipticity can arise if streamwise conduction of heat in the plate is allowed (which may be desirable in practice to reduce or possibly eliminate the need for an external supply of heat). When the temperature of the plate is below the adiabatic flame temperature, heat at first flows from the plate to the gas and then, after the flame beings to develop, flows from the gas to the plate. The distance along the plate from the point at which heating is initiated to the point at which heat begins to flow back into the plate is a measure of the length of a hot plate required to stabilize a flame. From the analysis of ignition of a material exposed to an elevated surface temperature, which is cited in Section 8.3, it may be inferred that the critical length so obtained will be somewhat less than that based on a criterion of thermal runaway in activation-energy asymptotics. The rapid thermal expansion associated with the sudden onset of heat release may cause the boundary-layer approximation to fail at the point of flame establishment, but nevertheless the approximation may yield reasonable estimates of stabilization lengths because of the abruptness of flame development after the ignition delay.

Limitations on temperatures of solid materials often cause the methods of stabilization by solid elements, discussed so far, to be impractical. In most applications of stabilization by solid elements the flame is attached in the wake behind the element, so that the solid is not fully exposed to the flame temperature. Representative examples are bluff-body flame stabilizers, such as the stabilizing rods or plates placed normal to the flow in ramjets and afterburners, which were mentioned in Sections 5.1.1 and 10.3.5. A distinctive feature of bluff-body flame stabilization is the presence of a recirculation zone behind the body. Unlike the alternate vortices shed from bluff bodies in cold flow over the Reynolds-number range of practical interest, a well-defined vortex, steady in the mean, is observed to exist just downstream from the stabilizer when combustion occurs. This is a toroidal vortex for an axisymmetric stabilizer or a pair of identical counterrotating line vortices for rodlike stabilizers. The reason for the drastic change in the flow pattern produced by combustion has not been properly clarified, although suggestions have been made that the increase in viscosity (by about a factor of 10 because of the temperature increase) may reduce the effective Reynolds number into a range too low for shedding to occur. The full explanation cannot be as simple as this, and further research on the effect of combustion on the flow is needed. For analyzing stabilization, the fortuitous occurrence of the recirculation zone may be accepted as an experimental fact.

The gases in the recirculation zone are hot combustion products that have been cooled only slightly by heat transfer to the stabilizer. These hot gases cause ignition by transferring heat to the cold combustible across the mixing layer that exists at the outer edge of the recirculation zone. Thus for these wake-stabilized flames, the stabilization mechanism in fact involves the interaction of two gas streams, and suitable stabilization analyses are derivable from those for stabilization by fluid streams. Therefore, before discussing bluff-body flame stabilization further, we shall mention various methods of stabilization by fluid streams.

An alternative to a bluff body is a recess in the wall of the combustor [56]. Recesses with leading edges that are adequately sharp produce flow separation to establish a recirculation zone within the recess. The stabilization mechanism then is much like that of the bluff body; hot recirculating gases ignite the combustible stream by heat transfer across the mixing layer. The length of the wall recess plays the same role as the length of the bluff-body recirculation zone in stabilization criteria unless this length is large enough for flow reattachment to occur within recess, in which case the vortex length depends instead of the depth of the recess. A design that corresponds to a limiting case involving reattachment is the *dump combustor*, in which a duct with a small cross-sectional area empties coaxially into a duct with a larger cross-sectional area at a right-angle step [57]. The flame is stabilized by the recirculation region that forms at the step. Designs of this type are becoming popular since they offer benefits such as reduced losses in total pressure; associated wall-heating problems are solvable, for example, by transpiration cooling if necessary.

Introduction of swirl can be beneficial in combustor designs, for example, as a means for tailoring locations and extents of recirculation regions [58], [59], [60]. Rotating fluid streams injected near the periphery of a combustor exert different effects from those of rotation introduced near the axis; influences of swirl can differ greatly in different configurations [58], [59]. There are numerous possible geometrical arrangements of combustion chambers with swirl, and the fluid mechanics of many of them are poorly understood.

Fluid streams may be used either to augment the performance of bluff-body stabilizers or to produce flame stabilization by themselves [2]. A practical example in which extensive use is made of fluid streams is the can-type stabilizer of turbojet engines [61], in which small contoured perforations in the can surrounding the fuel injectors direct air streams in a manner needed to provide desired performance. In a much simpler configuration, stabilization can be achieved by an axisymmetric reverse jet, in which fluid is injected in an upstream direction along the axis of a combustor without swirl [2]; here, too, recirculation plays a role in the stabilization mechanism since a toroidal vortex develops between the edge of the jet and the incoming combustible gas. To stabilize a flame, jets of fluid can be injected in other directions, for example, transverse to the main flow. The injected fluid in the stabilizing stream may be air, fuel, an inert gas, or a combustible mixture (for example, forming a pilot flame in a low-velocity jet). Stabilization analyses must differ for different injected fluids.

Recirculation is not essential for flame stabilization. A stream of hot fluid injected in the downstream direction can anchor a flame in a coflowing combustible. Heat transfer from the hot inert gas to the cool combustible gas across the mixing layer that separates the two gases can cause ignition. Although this method of stabilization usually requires excessive combustor lengths, making it impractical, basically the same mechanism of stabilization applies in stabilization by hot recirculating gases, for example behind bluff-body stabilizers. Therefore, there has been interest in developing detailed analyses of stabilization for this simple configuration.

A problem of this type that has been subjected to careful analysis beginning with the early work of Marble and Adamson [62] is illustrated schematically in Figure 12.4. At the point $x = 0$, a stream of cold (temperature T_1) combustible gas traveling at the velocity u_1 comes into contact with a stream of hot (temperature T_2) inert gas traveling at the velocity u_2. As

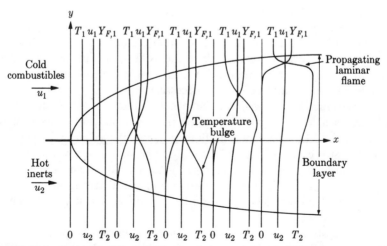

FIGURE 12.4. Schematic diagram of the development of the velocity, temperature, and fuel mass-fraction profiles in a planar mixing layer.

these two streams mix, the temperature of the combustible increases (by means of heat conduction and diffusion) until the reaction begins to take place at an appreciable rate. When T_2 is less than the adiabatic flame temperature, the heat released by the reaction produces a local temperature maximum in the mixing zone. A laminar flame then develops in the mixing layer and propagates into the combustible gas. In this manner, the laminar flame is ignited and stabilized by the laminar mixing region.

The objective of the analyses are to predict the velocity, temperature, and composition profiles as functions of x and thereby to obtain an estimate of the point of laminar flame attachment. The ratio of the length x required for flame development to a representative length of an experimental apparatus is a parameter determining whether a flame can be established by the laminar mixing process in the given apparatus. Boundary-layer equations are employed for describing the mixing and the initial development of the temperature bulge. With $F' = u/u_1$, the velocity profile is found to be described by equation (39) with $F'(\infty) = 1$ and $F'(-\infty) = u_2/u_1$ as boundary conditions; some solutions are given in [63]. By letting $C = 1$ and talking $\kappa_i = 0$ and the Lewis and Prandtl numbers to be unity, the problem may be reduced to one involving a single partial differential equation—for example, for the fuel mass fraction Y_F. Because of the chemical source term in this equation, self-similar solutions of the type found in the preceding section do not exist.

Near the point where the two streams first meet the chemical reaction rate is small and a self-similar frozen-flow solution for Y_F applies. This frozen solution has been used as the first term in a series expansion [62] or as the first approximation in an iterative approach [64]. An integral method also has been developed [62], in which ordinary differential equations are solved for the streamwise evolution of parameters that characterize profile shapes. The problem also is well suited for application of activation-energy asymptotics, as may be seen by analogy with [65]. The boundary-layer approximation fails in the downstream region of flame spreading unless the burning velocity is small compared with u_1; it may also fail near the point where the temperature bulge develops because of the rapid onset of heat release there.

A number of different specific criteria for flame attachment in the mixing zone may be identified. In the early work [62], [66]–[68], it was popular to define the streamwise distance from the beginning of the mixing zone to the point at which the temperature bulge first appears as the characteristic flame-development length. However, this length depends strongly on the parameters of the system, and its calculated value can be influenced appreciably by the types of approximations made in the analysis [64]. A better definition of a characteristic length would be the distance to a point of thermal runaway, within the context of activation-energy asymptotics. The asymptotic analysis that is needed for applying this definition has not been performed. A simpler but inherently less precise definition of a combustion development length can be based upon a comparison of a residence time and a reaction time, a Damköhler first similarity group, a concept upon which we elaborated in Section 4.1.

Neglecting influences of transport phenomena on the chemical reaction rate, we may hypothesize that the distance downstream at which the flame first appears equals the product of the chemical reaction time and the mean convective velocity for the element of the combustible mixture that first reacts. The mean convective velocity for this element is roughly $(u_1 + u_2)/2$. The reaction time can be estimated from the reaction rate w_F (mass per unit volume per unit time produced) for the principal reactant F; namely, the reaction time is roughly $\rho_1 Y_{F,1}/(-w_F)$. Employing the definition of the reaction rate ω given in equation (1-43), we then find that the flame development length is roughly

$$x_a = (u_1 + u_2)\rho_1 Y_{F,1}/[2W_F(v_F' - v_F'')\omega]. \tag{61}$$

Equation (61) is ambiguous until we state at precisely what temperature, pressure, and composition the reaction-rate function ω is to be evaluated. The ambient pressure p and the initial mass fractions $Y_{i,1}$ are reasonable first approximations for the pressure and composition. Since the principal profile changes (for example, the development of the local temperature maximum) occur well within the hot inert stream throughout most of the combustion development region, the most reasonable temperature to use in ω in equation (61) is T_2. Thus, the specification $\omega = \omega(T_2, p, Y_{i,1})$ serves to define the right-hand side of equation (61) unambiguously. For example, for the unimolecular reaction $F \to P$, it can be seen from equations (1-8) and (1-43) that equation (61) then becomes

$$x_a = \left(\frac{u_1 + u_2}{2}\right)\left(\frac{T_2}{T_1}\right)(B_1 T_2^{\alpha_1})^{-1} e^{E_1/(R^0 T_2)}. \tag{62}$$

The results given in equations (61) and (62) correctly yield the principal velocity, pressure, and temperature dependences of x_a. For representative values of the parameters, equation (62) is also in rough numerical agreement with values for the combustion development length obtained in other studies [64].

Equation (62) can be applied to derive an approximate criterion for flame stabilization by bluff-body stabilizers. Empirically, when the approach flow velocity u_1 in the combustor exceeds a critical **blowoff** velocity $u_{1,\text{max}}$, the flame is blown downstream and can no longer be stabilized by the bluff body. A knowledge of $u_{1,\text{max}}$ is essential in the design of ramjet and afterburner combustors. Although there are many differences between the mixing zone considered above and a bluff-body flame stabilization region, an approximate criterion for flame stabilization by a bluff body is the requirement that the length of the hot recirculation zone, l, be greater than or equal to the flame-attachment length x_a given in equation (61). The subscript 2 in equation (61) now identifies conditions inside the recirculation zone. Since the velocity in the recirculation zone is negligibly small ($u_2 \approx 0$) and the temperature in the recirculation zone is, empirically, very near the adiabatic flame temperature ($T_2 \approx T_\infty$) [2], the criterion for flame attachment becomes

$$x_a/l = u_1 \rho_1 Y_{F,1}/[2lW_F(v_F' - v_F'')\omega(T_\infty, p, Y_{i,1})] \leq 1. \tag{63}$$

This result can conveniently be expressed in terms of the laminar flame speed v_0 by employing equation (5-2); the result is

$$x_a/l = u_1(\lambda/c_p)/(2l\rho_1 v_0^2) \leq 1, \tag{64}$$

since w in equation (5-2) corresponds to $W_F(v_F' - v_F'')\omega(T_\infty, p, Y_{i,1})/Y_{F,1}$. Equation (64) shows that the blowoff velocity (the maximum value of u_1) is

$$u_{1,max} = l\rho_1 v_0^2[2/(\lambda/c_p)]. \tag{65}$$

A characteristic transverse dimension d of the flame holder can be measured more easily than the length l of the recirculation zone. The ratio l/d experimentally has a practically constant value between 5 and 10, independent of flow conditions for hot turbulent wakes. Hence, $l \sim d$ in equation (65), so that $u_{1,max} \sim d$. Of greater interest than the dependence of $u_{1,max}$ on the density ρ_1 is its dependence on the pressure p at constant initial temperature T_1. Since $\rho_1 \sim p$ at constant T_1, and λ/c_p may be expected to depend only on temperature, equation (65) yields

$$u_{1,max} \sim dpv_0^2. \tag{66}$$

The dependences of the blowoff velocity on size, d, pressure, p, and flame speed, v_0, given by equation (66) generally agree with experiment and provide reasonable correlations of empirical data [2], [69].* Although this reasoning is very rough, the complexity of the flow field about a bluff body necessitates the use of imprecise concepts such as this at the present time if simple correlation formulas are to be derived.†

12.4. PROCESSES OF FLAME SPREAD

Imprecise concepts are also helpful for describing the spread of flames in the gas phase, through condensed-phase fuels. A general approach, applicable in complex configurations, is first to identify a surface of fuel involvement that encloses the fuel. This surface, which flames approach at some

---

    * See also N. K. Rizk and A. H. Lefebvre, *AIAA Journal* **22**, 1444 (1984).
    † The concepts are imprecise or approximate in the sense that their application cannot be expected to produce precisely correct results; the concepts themselves are defined precisely.

but not all points, is the boundary of a control volume within which the virgin fuel is contained. An ignition stimulus of some kind must be transferred across the surface of fuel involvement if flame spread is to occur. In practically all problems, the stimulus involves heat in one way or another. Therefore, attention is focused on heat transfer across the surface of fuel involvement.

Let q denote an energy per unit area per unit time transferred across the surface. As a first approximation, let us assume that a critical enthalpy increase Δh per unit mass of fuel is needed for ignition to occur. If ρ is the fuel density and V is a spread velocity, then, according to an energy balance,

$$\rho V \Delta h = q. \tag{67}$$

This equation may be employed to calculate a spread velocity if q, ρ, and Δh can be estimated. The concept of a critical temperature T_c for ignition of the fuel in the presence of a flame may be introduced to provide an approximate expression for Δh, namely,

$$\Delta h = \int_{T_0}^{T_c} c_p \, dT, \tag{68}$$

where T_0 is the initial temperature of the fuel and c_p is its specific heat, interpreted to include appropriate delta functions if energetically nonneutral phase changes occur between T_0 and T_c. As may be seen from the discussion in Section 8.3, for example, the value of T_c varies with experimental conditions and may be considered to be a known property of the fuel only in a first approximation. If approximations for ρ, c_p, and T_c are available, then the problem of obtaining V from equation (67) is reduced to one of studying the heat-transfer mechanisms that determine q.

Equation (67) has been employed in analyzing many different processes of flame spread [3], [70], [71]. Its utility extends beyond the specific class of problems addressed here. It can provide an approximate description of the rate of normal flame propagation—for example, through a premixed combustible gas, the problem addressed in Chapter 5—and it has been used for analyzing fire spread, even for fires involving only smoldering combustion without gas-phase flames [3]. The choice of whether to adopt a description as simplified as the present one depends on the complexity of the problem; complicated analyses can be pursued with confidence only for sufficiently simple problems. Thus the problem of Chapter 5 was seen to be amenable to much more thorough analysis; for flame propagation through suspensions of particulate fuels, with significant contributions of radiant-energy transfer, descriptions based on equation (67) are more prevalent and better justified as first steps. For spread of forest fires, the complexities are sufficiently great to preclude detailed analysis of the overall problem and to make formulations based on principles like equation (67) extremely attractive [70]. Spread of flames over surfaces of solid propellants also can be complicated enough to make equation (67) useful [72].

Geometric aspects of spread processes must be taken into account in applying equation (67). As an example, consider one-dimensional flame spread through a porous bed of solid fuels, such as grass, sticks, pine needles, or a thicket of heavy brush burning in air. The bed then is not pure fuel but instead contains pockets of air, the heating of which does not contribute a resistance to flame spread. In this case, if each fuel element is heated throughout its interior prior to flame arrival, then in terms of the packing fraction f, the ratio of the volume of the bed occupied by fuel to the total bed volume, it is seen that the density in equation (67) is $\rho = f\rho_F$, where ρ_F is the density of the fuel. If the fuel elements are large enough and the spread rate is high enough, then the fuel elements are not heated completely prior to flame arrival. In equation (67) we must then use $\rho = \gamma f\rho_F$, where γ is the fraction of the volume of a typical fuel element heated prior to flame arrival. If the heating extends normal from the surface of fuel involvement to a distance h within the bed, then the heating time of a fuel element prior to flame arrival is h/V, and the depth of the fuel element heated will be of the order of $\sqrt{\alpha_F h/V}$, where α_F is the thermal diffusivity of the fuel. If this depth is greater than the thickness of a fuel element, then the elements are thermally thin and $\gamma \approx 1$; if it is less, then the elements are thermally thick in the spread process and $\gamma < 1$. In terms of the surface-to-volume ratio s of the fuel elements (for example, the reciprocal of the stick diameter for long cylindrical sticks), we then roughly have $\gamma = s\sqrt{\alpha_F h/V}$ for thermally thick fuels. Thus it is seen that in applications of equation (67) ρ depends upon V under certain circumstances. If the energy transfer into the bed occurs mainly by radiation with negligible influences of gas motion, then $h = 1/(fs)$ provides a rough approximation, and for thermally thick fuels, equation (67) gives

$$V = q^2/[fs\alpha_F \rho_F^2(\Delta h)^2]. \qquad (69)$$

Heat-transfer expressions for q also may involve velocities in some situations. To illustrate this as well as two-dimensional geometric aspects of the use of equation (67), let us consider the downward spread of flames along the flat vertical surfaces of solid fuels that burn in a diffusion flame with an oxidizing gas under conditions of negligible radiant-energy transfer, so that q involves only conduction and convection, and with velocities low enough that the flow remains laminar (as it often does in experiments of this type, at least in the flow region where the spread occurs). This configuration serves to demonstrate well the necessary distinction that often must be made between spread rates and normal burning rates; the velocity at which the fuel surface regresses normal to itself (that is, in the horizontal direction) along the portion bathed by flame is the normal burning rate and is entirely different from the vertical velocity at which the boundary of the burning area moves downward along the fuel sheet, the spread velocity. If there were an opposed (upward) forced-convective flow at a large-enough velocity to dominate natural convection and to assure applicability of the boundary-layer approximation, then the analysis of Section 12.2 could be employed to obtain the normal burning rate; in the absence of forced convection, corresponding free-convection boundary-layer analyses [73]–[76] can be used. Calculation of the spread rate is more difficult.

If the vertical sheet of fuel is burning on both faces, then its center plane is one of symmetry across which no heat transfer occurs. For this problem of two-sided burning, let $2L$ be the thickness of the fuel sheet; alternatively we may consider one-sided burning of a sheet of thickness L with a perfectly insulating wall at the back face of the sheet. The surface of fuel involvement may be drawn with a horizontal upper face at the elevation where vigorous fuel gasification begins and with vertical surfaces extending downward along the insulating plane and along the exposed face of the fuel. With this selection, since the spread is downward, q in equation (67) represents the total energy per unit time transmitted across the surface of fuel involvement divided by the area of the horizontal upper face of this surface. If the transfer is by conduction through the gas from the foot of the flame to the exposed face of the fuel, over a distance l down from the top of the surface of fuel involvement, and with the flame standing a distance d off the surface, then q is approximately $(l/L)\lambda_G(T_f - T_c)/d$, where λ_G is the thermal conductivity of the gas, T_f the flame temperature, and T_c the ignition temperature of the fuel*; the factor l/L here is the area ratio needed in the definition of q. For thermally thin fuels, and with the approximation $l = d$, use of this expression for q in equation (67) gives

$$V = [\lambda_G/(\rho_F c_F L)](T_f - T_c)/(T_c - T_0), \tag{70}$$

where c_F is the average heat capacity per unit mass for the fuel between temperatures T_0 and T_c. For thermally thick fuels $\gamma = \sqrt{\alpha_F l/V}/L$ and with the value of l estimated as a heat-conduction length for downward penetration of heat in the gas against an average upward gas velocity V_G—namely, $l = \alpha_G/V_G$, where α_G is the thermal diffusivity of the gas—we find that

$$V = V_G(\Gamma_G^2/\Gamma_F^2)(T_f - T_c)^2/(T_c - T_0)^2. \tag{71}$$

Here $\Gamma_G \equiv \sqrt{\rho_G c_G \lambda_G} = \lambda_G/\sqrt{\alpha_G}$ and $\Gamma_F \equiv \sqrt{\rho_F c_F \lambda_F} = (\rho_F c_F)\sqrt{\alpha_F}$ represent thermal inertias (or thermal responsivities) of the gas and fuel, respectively.

For the downward-spread problems that have been considered here and for the closely related problems of spread against an opposed forced-convective gas velocity, the configurations are sufficiently uncomplicated to

---

* For polymeric fuels, the analysis of Section 7.4 is relevant to the evaluation of T_c, therein denoted by T_i.

motivate detailed analyses of simplified models of the processes. Many analyses of this type have been completed in recent years [77]–[95] and have considerably improved our understanding of the spread mechanisms. Equation (70) (with an additional factor of $\sqrt{2}$ on the right-hand side) and equation (71) were derived first by de Ris [77], who addressed a model problem in which V_G is constant throughout the gas, solving the differential equations for heat conduction and species diffusion without a boundary-layer approximation by introducing Fourier-transform techniques that enabled a complex-variable method of the Wiener-Hopf type to be employed to factor a kernel, as needed for completing the solution. For configurations dominated by forced convection, he interpreted V_G in equation (71) as the externally applied gas velocity; for downward spread without forced convection, he later [78] (in a comment) identified V_G as a characteristic buoyant velocity, $[g\alpha_G(T_f - T_0)/T_0]^{1/3}$, where g is the acceleration of gravity. The interpretations for forced convection are modified somewhat if the effects of the actual nonuniform velocity in the gas are taken into account [92], [94]. For thermally thick fuels, if conductive heat transfer occurs mainly through the solid, then the dependences of ρ and q upon V are such that the spread velocity cancels from equation (67), and an alternative principle, a balance between the rate of energy release and the rate of energy transported away, is needed for obtaining the spread rate [93]. Further investigations of spread over thermally thin fuels have indicated [95] that equation (70) is better than the corresponding equation with the factor $\sqrt{2}$; a factor of 0.8 produces best agreement with experiment. The results in equations (70) and (71) apply only for a limit of infinite Damköhler number in which the rate of heat release in the gas phase is large, so that processes of heat and mass transfer control the spread rate. In experiments, finite-rate gas-phase chemistry often exerts a large effect on the spread rate and thereby necessitates development of more-complicated analyses; some steps in this direction have been taken (for example, [78], [82], [84], and [87]), and correlations of spread-rate data have been obtained on the basis of Damköhler numbers [88], [89], [92], [95]–[98].

There are processes of flame spread among discrete fuel elements in which the rate of downward spread along an element plays a role. For example, in a linear, horizontal array of vertically oriented matchsticks flame propagation from one stick to the next may occur through a mechanism in which the sticks successively ignite at the top and burn in a manner such that a flame propagates down each stick [99]. The horizontal velocity at which flames propagate through the array depends on the ignition time of a stick under convective heating by adjacent burning sticks and not on the rate of downward spread, but the critical stick spacing above which horizontal propagation does not occur may be defined by equating the ignition time to the burning time of a stick, the latter being approximately the time required for a flame to propagate from the top to the bottom of the stick. The top of an unignited stick exposed to an adjacent flame bends toward the flame because of preferential outgassing on the exposed side; after flame passage, the charred, extinguished sticks bend back in the opposite direction, again toward the flame, through continued outgassing by heating from the flame that has passed. In some experiments [99], these fuel motions affect flame spread negligibly, but in others they can be important [3], especially for polymeric materials that soften and flow when exposed to heat. Downward flow of burning viscous liquid controls rates of downward flame spread for some materials [3]. Related studies of flame spread through other regular arrays of small-scale fuel elements have been made ([100] and [101]).

In contrast to downward spread, upward spread of flames along vertical fuel surfaces usually is acceleratory. Rates of heat transfer to the fuel by radiation and conduction from the luminous flames of height l, that bathe the surface are so great that, in comparison, transfer elsewhere can be neglected in the first approximation. If $\bar{q}$ is the average normal energy flux from these flames to the surface, then from geometric considerations, $q = (l/L)\bar{q}$ for the q in equation (67), where L is the thickness of the fuel sheet. For thermally thick fuels, $\gamma = \sqrt{\alpha_F l/V}/L$ again, so that equation (67) gives

$$V = \bar{q}^2 l/[\Gamma_F^2 (T_c - T_0)^2]. \tag{72}$$

In equation (72) l depends on the total rate of mass loss of the fuel. In terms of the normal regression velocity r of the burning surface, this rate (per unit horizontal distance along the face of the sheet) is $M = \rho_F r x$, where x is the vertical length of the burning portion of the sheet. Geometrically, $x \approx L(V/r)$ for fully developed, steady spread. Various flame-height results for l as a function of M are available [71]; l is approximately proportional to a power of M less than unity [3], [71], so that substitutions into equation (72) enable V to be estimated for steady spread. Prior to attainment of a steady state, the relationship $V = dx/dt$ may be used in equation (72) to provide a differential equation that can be solved for $x(t)$ during the acceleratory period (which frequently persists throughout an experiment since the steady-state V often is too large to measure). Wind-aided spread along horizontal fuel surfaces possesses many similarities to upward spread. Detailed analyses of rates of upward and wind-aided spread, roughly along the lines sketched here but incorporating a number of refinements, have been developed [102]–[110].

Additional physical phenomena influence rates of horizontal spread of flames over surfaces of liquid fuels [111]–[120]. If the equilibrium vapor pressure of the liquid fuel at its initial temperature is high enough for the fuel concentration in the gas mixture at the liquid surface to lie above the lower flammability limit, then premixed flame propagation occurs in the gas above the liquid, and flame spread is rapid. If the liquid is cold and highly viscous, then spread mechanism resemble those for solids, and spread rates are slow. Between these two extremes, motion of the liquid usually plays an important role in the spread process. Two properties of liquids that contribute to this motion are thermal expansion and the variation of the surface tension with temperature. These two properties give rise to flows driven by buoyancy and to flows driven by surface-tension variations, respectively.

Estimates of spread velocities by the two mechanisms of liquid motion are readily developed. Consider buoyancy-driven flow first. The pressure difference in the horizontal direction between hot and cold liquid is of the order of $gh(T_c - T_0)(d\rho_F/dT)$, where h is the depth of the heated layer, T_c now is the temperature of the liquid surface beneath the flames, typically slightly less than the normal boiling point, and the temperature derivative of the liquid density is to be evaluated at constant pressure. If the resulting horizontal pressure gradient is balanced against viscous forces for steady flow at a liquid-surface velocity V, then it is found that

$$V = (gh^3/l)(T_c - T_0)(d\rho_F/dT)/\mu_F, \tag{73}$$

where l measures the horizontal distance ahead of the flames over which an elevated surface temperature exists, μ_F is the viscosity of the liquid, and the characteristic vertical distance over which nonnegligible horizontal liquid velocities persist has been taken to be the depth h. If heat transfer to the unignited liquid occurs mainly by radiation from vertical flames above the burning liquid, then l in equation (73) is approximately the flame height; transparency of the liquid to flame radiation or conduction of heat through metallic sides of the liquid container can introduce complications and modify this estimate. For liquids contained in shallow pans, the value of h in equation (73) may be roughly the depth of the liquid. For deeper pools, a boundary layer develops at the surface of the liquid, and h is then approximately the depth of this boundary layer. This depth may be estimated by balancing viscous and inertial forces in the horizontal direction for steady flow; since the horizontal velocity gradient in the liquid boundary layer is of order V/l, we find that $\rho_F V^2/l = \mu_F V/h^2$, approximately, so that $h = \sqrt{\nu_F l/V}$, where ν_F is the kinematic viscosity of the liquid. Use of this result in equation (73) enables V to be estimated and h to be calculated for sufficiently deep liquids; the calculations remain consistent if the resulting h is less than the liquid depth, that is, if V is less than the value obtained from equation (73) by using the liquid depth for h. Thus the deep-pool expressions—for example,

$$V = (g\sqrt{l\nu_F}\, l\rho'/\rho_F)^{2/5} \qquad \text{where} \quad \rho' \equiv (d\rho_F/dT)(T_c - T_0)/l$$

—apply when the actual depth of the liquid exceeds $[l\mu_F^2/(g\rho'\rho_F)]^{1/5}$, a value typically on the order of 1 mm to 1 cm.

Next consider flows driven by variations of surface tension. Associated with the temperature gradient $\partial T/\partial x$ along the surface of the liquid ahead of the flame is a gradient in the surface tension, σ, given by $\sigma' = (d\sigma/dT)(\partial T/\partial x)$. This surface-tension gradient causes a net surface force per unit area on an element of liquid at the surface, which may be balanced against a viscous force per unit area, estimated as $\mu_F V/h$, where the velocity gradient has been approximated as the ratio of the liquid velocity at the surface (the spread velocity) to the depth of the heated layer. With $\partial T/\partial x$ approximated as $(T_c - T_0)/l$, the resulting estimate for the spread velocity is

$$V = (h/l)(T_c - T_0)(d\sigma/dT)/\mu_F. \qquad (74)$$

The discussions in the preceding paragraph concerning l and h again apply, and it is found that the deep-pool spread-rate formula, $V = (\sigma'/\mu_F)(lv_F\mu_F/\sigma')^{1/3}$, may be used if the liquid depth exceeds $(lv_F\mu_F/\sigma')^{1/3}$, a value which again typically lies between 1 mm and 1 cm. The question of whether buoyancy forces or surface-tension forces are more important in the spread process may be addressed by evaluating V according to each mechanism separately; the mechanism giving the larger V predominates. Thus from equations (73) and (74) it is readily inferred that the surface-tension mechanism is of greater importance for sufficiently shallow pools. For deep pools it is found from the spread-velocity formulas that the effects of variations in surface tension predominate if

$$l^4 g^3 v_F^2 \mu_F^2 (d\rho_F/dT)^3/[(T_c - T_0)^2(d\sigma/dT)^5] < 1,$$

which often is satisfied for laboratory experiments but is violated if l becomes sufficiently large. In applications both mechanisms often contribute, and the sum of equations (73) and (74) then provides a rough estimate of the spread velocity, although it is both possible and desirable to pursue more careful analyses to obtain improved estimates.

REFERENCES

1. Th. von Kármán and G. Millán, *4th Symp.* (1953), 173–177.
2. F. A. Williams, "Flame Stabilization in Premixed Turbulent Gases", in *Applied Mechanics Surveys*, H. N. Abramson, H. Liebowitz, J. M. Crowley, and S. Juhasz, eds., Washington, D.C.: Spartan Books, 1966, 86–91.
3. F. A. Williams, *16th Symp.* (1977), 1281–1294.
4. H. Schlichting, *Boundary-Layer Theory*, New York: McGraw-Hill, 1960; see Chapter VII in particular.
5. B. P. Mullins and S. S. Penner, *Explosions, Detonations, Flammability and Ignition*, New York: Pergamon Press, 1959, 94–99.
6. L. Lees, *Third Combustion and Propulsion Colloquium*, AGARD, New York: Pergamon Press, 1959, 451–497.

7. P. M. Chung, "Chemically Reacting Nonequilibrium Boundary Layers," in *Advances in Heat Transfer*, vol. 2, New York: Academic Press, 1965, 109–270.

8. L. Rosenhead, ed., *Laminar Boundary Layers*, Oxford: Clarendon Press, 1963.

9. K. Stewartson, *The Theory of Laminar Boundary Layers in Compressible Fluids*, Oxford: Clarendon Press, 1964.

10. F. K. Moore, ed., *Theory of Laminar Flows*, vol. IV of *High Speed Aerodynamics and Jet Propulsion*, Princeton: Princeton University Press, 1964.

11. C. C. Lin, ed., *Turbulent Flows and Heat Transfer*, vol. V of *High Speed Aerodynamics and Jet Propulsion*, Princeton: Princeton University Press, 1959, 75–195, 288–427.

12. W. H. Dorrance, *Viscous Hypersonic Flow*, New York: McGraw-Hill, 1962.

13. W. D. Hayes and R. F. Probstein, *Hypersonic Flow Theory*, New York: Academic Press, 1959, Chapter 8.

14. W. Mangler, *Z. angew. Math. Mech.*, **28**, 97 (1948).

15. P. L. Howarth, *Proc. Roy. Soc. London*, **164A**, 547 (1938); **194A**, 16 (1948).

16. A. Dorodnitzyn, *Dokl. Akad. Nauk SSSR*, **34**, 213 (1942).

17. F. K. Moore, "Unsteady Laminar Boundary-Layer Flow," NACA Tech. Note No. 2471 (1951).

18. K. Stewartson, *Qurat. J. Mech. Appl. Math.* **4**, 182 (1951).

19. F. A. Williams, *AIAA Journal* **3**, 2112 (1965).

20. H. W. Emmons, *Z. angew. Math. Mech.*, **36**, 60 (1956).

21. F. A. Williams, *C & F* **5**, 207 (1961).

22. L. Roberts, *J. Fluid Mech.*, **4**, 505 (1958).

23. G. W. Sutton, *J. Aero. Sci.*, **25**, 29 (1958); **26**, 397 (1959).

24. S. M. Scala and G. W. Sutton, "The Two-Phase Hypersonic Boundary Layer—A Study of Surface Melting," *1958 Heat Transfer and Fluid Mechanics Institute*, Stanford: Stanford University Press, 1958, 231–240.

25. L. Lees, *ARS Journal*, **29**, 345 (1959).

26. H. A. Bethe and M. C. Adams, *J. Aero. Sci.*, **26**, 321 (1959); M. C. Adams, *ARS Journal*, **29**, 625 (1959).

27. G. A. Tirskii, *Prikl. Mat. Mekh.*, **25**, 291 (1961).

28. D. B. Spalding, *Aero. Quart.*, **12**, 237 (1961).

29. D. B. Spalding, *Convective Mass Transfer*, London: Arnolds Press, 1963.

30. D. B. Spalding, *Int. J. Heat Mass Transfer*, **2**, 15 (1961); D. B. Spalding and H. L. Evans, *Int. J. Heat Mass Transfer*, **2**, 199, 314 (1961).

31. M. Sibulkin, *J. Aero. Sci.*, **19**, 570 (1952).

32. L. Lees, *Jet Propulsion*, **26**, 259 (1956).

33. R. F. Probstein, *Jet Propulsion*, **26**, 497 (1956).

34. R. Bromberg, *J. Aero. Sci.*, **23**, 976 (1956).

35. M. Romig, *Jet Propulsion*, **26**, 1098 (1956); **27**, 1255 (1957).

36. J. A. Fay, F. R. Riddell, and N. H. Kemp, *Jet Propulsion*, **27**, 672 (1957).

37. J. A. Fay, and F. R. Riddell, *J. Aero. Sci.*, **25**, 73 (1958).

38. P. A. Libby, "Laminar Hypersonic Heat Transfer on a Blunt Body According to the Integral Method," *1958 Heat Transfer and Fluid Mechanics Institute*, Stanford: Stanford University Press, 1958, 216–230.

39. N. H. Kemp, P. H. Rose, and R. W. Detra, *J. Aero. Sci.*, **26**, 421 (1959).

40. I. P. Ginzburg, *Int. J. Heat Mass Transfer*, **4**, 89 (1961).

41. G. A. Marxman and M. Gilbert, *9th Symp.* (1963), 371–383.

42. F. A. Williams, "Grain Design and Throttling of Hybrid Rocket Motors," *Aerospace Chemical Engineering*, D. J. Simkin, L. Isenberg, and S. E. Stephanou, eds. vol. 62, Chemical Engineering Progress Symposium Series, New York: American Institute of Chemical Engineers, 1966, 86–91.

43. G. A. Marxman, *11th Symp.* (1967), 269–289.

44. P. A. Libby, *ARS Journal*, **32**, 388 (1962).

45. J. S. Evans and C. J. Schexnayder, Jr., *AIAA Journal* **18**, 188 (1980).

46. T. Ahmad and G. M. Faeth, *17th Symp.* (1979), 1149–1160.

47. D. A. Dooley, "Ignition in the Laminar Boundary Layer of a Heated Plate," *1957 Heat Transfer and Fluid Mechanics Institute*, Stanford: Stanford University Press, 1957, 321–342.

48. T. Y. Toong, *6th Symp.* (1957), 532–540.

49. O. P. Sharma and W. A. Sirignano, *CST* **1**, 481 (1970).

50. C. K. Law, *CST* **19**, 237 (1979).

51. C. K. Law and H. K. Law, *J. Fluid Mech.* **92**, 97 (1979).

52. C. Treviño and M. Sen, *18th Symp.* (1981), 1781–1789.

53. C. Treviño and M. Sen, *C & F* **43**, 121 (1981).

54. C. Treviño and A. C. Fernández-Pello, *CST* **26**, 245 (1981).

55. C. Treviño and A. C. Fernández-Pello, *C & F* **49**, 91 (1983).

56. P. R. Choudhury and A. B. Cambel, *8th Symp.* (1962), 963–970.

57. R. B. Edelman and P. T. Harsha, *Prog. Energy Combust. Sci.* **4**, 1 (1978).

58. N. A. Chigier and A. Chervinsky, *11th Symp.* (1967), 489–499.

59. D. G. Lilley, *AIAA Journal* **15**, 1063 (1977).

60. M. J. Oven, F. C. Gouldin, and W. J. McLean, *17th Symp.* (1979), 363–374.

61. W. S. Blazowski, *Prog. Energy Combust. Sci.* **4**, 177 (1978).

62. F. E. Marble and T. C. Adamson, Jr., *Jet Propulsion* **24**, 85 (1954).

63. R. C. Lock, *Quart. J. Mech. Appl. Math.* **4**, 42 (1951).

64. F. A. Williams, *C & F* **8**, 166 (1964).

65. A. Liñán and A. Crespo, *CST* **14**, 95 (1976).

66. S. I. Cheng and A. A. Kovitz, *6th Symp.* (1957), 418–427.

67. S. I. Cheng and A. A. Kovitz, *J. Fluid Mech.* **4**, 64 (1958).

68. S. I. Cheng and H. H. Chiu, *Int. J. Heat Mass Transfer* **1**, 280 (1960).

69. E. A. De Zubay, *Aero. Digest*, **61**, 54 (1950).

70. H. W. Emmons, *Fire Research Abstracts and Reviews* **5**, 163 (1963).

71. F. A. Williams, *Prog. Energy Combust. Sci.* **8**, 317 (1982).

72. F. A. Williams, M. Barrère, and N. C. Huang, *Fundamental Aspects of Solid Propellant Rockets*, AGARDograph no. 116, Slough, England: Techvision Services, 1969, 522–527.

73. F. J. Kosdon, F. A. Williams, and C. Buman, *12th Symp.* (1969), 253–264.

74. J. S. Kim, J. de Ris, and F. W. Kroesser, *13th Symp.* (1971), 949–961.

75. P. J. Pagni and T. M. Shih, *16th Symp.* (1977), 1329–1343.

76. T. Ahmad and G. M. Faeth, *Journal of Heat Transfer* **100**, 112 (1978).

77. J. de Ris, *12th Symp.* (1969), 241–252.

78. F. A. Lastrina, R. S. Magee, and R. F. McAlevy III, *13th Symp.* (1971), 935–948.

79. Y. Ohki and S. Tsugé, *CST* **9**, 1 (1974).

80. W. A. Sirignano, *Acta Astronautica* **1**, 1285 (1974).

81. A. Fernández-Pello, M. Kindelan, and F. A. Williams, *Ing. Aeron. y Astron.* **135**, 41 (1974).

82. A. Fernández-Pello and F. A. Williams, *15th Symp.* (1975), 217–231.

83. M. Sibulkin, J. Kim, and J. V. Creeden, Jr., *CST* **14**, 43 (1976).

84. A. Fernández-Pello and F. A. Williams, *C & F* **28**, 251 (1977).

85. C. C. Feng and W. A. Sirignano, *C & F* **29**, 247 (1977).

86. A. C. Fernández-Pello, S. R. Ray, and I. Glassman, *CST* **19**, 19 (1978).

87. A. E. Frey, Jr. and J. S. T'ien, *C & F* **36**, 263 (1979).

88. R. A. Altenkirch, R. Eichhorn, and P. C. Shang, *C & F* **37**, 71 (1980).

89. A. C. Fernández-Pello, S. R. Ray, and I. Glassman, *18th Symp.* (1981), 579–589.

90. R. A. Altenkirch, M. Rezayat, R. Eichhorn, and F. J. Rizzo, *Journal of Heat Transfer* **104**, 734 (1982).

91. P. H. Thomas, *CST* **28**, 173 (1982).

92. I. S. Wichman, F. A. Williams, and I. Glassman, *19th Symp.* (1982), 835–845.

93. I. S. Wichman and F. A. Williams, *CST* **32**, 91 (1983).

94. I. S. Wichman, *C & F* **50**, 287 (1983).

95. I. S. Wichman and F. A. Williams, *CST* **33**, 207 (1983).

96. A. C. Fernández-Pello and T. Hirano, *CST* **32**, 1 (1983).

97. S. R. Ray and I. Glassman, *CST* **32**, 33 (1983).

98. R. A. Altenkirch, R. Eichhorn, and A. R. Rizvi, *CST* **32**, 49 (1983).

99. M. Vogel and F. A. Williams, *CST* **1**, 429 (1970).

100. H. W. Emmons and T. Shen, *13th Symp.* (1971), 917–926.

101. F. R. Steward and K. N. Tennankore, *18th Symp.* (1981), 641–646.

102. G. H. Markstein and J. de Ris, *14th Symp.* (1973), 1085–1097.

103. L. Orloff, J. de Ris, and G. H. Markstein, *15th Symp.* (1975), 183–192.

104. A. C. Fernández-Pello, *CST* **17**, 87 (1977).

105. M. Sibulkin and J. Kim, *CST* **17**, 39 (1977).

106. A. C. Fernández-Pello, *C & F* **31**, 135 (1978).

107. A. C. Fernández-Pello, *C & F* **36**, 63 (1979).

108. K. Annamalai and M. Sibulkin, *CST* **19**, 185 (1979).

109. A. C. Fernández-Pello and C. P. Mao, *CST* **26**, 147 (1981).

110. G. Carrier, F. Fendell, and S. Fink, *CST* **32**, 161 (1983).

111. J. H. Burgoyne, A. F. Roberts, and P. G. Quinton, *Proc. Roy. Soc. London* **308A**, 39 (1968).

112. I. Glassman, J. G. Hansel, and T. Eklund, *C & F* **13**, 99 (1969).

113. R. Mackinven, J. G. Hansel, and I. Glassman, *CST* **1**, 293 (1970).

114. W. A. Sirignano and I. Glassman, *CST* **1**, 307 (1970).

115. K. E. Torrance, *CST* **3**, 133 (1971).

116. K. Akita, *14th Symp.* (1973), 1075–1083.

117. K. E. Torrance and R. L. Mahajan, *15th Symp.* (1975), 281–287.

118. K. E. Torrance and R. L. Mahajan, *CST* **10**, 125 (1975).

119. I. Glassman and F. L. Dryer, *Fire Safety Journal* **3**, 123 (1980).

120. T. Suzuki and T. Hirano, *19th Symp.* (1982), 877–884.

APPENDIX A

Summary of Applicable Results of Thermodynamics and Statistical Mechanics

Thermodynamics and statistical mechanics deal with systems in *equilibrium* and are therefore applicable to phenomena involving flow and irreversible chemical reactions only when departures from complete equilibrium are small. Fortunately this is often true in combustion problems, but occasionally thermodynamical concepts yield useful results even when their validity is questionable [for example, in the analysis of detonation structure (see Section 6.1.5) and in transition-state theory (see Section B.3.4)]. The presentation is restricted to chemical systems; appropriate independent thermodynamic coordinates are pressure, p, volume, V, and the total number of moles of a chemical species in a given phase, N_i. Moreover, results related to combustion theory are emphasized.

In Section A.1, the general laws of thermodynamics are stated. The results of statistical mechanics of ideal gases are summarized in Section A.2. Chemical equilibrium conditions for phase transitions and for reactions in gases (real and ideal) and in condensed phases (real and ideal) are derived in Section A.3, where methods for computing equilibrium compositions are indicated. In Section A.4 heats of reaction are defined, methods for obtaining heats of reaction are outlined, and adiabatic flame-temperature calculations are discussed. In the final section (Section A.5), which is concerned with condensed phases, the phase rule is derived, dependences of the vapor pressure and of the boiling point on composition in binary mixtures are analyzed, and properties related to osmotic pressure are discussed.

A.1. GENERAL THERMODYNAMICAL RESULTS

A.1.1. The laws of thermodynamics

In an open (size and composition not fixed) chemical system, the quantities p, V, N_i ($i = 1, \ldots, N$) constitute a convenient complete set of independent variables for specifying the thermodynamic state of the system. The pressure p is an intensive property, while the volume V and the mole number N_i ($i = 1, \ldots, N$) are extensive variables.* Here the numbers of moles of the same species present in different phases are distinguished by different values of the subscript i, so that N is the *sum* of the total number of species present in *each* phase.

The zeroth law of thermodynamics states that there exists an additional intensive variable, **temperature** $T = T(p, V, N_i)$, which has the same value for all systems in equilibrium with each other.

The first law of thermodynamics is conservation of energy, which states that there exists an extensive function $U = U(p, V, N_i)$, called the **internal energy**, having the property that for a closed system (one that does not exchange material with its surroundings) the heat added to the system in an infinitesimal process is

$$d\hat{Q} = dU + p \, dV.^\dagger \tag{1}$$

In the classical formulation, the second law of thermodynamics states that there exist an absolute scale for the temperature T and an extensive function $S(p, V, N_i)$, called the **entropy**, such that for an infinitesimal process in a closed system

$$T \, dS \geq d\hat{Q}, \tag{2}$$

where the equality holds for reversible[‡] processes and the inequality is valid for natural processes.

The particular result of statistical mechanics that is sometimes referred to as the third law of thermodynamics appears to be of little use in combustion.

---

* An *intensive property* may be defined as a property that is unchanged when the size of the system is increased by adding to it any number of systems that are identical to the original system. An *extensive property* is one that increases in proportion to the size (for example, volume) of the system in such a "process." Thus an intensive property may be formed from any extensive property through division by any other extensive property.

† Since we are developing the thermodynamics of equilibrium chemical systems, the only work term included here is $p \, dV$. The equation of energy conservation in a flow system is derived in Appendix C.

‡ A reversible process is a process performed in such a way that the system and its surroundings can both return to their initial states. Such a process must be carried out very slowly so that the system remains in an equilibrium state throughout the process.

A.1.2. Thermodynamic functions

From equations (1) and (2) it follows that $T = (\partial U/\partial S)_{V, N_i}$ and $p = -(\partial U/\partial V)_{S, N_i}$, whence by defining the **chemical potential** of a given species in a given phase as

$$\mu_i \equiv (\partial U/\partial N_i)_{S, V, N_{j(j \neq i)}},$$

it is found that

$$dU = T \, dS - p \, dV + \sum_{i=1}^{N} \mu_i \, dN_i. \tag{3}$$

From equation (3) it is seen that the **Helmholtz function** $A \equiv U - TS$, the **enthalpy** $H \equiv U + pV$, and the **Gibbs function** $G \equiv H - TS$ obey the relations*

$$dA = -S \, dT - p \, dV + \sum_{i=1}^{N} \mu_i \, dN_i, \tag{4}$$

$$dH = T \, dS + V \, dp + \sum_{i=1}^{N} \mu_i \, dN_i, \tag{5}$$

and

$$dG = -S \, dT + V \, dp + \sum_{i=1}^{N} \mu_i \, dN_i. \tag{6}$$

Since the partial molar value (with respect to i) of any extensive property χ is defined as $(\partial \chi/\partial N_i)_{p, T, N_{j(j \neq i)}}$, equation (6) shows that μ_i is the **partial molar Gibbs function**.

Equations (3)–(6) display a certain degree of symmetry and have been called the *fundamental equations* of chemical thermodynamics [2]. They are very useful in providing general relations between thermodynamic properties. For example, from equation (6) it follows that

$$(\partial S/\partial p)_{T, N_i} = -(\partial V/\partial T)_{p, N_i};$$

relations of this type are usually called *Maxwell equations*. Furthermore, if any one of the properties U, A, H, or G is known as a function of the independent variables appearing in its fundamental equation, then all the other thermodynamic properties can be evaluated as functions of these independent variables by using the appropriate fundamental equation and property definitions to relate the unknown properties to the known function and its derivatives.

An integrated relation of completely general validity between G and the μ_i's and N_i's can be obtained from equation (6) by using the fact that G

---

* The names given to these functions vary considerably in the literature; the nomenclature chosen here is that which appears to be least likely to lead to confusion (see [1]).

is an extensive property. In a "process" in which the size of the system is increased by adding systems with the same intensive properties, all intensive properties remain constant and all extensive properties increase proportionally. Hence $dT = 0$, $dp = 0$, and $d\mu_i = 0$ in such a process, which shows that equation (6) can readily be integrated from $G = 0$, $N_i = 0$ to G, N_i, yielding

$$G = \sum_{i=1}^{N} \mu_i N_i. \tag{7}$$

Similar integrations may be performed in equations (3)–(5), but no new relationships are obtained; the result is always equivalent to equation (7). On the other hand, precisely this same procedure may be used in conjunction with the general relation

$$d\chi = \left(\frac{\partial \chi}{\partial T}\right)_{p, N_i} dT + \left(\frac{\partial \chi}{\partial p}\right)_{T, N_i} dp + \sum_{i=1}^{N} \left(\frac{\partial \chi}{\partial N_i}\right)_{p, T, N_j(j \neq i)} dN_i$$

to show that any extensive property χ may be expressed in terms of its partial molar values, namely,

$$\chi = \sum_{i=1}^{N} \left(\frac{\partial \chi}{\partial N_i}\right)_{p, T, N_j(j \neq i)} N_i. \tag{7a}$$

Differentiating equation (7) and substituting the result into equation (6) gives

$$\sum_{i=1}^{N} N_i \, d\mu_i = -S \, dT + V \, dp, \tag{8}$$

a general result which is useful in discussions of phase equilibria.

A.2. PERTINENT RESULTS OF STATISTICAL MECHANICS

A.2.1. Background

Two important objectives of statistical mechanics are (1) to verify the laws of thermodynamics from a molecular viewpoint and (2) to make possible the calculation of thermodynamic properties from the molecular structure of the material composing the system. Since a thorough discussion of the foundations, postulates, and formal development of statistical mechanics is beyond the scope of this summary, we shall dispose of objective (1) by merely stating that for all cases in which statistical mechanics has successfully been developed, the laws quoted in the preceding section have been found to be valid. Furthermore, in discussing objective (2), we shall merely quote results; the reader is referred to the literature [3–7] for amplification.

Since conditions encountered in combustion problems are virtually always in the range where departures from Boltzmann statistics are negligibly small, the results of Fermi-Dirac and Bose-Einstein statistics will not be considered here.* Complications associated with strong many-body intermolecular forces (for example, liquids) will also be neglected; we shall restrict our attention to ideal gases, although many of the formulas will also be valid for solids.

The state of a gas is completely determined by specifying the states of all molecules composing the gas. In view of the large number of molecules involved (and the continual external perturbations present in any real system), one cannot hope to determine the state of a gas; the most information that can be obtained is a specification of some sort of average state. This average state is the thermodynamic state (that is, only the thermodynamic properties are specified), which might be expected to depend on all possible states of the molecules composing the gas.

Molecular states are determined by the solution of the steady-state Schrödinger equation [8–10] of quantum mechanics for the wave function of the system (the nuclei and electrons composing the molecule). For each molecular state (that is, for each wave function), an energy of the molecule is obtained as an eigenvalue of the Schrödinger equation. The different energy eigenvalues are called *energy levels* and will be labeled by the subscript α, which conventionally increases as the energy increases. Thus $\epsilon_{i,\alpha}$ will denote the α **energy level** of a molecule of kind i. The quantities α, the definitions of which are clearly somewhat arbitrary, are called **quantum numbers**.[†] It is often found that a number of different states (wave functions) have the same energy eigenvalue; the number of states with energy $\epsilon_{i,\alpha}$ is called the **degeneracy** of energy level α and will be denoted by $g_{i,\alpha}$ for molecules of species i.

For the purposes of statistical mechanics it is sufficient to know only the energy levels and their corresponding degeneracies for the molecules; a knowledge of the wave function itself is unnecessary. Therefore, the specification of $\epsilon_{i,\alpha}$ and $g_{i,\alpha}$ for all energy levels α will be termed a specification of the structure of a molecule of type i. Although, in principle, the structure of a molecule is determined by solving the appropriate Schrödinger equation, in practice this is prohibitively difficult except for the simplest molecules. More often, the structure is found by analyzing spectroscopically the light emitted and absorbed by the molecule (emission or absorption is accompanied by a transition of the molecule from one energy level to another,

---

* The kind of statistics obeyed by the system depends on the symmetry properties of the quantum-mechanical wave functions describing the molecules composing the system [3–7].

† For example, in some cases the α values may be taken as either integers $(0, 1, \ldots)$ or half-integers $(\frac{1}{2}, \frac{3}{2}, \ldots)$; the choice is based on the nature of the particular Schrödinger equation describing the molecule.

and the frequency v of the light is related to the energy difference of the levels, $\Delta\epsilon = hv$, h = Planck's constant). In many cases the spectroscopic data are correlated by fitting to functional forms derived from simple quantum-mechanical models of the molecular structure.

A.2.2. Summary of results

The results of Boltzmann (or "classical") statistics are conveniently expressed in terms of the **partition function** for species i,

$$Q_i \equiv \sum_\alpha g_{i,\alpha} e^{-\epsilon_{i,\alpha}/k^0 T}, \quad i = 1, \ldots, N, \tag{9}$$

where k^0 is Boltzmann's constant and the summation is over all energy levels α.* The primary prediction of Boltzmann statistics is that in thermodynamic equilibrium, the average number of moles of species i in energy level α is

$$N_{i,\alpha} = \frac{N_i g_{i,\alpha} e^{-\epsilon_{i,\alpha}/k^0 T}}{Q_i} = -N_i k^0 T \frac{\partial \ln Q_i}{\partial \epsilon_{i,\alpha}}, \quad i = 1, \ldots, N; \tag{10}$$

that is, $N_{i,\alpha}$ is proportional to $g_{i,\alpha} e^{-\epsilon_{i,\alpha}/k^0 T}$. The importance of the partition function arises mainly from the fact that all thermodynamic properties of the system can be expressed in terms of Q_i and its derivatives with respect to T and V. It can, in fact, be shown that [3], [4].

$$A = \sum_{i=1}^N N_i R^0 T [\ln(\mathscr{A} N_i / Q_i) - 1], \tag{11}$$

$$U = \sum_{i=1}^N N_i R^0 T^2 \frac{\partial \ln Q_i}{\partial T},^\dagger \tag{12}$$

$$S = \sum_{i=1}^N \frac{\partial}{\partial T} \{ N_i R^0 T [-\ln(\mathscr{A} N_i / Q_i) + 1] \}, \tag{13}$$

$$p = \sum_{i=1}^N N_i R^0 T \frac{\partial \ln Q_i}{\partial V},^\ddagger \tag{14}$$

---

* It may be noted that in nearly all cases α is really a set of numbers; more than one quantum number is used in identifying the energy levels. Furthermore, for certain systems some energy levels may be continuous and the corresponding summations become integrals.

† This provides a caloric equation of state for the system. It follows directly from the fact that U is the average energy of the system, that is,

$$U = \sum_{i=1}^N \sum_\alpha (\mathscr{A} N_{i,\alpha} \epsilon_{i,\alpha}),$$

where $N_{i,\alpha}$ is given by equation (10).

‡ Clearly this relation is equivalent to an equation of state (derived from statistical mechanics) for the system; it reduces to the ideal-gas equation of state in the present case. Equation (14) follows from the fact that pressure may in general be defined as the average value of the sum over all molecules of $-\partial\epsilon_{i\alpha}/\partial V$; that is, $p = -\sum_{i=1}^N \sum_\alpha (\mathscr{A} N_{i,\alpha} \partial\epsilon_{i,\alpha}/\partial V)$.

and

$$\mu_i = R^0 T \ln(\mathscr{A} N_i / Q_i), \quad i = 1, \ldots, N, \tag{15}$$

where $\mathscr{A}$ is Avogadro's number and $R^0 = \mathscr{A} k^0$ is the universal gas constant. From equations (11)–(15)* and from the definitions in the preceding section, all previously defined thermodynamic functions can be related to T, V, and N_i provided that Q_i is a known function of T and V. The problem of calculating thermodynamic properties of gases therefore reduces to the problem of evaluating Q_i in terms of T and V.

A.2.3. Evaluation of partition functions

Aside from electronic excitations, a molecule composed of n atoms has $3n$ degrees of freedom, one degree of freedom corresponding to each Cartesian coordinate determining the location of each atom. Three of these degrees of freedom are associated with translation of the center of mass of the molecule, while two (for linear molecules[†]) or three (for nonlinear molecules) may be associated with rotation of the molecule as a whole. The remaining $3n - 5$ or $3n - 6$ degrees of freedom generally involve vibrational motions of the molecule.

One important property of the partition function is that for degrees of freedom which are separable in the sense that the Schrödinger equation for the system may be simplified by applying the method of separation of variables to the coordinates representing these degrees of freedom, the partition function for the system equals the product of partition functions associated with each set of separable degrees of freedom. Explicitly, if the degrees of freedom of the molecule are separable into two groups, 1 and 2, then the Schrödinger equation shows that the energy of the molecule in any given state is $\epsilon_{i,\alpha} = \epsilon_{i,\beta} + \epsilon_{i,\gamma}$, where $\epsilon_{i,\beta}$ and $\epsilon_{i,\gamma}$ represent the energies associated with groups 1 and 2, respectively, and for each state of group 1 (or 2) there exist states of the system corresponding to all possible states of group 2 (or 1).[‡] Hence,

$$\sum_\alpha g_{i,\alpha} e^{-\epsilon_{i,\alpha}/k^0 T} = \sum_{\beta,\gamma} g_{i,\beta} g_{i,\gamma} e^{-(\epsilon_{i,\beta} + \epsilon_{i,\gamma})/k^0 T},$$

whence it follows from equation (9) that $Q_i = Q_{i,1} Q_{i,2}$, where

$$Q_{i,1} = \sum_\beta g_{i,\beta} e^{-\epsilon_{i,\beta}/k^0 T} \quad \text{and} \quad Q_{i,2} = \sum_\gamma g_{i,\gamma} e^{-\epsilon_{i,\gamma}/k^0 T}.$$

---

* Actually, equation (11) alone is sufficient; equations (12)–(15) may be derived from equation (11) by purely thermodynamical considerations.

† Linear molecules are molecules in which all of the atoms lie along a straight line in the state of lowest energy.

‡ Note that α consists of at least two quantum numbers (β and γ) in this case.

The importance of this property of the partition function arises from the fact that in ideal gases some molecular degrees of freedom (for example, translational) are rigorously separable while, with conventional approximations, other degrees of freedom (for example, rotational) also become separable.

After separation, the Schrödinger equations for most groups of degrees of freedom become identical with the Schrödinger equations for simple systems for which the quantum mechanical structure has been determined. Since the partition functions for these simple systems are easily evaluated from their structure, the partition function of the molecule often can be computed as a product of known partition functions. The partition functions *per degree of freedom* for these simple systems are listed below.

The function

$$\left(\frac{2\pi m k^0 T}{h^2}\right)^{1/2} V^{1/3}$$

represents a structureless particle of mass m in a box of volume V and corresponds to a translational degree of freedom. Factors from translational degrees of freedom are the only factors in the partition function that depend on V; all others are functions of temperature alone.

The function

$$\left(\frac{8\pi^2 I k^0 T}{h^2}\right)^{1/2}$$

represents a linear rigid rotator with moment of inertia I and corresponds to a rotational degree of freedom for a linear molecule, in which case $I = \sum_{j=1}^{n} m_j a_j^2$, where m_j is the mass of the jth atom of the molecule, and a_j is the distance of the jth atom from the center of mass of the molecule. The formula quoted here may become inaccurate below room temperature and requires minor modification (multiplication by a constant) if some of the atoms in the molecule are identical [4]. Partition functions for symmetrical and unsymmetrical tops have also been computed and (except for a constant factor near unity) agree with the formula quoted here if I is a principal moment of inertia; these systems would correspond to rotational degrees of freedom for nonlinear molecules.

The function

$$(1 - e^{-h\nu/k^0 T})^{-1}$$

represents a harmonic oscillator of natural frequency ν and corresponds to a vibrational degree of freedom.

Analyses of linear nonrigid rotators, anharmonic oscillators, and vibrating rotators, yielding first-order corrections for nonrigidity, anharmonicity, and vibration-rotation interaction (nonseparability of vibrational and rotational modes), respectively, have also been completed and are conventionally used in obtaining corrections (which are most important at elevated temperatures) to the simple product form of the molecular partition function. These (more complicated) formulas [4], [11] will not be quoted here.

Electronic excitations have not been considered above. Electronic systems for which the quantum mechanical structure has been theoretically determined, at least to a close approximation, include the hydrogen atom and hydrogenlike atoms (for example, sodium). Accurate experimental values for electronic energy levels are available for an extremely large number of atoms and molecules. For nearly all molecules the energy difference between successive electronic energy levels is so great that provided the temperature is not too large ($T \lesssim 2000$K), successive electronic terms in the sum in equation (9) decrease so rapidly that only the first term (or, possibly, the first few terms) in the sum must be retained. An electronic factor $g_{i,\alpha} e^{-\epsilon_{i,\alpha}/k^0 T}$ is therefore multiplied by the partition function computed by the separation procedure described above.* When chemical reactions are of importance, the value of $\epsilon_{i,\alpha}$ in the ground state cannot be chosen arbitrarily; it must include the energy of formation of the molecule.

Finally, nuclear spin also produces an additional constant factor in the partition function; a nucleus with spin quantum number s contributes a factor $2s + 1$ to Q_i.

In order to use the procedure described above to calculate the partition function of a given molecule as a function of T and V, one must know the appropriate I and v values for the various degrees of freedom of the molecule. A great deal of this type of data is tabulated for diatomic molecules in [11] and for polyatomic molecules in [12]. Updated information may be found in [13], [14], and [15].

A.3. CHEMICAL EQUILIBRIUM

A.3.1. General equilibrium condition

Although conventional derivations of the conditions for chemical equilibrium sometimes are restricted to isothermal, isobaric processes (possibly because often $dT = 0$ and $dp = 0$ in experimental equilibrium determinations), the general equilibrium conditions do not depend on these assumptions, as will be seen from the following development. By substituting equation (1) into equation (2) and using equation (3) to eliminate dU from the resulting expression, we find that

$$\sum_{i=1}^{N} \mu_i \, dN_i \leq 0 \qquad (16)$$

---

* In general, the I and v values for the nuclei change when the electronic energy levels change; if more than one electronic level must be considered, Q_i is the *sum* of terms involving perature for thermal explosions.

in a closed system. In equation (16), the inequality holds for natural processes and the equality is valid in equilibrium. A single arbitrary chemical reaction may be written in the form

$$\sum_{i=1}^{N} v_i' \mathfrak{M}_i \rightleftharpoons \sum_{i=1}^{N} v_i'' \mathfrak{M}_i, \tag{17}$$

where v_i' is the stoichiometric coefficient for species i appearing as a reactant, v_i'' is the stoichiometric coefficient for species i appearing as a product, and $\mathfrak{M}_i$ represents the chemical symbol for species i.* For a closed system in which only the reaction in (17) may take place, equation (17) implies that there is a relationship between the change in the number of moles of each species, namely,

$$dN_i/(v_i'' - v_i') = dN_j/(v_j'' - v_j') \tag{18}$$

for any two species i and j. From equation (18) it follows that a single **reaction progress variable** ϵ may be defined such that

$$dN_i = (v_i'' - v_i')\, d\epsilon, \quad i = 1, \ldots, N. \tag{19}$$

Substituting equation (19) into equation (16) yields

$$\left[\sum_{i=1}^{N} \mu_i(v_i'' - v_i') \right] d\epsilon \leq 0. \tag{20}$$

Equation (20) shows that in natural processes in closed systems, ϵ increases [the reaction in (17) goes in the forward direction, $d\epsilon > 0$] when

$$\sum_{i=1}^{N} \mu_i(v_i'' - v_i') < 0$$

and ϵ decreases ($d\epsilon < 0$) when

$$\sum_{i=1}^{N} \mu_i(v_i'' - v_i') > 0.$$

The equilibrium condition [the requirement that the equality in equation (20) be valid for an arbitrary small change $d\epsilon$] is clearly

$$\sum_{i=1}^{N} \mu_i(v_i'' - v_i') = 0. \tag{21}$$

Although, for brevity, the subscript i was previously said to identify a particular species, it will be recalled from the general definition in Section A.1 that i actually identifies a particular species in a particular phase. Hence equation (21) also expresses the condition for phase equilibrium. For example,

---

* Species which are not reactants have $v_i' = 0$, while those that do not appear as products have $v_i'' = 0$.

if equation (17) is taken to represent a simple phase transition of a given species from phase 1 to phase 2, then $v'_1 = 1$, $v'_2 = 0$, $v''_1 = 0$, $v''_2 = 1$, and equation (21) becomes $\mu_1 = \mu_2$, which is the equilibrium condition for any given material present in phases 1 and 2. In fact, this result and, more generally, the properties expressed by equations (20) and (21) provide the justification for the name chemical potential for μ_i; a reaction moves in the direction of decreasing chemical potential and reaches equilibrium only when the potential of the reactants equals the potential of the products.

A.3.2. Phase equilibria

If phases 1 and 2 each contain only one species, then the above expression $\mu_1 = \mu_2$ for this species may be used to obtain a relation between p and T at phase equilibrium. Differentiating $\mu_1 = \mu_2$ yields $d\mu_1 = d\mu_2$ for a small change which maintains equilibrium. Expressing $d\mu_1$ and $d\mu_2$ in terms of dT and dp by applying equation (8) to each phase separately, we then obtain*

$$\frac{1}{N_1}(-S_1\,dT + V_1\,dp) = \frac{1}{N_2}(-S_2\,dT + V_2\,dp),$$

which, after rearrangement, becomes

$$\frac{dp}{dT} = \frac{S_2/N_2 - S_1/N_1}{V_2/N_2 - V_1/N_1}. \tag{22}$$

Equation (22) is the well-known Clapeyron equation for phase equilibrium.

The enthalpy change of a mole of the species in passing from phase 1 to phase 2 is the **heat of transition** per mole, $L_{12} \equiv H_2/N_2 - H_1/N_1$. In equation (22) $S_2/N_2 - S_1/N_1$ is often replaced by L_{12}/T; their equivalence follows from the definition $G = H - TS$ and the result $G_1/N_1 = G_2/N_2$ [which is a direct consequence of equation (7) (applied to each phase) since $\mu_1 = \mu_2$].

For transitions in which 1 is a condensed phase and 2 is a gas, $V_2/N_2 \gg V_1/N_1$ since the volume per mole of a gas is considerably larger (usually by a factor of about 10^3) than that of a liquid or solid. If it is also assumed that the gas obeys the ideal-gas equation of state, $pV_2 = N_2 R^0 T$, then equation (22) reduces to

$$dp/dT = pL_{12}/R^0 T^2,$$

which is readily integrated, yielding the Clausius-Clapeyron equation

$$p = p_0 \exp\left[-\int_T^{T_0} \frac{L_{12}}{R^0 T^2}\,dT\right], \tag{23}$$

---

* Necessarily, $T_1 = T_2 \equiv T$ and $p_1 = p_2 \equiv p$ for thermal and mechanical equilibrium.

where p_0 is the value of p at the arbitrarily chosen temperature T_0. Equation (23) gives the temperature dependence of the equilibrium **vapor pressure** p; since $L_{12} \approx$ constant for a sufficiently small temperature range, equation (23) shows that roughly $p \sim e^{-L_{12}/(R^0 T)}$.

A.3.3. Ideal-gas reactions

For chemical reactions in ideal gases, it is possible to express the equilibrium condition in more convenient forms by relating μ_i to other properties of the gas mixture. Substituting equation (15) into equation (21) yields

$$\sum_{i=1}^{N} \ln \left[(\mathscr{A} N_i / Q_i)^{v_i'' - v_i'} \right] = 0,$$

which may be written in the form

$$\prod_{i=1}^{N} c_i^{v_i'' - v_i'} = \prod_{i=1}^{N} Q_i'^{v_i'' - v_i'} \equiv K_c, \tag{24}$$

where the molar **concentration** of species i has been defined as $c_i \equiv N_i / V$ and the partition function *per unit volume* is* $Q_i' \equiv Q_i / (\mathscr{A} V)$. It was seen in Section A.2 that for ideal gases Q_i is proportional to V, implying that Q_i' depends on temperature alone (in addition to physical and molecular constants). Hence K_c, which is called the **equilibrium constant for concentrations**, depends only on temperature, and equation (24) expresses a necessary relationship between reactant and product concentrations for equilibrium of the reaction of (17) in an ideal-gas mixture at a given T.

From the same kind of reasoning that was used in defining the total pressure p, the **partial pressure** of species i may be defined as

$$p_i = N_i R^0 T \frac{\partial \ln Q_i}{\partial V}, \quad i = 1, \ldots, N,$$

whence the fact that Q_i is proportional to V implies that

$$p_i = c_i R^0 T, \quad i = 1, \ldots, N, \tag{25}$$

and equation (14) implies that $p = \sum_{i=1}^{N} p_i$. By using equation (25) to eliminate c_i from equation (24), we find that

$$\prod_{i=1}^{N} p_i^{v_i'' - v_i'} = K_c (R^0 T)^{\sum_{i=1}^{N} (v_i'' - v_i')} \equiv K_p, \tag{26}$$

where K_p, which is the **equilibrium constant for partial pressures**, is seen to be a function of temperature alone.

---

* The factor $\mathscr{A}$ is included here because we desire an expression for molar (instead of molecular) concentrations.

The **mole fraction** of species i is defined as $X_i \equiv N_i / \sum_{i=1}^{N} N_i = c_i/c$, where $c \equiv \sum_{i=1}^{N} c_i$ is the total number of moles per unit volume. Since it follows from equation (25) that $X_i = p_i/p$, equation (26) shows that in equilibrium

$$\prod_{i=1}^{N} X_i^{v_i'' - v_i'} = K_p p^{-\sum_{i=1}^{N} (v_i'' - v_i')} \equiv K_X, \tag{27}$$

where K_X is the **equilibrium constant for mole fractions**. Since equation (27) shows that K_X depends on p as well as on T, it is more reasonable to tabulate K_p or K_c than K_X; tables of experimental values of K_p for many reactions may be found in [13], [16], and [17]. Equations (24), (26), and (27) are alternative forms of the ideal-gas chemical-equilibrium condition.

A.3.4. Non-ideal-gas reactions

For nonideal gases the above simplifications are not applicable and it would appear to be best to use equation (21) directly as the chemical equilibrium condition. However, it is conventional when dealing with non-ideal gases to replace μ_i by the **fugacity** f_i defined below. By solving equation (15) for N_i and using equation (25) to express N_i in terms of p_i, it is found that

$$p_i = R^0 T Q_i' e^{\mu_i / R^0 T}, \qquad i = 1, \ldots, N, \tag{28}$$

for an ideal gas. For nonideal gases, the fugacity is defined as

$$f_i = R^0 T Q_i' e^{\mu_i / R^0 T}, \qquad i = 1, \ldots, N, \tag{29}$$

where $Q_i' = Q_i'(T)$ is the previously defined partition function per unit volume, computed for species i in the limit $p \rightarrow 0$ (that is, when intermolecular forces become negligible and the gas becomes ideal). From equations (28) and (29), we see that f_i approaches the partial pressure when the gas becomes ideal, that is, $\lim_{p \rightarrow 0}(f_i/p_i) = 1$.

By using equations (29) in equation (21), it can be shown that the condition for chemical equilibrium in nonideal gases is

$$\prod_{i=1}^{N} f_i^{v_i'' - v_i'} = \prod_{i=1}^{N} (Q_i' R^0 T)^{v_i'' - v_i'} \equiv K_f, \tag{30}$$

where K_f, the **equilibrium constant for fugacities**, is a function of T alone and, in fact, is equal to the K_p for the same reaction in an ideal gas mixture. The only difficulty in non-ideal-gas equilibrium calculations is therefore in determining the fugacities.

A.3.5. Reactions in condensed phases

The equilibrium conditions for a reaction in a single, homogeneous, condensed phase (liquid or solid) can be related to the equilibrium conditions for the same reaction occurring in a gas mixture which is in (possibly meta-stable) equilibrium with the condensed phase by using the phase-equilibrium conditions $\mu_{i(\text{cond phase})} = \mu_{i(\text{gas})}$ for all species i. Thus equation (21) for the reaction in the condensed phase becomes identical with equation (21) for the same reaction in the gas.

An **ideal solution** [2], [14], [18] may be defined as any multicomponent condensed phase in which the mole fraction of each species is given by

$$X_i = \alpha_i f_i, \quad i = 1, \ldots, N, \tag{31}$$

where f_i is the fugacity of species i in the gas which would be in equilibrium with the given condensed phase, and the constants of proportionality α_i may depend on p and T, but must be independent of composition, X_i. This concept of an ideal solution is of value because it represents the simplest kind of condensed mixture that has any pretense to physical reality; although most solutions are not ideal [by the definition in equation (31)], there exist some real mixtures which are ideal, and many other solutions approach ideal behavior as they become dilute. In most cases the constants α_i in equation (31) are empirically found to have little, if any, pressure dependence, $\alpha_i \approx \alpha_i(T)$. When the gas in equilibrium with the condensed phase is ideal ($f_i = p_i$), equation (31) reduces to Raoult's law, $X_i \sim p_i$.

From equations (30) and (31) the condition for equilibrium of the reaction of (17) in an ideal solution is found to be

$$\prod_{i=1}^{N} X_i^{v_i'' - v_i'} = K_f \prod_{i=1}^{N} \alpha_i^{v_i'' - v_i'} \equiv K_X, \tag{32}$$

where, in addition to having a usually strong T dependence, K_X may now also depend weakly on p through α_i. In practice, equation (32) is often replaced by an equilibrium equation for concentrations, c_i, but the result will not be given here because, except in special cases, the equilibrium "constant" in the new equation depends on composition (for example, c_i), as well as on p and T, even for ideal solutions.

Equilibrium conditions for chemical reactions in nonideal solutions are conventionally expressed in terms of **activities**, defined as

$$a_i = \beta_i e^{\mu_i/R^0 T}, \quad i = 1, \ldots, N, \tag{33}$$

where the coefficient β_i, which may depend on p and T but is independent of all compositions, X_i, is defined by assigning a definite value to each a_i in a mixture at some chosen composition (all X_i given). The standard composition for assigning values to a_i depends on the system under consideration; for example, for a solvent or a liquid mixture β_i is defined by $\lim_{X_i \to 1} (a_i/X_i) = 1$, while in a binary system involving a solid solute i, β_i is defined by $\lim_{X_i \to 0} (a_i/X_i) = 1$. In these two examples, as nearly always, the choice of the standard composition is governed by the observation that the mixture becomes ideal at the limiting standard composition.

By solving equation (33) for μ_i and substituting the result into equation (21), we find that the equilibrium condition in nonideal mixtures becomes

$$\prod_{i=1}^{N} a_i^{v_i'' - v_i'} = \prod_{i=1}^{N} \beta_i^{v_i'' - v_i'} \equiv K_a, \tag{34}$$

where the **equilibrium constant for activities,** K_a, depends on p and T but is independent of composition. As with equation (30), the greatest difficulty in using equation (34) is in determining a_i, not K_a (K_a is simply K_X in the limit of ideality). The reader should be cautioned that many different quantities similar to a_i are used in practice; to name a few, there are activities defined in terms of concentrations and molalities instead of mole fractions, and there are activity coefficients and osmotic coefficients defined differently.

A.3.6. Heterogeneous reactions

Equilibria for surface processes and for heterogeneous processes more complicated than simple phase transitions should perhaps also be considered here. Much is known about the thermodynamic properties of surface phases and about equilibria in multiphase systems [2]. However, for our purposes it appears to be sufficient to consider restricted classes of these phenomena, and it will be of greater value to treat these classes as limiting cases of heterogeneous rate processes (see Section B.4).

A.3.7. Calculation of equilibrium compositions

We have been restricting our attention to the equilibrium condition for a single reaction, (17). It is also worthwhile to investigate the problem of computing equilibrium compositions for systems in which many reactions may occur simultaneously. Explicitly, the problem is the following: given the pressure, the temperature, and the total number of moles of atoms* in the system (irrespective of the number of moles of the chemical compounds in which these atoms may appear), find the number of moles of all chemical species (compounds and free atoms) in the system at equilibrium.

Let us consider an ideal homogeneous mixture (either a gas or a condensed phase) containing N different chemical species and admitting the M independent chemical reactions

$$\sum_{i=1}^{N} v_{i,k}' \, \mathfrak{M}_i \rightleftharpoons \sum_{i=1}^{N} v_{i,k}'' \, \mathfrak{M}_i, \qquad k = 1, \ldots, M. \tag{35}$$

---

* In typical practical problems, these numbers are known from the amounts of the various chemical species initially put into the system.

The reactions in (35) will be independent if and only if the corresponding algebraic equations for $\mathfrak{M}_i$,

$$\sum_{i=1}^{N} (v''_{i,k} - v'_{i,k})\mathfrak{M}_i = 0, \qquad k = 1, \ldots, M,$$

are linearly independent. The conditions for linear independence of such equations are well known; they may be investigated in terms of the corresponding matrix equation

$$\mathbf{v}\mathfrak{M} = 0,$$

where the components of the rectangular matrix $\mathbf{v}$ are $(v''_{i,k} - v'_{i,k})$ and the components of the vector $\mathfrak{M}$ are $\mathfrak{M}_i$. Since $X_i = N_i / \sum_{i=1}^{N} N_i$, either equation (27) or equation (32) for each of the M reactions in equation (35) will provide a total of M independent equilibrium equations relating the N unknowns N_i. Independence is an important restriction here because, if any of the reactions in equation (35) is not independent of the others, then the equilibrium equation [for example, equation (27) or equation (32)] for this reaction will not be independent of the equilibrium equations for the other reactions.

A consequence of the independence condition is that in general $M \leq N$; this is to be expected because, in addition to the equilibrium equations, it will be necessary in computing equilibrium compositions to use equations expressing the fact that atoms are neither created nor destroyed in chemical reactions. If $v_i^{(j)}$ denotes the number of atoms of kind j in a molecule of species i, then these atom-conservation equations may be written in the form

$$\sum_{i=1}^{N} v_i^{(j)} N_i = \mathfrak{N}_j, \quad j = 1, \ldots, L, \tag{36}$$

where $\mathfrak{N}_j$ is the known total number of moles of atom j in the system, and L is the number of different kinds of atoms present. Actually, some of the relations in equation (36) may not be independent of the others; this would occur, for example, if two atoms always appeared in the same combination (for example, in the two-component system experiencing only the reaction $N_2O_4 \rightleftharpoons 2NO_2$, nitrogen and oxygen always appear in the combination NO_2). Thus not all the L relations in equation (36) are useful in computing the equilibrium composition; only a subset composed of L' ($\leq L$) linearly independent relations is to be included ($L' \equiv$ largest number of linearly independent relations).

It is always found that $M + L' = N$, whence M (generally nonlinear) equilibrium equations and L' (linear) atom-conservation equations determine the equilibrium composition (that is, may be solved for all N_i). In system of practical importance (for example, for products of the combustion of hydrocarbon fuels in air), the complete set of N equations is quite complicated and usually must be solved iteratively. Methods of solution that have been found to be efficient may be divided into two broad classes [19], simplified methods for particular systems and general methods for arbitrary systems. The first class comprises techniques for which the computations often can be performed by hand [17], [19], [20], [21]; the second class is composed of methods that necessitate utilization of electronic computers [19], [21]–[25]. Although hand computations provide a good feeling for the relative importance of various reactions, the complexities of many systems of interest are so great that computer methods are essential in complete calculations. Now, packaged computer routines are available (for example, [25]) for calculating equilibrium compositions accurately in practically all chemical systems of interest. The limitations on accuracies of calculations stem primarily from uncertainties in thermochemical data [13], seldom from deficiencies in computational abilities [19], [21]. Working computer routines for calculating equilibria are rapidly becoming standard tools in combustion laboratories; exercising such routines frequently proves helpful in combustion research.

In combustion applications, the ideal homogeneous systems of interest are gaseous. Condensed phases, notably carbon and oxides of metals, are often present in high-temperature equilibria. Therefore, multiphase equilibria also are important in combustion. If a species i, which is present in the ideal-gas mixture, condenses to a liquid or solid, then its partial pressure remains equal to its equilibrium vapor pressure, $p_i(T)$, a unique function of temperature given by equation (23). The K_p for gaseous equilibria involving species that condense are often written with the factor $p_i(T)$ included in a modified definition of K_p for computational convenience [13], [17]. With p and T specified, the mole fraction $X_i = p_i/p$ in the gas for a species that condenses becomes known in advance, thereby reducing by one the number of unknowns to be found in calculating the gaseous equilibrium. Associated with this reduction is the loss for the gas of one independent equation of the type given in equation (36); the condensed phase acts as a reservoir for the condensible species and contributes any amount of that material needed by the gas in attaining equilibrium. After the gaseous equilibrium has been calculated, the amount of the material condensed may be evaluated from the appropriate balance equation that had previously been deleted from equation (36). Existing computer routines [25] automatically account for condensed phases; they calculate multiphase equilibria. The development of efficient methods for including condensed phases in general procedures for computation of chemical equilibria was a major problem that has been solved by mathematical analysis and the invention of novel algorithms [19], [23], [25].

A.4. HEATS OF REACTION

A.4.1. Definition of heat of reaction

There are a number of different ways in which a heat of reaction may be defined. One of the most general definitions is the following. If a closed system containing a given number of moles N_i of N different species at a given T and p is caused to undergo an isobaric process in which the N_i are changed to prescribed final values and in which the initial and final values of T are the same, then the heat liberated by the system, $-\hat{Q} \equiv -\int d\hat{Q}$, is the **heat of reaction** for this process.* From equation (1) and the definition of the enthalpy H, it follows that $d\hat{Q} = dH - V\,dp$, which implies that $\hat{Q} = \int dH \equiv \Delta H$ for isobaric processes. Hence, according to the above definition, the heat of reaction is the negative of the enthalpy increase of the system during the reaction and does not depend on the particular kind of isobaric process occurring.

It will be observed that in addition to p and T, the heat of reaction $-\Delta H$ depends on the specified initial and final compositions. To make the definition more precise when investigating a single (general) reaction as given in equation (17), it is often stipulated that the system must initially be composed of v_i' moles of each species $\mathfrak{M}_i$ and finally should contain v_i'' moles of each species $\mathfrak{M}_i$. The heat of reaction defined in this manner is, in general, different from $-\Delta H$ for the same process occurring in a system of different composition (for example, a system in which $v_i' + 1$ moles were transformed to $v_i'' + 1$ moles or one in which the reaction took place in a dilute solution containing a great excess of a species not entering into the reaction). Heats of reaction under other conditions (for example, in dilute solutions) are, however, sometimes also defined.

Since equations (12) and (14) and the form of the partition functions Q_i show that for ideal-gas mixtures, $H(=U + pV)$ may be expressed in the form

$$H = \sum_{i=1}^{N} N_i H_i, \tag{37}$$

where H_i, which depends only on T, is the enthalpy of one mole of pure gaseous species i at temperature T, it is clear that in the special case of ideal-gas reactions,

$$-\Delta H = -\sum_{i=1}^{N} (v_i'' - v_i') H_i. \tag{38}$$

---

* Most of the more recent scientific texts, with good justification, define the heat of reaction as the heat adsorbed ($+\hat{Q}$). Our reason for the definition used here is that in combustion, the overall reactions are exothermic ($\hat{Q}$ is negative), and it would be somewhat inconvenient to work continually with negative heats of reaction.

In other words, $-\Delta H$ for ideal-gas reactions is independent of both the pressure and the composition of the mixture in which the reaction occurs.*

A.4.2. Differential heat of reaction

An alternative definition of the heat of reaction, which is more useful for some problems than the definition given above, may be obtained by considering an infinitesimal, isothermal, isobaric process. Differentiation of equation (7a) with $\chi = H$ yields

$$dH = \sum_{i=1}^{N} \left(\frac{\partial H}{\partial N_i}\right)_{p,\,T,\,N_j(j \neq i)} dN_i + \sum_{i=1}^{N} N_i d\left(\frac{\partial H}{\partial N_i}\right)_{p,\,T,\,N_j(j \neq i)},$$

which may be substituted into the general relation preceding equation (7a) (with $\chi = H$) to show that

$$\sum_{i=1}^{N} N_i d\left(\frac{\partial H}{\partial N_i}\right)_{p,\,T,\,N_j(j \neq i)} = 0$$

when $dT = 0$ and $dp = 0$. Hence

$$dH = \sum_{i=1}^{N} \left(\frac{\partial H}{\partial N_i}\right)_{p,\,T,\,N_j(j \neq i)} dN_i \qquad (39)$$

for isothermal, isobaric processes.

Considering the reaction in (17) and introducing the corresponding reaction-progress variable, we find from equations (19) and (39) that

$$dH = \left[\sum_{i=1}^{N} (v_i'' - v_i')\left(\frac{\partial H}{\partial N_i}\right)_{p,\,T,\,N_j(j \neq i)}\right] d\epsilon. \qquad (40)$$

The heat of reaction (sometimes called the **differential** or **partial heat of reaction** to distinguish it from that defined previously) may then be defined as the heat evolved in this infinitesimal process per unit change of ϵ, whence the relation $d\hat{Q} = dH$ and equation (40) show that the heat of reaction may be written as

$$-\frac{dH}{d\epsilon} = -\sum_{i=1}^{N} (v_i'' - v_i')\left(\frac{\partial H}{\partial N_i}\right)_{p,\,T,\,N_j(j \neq i)}. \qquad (41)$$

According to this definition, the heat of reaction $-dH/d\epsilon$ for any reaction in any given mixture depends on the values of p, T and N_i in that mixture and may be evaluated in terms of the partial molar enthalpies of the reacting species.

---

* For reactions occurring in any homogeneous phase, $-\Delta H$ seldom ever depends strongly on p.

Although $-dH/d\epsilon \neq -\Delta H$ in general, for ideal-gas reactions the substitution of equation (37) into the right-hand side of equation (41) yields*

$$-\frac{dH}{d\epsilon} = -\sum_{i=1}^{N} (v_i'' - v_i')H_i, \qquad (42)$$

which shows by comparison with equation (38) that the definitions are equivalent $(-dH/d\epsilon = -\Delta H)$ for ideal gases.

A.4.3. Heat of formation and other properties

Heats of reaction for particular kinds of chemical processes are often identified by special names. The heat liberated, $-\Delta H$, when a hydrocarbon or a compound containing carbon, hydrogen, and oxygen combines with oxygen to yield H_2O and CO_2 as products is called the **heat of combustion** of the substance. The heat required, $+\Delta H$, per mole to break a diatomic molecule into its constituent atoms is the **heat of dissociation** of the molecule. When solvent is added to a given solution the heat absorbed, $+\Delta H$, is the **heat of dilution**. Of particular utility is the **heat of formation** (denoted by ΔH_f) of a substance, which is defined as the negative of the heat of reaction (that is, $+\Delta H$) for the reaction in which one mole of the material is formed from its elements in their standard states. The **standard state** of an element is the form that is stable at room temperature and atmospheric pressure (for example, A(g) for argon (g $\equiv$ gas), O_2(g) for oxygen, and C(s) for carbon (s $\equiv$ solid, graphite in this case)].

The importance of the heat of formation arises from the fact that $-\Delta H$ for any reaction can be evaluated in terms of the ΔH_f values for the reactant and product molecules. This follows from the additive property of heats of reaction, often referred to as Hess's law. If any given process is broken into a number of parts, then from the fact that H is a state function it is clear that ΔH for the given process is the sum of the ΔH values for each part. Since an arbitrary reaction [for example, equation (17)] may be broken into reactions in which each reactant (in succession) is decomposed into its elements (in their standard states) followed by reactions in which the elements are recombined to form the products, it follows that

$$-\Delta H = -\sum_{i=1}^{N} (v_i'' - v_i') \Delta H_{f,i}, \qquad (43)$$

where $\Delta H_{f,i}$ is ΔH_f for species i in its appropriate state.[†] Thus once the

---

* Alternatively, equations (7a) and (37) show that, for each species in an ideal-gas mixture, the molar enthalpy is equal to the partial molar enthalpy.

† If a given species occurs in different states (for example, different phases) in the reaction, then in equation (43) a different subscript must, in general, be assigned to the species in each of its states because its ΔH_f value may be different in each state.

ΔH_f values are known for a set of compounds, $-\Delta H$ for *any* reaction among these compounds may easily be computed. It is therefore conventional to tabulate ΔH_f for all compounds and to use equation (43) to compute heats of reaction from the tables.

The **standard heat of formation** of a material is its value of ΔH_f at one atmosphere and 25°C. Values of standard heats of formation may be found in [13], [15]–[18], and [26]–[29], for example; [13] and [29] are particularly extensive. Estimates of standard heats of formation for materials not appearing in tables may be derived by the bond-energy procedure described in [17].

A.4.4 The equations of Kirchhoff and van't Hoff

In order to obtain $-\Delta H$ at conditions other than 1 atm and 25°C from equation (43) and tables of standard heats of formation, it is necessary to compute the enthalpy change of the reactant mixture and of the product mixture in going from 1 atm and 25°C to the given p and T; additional tables are available to facilitate these computations for a number of materials [13], [15]–[17], [26]–[28]. Usually the pressure dependence of $-\Delta H$ is negligible, and from equation (38) it follows that for ideal-gas reactions,

$$\left[\frac{\partial(-\Delta H)}{\partial T}\right]_p = \frac{d(-\Delta H)}{dT} = -\sum_{i=1}^{N}(v_i'' - v_i')\frac{dH_i}{dT} = -\sum_{i=1}^{N}(v_i'' - v_i')C_{p,i} \tag{44}$$

(Kirchhoff's equation), where $C_{p,i}$ is the molar heat capacity at constant pressure for species i at temperature T.

A thermodynamic relationship between the differential heat of reaction, $-dH/d\epsilon$, and the temperature dependence of the equilibrium condition for the reaction can be derived in the following way. The relation

$$\left(\frac{\partial H}{\partial N_i}\right)_{p,T,N_j(j\neq i)} = \left(\frac{\partial H}{\partial N_i}\right)_{S,p,N_j(j\neq i)} + \left(\frac{\partial S}{\partial N_i}\right)_{T,p,N_j(j\neq i)}\left(\frac{\partial H}{\partial S}\right)_{p,N_j(\text{all }j)} \tag{45}$$

is a mathematical identity. From equation (5) it follows that

$$\left(\frac{\partial H}{\partial N_i}\right)_{S,p,N_j(j\neq i)} = \mu_i \quad \text{and} \quad \left(\frac{\partial H}{\partial S}\right)_{p,N_j(\text{all }j)} = T,$$

while one of the Maxwell relations for equation (6) is

$$\left(\frac{\partial S}{\partial N_i}\right)_{T,p,N_j(j\neq i)} = -\left(\frac{\partial \mu_i}{\partial T}\right)_{p,N_j(\text{all }j)}.$$

Hence equation (45) becomes

$$\left(\frac{\partial H}{\partial N_i}\right)_{p,T,N_j(j\neq i)} = \mu_i - T\left(\frac{\partial \mu_i}{\partial T}\right)_{p,N_j(\text{all }j)} = -T^2\left(\frac{\partial}{\partial T}\frac{\mu_i}{T}\right)_{p,N_j(\text{all }j)}. \tag{46}$$

Multiplying equation (46) by $v_i'' - v_i'$, summing over i, and using (41), we then obtain

$$-\frac{dH}{d\epsilon} = T^2 \left\{ \frac{\partial}{\partial T} \left[\frac{1}{T} \sum_{i=1}^{N} (v_i'' - v_i')\mu_i \right] \right\}_{p, N_j(\text{all } j)}, \tag{47}$$

which is the desired relationship between $-dH/d\epsilon$ and the T derivative of the equilibrium condition.

For ideal-gas reactions, equation (47) determines the temperature dependence of the equilibrium constant. Since $p_i = pN_i/\sum_{i=1}^{N} N_i$, it follows from equation (28) that

$$\left(\frac{\partial}{\partial T} \frac{\mu_i}{T} \right)_{p, N_j(\text{all } j)} = -R^0 \left[\frac{\partial}{\partial T} \ln (R^0 T Q_i') \right]_{p, N_j(\text{all } j)}$$

$$= -R^0 \frac{d}{dT} \ln (R^0 T Q_i'),$$

where the last equality is a consequence of the fact that Q_i' depends only on T. Multiplying this expression by $v_i'' - v_i'$, summing over i, and substituting the result into equation (47) yields

$$-\frac{dH}{d\epsilon} = -R^0 T^2 \frac{d}{dT} \left\{ \ln \left[(R^0 T)^{\sum_{i=1}^{N} (v_i'' - v_i')} \prod_{i=1}^{N} Q_i'^{v_i'' - v_i'} \right] \right\}. \tag{48}$$

Since equations (24) and (26) show that

$$K_p = (R^0 T)^{\sum_{i=1}^{N} (v_i'' - v_i')} \prod_{i=1}^{N} Q_i'^{(v_i'' - v_i')},$$

equation (48) reduces to

$$-\frac{dH}{d\epsilon} = -R^0 T^2 \frac{d(\ln K_p)}{dT} = R^0 \frac{d(\ln K_p)}{d(1/T)}; \tag{49}$$

the slope of the curve of $\ln K_p$ versus $1/T$ is the heat of reaction (in units of R^0).

In equation (49), which is the van't Hoff equation, $-dH/d\epsilon$ may be replaced by $-\Delta H$, since these two quantities are equal for ideal-gas reactions. Relationships analogous to equation (49) may be derived for each of the equilibrium constants defined in Section A.3, but for reactions in systems other than ideal-gas mixtures, $-\Delta H$ and $-dH/d\epsilon$ may not, in general, be equated in these expressions. Heats of reaction can be determined directly either by spectroscopic measurements followed by the application of statistical mechanics (for ideal-gas reactions) or by calorimetric measurements of $\hat{Q}$ (for arbitrary reactions). Since the measurement of equilibrium compositions may be simpler than either of the above procedures, in practice equation (49) is often used to obtain heats of reaction from experimental values of K_p at neighboring temperatures.

A.4.5. The adiabatic flame temperature

A thermodynamic quantity of considerable importance in many combustion problems is the **adiabatic flame temperature**. If a given combustible mixture (a closed system) at a specified initial T and p is allowed to approach chemical equilibrium by means of an isobaric, adiabatic process, then the final temperature attained by the system is the adiabatic flame temperature T_f. Clearly T_f depends on the pressure, the initial temperature and the initial composition of the system. The equations governing the process are $p =$ constant (isobaric), $H =$ constant (adiabatic, isobaric) and the atom-conservation equations; combining these with the chemical-equilibrium equations (at p, T_f) determines all final conditions (and therefore, in particular, T_f). Detailed procedures for solving the governing equations to obtain T_f are described in [17], [19], [27], and [30], for example. Essentially, a value of T_f is assumed, the atom-conservation equations and equilibrium equations are solved as indicated at the end of Section A.3, the final enthalpy is computed and compared with the initial enthalpy, and the entire process is repeated for other values of T_f until the initial and final enthalpies agree.

Knowledge of the enthalpies of the various mixtures as functions of T and p is required in carrying out the computation. Usually equation (37) is an excellent approximation for H, and the pressure dependence of H_i is often negligible (these results are exactly true for ideal gases); even in the few cases where these approximations are not accurate, insufficient information is available to merit using more complicated equations. In equation (37), the temperature dependence of H_i is then given by

$$H_i(T) = \Delta H_{f,i}(T^0) + \int_{T^0}^{T} C_{p,i}\, dT, \tag{50}$$

where T^0 is the standard reference temperature (usually 25°C), $C_{p,i}(T)$ is the molar heat capacity of species i in the phase that is stable at atmospheric pressure and temperature T, and it is understood that when more than one phase is stable at 1 atm between T^0 and T, then the heats of transition (at the transition temperatures at 1 atm) are also to be included on the right-hand side of equation (50). References cited in the preceding subsection provide data for use in equation (50).

Except under special circumstances—for example, for systems having flame temperatures so low that dissociation is negligible in product gases, thereby causing combustion reactions to proceed essentially to completion—manual computation of adiabatic flame temperatures is very tedious. Fortunately, programs for electronic computers are now available for calculating adiabatic flame temperatures in practically all systems of interest (for example, [25]). These programs include the needed thermodynamic data and provide information on additional equilibrium properties such as compositions at the final state. The availability of such programs is highly beneficial to combustion research. Tabulations of results of various calculations of adiabatic flame temperatures may be found in [17], [27], [28], and [30], for example. Most adiabatic flame temperatures lie between 1000K and 5000K and average 2200K for typical hydrocarbon-air mixtures.

A.5. CONDENSED PHASES

A.5.1. The phase rule

The framework within which condensed phases are studied is provided by the phase rule, which gives the number of independent thermodynamic variables required to determine completely the state of the system. This number, called the **number of degrees of freedom** or the **variance** of the system, will be denoted by F. The number of phases present (that is, the number of homogeneous, physically distinct parts) will be denoted by P, and the number of independently variable chemical constituents will be called C. By *independently variable* constituents, we mean those whose concentrations are not determined by the concentrations of other constituents through chemical-equilibrium equations or other subsidiary conditions. The phase rule states that

$$F = 2 + C - P. \tag{51}$$

In order to prove equation (51), it is convenient to consider first the case in which N chemical components are present in each of the P phases and there are no chemical reactions or subsidiary conditions. In writing the thermodynamic properties, the phase will be identified by superscripts and the species will be identified by subscripts. If p, T, and $X_i^{(j)}$ are known for $i = 1, \ldots, N$ and for $j = 1, \ldots, P$, then the state of the system is completely determined (except for the total mass of material in each phase, knowledge of which is seldom wanted). Since

$$\sum_{i=1}^{N} X_i^{(j)} = 1, \qquad j = 1, \ldots, P$$

by definition of $X_i^{(j)}$, the total number of fundamentally independent parameters appearing here is $2 + P(N - 1)$. However, as was shown near the beginning of Section A.3, each species must obey the phase equilibrium condition $\mu_i^{(j)} = \mu_i^{(k)}$, $i = 1, \ldots, N$, for every pair of phases j and k. If these equations are written in a rectangular array, it immediately becomes evident that there are $N(P - 1)$ different equations of this kind. Hence, the net number of independent thermodynamic variables is

$$F = 2 + P(N - 1) - N(P - 1) = 2 + N - P = 2 + C - P$$

since $N = C$ in this case. If substance 1 were not present in phase 1, then the single relation $\mu_1^{(1)} = \mu_1^{(j)}$ would be lost, but there would also be one less variable to determine, since $X_1^{(1)} = 0$, so that the result for F would be unchanged. It is therefore clear that equation (51) does not depend on the restriction that all N species be present in each phase. If chemical reactions can occur, then C is decreased by one for each independent reaction, but an additional independent equilibrium equation [equation (21)] is obtained for each reaction. Therefore, F also decreases in such a way that equation (51) remains valid. The same remark clearly applies to other kinds of restrictive conditions (for example, conditions for electrical neutrality). Thus $C = N - R$ in general, where R is the total number of independent chemical reactions and subsidiary conditions, and equation (51) is of general validity.

Systems involving condensed phases are conventionally investigated by using state diagrams. The phase rule determines the number of dimensions required for such a diagram and predicts whether a given set of states will lie on a point, line, or area, for example, in the diagram. For three-component systems, triangular diagrams in which each vertex corresponds to one pure species are popular. In binary systems p-T, p-X, and T-X diagrams are useful. The state of a pure substance ($C = 1$, $F = 3 - P$) can always be represented on a two-dimensional diagram; this provides the basis for charts of thermodynamic properties (for example, Mollier diagrams). We shall briefly study completely miscible binary liquid (or solid) mixtures in order to illustrate the analytical procedures.

A.5.2. Vapor pressures of binary mixtures

Let us consider vapor-liquid (or vapor-solid) equilibria for binary mixtures. For the sake of simplicity it will be assumed that all gases are ideal. In addition to the vapors of each component of the condensed phase, the gas will be assumed to contain a completely insoluble constituent, the partial pressure p_3 of which may be adjusted so that the total pressure of the system, p, assumes a prescribed value. Therefore, $C = 3$, $P = 2$, and, according to equation (51), $F = 3$. Let us study the dependence of the equilibrium vapor pressures of the two soluble species p_1 and p_2 on their respective mass fractions in the condensed phase X_1 and X_2 at constant temperature and at constant total pressure. Since it is thus agreed that T and p are fixed, only one remaining variable [say $X_1 (= 1 - X_2)$] is at our disposal; p_1, p_2 and the total vapor pressure $p' \equiv p_1 + p_2$ will depend only on X_1.

The functional relationship between p_i and X_1 depends on the nature of the material. If the condensed mixture is ideal with respect to species 1, then equation (31) is valid for species 1. Since $f_i = p_i$ for ideal gases, equation (31) reduces to **Raoult's law,**[*]

$$X_1 = \alpha_1 p_1, \tag{52}$$

---

[*] Equation (52) is also, in essence, equivalent to Henry's law for the solubility of a normally gaseous species 1 in a liquid 2.

where α_1 is a constant since p and T are fixed. Since p_1 approaches the vapor pressure p_1^0 of pure species 1 as $X_1 \rightarrow 1$, equation (52) implies that $\alpha_1 = 1/p_1^0$ and may be written in the form

$$p_1 = X_1 p_1^0. \tag{52a}$$

If p_1 exceeds the value given by equation (52a), then the p_1-X_1 vapor-pressure curve is said to exhibit a *positive deviation*; on the other hand, cases in which $p_1 < X_1 p_1^0$ are described as *negative deviations*. Molecular interpretations for the causes of positive and negative deviations are discussed in [18]. Typical vapor-pressure curves exhibiting positive deviations are shown in Figure A.1.

It can be shown thermodynamically that the p_1-X_1 curve determines the shape of the p_2-X_1 curve (and therefore the p'-X_1 curve). Since p and T are constant, if equation (8) is applied to the (binary) condensed phase, it is found that

$$N_1 \, d\mu_1 + N_2 \, d\mu_2 = 0, \tag{53}$$

which is often called the **Gibbs-Duhem equation**. Dividing equation (53) by $N_1 + N_2$, using the fact that μ_i in the condensed phase must equal μ_i in the gas for equilibrium, and employing equation (28) to evaluate μ_i in the ideal gas, we find that

$$X_1 \, d \ln p_1 + X_2 \, d \ln p_2 = 0, \tag{54}$$

since T is constant (and Q_i' depends only on T). In view of the identity $dX_2 = -dX_1$ (which follows from $X_2 = 1 - X_1$), equation (54) may be written in the form

$$\frac{d \ln p_2}{d \ln X_2} = \frac{d \ln p_1}{d \ln X_1}, \tag{55}$$

which is called the **Duhem-Margules relation**. From equation (55) it is clear that apart from a constant of integration, the p_2-X_1 curve is uniquely determined by $p_1(X_1)$. From the fact that most liquids approach ideal behavior in the limit in which their mole fraction approaches unity, it follows that equation (52a) is usually valid near $X_1 = 1$, whence equation (55) implies that $d \ln p_2/d \ln X_2 = 1$ (that is, $p_2 \sim X_2$) near $X_2 = 0$. This linear dependence is illustrated in Figure A.1, where it may be seen that the constant of proportionality between p_2 and X_2 near $X_2 = 0$ in general differs from that near $X_2 = 1$. In the special case when species 1 obeys equation (52a) over the entire range of concentration, the constant of proportionality in the relation $p_2 \sim X_2$ may be evaluated at $X_2 = 1$, where $p_2 = p_2^0$ (the equilibrium vapor pressure of pure species 2), yielding $p_2 = X_2 p_2^0$; that is, if species 1 obeys Raoult's over the entire composition range, then so does species 2. Another consequence of equation (55) is that positive deviations of species 1 must be accompanied by positive deviations of species 2 (see Figure A.1).

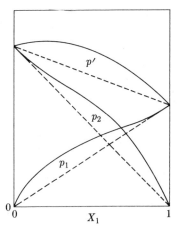

FIGURE A.1. Vapor pressures versus condensed-phase composition for completely miscible binary systems.

If both species 1 and 2 obey Raoult's law, then $p' = p_1 + p_2$ is a linear function of X_1, but if species 1 and 2 deviate positively (or negatively), then a maximum (or minimum) may develop in the $p'-X_1$ curve, as illustrated in Figure A.1. We can show that the difference between the relative concentration of the species in the gas and in the liquid, $p_1/p_2 - X_1/X_2$, depends on the slope of the $p'-X_1$ curve. Equation (55) can be written in the form

$$\frac{X_2}{p_2}\frac{dp_2}{dX_1} = -\frac{X_1}{p_1}\frac{dp_1}{dX_1},$$

from which it follows that

$$\frac{dp'}{dX_1} = \frac{dp_1}{dX_1} + \frac{dp_2}{dX_1} = \left[\frac{p_2}{p_1}\frac{dp_1}{dX_1}\right]\left(\frac{p_1}{p_2} - \frac{X_1}{X_2}\right). \tag{56}$$

Since the quantity in brackets in equation (56) is always positive, the sign of $(p_1/p_2 - X_1/X_2)$ is the same as the sign of dp'/dX_1. Hence, the vapor is richer in species 1 in regions where $dp'/dX_1 > 0$, and the condensed phase is richer in species 1 where $dp'/dX_1 < 0$. The vapor and condensed phase have the same composition where $dp'/dX_1 = 0$, a condition under which the mixture is said to be **azeotropic**.

A.5.3. Boiling points of binary mixtures

It is possible to infer the shape of curves of the normal boiling (or sublimation) point T_B (that is, the temperature at which $p' = 1$ atmosphere) versus concentration from the vapor-pressure curves described above. Since vapor pressures increase as T increases [see equation (23)], a typical graph of p' versus X_1 at successively higher T values would show a series of curves with shapes similar to the $p' - X_1$ curves of Figure A.1 but at continually

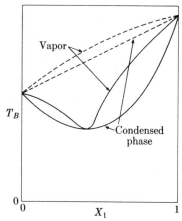

FIGURE A.2. Normal boiling point versus condensed-phase and gas compositions for completely miscible binary systems.

increasing p' levels. Curves formed by plotting versus X_1 the T values for the p' curves successively intersected by the horizontal line $p' = 1$ atmosphere constitute the desired boiling-point graphs. From the shape of the p'-X_1 curves in Figure A.1, it is clear that the resulting T_B-X_1 curves will appear as illustrated in Figure A.2; the dashed and solid curves in Figure A.2 correspond to the dashed and solid curves of Figure A.1, respectively. Also plotted in Figure A.2 are curves of T_B versus the composition p_1/p' of the corresponding binary vapor,* obtained from the p_1 and p' values at the corresponding X_1 on the vapor-pressure curve at the appropriate T; the location of these curves relative to the T_B-X_1 (condensed-phase) curves is qualitatively explained by the discussion in the preceding paragraph. We see from Figure A.2 that the existence of a maximum (minimum) in the p'-X_1 curve implies that a minimum (maximum) exists in the T_B-X_1 curve. Azeotropic mixtures are seen to correspond to extrema in the T_B-X_1 curves. Figure A.2 also directly shows the behavior of binary mixtures during distillation and suggests fractional distillation [18] as a means of separating the constituents. Much additional information of this type may be found in [2], [7], and [18].

A.5.4. Temperature dependence of vapor pressures of binary mixtures

We can readily derive an accurate, analytical, thermodynamic expression for the temperature dependence of the vapor pressure of binary

---

* The curves in Figure A.2 rigorously represent the conventional boiling point only if the total-pressure dependence of p_1 and p_2 (at given X_1 and T) is so slight that p_1 and p_2 change negligibly in passing to the limit $p_3 \to 0$ ($p' \to p$). This condition is satisfied in nearly all binary mixtures.

mixtures, analogous to equation (22) for a pure substance. Let us identify properties in the condensed phase by the superscript c and properties in the gas by the superscript g. Since $\mu_i^{(c)} = \mu_i^{(g)} \equiv \mu_i$ in equilibrium, equation (8) for either phase j ($j = c$ or g) reduces to

$$N_1^{(j)} d\mu_1 + N_2^{(j)} d\mu_2 = -S^{(j)} dT + V^{(j)} dp. \tag{57}$$

In order to eliminate $d\mu_2$, we may divide equation (57) by $N_2^{(j)}$ and subtract the resulting expression with $j = c$ from that with $j = g$. If the identity $S^{(j)} = H^{(j)}/T - G^{(j)}/T = H^{(j)}/T - (\mu_1 N_1^{(j)} + \mu_2 N_2^{(j)})/T$, which follows from the definition of G and equation (7), is utilized in the result, then we find that

$$\left(\frac{N_1^{(g)}}{N_2^{(g)}} - \frac{N_1^{(c)}}{N_2^{(c)}}\right) Td\left(\frac{\mu_1}{T}\right) = \left(\frac{V^{(g)}}{N_2^{(g)}} - \frac{V^{(c)}}{N_2^{(c)}}\right) dp - \left(\frac{H^{(g)}}{N_2^{(g)}} - \frac{H^{(c)}}{N_2^{(c)}}\right) \frac{dT}{T}. \tag{58}$$

For a change in which the composition of the condensed phase remains fixed ($dN_i^{(c)} = 0$), the left-hand side of equation (58) may be evaluated by using the mathematical identity

$$Td\left(\frac{\mu_1}{T}\right) = T\left[\frac{\partial}{\partial T}\left(\frac{\mu_1}{T}\right)\right]_{p, N_j^{(c)} (\text{all } j)} dT + T\left[\frac{\partial}{\partial p}\left(\frac{\mu_1}{T}\right)\right]_{T, N_j^{(c)} (\text{all } j)} dp.$$

Since a Maxwell relation for equation (6) shows that the partial molar volume of the condensed phase is

$$\left(\frac{\partial V^{(c)}}{\partial N_1^{(c)}}\right)_{p, T, N_j^{(c)} (j \neq 1)} = \left(\frac{\partial \mu_1}{\partial p}\right)_{T, N_j^{(c)} (\text{all } j)} = T\left(\frac{\partial}{\partial p}\frac{\mu_1}{T}\right)_{T, N_j^{(c)} (\text{all } j)},$$

we see by using equation (46) for species 1 in the condensed phase that the above identity can be written in the form

$$Td\left(\frac{\mu_1}{T}\right) = -\left(\frac{\partial H^{(c)}}{\partial N_1^{(c)}}\right)_{p, T, N_2^{(c)}} \frac{dT}{T} + \left(\frac{\partial V^{(c)}}{\partial N_1^{(c)}}\right)_{p, T, N_2^{(c)}} dp \tag{59}$$

($dN_i^{(c)} = 0$). Substituting equation (59) into equation (58) and solving for dp yields

$$\frac{dp}{dT} = \frac{1}{T}\left[\frac{X_1^{(g)}(\hat{H}_1^{(g)} - \hat{H}_1^{(c)}) + X_2^{(g)}(\hat{H}_2^{(g)} - \hat{H}_2^{(c)})}{X_1^{(g)}(\hat{V}_1^{(g)} - \hat{V}_1^{(c)}) + X_2^{(g)}(\hat{V}_2^{(g)} - \hat{V}_2^{(c)})}\right], \tag{60}$$

where use has been made of equation (7a) for $H^{(c)}$, $V^{(c)}$, $H^{(g)}$, and $V^{(g)}$ in order to simplify the right-hand side. In equation (60), for brevity, partial molar quantities have been denoted by $\hat{\chi}_i^{(j)} = (\partial \chi^{(j)}/\partial N_i^{(j)})_{p, T, N_k^{(j)} (k \neq i)}$, with $\chi = H$ and $\chi = V$. Equation (60) determines how the equilibrium total vapor-pressure of a binary mixtures varies with temperature when the composition of the condensed phase is kept fixed ($dX_i^{(c)} = 0$).

No approximations are involved in the derivation of equation (60). For nearly all systems the molar volume of the gas ($X_1^{(g)}\hat{V}_1^{(g)} + X_2^{(g)}\hat{V}_2^{(g)}$) is very much greater than the molar volume or partial molar volumes of the condensed phase, and therefore the terms $\hat{V}_i^{(c)}$ in equation (60) are negligibly small. If it is also assumed that the gas obeys the ideal-gas equation of state, then equation (60) may be written in the form

$$d(\ln p)/d(1/T) = -[X_1^{(g)}(\hat{H}_1^{(g)} - \hat{H}_1^{(c)}) + X_2^{(g)}(\hat{H}_2^{(g)} - \hat{H}_2^{(c)})]/R^0, \quad (60a)$$

the integral of which is similar to equation (23). It is apparent that in equation (60a) the quantity in the brackets, which is the differential heat of transition for the binary mixture, plays a role equivalent to that of the heat of transition L_{12} in equation (23).

A.5.5. Colligative properties of solutions

The difference between the physico-chemical properties of a pure species and of a mixture is perhaps most easily illustrated by considering a liquid solvent 1 in which an involatile solute 2 is dissolved. The equilibrium vapor pressure of the solute is assumed to be so low that the solute is present to a negligible extent in the gaseous phase ($X_2^{(g)} \approx 0$). The phenomena exhibited by such systems are often referred to as the **colligative properties** of solutions. One such property is the fact that addition of the solute raises the boiling point of the mixture above that of the pure solvent; this is clear from the reasoning by which Figure A.2 was deduced from Figure A.1. If the vapor of the solvent is an ideal gas and the solution is ideal, then it is possible to derive an explicit expression for the increase in the boiling temperature.

Since $X_2^{(g)} \approx 0$ ($p \approx p_1$), equation (60a) reduces to

$$d(\ln p_1)/d(1/T) = -(\hat{H}_1^{(g)} - \hat{H}_1^{(c)})/R^0, \quad (61)$$

($dX_1^{(c)} = 0$). From the condition for ideality of the solution it can be shown that the differential heat of vaporization of the solution

$$(\hat{H}_1^{(g)} - \hat{H}_1^{(c)})$$

appearing in equation (61) must equal the heat of vaporization of the pure solvent, L; the proof essentially involves noticing that at constant X_1, equation (52a) implies that the $1/T$ derivative of $\ln p_1$ equals the $1/T$ derivative of $\ln p_1^0$ and using the fact that p_1^0 obeys the Clapeyron equation for the pure solvent. Assuming that L is constant between the boiling temperature of the pure solvent T_B^0 and the boiling temperature of the solution T_B, we may integrate equation (61) from T_B^0 to T_B, obtaining

$$\ln(p_0) - \ln(X_1 p_0) = \left(\frac{L}{R^0}\right)\left(\frac{1}{T_B^0} - \frac{1}{T_B}\right), \quad (62)$$

where p_0 is atmospheric pressure and use has been made of the facts that $p_1 = p_0$ at $T = T_B$ and $p_1^0 = p_0$ (while $p_1 = X_1 p_1^0$, $X_1 \equiv X_1^{(c)}$) at $T = T_B^0$ by definition of the boiling temperatures. Since $X_1 = 1 - X_2$, where X_2

is the mole fraction of solute in the solution, equation (62) may be written in the form

$$\frac{T_B - T_B^0}{T_B} = -\left(\frac{R^0 T_B^0}{L}\right) \ln(1 - X_2). \tag{63}$$

If the solution is dilute, then X_2 and $(T_B - T_B^0)/T_B$ are small compared with unity, and an expansion of equation (63) in powers of those two small quantities yields

$$(T_B - T_B^0) = X_2 R^0 T_B^{0^2}/L. \tag{64}$$

Equation (64) shows that the elevation of the boiling point is proportional to the concentration of the solute; the prediction in equation (64) is in approximate agreement with experiment for a number of dilute solutions.

Other colligative properties include the lowering of the freezing point by a solute and osmotic pressure. By a procedure similar to that described above, it can be shown that the difference between the normal freezing temperature of a pure solvent, T_F^0, and that of a solution with an involatile solute, T_F, is roughly given by the expression

$$(T_F^0 - T_F) = X_2(1 - k)R^0 T_F^{0^2}/L_F, \tag{65}$$

where L_F is the heat of fusion of the solvent, and $k = k(p, T)$, which is called a **distribution coefficient**, is the equilibrium ratio of the mole fraction of the solute in the solid solution to the mole fraction of the solute in the liquid solution, X_2. In most cases k is small compared with unity; often solid solutions do not occur and $k \equiv 0$. However, there exist examples of systems in which $k > 1$, and the freezing point actually increases upon the addition of solute.

Membranes which are permeable to a solvent but are almost completely impermeable to a solute dissolved in this solvent have been constructed; they are called **semipermeable membranes**. It is observed that if a solution is separated from its solvent by a semipermeable membrane, then the pressure in the solution must exceed that in the solvent in order to prevent the solvent from passing through the membrane and going into solution. The pressure difference required to maintain equilibrium is called the **osmotic pressure**, π, and is found in a very rough approximation to obey the equation

$$\pi = c_2 R^0 T, \tag{66}$$

where c_2 is the concentration (moles per unit volume) of the solute in the solution. Equation (66), which bears a striking resemblance to the ideal-gas law, can be derived by procedures somewhat analogous to those used in obtaining equation (64). The physics of ideal gases differs greatly from that underlying equation (66).

Our purpose in Section A.5 has merely been to illustrate the methods used in analyzing condensed phases and to state some of the more important results. A wealth of related material may be found in [2], [7], [14], and [18]. Although there are many combustion problems in which properties of condensed phases are unimportant, occasionally nearly all the phenomena discussed in these references must be considered (note, for example, Section 3.3.6).

REFERENCES

1. M. W. Zemansky, *Heat and Thermodynamics*, New York: McGraw-Hill, 1957.
2. E. A. Guggenheim, *Thermodynamics*, New York: Interscience, 1957.
3. R. H. Fowler and E. A. Guggenheim, *Statistical Thermodynamics*, New York: Macmillan, 1939.
4. J. Mayer and M. Mayer, *Statistical Mechanics*, New York: Wiley, 1940.
5. R. C. Tolman, *The Principles of Statistical Mechanics*, Oxford: Clarendon Press, 1938.
6. T. L. Hill, *An Introduction to Statistical Thermodynamics*, Reading, Mass.: Addison-Wesley, 1960.
7. L. D. Landau and E. M. Lifshitz, *Statistical Physics*, Reading, Mass.: Addison-Wesley, 1969.
8. D. Bohm, *Quantum Theory*, New York: Prentice-Hall, 1951.
9. L. D. Landau and E. M. Lifshitz, *Quantum Mechanics*, Reading, Mass.: Addison-Wesley, 1958.
10. L. I. Schiff, *Quantum Mechanics*, New York: McGraw-Hill, 1968.
11. G. Herzberg, *Molecular Spectra and Molecular Structure I. Spectra of Diatomic Molecules*, New York: D. Van Nostrand, 1950.
12. G. Herzberg, *Molecular Spectra and Molecular Structure II. Infrared and Raman Spectra of Polyatomic Molecules*, New York: D. Van Nostrand, 1945.
13. D. R. Stull and H. Prophet, *JANAF Thermochemical Tables*, 2nd ed., Washington: National Bureau of Standards, NSRDS-NBS 37 (June 1971).
14. E. A. Moelwyn-Hughes, *Physical Chemistry*, 2nd ed., New York: Pergamon Press, 1961.
15. *Journal of Physical and Chemical Reference Data* 1, (1972)–present.
16. F. D. Rossini et al., *Selected Values of Properties of Hydrocarbons*, Washington: National Bureau of Standards, Circular C461 (November 1947).
17. S. S. Penner, *Chemistry Problems in Jet Propulsion*, New York: Pergamon Press, 1958.
18. S. Glasstone, *Textbook of Physical Chemistry*, New York: D. Van Nostrand, 1946.
19. F. A. Williams, M. Barrère, and N. C. Huang, *Fundamental Aspects of Solid Propellant Rockets*, AGARDorgraph No. 116, Slough, England: Technivision Services, 1969, 122–165.
20. J. M. Carter and D. Altman, "High Temperature Equilibrium," in *Combustion Processes*, vol. II of *High Speed Aerodynamics and Jet Propulsion*, B. Lewis, R. N. Pease, and H. S. Taylor, eds., Princeton: Princeton University Press, 1956, 3–25.
21. F. von Zeggeren and S. H. Storey, *The Computation of Chemical Equilibria*, Cambridge: Cambridge University Press, 1970.

22. V. N. Huff, S. Gordon, and V. E. Morell, *Basic Considerations in the Combustion of Hydrocarbon Fuels with Air*, NACA Rept. No. 1037 (1951).
23. S. R. Brinkley, "Computational Methods in Combustion Calculations," in *Combustion Processes*, vol. 2 of *High Speed Aerodynamics and Jet Propulsion*, B. Lewis, R. N. Pease and H. S. Taylor, eds., Princeton: Princeton University Press. 1956, 64–98.
24. W. B. White, S. M. Johnson, and G. B. Dantzig, *Chemical Equilibrium in Complex Mixtures*, Rept. No. P1059, Rand Corporation (1957); *J. Chem. Phys.* **28**, 751 (1958).
25. S. Gordon and B. J. McBride, *Computer Program for Calculation of Complex Chemical Equilibrium Composition, Rocket Performance, Incident and Reflected Shocks and Chapman-Jouguet Detonations*, NASA SP-273 (1971).
26. R. C. Weast, *Handbook of Physics and Chemistry*, 54th Edition, Cleveland, Ohio: Chemical Rubber Co., 1973.
27. A. M. Kanury, *Introduction to Combustion Phenomena*, New York: Gordon and Breach, 1975.
28. H. C. Barnett, et al., *Basic Considerations in the Combustion of Hydrocarbon Fuels with Air*, NACA Rept. No. 1300 (1959).
29. J. S. Rockenfeller and F. D. Rossini, *J. Phys. Chem.* **65** 267 (1961); D. D. Wagman et al., *Selected Values of Chemical Thermodynamic Properties*, Washington: National Bureau of Standards, Tech. Notes 270-3 (January 1968), 270-4 (May 1969), 270-5 (March 1971), 270-6 (November 1971).
30. I. Glassman, *Combustion*, New York: Academic Press, 1977.

APPENDIX B

Review of Chemical Kinetics

The necessity of considering chemical reactions that proceed at finite rates distinguishes combustion theory from other extensions of fluid dynamics. Concepts of chemical kinetics therefore comprise an integral part of the subject. The phenomenological laws for rates of chemical reactions are presented in Section B.1. Various mechanisms for chemical reactions are considered in Section B.2, which includes discussion of recent work in explosion theory. This section contains material specifically related to combustion that is seldom found in basic texts on chemical kinetics. Theoretical predictions of reaction-rate functions for homogeneous and heterogeneous processes are addressed in Sections B.3 and B.4, respectively. References [1]–[4] are textbooks of a basic nature on chemical kinetics; [5]–[12] contain, in addition, material more directly applicable in combustion.

B.1. THE LAW OF MASS ACTION

B.1.1. Statement of the law

An arbitrary chemical reaction may be written in the form

$$\sum_{i=1}^{N} v_i' \mathfrak{M}_i \rightarrow \sum_{i=1}^{N} v_i'' \mathfrak{M}_i, \tag{1}$$

where v_i' and v_i'' are the stoichiometric coefficients for species i appearing as a reactant and as a product, respectively, and $\mathfrak{M}_i$ is the chemical symbol for species i. For the reaction in (1) there exists a relationship among changes in the concentrations c_i (moles per unit volume) of all species i. If $\hat{\omega}_i$ (moles per unit volume per second) denotes the time rate of increase of the concentration of species i, then equation (1) states that

$$\hat{\omega}_i/(v_i'' - v_i') = \hat{\omega}_j/(v_j'' - v_j') \tag{2}$$

for any pair of species i and j [compare equation (A-18)]. We may therefore define a reaction rate for the reaction in (1) as $\omega \equiv \hat{\omega}_1/(v_1'' - v_1')$, where species 1 may be taken to be any species for which $v_1'' - v_1' \neq 0$. It will be noted that ω is the time rate of production (moles per unit volume per second) of any species for which $v_i'' - v_i' = 1$. Equation (2) implies that in reaction (1),

$$\hat{\omega}_i = (v_i'' - v_i')\omega, \quad i = 1, \ldots, N, \tag{3}$$

for every species.

In the form usually quoted in chemical kinetics, the phenomenological law of mass action states that the rate of a reaction is proportional to the product of the concentrations of the reactants. For the general reaction given in equation (1), we may therefore write

$$\omega = k \prod_{i=1}^{N} c_i^{v_i'}, \tag{4}$$

where the proportionality factor k is called the **specific reaction-rate constant** for reaction (1). In virtually all systems, k depends mainly on temperature; the value of k often increases rapidly with increasing temperature.

Derivations of equation (4) involve a microscopic viewpoint. The reasoning, in its simplest form, is that the reaction rate is proportional to the collision rate between appropriate molecules, and the collision rate is proportional to the product of the concentrations. Implicit in this picture is the idea that equation (4) will be valid only if equation (1) represents a process that actually occurs at the molecular level. Equation (1) must be an *elementary* reaction step, with v_i' molecules of each molecular species i interacting in the microscopic process; equation (4) will not be meaningful if equation (1) is the overall methane-oxidation reaction $CH_4 + 2O_2 \rightarrow CO_2 + 2H_2O$, for example. Thus, there are two basic problems in chemical kinetics; the first is to determine the reaction mechanism, that is, to find the elementary steps by which the given reaction proceeds, and the second is to determine the specific rate constant k for each of these steps. These two problems are discussed in Sections B.2 and B.3, respectively.

B.1.2. Multiple reactions; equilibrium constant

Since most reactions involve a number of simultaneous elementary steps, the net rate of production of a species i usually equals a sum of terms, each term of which corresponds to the production rate in one of the reaction steps. An arbitrary number M of simultaneous elementary steps may be represented by the equation

$$\sum_{i=1}^{N} v'_{i,k} \mathfrak{M}_i \rightarrow \sum_{i=1}^{N} v''_{i,k} \mathfrak{M}_i, \quad k = 1, \ldots, M, \tag{5}$$

where the symbols are the same as in equation (1), except that the subscript k on the stoichiometric coefficients identifies the reaction step. If the rate of production of species i in reaction k is defined as $\hat{\omega}_{i,k}$ (moles per unit volume per second), then the net rate of production of species i in all chemical reactions will be given by

$$\hat{\omega}_i = \sum_{k=1}^{M} \hat{\omega}_{i,k}, \quad i = 1, \ldots, N. \tag{6}$$

As with equation (1), equation (5) implies that a reaction rate ω_k for the reaction step k may be defined in such a way that

$$\hat{\omega}_{i,k} = (v''_{i,k} - v'_{i,k})\omega_k, \quad i = 1, \ldots, N, \quad k = 1, \ldots, M. \tag{7}$$

The law of mass action yields

$$\omega_k = k_k \prod_{i=1}^{N} c_i^{v_{i,k}}, \quad k = 1, \ldots, M, \tag{8}$$

where k_k is the specific rate constant for the step k.

With many reaction schemes, particularly in systems which are near equilibrium, for each reaction in equation (5) the reverse reaction

$$\sum_{i=1}^{N} v''_{i,k} \mathfrak{M}_i \rightarrow \sum_{i=1}^{N} v'_{i,k} \mathfrak{M}_i$$

also appears as a member of the set. In such cases it is often convenient to replace set (5) by the equivalent equations

$$\sum_{i=1}^{N} v'_{i,k} \mathfrak{M}_i \rightleftharpoons \sum_{i=1}^{N} v''_{i,k} \mathfrak{M}_i, \quad k = 1, \ldots, M, \tag{9}$$

where M is half of the M in equation (5). Corresponding forward and backward reactions are grouped together in equation (9). With this convention, equations (7) and (8) are replaced by

$$\hat{\omega}_{i,k} = (v''_{i,k} - v'_{i,k})\omega'_k, \quad i = 1, \ldots, N, \quad k = 1, \ldots, M \tag{10}$$

and

$$\omega'_k = k_{f,k} \prod_{i=1}^{N} c_i^{v_{i,k}} - k_{b,k} \prod_{i=1}^{N} c_i^{v''_{i,k}}, \quad k = 1, \ldots, M, \tag{11}$$

where $k_{f,k}$ and $k_{b,k}$ are the specific reaction-rate constants for the forward and backward reactions, respectively. The quantity ω_k' is seen to be the net forward rate of the kth reaction of equation (9).

An expression for the equilibrium constant in terms of the forward and backward specific rate constants for a given reaction can be derived from equations (9)–(11). Let us consider a single reaction [$M = 1$ in equations (9)–(11)] and omit the subscript k. At equilibrium, the net rate of production of each species is zero ($\hat{\omega}_i = 0$ for all i). Hence equation (10) implies that $\omega' = 0$ at equilibrium for the reaction under consideration. In view of equation (11), this requires that

$$k_f/k_b = \prod_{i=1}^{N} c_i^{v_i'' - v_i'}.$$ (12)

But, at equilibrium, the right-hand side of equation (12) is the equilibrium constant for concentrations K_c [see equation (A-24)]. Hence,

$$k_f/k_b = K_c.$$ (13)

The kinetic view of the equilibrium condition, which emerges from this reasoning, is that the forward rates become equal to the backward rates (not that both forward and backward rates go to zero). Since equilibrium constants can be measured much more accurately than can specific reaction-rate constants (whose uncertainties often exceed a factor of 10), equation (13) is generally used to calculate one specific rate constant from the equilibrium constant and the other rate constant.

B.1.3. Reaction order and molecularity

If no processes (such as expansion or diffusion) other than chemical reactions cause the concentrations c_i to change, then, from the definition of $\hat{\omega}_i$,

$$dc_i/dt = \hat{\omega}_i, \quad i = 1, \ldots, N.$$ (14)

For a single, elementary, irreversible, reaction step, $\hat{\omega}_i$ is given by equation (3) and equations (4) and (14) imply that

$$dc_i/dt = [(v_i'' - v_i')k] \prod_{j=1}^{N} c_j^{v_j'},$$ (15)

where $(v_i'' - v_i')k$ is constant (at least to a good approximation) in isothermal systems. Under isothermal conditions with all concentrations held fixed except one, that of species j, empirical measurements often yield $dc_i/dt \sim c_j^{n_j}$, where the exponent n_j is constant. In all such cases, the number n_j is called the **order** of the reaction with respect to species j. Furthermore, if under isothermal conditions the concentrations of all species are changed proportionally—for example, by changing the total pressure at fixed relative concentrations—then empirical measurements often show that $dc_i/dt \sim c^n$, where $c \equiv \sum_{i=1}^{N} c_i$, and the exponent n is constant. In all such cases, the number n is called the **overall order** (or simply the order or sometimes the pressure exponent) of the reaction. From equation (15) it is clear that if the reaction is a single, elementary step, then $n_i = v_i'$, and $n = \sum_{i=1}^{N} n_i$. In this case, n_i is the molecularity of the reaction with respect to species i, and n is the (overall) molecularity of the reaction; the **molecularity** of a reaction is thus defined as the number of molecules that interact in the process. When multiple steps are involved and $\hat{\omega}_i$ is determined by equations (6)–(8), it is not possible to write equation (15) and, in general, neither $n_i = v_i'$ nor $n = \sum_{i=1}^{N} n_i$. In such cases the reaction order is not simply related to any molecularity, and the n_i may assume nonintegral values.

Slight variations in the definition of reaction order may be found in the literature. For example, for some reactions there may be reason to believe that one species (for example, nitrogen) is inert, in the sense that changes in its concentration do not influence the rate of the reaction at constant pressure, and in such cases n_j may be defined by considering simultaneous changes in the concentrations of species j and of the inert, with the total pressure held constant. In general, the resulting value of n_j differs from that defined above; therefore it is important to ascertain the specific definition employed. An empirical formula that is often useful—for example, in the presence of an inert—is

$$dc_i/dt \sim c^n \prod_{j=1}^{N} X_j^{n_j}, \tag{16}$$

where $X_j \equiv c_j/c$ and $n_j = 0$ for the inert. Here the overall order n is defined as above, while the orders n_i are defined in terms of changes in which fixed values are maintained for c and for all X_j except that of the inert and of the species i whose order is being measured.

B.2. REACTION MECHANISMS

B.2.1. General methods

If the rates of all elementary steps that could conceivably occur in a given overall reaction were known, then (in principle) the actual mechanism by which the major part of the reaction takes place could be computed by straightforward procedures for any given experimental conditions. Unfortunately, such detailed rate information is seldom available. Therefore, the only feasible approach to the problem of determining the mechanism of a given reaction is to measure the dependence of the overall reaction rate on the concentrations of as many species as possible, and then to search for a reaction mechanism that is in agreement with this dependence.* The search is, of course, guided by past experience with similar reactions and by other known chemical facts and principles. However, there are few general rules that can be stated for unraveling reaction mechanisms; nearly every case must be treated individually. Furthermore, the mechanism depends on the experimental conditions (pressure, temperature, etc.); for example, the dependence of the rate constants on temperature will cause some elementary steps to become negligibly slow and other new ones to become important as the temperature changes. In view of these difficulties, in this section we can do little more than introduce nomenclature, state a few guiding principles, and give some examples of reaction mechanisms that have been discovered. In addition, an outline of explosion theory will be given in the section on chain reactions.

B.2.2. First-order reactions and unimolecular reactions

The simplest chemical process is the first-order reaction. If $i = 1$ denotes the reactant, then $n = 1$, $n_1 = 1$, $n_j = 0$ for $j \neq 1$, and equation (16) reduces to

$$dc_1/dt = -kc_1, \tag{17}$$

where k (s^{-1}) is a rate constant (and $k > 0$). Integrating equation (17) shows that c_1 decays exponentially with time and that

$$k = (1/t) \ln (c_{10}/c_1) \tag{18}$$

when $c_1 = c_{10}$ at $t = 0$. Equation (18) shows that a plot of $\ln c_1$ versus t is a straight line; such graphs are often used to test for first-order reactions. Equation (18) also shows that the negative of the slope of the graph is the rate constant k.

The simplest type of system that obeys equation (17) is the unimolecular process $\mathfrak{M}_1 \to$ products. Since a stable molecule $\mathfrak{M}_1$ should not spontaneously break up into reaction products, the mechanism by which the unimolecular process occurs must be explained. Many unimolecular reactions are believed to follow the mechanism proposed by Lindemann, namely,

$$\mathfrak{M}_1 + X \underset{k_2}{\overset{k_1}{\rightleftharpoons}} \mathfrak{M}_1^* + X,$$

$$\mathfrak{M}_1^* \overset{k_3}{\longrightarrow} \text{products},$$

where X stands for any molecule, the superscript $*$ on $\mathfrak{M}_1$ signifies that the molecule is in an excited (in this case, unstable) state, and the specific rate constants for the reaction steps are written above (or below) the arrows.

---

* It is seldom possible to prove that an acceptable mechanism is unique.

The notation indicates that the $\mathfrak{M}_1$ molecules acquire sufficient excitation energy to react through collisions with other molecules. For this mechanism, equation (14) and the general rate expressions given in equations (6)–(8) yield

$$dc_P/dt = k_3 c_{\mathfrak{M}_1^*}$$

and

$$dc_{\mathfrak{M}_1^*}/dt = k_1 c_{\mathfrak{M}_1} c_X - k_2 c_{\mathfrak{M}_1^*} c_X - k_3 c_{\mathfrak{M}_1^*},$$

where c_P is the concentration of a typical reaction product. Often the time variation of $c_{\mathfrak{M}_1^*}$ is so slight that each term on the right-hand side of this last equation greatly exceeds the left-hand side. Then $dc_{\mathfrak{M}_1^*}/dt$ may be neglected, and the solution for $\mathfrak{M}_1^*$ becomes

$$c_{\mathfrak{M}_1^*} = k_1 c_{\mathfrak{M}_1} c_X/(k_2 c_X + k_3).$$

Substituting this result into the equation for dc_P/dt, we find that

$$dc_P/dt = k_1 c_{\mathfrak{M}_1} c_X/[1 + (k_2/k_3)c_X]. \tag{19}$$

If c_X (the total number of moles per unit volume) is large enough (the pressure is high enough), or if k_3/k_2 is small enough (the decay reaction is slow enough), then the second term in the denominator of equation (19) is dominant, $(k_2/k_3)c_X \gg 1$, and the equation reduces to

$$dc_P/dt = (k_1 k_3/k_2)c_{\mathfrak{M}_1} = \text{constant } c_{\mathfrak{M}_1}, \tag{20}$$

which is the required first-order rate expression. However, at sufficiently low pressures, c_X may be reduced to a point where $(k_2/k_3)c_X \ll 1$, whence the rate expression, equation (19), becomes

$$dc_P/dt = k_1 c_{\mathfrak{M}_1} c_X,$$

which is of second order in the overall sense ($n = 2$). Experiments have shown that a number of unimolecular reactions become second-order processes at low pressures, thus supporting the Lindemann mechanism.

Conditions necessary for neglecting $dc_{\mathfrak{M}_1^*}/dt$ in the manner employed above may be investigated through formal approximations in reaction-rate theory. This will be considered further, with application to the Lindemann mechanism, in Section B.2.5. The mechanism itself generally contains fundamental inaccuracies and is best viewed as a simplified approximation to more-complex mechanisms. In particular, molecules capable of experiencing unimolecular decomposition or isomerization may exist in many different vibrationally excited states, and the rate constant for the reaction may differ in each state. Approximate means for summing over states to obtain average rate constants have been developed; an introduction to these considerations may be found in [3].

B.2.3. Higher-order reactions

Expressions similar to equation (17) may easily be derived for various second-, third-, and higher-order reactions. These expressions are readily integrated for all second-order reactions and for many third- and higher-order reactions, yielding (in many cases) relations analogous to equation (18), which define useful concentration-time graphs. The dimensions of the rate constant k for an nth order reaction are $(concentration)^{-(n-1)} (time)^{-1}$.

If some of the reactant or product species are present in excessive quantities, then the fractional changes in their concentrations over the entire duration of the reaction may be immeasurably small. In such cases the concentrations of the reactants present in excess remain approximately constant and may be absorbed into the rate constant k. A measurement of the order of the reaction from concentration-time plots then does not reveal the dependence of the rate on the concentrations of the overabundant species; the measurement yields the **pseudo molecularity** of the reaction, that is, the sum of the orders with respect to the species that are not present in excess. Thus a number of higher-order reactions are found to be pseudounimolecular under certain conditions. This observation provides the basis for the **isolation method** of determining the order of a complex reaction with respect to a particular reactant; in this method, the apparent overall order (pseudomolecularity) of the reaction is measured under conditions in which all of the reactants except the one of interest are present in excess.

Although examples of reactions of very high orders (greater than 5) have been reported in the literature, it appears unlikely that reaction steps with molecularity greater than 3 are of any importance in reaction mechanisms because molecular considerations (see Section B.3) of the probabilities of many-body collisions indicate that reactions of higher molecularity will proceed extremely slowly. Trimolecular reactions in gases often are important only at moderate or high pressures (on the order of or greater than atmospheric pressure). Since unimolecular reactions will occur only for relatively unstable reactants, we may conclude that most elementary reaction steps in gases will be bimolecular.

B.2.4. Opposing reactions

Perhaps the simplest example of a reaction mechanism which is more complicated than that of equation (1) is the one-step reversible reaction

$$\sum_{i=1}^{N} v_i' \mathfrak{M}_i \quad \rightleftharpoons \quad \sum_{i=1}^{N} v_i'' \mathfrak{M}_i. \tag{21}$$

For reaction (21), equations (10), (11), and (14) (with the subscript k omitted) imply that

$$dc_i/dt = (v_i'' - v_i') \left(k_f \prod_{j=1}^{N} c_j^{v_j'} - k_b \prod_{j=1}^{N} c_j^{v_j''} \right), \quad i = 1, \ldots, N. \tag{22}$$

If $c_i \equiv c_{i0}$ at $t = 0$ for all i, then equation (22) shows (through appropriate integrations) that

$$(c_j - c_{j0})/(v_j'' - v_j') = (c_i - c_{i0})/(v_i'' - v_i')$$

for all j with any chosen i. Substituting this relation into equation (22), we obtain (for any chosen i)

$$\int_{c_{i0}}^{c_i} [dc_i/f(c_i)] = (v_i'' - v_i')k_f t, \tag{23}$$

where

$$f(c_i) \equiv \prod_{j=1}^{N} \left[c_{j0} + \left(\frac{v_j'' - v_j'}{v_i'' - v_i'} \right)(c_i - c_{i0}) \right]^{v_i'}$$

$$- \left(\frac{k_b}{k_f} \right) \prod_{j=1}^{N} \left[c_{j0} + \left(\frac{v_j'' - v_j'}{v_i'' - v_i'} \right)(c_i - c_{i0}) \right]^{v_i''}. \tag{24}$$

Thus the problem of determining the time dependence of the concentration of any species reduces to the problem of evaluating the quadrature on the left-hand side of equation (23).

For the unimolecular case $\mathfrak{M}_1 \rightleftharpoons \mathfrak{M}_2$, the integral in equation (23) is easily evaluated, giving

$$\ln \left[\frac{(k_f + k_b)c_1 - k_b(c_{10} + c_{20})}{k_f c_{10} - k_b c_{20}} \right] = -(k_f + k_b)t. \tag{25}$$

Since $f(c_1) = 0$ at equilibrium (that is, at $c_1 \equiv c_{1e}$), equation (24) shows that $(k_f + k_b)c_{1e} = k_b(c_{10} + c_{20})$, whence equation (25) can be written as

$$(k_f + k_b) = (1/t) \ln [\text{constant}/(c_1 - c_{1e})], \tag{26}$$

where the constant depends on the initial concentrations and the rate constants. Comparison of equations (18) and (26) shows that the remarks in the first paragraph of Section B.2.2 also apply to unimolecular reversible reactions, provided that the rate constant is replaced by the sum of the forward and backward rate constants and the reactant concentration is replaced by the departure of the reactant concentration from its equilibrium value.

A classical example of a process that follows the mechanism of (21) is the hydrogen-iodine reaction $H_2 + I_2 \rightleftharpoons 2HI$.[*] · If $c_{HI} = 0$ and $c_{H_2} = c_{I_2} \equiv c_0$ at $t = 0$ for this bimolecular process, then equation (22) becomes

$$dc_{HI}/dt = (k_f/2)(2c_0 - c_{HI})^2 - (2k_b)c_{HI}^2, \tag{27}$$

---

[*] For all specific examples quoted, it must be remembered that the given mechanism is valid only for a certain range of pressure and temperature.

the solution to which can be shown to give

$$\sqrt{k_f k_b} = \frac{1}{4c_0 t} \ln\left[\frac{2c_0 + c_{HI}(2\sqrt{k_b/k_f} - 1)}{2c_0 - c_{HI}(2\sqrt{k_b/k_f} + 1)}\right]. \tag{28}$$

If equation (13) is used for k_b/k_f and the empirical value of the equilibrium constant is employed, then the c_{HI}-t relationship given by equation (28) agrees with experimental results.

　　Other reaction mechanisms that are simple enough for their concentration-time curves to be computed analytically, without making approximations, include consecutive reactions (for example, $\mathfrak{M}_1 \to \mathfrak{M}_2 \to \mathfrak{M}_3 \to \cdots$) and parallel (side) reactions (for example, $\mathfrak{M}_1 \to \mathfrak{M}_2$, $\mathfrak{M}_1 \to \mathfrak{M}_3, \ldots$). Reaction mechanisms in combustion processes usually are considerably more complex than these.

B.2.5. Chain reactions and related processes

B.2.5.1. Initiation, propagation, and termination steps.*

　　The hydrogen-chlorine reaction $H_2 + Cl_2 \to 2HCl$ is known to involve the steps

$$Cl_2 + h\nu \longrightarrow 2Cl, \tag{29}$$

$$Cl + H_2 \longrightarrow HCl + H, \tag{30}$$

$$H + Cl_2 \longrightarrow HCl + Cl, \tag{31}$$

and

$$2Cl + X \longrightarrow Cl_2 + X^*, \tag{32}$$

in addition to some other steps. Equations (29)–(32) are illustrative of a variety of types of elementary reaction steps.

　　In equation (29) the symbol $h\nu$ denotes the energy of a quantum of light of frequency ν. The notation signifies that when the H_2-Cl_2 mixture is irradiated by light of an appropriate frequency, light quanta are absorbed by chlorine molecules, causing them to dissociate. Any process of this kind, where radiation induces a reaction in an otherwise metastable system, is termed **photosensitization**. There are also many examples of processes in which reaction intermediaries or products are formed in excited states and deactivate by emitting nonthermal radiation; this phenomenon is called

---

*The clear and concise text by Dainton [13] is devoted solely to chain reactions. Background material on this subject also may be found in [6]–[11], for example.

chemiluminescence*. All chemical reactions involving light are classified as **photochemical** processes. Semiquantitative explanations of these phenomena exist [2].[†]

The reactions of (30) and (31) both occur rapidly, and their net effect is that one H_2 molecule and one Cl_2 molecule have been converted into two HCl molecules. A Cl atom reappears and may therefore initiate the process (30), (31) again. The presence of a few free Cl atoms thus causes the overall reaction to proceed very rapidly. The overall reaction is therefore called a **chain reaction**, and the H atoms and Cl atoms, which are continually re-generated, are said to be **chain carriers**. Reactions (30) and (31) are **chain-carrying** or **propagation** steps.

Since the chain carrier Cl is generated by reaction (29), this process is said to be a **chain-initiation** step. The energy required to dissociate the Cl_2 molecule might also be acquired by collision with another molecule or with a hot solid object (such as, a wall) in contact with the gas; chain-initiation processes need not be photochemical.

In reaction (32), the chain carrier Cl is recombined to form Cl_2; this is a **chain-breaking** or **termination** step. Equation (32) describes a three-body collision in which the recombination energy is carried away in the form of excitation energy of the third molecule X. It can easily be shown that the two-body recombination process $2Cl \rightarrow Cl_2$ is forbidden because it cannot satisfy the laws of the conservation of both energy and momentum. In order to conserve momentum, the relative kinetic energy of the colliding molecules (as well as the dissociation energy) would have to be carried away as internal excitation energy of the Cl_2 molecule; but then the Cl_2 molecule would be unstable and would re-dissociate. For a two-body process of the type $A + B \rightarrow C$ to occur, the product C must be a relatively large molecule having a sufficient number of internal degrees of freedom to retain the energy of association for times on the order of bimolecular collision times; low temperatures and high pressures favor such processes. Instead of the third molecule X, radiation (chemiluminescence) or a solid wall (for collisions at a wall) may carry away the necessary energy.

The hydrogen-oxygen reaction $2H_2 + O_2 \rightarrow 2H_2O$ involves a chain mechanism in which the atoms H and O and the hydroxyl radical OH act as chain carriers.[‡] One of the elementary steps in the mechanism is

$$H + O_2 \longrightarrow OH + O. \tag{33}$$

In this step, one chain carrier (H) is consumed and two (OH and O) are

---

* Chemiluminescent processes occur in most combustion reactions, giving flames many of their characteristic colors. However, hot solid carbon particles in flames emit (usually yellow) thermal radiation in an equilibrium radiative process which therefore is not chemi-luminescence.

† Also see Chapter 10 of [11] and Chapter III of [14], for example.

‡ See, for example, [5], [7], [8], [12], [14]–[16].

produced. Therefore, reaction (33) is a **chain-branching** step. Chain-branching reactions can lead to a rapid production of relatively large concentrations of chain carriers, which, in turn, cause the overall reaction to proceed extremely rapidly, resulting in a **branched-chain explosion**.

B.2.5.2. The steady-state and partial-equilibrium approximations

If the reaction mechanism is too complicated for the set defined by equation (14) to be integrated analytically, then accurate concentration histories, $c_i(t)$, may be sought by numerical integration. Attempts to perform such integrations often encounter difficulties stemming from terms on the right-hand side of equation (6) being large, opposite in sign, and nearly equal in magnitude. Under this condition the system has been called *stiff* [17]; in recent years much attention has been devoted to the integration of stiff systems. A significant degree of success has been obtained (for example, [18], [19], [20]), and more recently effort has been placed on ascertaining sensitivities of solutions to changes in values of rate constants [21]–[24]. A motivation for this emphasis is that rate constants are not often known accurately. If predicted histories $c_i(t)$ are sufficiently sensitive to values of poorly known rate constants, then closer estimates for them may be obtained by comparing measured histories with those calculated through numerical integrations performed for various values of the uncertain constants.

Although numerical integrations of equation (14) usually can be performed accurately now, some systems remain that are too complex or too stiff for the computational methods to succeed. To analyze such systems, as well as to understand better the behaviors of various systems amenable to accurate numerical integrations, rational approximations producing simplifications are wanted. Two important approximations of this type are the steady-state and partial-equilibrium approximations.

The **steady-state approximation** is a simplification applied to the description of reaction intermediaries. It has often been successful for chain reactions with sufficiently low degrees of branching as well as for straight chain reactions and for many reactions that cannot be classified as chain reactions. In this approach, given a reaction mechanism, identification is made of the reaction intermediaries—such as chain carriers—which are neither initial reactants nor principal products, and the approximation $\hat{\omega}_i = 0$ is employed in equation (6) for one or more of these intermediaries. This algebraic equation is solved for the concentration of intermediary i in terms of the concentrations of other species. The full kinetic expressions, equations (6), (7), (8), and (14), are employed in computing dc_i/dt for species to which the steady-state approximation is not applied, with the concentrations of the other species having been eliminated through solution of the set of algebraic equations $\hat{\omega}_i = 0$. The order of the nonlinear system of differential equations is thereby reduced. If the steady-state approximation is applied to all reaction intermediaries, then the only equations requiring integration are those for dc_i/dt of the reactants and products; often the problem may be reduced to that of evaluating a single integral of the type of equation (23), with $f(c_i)$ given by an expression differing from equation (24).

A general criterion for applicability of the steady-state approximation to species i may be stated by writing equation (6) in the form $\hat{\omega}_i = \hat{\omega}_{i+} - \hat{\omega}_{i-}$, where $\hat{\omega}_{i+}$ is the sum of the positive terms on the right-hand side of the equation and $\hat{\omega}_{i-}$ denotes the sum of the magnitudes of the negative terms. The approximation is $\hat{\omega}_{i+} = \hat{\omega}_{i-}$, which—according to equation (14)— will be good if $|dc_i/dt| \ll \hat{\omega}_{i+}$. It should be realized that the approximation does *not* state that $dc_i/dt = 0$ but instead is based on $\hat{\omega}_{i+}$ and $\hat{\omega}_{i-}$ both being large compared with dc_i/dt. The validity of the approximation may be checked a posteriori by an iterative scheme in which all $c_i(t)$ are obtained by use of the approximation, dc_i/dt is then calculated for the intermediary in question by differentiation, and the magnitude of this quantity is compared with $\hat{\omega}_{i+}(t)$ or $\hat{\omega}_{i-}(t)$, the latter having been calculated by use of the steady-state functions $c_i(t)$ in equations (7) and (8). If the approximation is marginal, then improvement may be achieved by putting $\hat{\omega}_i(t) = dc_i/dt$ for the intermediary, the derivative being calculated from the steady-state $c_i(t)$, and solving the algebraic equation $\hat{\omega}_i = \hat{\omega}_i(t)$ instead of $\hat{\omega}_i = 0$ for c_i in order to obtain a better approximation for the intermediary concentration. By such methods it is often found that the steady-state approximation for chain carriers is inaccurate in the initial or final stages of a chain reaction, during which chain carriers are being produced or destroyed relatively rapidly through the predominance of initiation or termination steps. However, the rates of propagation steps often exceed those of initiation and termination so greatly during the major part of straight-chain reactions that the steady-state approximation is quite accurate for most of the reaction history.

The classical example of a complex straight-chain reaction for which the results of the steady-state approximation agree with experimental measurements is the hydrogen-bromine reaction $H_2 + Br_2 \rightarrow 2\,HBr$ [5]. The inferred mechanism is

$$Br_2 + X \xrightarrow{k_1} 2\,Br + X \qquad \text{(chain initiating)}$$

(H_2 dissociation is negligibly slow),

$$Br + H_2 \xrightarrow{k_2} HBr + H \qquad \text{(chain carrying)},$$

$$H + Br_2 \xrightarrow{k_3} HBr + Br \qquad \text{(chain carrying)}$$

(the reverse of this reaction is negligibly slow),

$$H + HBr \xrightarrow{k_4} H_2 + Br \qquad \text{(chain carrying)},$$

and

$$2\,Br + X \xrightarrow{k_5} Br_2 + X^* \qquad \text{(chain breaking)}$$

(low concentrations of H atoms make H recombination negligibly slow). With this mechanism, the steady-state approximations $\hat{\omega}_H = 0$ and $\hat{\omega}_{Br} = 0$ can easily be shown [5] to lead to $c_{Br} = \sqrt{k_1/k_5}\sqrt{c_{Br_2}}$ and

$$c_H = k_2 c_{H_2}\sqrt{k_1/k_5}\ \sqrt{c_{Br_2}}/(k_3 c_{Br_2} + k_4 c_{HBr}).$$

The rate of production of HBr consequently is given by

$$\begin{aligned}
dc_{HBr}/dt &= k_2 c_{Br} c_{H2} + k_3 c_H c_{Br_2} - k_4 c_H c_{HBr} \\
&= 2k_2\sqrt{k_1/k_5}\ c_{H_2}\sqrt{c_{Br_2}}/[1 + (k_4/k_3)(c_{HBr}/c_{Br_2})],
\end{aligned} \qquad (34)$$

with

$$dc_{H_2}/dt = dc_{Br_2}/dt = -dc_{HBr}/dt/2.$$

The last relation in equation (34) shows that the fairly complicated reaction-rate expression [constant $c_{H_2}\sqrt{c_{Br_2}}/(1 + \text{constant } c_{HBr}/c_{Br_2})$], which is correctly predicted by the steady-state approximation, has an overall order of 1.5 (even though none of the significant reaction steps have molecularities less than 2) and exhibits inhibition by products (an explicit inverse dependence on the concentration of HBr). Applicability estimates indicate that at atmospheric pressure the steady-state approximation for Br fails at temperatures above 1000 K because the initiation and propagation steps become too rapid in comparison with Br recombination, thereby causing $\hat{\omega}_{Br^-}$ to cease to be large in comparison with dc_{Br}/dt. At higher temperatures, $\hat{\omega}_H = 0$ may be retained to show that $c_{HBr}(t)$ may be calculated by solving two simultaneous first-order differential equations.

The **partial-equilibrium approximation** differs from the steady-state approximation in that it refers to a particular reaction instead of to a particular species. The mechanism must include the forward and backward steps of any reaction that maintains partial equilibrium, and the approximation for a reaction k is then expressed by setting $\omega'_k = 0$ in equation (11). It is not always proper to conclude from this that when equations (6), (10), and (11) are employed in equation (14), the terms $\hat{\omega}_{i,k}$ may be set equal to zero for each k that maintains partial equilibrium; partial equilibria occur when the forward and backward rates are both "large," and a small fractional difference of these two large quantities may contribute significantly to dc_i/dt. The criterion for validity of the approximation is that ω'_k be small compared with the forward or backward rate.

To apply a partial-equilibrium approximation for a reaction $k = 1$, identify a species, say $i = 1$, for which $v''_{i,1} - v'_{i,1} \neq 0$, and solve equation (11) with $\omega'_1 = 0$ for c_1 as a function of the other concentrations c_i. Also write equation (14) for $i \neq 1$ in the form

$$\frac{dc_i}{dt} = \sum_{k=2}^{M} \hat{\omega}_{i,k} + \left(\frac{v''_{i,1} - v'_{i,1}}{v''_{1,1} - v'_{1,1}}\right)\left[\frac{dc_1}{dt} - \sum_{k=2}^{M} \hat{\omega}_{1,k}\right], \quad i = 2, \ldots, N, \qquad (35)$$

which may be derived by use of equations (6) and (10). Since c_1 is a known function of the c_i for $i \neq 1$, equation (35) constitutes a set of $N - 1$ simultaneous first-order equations for the $N - 1$ concentrations c_i, $i = 2, \ldots, N$. In this manner, the partial-equilibrium approximation has reduced the order of the system of differential equations that must be solved. Introduction of additional partial-equilibrium approximations reduces the order further by sequential application of the same type of elimination procedure. This procedure is not straightforward enough to implement for calculation by electronic computers, and development of efficient methods for use of partial-equilibrium approximations remains a topic of current research [24], [25].

The same type of difficulty that is resolved by use of equation (35) for the partial-equilibrium approximation may also arise in connection with the steady-state approximation. For example, part of the sum of terms that contribute to the production rate of a primary species, to which the steady-state approximation is not applied, may be a constant multiple of $\hat{\omega}_i$ for an intermediary that is subject to the steady-state approximation, and the remaining terms in the production rate may be smaller than $\hat{\omega}_i$ even though $\hat{\omega}_i$ is small compared with $\hat{\omega}_{i+}$. Under this condition, inaccurate results for the concentration history of the primary species will be obtained by use of the steady-state approximation for the intermediary unless a substitution analogous to equation (35) is employed. In the absence of appropriate substitutions, the previously stated condition for validity of the steady-state approximation, although necessary, is not sufficient in all respects. Often less complicated criteria are stated for the validity of steady-state approximations; for example, it is frequently suggested that the steady state for an intermediary will be acceptable if its concentration is small compared with the concentrations of the major species. These simple criteria, which are useful for obtaining insights and estimates, usually are necessary but not sufficient. Especially when combinations of steady-state and partial-equilibrium approximations are employed, correct specification of necessary and sufficient conditions can become complex.

The Lindemann mechanism for unimolecular reactions, discussed in Section B.2.2, provides a convenient vehicle for illustrating partial-equilibrium approximations and for comparing them with steady-state approximations, even though this mechanism is not a chain reaction. To use the partial-equilibrium approximation for the two-body production of $\mathfrak{M}_1^*$, select $\mathfrak{M}_1$, for example, as the species whose concentration is to be determined by partial equilibrium and use

$$dc_{\mathfrak{M}_1}/dt = k_2 c_{\mathfrak{M}_1^*} c_X - k_1 c_{\mathfrak{M}_1} c_X \qquad (36)$$

to eliminate the equilibrium terms from the differential equation preceding equation (19), obtaining

$$dc_{\mathfrak{M}_1^*}/dt + dc_{\mathfrak{M}_1}/dt = -k_3 c_{\mathfrak{M}_1^*}, \qquad (37)$$

which is exact. The partial-equilibrium approximation,

$$c_{\mathfrak{M}_1} = (k_2/k_1)c_{\mathfrak{M}_i^*}, \tag{38}$$

may be substituted into equation (37), which may then be integrated to give

$$c_{\mathfrak{M}_i^*} = (k_1/k_2)c_{\mathfrak{M}_1}(0)e^{-\alpha t} \tag{39}$$

where $c_{\mathfrak{M}_1}(0)$ is the concentration of $\mathfrak{M}_1$ at time zero and $\alpha = k_1 k_3/(k_1 + k_2)$. Substitution of equation (39) into the equation for dc_p/dt, preceding equation (19), enables another integration to be performed, yielding

$$c_P = c_P(0) + [(k_1 + k_2)/k_2]c_{\mathfrak{M}_1}(0)(1 - e^{-\alpha t}), \tag{40}$$

where $c_P(0)$ is the initial product concentration. The rate formulas show that the reaction is of first order whenever partial equilibrium prevails. Freedom to specify $c_{\mathfrak{M}_i^*}(0)$ independently of $c_{\mathfrak{M}_1}(0)$ is lost by virtue of equation (38); if equation (38) is not satisfied initially, then there must be an initial stage in which partial equilibrium is invalid. The criterion for applicability of partial equilibrium after this initial stage is that the left-hand side of equation (36) be small compared with either term on the right. Use of equations (38) and (39) to evaluate these quantities shows that $\alpha \ll k_1 c_X$, that is,

$$k_3 \ll (k_1 + k_2)c_X, \tag{41}$$

is required for partial equilibrium.

If, erroneously, equation (37) were not derived before introduction of equation (38), then $dc_{\mathfrak{M}_i^*}/dt = -k_3 c_{\mathfrak{M}_i^*}$ would be obtained, resulting incorrectly in $c_{\mathfrak{M}_i^*} = (k_1/k_2)c_{\mathfrak{M}_1}(0)e^{-k_3 t}$, in place of equation (39). This result, in fact, corresponds not to partial equilibrium but rather to introduction of a steady-state approximation for $\mathfrak{M}_1$, and it may be used with equation (38) in the criterion $|dc_{\mathfrak{M}_1}/dt| \ll k_1 c_{\mathfrak{M}_1}c_X$ for applicability of this steady state to show that $k_3 \ll k_1 c_X$ is needed. This condition is more stringent than equation (41) and becomes equivalent to equation (41) if $k_2 \ll k_1$, under which condition equation (39) reduces to the result derived here. Thus the partial equilibrium has broader applicability than the steady-state approximation for $\mathfrak{M}_1$. Since $\mathfrak{M}_1$ is a reactant, it is not surprising that a steady-state approximation for it is of limited utility.

A more reasonable steady-state approximation would be that for the intermediary $\mathfrak{M}_i^*$. With this approximation, the development leading to equation (19) follows, and equation (36) becomes

$$dc_{\mathfrak{M}_1}/dt = -k_1 k_3 c_X c_{\mathfrak{M}_1}/(k_2 c_X + k_3), \tag{42}$$

the integral of which is $c_{\mathfrak{M}_1} = c_{\mathfrak{M}_1}(0)e^{-\beta t}$, where $\beta = k_1 k_3 c_X/(k_2 c_X + k_3)$. Use of this result in equation (19) enables $c_P(t)$ to be obtained by a further integration. The equation preceding equation (19) gives $c_{\mathfrak{M}_i^*}(t)$ from $c_{\mathfrak{M}_1}(t)$;

again, there must be an initial stage in which the approximation is invalid if $c_{\mathfrak{M}_i}(0)$ is not consistent with this relationship. The criterion for applicability of this steady-state approximation after the initial stage is $|dc_{\mathfrak{M}_1^*}/dt| \ll k_1 c_{\mathfrak{M}_1} c_X$, which is $\hat{\omega}_{i+}$ in the general notation given earlier. Use of equation (42) and the relationship preceding equation (19) then shows that

$$k_3 k_1 c_X \ll (k_2 c_X + k_3)^2 \tag{43}$$

is required if $\mathfrak{M}_1^*$ is to achieve a steady state. Equations (43) and (41) define different but partially overlapping regimes. If $k_3 \gg (k_1 + k_2)c_X$, then equation (41) is violated but equation (43) is satisfied. If $k_2 c_X \ll k_3 \ll k_1 c_X$, then equation (43) is violated but equation (41) is satisfied. If $k_1 \ll k_2$ and $k_3 \ll k_2 c_X$, then equations (41) and (43) are both satisfied and $\beta \approx \alpha$, so that the two approximations lead to the same predictions. Typically, $k_1 \ll k_2$, a condition under which satisfaction of equation (41) implies satisfaction of equation (43), but not conversely; thus the steady-state approximation for $\mathfrak{M}_1^*$ has wider applicability than the partial-equilibrium approximation if $k_1 \ll k_2$. These observations suggest that prior knowledge of relative rates of various steps is helpful in deciding whether to try to employ steady-state or partial-equilibrium approximations.

B.2.5.3. Branched-chain explosions

Explosions in approximately homogeneous chemical systems have been studied experimentally by introducing reactants into a preheated vessel, by use of a piston-driven adiabatic compression to achieve a rapid increase of temperature, and by shock-tube techniques in which a shock wave produces compression. In these experiments the temperature increase initiates measurable rates of reaction, and reaction histories are recorded with the objective of ascertaining, for example, whether the heat-release rate eventually experiences an abrupt increase that would be classified as an explosion, and—if so—the time that elapses before this increase occurs (the induction time). A simplified kinetic model may be employed to illustrate how chain branching may be responsible for such explosions. Let R signify reactant, P signify product, and C signify chain carrier, and write the initiation, propagation, and termination steps, respectively, as $R \to C$, $C + R \to P + \alpha C$ and $C \to P$, where α is a branching constant, equal to unity for straight-chain reactions [for example, equations (30) and (31)]. With k_i, k_p, and k_t representing the rate constants for these three steps, equation (14) gives

$$dc_R/dt = -k_i c_R - k_p c_R c_C, \qquad dc_P/dt = k_p c_R c_C + k_t c_C, \tag{44}$$

and

$$dc_C/dt = k_i c_R + (\alpha - 1)k_p c_R c_C - k_t c_C. \tag{45}$$

Although the nonlinearity $c_R c_C$ in the propagation prevents an exact solution to this system from being obtained analytically, the qualitative behavior of the solution may be deduced from the differential equations.

From equation (44) we see that the reactant concentration decreases monotonically with time from its initial value $c_R(0)$, while that of the product increases monotonically. The concentration of the chain carrier exhibits a more complicated behavior, according to equation (45). If $c_C = 0$ initially, then there is a linear increase of c_C with time at early times by initiation. If $k_t > (\alpha - 1)k_p c_R(0)$ (which applies, for example, to straight-chain reactions), then dc_C/dt decreases continually with time, and after a sufficient amount of reactant depletion and carrier buildup, c_C begins to decrease and eventually decays exponentially through termination with a time constant k_t^{-1}, that is, $c_C \sim e^{-k_t t}$. However, there are other ranges of parameters in which c_C behaves differently; for example, it may exhibit a tendency to oscillate if $\alpha = 2$ and $k_p c_R(0) \geq k_i + k_t$. Often both $k_t \ll (\alpha - 1)k_p c_R(0)$ and $k_i \ll (\alpha - 1)k_p c_R(0)$, under which conditions the linear increase in c_C is followed by a stage of exponential increase with t, during which propagation dominates and produces a short time constant of approximately $[(\alpha - 1)k_p c_R]^{-1}$. The exponential growth ceases approximately when a propagation-termination balance is established, $c_R = k_t/[(\alpha - 1)k_p]$, and a maximum of c_C is then reached and is followed by a decrease, eventually exponential. The occurrence of the stage of exponential growth is typical of branched-chain explosions.

If substantial heat release is associated with propagation alone, then the onset time for explosion is roughly equal to the time needed for the propagation rate to become equal to the initiation rate, approximately $[(\alpha - 1)k_p c_R(0)]^{-1}$ [since $c_C \approx k_i c_R(0)t$ during initiation]. Often propagation involves negligible heat release, and termination steps produce the major heat release. Under such conditions, the induction time may be interpreted as the time required for the termination rate to become equal to the propagation rate, a time usually large compared with $[(\alpha - 1)k_p c_R(0)]^{-1}$. To estimate this time, use

$$c_c \approx \{k_i/[(\alpha - 1)k_p]\}\exp[(\alpha - 1)k_p c_R(0)t]$$

as an approximate solution to equation (45) during exponential growth, where reactant depletion has been neglected and the initial condition has been evaluated from equality of initiation and propagation rates. This equation may be used to integrate equation (44) for c_R with initiation neglected, giving

$$c_R = c_R(0)\exp\{-e^{(\alpha - 1)k_p c_R(0)t}k_i/[(\alpha - 1)^2 k_p c_R(0)]\}.$$

Since this equals $k_t/[(\alpha - 1)k_p]$ at the propagation-termination balance, the approximate formula for the induction time is

$$t = [(\alpha - 1)k_p c_R(0)]^{-1} \ln\{[(\alpha - 1)^2 k_p c_R(0)/k_i] \ln[(\alpha - 1)k_p c_R(0)/k_t]\}. \tag{46}$$

It may be noted from this formula that an increase in $(\alpha - 1)k_p c_R(0)$, in k_i (less strongly), or in k_t (much less strongly) produces a decrease of the induction time. There are elements of arbitrariness in the definition of induction time as well as in judgments of whether explosion occurs.

Since it has been seen that dc_C/dt decreases with t if $k_t > (\alpha - 1)k_p c_R(0)$ but may exhibit an explosionlike growth if $k_t < (\alpha - 1)k_p p_R(0)$, it may be reasonable to identify the equality $k_t = (\alpha - 1)k_p c_R(0)$ with the boundary of explosion. Alternative reasoning that leads to this same result makes use of the steady-state approximation for the intermediary C. Vanishing of the right-hand side of equation (45) yields

$$c_C = (k_i/k_p)/[\beta - (\alpha - 1)], \tag{47}$$

where $\beta = k_t/(k_p c_R)$ measures the number of termination steps per propagation step experienced by C. Substitution of equation (47) into equation (44) yields

$$dc_P/dt = k_i(c_R + k_t/k_p)/[\beta - (\alpha - 1)] \tag{48}$$

for the overall rate of production of products. If $\beta > \alpha - 1$, then equations (47) and (48) provide positive values for c_C and the rate, and the steady-state approximation *may* be acceptable (to ascertain whether it is would require further study). If $\beta \leq \alpha - 1$, then equation (47) yields an infinite or negative value of c_C, and therefore the steady-state approximation is clearly not acceptable under this condition. The limit in which the steady state predicts infinite c_C and rate has been identified as the explosion limit and agrees with the estimate obtained above. Hence,

$$\alpha \geq 1 + \beta \tag{49}$$

may be employed as a criterion for the occurrence of a branched-chain explosion. Alternative approaches to derivations that lead to results such as those given in equations (48) and (49) include phenomenological reasoning and approximate analyses of different simplified models for the reacting system.*

It may be noted from equation (48) that for $\alpha = 1$, the overall rate is proportional to β^{-1}, the number of propagation steps per termination step, which may be termed the **chain length**. Since chain reactions typically have

---

* See, for example, [7], 111–283, [13], 76–142, and [15], 17–40.

long chain lengths, β usually is small. Therefore, relatively small amounts of branching (α slightly in excess of unity) typically are sufficient to satisfy equation (49) and to lead to a branched-chain explosion.

In many systems, the quantity β can reasonably be expressed as the sum of two terms

$$\beta = \beta_s + \beta_c, \tag{50}$$

where β_s is the number of heterogeneous termination reactions (at the walls or other solid surfaces) per propagation step and β_c is the number of homogeneous termination reactions (generally by means of three-body collisions) per propagation step. Typically, β_s decreases as the pressure p increases, while $\beta_c \sim p$; therefore, a graph of $\beta(p)$ exhibits a minimum, leading to the possibility of an explosion peninsula for which equation (49) is satisfied only over an intermediate range of p.

To understand these pressure dependences, write $\beta_s = k_s/(k_p c_R)$ and $\beta_c = k_c c_A c_B/(k_p c_R)$, where $k_t = k_s + k_c c_A c_B$. Here k_s is the rate constant for heterogeneous termination, while k_c is that for three-body recombination, c_A and c_B being the concentrations of the two species A and B that collide with C in the homogeneous termination. If diffusion rates are fast compared with chemical rates of heterogeneous termination, then the considerations to be presented in Section B.4 describe the functional dependences of k_s; it is more often true that diffusion is slow and controls the heterogeneous rate, under which conditions k_s equals the ratio of a diffusion coefficient to the square of a characteristic dimension of the vessel containing the reacting gases. Since diffusion coefficients vary inversely with p (Appendix E), it then found that $k_s \sim p^{-1}$, whence $\beta_s \sim p^{-2}$ (since $c_R \sim p$). The most representative process of homogeneous termination is $2C + X \rightarrow C_2 + X$,* where X denotes any third molecule; this process would be described by setting $A = C$ and $B = X$, although use of $A = B = X$ leads to results that are qualitatively the same and avoids introducing an additional nonlinearity into equation (45). Since $c_A \sim p$, $c_B \sim p$ and $c_R \sim p$, it is seen that $\beta_c \sim p$. Unlike β_s, which decreases as the size of the vessel increases, the term β_c is independent of the characteristic dimension. The temperature dependences of the diffusion coefficient and of k_c are relatively weak, but k_p typically increases rapidly with increasing temperature. Hence β_s and β_c both decrease with increasing temperature, producing a widening of the explosion peninsula as temperature increases, according to equation (49).

The model scheme described by equations (44) and (45) represents a simplification of mechanisms that apply to real branched-chain explosions. Initiation steps are seldom ever unimolecular; $R + X \rightarrow 2C + X$ or $2R \rightarrow$ carriers would be more realistic, and the latter would introduce further nonlinearity. The stable species produced in the termination step are not all products but instead include substantial amounts of other molecules, such as reactants. Finally, there are generally more than one reactant,

product and chain carrier involved in realistic mechanisms. An example is
the hydrogen-oxygen explosion, which involves the steps

$$H_2 + X \longrightarrow 2H + X \qquad \text{(initiation)},$$

$$H + O_2 \longrightarrow OH + O \qquad \text{(branching)},$$

$$O + H_2 \longrightarrow OH + H \qquad \text{(branching)},$$

$$OH + H_2 \longrightarrow H_2O + H \qquad \text{(propagation)},$$

$$H \longrightarrow \text{walls} \qquad \text{(termination)},$$

$$2H + X \longrightarrow H_2 + X \qquad \text{(termination)},$$

$$H + OH + X \longrightarrow H_2O + X \qquad \text{(termination)},$$

$$H + O_2 + X \longrightarrow HO_2 + X.$$

The first of these steps is the simplest of the possible initiation reactions,
the second and third are chain-branching, the fourth a straight-chain
propagation, the fifth heterogeneous termination, and the last three homogeneous termination, with the final one dominant because of low radical
concentrations. The explosion-limit curve for hydrogen-oxygen mixtures is
sketched schematically in Figure B.1. The explosion peninsula in the lower
part of this curve is explained qualitatively by the analysis that has been
given for the simplified model. The third explosion limit shown is believed
to involve a change in the character of HO_2, which at low pressures is stable

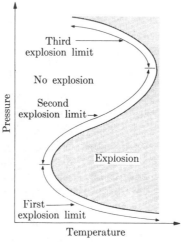

FIGURE B.1. Schematic representation of the dependence of the explosion limit for
hydrogen-oxygen mixtures on pressure and temperature for a fixed initial composition and
in a given container.

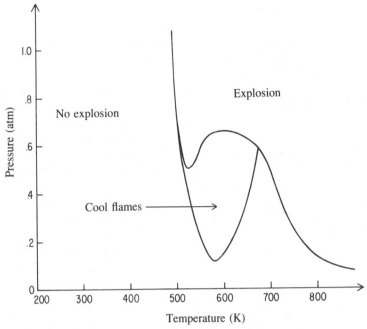

FIGURE B.2. A measured dependence of the explosion limit for a propane-oxygen mixture on pressure and temperature in a given vessel (adapted from information given in [8]).

enough to diffuse to the walls without reacting but may regenerate chain carriers through $H_2 + HO_2 \rightarrow H_2O_2 + H$ at elevated pressures [7], [8], [12], [13], [16]. Thermal explosion processes (see Section B.2.5.4) may also be involved in the third limit [7], [8].

Explosion-limit curves exhibiting the explosion peninsula (without the third limit) have also been observed for other systems, such as CO oxidation [7], [8], [12], [13], [16], and results of the model in equations (44) and (45) are in qualitative agreement with these observations as well. Explosion-limit measurements help in clarifying reaction mechanisms. Hydrocarbon oxidation exhibits a different type of branched-chain explosion behavior [8], [12], [16]. A representative limit curve for a propane-oxygen mixture is shown in Figure B.2. In the region marked "cool flames," propagation of a luminous reaction front through the vessel is observed to occur one to six times in the course of the slow reaction. Each such front, called a **cool flame**, is believed to be initiated by buildup of chain carriers and to consume these intermediaries as it propagates, thereby preventing explosion. Cool flames have been stabilized by flowing the reactant mixture through a tube whose wall temperature is controlled carefully by heating elements [26]. Model kinetic mechanisms are known that produce periodicities resembling, in some respects, the repeated reactions observed in cool flames [27].

B.2.5.4. Thermal explosions

Explosion is a general term that connotes noise and destruction associated with a sudden occurrence of pressure-wave propagation. In combustion, the definition has been specialized to describe the spontaneous development of rapid rates of heat release by chemical reactions in initially nearly homogeneous systems. Pressure-wave generation usually follows but is not germane to the definition. Within the restricted definition employed here, there are two types of explosions. In addition to the branched-chain explosions just discussed, there are thermal explosions, which do not require complex kinetics for their explanation but instead can be understood on the basis of a one-step approximation for the chemical reaction. Since overall rates of chemical reactions generally increase with increasing temperature, the heat released from an exothermic process may accelerate the rate by raising the temperature of the reacting mixture. The increased rate of reaction further increases the rate of heat release, leading to a rapid buildup of thermal energy in the system. The importance of the dependence of the rate upon temperature distinguishes thermal explosions from branched-chain explosions, which may occur in isothermal systems.

Thermal explosions have been discussed in various books and review articles (for example, [7], [28]–[31]). The presentations in [28] and [30] are especially extensive. These explosions have been studied for both gaseous and condensed-phase reactants. In addition to being of fundamental interest to chemical kinetics, they are of practical concern in wide-ranging problems of fire and explosion safety; for example, a hazard in storage and transportation of crystalline NH_4NO_3, used in fertilizers, is the possible occurrence of a thermal explosion if suitable packing precautions are not applied.

Statement of an appropriate model for describing thermal explosions necessitates introduction of an energy balance. For purposes of discussion, a balance for the concentration c_F of the fuel also will be included here. The latter is given by equation (14) only if loss processes, such as diffusion to walls where catalytic destruction may occur, are neglected. The accurate inclusion of these losses violates homogeneity and necessitates introduction of spatial derivatives into the equations (Appendix E). As a basis for a simplified model that preserves homogeneity, the loss effects will be approximated by use of a mass-transfer coefficient h_F in a negative term added to the right-hand side of equation (14). Similarly, a phenomenological heat-transfer coefficient h_T will be employed in the energy balance to maintain homogeneity for the temperature T.

Consider a vessel of volume V with internal surface area S, which initially contains a reactant mixture with uniform fuel concentration c_{F0} at a uniform temperature T_0, which may differ from the constant wall temperature T_w. Approximate $\hat{\omega}_F$ in equation (14) by $\hat{\omega}_F = -kf$, where f describes the dependence of the rate on c_F, and k its dependence on T. The Arrhenius-type approximation $k = B_0 e^{-E/R^0 T}$ will be employed (see Section B.3.1) with B_0 and E/R^0 constants. By introducing a dependence of B_0 on the initial concentrations, the value of f is selected to be unity when $c_F = c_{F0}$; it decreases as the reaction progresses, accounting for the effect of reactant depletion on the rate, and it approaches zero as the reaction goes to completion. The generalization of equation (14) for fuel is

$$V dc_F/dt = -VB_0 f e^{-E/R^0T} - Sh_F c_F, \tag{51}$$

the negative of which states that the time rate of decrease of the number of fuel molecules in the vessel is the sum of the rate at which those molecules are consumed chemically and the rate at which they are lost to the walls. If c_v denotes the specific heat at constant volume for the reacting system and ρ its density, then an analogous energy balance for the system is

$$V \rho c_v dT/dt = VQ_F B_0 f e^{-E/R^0T} - Sh_T(T - T_w), \tag{52}$$

where Q_F represents the energy released in the reaction per mole of fuel consumed. This equation states that the rate of energy buildup in the system is the difference between the rate of energy production and the rate of energy loss to the walls. Linear approximations have been introduced for the loss terms in these equations to achieve simplicity.

Research [28] has identified efficient nondimensional variables for use in equations (51) and (52). Let $\tau = t/t_C$, where $t_C = V \rho c_v/(Sh_T)$ is a cooling time proportional to the ratio of the volume to the surface area of the container. Introduce $\varphi = c_F/c_{F0}$ and $\theta = (T - T_w)E/(R^0 T_w^2)$ as nondimensional concentration and temperature variables. Then equations (51) and (52) become

$$d\varphi/d\tau = -\epsilon\gamma\delta f e^{\theta/(1+\epsilon\theta)} - \alpha\varphi \tag{53}$$

and

$$d\theta/d\tau = \delta f e^{\theta/(1+\epsilon\theta)} - \theta \tag{54}$$

where $\alpha = \rho c_v h_F/h_T$, $\epsilon = R^0 T_w/E$, $\gamma = \rho c_v T_w/(Q_F c_{F0})$, and

$$\delta = \left(\frac{Q_F c_{F0}}{\rho c_v T_w}\right)\left(\frac{E}{R_0 T_w}\right)\left(\frac{V \rho c_v}{Sh_T}\right)\left(\frac{B_0}{c_{F0}}\right)e^{-E/R^0 T_w}. \tag{55}$$

The initial conditions are $\varphi = 1$ and $\theta = \theta_0 \equiv (T_0 - T_w)E/(R^0 T_w^2)$ at $\tau = 0$. The simplest approximation to the function $f(\varphi)$ is $f = \varphi$. Here α is the ratio of a transfer coefficient for fuel to that for heat, ϵ is a ratio of the thermal energy at wall temperature to the activation energy (see Section B.3), γ is a ratio of the thermal energy at wall temperature to the total energy released by the reaction, and $\epsilon\gamma\delta$ is the ratio of the cooling time to the characteristic time of chemical reaction at the wall temperature. Frank-Kamenetskii [28] has emphasized that in combustion, the parameters ϵ and γ are small. He also introduced the parameter δ, defined in equation (55), as occupying a role of central importance in thermal explosions.

If losses are negligible, then addition of $\epsilon\gamma$ times equation (54) to equation (53) reveals that $\epsilon\gamma\theta + \varphi$ remains constant, that is, $\varphi = 1 - \epsilon\gamma(\theta - \theta_0)$, which may be used in f in equation (54) to obtain a single differential equation for $\theta(\tau)$ under these conditions. Since $\gamma\epsilon \ll 1$, often reactant depletion is negligible except in late stages of the process, and $f = 1$ may be employed in equation (54). Since $\epsilon \ll 1$, in a first approximation, equation (54) with $f = 1$ becomes

$$d\theta/d\tau = \delta e^{\theta} - \theta, \tag{56}$$

which defines the simplest model for a thermal explosion if $\theta = 0$ at $\tau = 0$ (that is, $T_0 = T_w$, the initial temperature of the reacting medium is the same as the wall temperature).

Graphs of the heat-generation and heat-loss terms on the right-hand side of equation (54) or (56) as functions of θ are useful in defining necessary conditions for the occurrence of thermal explosions. A plot of this type is is shown schematically in Figure B.3. Steady-state conditions of constant temperature correspond to equality of rates of generation and loss, intersections of the solid and broken curves in Figure B.3. Consider first the generation curve labeled e^{θ}, which applies under the approxmations leading to equation (56). If δ is small enough, then there are two intersections of the loss line with this generation curve. The lower intersection defines a stable,

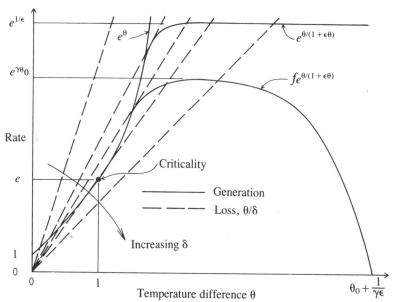

FIGURE B.3. Schematic illustration of nondimensional rates for heat generation and heat loss as functions of the nondimensional difference between bulk temperature and wall temperature for thermal explosions.

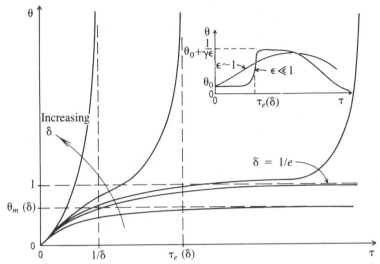

FIGURE B.4. Schematic illustration of nondimensional temperature-time histories for thermal explosions for the simplified Frank-Kamenetskii model (main graph) and for a more complete model that includes reactant depletion (inset).

slow reaction, while the upper intersection is not significant because it corresponds to an unstable steady state. The instability is evident from the fact that a small increase in temperature causes the generation to exceed the loss, and therefore θ will increase even more according to equation (56); similarly, a decrease in θ causes the loss to exceed the generation, so that θ decreases further. Thus at sufficiently small δ, there exists one stable state, and a steady, slow reaction is possible. If δ is large enough, then there is no intersection; the generation rate then always exceeds the loss rate, causing θ to increase continually with τ, as is qualitatively anticipated during a thermal explosion. The critical value of δ for the occurrence of thermal explosion may therefore be identified as the value for which the dashed line is tangent to the solid curve. At values of δ below this critical value, a steady state may occur, but at values above this value, it cannot.*

Mathematical identification of the tangency condition is straightforward when equation (56) applies. Equality of generation and loss rates requires $\delta e^\theta = \theta$, and equality of the slopes requires $\delta e^\theta = 1$. Hence $\theta = 1$ at the point marked "criticality" in Figure B.3, and $1/e$ is the critical value of δ. Solutions of equation (56) for various values of δ, shown schematically in Figure B.4, confirm that $\delta = 1/e$ can properly be interpreted as the critical measure of the ratio of the generation rate to the loss rate, above which

---

* Even if δ is below the critical value, thermal explosion may occur if the initial temperature is higher than the second intersection, but seldom are experiments performed with initial temperatures so high.

thermal explosion occurs. For $\delta < 1/e$, $d\theta/d\tau$ decreases continually with increasing τ, and θ increases from zero, approaching a limiting value $\theta_m(\delta)$ as $\tau \to \infty$; this limit is given by the smaller solution to $\theta_m = \delta e^{\theta_m}$, at which value $d\theta/d\tau = 0$, giving a steady, slow reaction. For $\delta > 1/e$, the behavior is markedly different, there being a maximum value of τ, called $\tau_e(\delta)$ in Figure B.4, at which—according to equation (56)—θ approaches infinity. A predicted temperature of infinity has been termed a **thermal runaway** and the time at which it occurs has been called an **explosion time**. The nondimensional explosion time $\tau_e(\delta)$ may be quite large if δ is near $1/e$. For $\delta \gg 1/e$, the explosion time can be obtained with ease from equation (56), because the loss term θ always remains negligible in comparison with the generation term δe^θ. The integral of $d\theta/d\tau = \delta e^\theta$ with $\theta(0) = 0$ is simply $\theta = -\ln(1 - \delta\tau)$, whence $\tau_e(\delta) = 1/\delta$ when δ is large; in dimensional variables the explosion time becomes

$$t_e = \rho c_v R^0 T_w^2 e^{E/R^0 T_w}/(E Q_F B_0), \tag{57}$$

which is the chemical reaction time diminished by heat-release and activation-energy factors. Calculation of the explosion time for δ near $1/e$ involves a more complicated procedure, which results in $\tau_e(\delta) = \sqrt{2\pi^2/(e\delta - 1)}$.

The infinite θ at a finite τ is contradictory to equation (54) and is a consequence of the approximations that lead to equation (56). Even if reactant depletion is neglected, retention of the factor $(1 + \epsilon\theta)^{-1}$ in the exponential in the rate term of equation (56) would cause the generation rate to level off at a value $\delta e^{1/\epsilon}$ for very large values of θ, as illustrated schematically in Figure B.3 by the curve labeled $e^{\theta/(1 + \epsilon\theta)}$. This curve is drawn for illustrative purposes with an incorrect scale so that the leveling fits on the graph; in in reality, the leveling occurs at values of θ too large to be of realistic interest. Nevertheless, it is seen that for a range of values of δ, the leveling produces a third intersection of the generation and loss curves. This upper intersection is stable and exists for all values of δ greater than a minimum value (corresponding to the upper tangent line in the illustration), below which thermal explosion cannot occur, irrespective of the value of θ_0. If conditions are such that thermal explosion occurs, then at large values of τ, θ approaches a constant value corresponding to a generation-loss balance at the upper intersection. The approximation e^θ to the Arrhenius rate expression has been termed the **Frank-Kamenetskii rate expression**; a fundamental difference between the two is the absence of the upper intersection for e^θ. If ϵ is small, then at the time $\tau_e(\delta)$—obtained from equation (56)—a rapid increase in θ occurs, as illustrated in the inset in Figure B.4. Since this increase would be interpreted as an explosion, the explosion time obtained from the study of equation (56) remains valid, except for corrections that are small when ϵ is small. However, the thermal runaway does not proceed to infinite θ. If ϵ is not small, then a gradual change of θ occurs, like that sketched in the curve labeled $\epsilon \sim 1$ in the inset in Figure B.4, and a well-defined explosion event, to which an explosion time can be ascribed, does not exist. Thus thermal explosions require small ϵ—that is, a heat-release rate that depends strongly on temperature.

The reasoning that has just been outlined neglects reactant depletion. If reactant depletion is taken into account in equation (54), then the generation curve labeled $f e^{\theta/(1 + \varepsilon\theta)}$ in Figure B.3 is obtained. The maximum temperature then is limited by the adiabatic flame temperature, and the generation rate is restricted to a value well below $\delta e^{1/\varepsilon}$. Moreover, the generation rate approaches zero at a value of θ corresponding to complete depletion of reactants; eventually θ itself approaches zero, as illustrated in the inset in Figure B.4, due to continued heat loss to the walls. In Figure B.3 it is seen that an upper intersection exists, the situation being entirely analogous to that discussed for the generation curve labeled $e^{\theta/(1 + \varepsilon\theta)}$. However, the values of θ and of the generation rate at the upper intersection are much less than those for the curve that neglects reactant depletion, and it is therefore essential to take reactant depletion into account if post-explosion phenomena are to be described with any accuracy.

Continuing studies in the theory of thermal explosions have been directed along a number of lines (for example, [32]–[34]). Usually the objective has been to consider influences of phenomena beyond those contained in equation (56), such as the simultaneous presence of two exothermic reactions [32], but occasionally models leading to even greater simplification have been sought [33]. Especially notable is the recent emergence of asymptotic methods in thermal-explosion theory [35]–[38], since the smallness of ϵ and γ make the problems ideally suited for application of such techniques. These studies have, for example, shed light on the influence of reactant depletion on explosion [38]. There have also been many studies of partial differential equations that avoid the homogeneity approximation which necessitates introducing h_T; these studies help to clarify influences of vessel shape on thermal explosions.*

B.2.5.5. Kinetics of hydrocarbon combustion

Detailed kinetic mechanisms for the combustion of hydrocarbon fuels have been subjected to relatively intensive study in recent years. The mechanisms are complex and involve a variety of chain carriers. Fundamental to such studies are data on rates of elementary steps. Extensive compilations of rate information are becoming available [12], [39]–[47]. Users should realize that uncertainties remain in rates of various elementary steps. Since these uncertainties sometimes exceed an order of magnitude, studies of

---

* Much of this work is presented by Y. B. Zel'dovich, G. I. Barenblatt, V. B. Librovich, and G. M. Makhviladze in *The Mathematical Theory of Combustion and Explosion*, Moscow: Nauka, 1980.

accuracies and of sensitivities of predictions are desirable in applications. Dominant mechanisms are dependent upon pressure, temperature and reactant mixtures employed. As an example, consider oxidation of methane in air at atmospheric pressure and near the adiabatic flame temperature achieved for a stoichiometric mixture initially at room temperature.

The initiation step is believed to be either

$$CH_4 + X \longrightarrow CH_3 + H + X$$

or

$$CH_4 + O_2 \longrightarrow CH_3 + HO_2,$$

the latter being more important at lower temperatures. The thermal scission described by the former is slower for CH_4 than for other hydrocarbons because of the greater strength of the C—H bond in CH_4. At these high temperatures, the H or HO_2 quickly provide a "radical pool" of chain carriers OH, O, and H by the chain mechanisms of the hydrogen-oxygen reaction (see Section B.2.5.3). The fuel is attacked by these radicals; for example,

$$CH_4 + OH \longrightarrow CH_3 + H_2O.$$

The methyl radicals may undergo the sequence

$$CH_3 + O_2 \longrightarrow CH_3O + O,$$

$$CH_3O + O_2 \longrightarrow CH_2O + HO_2,$$

producing formaldehyde, which is oxidized in steps such as

$$CH_2O + OH \longrightarrow HCO + H_2O,$$

followed by

$$HCO + X \longrightarrow H + CO + X.$$

If methyl concentrations become sufficiently high, then

$$CH_3 + O \longrightarrow CH_2O + H$$

replaces the two-step formaldehyde production given above. In addition,

$$CH_3 + CH_3 \longrightarrow C_2H_6$$

may occur [42], introducing a number of additional reactions associated with ethane oxidation.

The mechanism described here results in H_2O and CO as reaction products. The oxidation of CO to CO_2 occurs largely by

$$OH + CO \longrightarrow CO_2 + H.$$

Attempts to simplify the scheme outline here have employed a two-step process, the first typically being combustion to form H_2O and CO and the second the oxidation of CO to CO_2 (for example, [48]). Such endeavors have resulted in controversies concerning suitable phenomenological expressions for the overall rates of each step. Experiments or detailed kinetic computations may be employed to provide overall rate parameters for one-step approximations to the complex mechanism more easily than to provide parameters for two-step approximations. Nevertheless, intermediate approximations, such as two-step mechanisms, appear to be worthy of further pursuit, since extensive calculations by electronic computers with complete mechanisms are time-consuming and often experience difficulties in attempting to exhibit the key reaction steps of the mechanism and to develop an understanding of the kinetics.

Accurate specifications of kinetic mechanisms for combustion often are less critical to calculation of overall rates of heat release than to estimation of amounts of pollutants produced. Pollutants of primary interest are oxides of nitrogen, oxides of sulfur, unburned hydrocarbons, and particulates such as smoke or soot [12]. A primary mechanism by which nitric oxide is formed in flames is that attributed to Zel'dovich, namely,

$$O + N_2 \longrightarrow NO + N,$$

$$N + O_2 \longrightarrow NO + O,$$

The rates of these two steps are known accurately, and the second is fast compared with the first. Hence the rate at which NO is produced depends strongly on the concentration of O atoms, which in turn is highly sensitive to the kinetics of the combustion process. Since O concentrations often exceed values for equilibrium of $O_2 \rightleftarrows 2O$ in flames, rates of production of NO may exceed those calculated from an O-equilibrium assumption. Another mechanism that may be of importance under fuel-rich conditions is $HCN + O \rightarrow CH + NO$, the HCN having been produced either by $CH + N_2 \rightarrow HCN + N$ or from the fuel itself if the fuel contains sufficient amounts of bound nitrogen. Oxidation of NH, NH_2, and NH_3 may also contribute to NO production for fuels containing nitrogen—for example, $NH + O_2 \rightarrow HNO + O$, $HNO + O_2 \rightarrow HO_2 + NO$. Extensive discussions of mechanisms for production of NO and SO_2 may be found in the literature (see [12], for example).

Production of soot in hydrocarbon combustion is a complicated kinetic process involving transformations of fuel molecules in fuel-rich zones under the influence of heat (called **pyrolysis**) to produce new fuel molecules that contain more C and less H than the original molecules. The elementary steps involved are so numerous that simplifying approximations are essential for both efficient computation of soot production and development of understanding of the kinetic mechanisms. Progress is being made toward discovery of suitable simplications [49]. It is known that small amounts of O_2 can sometimes increase the rate of the process of soot production (while larger amounts may reduce particulates by oxidizing soot);

consideration therefore needs to be given to *oxygen-catalyzed pyrolysis* in the development of simplified models.

B.2.6. Catalysis

Fundamentally a **catalyst** is a substance that alters the rate of reaction, but that itself suffers no net change in properties or concentration during the process.* There are a number of these materials which, when present in very small amounts, exert profound influences on the rates of certain reactions (occasionally, a catalyst can cause an explosion in a system exhibiting no observable reaction). Since its concentration remains unchanged during the reaction, a catalyst cannot affect the equilibrium composition of the system. This result implies that a good catalyst for a given forward reaction also can function as a catalyst for the corresponding backward reaction. Some reactions are known to be **autocatalytic** (a product of the reaction acts as a catalyst) and consequently exhibit peculiar concentration-time curves. Homogeneous catalysts generally change the homogeneous reaction mechanism by providing an alternative reaction path along which the reaction can proceed more rapidly. Heterogeneous catalysts (that is, certain solid surfaces) usually adsorb the reactants and then promote critical reaction steps at **active centers** (certain adsorption sites), thereby increasing the rate of the overall reaction. Some substances are also known to act as **negative catalysts** (or, **inhibitors**), that is, to reduce certain reaction rates without themselves being altered; inhibitors of homogeneous reactions might combine readily with chain carriers and give up these chain carriers slowly, while inhibitors of heterogeneous reactions may occupy most of the active centers through (perhaps preferential) adsorption.

Examples of catalysts in combustion reaction include the effect of H_2O on the carbon monoxide oxidation reaction $CO + \frac{1}{2}O_2 \rightarrow CO_2$.[†] Nitric oxide also catalyzes CO oxidation through the mechanism $2NO + O_2 \rightarrow 2NO_2$ (overall) and $NO_2 + CO \rightarrow NO + CO_2$. In both of these examples, an intermediate compound (for example, NO_2) is formed and then destroyed. The addition of a small amount of NO_2 to an $H_2 - O_2$ mixture leads to a branched-chain explosion by introducing the relatively rapid initiation step $NO_2 + X \rightarrow NO + O + X$, with the O atoms so produced generating the usual $H_2 - O_2$ chain. The NO_2 also participates in the efficient termination step $NO_2 + O \rightarrow NO + O_2$, which is sufficiently important at large concentrations of NO_2 to cause a slow reaction to be

---

* Use of the term *catalyst* is commonly broadened to include substances that generate catalytic species. Often a so-called catalyst is consumed during the reaction, having produced species that first exert catalytic influences then emerge in altered forms.

[†] See [12] and [50] for discussions of the mechanism of this reaction.

observed instead of an explosion. Important negative catalysts for the pre-burning process in spark-ignition engines are **antiknock** compounds, such as tetraethyl lead. Inside the engine cylinder these materials form lead peroxide, which presumably combines with chain carriers, thus decreasing the concentration of chain carriers and preventing branched-chain ex-plosions. Bromine-containing compounds similarly reduce chain-carrier concentrations in hydrocarbon oxidation, thereby inhibiting combustion and causing some such materials to be attractive fire suppressants.

B.3. DETERMINATION OF THE SPECIFIC REACTION-RATE CONSTANT

B.3.1. The Arrhenius law

In discussing reaction-rate constants, we shall focus our attention on a single reaction step, equation (1). For any such process, the temperature T dependence of the specific reaction-rate constant k that appears in equa-tion (4) is given empirically by the Arrhenius expression

$$k = Ae^{-E/R^0T}, \tag{58}$$

where R^0 is the universal gas constant, E represents the **activation energy**, and A is the **frequency factor** for the reaction step. In equation (58), E is a constant but A may depend weakly on T. Typically, the approximation

$$A = BT^\alpha, \tag{59}$$

where B = constant, α = constant, and $-1 < \alpha \lesssim 2$, is well within experi-mental error.

Experimentally, E is found by plotting measured values of $\ln k$ against $1/T$ and computing the slope of the best-fit straight line through the data points (slope = $-E/R^0$). After E has been obtained, A may be calculated from equation (58) and the measured values of k. Often E can be determined within a few percent; however, difficulties in measuring the absolute value of the reaction-rate constant often produce uncertainties in A that exceed a factor of 2. If the reaction were not an elementary step then E obtained in this way would be called an **overall activation energy**.

B.3.2. The activation energy

For the sake of simplicity, the theoretical basis of equation (58) will be considered only for homogeneous reactions in ideal gas mixtures, although the principal concepts presented require only small modifications to be applied to other homogeneous reactions and to heterogeneous processes.

At the molecular level, a homogeneous reaction in an ideal gas in-volves a collision between two or three appropriate reactant molecules.

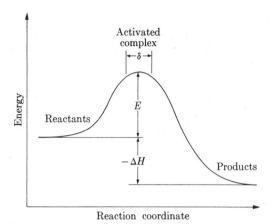

FIGURE B.5.   Schematic illustration of potential-energy surface.

This collision process is governed by the Schrödinger equation of quantum mechanics, in which the coordinates of all of the electrons and nuclei composing the colliding molecules appear as independent variables. If the collision process is slow enough to ensure that the solution for the time-dependent Schrödinger equation differs negligibly from the solution for the appropriate steady-state Schrödinger equation and that the kinetic energy of each nucleus is negligible compared with that of the electrons,* then the nuclei may be treated as moving about on a potential-energy "surface," the energy of which is determined by the motion of the electrons corresponding to the instantaneous locations of the nuclei.

The number of dimensions of this surface depends on the number and nature of the colliding nuclei. In a certain region of this surface the energy assumes a local minimum value and the relative positions of the nuclei correspond to a system of reactant molecules. In another region of local energy minimum the system may be identified as reaction products. A potential-energy barrier on the surface generally separates these two regions. At a saddle point on this barrier, the height of the surface (above the energy of the reactant region) assumes a minimum value. The collision of the reactant molecules can produce products only if the energy of the reactants (for example relative kinetic energy) exceeds this minimum height.[†] The minimum barrier height therefore may be termed an **activation energy** E which reactants must acquire before they can react.

This situation is illustrated schematically for a one-dimensional potential-energy surface in Figure B.5, from which it can be inferred that

---

* Reactions that are extremely rapid in a macroscopic sense may still be slow according to these criteria.

† Quantum-mechanical barrier-penetration effects usually are negligible for chemical reactions.

$E \geq 0$ for exothermic reactions and $E \geq \Delta H (-\Delta H \equiv$ heat of reaction) for endothermic processes. Typically $E \sim 30\,\text{kcal/mol}$, although values may range from 0 for three-body radical recombinations to magnitudes in excess of 80 kcal/mol in unusual instances.

The Boltzmann energy-distribution law states that the probability that a molecule possesses energy E is proportional to e^{-E/R^0T}. More-elaborate computations, based on Boltzmann statistics, of the probability that reactants possess energies greater than or equal to the activation energy E yield essentially (sometimes precisely) this same proportionality factor. Hence the factor e^{-E/R^0T} in equation (58) represents the fraction of collisions between reactant molecules in which reaction products can be formed, and the empirical actication energy defined in Section B.3.1 is essentially the same as the activation energy that we have been discussing in this section.

It may be emphasized that the viewpoint described above provides a basis for a theoretical calculation of the activation energy; namely, the shape of the potential-energy surface is computed, the saddle point is located, and its energy is noted. The shape of the potential-energy surface is determined by solving the Schrödinger equation for the electrons, with various choices of nuclear coordinates. In this equation the required potential energy appears as an eigenvalue. Variational procedures have been developed for finding the eigenvalue, but the task is so complicated that semiempirical approximations must be used for most reactions. Therefore, the experimental procedure described in Section B.3.1 remains the most dependable method for obtaining the activation energy.

B.3.3. Collision reaction-rate theory [2], [3], [11]

In order to predict the value of the frequency factor, one may assume that all collisions between reactant molecules with sufficient activation energy result in the instantaneous formation of the reaction products. With this simple hypothesis (collision theory), if the activation energy is known, then the problem of computing the reaction rate reduces to the problem of computing the rate of collision between the appropriate reactant molecules in the ideal gas mixture. This last problem is easily solved by the elementary kinetic theory of gases.

We shall illustrate the reasoning by considering reactions involving binary collisions between molecules of two species (identified by subscripts 1 and 2). If v_{12} represents the number of collisions per unit volume per second between molecules of kinds 1 and 2, then the collision-theory hypothesis states that

$$\mathscr{A}\omega = v_{12}e^{-E/R^0T}, \tag{60}$$

where Avogadro's number $\mathscr{A}$ converts the number of moles (in ω) to the number of molecules. According to equation (60), the number of product molecules (with $v_i'' - v_i' = 1$) produced per unit volume per second equals the number of collisions between appropriate reactant molecules per unit volume per second multiplied by the fraction of these collisions involving sufficient activation energy. In view of equations (4) and (58), equation (60) implies that the frequency factor is

$$A = v_{12}/(\mathscr{A} c_1 c_2). \qquad (61)$$

An expression for v_{12} is derived from elementary kinetic theory in Section E.2.3. By using this result in equation (61), we find that

$$A = \mathscr{A} \sigma_{12} \bar{v}_{12} = \mathscr{A} \sigma_{12} (8k^0 T/\pi \mu_{12})^{1/2}, \qquad (62)$$

where σ_{12} is the collision cross section for molecules of types 1 and 2, $\bar{v}_{12}$ is the average relative velocity of molecules of kinds 1 and 2, μ_{12} is the reduced mass of molecules of types 1 and 2 $[\mu_{12} = W_1 W_2/(W_1 + W_2)\mathscr{A}]$, and k^0 is Boltzmann's constant $(k^0 = R^0/\mathscr{A})$.*

Equation (62) implies that the frequency factor depends on the single molecular parameter σ_{12}, which can be deduced from data on transport properties or from molecular-beam experiments (see Appendix E). Equation (62) predicts a weak temperature dependence of A, namely,

$$A \sim \sigma_{12}\sqrt{T} \sim T^\alpha, \quad 0 \lesssim \alpha \lesssim 1/2.$$

Examples are known in which equation (62) agrees closely with experiment (for example, the reaction $2\,HI \rightarrow H_2 + I_2$). However, experimentally determined frequency factors often differ considerably from the values given by equation (62) (a difference of a factor of 10^6 is not uncommon). Since the experimental rates usually are lower than the rates predicted by collision theory, equation (62) is conventionally corrected by introducing a **steric factor** P, which originally was interpreted as accounting for the fact that activated collisions lead to reaction only if the incident molecules have the correct relative geometrical orientation (or, alternatively, only if the activation energy is in the proper modes). Thus, in place of equation (62), use is made of the expression

$$A = PZ, \qquad (63)$$

where the *collision number* is defined as

$$Z \equiv \mathscr{A} \sigma_{12} (8k^0 T/\pi \mu_{12})^{1/2}. \qquad (64)$$

Equation (63) is precisely correct for bimolecular reactions by definition, but the unknown steric factor P must be determined before

---

* A more detailed derivation of this result, on the basis of the kinetic theory of gases, with classical mechanics employed and with reactive collisions presumed to occur whenever a pair of colliding molecules reaches the top of the potential-energy barrier, has been given by R. D. Present, *J. Chem. Phys.* **31**, 747 (1959).

equation (63) can be used. Transition-state theory provides an alternative and complementary viewpoint which is capable of predicting values for P.

B.3.4. Transition-state theory [2], [3], [11], [51], [52]

If colliding reactant molecules are to form products, they must first reach the top of the potential-energy barrier illustrated in Figure B.5. Transition-state theory assumes that the reacting system at the top of the barrier is a molecule (to which thermodynamics may be applied) and that this molecule, which is called the **activated complex**, is in chemical equilibrium with the reactants. The rate at which the activated complex decays to products then equals the reaction rate.

The activated complex always is identified by the symbol $\neq$. It is an ordinary molecule in every respect except that one ordinarily vibrational degree of freedom (that in the direction of the **reaction coordinate** shown in Figure B.5) is replaced by a translational degree of freedom. This conclusion is a consequence of the observation that in the neighborhood of the saddle point on the potential-energy surface, the potential energy increases (leading to ordinary vibrations) in all coordinate directions except that of the re- reaction coordinate. The activated complex thus has four translational degrees of freedom and one less vibrational degree of freedom than would otherwise be expected.

In order to derive an equation for the reaction rate, let τ denote the average time for an activated complex to decay to products. Then, from the definition of ω,

$$\omega = c_{\neq}/\tau \tag{65}$$

($c_{\neq} \equiv$ concentration of the complex). Equation (4) therefore implies that

$$k = \left(c_{\neq} \middle/ \prod_{i=1}^{N} c_i^{v_i} \right) \frac{1}{\tau}. \tag{66}$$

The assumption of equilibrium between the complex and reactants implies that

$$c_{\neq} \middle/ \prod_{i=1}^{N} c_i^{v_i} = K_c^{\neq},$$

the equilibrium constant for concentrations for the reaction in which the complex is formed. Assuming that the activated state exists for a distance δ in the reaction coordinate at the top of the barrier, we may expect that $\tau = \delta/\bar{v}$, where $\bar{v}$ is the average velocity in the positive direction of the reaction coordinate at the top of the barrier. Hence

$$k = K_c^{\neq}(\bar{v}/\delta). \tag{67}$$

If the velocity distribution in the reaction coordinate is assumed to be Maxwellian (because of the hypothesis of thermodynamic equilibrium for the complex), then a simple integration shows that $\bar{v} = (k^0 T/2\pi m_{\neq})^{1/2}$, where $m_{\neq}$ is an appropriate reduced mass of the reacting components of the activated-complex molecule. The use of this result and of equation (A-24) for $K_c^{\neq}$ in equation (67) yields

$$k = (Q'_{\neq}/\delta)(k^0 T/2\pi m_{\neq})^{1/2} \Big/ \prod_{i=1}^{N} Q'^{v_i}_i, \tag{68}$$

where Q' represents the partition function per unit volume. Since methods for computing partition functions have been summarized in Appendix A, the only troublesome factor in equation (68) is $Q'_{\neq}/\delta$. This factor can be simplified by extracting the contribution to the partition function of the activated complex caused by translation in the reaction coordinate. According to Section A.2.3, this contribution is $(2\pi m_{\neq} k^0 T/h^2)^{1/2}\delta$. Hence equation (68) reduces to

$$k = (k^0 T/h)\hat{Q}'_{\neq} \Big/ \prod_{i=1}^{N} Q'^{v_i}_i, \tag{69}$$

where $\hat{Q}'_{\neq}$ is the partition function per unit volume for an ordinary molecule, but with one vibrational degree of freedom omitted. Therefore, $\hat{Q}'_{\neq}$ may easily be estimated by hypothesizing a reasonable structure for the activated complex, and k may be calculated from equation (69) by using the methods discussed in Section A.2.

From the preceding paragraph, the reader will note that many assumptions are involved in transition-state theory. Alternative derivations exhibit differing hypotheses. In a quicker but perhaps less intuitive derivation, translation in the reaction coordinate is treated formally as the low-frequency limit of a vibrational mode. Expansion of the vibrational partition function given in Section A.2.3 then yields $Q'_{\neq} = \hat{Q}'_{\neq}(k^0 T/hv)$, which is substituted into equation (A-24), to be used directly in equation (66), thereby producing equation (69) when $v = 1/\tau$. The decay time thus is identified as the reciprocal of the small frequency of vibration in the direction of the reaction coordinate.

Although equation (69) does not explicitly exhibit the factor $e^{-E/R^0 T}$, this factor will arise in computing the partition functions because of the difference between the ground-state energy of the activated complex and the ground-state energy of the reactants. If q_i denotes the partition function per unit volume (for species i) from which the ground-state energy factor has been removed and $q_{\neq}$ denotes the corresponding quantity for the complex without the degree of freedom corresponding to the reaction coordinate, then equation (69) becomes

$$k = (k^0 T/h)\left(q_{\neq} \Big/ \prod_{i=1}^{N} q_i^{v_i}\right)e^{-E/R^0 T},$$

which implies that the frequency factor is given by

$$A = (k^0 T/h) q_{\neq} \left/ \prod_{i=1}^{N} q_i^{\nu_i} \right.$$ (70)

[see equation (58)].* The partition-function ratio in equation (70) can be interpreted in terms of an entropy of activation; this leads to an approximate interpretation of the steric factor as a thermodynamic probability of activation and to a simple physical explanation of the observation that $P < 1$ [2].

B.3.5. Comparison between transition-state theory and collisional theory

In order to illustrate the consequences of equation (70), it will be assumed that the partition functions for the reactants and the complex can be expressed as products of the appropriate numbers of translational, rotational and vibrational partition functions.† For simplicity we shall also neglect factors associated with nuclear spin and electronic excitation. If $s_i \equiv$ total number of atoms in a molecule of species i and $u_i \equiv 0$ for nonlinear molecules, 1 for linear molecules, and 3 for monatomic molecules, then the correct numbers of the various kinds of degrees of freedom are obtained in equation (70) by letting

$$q_i = \prod_{t=1}^{3} q_{i,t} \prod_{r=1}^{3-u_i} q_{i,r} \prod_{v=1}^{3s_i-6+u_i} q_{i,v},$$ (71)

where the indices t, r, and v identify partition functions per unit volume for single translational, rotational and vibrational degrees of freedom, respectively. Here the "per unit volume" part of q_i is included in $q_{i,t}$; that is, each $q_{i,t}$ may be assumed to be $(\mathscr{A}V)^{-1/3}$ times the partition function per translational degree of freedom listed in Section A.2.3, and the remaining $q_{i,r}$ and $q_{i,v}$ factors are equal to the appropriate unmodified partition functions per degree of freedom given in Section A.2.3. Equation (71) may also be used for $q_{\neq}$, provided that the number of terms in the vibrational product is replaced by $3s_{\neq} - 7 + u_{\neq}$ to account for the absence of one vibrational mode.

The predictions of transition-state theory are most easily compared with those of collision theory for bimolecular reactions. The comparison may be facilitated by expressing Z in terms of a ratio of partition functions. The definition of Z [equation (64)] may be rewritten as

$$Z = \left(\frac{k^0 T}{h} \right) \left[\frac{(2\pi\mu_{12} k^0 T)^{1/2}}{\mathscr{A}^{1/3} h} \right]^{-3} \left[\frac{(8\pi\mu_{12}\sigma_{12} k^0 T)^{1/2}}{h} \right]^2 .$$

---

* The quantity q_i may be computed more easily than Q_i'.
† This assumption is made in nearly every application of transition-state theory.

The second factor here, which will be denoted by q_T^{-3}, is the reciprocal of the translational partition function per unit volume for a molecule of mass μ_{12} (see Section A.2.3). Since the moment of inertia of a system composed of the masses m_1 and m_2 separated by a distance d such that $\pi d^2 = \sigma_{12}$ is $I = \mu_{12}\sigma_{12}/\pi$, $\mu_{12} \equiv m_1 m_2/(m_1 + m_2)$, the last factor in the above expression, which will be denoted by q_R^2, is the rotational partition function for this system (see Section A.2.3). Thus in terms of rotational and translational partition functions per unit volume,

$$Z = (k^0 T/h)(q_R^2/q_T^3). \tag{72}$$

Equations (63), (70), and (72) yield

$$P = (q_T^3/q_R^2)q_{\neq}/q_1 q_2 \tag{73}$$

for a bimolecular reaction involving the collision of molecules of species 1 and 2. If use is made of the approximation given in equation (71), then this expression for the steric factor, derived from transition-state theory, becomes

$$P = \frac{q_T^3 \displaystyle\prod_{t=1}^{3} q_{\neq,t} \prod_{r=1}^{3-u_{\neq}} q_{\neq,r} \prod_{v=1}^{3s_{\neq}-7+u_{\neq}} q_{\neq,v}}{q_R^2 \displaystyle\prod_{t=1}^{3} q_{1,t} \prod_{r=1}^{3-u_1} q_{1,r} \prod_{v=1}^{3s_1-6+u_1} q_{1,v} \prod_{t=1}^{3} q_{2,t} \prod_{r=1}^{3-u_2} q_{2,r} \prod_{v=1}^{3s_2-6+u_2} q_{2,v}}. \tag{74}$$

The order of magnitude of P may be inferred from equation (74) by assuming that all translational partition functions are roughly equal, all rotational partition functions per degree of freedom are approximately equal to q_R (which is somewhere between 3 and 100 depending on the moment of inertia and T), and all vibrational partition functions are roughly equal to an average vibrational factor q_v (which will lie between 1 and 10 depending on the characteristic vibrational frequency and T). These approximations will be used only for the purpose of discussion; the resulting error should seldom exceed a factor of 10. Equation (74) reduces to

$$P \approx (q_V/q_R)^{5+u_{\neq}-u_1-u_2}, \tag{75}$$

where q_V/q_R lies roughly between 1 and 10^{-2}, depending on T and the molecular structures of the reactants and the complex. For reactions between two monatomic molecules (which rarely occur), $u_1 = u_2 = 3$ and $u_{\neq} = 1$; hence $P \approx 1$ according to equation (75), and the collision theory in its simplest form agrees with transition-state theory. For reactions between two diatomic molecules, typically $q_V/q_R \sim 0.5$ at room temperature, $u_1 = u_2 = 1$ and (probably) $u_{\neq} = 0$; hence $P \approx (q_V/q_R)^3 \sim 10^{-1}$, and the simplest collision theory will sometimes yield a good approximation to the results of transition-state theory. For reactions between nonlinear poly-atomic molecules, $u_1 = u_2 = u_\neq = 0$ and typically $q_V/q_R \sim 0.1$ at room temperature; hence $P \approx (q_V/q_R)^5 \sim 10^{-5}$, and the simplest collision theory is generally in poor agreement with transition-state theory.

It should be emphasized that a disagreement between the results of the two rate theories under discussion does not imply that one is right and the other is wrong. Each theory involves different assumptions which may be correct for some reactions and incorrect for others. The advantage of trans-ition-state theory is that the predicted reaction rate depends on a greater number of molecular parameters, thus affording the possibility of describing a greater variety of reactions.

If E is near zero, then—because of the presence of vibrational partition functions in both the numerator and the denominator of equation (70)—the transition-state theory may predict that k varies nonmonotonically with T[53]. The collisional theory does not possess this flexibility.

B.3.6. Other applications of transition-state theory

The application of transition-state theory to unimolecular reactions is very straightforward. With the subscript 1 identifying the reactant, equation (70) reduces to

$$A = (k^0 T/h)(q_\neq/q_1). \tag{76}$$

The activated complex should differ from the reactant primarily in having one vibrational bond broken. If q_V denotes the partition function for this vibrational mode, then equation (76) becomes, approximately,

$$A = (k^0 T/h)/q_V, \tag{77}$$

from which the frequency factor is easily computed.

Transition-state theory can be developed for reactions occurring on surfaces by a procedure paralleling that given in Section B.3.4 [2]. The only modifications required are that volume concentrations must be replaced by surface concentrations and that a factor representing the concentration (number per unit surface area) of active surface sites at which the reaction may occur must be introduced. The viewpoint of transition-state theory can also easily be extended to include rate predictions for vaporization (for example, [54]) and for other heterogeneous processes involving phase changes.

B.3.7. Modern developments in reaction-rate theory

There are more-recent developments in reaction-rate theory that yield a deeper understanding of certain reactions than can be obtained with either of the theories outlined above. The newer analyses provide alternative viewpoints for studying some reactions and also modify and amplify preceding results. We shall not discuss these newer developments because they are somewhat limited in scope and do not invalidate the ideas presented above. However, certain innovations deserve to be mentioned.

The importance of electronically excited states in reaction kinetics is well established [14]. Electronic excitation leads to qualitative as well as to quantitative modifications in the preceding theories, especially in connection with the intersection of the potential-energy surfaces corresponding to different electronic states. Structures of electronically excited activated complexes have been studied (for example, [55]) and have been used in postulating kinetic mechanisms for the production of nonequilibrium excited species that have been observed (for example, [56]) in hydrogen-oxygen flames.

Increased attention has been focused on vibrational, rotational, and translational nonequilibria in reacting systems as well. To account for these nonequilibrium effects, it is becoming increasingly traditional to express specific reaction-rate constants in terms of sums or integrals of reaction cross-sections over states or energy levels of the reactants involved [3], [11]. This approach helps to relate the microscopic and macroscopic aspects of rate processes and facilitates the use of fundamental experimental information, such as that obtained from molecular-beam studies [57], in calculation of macroscopic rate constants. Proceeding from measurements at the molecular level to obtain the rate constant defined in equation (4) remains a large and ambitious task.

There is considerable interest in nonequilibria encountered in reacting systems from the viewpoint of their relationships to photon propagation. The numerous existing examples of chemiluminescence suggest that chemical reactions are promising for affording population inversions needed in lasers. Prospects for developing relatively high-power chemical lasers through combustion have prompted many new investigations [58]–[60]. These studies help to improve knowledge of reaction rates, for example by engendering more thorough descriptions of collisional activation and deactivation rates of excited vibrational states.

Some of the continuing approaches to reaction-rate theory that differ from either the simple collisional theory or the transition-state theory discussed here are cited on pages 98–112 of [4]. Examples of differing approaches may be found in particular in theories for rates of three-body radical-recombination processes [61]. Advances in methods for calculating rate constants relevant to the Lindemann view of unimolecular processes also are providing new information relevant to unimolecular and bimolecular rates. Future work may be expected to produce further results of use in combustion problems.

B.4. THE RATES OF HETEROGENEOUS PROCESSES

B.4.1. Description of the heterogeneous processes to be discussed

In this section consideration is restricted to reactions between gaseous species adsorbed on a solid or liquid surface. Principally, we shall consider the adsorption process, adopting the kinetic viewpoint originally developed by Langmuir. The method is illustrative of kinetic procedures that can also be used to analyze processes involving gas-liquid and gas-solid phase changes. We shall not discuss reactions occurring at surfaces with steps that do not involve adsorption.

A reaction between gaseous species occurs in an adsorbed gas layer on a surface by means of the following sequence of events: (1) Reactant molecules in the gas migrate to the surface, (2) these reactant molecules are adsorbed on the surface, (3) the adsorbed reactants react to form adsorbed products, (4) the product molecules leave the surface, and (5) the product molecules migrate to the bulk of the gas. Since these processes occur in series, the rate of the slowest step determines the overall reaction rate.

B.4.2. Gas transport rate-controlling

Unless the pressure is so low that the molecular mean free path (Appendix E) is comparable to the dimensions of the chamber enclosing the gas,* the rates of the first and last steps, 1 and 5, are governed by molecular diffusion rates (or by rates of mass transfer through a boundary layer, if the gas is in motion). Although these macroscopic transfer processes, which lie within the realm of fluid mechanics (Appendixes C–E), often are the rate-determining steps [62] (see also Chapter 12), we shall assume here that they are so fast that they can be neglected.

B.4.3. Adsorption or desorption rate-controlling

We shall now consider systems in which steps 2 or 4 are rate-controlling and later investigate systems in which step 3 determines the overall rate. Much additional information on these topics is available [63]–[66].

The rate of adsorption of a molecular species depends on the nature of the adsorption process. Gas molecules may be held on a surface by the weak van der Waal attractive forces which exist between all molecules or by stronger chemical types of bonds (chemisorption) which depend on the

---

* At these low pressures, the rates of steps 1 and 5, which are then governed by the kinetic theory of gases, generally become so fast that they do not affect the overall rate.

chemical characteristics of the adsorbed molecule and the surface. Although van der Waal adsorption may promote reactions by bringing molecules close together for a longer time than in the bulk of the gas, chemisorption usually should be more effective in promoting reactions because of the consequent modification in the chemical structure of the adsorbed molecule. For some chemisorption processes it will be reasonable to assume that the nature of the chemical bonding causes, at most, a *monomolecular* layer to be adsorbed on the surface. The following analysis of the rate of adsorption for these processes was first developed by Langmuir.

For ideal gases, the number of moles of chemical species i in the gas striking the surface per unit area per second is

$$\beta_i = c_i \sqrt{k^0 T / 2\pi m_i}, \quad i = 1, \ldots, N, \tag{78}$$

where c_i is the concentration (moles per unit volume) of species i in the gas, the square-root factor (in which m_i is the mass of a molecule of species i) is the average velocity normal to the surface for molecules of species i [see equation (68)], and N is the total number of chemical species in the gas. If θ_i denotes the fraction of the surface area covered by adsorbed molecules of species i, then for adsorption of a monomolecular layer

$$\theta = \sum_{i=1}^{N} \theta_i \tag{79}$$

is the total fraction of the surface area covered by adsorbed molecules of any kind and $1 - \theta$ is the fraction of the surface area available for adsorption.* Therefore, the number of moles of species i being adsorbed from the gas per unit surface area per second is

$$b_i' = \beta_i (1 - \theta)\alpha_i, \quad i = 1, \ldots, N, \tag{80}$$

where α_i, which is called a surface **accommodation coefficient** for molecules of species i, is the probability that a molecule of species i, which strikes a free adsorption site, will adhere to the surface.

Adsorbed molecules will also return to the gas because of thermal agitation. If μ_i denotes the number of moles of adsorbed molecules of species i per unit area per second which would return to the gas from a surface completely covered by molecules of species i, then the number of moles of species i per unit surface area per second returning to the gas is

$$b_i'' = \mu_i \theta_i, \quad i = 1, \ldots, N. \tag{81}$$

---

* Throughout this analysis, it is assumed that all molecular species are adsorbed on the same sites, so that an area covered by molecules of one species will not admit molecules of another species. Seldom is sufficient information available concerning the properties of the surface to justify any other assumption.

Hence, the net rate of adsorption (moles per unit area per second) of species i is

$$b_i = b'_i - b''_i = \beta_i(1 - \theta)\alpha_i - \mu_i\theta_i, \quad i = 1, \ldots, N. \tag{82}$$

In equation (82), the phenomenological coefficients α_i and μ_i, which according to the preceding development may depend on temperature but are independent of the pressure and of the compositions of the gas and adsorbed phases, can be predicted only by more detailed theories such as transition-state theory.

When steps 2 or 4 are rate-controlling, equation (82) determines the overall rate of the surface reaction, and the forward and backward reactions for step 3 may be assumed to be so fast that equilibrium is maintained for this step. If the surface reaction is given by equation (1), then the phenomenological condition for equilibrium of step 3 may be written as

$$\prod_{i=1}^{N} \gamma_i^{v''_i - v_i} = K_\gamma, \tag{83}$$

where γ_i is the surface concentration (moles per unit area) of species i and K_γ (which is expected to depend strongly on T but only weakly, if at all, on pressure or the γ_i) is the equilibrium constant for the surface reaction. The quantity γ_i is simply related to θ_i: If γ_0 is the number of moles of adsorption sites per unit surface area, that is, the maximum surface concentration, then

$$\gamma_i = \gamma_0\theta_i, \quad i = 1, \ldots, N. \tag{84}$$

Therefore, equation (83) provides a relationship among the values of θ_i appearing in equation (82).

Equations (82) and (83) are sufficient for determining the overall reaction rate when either step 2 or 4 controls the rate. However, since equation (82) contains N rate equations, the result cannot be expressed simply in terms of gas-phase concentrations. In order to obtain simpler formulas and to see the significance of the result, one may investigate the case in which all adsorption processes except one (say, that for species 1) are so fast that they may be assumed to be in equilibrium. In this case, $b_i = 0$ $i = 2, 3, \ldots, N$, whence [from equation (82)]

$$\beta_i(1 - \theta)\alpha_i = \mu_i\theta_i, \quad i = 2, \ldots, N. \tag{85}$$

The overall reaction rate is then most easily measured in terms of the rate of production of species 1; if the total surface area is a,* then the net number of moles per second of species 1 added to the gas is

$$-ab_1 = a\mu_1\theta_1 - a\beta_1(1 - \theta)\alpha_1. \tag{86}$$

---

* The quantity a is often difficult to determine; it is usually considerably greater than the geometrical area of the surface because of surface irregularities at the molecular level.

When species 1 is a reaction product, then the second term on the right-hand side of this equation often will be negligibly small; the first term on the right-hand side often will be negligible when species 1 is a reactant. In equation (86), c_1 appears through β_1 [see equation (78)], and the only other composition-dependent factors are the surface concentration fractions θ_1 and θ. Equation (79) gives θ in terms of θ_i, and, in view of equation (84), equations (83) and (85) provide N independent expressions relating the quantities θ_i to the volume concentrations c_i (through β_i) and temperature. Clearly, the explicit dependence of the overall reaction rate on the gas-phase concentrations may be quite complicated even in this simple case. The reader may investigate the nature of the result by considering particular reactions (for example, $\mathfrak{M}_1 \to \mathfrak{M}_2$ with $N = 2$). Instead of proceeding with such an analysis, we shall turn to the case in which step 3 controls the overall reaction rate.

B.4.4. Surface reaction rate-controlling

For many heterogeneous reactions, the surface reaction rate is slow compared with rates of the adsorption processes (steps 2 and 4). The overall reaction rate may then be determined by setting $b_i = 0$ in equation (82) for all species i; thus

$$\beta_i(1 - \theta)\alpha_i = \mu_i\theta_i, \quad i = 1, \ldots, N. \tag{87}$$

For simplicity we shall neglect the reverse surface reaction and shall employ the phenomenological expression

$$b = k \prod_{i=1}^{N} \gamma_i^{v_i} \tag{88}$$

for the rate (moles of a principal product produced per unit surface area per second) of the reaction given in equation (1). The factor k in equation (88) [(moles per unit area)$^{1-m}$(time)$^{-1}$, $m \equiv \sum_{i=1}^{N} v_i'$] is the specific reaction-rate constant for the surface reaction; it is expected to be influenced most strongly by temperature and might be computed theoretically from transition-state analyses. Equation (87) can be used to express the reaction rate given in equation (88) in terms of gas concentrations.

In view of equation (79), equations (87) provides N linear relations for the N quantities θ_i. Since equation (87) can be rewritten as

$$\theta = 1 - (\mu_i\theta_i/\alpha_i\beta_i), \quad i = 1, \ldots, N, \tag{89}$$

the quantity $\mu_i\theta_i/\alpha_i\beta_i$ must be the same for all species i. Hence

$$\theta_i = (\alpha_i\beta_i/\mu_i)(\mu_1\theta_1/\alpha_1\beta_1), \quad i = 1, \ldots, N. \tag{90}$$

The substitution of equation (90) into equation (79) relates θ to θ_1. This result may be substituted into equation (89) with $i = 1$ to show that

$$\sum_{i=1}^{N} (\alpha_i\beta_i/\mu_i)(\mu_1\theta_1/\alpha_1\beta_1) = 1 - (\mu_1\theta_1/\alpha_1\beta_1).$$

The solution of this equation for θ_1 yields

$$\theta_j = (\alpha_j\beta_j/\mu_j)\bigg/\left[1 + \sum_{i=1}^{N} (\alpha_i\beta_i/\mu_i)\right], \quad j = 1, \ldots, N, \tag{91}$$

since the choice of species 1 was arbitrary. The surface and gas-phase concentrations may be introduced explicitly into equation (91) by using equations (78) and (84); the result is

$$\gamma_j = \gamma_0\kappa_j c_j\bigg/\left[1 + \sum_{i=1}^{N} (\kappa_i c_i)\right], \quad j = 1, \ldots, N, \tag{92}$$

where

$$\kappa_i \equiv (\alpha_i/\mu_i)\sqrt{k^0 T/2\pi m_i}, \quad i = 1, \ldots, N, \tag{93}$$

depends only on T.

Before utilizing equation (92) in equation (88), let us consider the equilibrium surface concentration of an adsorbed gas as predicted by equation (92). Suppose that $N = 1$ and no surface reaction occurs. Equation (92) then reduces to

$$\gamma_1 = \gamma_0\kappa_1 c_1/(1 + \kappa_1 c_1). \tag{94}$$

Equation (94) shows that, at constant temperature, $\gamma_1 \sim c_1$ at low values of $c_1(\kappa_1 c_1 \ll 1)$ and $\gamma_1 \approx \gamma_0 = $ constant at high values of $c_1(\kappa_1 c_1 \gg 1)$. For intermediate values of c_1, $\gamma_1 \sim c_1^n$ over short ranges of c_1, where $0 \leq n \leq 1$. This dependence often is observed, and equation (94) has been verified experimentally for a number of systems.

Instead of c_1, the partial pressure of species 1 (or the total pressure if only one species is present) is generally measured in experiments and is used in equation (94). Since the proportionality factor between partial pressures and c_i depends only on T, this factor may be absorbed into κ_i, and equations (92) and (94) remain valid when concentrations are replaced by partial pressures. When written in terms of partial pressures, equation (94) is the **Langmuir adsorption isotherm**.

At constant c_1 as the temperature increases the rate of return of adsorbed molecules to the gas, μ_1, usually increases rapidly, thus causing κ_1 and therefore γ_1, as given by equation (94) to decrease rapidly. This agrees with experimental observations. Similar reasoning shows that a high surface accommodation coefficient α_1 favors strong adsorption.

A number of adsorption isotherms with concentration dependences differing from that given in equation (94) have been observed. These can be explained by extensions of the Langmuir analysis which account for various types of *multimolecular* adsorbed layers.

Let us return now to the problem of surface reactions. An explicit expression for the reaction rate in terms of gas-phase concentrations can be obtained by substituting equation (92) into equation (88):

$$b = k \left[\prod_{i=1}^{N} (\gamma_0 \kappa_i c_i)^{v_i} \right] \left[1 + \sum_{i=1}^{N} (\kappa_i c_i) \right]^{-\Sigma_{i=1}^{N} v_i}. \tag{95}$$

The implications of equation (95) are most easily seen by investigating special cases.

For the unimolecular reaction $\mathfrak{M}_1 \to \mathfrak{M}_2$ with $N = 2$, equation (95) reduces to

$$b = k\gamma_0 \kappa_1 c_1 / (1 + \kappa_1 c_1 + \kappa_2 c_2). \tag{96}$$

Even when the reaction product is very weakly adsorbed ($\kappa_2 c_2 \ll 1$), equation (96) may differ from the first-order expression ($b \sim c_1$ at constant T) obtained for homogeneous reactions. Equation (96) yields $b \sim c_1$ at constant T only when both the reactant and the product are weakly adsorbed ($\kappa_1 c_1 \ll 1$ and $\kappa_2 c_2 \ll 1$). When $\kappa_2 c_2 \ll 1$ and the reactant is strongly adsorbed ($\kappa_1 c_1 \gg 1$), equation (96) gives a zero-order kinetic expression ($b = k\gamma_0 =$ constant at constant T) for the unimolecular surface process. With $\kappa_2 c_2 \ll 1$, the apparent order of the reaction decreases from 1 to 0 as c_1 (for example, the pressure) is increased. If the reaction product is strongly adsorbed ($\kappa_2 c_2 \gg 1$) and is also adsorbed more strongly than the reactant ($\kappa_2 c_2 \gg \kappa_1 c_1$), then equation (96) becomes

$$b = (k\gamma_0 \kappa_1 / \kappa_2)(c_1/c_2) \sim (c_1/c_2),$$

which shows that the overall order of the reaction is zero, the order of the reaction with respect to the product is -1, and the reaction rate is retarded by the formation of reaction products. This result is a consequence of the fact that strongly adsorbed products occupy nearly all the adsorption sites, thus preventing the reactant from being adsorbed.

For the biomolecular reaction $\mathfrak{M}_1 + \mathfrak{M}_2 \to$ products in which none of the products are appreciably adsorbed, equation (95) reduces to

$$b = k\gamma_0^2 \kappa_1 \kappa_2 c_1 c_2 / (1 + \kappa_1 c_1 + \kappa_2 c_2)^2. \tag{97}$$

Equation (97) yields the usual second-order expression

$$b = (k\gamma_0^2 \kappa_1 \kappa_2) c_1 c_2 \sim c_1 c_2$$

at constant T when both reactants are weakly adsorbed ($\kappa_1 c_1 \ll 1$ and $\kappa_2 c_2 \ll 1$). Of greater interest is the case in which one reactant (say, species 2) is strongly adsorbed ($\kappa_2 c_2 \gg 1$) and is also adsorbed much more strongly than is the other reactant ($\kappa_2 c_2 \gg \kappa_1 c_1$). Equation (97) then reduces to

$$b = (k\gamma_0^2 \kappa_1 / \kappa_2)(c_1/c_2) \sim (c_1/c_2),$$

from which it is seen that the presence of the reactant species 2 in the gas retards the overall reaction rate. The physical explanation of this effect should be apparent from the preceding development.

REFERENCES

1. C. N. Hinshelwood, *The Kinetics of Chemical Change*, Oxford: Clarendon Press, 1940.
2. S. Glasstone, *Textbook of Physical Chemistry*, New York: D. Van Nostrand, 1946, Chapter 13.
3. S. W. Benson, *The Foundations of Chemical Kinetics*, New York: McGraw-Hill, 1960.
4. K. J. Laidler, *Chemical Kinetics*, New York: McGraw-Hill, 1965.
5. S. S. Penner, *Chemistry Problems in Jet Propulsion*, New York: Pergamon Press, 1958, Chapter 17.
6. N. N. Semenov, *Chemical Kinetics and Chain Reactions*, London: Oxford University Press, 1935.
7. N. N. Semenov, *Some Problems in Chemical Kinetics and Reactivity*, vols. I and II, Princeton: Princeton University Press, 1958, 1959.
8. G. J. Minkoff and C. F. H. Tipper, *Chemistry of Combustion Reactions*, London: Butterworths Scientific Publications, 1962.
9. V. N. Kondratiev, *Chemical Kinetics of Gas Reactions*, New York: Pergamon Press, 1964.
10. A. van Tiggelen et al., *Oxidations et Combustions*, vols. I and II Paris: l'institut Français du pétrole, 1968.
11. W. C. Gardiner, Jr., *Rates and Mechanisms of Chemical Reactions*, New York: W. A. Benjamin, 1969.
12. I. Glassman, *Combustion*, New York: Academic Press, 1977.
13. F. S. Dainton, *Chain Reactions*, New York: Wiley, 1956.
14. K. J. Laidler, *The Chemical Kinetics of Excited States*, Oxford: Clarendon Press, 1955.
15. B. P. Mullins and S. S. Penner, *Explosions, Detonations, Flammability and Ignition*, New York: Pergamon Press, 1959, 17–40.
16. B. Lewis and G. von Elbe, *Combustion, Flames and Explosion of Gases*, 1st ed., New York: Academic Press, 1951.
17. C. F. Curtiss and J. O. Hirschfelder, *Proc. Nat. Acad. Sci.* **38**, 235 (1952).
18. C. W. Gear, *Numerical Initial Value Problems in Ordinary Differential Equations*, Engelwood Cliffs, N.J.: Prentice-Hall, 1971.
19. R. J. Gelinas, *J. Comp. Phys.* **9**, 222 (1972).
20. L. F. Shampine and C. W. Gear, *SIAM Review* **21**, 1 (1979).
21. R. P. Dickinson and R. J. Gelinas, *J. Comp. Phys.* **21**, 123 (1976).
22. R. I. Cukier, H. B. Levine, and K. E. Shuler, *J. Comp. Phys.* **26**, 1 (1978).
23. E. P. Dougherty, J. T. Huang, and H. Rabitz, *J. Chem. Phys.* **71**, 1794 (1979).
24. S. H. Lam, "Singular Perturbations for Stiff Equations using Numerical Methods," in *Recent Advances in Aerospace Sciences*, C. Casci, ed., New York: Plenum Publishing Corp., 1985, 3–19.

25. J. D. Ramshaw, *Phys. Fluids* **23**, 675 (1980).
26. F. W. Williams and R. S. Sheinson, *CST* **7**, 85 (1973).
27. S. P. Hastings and J. D. Murray, *SIAM J. Appl. Math.* **28**, 678 (1975).
28. D. A. Frank-Kamenetskii, *Diffusion and Heat Transfer in Chemical Kinetics*, New York: Plenum Press, 1969.
29. P. H. Thomas, *C & F* **9**, 369 (1965).
30. P. Gray and P. R. Lee, *Oxidation and Combustion Reviews* **2**, 1 (1967).
31. B. F. Gray and C. H. Yang, *11th Symp.* (1967), 1057–1061.
32. P. C. Bowes, *C & F* **13**, 521 (1969).
33. B. F. Gray, *C & F* **21**, 313 (1973).
34. T. Boddington, P. Gray, and G. Wake, *Proc. Roy. Soc. London* **357A**, 403 (1977).
35. D. R. Kassoy, *CST* **10**, 27 (1975).
36. C. E. Hermance, *CST* **10**, 261 (1975).
37. D. R. Kassoy, *Quart. J. Mech. Appl. Math.* **28**, 63 (1975).
38. D. R. Kassoy and A. Liñán, *Quart. J. Mech. Appl. Math.* **31**, 99 (1978).
39. D. Seery and C. T. Bowman, *C & F* **14**, 37 (1970).
40. D. L. Baulch, D. D. Drysdale, D. G. Horne, and A. C. Lloyd, *Evaluated Kinetic Data for High Temperature Reactions*, London: Butterworths Scientific Publications, 1973.
41. D. E. Jensen and G. A. Jones, *C & F* **32**, 1 (1978).
42. C. K. Westbrook, *CST* **20**, 5 (1979).
43. L. A. Lovachev and L. N. Lovachev, *CST* **19**, 195 (1979).
44. G. Dixon-Lewis, *C & F* **36**, 1 (1979).
45. C. K. Westbrook and F. L. Dryer, *CST* **20**, 125 (1979).
46. C. K. Westbrook, *CST* **23**, 191 (1980).
47. F. L. Dryer and C. K. Westbrook, *18th Symp.* (1981), 749–767.
48. R. B. Edelman and P. T. Harsha, *Prog. Energy Combust. Sci.* **4**, 1 (1978).
49. I. Glassman, "Phenomenological Models of Soot Processes in Combustion Systems," Report no. 1450, Department of Mechanical and Aerospace Engineering, Princeton University, Princeton (1979).
50. K. G. P. Sulzmann, B. F. Myers, and E. R. Bartle, *J. Chem. Phys.* **42**, 3969 (1965).
51. S. Glasstone, K. J. Laidler, and H. Eyring, *The Theory of Rate Processes*, New York: McGraw-Hill, 1941.
52. H. S. Johnston, *Gas Phase Reaction Rate Theory*, New York: Ronald Press, 1966.
53. F. L. Dryer, D. W. Naegeli, and I. Glassman, *C & F* **17**, 270 (1971).
54. S. S. Penner, *J. Phys. Chem.* **56**, 475 (1952).
55. K. E. Shuler, *J. Chem. Phys.* **21**, 624 (1953).
56. K. E. Shuler, *J. Chem. Phys.* **18**, 1221 (1950); **19**, 888 (1951).
57. R. D. Levine and R. B. Bernstein, *Molecular Reaction Dynamics*, London: Oxford University Press, 1974.
58. M. S. Dzhidzhoev, V. T. Platonenko, and R. V. Khokhlov, *Uspekhi Fizika Nauk* **100**, 641 (1970).
59. M. L. Zwillenberg, D. W. Naegeli, and I. Glassman, *CST* **8**, 237 (1974).
60. W. R. Warren, Jr., *Astronautics and Aeronautics* **36** (April, 1975).
61. W. G. Vincenti and C. H. Kruger, Jr., *Introduction to Physical Gas Dynamics*, New York: Wiley, 1965.
62. D. E. Rosner, *A.I.Ch.E. Journal*, **9**, 321 (1963); *AIAA Journal* **2**, 593 (1964); *Chem. Eng. Sci.* **19**, 1 (1964).

63. P. A. Libby and F. A. Williams, *AIAA Journal* **3**, 1152 (1965).
64. D. O. Hayward and B. M. W. Trapnell, *Chemisorption*, London: Butterworths Scientific Publications, 1964.
65. G. R. Belton and W. L. Worrell, eds., *Heterogeneous Kinetics at Elevated Temperatures*, New York: Plenum Press, 1970.
66. D. E. Rosner, *Annual Review of Materials Science* **2**, 573 (1972).

APPENDIX C

Continuum Derivation of the Conservation Equations

The equations for conservation of mass, momentum, and energy for a one-component continuum are well known and are derived in standard treatises on fluid mechanics [1]–[3]. On the other hand, the conservation equations for reacting, multicomponent gas mixtures are generally obtained as the equations of change for the summational invariants arising in the solution of the Boltzmann equation (see Appendix D and [4] and [5]). One of several exceptions to the last statement is the analysis of von Kármán [6], whose results are quoted in [7] and are extended in a more recent publication [8] to a point where the equivalence of the continuum-theory and kinetic-theory results becomes apparent [9]. This appendix is based on material in [8].

The objective is to derive the conservation laws for multicomponent, reacting gas mixtures. To this end we invent a physical model consistent with continuum theory. The model involves the idea of a multicomponent continuum composed of coexistent continua, each obeying the laws of dynamics and thermodynamics, a notion which was first introduced by Stefan in 1871.* For an N-component gas mixture we presume the existence of N distinct continua within any arbitrary volume, continuum K corresponding to the chemical species K. The terms continuum K, species K and component

---

* We arrive at the model of simultaneous coexistent continua as the logical transcription to continuum theory of the fact that the entire volume is accessible to all of the different molecules in a gas mixture.

K will be used interchangeably, it being understood that each of these phrases refers to continuum K of the coexistent continua so long as we are following the derivation of conservation laws from continuum theory. It is apparent that each space point in the multicomponent continuum has N velocities v_i^K $(K = 1, 2, \ldots, N)$, one velocity for each of the coexistent continua.

Section C.1 contains relevant definitions and basic mathematical relations, which will be used in subsequent sections. In Sections C.2, C.3, and C.4 we treat, respectively, the equations for conservation of mass, momentum and energy. The results are shown to be equivalent to the relations obtained from the kinetic theory of nonuniform gas mixtures in Section C.5.

C.1. DEFINITIONS AND BASIC MATHEMATICAL RELATIONS

The multicomponent continuum is considered to be defined in regions of space, every point in a region being an interior point of the region. All properties of the N continua, including the velocities v_i^K $(K = 1, 2, \ldots, N)$, are assumed to be described by functions continuously differentiably in all variables within the region. This statement will be said to define **continuous flow** for the multicomponent continuum.

The conservation equations for continuous flow of species K will be derived by using the idea of a control volume $\tau^K(t)$ enclosed by its control surface $\sigma^K(t)$ and lying wholly within a region occupied by the continuum; here t denotes the time. In this appendix only, the notation of Cartesian tensors will be used.* Let x_i $(i = 1, 2, 3)$ denote the Cartesian coordinates of a point in space. In Cartesian tensor notation, the divergence theorem for any scalar function belonging to the Kth continuum $\alpha^K(x_i, t)$, becomes

$$\int_{\sigma^K} \alpha^K n_i^K \, d\sigma = \int_{\tau^K} \alpha_{,i}^K \, d\tau, \tag{1}$$

where n_i^K denotes the outward normal to the surface σ^K and $\alpha_{,i}^K$ represents the gradient of the scalar α^K. For any vector function belonging to the Kth continuum, $u_i^K(x_i, t)$, we have

$$\int_{\sigma^K} u_i^K n_i^K \, d\sigma = \int_{\tau^K} u_{i,i}^K \, d\tau, \tag{2}$$

with $u_{i,i}^K$ denoting the divergence of the vector u_i^K.

---

* Repeated subscript indices imply summation over all allowed values of the indices. The practical necessity of employing cartesian tensor notation here forces us to adopt a notation differing in some other respects from that employed elsewhere in the book (for example, the use of superscript K to identify species).

Consider that some property of the Kth continuum has a density (per unit volume) equal to $\alpha^K(x_i, t)$, and let $A^K(t)$ be the amount of this property contained within the control volume τ^K. Thus

$$A^K(t) = \int_{\tau^K} \alpha^K(x_i, t) \, d\tau. \tag{3}$$

For example, if $\alpha^K = \rho^K = $ the density of mass of species K, then A^K is the total mass of species K contained within τ^K. The property α^K has a density per unit mass of mixture equal to $\beta^K(x_i, t)$, where

$$\alpha^K = \rho\beta^K, \tag{4}$$

with $\rho = \sum_{K=1}^N \rho^K$ representing the density of mass for the fluid mixture.

The derivative dA^K/dt is defined to mean the time rate of change of A^K as the volume τ^K and its surface σ^K move with the flow of species K. Consider that equation (3) holds at a time t_0; at time $t_0 + \Delta t$ the particles in τ^K at x_i will have been displaced to new positions x_i' and will be contained within some new volume $\tau^K(t_0 + \Delta t) = \tau'^K$ enclosed by a surface $\sigma^K(t_0 + \Delta t) = \sigma'^K$; in general τ'^K and σ'^K are different from τ^K and σ^K. Therefore,

$$A^K(t_0 + \Delta t) = \int_{\tau'^K} \alpha^K(x_i', t_0 + \Delta t) \, d\tau$$

and

$$\left(\frac{dA^K}{dt}\right)_{t=t_0} = \lim_{\Delta t \to 0} \left\{ \frac{1}{\Delta t} \left[\int_{\tau'^K} \alpha^K(x_i', t_0 + \Delta t) \, d\tau - \int_{\tau^K} \alpha^K(x_i, t_0) \, d\tau \right] \right\}. \tag{5}$$

It is demonstrated in Section C.6 that equation (5) is equivalent to the relation

$$\frac{dA^K}{dt} = \int_{\tau^K} \left[\frac{\partial\alpha^K}{\partial t} + (\alpha^K v_i^K)_{,i} \right] d\tau. \tag{6}$$

Hence, using the divergence theorem given in equation (2), it is found that

$$\frac{dA^K}{dt} = \int_{\tau^K} \frac{\partial\alpha^K}{\partial t} \, d\tau + \int_{\sigma^K} \alpha^K v_i^K n_i \, d\sigma. \tag{7}$$

Equation (7) expresses the idea that the time rate of change of A^K in a flow, for an arbitrary volume τ^K bounded by a surface σ^K, is equal to the stationary rate of change of A^K in the interior of τ^K plus the rate of change of A^K due to the movement of τ^K and σ^K.

Equation (6) may be rewritten in the equivalent form

$$\frac{d}{dt}\left(\int_{\tau^K} \rho\beta^K \, d\tau \right) = \int_{\tau^K} \left[\frac{\partial(\rho\beta^K)}{\partial t} + (\rho\beta^K v_i^K)_{,i} \right] d\tau. \tag{8}$$

The overall transport equation for the multicomponent continuum is then obtained by summing over components, a procedure which is in accord with the idea of independent coexistent continua. We choose at the arbitrary time t all of the control volumes τ^K to be coexistent, that is, $\tau^K \equiv \tau$ for all K. We henceforth refer to a volume τ thus defined as being *of the multicomponent continuum* at time t. After summation, equation (8) becomes

$$\sum_K \left(\frac{d}{dt} \int_{\tau^K} \rho \beta^K \, d\tau \right)_{\tau^K = \tau} = \int_\tau \left[\frac{\partial}{\partial t} \left(\sum_K \rho \beta^K \right) + \sum_K (\rho \beta^K v_i^K)_{,\,i} \right] d\tau. \qquad (9)$$

C.2. CONTINUITY EQUATIONS

Denote by w^K the net production of mass of species K per unit volume per unit time. Since mass is neither created nor destroyed by chemical reactions but only converted from one species to another, it follows that

$$\sum_K w^K = 0. \qquad (10)$$

The conservation of the mass of species K in an arbitrary volume τ^K is expressed by the equation

$$\frac{d}{dt} \left(\int_{\tau^K} \rho Y^K \, d\tau \right) = \int_{\tau^K} w^K \, d\tau, \qquad (11)$$

where Y^K is the mass or weight fraction of species K (that is, $\rho^K = \rho Y^K$, and Y^K equals the mass of species K in unit mass of mixture). Let $\beta^K = Y^K$ in equation (8); then equation (11) becomes

$$\int_{\tau^K} \left[\frac{\partial (\rho Y^K)}{\partial t} + (\rho Y^K v_i^K)_{,\,i} - w^K \right] d\tau = 0,$$

and, since τ^K is arbitrary,

$$w^K = \frac{\partial (\rho Y^K)}{\partial t} + (\rho Y^K v_i^K)_{,\,i}. \qquad (12)$$

Now let v_i^K, the flow velocity for species K, be represented as

$$v_i^K = v_i' + V_i^K, \qquad (13)$$

where

$$v_i' \equiv \sum_K Y^K v_i^K. \qquad (14)$$

The summation in equation (14) is extended over all N distinct chemical components. Thus v_i' is the mass-weighted average velocity of the fluid mixture, and V_i^K is said to denote the diffusion velocity of species K. Since

$$\sum_k Y^k = 1, \qquad (15)$$

it follows from equations (13) and (14) that

$$\sum_K Y^K V_i^K = 0. \tag{16}$$

Introducing equation (13) into equation (12) leads to the following equation for continuity of species K:

$$w^K = \frac{D}{Dt}(\rho Y^K) + \rho Y^K v'_{i,i} + (\rho Y^K V_i^K)_{,i}, \tag{17}$$

where

$$\frac{D}{Dt}(\ ) \equiv \frac{\partial}{\partial t}(\ ) + v'_i(\ )_{,i} \tag{18}$$

is the Euler total time derivative following the mass-weighted average motion of the multicomponent continuum. Summing equation (17) over all distinct components, in view of equations (10), (15), and (16), leads to the overall continuity equation

$$\frac{D\rho}{Dt} + \rho v'_{i,i} = 0. \tag{19}$$

Equation (19) is evidently also the correct form of the continuity equation for a one-component system.

We may now transform equation (9) by using equations (18) and (19) to obtain a form which is useful for the derivation of the differential equations expressing conservation of momentum and energy:

$$\sum_K \left(\frac{d}{dt} \int_{\tau K} \rho \beta^K \, d\tau \right)_{\tau K = \tau} = \int_\tau \left[\rho \frac{D\beta}{Dt} + \left(\rho \sum_K \beta^K V_i^K \right)_{,i} \right] d\tau, \tag{20}$$

where

$$\beta \equiv \sum_K \beta^K.$$

C.3. MOMENTUM EQUATION

For an arbitrary volume τ of the multicomponent continuum, the total rate of change of linear momentum in the jth coordinate direction must equal the sum of the following: (1) the surface integral of the stress vector $\sum_K \sigma_{ij}^K n_i$, where σ_{ij}^K equals* the component in the direction x_j of the stress vector acting on that face of an elemental parallelepiped of species K which has an

---

* The species vectors $\sigma_{ij}^K n_i$ and ρf_j^K represent the sums of all forces which act upon species K and which move with the velocity of species K in the mixture. These definitions are used in Section C.5 to identify our results with the results obtained from kinetic theory.

outward normal in the direction x_i; (2) the volume integral of the total vector body force $\sum_K \rho^K f_j^K$ acting on unit volume of mixture, where f_j^K is the vector body force per unit mass of species K; and (3) the volume integral of the total rate of generation of momentum in unit volume through production of species. Let the rate of generation of momentum in unit volume for species K be $w^K m_j^K$, where m_j^K is the average momentum of the generated mass of species K per unit mass of species K. We postulate that, overall, linear momentum is neither created nor destroyed by chemical reactions; the consequent conservation principle states that the total rate of generation of linear momentum per unit volume by chemical production of species is zero:

$$\sum_K w^K m_j^K = 0. \tag{21}$$

The total rate of change of linear momentum is then expressed mathematically by

$$\sum_K \left(\frac{d}{dt} \int_{\tau^K} \rho Y^K v_j^K \, d\tau \right)_{\tau^K = \tau} = \int_\sigma \sum_K \sigma_{ij}^K n_i \, d\sigma + \int_\tau \sum_K \rho^K f_j^K \, d\tau. \tag{22}$$

In view of the divergence theorem, equation (1), and the transport relation given in equation (20), with

$$\beta^K = Y^K v_j^K, \qquad \beta = v_j',$$

equation (22) becomes

$$\int_\tau \left[\rho \frac{Dv_j'}{Dt} + \left(\rho \sum_K Y^K V_j^K V_i^K \right)_{,i} \right] d\tau = \int_\tau \sum_K (\sigma_{ij,i}^K + \rho^K f_j^K) \, d\tau. \tag{23}$$

We now define σ_{ij}^D, the diffusion stress tensor, as

$$\sigma_{ij}^D \equiv -\rho \sum_K Y^K V_i^K V_j^K \tag{24}$$

and f_j, the vector body force per unit mass of mixture, as

$$f_j = \sum_K f_j^K Y^K. \tag{25}$$

Since τ is arbitrary, equation (23) then leads to an expression for overall conservation of momentum, namely,

$$\rho \frac{Dv_j'}{Dt} \equiv \rho \frac{\partial v_j'}{\partial t} + \rho v_i' v_{j,i}' = \sum_K \sigma_{ij,i}^K + \sigma_{ij,i}^D + \rho f_j. \tag{26}$$

If we define σ_{ij} as

$$\sigma_{ij} \equiv \sum_K \sigma_{ij}^K + \sigma_{ij}^D, \tag{27}$$

then equation (26) is reduced to the well-known form of the momentum equation for one-component systems. Therefore, σ_{ij} is the stress tensor and

ρf_j is the body force acting on an elemental parallelepiped which is moving with the mass-weighted average velocity v'_j. Furthermore, we can express σ_{ij} as a sum of partial stress tensors $\sigma_{ij}^{*,K}$

$$\sigma_{ij} \equiv \sum_K \sigma_{ij}^{*,K}, \tag{28}$$

where, from equations (24) and (27)

$$\sigma_{ij}^{*,K} \equiv \sigma_{ij}^K - \rho Y^K V_i^K V_j^K.$$

Each stress tensor can then always be expressed as the sum of a mean pressure tensor, a viscous stress tensor, and a viscous diffusion-stress tensor; thus,

$$\sigma_{ij}^{*,K} \equiv -p^K \delta_{ij} + \tau_{ij}^{V,K} + \tau_{ij}^{D,K}, \tag{29}$$

where

$$p^K \equiv -\tfrac{1}{3}\sigma_{ii}^{*,K},$$

and

$$\sigma_{ij} \equiv -p\delta_{ij} + \tau_{ij}^V + \tau_{ij}^D, \tag{30}$$

where

$$p \equiv -\tfrac{1}{3}\sigma_{ii}.$$

The total pressure p is the sum of the partial pressures p^K for the different species, that is,

$$p = \sum_K p^K, \tag{31}$$

and so, in view of equation (28), it follows now that

$$\tau_{ij}^V = \sum_K \tau_{ij}^{V,K}, \tag{32}$$

$$\tau_{ij}^D = \sum_K \tau_{ij}^{D,K}. \tag{33}$$

The equations of von Kármán [6], [7] are obtained by using equation (30) in equation (26), namely,

$$\rho \frac{Dv'_j}{Dt} = -p_{,j} + (\tau_{ij}^V + \tau_{ij}^D)_{,i} + \rho f_j. \tag{34}$$

C.4. ENERGY EQUATION

For an arbitrary volume τ of the multicomponent continuum, the first law of thermodynamics states that

> Rate of increase of (internal plus kinetic) energy = rate at which work is done on τ (by body forces plus surface stresses) + rate of inward transport of heat by radiation, thermal conduction, and other transport process through the surface σ enclosing τ + rate of generation of energy through production of species within τ + rate at which work is done on material produced within τ.

Let u^K denote the absolute internal energy of species K per unit mass of species K and let u denote the absolute internal energy per unit mass of mixture. Then

$$u \equiv \sum_K Y^K u^K. \tag{35}$$

The kinetic energy of species K per unit mass of species K is $\frac{1}{2}v_j^K v_j^K$. The total rate at which work is done on τ by surface stresses and body forces is represented as the superposition of the rate of work done on the individual continua by their own surface stresses and body forces.

For the mass w^K of species K, which is generated by chemical reaction in unit volume per unit time, the sum of (1) the internal and kinetic energy carried by this mass, and (2) the work done on this mass in unit time, is $w^K(\eta^K + \frac{1}{2}m_j^K m_j^K)$, where η^K is the average specific enthalpy of generated mass of species K.* We postulate that, overall, energy is redistributed among various states but is neither created nor destroyed by chemical reaction; the consequent conservation principle states that the total rate of generation of (absolute enthalpy plus kinetic energy) per unit volume by chemical production of species is zero:

$$\sum_K w^K(\eta^K + \frac{1}{2}m_j^K m_j^K) = 0. \tag{36}$$

The analytical expression of the first law of thermodynamics, subject to the fundamental postulates of independence and conservation, is therefore

$$\sum_K \left[\frac{d}{dt} \int_{\tau K} \rho Y^K(u^K + \frac{1}{2}v_j^K v_j^K)\, d\tau\right]_{\tau K = \tau}$$
$$= \sum_K \left[\int_\sigma \sigma_{ij}^K n_i v_j^K\, d\sigma + \int_\tau \rho^K f_j^K v_j^K\, d\tau\right] - \int_\sigma \sum_K q_j^K n_j\, d\sigma. \tag{37}$$

Here q_j^K is the heat flux vector for species K, taken as positive for outward heat transport. Equation (37) can be transformed by the use of equation (20) with

$$\beta^K = Y^K u^K + \frac{1}{2}Y^K v_j^K v_j^K$$

and

$$\beta = u + \frac{1}{2}v_j' v_j' + \frac{1}{2}\sum_K Y^K V_j^K V_j^K. \tag{38}$$

Since τ is arbitrary, the following differential equation for overall conservation of energy then results:

$$\rho\frac{D}{Dt}(u + \frac{1}{2}v_j' v_j') + \frac{1}{2}\rho\frac{D}{Dt}\left(\sum_K Y^K V_j^K V_j^K\right) + \left[\rho\sum_K (Y^K u^K V_i^K + \frac{1}{2}Y^K v_j^K v_j^K V_i^K)\right]_{,i}$$
$$= \rho\sum_K Y^K f_j^K v_j^K + \sum_K (\sigma_{ij}^K v_j^K)_{,i} - \sum_K q_{j,j}^K. \tag{39}$$

---

* The quantity η^K should not be confused with the total (average) specific enthalpy of species K, which is denoted by h^K, as in equation (40).

Let h^K denote the absolute specific enthalpy of species K, which is defined as

$$Y^K h^K = Y^K u^K + \frac{p^K}{\rho}. \tag{40}$$

The absolute specific enthalpy of the mixture is then $h = \sum_K Y^K h^K = u + p/\rho$. The total heat-flux vector is defined as

$$q_j = \sum_K q_j^K. \tag{41}$$

In view of the definitions given by equations (29), (30), (40), and (41), equation (39) can be written in a desired form:

$$\begin{aligned}
\rho \frac{D}{Dt} &(u + \tfrac{1}{2} v_j' v_j') + \left(\rho \sum_K Y^K h^K V_j^K \right)_{,j} \\
&= \rho f_j v_j' - (p v_j')_{,j} \\
&\quad + [(\tau_{ij}^V + \tau_{ij}^D) v_j']_{,i} - q_{j,j} \\
&\quad + \rho \sum_K Y^K f_j^K V_j^K + \sum_K [(\tau_{ij}^{V,K} + \tau_{ij}^{D,K}) V_j^K]_{,i} \\
&\quad + \tfrac{1}{2} \left(\rho \sum_K Y^K V_i^K V_j^K V_j^K \right)_{,i} - \rho \frac{D}{Dt} \left(\tfrac{1}{2} \sum_K Y^K V_j^K V_j^K \right). \tag{42}
\end{aligned}$$

C.5. COMPARISON BETWEEN THE CONSERVATION LAWS DERIVED FOR INDEPENDENT COEXISTENT CONTINUA AND THE KINETIC-THEORY RESULTS FOR MULTICOMPONENT GAS MIXTURES

In order to show that the model of independent, coexistent continua represents correctly a real mixture of gases composed of different chemical species, we must compare the results obtained from this model with those of the kinetic theory of nonuniform gas mixtures (see Appendix D). Quantities such as the density ρ, the mass-weighted average velocity v_j', and the body force f_j have obviously analogous meanings in both the kinetic theory and the coexistent-continua model. On the other hand, the precise kinetic-theory meaning of terms such as the stress tensor σ_{ij}^K, the absolute internal energy per unit mass u^K, and the heat-flux vector q_j^K is not immediately apparent. In view of the known success of continuum theory for one-component systems, we shall identify the continuum-theory properties σ_{ij}^K, u^K and q_j^K for species K with their kinetic-theory counterparts. The proof then involves a comparison of the conservation equations obtained from multicomponent-continuum theory (replacing continuum properties for each species by their kinetic-theory definitions) with the conservation equations obtained from the kinetic theory of nonuniform gas mixtures. For readers not familiar with the kinetic-theory approach, to obtain a more complete understanding of this section it is advisable to read Appendix D first.

C.5.1. Definitions of kinetic theory

Let $c_j^{K,m}$ be the velocity of a particular molecule m of species K, and let $V_j^{\prime K,m}$ be the velocity of this molecule in excess of the velocity v_j^K which is identified in kinetic theory as the mean of velocity of all molecules of species K. Then

$$c_j^{K,m} = v_j^K + V_j^{\prime K,m} \equiv v_j^\prime + V_j^K + V_j^{\prime K,m},$$

$$\langle c_j^{K,m} \rangle = v_j^K \quad \text{and} \quad \langle V_j^{\prime K,m} \rangle = 0,$$

where the angle brackets indicate an average over all molecules of species K taken with respect to a distribution function appropriate for the mixture.

From kinetic theory, the following definitions* for the properties of species K in the mixture are obtained:

$$\sigma_{ij}^K \equiv -\rho^K \langle V_i^{\prime K,m} V_j^{\prime K,m} \rangle, \tag{43}$$

$$u^K \equiv \langle \tfrac{1}{2} V_j^{\prime K,m} V_j^{\prime K,m} + i^{K,m} \rangle \equiv \tfrac{1}{2} \langle V_j^{\prime K,m} V_j^{\prime K,m} \rangle + i^K, \tag{44}$$

$$q_j^K \equiv \rho^K \langle (\tfrac{1}{2} V_i^{\prime K,m} V_i^{\prime K,m} + i^{K,m}) V_j^{\prime K,m} \rangle$$
$$\equiv \rho^K (\langle \tfrac{1}{2} V_j^{\prime K,m} V_i^{\prime K,m} V_i^{\prime K,m} \rangle + \langle i^{K,m} V_j^{\prime K,m} \rangle). \tag{45}$$

In equations (44) and (45), $u^{K,m}$, the total internal energy per unit mass of a molecule of species K, is expressed as the sum of $\tfrac{1}{2} V_j^{\prime K,m} V_j^{\prime K,m}$, the molecular translatory kinetic energy per unit mass, and $i^{K,m}$, the contribution of additional internal energy terms (rotational, vibrational, etc.) per unit mass. We have then defined $u^K \equiv \langle u^{K,m} \rangle$ and $i^K \equiv \langle i^{K,m} \rangle$.

The corresponding definitions in the kinetic theory for the properties of the gas mixture will be denoted here by the superscript T; these are[†] the mixture stress tensor σ_{ij}^T, where

$$\sigma_{ij}^T \equiv -\sum_K \rho^K \langle (V_i^{\prime K,m} + V_i^K)(V_j^{\prime K,m} + V_j^K) \rangle$$

$$\equiv -\sum_K \rho^K \langle V_i^{\prime K,m} V_j^{\prime K,m} \rangle - \sum_K \rho^K V_i^K V_j^K, \tag{46}$$

the internal energy per unit mass of mixture u^T, where

$$\rho u^T \equiv \sum_K [\tfrac{1}{2} \rho^K \langle (V_j^{\prime K,m} + V_j^K)(V_j^{\prime K,m} + V_j^K) \rangle + \rho^K i^K]$$

$$\equiv \tfrac{1}{2} \sum_K \rho^K \langle V_j^{\prime K,m} V_j^{\prime K,m} \rangle + \tfrac{1}{2} \sum_K \rho^K V_j^K V_j^K + \sum_K \rho^K i^K, \tag{47}$$

---

* See, for example, equations (18), (23), (24), (25), and (28) in Section D.2.
† See, for example, equations (21), (27), and (29) in Section D.2.

and the heat-flux vector for the mixture q_j^T, where

$$q_j^T \equiv \sum_K \rho^K \langle [\tfrac{1}{2}(V_i'^{K,m} + V_i^K)(V_i'^{K,m} + V_i^K) + i^{K,m}](V_j'^{K,m} + V_j^K) \rangle$$

$$\equiv \sum_K \rho^K \{ \langle \tfrac{1}{2} V_i'^{K,m} V_i'^{K,m} V_j'^{K,m} \rangle + \tfrac{1}{2} V_j^K \langle V_i'^{K,m} V_i'^{K,m} \rangle$$

$$+ V_i^K \langle V_i'^{K,m} V_j'^{K,m} \rangle + \tfrac{1}{2} V_i^K V_i^K V_j^K$$

$$+ \langle i^{K,m} V_j'^{K,m} \rangle + i^K V_j^K \}. \tag{48}$$

When use is made of equations (43), (44), and (45) in equations (46), (47), and (48), the following identities are obtained between the properties σ_{ij}^T, ρu^T and q_j^T of the gas mixture and the properties σ_{ij}^K, $\rho^K u^K$ and q_j^K of the individual species:

$$\sigma_{ij}^T \equiv \sum_K \sigma_{ij}^K - \sum_K \rho^K V_i^K V_j^K, \tag{49}$$

$$\rho u^T \equiv \sum_K \rho^K u^K + \tfrac{1}{2} \sum_K \rho^K V_j^K V_j^K, \tag{50}$$

$$q_j^T \equiv \sum_K q_j^K - \sum_K \sigma_{ij}^K V_i^K + \sum_K \rho^K u^K V_j^K + \tfrac{1}{2} \sum_K \rho^K V_i^K V_i^K V_j^K. \tag{51}$$

In each of the above relations, the property for the mixture is equal to the sum, over all species, of the corresponding property for the components plus various diffusion terms. The diffusion terms arise because the reference coordinate system for species K is taken to move with velocity $v_j' + V_j^K$, which is the mass-weighted average velocity for molecules in species K alone; the reference coordinate system for the mixture on the other hand is taken to move with velocity v_j', which is the mass-weighted average velocity for all molecules in the mixture. The latter is a natural coordinate system in that it is consistent with equations (19) and (34).

Derivations of conservation equations from the viewpoint of kinetic theory usually do not exhibit explicitly the diffusion terms, such as diffusion stresses, that appear on the right-hand sides of equations (49), (50), and (51), since it is unnecessary to introduce quantities such as σ_{ij}^K specifically in these derivations. Kinetic-theory developments work directly with the left-hand sides of equations (49), (50), and (51). Transport coefficients (Appendix E) are defined only in terms of these kinetic-theory quantities because prescriptions for calculating the individual continua transports, σ_{ij}^K and q_j^K, are unduly complex. Moreover, measurement of diffusion stresses is feasible only by direct measurement of diffusion velocities, followed by use of equation (24). Therefore, it has not been fruitful to study the diffusion terms which, in a sense, may be viewed as artifacts of the continuum approach.

C.5.2. Comparison of conservation equations

The conservation equation for species K, as given by equation (17), is readily seen to be identical with the corresponding relation in the kinetic theory for multicomponent gas mixtures if w^K is the net mass rate of production of species K per unit volume by chemical reaction.[*] Explicit evaluation of w^K requires the introduction of the laws of chemical kinetics.[†]

The expression for overall conservation of momentum, equation (26), is also identical with the corresponding relation in the kinetic theory,[‡] since a comparison of equations (27) and (49) shows that $\sigma_{ij} \equiv \sigma_{ij}^T$.

To demonstrate the equivalence of the energy conservation equations, rewrite equation (39) in the form

$$\rho \frac{D}{Dt}\left[\sum_K (Y^K u^K + \tfrac{1}{2}Y^K V_j^K V_j^K) + \tfrac{1}{2}v_j' v_j'\right] + \left(\rho \sum_K Y^K u^K V_j^K\right)_{,j}$$

$$= \rho f_j v_j' + \rho \sum_K Y^K f_j^K V_j^K + \left(v_j' \sum_K \sigma_{ij}^K\right)_{,i} + \sum_K (\sigma_{ij}^K V_j^K)_{,i}$$

$$- \sum_K q_{j,j}^K - \left(\rho v_j' \sum_K Y^K V_i^K V_j^K\right)_{,i} - \tfrac{1}{2}\left(\rho \sum_K Y^K V_j^K V_j^K V_i^K\right)_{,i}. \quad (52)$$

Replacing $\sum_K \sigma_{ij}^K$, $\sum_K Y^K u^K$ and $\sum_K q_j^K$ by their kinetic-theory equivalents, as given by equations (49), (50), and (51), reduces equation (52) to

$$\rho \frac{D}{Dt}[u^T + \tfrac{1}{2}v_j' v_j'] = \rho f_j v_j' + \rho \sum_K Y^K f_j^K V_j^K - q_{j,j}^T + (\sigma_{ij}^T v_j')_{,i}. \quad (53)$$

Multiplying the momentum conservation equations by v_j' and contracting produces the scalar equation

$$\tfrac{1}{2}\rho \frac{D}{Dt}(v_j' v_j') = \sigma_{ij,i}^T v_j' + \rho f_j v_j'. \quad (54)$$

With this relation, equation (53) can be placed in a form which is identical with the usual form of the result obtained from kinetic theory,[§] namely,

$$\rho \frac{D}{Dt}(u^T) = \rho \sum_K Y^K f_j^K V_j^K - q_{j,j}^T + \sigma_{ij}^T v_{j,i}'. \quad (55)$$

C.6. PROOF OF EQUATION (6)

In order to prove equation (6), it is more convenient to work from equation (3) than from the limiting relation given in equation (5) and also to introduce the Lagrangian representation [10]. For any continuum K, let the three parameters a_i^K identify the individual point particles of continuum K; for

---

[*] See, for example, equation (D-39).
[†] See Appendix B or Chapter 1.
[‡] See, for example, equation (D-35).
[§] See, for example, equation (D-37).

definiteness, suppose that a_i^K are the spatial coordinates of the particles of continuum K at some fixed time t_0. The spatial coordinates x_i for any particle a_j^K at any time $t, t \geq t_0$, are then assumed to be given by the functions $x_i^{*K}(a_j^K, t)^*$ which are taken to be single-valued and at least twice continuously differentiable with respect to each of their variables:

$$x_i = x_i^{*K}(a_j^K, t), \quad t \geq t_0, \tag{56}$$

and $a_i^K \equiv x_i^{*K}(a_j^K, t_0)$. The transformations are assumed to be one-to-one, so that the inverse transformations $a_i^K(x_j, t)$ also exist and are twice continuously differentiable. The flow velocities, or "particle velocities," for continuum K, $v_i^{*K}(a_j^K, t)$, are then defined as

$$v_i^{*K}(a_j^K, t) \equiv \frac{\partial x_i^{*K}}{\partial t} = v_i^K(x_j, t),$$

where the v_i^K are defined by the inverse transformation. Similarly, the Jacobian of equation (56) is given as Δ^{*K} or Δ^K:

$$\Delta^{*K}(a_j^K, t) \equiv \det\left\{\frac{\partial x_i^{*K}}{\partial a_j^K}\right\} = \Delta^K(x_j, t). \tag{57}$$

If the integral of equation (3) is changed with the use of equation (56) to an integration at time $t = t_0$ over the volume τ_0^K, then

$$A^K(t) = \int_{\tau_0^K} \alpha^{*K}(a_j^K, t) \Delta^{*K} \, d\tau_0^K. \tag{58}$$

The definition of the time derivative given in equation (5) is therefore equivalent to

$$\frac{dA^K}{dt} = \int_{\tau_0^K} \left(\frac{\partial \alpha^{*K}}{\partial t} \Delta^{*K} + \alpha^{*K} \frac{\partial \Delta^{*K}}{\partial t}\right) d\tau_0^K. \tag{59}$$

But it is readily shown that[†]

$$\frac{\partial \Delta^{*K}}{\partial t} = \frac{\partial v_i^K}{\partial x_i} \Delta^{*K} \equiv v_{i,i}^K \Delta^{*K}$$

and, therefore, equation (59) may be written as

$$\frac{dA^K}{dt} = \int_{\tau_0^K} \left(\frac{\partial \alpha^{*K}}{\partial t} + v_{i,i}^K \alpha^{*K}\right) \Delta^{*K} \, d\tau_0^K. \tag{60}$$

Transformation of equation (60) to spatial coordinates leads to equation (6).

---

    * In this section only, an asterisk on any function indicates that its variables are a_i, t; functions without asterisks have the independent variables x_i, t.
    † See [10], equation (7.07).

REFERENCES

1. L. Prandtl, *Essentials of Fluid Dynamics*, Glasgow: Blackie and Sons, Ltd., 1952, Chapter II.
2. H. W. Liepmann, and A. Roshko, *Elements of Gasdynamics*, New York: Wiley, 1957, Chapter 7.
3. G. K. Batchelor, *An Introduction to Fluid Mechanics*, Cambridge: Cambridge University Press, 1967, Chapter 3.
4. S. Chapman and T. G. Cowling, *The Mathematical Theory of Non-Uniform Gases*, Cambridge: Cambridge University Press, 1953, Chapter 3.
5. J. O. Hirschfelder, C. F. Curtiss, and R. B. Bird, *Molecular Theory of Gases and Liquids*, New York: Wiley, 1954, Chapter 7.
6. Th. von Kármán, Sorbonne Lectures, Paris (1950–51).
7. S. S. Penner, *Introduction to the Study of Chemical Reactions in Flow Systems*, London: Butterworths Scientific Publications, 1955, Chapter 2.
8. W. Nachbar, F. Williams, and S. S. Penner, *Quart. Appl. Math.*, **17**, 43–54 (1959).
9. For related treatments, see the following: J. Stefan, *Sitzungsber. Akad. Wiss. Wien*, **63**^{2}, 63 (1871); C. Eckart, *Phys. Rev.*, **58**, 267 (1940); C. Truesdell, *J. Rat. Mech. Anal.*, **1**, 125 (1952).
10. R. Courant and K. O. Friedrichs, *Supersonic Flow and Shock Waves*, New York: Interscience, 1948, Section 7.

APPENDIX D

Molecular Derivation of the Conservation Equations

The rigorous approach to a kinetic-theory derivation of the fluid-dynamical conservation equations,* which begins with the Liouville equation and involves a number of subtle assumptions, will be omitted here because of its complexity. The same result will be obtained in a simpler manner from a physical derivation of the Boltzmann equation,[†] followed by the identification of the hydrodynamic variables and the development of the equations of change. For additional details the reader may consult [1] and [2].

D.1. THE VELOCITY DISTRIBUTION FUNCTION AND THE BOLTZMANN EQUATION

Let us focus our attention on a region of space containing molecules of N different chemical species. The viewpoint is versatile, in that molecules of the same kind in different energy levels may be considered to be different species. In defining the velocity distribution function for molecules of species i,

$$f_i(\mathbf{x}, \mathbf{v}, t)\, d\mathbf{x}\, d\mathbf{v}, \quad i = 1, \dots, N,$$

---

* See, for example, 449–452 of [1].
[†] See, for example, 444–449 of [1]

will denote the probable number of molecules of type i in the position range $dx \equiv dx\,dy\,dz$ about the spatial position $\mathbf{x}$ and with velocities in the range $d\mathbf{v} \equiv dv_x\,dv_y\,dv_z$ about the velocity $\mathbf{v}$ at time t; the six-dimensional space consisting of the coordinates $(\mathbf{x}, \mathbf{v})$ may be called **phase space**.

Consider a molecule of kind i at position $\mathbf{x}$ with velocity $\mathbf{v}$ at time t. If there were no intermolecular collisions, unimolecular reactions, radiative transitions, and so on, then this molecule would move in such a way that at a short time later $t + dt$ its position would be $\mathbf{x} + \mathbf{v}\,dt$ and its velocity would be $\mathbf{v} + \mathbf{f}_i\,dt$, where $\mathbf{f}_i(\mathbf{x}, \mathbf{v}, t)$ is the external force (for example, gravitational or electromagnetic) on molecules of kind i per unit mass of molecules of kind i (that is, $\mathbf{f}_i = \mathbf{F}_i/m_i =$ acceleration, where $\mathbf{F}_i$ is the external force and m_i is the molecular mass). Therefore, the only i molecules arriving at the phase-space position $(\mathbf{x} + \mathbf{v}\,dt, \mathbf{v} + \mathbf{f}_i\,dt)$ at time $t + dt$ would be those at $(\mathbf{x}, \mathbf{v})$ at time t, and hence, counting all molecules of kind i in the phase-space element $(d\mathbf{x}, d\mathbf{v})$,*

$$f_i(\mathbf{x} + \mathbf{v}\,dt, \mathbf{v} + \mathbf{f}_i\,dt, t + dt)\,d\mathbf{x}\,d\mathbf{v} = f_i(\mathbf{x}, \mathbf{v}, t)\,d\mathbf{x}\,d\mathbf{v}.$$

Since collisions and other similar events cause some molecules of type i in the range $(d\mathbf{x}, d\mathbf{v})$ about $(\mathbf{x}, \mathbf{v})$ at time t to arrive at a phase space position outside of the range $(d\mathbf{x}, d\mathbf{v})$ about $(\mathbf{x} + \mathbf{v}\,dt, \mathbf{v} + \mathbf{f}_i\,dt)$ at time $t + dt$ [let $(\delta f_i/\delta t)^{(-)}\,d\mathbf{x}\,d\mathbf{v}\,dt$ denote the number of these molecules], and these events also cause some i molecules not in $(d\mathbf{x}, d\mathbf{v})$ about $(\mathbf{x}, \mathbf{v})$ at t to arrive in $(d\mathbf{x}, d\mathbf{v})$ about $(\mathbf{x} + \mathbf{v}\,dt, \mathbf{v} + \mathbf{f}_i\,dt)$ at $t + dt$ [let $(\delta f_i/\delta t)^{(+)}\,d\mathbf{x}\,d\mathbf{v}\,dt$ denote the number of these molecules], the preceding equation is not valid and must be replaced by

$$f_i(\mathbf{x} + \mathbf{v}\,dt, \mathbf{v} + \mathbf{f}_i\,dt, t + dt)\,d\mathbf{x}\,d\mathbf{v}$$
$$= f_i(\mathbf{x}, \mathbf{v}, t)\,d\mathbf{x}\,d\mathbf{v} + (\delta f_i/\delta t)\,d\mathbf{x}\,d\mathbf{v}\,dt, \quad i = 1, \ldots, N, \qquad (1)$$

where

$$\delta f_i/\delta t \equiv (\delta f_i/\delta t)^{(+)} - (\delta f_i/\delta t)^{(-)}, \quad i = 1, \ldots, N,$$

is the net time rate of change of f_i caused by collisions and other molecular processes. Passing to the limit $dt \to 0$ in equation (1) shows that

$$\partial f_i/\partial t + \mathbf{v} \cdot \nabla_x f_i + \mathbf{f}_i \cdot \nabla_v f_i \equiv \delta f_i/\delta t, \quad i = 1, \ldots, N, \qquad (2)$$

in which the subscripts x and v on the gradient operator distinguish derivatives with respect to spatial and velocity coordinates. Equation (2) is a generalization of the **Boltzmann equation**.[†]

---

* Except for special cases such as that of electromagnetic forces, it must actually be assumed that $\mathbf{f}_i$ is independent of $\mathbf{v}$ for the volume elements on both sides of this equation to be equal (see [2], 322–324).

† Boltzmann's name is best reserved for the form that this equation takes when only binary collisions contribute to $\delta f_i/\delta t$, an explicit, quadratically nonlinear, integrodifferential equation.

D.2. DEFINITIONS OF FLUID-DYNAMICAL VARIABLES

The conventional variables of fluid dynamics may be defined as suitable integrals over velocity space, weighted by the distribution function.* For example, the total number of molecules of kind i per unit spatial volume at $(\mathbf{x}, t)$ is clearly

$$n_i(\mathbf{x}, t) = \int_{-\infty}^{\infty} \int_{-\infty}^{\infty} \int_{-\infty}^{\infty} f_i(\mathbf{x}, \mathbf{v}, t) \, dv_x \, dv_y \, dv_z \equiv \int f_i \, d\mathbf{v}, \quad i = 1, \ldots, N, \quad (3)$$

and therefore the molar concentration of species i is

$$c_i = n_i/\mathscr{A} = \int f_i \, d\mathbf{v}/\mathscr{A}, \quad i = 1, \ldots, N, \tag{4}$$

where $\mathscr{A}$ is Avogadro's number. Other related fluid variables are the density of species i (mass per unit volume),

$$\rho_i = m_i n_i = W_i c_i, \quad i = 1, \ldots, N, \tag{5}$$

where $W_i = \mathscr{A} m_i$ is the molecular weight of species i, the total (local) density,

$$\rho = \sum_{i=1}^{N} \rho_i, \tag{6}$$

the mass fraction of species i,

$$Y_i = \rho_i/\rho, \quad i = 1, \ldots, N, \tag{7}$$

the total number of moles per unit volume,

$$c = \sum_{i=1}^{N} c_i = n/\mathscr{A}, \tag{8}$$

where $n = \sum_{i=1}^{N} n_i$ is the total number of molecules per unit volume, and the mole fraction of species i,

$$X_i = c_i/c, \quad i = 1, \ldots, N. \tag{9}$$

In terms of n_i, the local average value of any property G of species i can be defined as

$$\bar{G}_i(\mathbf{x}, t) = \int G_i(\mathbf{x}, \mathbf{v}, t) f_i(\mathbf{x}, \mathbf{v}, t) \, d\mathbf{v}/n_i, \quad i = 1, \ldots, N. \tag{10}$$

Thus the average velocity of molecules of type i is

$$\bar{\mathbf{v}}_i(\mathbf{x}, t) = \int \mathbf{v} f_i \, d\mathbf{v}/n_i, \quad i = 1, \ldots, N. \tag{11}$$

---

* Some of these definitions are physically meaningful only when the time between collisions and other microscopic molecular processes is greatly in excess of the duration time of these processes.

The mass-weighted average velocity of the mixture is

$$\mathbf{v}_0 = \sum_{i=1}^{N} Y_i \bar{\mathbf{v}}_i, \tag{12}$$

which is the ordinary flow velocity of fluid dynamics. In special problems, mixture velocities other than $\mathbf{v}_0$ are useful (for example, the number-weighted average velocity $\sum_{i=1}^{N} c_i \bar{\mathbf{v}}_i / c$), but $\bar{\mathbf{v}}_0$ is usually of such predominant importance that in the applications (namely, after the molecular velocity $\mathbf{v}$ has disappeared from the development) the subscript 0 on $\mathbf{v}_0$ has been omitted. The difference between the molecular velocity and $\mathbf{v}_0$ is

$$\mathbf{V}(\mathbf{x}, \mathbf{v}, t) = \mathbf{v} - \mathbf{v}_0, \tag{13}$$

and the average value of $\mathbf{V}$ for species i is the diffusion velocity of species i,

$$\bar{\mathbf{V}}_i(\mathbf{x}, t) = \int \mathbf{V} f_i \, d\mathbf{v}/n_i, \quad i = 1, \ldots, N. \tag{14}$$

It may be seen from equations (11) and (13) that $\bar{\mathbf{V}}_i = \bar{\mathbf{v}}_i - \mathbf{v}_0$ and from equation (12) that $\sum_{i=1}^{N} Y_i \bar{\mathbf{V}}_i = 0$.

The translational temperature of species i, $T_i(\mathbf{x}, t)$, may be defined by

$$3 R^0 T_i / 2 = n_i (m_i \overline{V_i^2/2})/c_i = (m_i/2) \int V^2 f_i \, d\mathbf{v}/c_i, \quad i = 1, \ldots, N, \tag{15}$$

and the translational temperature of the mixture $T(x, t)$ is determined by the equation

$$3 R^0 T/2 = \sum_{i=1}^{N} X_i \, 3 R^0 T_i / 2 = \sum_{i=1}^{N} n_i (m_i \overline{V_i^2/2})/c$$

$$= \sum_{i=1}^{N} (m_i/2) \int V^2 f_i \, d\mathbf{v}/c, \tag{16}$$

where $V^2 = \mathbf{V} \cdot \mathbf{V}$ and R^0 is the universal gas constant ($R^0 = \mathscr{A} k^0$ if k^0 is Boltzmann's constant). For gases in thermodynamic equilibrium f_i reduces to the Maxwell velocity distribution,*

$$f_i \to n_i [m_i/(2\pi k^0 T)]^{3/2} \exp[-(m_i V^2)/(2k^0 T)], \quad i = 1, \ldots, N, \tag{17}$$

from which it may be seen [by substituting equation (17) into equations (15) and (16) and performing the integrations] that the translational temperatures in equations (15) and (16) reduce to the thermodynamic temperature appearing in equation (17) at equilibrium. In nonequilibrium systems the translational temperature defined in equation (16) may differ from temperatures associated with internal degrees of freedom and from translational temperatures associated with individual species.

---

* See, for example, 69–74 of [2].

The stress tensor for species i, $\mathbf{P}_i(\mathbf{x}, t)$, is defined as

$$\mathbf{P}_i = n_i m_i \overline{(\mathbf{VV})}_i = \int m_i(\mathbf{VV}) f_i \, d\mathbf{v}, \quad i = 1, \ldots, N, \tag{18}$$

where $(\mathbf{VV})$ is a dyadic product. It is easily seen* that $\mathbf{P}_i$ has the physical meaning of a stress tensor for gases; for example, the total momentum per second of i molecules in the $\mathbf{m}$ direction transported in the $\mathbf{n}$ direction across a surface of unit area normal to the $\mathbf{n}$ direction and moving with velocity $\mathbf{v}_0$ is $\mathbf{m} \cdot \mathbf{P}_i \cdot \mathbf{n}$. It may be noted that $\mathbf{P}_i$ is symmetric, that its diagonal elements are normal stresses and that its off-diagonal elements are shear stresses. In thermodynamic equilibrium, it is found from equations (17) and (18) that

$$\mathbf{P}_i \rightarrow c_i R^0 T \mathbf{U}, \quad i = 1, \ldots, N, \tag{19}$$

where $\mathbf{U}$ is the unit tensor and $c_i R^0 T$ clearly equals the equilibrium partial pressure of species i in the gas. In nonequilibrium systems, the partial pressure of species i may be defined as

$$p_i = (\text{tr } \mathbf{P}_i)/3 = c_i R^0 T_i, \quad i = 1, \ldots, N, \tag{20}$$

where tr denotes the trace of the tensor (tr $\mathbf{P}_i$ is the sum of the diagonal elements of $\mathbf{P}_i$) and where the last equality follows from equation (15). The total stress tensor $\mathbf{P}$ is defined as

$$\mathbf{P} = \sum_{i=1}^{N} \mathbf{P}_i, \tag{21}$$

and the (nonequilibrium) hydrostatic pressure is

$$p = (\text{tr } \mathbf{P})/3 = c R^0 T, \tag{22}$$

in which the last equality is a consequence of equation (16). It will be noted from equation (22) that essentially by definition, the ideal gas law is valid even for nonequilibrium systems.

The symbol $U_i(\mathbf{x}, \mathbf{v}, t)$ will denote the total absolute internal energy of a molecule of type i traveling with velocity $\mathbf{v}$ and is given by

$$U_i = m_i V^2/2 + U_i^+, \quad i = 1, \ldots, N, \tag{23}$$

where U_i^+ is the contribution to U_i from internal (rotational, vibrational, and electronic) degrees of freedom. When molecules of species i may be in different internal energy levels, U_i^+ must be interpreted as the average contribution to U_i from internal degrees of freedom at $(\mathbf{x}, \mathbf{v}, t)$. The reasonable

---

* See, for example, pp. 28–36 of [2]. Unlike solids, in which transmission of forces between molecules is the dominant contributor to surface stresses, gases experience surface stresses primarily by migration of molecules with differing momenta across surfaces.

assumption that U_i^+ is independent of $\mathbf{x}$, $\mathbf{v}$, and t will be made subsequently.*
The average internal energy per mole of molecules of type i at $(\mathbf{x}, t)$ is

$$\mathscr{A}\bar{U}_i \equiv \int U_i f_i \, d\mathbf{v}/c_i = 3R^0 T_i/2 + \mathscr{A}\bar{U}_i^+, \quad i = 1, \ldots, N, \qquad (24)$$

where $\bar{U}_i^+ \equiv \int U_i^+ f_i \, d\mathbf{v}/n_i$ and where use has been made of equation (15).
The internal energy per unit mass of species i is

$$u_i = \bar{U}_i/m_i, \quad i = 1, \ldots, N; \qquad (25)$$

the average molar internal energy of the mixture is

$$\mathscr{A}\bar{U} = \mathscr{A} \sum_{i=1}^{N} c_i \bar{U}_i/c = 3R^0 T/2 + \mathscr{A}\bar{U}^+, \qquad (26)$$

where $\bar{U}^+ \equiv \sum_{i=1}^{N} n_i \bar{U}_i^+/n$ and where use has been made of equation (16),
and the average internal energy per unit mass of the mixture is

$$u \equiv \sum_{i=1}^{N} Y_i u_i = \mathscr{A}\bar{U}c/\rho. \qquad (27)$$

The net total energy per unit area per second carried by molecules of
species i across a surface with normal $\mathbf{n}$ and moving with velocity $\mathbf{v}_0$ is
$\mathbf{q}_i \cdot \mathbf{n}$, where

$$\mathbf{q}_i \equiv n_i(\overline{U_i \mathbf{V}})_i = \int U_i \mathbf{V} f_i \, d\mathbf{v}, \quad i = 1, \ldots, N, \qquad (28)$$

is called the heat-flux vector for species i. The total heat-flux vector is

$$\mathbf{q} = \sum_{i=1}^{N} \mathbf{q}_i + \mathbf{q}_R, \qquad (29)$$

where $\mathbf{q}_R$ is the radiant heat-flux vector.[†] From equations (17) and (28) it is
readily seen that $\mathbf{q}_i = 0$ at thermodynamic equilibrium.

The final variable needed in the fluid equations is the mass of species i
per unit volume per second produced by chemical reactions, w_i. From the
definition of $\delta f_i/\delta t$, it is clear that the number of molecules of species i per
unit spatial volume per second produced by chemical processes is $\int (\delta f_i/\delta t) \, dv$,
whence it follows that

$$w_i = m_i \int (\delta f_i/\delta t) \, dv, \quad i = 1, \ldots, N. \qquad (30)$$

---

* This, of course, does not imply that $\bar{U}_i^+$ is independent of $\mathbf{x}$ and t.
† The net energy per unit area per second carried by radiation across a surface with
normal $\mathbf{n}$ and moving with velocity $\mathbf{v}_0$ is $\mathbf{q}_R \cdot \mathbf{n}$. It is often important to account for this radiant
flux even though the contribution of the radiation density to the internal energy u is negligible.

The phenomenological expressions of chemical kinetics for w_i are discussed in Appendix B and are quoted in Chapter 1.

D.3. THE EQUATION OF CHANGE

The fluid equations governing the time and space dependences of the variables defined in the preceding section are obtained by multiplying equation (2) by suitable factors and integrating over velocity space. A systematic derivation of these equations is therefore facilitated by obtaining a general expression for the integral over all $\mathbf{v}$ of the product of equation (2) and an arbitrary function of velocity for species i, $\psi_i(\mathbf{v})$. Bearing in mind that $\mathbf{x}$, $\mathbf{v}$, and t are all independent variables in equation (2), we readily obtain

$$\frac{\partial}{\partial t}(n_i \bar{\psi}_i) + \mathbf{V}_x \cdot [n_i(\overline{\psi_i \mathbf{v}})_i] - n_i \mathbf{f}_i \cdot (\overline{\mathbf{V}_v \psi_i})_i = \int \psi_i \frac{\delta f_i}{\delta t} d\mathbf{v}, \quad i = 1, \ldots, N,$$

(31)

where the bars denote averages of the type defined in equation (10), $\mathbf{f}_i$ has been assumed to be independent of $\mathbf{v}$, the identity $\mathbf{v} \cdot \mathbf{V}_x f_i = \mathbf{V}_x \cdot (\mathbf{v} f_i)$ has been employed, the last term on the left-hand side has been integrated by parts, and it has been assumed that $|\mathbf{v}|^2 \psi_i f_i \to 0$ as $|\mathbf{v}| \to \infty$ (which is true for all ψ_i used). Equation (31) is called the equation of change for $\psi_i(\mathbf{v})$.

D.4. SUMMATIONAL INVARIANTS

For all microscopic processes (collisions, radiative transitions, and so forth, in which $\mathbf{v}$ may change but the macroscopic $\mathbf{x}$ and t are fixed for the molecules experiencing the process), certain quantities may be conserved. If ψ_i is such a quantity, then it is called a summational invariant because

$$\sum_{i=1}^{N} \psi_i (\delta f_i / \delta t) \, d\mathbf{v} = 0,$$

(32)

in this case, according to the physical interpretation of the integral (the integral is the change in the total ψ_i per unit volume per second due to microscopic processes). If the mass, momentum, and energy removed by radiation in collisions and by purely radiative transitions are neglected, then the conservation of mass, momentum, and energy in molecular processes implies that equation (32) will be valid for

$$\psi_i = m_i, \quad \psi_i = m_i \mathbf{v}, \quad \text{and} \quad \psi_i = \tfrac{1}{2} m_i v^2 + U_i^+,$$

respectively. If energy transfer associated with radiation is included, then an energy balance for the radiation implies that for the last of these ψ_i,

$$\sum_{i=1}^{N} \int \psi_i (\delta f_i / \delta t) \, dv = -\mathbf{V}_x \cdot \mathbf{q}_R.$$

D.5. MACROSCOPIC CONSERVATION EQUATIONS

D.5.1. Overall continuity

Setting $\psi_i = m_i$ in equation (31) and summing over all i yields

$$\sum_{i=1}^{N} \left[\frac{\partial}{\partial t} (n_i m_i) + \mathbf{V}_x \cdot (n_i m_i \bar{\mathbf{v}}_i) \right] = \sum_{i=1}^{N} \int m_i \frac{\delta f_i}{\delta t} \, dv = 0,$$

since equation (32) is valid for $\psi_i = m_i$. From equations (5), (6), (7), and (12), we see that this expression may be written in the form

$$\partial \rho / \partial t + \mathbf{V}_x \cdot (\rho \mathbf{v}_0) = 0, \tag{33}$$

which is the familiar continuity equation of fluid dynamics.

D.5.2. Momentum conservation

Since $\overline{[\mathbf{V}_r(m_i \mathbf{v})]}_i = m_i \mathbf{U}$ and

$$\overline{[(m_i \mathbf{v})\mathbf{v}]}_i = m_i \overline{[(\mathbf{v}_0 + \mathbf{V})(\mathbf{v}_0 + \mathbf{V})]}_i$$
$$= m_i [(\mathbf{v}_0 \mathbf{v}_0) + (\mathbf{v}_0 \bar{\mathbf{V}}_i) + (\bar{\mathbf{V}}_i \mathbf{v}_0) + \mathbf{P}_i / (m_i n_i)]$$

[see equation (18)], letting $\psi_i = m_i \mathbf{v}$ in equation (31) and summing over i gives

$$\sum_{i=1}^{N} \left\{ \frac{\partial}{\partial t} (n_i m_i \bar{\mathbf{v}}_i) + \mathbf{V}_x \cdot (n_i m_i [(\mathbf{v}_0 \mathbf{v}_0) + (\mathbf{v}_0 \bar{\mathbf{V}}_i) + (\bar{\mathbf{V}}_i \mathbf{v}_0) + \mathbf{P}_i / (m_i n_i)]) \right.$$
$$\left. - n_i \mathbf{f}_i \cdot (m_i \mathbf{U}) \right\} = 0$$

because $\psi_i = m_i \mathbf{v}$ satisfies equation (32). Using equations (5), (6), (7), (12), (14), and (21), we find that the preceding equation reduces to

$$\frac{\partial}{\partial t} (\rho \mathbf{v}_0) + \mathbf{V}_x \cdot [\rho (\mathbf{v}_0 \mathbf{v}_0)] + \mathbf{V}_x \cdot \mathbf{P} - \sum_{i=1}^{N} \rho_i \mathbf{f}_i = 0. \tag{34}$$

In view of equation (33) and the identity $\mathbf{V}_x \cdot [\rho (\mathbf{v}_0 \mathbf{v}_0)] = [\mathbf{V}_x \cdot (\rho \mathbf{v}_0)] \mathbf{v}_0 + \rho (\mathbf{v}_0 \cdot \mathbf{V}_x) \mathbf{v}_0$, equation (34) becomes

$$\partial \mathbf{v}_0 / \partial t + (\mathbf{v}_0 \cdot \mathbf{V}_x) \mathbf{v}_0 = -(\mathbf{V}_x \cdot \mathbf{P}) / \rho + \mathbf{f}, \tag{35}$$

where $\mathbf{f} \equiv \sum_{i=1}^{N} Y_i \mathbf{f}_i$ is the average body force per unit mass acting on the mixture. Equation (35) is the usual form of the momentum equation of fluid dynamics.

D.5.3. Energy conservation

In order to derive the energy equation, ψ_i may be set equal to $m_i v^2/2 + U_i^+$ in equation (31), U_i^+ is assumed to be independent of $\mathbf{v}$, $\mathbf{x}$, and t, equation (31) is summed over i, and radiant energy transfer is included in evaluating the right-hand side. Since $\overline{\psi}_i = m_i v_0^2/2 + m_i \mathbf{v}_0 \cdot \overline{\mathbf{V}}_i + \overline{U}_i$ from equations (11)–(14), (23), and (24), $(\overline{\psi_i \mathbf{v}})_i = m_i v_0^2 \mathbf{v}_0/2 + m_i v_0^2 \, \overline{\mathbf{V}}_i/2 + m_i(\mathbf{v}_0 \cdot \overline{\mathbf{V}}_i)\mathbf{v}_0 + m_i \mathbf{v}_0 \cdot (\mathbf{P}_i/m_i n_i) + \overline{U}_i \mathbf{v}_0 + \mathbf{q}_i/n_i$ from equations (11)–(14), (18), (23), (24), and (28) and $(\overline{\mathbf{V}_v \psi_i})_i = m_i(\mathbf{v}_0 + \overline{\mathbf{V}}_i)$ from equations (11), (13), and (14), the result is

$$\frac{\partial}{\partial t}(\tfrac{1}{2}\rho v_0^2 + \rho u) + \mathbf{V}_x \cdot (\tfrac{1}{2}\rho v_0^2 \mathbf{v}_0 + \rho u \mathbf{v}_0 + \mathbf{v}_0 \cdot \mathbf{P} + \mathbf{q} - \mathbf{q}_R)$$

$$- \sum_{i=1}^{N} \rho_i \mathbf{f}_i \cdot (\mathbf{v}_0 + \overline{\mathbf{V}}_i) = -\mathbf{V}_x \cdot \mathbf{q}_R, \quad (36)$$

where use has been made of the identity $\sum_{i=1}^{N} Y_i \overline{\mathbf{V}}_i = 0$ and of equations (5), (6), (7), (21), (25), (27), (29). Taking the dot product of $\rho \mathbf{v}_0$ with equation (35), using the identities $\mathbf{v}_0 \cdot \partial \mathbf{v}_0/\partial t = \partial/\partial t(v_0^2/2)$, $\mathbf{v}_0 \cdot [(\mathbf{v}_0 \cdot \mathbf{V}_x)\mathbf{v}_0] = \mathbf{v}_0 \cdot \mathbf{V}_x(v_0^2/2)$ and $\mathbf{v}_0 \cdot (\mathbf{V}_x \cdot \mathbf{P}) = \mathbf{V}_x \cdot (\mathbf{v}_0 \cdot \mathbf{P}) - \mathbf{P}:(\mathbf{V}_x \mathbf{v}_0)$ where the two dots ($:$) indicate that the tensors are to be contracted twice to form a scalar, and utilizing the resulting equation and (33) in equation (36), we see that

$$\rho \partial u/\partial t + \rho \mathbf{v}_0 \cdot \mathbf{V}_x u = -\mathbf{V}_x \cdot \mathbf{q} - \mathbf{P}:(\mathbf{V}_x \mathbf{v}_0) + \rho \sum_{i=1}^{N} Y_i \mathbf{f}_i \cdot \mathbf{V}_i, \quad (37)$$

which is the final form of the energy equation. The terms on the right-hand side of equation (37) may be viewed physically as an enumeration of the ways in which u may change, following a fluid element.

D.5.4. Species conservation

In deriving the species conservation equations, ψ_i will be set equal to unity, which is not a summational invariant, since the number of molecules need not be conserved in chemical reactions. With $\psi_i = 1$, equation (31) reduces to

$$\partial n_i/\partial t + \mathbf{V}_x \cdot (n_i \overline{\mathbf{v}}_i) = \int (\delta f_i/\delta t) \, d\mathbf{v}, \quad i = 1, \ldots, N, \quad (38)$$

which may be multiplied by m_i to show that

$$\frac{\partial}{\partial t}(\rho Y_i) + \mathbf{V}_x \cdot [\rho Y_i(\mathbf{v}_0 + \overline{\mathbf{V}}_i)] = w_i, \quad i = 1, \ldots, N, \quad (39)$$

where use has been made of equations (5)–(7), (11)–(14), and (30). Substituting equation (33) into equation (39) yields the form

$$\partial Y_i/\partial t + \mathbf{v}_0 \cdot \mathbf{V}_x Y_i = w_i/\rho - [\mathbf{V}_x \cdot (\rho Y_i \bar{\mathbf{V}}_i)]/\rho, \quad i = 1, \ldots, N, \quad (40)$$

for the conservation of mass of chemical species i. Equation (40) states that following the mass-average motion of the fluid, Y_i can change only by chemical reactions or diffusion. Only $N - 1$ of the relations given in equation (40) are independent, since summing equation (40) over all i gives the identity $0 = 0$; this result corresponds to the fact that only $N - 1$ of the Y_i are independent.

D.5.5. Summary

Equations (33), (35), (37), and (40) comprise $N + 5$ equations in the $N + 5$ unknowns, Y_i, ρ, u, and $\mathbf{v}_0$. These equations contain the functions $\mathbf{f}_i, \mathbf{P}, \mathbf{q}, \bar{\mathbf{V}}_i$, and w_i, which must be related to the other dependent and independent variables if the system is to form a closed set of equations. The $\mathbf{f}_i$ are specified by the nature of the external force field, if any, the w_i are determined by the chemical kinetics, and $\mathbf{q}, \mathbf{p}$ and $\bar{\mathbf{V}}_i$ are the transport properties investigated in Appendix E. It will be seen in Appendix E that the transport properties can rigorously be related to Y_i, ρ, u, and $\mathbf{v}_0$ only for near-equilibrium flows. In quoting equations (33), (35), (37), and (40) in Chapter 1, the subscript 0 is omitted from $\mathbf{v}_0$, the bar is omitted from $\bar{\mathbf{V}}_i$, and the subscript x is omitted from $\mathbf{V}_x$, since the molecular velocity never appears as an independent variable in the applications.

Most textbooks that present the conservation equations without giving kinetic-theory derivations lack the generality needed for application to combustion. A notable exception, recommended for study, is [3], which introduces some of the principles of combustion as well.

REFERENCES

1. J. O. Hirschfelder, C. F. Curtiss, and R. B. Bird, *Molecular Theory of Gases and Liquids*, New York: Wiley, 1954.
2. S. Chapman and T. G. Cowling, *The Mathematical Theory of Non-Uniform Gases*, Cambridge: Cambridge University Press, 1953.
3. L. D. Landau and E. M. Lifshitz, *Fluid Mechanics*, Reading, Mass.: Addison-Wesley, 1959.

APPENDIX E

Transport Properties

The diffusion velocities, viscous stresses, and heat fluxes appearing in the conservation equations of fluid dynamics [see Appendix C or D], which are determined by the molecular transport of mass, momentum, and energy, respectively, must be calculable in order to render the conservation equations soluble. The fact that these quantities cannot generally be directly related to the other variables appearing in the conservation equations is readily apparent, since the transport properties involve higher velocity moments of the distribution function [see, for example, equation (D-28)]. For near-equilibrium systems, Enskog has employed a series expansion of the velocity-distribution function about the Maxwellian distribution in order to obtain from the Boltzmann equation explicit expressions for the transport vectors (and tensor) in terms of the gradients of the dependent variables of fluid dynamics. The resulting closed set of equations constitute the Navier-Stokes equations, which have been found to be applicable for considerably large deviations from equilibrium.* Since Enskog's rigorous derivation of the Navier-Stokes equations necessarily is very long, a physical discussion somewhat analogous to those in [1]-[3] is presented here, and the reader is referred to [3]-[5] for the exact treatment. In view of the fact that simplified

---

* Higher-order approximations have not been proven to be useful; when the Navier-Stokes equations are invalid, the most satisfactory procedure is to solve the Boltzmann equation by methods that do not rely on the assumption of near-equilibrium flow.

analyses, while possibly giving more insight into the problem, lead to inaccurate expressions for the transport coefficients (the constants of proportionality between the fluxes and the gradients), the results of the exact theory also will be quoted here. Reference [6] may be recommended as a thorough textbook on transport phenomena.

In the following section molecular collisions are discussed briefly in order to define the notation appearing in the exact expressions for the transport coefficients. Diffusion is treated separately from the other transport properties in Section E.2 because it has been found [7] that closer agreement with the exact theory is obtained by utilizing a different viewpoint in this case. Next, a general mean-free-path description of molecular transport is presented, which is specialized to the cases of viscosity and heat conduction in Sections E.4 and E.5. Finally, dimensionless ratios of transport coefficients, often appearing in combustion problems, are defined and discussed. The notation throughout this appendix is the same as that in Appendix D.

E.1. COLLISION INTEGRALS

The exact kinetic theory of dilute gases leads to expressions for transport coefficients in terms of certain quantities called **collision integrals**, which depend on the dynamics of binary intermolecular collisions.* These integrals are defined in this section.

Since the collisions of predominant importance in transport processes are elastic and do not involve chemical reactions, a potential φ may be defined, the negative gradient of which is the force between the two interacting molecules. The interaction may accurately be treated classically for all molecules at room temperature and above. Consideration is restricted to central forces, for which φ depends only on r, the distance between the mass centers of the molecules [$\varphi = \varphi(r)$]. Useful results for more general potentials (which, rigorously, are required to describe interactions of polar molecules) have not been obtained. The arbitrary constant in the potential is defined by $\varphi(\infty) = 0$.

It is easily seen from classical mechanics[†] that the binary collision problem is mathematically equivalent to a one-body problem in which a body with the reduced mass

$$\mu = \mu_{ij} \equiv m_i m_j/(m_i + m_j) \tag{1}$$

(where m_i and m_j are the masses of the colliding molecules) moves in the

---

    * In dilute gases (those obeying the ideal gas law), ternary collisions occur so seldom that they are of no consequence in transport.

    † See, for example, Chapter 3 of [8] or Chapter 1 of [5].

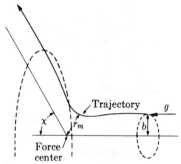

FIGURE E.1. Schematic diagram of the equivalent one-body collision.

fixed potential $\varphi(r)$. A schematic diagram of the collision in this representation is shown in Figure E.1. Here r_m is the distance of closest approach of the molecules, χ is the deflection angle in the collision, b is the impact parameter (the distance of closest approach if there were no potential), and g is the relative approach velocity of the two molecules when they are infinitely far apart. From an analysis of the collision [5], [8], it is found that $\chi(b, g)$ is given by

$$\chi = \pi - 2b \int_{r_m}^{\infty} \left(\frac{1}{r^2}\right) \left[1 - \left(\frac{\varphi}{\mu g^2/2}\right) - \left(\frac{b^2}{r^2}\right)\right]^{-1/2} dr, \qquad (2)$$

where r_m is determined by the solution of the equation

$$\varphi(r_m) = (\mu g^2/2)[1 - (b^2/r_m^2)]. \qquad (3)$$

The collision integrals are defined as

$$\Omega_{ij}^{(k, l)} = \sqrt{\frac{2\pi k^0 T}{\mu}} \int_{0}^{\infty} \int_{0}^{\infty} e^{-\hat{g}^2} \hat{g}^{2l+3}[1 - (\cos \chi)^k] b \, db \, d\hat{g}, \qquad (4)$$

where the dimensionless velocity $\hat{g}$ is

$$\hat{g} \equiv g\sqrt{\mu/2k^0 T}. \qquad (5)$$

From equations (1)–(5) it is apparent that in addition to the powers k and l, the $\Omega_{ij}^{(k, l)}$ depend on m_i, m_j, T and the form of the potential $\varphi(r)$. Therefore, only $\varphi(r)$ remains to be found in order to determine completely the collision integrals.

Theoretical calculations of $\varphi(r)$ from molecular structure are difficult and necessarily quite approximate in nature. It has been found to be more useful to assume a reasonable functional form for $\varphi(r)$ with adjustable constants and to choose these constants to fit experimental data. Experiments yielding information about $\varphi(r)$ are (1) measurements of transport coefficients [giving $\varphi(r)$ at low energies (large r)] and (2) cross-section measurements by molecular-beam scattering [giving $\varphi(r)$ at high energies (small r)]. A number of different functional forms $\varphi(r)$ have been employed, most of which are described in [5].

The Lennard-Jones potential [4], [5], [9] (sixth-power attraction, twelfth-power repulsion) is quite realistic and appears to be the one most commonly used in practice. This potential contains two adjustable parameters (a "size" and a "strength"), which are defined and listed for various chemical compounds in [5], [6], and [9]. The collision integrals appearing in the first approximations to the transport properties are tabulated as functions of useful dimensionless forms of these two parameters in [5], [6], and [9]. Similar tabulations for other potentials may also be found in [5].

E.2. DIFFUSION

E.2.1. Physical derivation of the multicomponent diffusion equation [10]

An equation determining the diffusion velocities $\bar{\mathbf{V}}_i$ for gas mixtures may be derived by the following physical arguments. Since the total momentum is conserved in collisions, the momentum of molecules of type i can be changed only by collisions with molecules of other types ($j \neq i$) and by body forces acting on species i. It is to be expected that on the average, the momentum transferred to an i molecule in a collision between molecules i and j is approximately $\mu_{ij}(\bar{\mathbf{V}}_j - \bar{\mathbf{V}}_i)$ because $(\bar{\mathbf{V}}_j - \bar{\mathbf{V}}_i)$ is the average relative velocity of molecules of species i and j. The fact that the average momentum transferred from i to j must equal the negative of that transferred from j to i implies that the average momentum transferred is symmetrical in i and j, whence the reduced mass μ_{ij} enters the expression. The body force acting on molecules of type i in a unit volume element is clearly $\rho Y_i \mathbf{f}_i$. Hence, the net rate of change of momentum of molecules of type i per unit volume is

$$\boldsymbol{\Gamma}_i = \sum_{j=1}^{N} \mu_{ij} v_{ij} (\bar{\mathbf{V}}_j - \bar{\mathbf{V}}_i) + \rho Y_i \mathbf{f}_i, \quad i = 1, \ldots, N, \tag{6}$$

where v_{ij} is the total number of collisions per unit volume per second between molecules of kinds i and j.

This change in momentum of species i is manifest in changes in both the random velocity and the ordered velocity of the molecules. The rate of change of the ordered momentum is $\rho Y_i D\mathbf{v}_0/Dt$, where the substantial derivative with respect to time is

$$D/Dt = \partial/\partial t + \mathbf{v}_0 \cdot \boldsymbol{\nabla}_x.$$

Since (neglecting the off-diagonal elements of the pressure tensor*) the partial

---

* These assumptions, which are usually tacit in elementary treatments, are required in the order of approximation used here, as may be seen, for example, from Section 7.3, 468–469 of [5].

pressure of species i, p_i is physically the momentum of molecules of type i transported per second across a surface of unit area traveling with the mass-average velocity of the fluid [see, for example, equations (D-15) and (D-20)], it follows that $\mathbf{V}_x p_i$ is the rate of change of momentum of random motion of molecules of type i per unit volume. Therefore, the quantity Γ_i also is given by the relation

$$\Gamma_i = \rho Y_i D\mathbf{v}_0/Dt + \mathbf{V}_x p_i, \quad i = 1, \ldots, N. \tag{7}$$

Under the assumption that the translational temperatures of all species, are the same,* equations (D-9), (D-20), and (D-22) show that Dalton's law of partial pressures is valid, that is,

$$p_i = X_i p, \quad i = 1, \ldots, N, \tag{8}$$

whence

$$\mathbf{V}_x p_i = p\mathbf{V}_x X_i + X_i \mathbf{V}_x p, \quad i = 1, \ldots, N. \tag{9}$$

When the off-diagonal elements of the pressure tensor are neglected as above,* the momentum equation, equation (D-35), becomes

$$\rho D\mathbf{v}_0/Dt = -\mathbf{V}_x p + \rho \sum_{j=1}^{N} Y_j \mathbf{f}_j. \tag{10}$$

Substituting equations (6), (9), and (10) into equation (7) yields

$$\mathbf{V}_x X_i = \sum_{j=1}^{N} \left(\frac{\mu_{ij} v_{ij}}{p} \right)(\mathbf{V}_j - \mathbf{V}_i) + (Y_i - X_i)\left(\frac{\mathbf{V}_x p}{p} \right)$$
$$+ \left(\frac{\rho}{p} \right) \sum_{j=1}^{N} Y_i Y_j (\mathbf{f}_i - \mathbf{f}_j), \quad i = 1, \ldots, N, \tag{11}$$

where use has been made of the identity $\sum_{j=1}^{N} Y_j = 1$. Equation (11) is essentially the desired diffusion equation.

The product $\mu_{ij} v_{ij}$ in equation (11) may be related to the binary diffusion coefficients by considering the limiting case of a constant-pressure process in a two-component system with no body forces, for which equation (11) reduces to

$$\mathbf{V}_x X_1 = (\mu_{12} v_{12}/p)(\bar{\mathbf{V}}_2 - \bar{\mathbf{V}}_1), \tag{12}$$

where the subscripts 1 and 2 identify the two components. The diffusion velocity $\bar{\mathbf{V}}'_1$ of species 1 with respect to the *number-weighted* average velocity of the mixture is

$$\bar{\mathbf{V}}'_1 \equiv (\mathbf{v}_0 + \bar{\mathbf{V}}_1) - [X_1(\mathbf{v}_0 + \bar{\mathbf{V}}_1) + X_2(\mathbf{v}_0 + \bar{\mathbf{V}}_2)] = X_2(\bar{\mathbf{V}}_1 - \bar{\mathbf{V}}_2), \tag{13}$$

---

* See previous footnote.

in which the last equality follows from the identity $(1 - X_1) = X_2$. Substituting equation (13) into equation (12) leads to the result

$$\mathbf{V}_x X_1 = -(\mu_{12} v_{12}/X_2 p)\bar{\mathbf{V}}_1'. \tag{14}$$

The binary diffusion coefficient D_{12} for species 1 and 2 is often defined by the equation

$$D_{12}\mathbf{V}_x X_1 = -X_1\bar{\mathbf{V}}_1' \tag{15}$$

under the present conditions; that is, it is the constant of proportionality between the mole-fraction gradient and the number flux with respect to the number-weighted average velocity. A comparison of equations (14) and (15) then shows that

$$D_{12} = X_1 X_2 p/\mu_{12} v_{12},$$

which implies that, in general,

$$D_{ij} = X_i X_j p/\mu_{ij} v_{ij}. \tag{16}$$

Utilizing equation (16) in equation (11) yields

$$\mathbf{V}_x X_i = \sum_{j=1}^{N} \left(\frac{X_i X_j}{D_{ij}}\right)(\bar{\mathbf{V}}_j - \bar{\mathbf{V}}_i) + (Y_i - X_i)\left(\frac{\mathbf{V}_x p}{p}\right)$$
$$+ \left(\frac{\rho}{p}\right)\sum_{j=1}^{N} Y_i Y_j(\mathbf{f}_i - \mathbf{f}_j), \quad i = 1, \ldots, N, \tag{17}$$

as the final form of the multicomponent diffusion equation. Equation (17) states that concentration gradients may be supported by diffusion velocities, pressure gradients (when the mass fractions differ from the mole fractions) and differences in the body force per unit mass on molecules of different species. It can be shown from the rigorous kinetic theory that equation (17) is valid to first order in a Sonine polynomial expansion of f_i when thermal diffusion (diffusion resulting from temperature gradients) is negligible [5]. The multicomponent diffusion equation obtained from the complete kinetic theory is [5]

$$\mathbf{V}_x X_i = \sum_{j=1}^{N} \left(\frac{X_i X_j}{D_{ij}}\right)(\bar{\mathbf{V}}_j - \bar{\mathbf{V}}_i) + (Y_i - X_i)\left(\frac{\mathbf{V}_x p}{p}\right)$$
$$+ \left(\frac{\rho}{p}\right)\sum_{j=1}^{N} Y_i Y_j(\mathbf{f}_i - \mathbf{f}_j) + \sum_{j=1}^{N}\left[\left(\frac{X_i X_j}{\rho D_{ij}}\right)\left(\frac{D_{T,j}}{Y_j} - \frac{D_{T,i}}{Y_i}\right)\right]\left(\frac{\mathbf{V}_x T}{T}\right),$$
$$i = 1, \ldots, N, \tag{18}$$

where $D_{T,i}$ is the thermal diffusion coefficient of species i in the multicomponent mixture. The physical viewpoint presented here is easily extended to show how thermal diffusion (the Soret effect) may arise, and it even enables one to determine the signs of the thermal diffusion coefficients [7]. In quoting equation (18) in Chapter 1, the bars over the diffusion velocities and the subscript x in $\mathbf{V}_x$ are omitted.

E.2.2. Simplified diffusion equations

For binary mixtures it may be shown that equation (18) reduces to

$$Y_1 \bar{\mathbf{V}}_1 = -D_{12}\left[\mathbf{V}_x Y_1 - \left(\frac{Y_1 Y_2}{X_1 X_2}\right)(Y_1 - X_1)\left(\frac{\mathbf{V}_x p}{p}\right)\right.$$
$$\left. - \frac{(Y_1 Y_2)^2}{(X_1 X_2)}\left(\frac{\rho}{p}\right)(\mathbf{f}_1 - \mathbf{f}_2) + \left(\frac{D_{T,1}}{\rho D_{12}}\right)\left(\frac{\mathbf{V}_x T}{T}\right)\right]. \tag{19}$$

From equation (19), we see that the familiar **Fick's law** of diffusion

$$Y_1 \bar{\mathbf{V}}_1 = -D_{12}\mathbf{V}_x Y_1, \tag{20}$$

is rigorously valid when (1) the mixture is a binary mixture, (2) thermal diffusion is negligible, (3) the body force per unit mass is the same on each species ($\mathbf{f}_1 = \mathbf{f}_2$), and (4) either pressure is constant ($\mathbf{V}_x p = 0$) or the molecular weights of both species are the same ($Y_1 = X_1$). Because of the complex forms of equations (18) and (19), in many of the applications discussed in this book equation (20) is utilized in order to make the governing equations tractable. The use of equation (20) in combustion problems is partially justified by the facts that conditions 2 and 3 are almost always valid approximations and $\mathbf{V}_x p/p$ usually is sufficiently small that conditions 4 is satisfied. Requirement 1 may be replaced by somewhat less stringent conditions (for example, that the binary diffusion coefficients of all pairs of species in the multicomponent mixture are equal), but, of the four conditions listed above, 1 (or its equivalent) usually is the most difficult to justify in combustion problems.

If the condition 1 is not imposed, then with approximations 2, 3, and 4, equation (18) reduces to

$$\mathbf{V}_x X_i = \sum_{j=1}^{N} \frac{X_i X_j}{D_{ij}}(\bar{\mathbf{V}}_j - \bar{\mathbf{V}}_i), \quad i = 1, \ldots, N, \tag{21}$$

which has been called the **Stefan-Maxwell equation** [6]. Equation (21) is more complicated to employ than equation (20), mainly because, in general, it is not readily soluble for diffusion velocities in terms of concentration gradients; it provides concentration gradients in terms of diffusion velocities. However, there are some combustion problems for which use of equation (21) will not introduce excessive complexity in seeking analytical solutions. Moreover, when electronic computers are employed to calculate numerical solutions to the differential equations of combustion, use of equation (21) should not be appreciably more complicated than use of equations involving multicomponent diffusion coefficients (see Section E.2.4) particularly in one-dimensional problems. Nevertheless, such calculations seldom have been based on equation (21).

E.2.3. Binary diffusion coefficients

Equation (16) constitutes a physically derived expression for the binary diffusion coefficients D_{ij}. This equation may be written in a more useful form by expressing the number of i-j collisions per unit volume per second (v_{ij}) in terms of more basic molecular parameters. Since there are n_i molecules of type i per unit volume, $v_{ij} = n_i/\tau_{ij}$, where τ_{ij} is the average time between collisions of a given molecule of kind i with molecules of kind j. Letting σ_{ij} denote the collision cross section for molecules of types i and j [for example, $\sigma_{ij} = \pi(r_i + r_j)^2$ for hard spheres with radii r_i and r_j] and $\bar{v}_{ij}$ denote the average relative velocity of molecules of kinds i and j, we may consider the given molecule of kind i to sweep out a volume $\sigma_{ij}\bar{v}_{ij}$ per second, such that the given i molecule will collide with any j molecule in this volume. Since, on the average the i molecule will collide with a j molecule after it has swept out a volume $1/n_j$, where n_j is the number of molecules of type j per unit volume, it is clear that the number of collisions of the given i molecule with j molecules per second is $1/\tau_{ij} = n_j(\sigma_{ij}\bar{v}_{ij})$. Hence

$$v_{ij} = n_i n_j \sigma_{ij}\bar{v}_{ij} = X_i X_j n^2 \sigma_{ij}\bar{v}_{ij}, \tag{22}$$

where use has been made of the definition of X_i. Substituting equation (22) and the ideal gas law, equation (D-22), into equation (16) yields

$$D_{ij} = k^0 T/n\mu_{ij}\sigma_{ij}\bar{v}_{ij} \tag{23}$$

as the equation for the diffusion coefficient.

The conventionally quoted result of the exact kinetic theory [5], [9] is that obtained from a first-order expansion in Sonine polynomials, namely,

$$D_{ij} = 3k^0 T/16n\mu_{ij}\Omega_{ij}^{(1,1)}, \tag{24}$$

where the collision integral $\Omega_{ij}^{(1,1)}$ is defined in equation (4). It may be shown [5] that $\Omega_{ij}^{(1,1)} = \sigma_{ij}\bar{v}_{ij}/4$ for hard-sphere molecules, in which case it follows that equation (23) is in error only by the constant factor of $\frac{3}{4}$. Since it is easily shown from the Maxwellian velocity distribution that

$$\bar{v}_{ij} = \sqrt{8k^0 T/\pi\mu_{ij}},$$

it follows from equation (23) that (since $n \sim p/T$), the pressure and temperature dependence of D_{ij} is $D_{ij} \sim T^{3/2}/p$. For realistic intermolecular force potentials, equation (24) gives $D_{ij} \sim T^\alpha/p$, where, roughly, $\frac{3}{2} \le \alpha \le 2$. Values of D_{ij} typically lie between 10^{-2} cm²/s and 10 cm²/s at atmospheric pressure. Equations (23) and (24) both predict that D_{ij} is independent of the relative concentrations X_i and X_j and that D_{ij} increases as the molecular weight decreases ($D_{ij} \sim 1/\sqrt{\mu_{ij}}$). The product

$$\rho D_{ij} = 3\overline{W}k^0 T/16\mu_{ij}\Omega_{ij}^{(1,1)},$$

where $\overline{W} \equiv \sum_{j=1}^{N} X_j W_j$ is the average molecular weight of the mixture [see equations (D-5)–(D-9)], often appears in combustion problems and may be called the mass diffusivity. References [5], [6], and [9] and references quoted therein may be consulted for tabulations of binary diffusion coefficients for gases. Sources of newer experimental information are quoted in [11] and data are updated continually in a useful periodical [12]. Binary diffusion coefficients for liquids and solids are not given by equation (24); the reader may consult [6], [13], and [14] for information on the former and [15] for information on the latter. Additional sources of data on these and other transport properties appear in [16]–[21].

E.2.4. Multicomponent diffusion coefficients

Diffusion equations often are written differently from those given in Section E.2.1 [6]. In particular, multicomponent diffusion coefficients differing from D_{ij} often are introduced so that diffusion velocities may be expressed directly as linear combinations of gradients. The multicomponent diffusion coefficients are defined so that they reduce to D_{ij} for binary mixtures [6]. Use of diffusion equations involving multicomponent diffusion coefficients is being made increasingly frequently.

For liquids and solids, the molecular theory of diffusion is less highly developed than for gases, and, therefore, fundamental reasons for preferring one correct form of the diffusion equation over another do not exist for condensed phases. However, the theory of diffusion for gases rests on firmer grounds, which enable better assessments to be made of advantages of differing formulations. In particular, it is known that correct statements of multicomponent diffusion equations for gas mixtures involve multicomponent diffusion coefficients that depend on mole fractions in a complex manner and that are not necessarily symmetric in the indices i and j [6]. In contrast, the D_{ij} in equation (24) are symmetric in i and j, and the product nD_{ij} is independent of mole fractions. This affords greater economy in tabulation for binary diffusion coefficients and also may enable greater accuracy in prediction to be achieved if use is made of equation (18) or (21). Therefore, there exist fundamental reasons for preferring not to introduce multicomponent diffusion coefficients for gas mixtures. Notwithstanding this advantage of the formulation in equation (18), there may exist problems whose solutions are facilitated by use of different forms of multicomponent diffusion equations, even if the greater complication in functional dependences of diffusion coefficients is retained. At the current state of development of computational capabilities, problems are not solved with the full equations including precise variations of coefficients; therefore, with few exceptions [22], tests of relative efficiencies of differing precise formulations are unavailable.

A common approximation that has been employed is to introduce an effective diffusion coefficient D_i for each species i in the mixture, such that either

$$Y_i \bar{\mathbf{V}}_i = -D_i \nabla_x Y_i, \quad i = 1, \dots, N, \tag{25}$$

or $X_i \bar{V}_i' = -D_i \nabla_x X_i$, $i = 1, \ldots, N$ [compare equations (15) and (20)]. It is important to realize that if an approximation of this kind is employed, then only $N - 1$ of the quantities D_i can be specified independently; the identities $\sum_{i=1}^{N} Y_i \bar{V}_i = 0$ and $\sum_{i=1}^{N} Y_i = 1$ may be used in equation (25) to show that $D_i = \sum_{j \neq i} D_j \nabla_x Y_j / \sum_{j \neq i} \nabla_x Y_j$. Thus one species must be treated differently from the others if equation (25) is to be applied consistently. A situation in which equation (25) can be derived logically is that in which all species except one, say $i = N$, are present only in trace amounts, so that $1 - X_N \ll 1$. In this case equation (25) is to be used only for $i = 1, \ldots, N - 1$, and it is found that $D_i = D_{iN}$, the binary diffusion coefficient for diffusion through the species present in excess. In more general situations, the formula

$$D_i = (1 - X_i) / \sum_{j \neq i} (X_j/D_{ij}), \quad i = 1, \ldots, N \tag{26}$$

is qualitatively correct but, as emphasized above, can be used properly at best for only $N - 1$ species. The diffusion velocity of the component for which equation (25) is not used must be obtained from the identity

$$\sum_{i=1}^{N} Y_i \bar{V}_i = 0.$$

In general, if equation (25) is introduced for all i, then the diffusion coefficients D_i must depend on the diffusion velocities, that is, on the solution to the problem, and the formulation then encounters complexities that prevent it from being very useful. Calculations have been performed in which equations (25) and (26) are employed for all species and the fundamental requirement $\sum_{i=1}^{N} Y_i \bar{V}_i = 0$ is violated; the results obtained may not be too bad if the degree of violation is not too great.

E.2.5. Thermal diffusion coefficients

The thermal diffusion coefficients $D_{T,i}$ appearing in equations (18) and (19) in general may be positive or negative and depend on pressure, temperature, and concentrations. It is almost always found that the dimensionless ratio $D_{T,i}/(\rho D_{ij})$ is less than $\frac{1}{10}$ for all pairs of gaseous species i and j, which implies that thermal diffusion usually is negligible in comparison with the ordinary concentration-gradient diffusion. The complicated expressions for the coefficients of thermal diffusion will therefore be omitted here; the reader is referred to [5] and [9] for theoretical results and for useful empirical formulas.

Thermal diffusion coefficients should not be confused with the thermal diffusivity, a quantity defined in terms of the thermal conductivity and referring to conduction of heat (see Section E.5). **Thermal diffusion** one of the cross-transport effects, is a physical process entirely separate from heat conduction. It tends to draw light molecules to hot regions and to drive heavy molecules to cold regions of the gas. Hydrogen is a species that is likely to be relatively strongly influenced by thermal diffusion. Fine particles with sizes small compared to the mean free path (see Section E.3) behave like large molecules and migrate toward cold regions under the influence of thermal diffusion; for such particles, as well as for larger particles, this effect often is called **thermophoresis**. Combustion problems exist, involving hydrogen or fine particles, in which thermal diffusion may be of importance. The phenomenon also may be of significance concerning certain flame instabilities [23]. Numerical solutions of differential equations of combustion by electronic computers occasionally have been obtained with thermal diffusion included. Analytical methods in which the effect is taken into account only in regions of high temperature gradients appear feasible for some combustion problems but currently are unexplored. References to literature or alternative ways to formulate equation (18) with thermal diffusion included may be found in [6]. Quantities related to $D_{T,i}$ and sometimes used in place of it include the *thermal diffusion ratio*, the *thermal diffusion factor*, and the *Soret coefficient* (see [6] for the definitions). In condensed phases, $D_{T,i}/(\rho D_{ij})$ tends to be larger than in gases [6].

E.3. UNIFIED ELEMENTARY TREATMENT OF TRANSPORT PROCESSES

The rough mean-free path development to be given in this section will be used subsequently for viscous stresses and for thermal conduction; it may also be applied to diffusion, but the results in this case are not so satisfactory as are those of the preceding section.

The essential features of the results, including all those needed for the subsequent applications, may be obtained by limiting our attention to a one-component gas. The symbol Q will denote any property of a molecule that can be changed by collisions (Q will later be set equal to momentum and energy), and $\bar{Q}$ will denote the average value of Q for all molecules. For simplicity we shall consider a geometrical configuration in which $\partial\bar{Q}/\partial x > 0$, $\partial\bar{Q}/\partial y = \partial\bar{Q}/\partial z = 0$, whence molecular motion will tend to transport $\bar{Q}$ in the negative x direction. Let F denote the flux of Q in the $+x$ direction, that is, the net amount of Q per unit area per second transported across a surface normal to the x axis by molecular motion. Our aim will be to compute F_0 in terms of $(d\bar{Q}/dx)_0$, where the subscript 0 identifies conditions at $x = 0$.

It will be assumed that at each collision, both colliding molecules exchange their property Q, equalizing it between them,* and that their resulting Q equals the $\bar{Q}$ appropriate to the level at which they collide. Both of these assumptions are, at best, approximations that may be valid on the

---

* "Persistence of velocity" corrections constitute an example of methods of accounting for deviations from this approximation.

average. As a consequence of these assumptions, F_0 is determined by the location of the last collision of a molecule before it crosses the plane $x = 0$. This location depends on the average distance that a molecule travels between collisions, which is called the **mean free path** l.

An estimate of the mean free path l may easily by obtained by a slight extension of the reasoning given at the beginning of Section E.2.3. Letting $j = i$ (where i will now identify the only species present in the one-component system), we find that the average time between collisions of a given i molecule with another i molecule is (see Section E.2.3) $\tau_{ii} = 1/(n_i \sigma_{ii} \bar{v}_{ii})$. If the average velocity of i molecules is $\bar{v}_i$, then clearly $l = \tau_{ii} \bar{v}_i = (1/n_i \sigma_{ii})(\bar{v}_i/\bar{v}_{ii})$. A rough estimate of the last factor in this expression may be obtained by assuming that $\bar{v}_i/\bar{v}_{ii} \approx (v_i^2/v_{ii}^2)^{1/2}$ and by adopting a statistically independent pair-velocity distribution (that is, by considering near-equilibrium conditions). The last condition implies that $\overline{v_{ii}^2} = 2\overline{v_i^2}$, whence $\bar{v}_i/\bar{v}_{ii} \approx 1/\sqrt{2}$ and

$$l = 1/\sqrt{2}\, n\sigma, \tag{27}$$

where $n = n_i$ and the subscripts on σ have been omitted in the one-component mixture.

It may be expected that, on the average, molecules crossing the plane $x = 0$ from below will have $Q = \bar{Q}(-l)$ (that is a value of Q appropriate to the position $x = -l$), and those crossing from above will have $Q = \bar{Q}(+l)$. Hence

$$F_0 = J_- \bar{Q}(-l) - J_+ \bar{Q}(+l), \tag{28}$$

where J_- is the number of molecules per unit area per second crossing the plane $x = 0$ from below and J_+ is the number per unit area per second crossing from above. Assuming that the system is sufficiently near equilibrium that the velocity distribution may be taken to be isotropic, we find that $J_+ = J_- \equiv J$, where $J = n\bar{v}_x$, in which $\bar{v}_x$ is the x component of velocity averaged only over those molecules traveling in the $+x$ direction. It is easily proven by integrating the (isotropic) velocity distribution over a hemisphere that in terms of the average velocity of a molecule $\bar{v}_i \equiv \bar{v}$, the relation $\bar{v}_x = \bar{v}/4$ is valid, whence

$$F_0 = (n\bar{v}/4)[\bar{Q}(-l) - \bar{Q}(+l)]. \tag{29}$$

When the mean free path is small compared with the distance over which $\bar{Q}$ changes appreciably, the quantities $\bar{Q}(-l)$ and $\bar{Q}(+l)$ in equation (29) may be expanded in a Taylor series about $x = 0$, and terms of order higher than the first may be neglected. The result is

$$F_0 = -(n\bar{v}l/2)(d\bar{Q}/dx)_0. \tag{30}$$

The final expression for the flux of Q in terms of the gradient of its average value is obtained by substituting equation (27) into equation (30), omitting the subscript 0 (since the choice of the position $x = 0$ was arbitrary) and generalizing to the case in which $\partial \bar{Q}/\partial y \neq 0$ and $\partial \bar{Q}/\partial z \neq 0$ (by assuming that the medium is isotropic), namely,

$$\mathbf{F} = -(\bar{v}/2\sqrt{2}\,\sigma)\,\mathbf{V}_x \bar{Q}. \tag{31}$$

From equation (31) it is seen that the constant of proportionality between $\mathbf{F}$ and $\mathbf{V}_x \bar{Q}$ is independent of the number density and inversely proportional to the collision cross section; it is independent of pressure and its temperature dependence is $\sim \sqrt{T/\sigma} \sim T^\alpha$, where $\frac{1}{2} \leq \alpha \leq 1$, usually. Equation (31) is employed in the next two sections.

E.4. VISCOSITY

E.4.1. Coefficient of viscosity

Let $v_{0,y}$ be the average velocity of the molecules in the y direction, and let us investigate the transport of momentum in the y direction across planes normal to the x direction. If m is the mass of a molecule, the appropriate value of Q is $Q = mv_y$, and F becomes the shear stress in the y direction on a surface with normal in the x direction [see Section D.2]. In this case F is usually denoted by τ_{xy}, and equation (31) reduces to

$$\tau_{xy} = -(m\bar{v}/2\sqrt{2}\,\sigma)\,dv_{0,y}/dx. \tag{32}$$

The coefficient of viscosity μ may be defined conventionally as the constant of proportionality between τ_{xy} and $-dv_{0,y}/dx$ in this geometry; that is,

$$\tau_{xy} = -\mu\,dv_{0,y}/dx, \tag{33}$$

whence it follows that

$$\mu = m\bar{v}/2\sqrt{2}\,\sigma. \tag{34}$$

Equation (34) implies that μ is independent of pressure and that the temperature dependence of μ is $\mu \sim \sqrt{T/\sigma} \sim T^\alpha$, where $\frac{1}{2} \leq \alpha \leq 1$.

The accurate kinetic theory for a one-component system yields [5]

$$\mu = 5k^0 T/8\Omega_{ii}^{(2,2)}, \tag{35}$$

which exhibits essentially the same pressure and temperature dependence predicted above. For hard-sphere molecules, equation (35) is found to reduce to

$$\mu = (5\pi/16)m\bar{v}/2\sqrt{2}\,\sigma, \tag{36}$$

with which the approximate result in equation (34) fortuitously agrees within 2%. Tables giving viscosity coefficients may be found in [5], [6], and [18], for example.

For binary and multicomponent mixtures of gases the viscosity coefficient depends on the concentrations, and the results of the accurate kinetic theory are quite complicated. In terms of the viscosities of the pure components at the same pressure and temperature, μ_i, a useful empirical formula is [5], [9].

$$\mu = \sum_{i=1}^{N} \left\{ X_i \middle/ \left[\frac{X_i}{\mu_i} + 1.385 \sum_{\substack{j=1 \\ (j \neq i)}}^{N} (X_j k^0 T / p m_i D_{ij}) \right] \right\}, \tag{37}$$

which (except for a factor of 2 in place of 1.385) has some theoretical justification [5]. For most of the combustion applications discussed in this book, it is seldom practical to account for the concentration dependence of μ.

The kinematic viscosity, μ/ρ, has the same units as binary diffusion coefficients and the same type of pressure and temperature dependence, $\mu/\rho \sim T^\alpha/p$, with $\frac{3}{2} \leq \alpha \leq 2$. At atmospheric pressure, its values typically range from 10^{-1} cm^2/s to 1 cm^2/s for gases in combustion problems. The increase in viscosity with increasing temperature can be an important effect in the fluid dynamics of combustion.

E.4.2. The pressure tensor

The appropriate generalization of equation (33) to arbitrary geometries in multicomponent mixtures can be shown from kinetic theory [5] or from a reasonable continuum treatment to be

$$\mathbf{T} = -\mu[(\nabla_x \mathbf{v}_0) + (\nabla_x \mathbf{v}_0)^T] + \eta(\nabla_x \cdot \mathbf{v}_0)\mathbf{U}, \tag{38}$$

where η is a second viscosity coefficient, the superscript T denotes the transpose of the tensor, and the (symmetric) shear tensor $\mathbf{T}$ is related to the pressure tensor $\mathbf{P}$ [equation (D-21)] and to the hydrostatic pressure p by the equation

$$\mathbf{P} = p\mathbf{U} + \mathbf{T}. \tag{39}$$

The shear stress τ_{xy} appearing in equation (33) is the xy component of $\mathbf{T}$. Equation (39) clearly reduces to $\mathbf{P} = p\mathbf{U}$ in equilibrium as is necessary (see Section D-2).

In quoting equations (38) and (39) in Chapter 1, the subscript 0 is omitted from $\mathbf{v}_0$ and the second viscosity coefficient is replaced according to the expression

$$\eta = \tfrac{2}{3}\mu - \kappa, \tag{40}$$

where $\kappa \geq 0$ is the coefficient of bulk viscosity. For monatomic gas mixtures, kinetic theory shows that $\eta = \tfrac{2}{3}\mu$ ($\kappa = 0$). For polyatomic gases, relaxation effects between the translational motion and the various internal degrees of freedom may in some cases lead to a positive value of κ,* which is essentially proportional to relaxation time for small relaxation times and is zero for large relaxation times [5]. Few theoretical numerical calculations or reliable experimental measurements of κ yet exist, and bulk viscosity usually is negligible in combustion processes.

E.5. HEAT FLUX

E.5.1. Thermal conductivity

In order to treat thermal conduction in one-component systems, we may let $Q = U_i \equiv U$, the total energy of the molecule defined in equation (D-23). Then $\mathbf{F}$ becomes the energy per second transported across a surface of unit area; that is, $\mathbf{F}$ is the heat-flux vector $\mathbf{q}$ [see equations (D-28) and (D-29) with $\mathbf{q}_R$ neglected]. Hence equation (31) becomes

$$\mathbf{q} = -(\bar{v}/2\sqrt{2}\,\sigma)\nabla_x \overline{U} = -(mc_v \bar{v}/2\sqrt{2}\,\sigma)\nabla_x T, \tag{41}$$

where c_v is the specific heat at constant volume (that is, $d\overline{U}/dT = mc_v$).

The thermal conductivity λ of a one-component gas is defined as the constant of proportionality between $\mathbf{q}$ and $-\nabla_x T$; namely,

$$\mathbf{q} = -\lambda \nabla_x T. \tag{42}$$

Equations (41) and (42) then show that

$$\lambda = mc_v \bar{v}/2\sqrt{2}\,\sigma = \mu c_v, \tag{43}$$

where use has been made of equation (34) in the last equality. Since c_v is independent of pressure and usually depends only weakly on temperature, equation (43) implies that λ is independent of pressure and that, roughly, $\lambda \sim T^\alpha, \frac{1}{2} \leq \alpha \leq 1$.

The accurate kinetic theory for a pure monatomic ideal gas [$c_v = \frac{3}{2}(k^0/m)$] gives [5]

$$\lambda = \tfrac{25}{16}(k^0 T c_v/\Omega_{ii}^{(2,\,2)}) = \tfrac{5}{2}\mu c_v, \tag{44}$$

which agrees with equation (43) in the pressure and temperature dependence of λ but differs in magnitude by a factor of 2.5, thus emphasizing the approximate nature of the treatment in Section E.3. For pure polyatomic gases a rigorous kinetic theory expression for λ has not yet provided useful numerical results [24]. The Eucken formula [25], based on the physical reasoning

---

* The pressure p appearing in equation (39) is not given by equation (D-22) when $\kappa \neq 0$, since equations (38) and (39) imply (tr $\mathbf{P}$)/3 $= p - \kappa(\nabla_x \cdot \mathbf{v}_0)$, reflecting the fact that when translational-internal relaxation processes are of importance, the translational temperature is not the appropriate temperature to associate with the hydrostatic pressure [3].

outlined below, is, therefore still the simplest way to compute thermal conductivities of polyatomic gases, although good experimental data are preferable when available.

Since the translational energy is proportional to the square of the velocity and the energy associated with internal degrees of freedom is approximately independent of the velocity of the molecule (see Section D.2), the translational and internal energies should be transported differently, leading to different constants of proportionality between λ and μc_v for each contribution. Equation (44) should be valid for the translational contribution, but the internal part will be transported more like momentum, whence $\lambda = \mu c_v$ may be approximately true for the internal contribution.* Adding the two contributions therefore yields

$$\lambda = \frac{5}{2}\left(\frac{3}{2}\frac{k^0}{m}\right)\mu + \left(c_v - \frac{3}{2}\frac{k^0}{m}\right)\mu,$$

which may be written as

$$\lambda = \frac{\mu c_v(9\gamma - 5)}{4} \qquad (45)$$

by utilizing the fact that the specific heat at constant pressure is $c_p = c_v + k^0/m$ and the definition of the heat-capacity ratio $\gamma \equiv c_p/c_v$.

For binary and multicomponent mixtures, the thermal conductivity depends on the concentrations as well as on temperature, and the formulas of the accurate kinetic theory are quite complicated [5]. Empirical expressions for λ are therefore more useful for both binary [9] and ternary [6], [26] mixtures, although few data exist for ternary mixtures. Tabulations of available experimental and theoretical results for thermal conductivities may be found in [5], [6], [13], and [18]–[21], for example. The thermal diffusivity, defined as $\lambda/\rho c_p$, often arises in combustion problems; its pressure and temperature dependences in gases are $\lambda/\rho c_p \sim T^\alpha/p$ ($\frac{3}{2} \leq \alpha \leq 2$), and its typical values in combustion lie between 10^{-1} cm²/s and 1 cm²/s at atmospheric pressure.

E.5.2. The heat-flux vector

While equation (42) is valid for one-component systems without radiant transport, for binary and multicomponent mixtures there are other effects besides thermal conduction that contribute to the heat flux q.

---

* Although this might not be expected to be exactly true because the momentum is proportional to the first, not zeroth, power of the velocity, the derivation in Section E.4.1 fundamentally involved the component of momentum in the y direction and the component of velocity in the x direction; these are independent in a first approximation, and therefore the analogy is reasonable.

When the average velocity of any component i differs from the mass-average velocity of the mixture, then $n_i \bar{\mathbf{V}}_i$ molecules of type i per unit area per second flow across a surface moving with the mass-average velocity of the gas mixture. If the enthalpy associated with an i molecule is denoted by $H_i = U_i + k^0 T$, these molecules will carry across the surface an average enthalpy per unit area per second equal to

$$\bar{H}_i n_i \bar{\mathbf{V}}_i, \quad \text{where} \quad \bar{H}_i = \bar{U}_i + k^0 T$$

and $\bar{U}_i$ is given by equation (D-24). Denoting the average enthalpy per unit mass of species i by $h_i = \bar{H}_i/m_i$, we find from equations (D-5), (D-6), and (D-7) that $\bar{H}_i n_i \bar{\mathbf{V}}_i = \rho h_i Y_i \bar{\mathbf{V}}_i$, whence the total enthalpy (of all species) per unit area per second flowing relative to the mass-average motion of the mixture is

$$\rho \sum_{i=1}^{N} h_i Y_i \bar{\mathbf{V}}_i.$$

This term constitutes an additional contribution to $\mathbf{q}$ in binary and multi-component systems.

Onsager's reciprocal relations of irreversible thermodynamics [27–30] imply that if temperature gradients give rise to diffusion velocities (thermal diffusion), then concentration gradients must produce a heat flux. This reciprocal cross-transport process, known as the Dufour effect, provides another additive contribution to $\mathbf{q}$. It is conventional to express the concentration gradients in terms of differences in diffusion velocities by using the diffusion equation, after which it is found that the Dufour heat flux is [5].

$$R^0 T \sum_{i=1}^{N} \sum_{j=1}^{N} (X_j D_{T,i}/W_i D_{ij})(\bar{\mathbf{V}}_i - \bar{\mathbf{V}}_j).$$

In most cases the Dufour effect is so small that it apparently often is negligible even when thermal diffusion is not negligible. Although it is omitted in the applications, this term is retained in the general equations for completeness.

Inserting the radiation flux and the terms discussed in the preceding paragraphs into equation (42) yields

$$\mathbf{q} = -\lambda \nabla_x T + \rho \sum_{i=1}^{N} h_i Y_i \bar{\mathbf{V}}_i + R^0 T \sum_{i=1}^{N} \sum_{j=1}^{N} \left(\frac{X_j D_{T,i}}{W_i D_{ij}} \right)(\bar{\mathbf{V}}_i - \bar{\mathbf{V}}_j) + \mathbf{q}_R$$

$$(46)$$

as the complete expression for the heat-flux vector. The bars on $\bar{\mathbf{V}}_i$ and the subscript x on $\mathbf{V}_x$ are omitted when equation (46) is quoted in Chapter 1.

Calculation of $\mathbf{q}_R$ necessitates consideration of radiation transport [31]–[33]. The spectral intensity $I_\nu(\mathbf{x}, \mathbf{\Omega}, t)$ is defined as the radiant energy per unit area per second, traveling in a direction defined by the unit vector $\mathbf{\Omega}$, per unit solid angle about that direction, per unit frequency range about the frequency v, at position $\mathbf{x}$ and time t. The relationship between $\mathbf{q}_R$ and I_v is

$$\mathbf{q}_R = \int_0^\infty \oiint I_v \mathbf{\Omega} \, d\Omega \, dv, \tag{47}$$

where dv and $d\Omega$ are elements of frequency and of solid angle, respectively, and the integration in $\mathbf{\Omega}$ space extends over the entire solid angle 4π. Statement of an equation for I_v necessitates introduction of absorption and scattering coefficients. The mass-based spectral absorption coefficient κ_v is defined such that $\rho\kappa_v$ is the fractional decrease in radiant intensity per unit length of beam travel, due to absorption. The mass-based spectral scattering function σ_v is defined such that the increase of I_v per unit length of ray travel due to scattering of radiation from rays traveling in other directions, defined by unit vectors $\mathbf{\Omega}'$, is $\rho \oiint \sigma_v I_v' \, d\Omega'$, where $I_v' \equiv I_v(\mathbf{x}, \mathbf{\Omega}', t)$.* With these definitions, the equation of radiation transport may be written as

$$\mathbf{\Omega} \cdot \nabla_x I_v = \rho\kappa_v(B_v - I_v) + \rho \oiint \sigma_v(I_v' - I_v) \, d\Omega' \tag{48}$$

where the Planck function is

$$B_v = (2hv^3/c^2)/(e^{hv/k^0 T} - 1), \tag{49}$$

in which h and c are Planck's constant and the velocity of light, respectively. Equation (48) states that, in the direction of ray travel, I_v changes because of emission $(\rho\kappa_v B_v)$, absorption $(-\rho\kappa_v I_v)$, and scattering (the last term). The scattering term exhibits both a gain from rays traveling in other directions and a loss through scattering of the ray under consideration. Since equation (48) is an integrodifferential equation when scattering is important, correct inclusion of $\mathbf{q}_R$ in equation (46) through use of equation (47) sometimes may prevent combustion problems from being formulated in terms of partial differential equations.

In view of the complexity associated with equation (48), approximate methods are needed for applications. References [6] and [33]–[38] may be consulted for these approximations. While scattering may be important in combustion situations involving large numbers of small condensed-phase particles, often the effects of scattering may be approximated as additional contributions to emission and absorption, thereby eliminating the integral term. Two classical limits in radiation-transport theory are those of optically thick and optically thin media; the former limit seldom is applicable in combustion, while the latter often is. In the optically thin limit, gas-phase

---

* Note that $(\rho\kappa_v)^{-1}$ and $(\rho\sigma_v)^{-1}$ represent characteristic lengths for absorption and for scattering, respectively.

absorption is small; if this and scattering are neglected in equation (48), then equation (47) may easily be shown to imply that

$$\mathbf{V}_x \cdot \mathbf{q}_R = 4\pi\rho \int_0^\infty \kappa_\nu B_\nu \, d\nu \equiv 4\sigma T^4/l_p, \tag{50}$$

where $\sigma = 2\pi^5 k^{04}/15c^2 h^3$ is the Stefan-Boltzmann constant and l_p is the Planck-mean absorption length, which is defined by equation (50). Integration of equation (50) shows that at the surface of a slab of thickness L, having uniform properties inside and no radiant flux incident on the back face, the component of $\mathbf{q}_R$ normal to the surface of the slab is

$$q_R = 4\sigma T^4 L/l_p \equiv \epsilon\sigma T^4, \tag{51}$$

where $\epsilon = 4L/l_p$ is the engineering emissivity of the slab. More generally, it is found that if absorption is included as well in this geometry, then

$$\epsilon = 1 - e^{-4L/lp}. \tag{52}$$

Data in references quoted above may be employed in equation (50) or (51) to estimate effects of radiation in combustion problems.

Often equation (51) is used to obtain radiant fluxes emitted from reaction regions; accompanied by geometrical considerations, it then provides estimates of radiant fluxes incident on solid or liquid surfaces near reaction regions. In many problems, the only significant effects of radiation are those concerning radiant transfer to and from solid surfaces. The normal component of radiant flux emitted from a solid surface is given by equation (51), with ϵ representing the surface emissivity. Absorption lengths often are small for solids, so that their surfaces may be characterized by an absorptivity (which may be shown to equal ϵ) and a reflectivity, $1 - \epsilon$, having properties such that the fraction ϵ of the incident energy flux is absorbed and the fraction $1 - \epsilon$ reflected, transmission being negligible. Improved descriptions of interaction of radiation with surfaces need to consider transmission, spectral variations and influences of incident angle (see references quoted above). There are a number of combustion problems in which it is important to employ equation (51) to calculate radiant losses from solid surfaces while all other influences of radiation remain negligible.

E.6. DIMENSIONLESS RATIOS OF TRANSPORT COEFFICIENTS

In flow problems, the conservation equations often can be written in forms involving dimensionless ratios of various transport coefficients. These ratios are defined in the present section. They refer to molecular transport, not to corresponding quantities that often have been defined for turbulent flows.

The Prandtl number, which is a rough measure of the relative importance of momentum transfer and heat transfer, is defined as

$$Pr = \mu c_p / \lambda, \tag{53}$$

which is given by

$$Pr = 4\gamma / (9\gamma - 5) \tag{54}$$

for a pure polyatomic gas according to equation (45). From equation (54) we see that Pr, which is approximately 0.74 for air ($\gamma = 1.4$), varies from $\frac{2}{3}$ for monatomic gases ($\gamma = \frac{5}{3}$) to unity as $\gamma \rightarrow 1$. In some systems, the Eucken formula is not accurate and Pr can differ considerably from unity (for example, it is large compared with unity for liquids).

The Schmidt number, a rough measure of the relative importance of momentum transfer and mass transfer, is defined by

$$Sc = \mu / \rho D_{12} \tag{55}$$

in a binary mixture. In multicomponent systems, Schmidt numbers may be defined for each pair of species. As in the case of the Prandtl number, Sc often is somewhat less than unity for gases; for liquids typically $Sc \gg Pr$.

The Lewis number (or Lewis-Semenov number) is a measure of the ratio of the energy transported by conduction to that transported by diffusion; namely,

$$Le = \lambda / \rho D_{12} c_p \tag{56}$$

in a binary mixture. As with the Schmidt number, Lewis numbers may be defined for each pair of species in multicomponent mixtures. From equations (53), (55) and (56) it is apparent that

$$Le = Sc/Pr. \tag{57}$$

In many gases, Le is very nearly unity; it is often slightly less than unity in combustible gas mixtures. The approximation $Le = 1$ is frequently helpful in theoretical combustion analyses. However, small departures of Le from unity often can produce new phenomena in combustion. Therefore, a number of studies have been made of influences of the value of Le.

There are many other dimensionless groups that arise in applications involving transport processes [6], [39], [40]. Such quantities, a number of which are of practical importance in combustion [41], are introduced in the main text only to the extent that they are needed in the presentation.

REFERENCES

1. J. Jeans, *Dynamical Theory of Gases*, Cambridge: Cambridge University Press, 1925.
2. E. H. Kennard, *Kinetic Theory of Gases*, New York: McGraw-Hill, 1938.

3. W. G. Vincenti and C. H. Kruger, Jr., *Introduction to Physical Gas Dynamics*, New York: Wiley, 1965.

4. S. Chapman and T. G. Cowling, *The Mathematical Theory of Non-Uniform Gases*, Cambridge, Cambridge University Press, 1953.

5. J. O. Hirschfelder, C. F. Curtiss, and R. B. Bird, *Molecular Theory of Gases and Liquids*, New York: Wiley, 1954.

6. R. B. Bird, W. E. Stewart, and E. N. Lightfoot, *Transport Phenomena*, New York: Wiley, 1960.

7. W. H. Furry, *Am. J. Phys.* **16**, 63 (1948).

8. H. Goldstein, *Classical Mechanics*, Reading, Mass.: Addison-Wesley, 1950, Chapter 3.

9. S. S. Penner, *Chemistry Problems in Jet Propulsion*, New York: Pergamon Press, 1957.

10. F. A. Williams, *Am. J, Phys.* **26**, 467 (1958).

11. M. Gordon, "References to Experimental Data on Diffusion Coefficients of Binary Gas Mixtures," *National Engineering Laboratory Rept. No. 647*, Department of Industry, Glasgow, Scotland (1977).

12. *Journal of Physical and Chemical Reference Data*, **1** (1972)–present.

13. R. C. Reid and T. K. Sherwood, *The Properties of Gases and Liquids*, New York: McGraw-Hill, 1958.

14. J. Frenkel, *Kinetic Theory of Liquids*, New York: Dover, 1955.

15. C. Kittel, *Introduction to Solid State Physics*. New York: Wiley, 1976.

16. *International Critical Tables*, New York: McGraw-Hill, 1929.

17. L. S. Marks, *Mechanical Engineers' Handbook*, 5th ed., New York: McGraw-Hill, 1951.

18. *Handbook of Chemistry and Physics*, 36th ed., Cleveland, Ohio: Chemical Rubber Publ. Co., 1954.

19. R. M. Fristrom and A. A. Westenberg, *Fire Research Abstracts and Reviews* **8**, 155 (1966).

20. R. W. Gallant, *Physical Properties of Hydrocarbons*, Houston: Gulf Publ. Co., 1968.

21. Y. S. Touloukian, *Thermophysical Properties of Matter*, New York: Plenum Press, 1970.

22. T. P. Coffee and J. M. Heimerl, *C & F* **43**, 273 (1981).

23. P. L. García-Ybarra and P. Clavin, "Cross Transport Effects in Nonadiabatic Premixed Flames," in *Combustion in Reactive Systems*, J. R. Bowen, N. Manson, A. K. Oppenheim, and R. I. Soloukhin, eds., vol. 76 of *Progress in Astronautics and Aeronautics*, New York: American Institute of Aeronautics and Astronautics, 1981, 463–481.

24. L. Monchick and E. A. Mason, *J. Chem. Phys.* **35**, 1676 (1961).

25. A. Eucken, *Physik. Zeit.* **14**, 324 (1913).

26. E. A. Mason and S. C. Saxena, *Phys. Fluids* **1**, 361 (1958).

27. L. Onsager, *Phys. Rev.* **38**, 2265 (1931).

28. S. R. deGroot, *Thermodynamics of Irreversible Processes*, Amsterdam: North-Holland, 1951.

29. I. Prigogine, *Thermodynamics of Irreversible Processes*. Springfield, Ill.: Thomas, 1955.

30. S. R. deGroot and E. P. Mazur, *Non-Equilibrium Thermodynamics*, New York: Wiley, 1962.

31. S. Chandrasekhar, *Radiative Transfer*, New York: Dover, 1960.
32. V. Kourganoff, *Basic Methods in Transfer Processes, Radiative Equilibrium and Neutron Diffusion*, New York: Dover, 1963.
33. S. S. Penner and D. B. Olfe, *Radiation and Reentry*, New York: Academic Press, 1968.
34. W. H. McAdams, *Heat Transmission*, New York: McGraw-Hill, 1954.
35. E. M. Sparrow and R. D. Cess, *Radiation Heat Transfer*, Belmont, Calif.: Brooks/Cole, 1967.
36. H. C. Hottel and A. F. Sarofim, *Radiative Transfer*, New York: McGraw-Hill, 1967.
37. F. R. Steward, "Radiative Heat Transfer Associated with Fire Problems," Part IV of *Heat Transfer in Fires: thermophysics, social aspects, economic impact*, P. L. Blackshear, ed., Scripta Book Co., New York: Wiley, 1974, 273–486.
38. J. deRis, *17th Symp.* (1977), 1003–1016.
39. F. A. Williams, *Fire Research Abstracts and Reviews* **11**, 1 (1969).
40. F. A. Williams, "Current Problems in Combustion Research," in *Dynamics and Modelling of Reactive Systems*, W. E. Stewart, W. H. Ray, and C. C. Conley, eds., New York: Academic Press, 1980, 293–314.
41. A. M. Kanury, *Introduction to Combustion Phenonema*, New York: Gordon and Breach, 1975.

Subject Index

Author Index